# MODELING AND MEASUREMENT METHODS FOR ACOUSTIC WAVES AND FOR ACOUSTIC MICRODEVICES

Edited by **Marco G. Beghi**

**Modeling and Measurement Methods for Acoustic Waves and for Acoustic Microdevices**
http://dx.doi.org/10.5772/2581
Edited by Marco G. Beghi

## Contributors

V.I. Alshits, V.N. Lyubimov, A. Radowicz, Kuo-Ming Lee, E. Raitman, V. Gavrilov, Ju. Ekmanis, Bo Hou, Weijia Wen, Marian Urbańczyk, Tadeusz Pustelny, Zine El Abiddine Fellah, Mohamed Fellah, Claude Depollier, Mohammad Hosseini Fouladi, Mohamed H. Nassir, Masomeh Ghassem, Marwan Shamel, Sim Yeng Peng, Sin Yi Wen, Pang Zong Xin, Mohd Jailani Mohd Nor, Noorsuhada Md Nor, Norazura Muhamad Bunnori, Azmi Ibrahim, Hamidah Mohd Saman, Shahiron Shahidan, Soffian Noor Mat Saliah, Fabio Di Fonzo, Francisco García Ferré, Dario Gastaldi, Pasquale Vena, Marco G. Beghi, Boris Sveshnikov, Sergey Nikitov, Sergey Suchkov, T. Laroche, S. Ballandras, Yongqiang Guo, Weiqiu Chen, T. Baron, E. Lebrasseur, F. Bassignot, G. Martin, V. Pétrini, S. Ballandras, Matthieu Chatras , Stéphane Bila, Sylvain Giraud, Lise Catherinot, Ji Fan, Dominique Cros, Michel Aubourg, Axel Flament, Antoine Frappé, Bruno Stefanelli, Andreas Kaiser, Andreia Cathelin, Jean Baptiste David, Alexandre Reinhardt, Laurent Leyssenne, Eric Kerhervé, Bodong Li, Hommood Al Rowais, Jürgen Kosel, Sergey E. Babkin, Abhilash Mandloi, Vivekanand Mishra, M. El Hassan, E. Kerherve, Y. Deval, K. Baraka, J.B. David, D. Belot, Jerzy Filipiak, Grzegorz Steczko, Rogério Pirk, Carlos d´Andrade Souto, Gustavo Paulinelli Guimarães, Luiz Carlos Sandoval Góes, J. K. Luo, Y. Q. Fu, W. I. Milne, Emerson de Sousa Costa, Eduardo Bauzer Medeiros, João Batista Carvalho Filardi, Jens M. Hovem

## Published by InTech

Janeza Trdine 9, 51000 Rijeka, Croatia

## Copyright © 2013 InTech



## Notice



**Publishing Process Manager** Marina Jozipovic
**Technical Editor** InTech Prepress, Novi Sad
**Cover** InTech Design team

Additional hard copies can be obtained from orders@intechopen.com

Modeling and Measurement Methods for Acoustic Waves and for Acoustic Microdevices,
Edited by Marco G. Beghi
p. cm.
ISBN 978-953-51-1189-4

# Meet the editor

Dr. Marco G. Beghi graduated in 1979 in Nuclear Engineering at Politecnico di Milano, Italy. He spent one year at University of California, Los Angeles. In 1984 he became research fellow at the Department of Nuclear Engineering, Politecnico di Milano, and in 2003 associate professor of Condensed Matter Physics. He was member of government bodies of Politecnico: Board of Administrators and Academic Senate. Presently he is in the Micro- and Nanostructured Materials Laboratory of the Department of Energy, Politecnico. Since 1991 he has been teaching Experimental Physics, Condensed Matter Physics, and Technology of Nuclear Materials, to undergraduate and graduate students of Nuclear and Materials Engineering. His experimental research concerns the physics of materials. He analyzed the mechanical behavior of metals in terms of dislocation dynamics and irreversible thermodynamics. He then worked on thin films and their properties, exploiting Brillouin spectrometry to measure the elastic properties. He co-authored over seventy peer reviewed publications.

# Contents

# Preface

Acoustics is a mature field which also enjoys a never ending youth. After centuries of investigations, innovations and new developments either come from the endless search for a deeper understanding, and for aspects or phenomena which have not yet been put in light, or are triggered by technological innovations. New technologies can involve phenomena which are known, but which happen in unprecedented conditions, or which need a description of unprecedented accuracy and precision. In particular, the availability of new production technologies leads to the introduction of new devices. In the last years the development of micro-fabrication technologies marked a major step. A whole new class of microdevices, which exploit acoustic waves for various tasks, was introduced and is rapidly growing. Acoustics based microdevices are being developed, in particular, for information processing purposes and for sensing purposes, and have become a key technology in fields like telecommunications. This technological breakthrough challenges the methods for modelling the behavior of acoustic waves: the quest for devices of ever improving performances can be faced only by modelling tools which become more accurate and more detailed, and able to cope with more complex configurations. At the same time, technological developments are sought for the fabrication of these devices, and for their integration in technological systems. Several chapters of this book deal with the modelling and fabrication techniques for microdevices, including the exploration of unconventional phenomena, configurations and applications.

But the microdevices field is only one of the research lines in acoustics. The present book offers a sampling which, far from being exhaustive, allows to appreciate how diverse and pervasive the applications of acoustic waves can be. Theoretical analyses are presented, as well as modelling techniques, resulting in a deeper understanding of phenomena or features ranging from the detection of cracks to the acoustics of the oceans. And measurement methods are discussed, which exploit the potential of acoustic waves as probes to investigate the properties of widely different materials and systems. Overall, it is possible to appreciate the huge range of involved length and time scales, and of media supporting acoustic excitations: from the sub-micrometric layers exploited in microdevices to propagation in the oceans, and from materials and configurations aiming at enhancing resonances, to those aiming instead at a damping behavior.

The first section of the book aims at a detailed theoretical exploration, also by sophisticated mathematical analyses, of various aspects of phenomena whose basic laws are well established. Alshits and co-workers present a very deep analysis of the configuration of the vector fields which describe the electrical state in piezoelectric media. In particular, they investigate the fields associated to bulk acoustic wave, showing the existence of non piezoactive directions and analyzing the effect of symmetries. Lee undertakes a rigorous analysis of a scattering phenomenon which is the basis of a widespread non destructive examination technique: the scattering of an acoustic wave by a crack. Lee considers the direct problem, the computation of the acoustic field scattered by a crack of known size and shape, then investigates the inverse problem: how accurately can a crack be reconstructed, from the acoustic waves scattered by it? Raitman et al. exploit the dynamical scattering theory to investigate the effect of the acoustic waves on the observed diffraction patterns. They achieve a detailed interpretation of various features, experimentally observed. Hou & Wen present an analysis, both theoretical and experimental, of the peculiar transmission properties of a plate containing sub wavelength holes, or arrays of holes. They give a detailed picture of various regimes, and of underlying phenomena like evanescent waves.

The second section presents the exploitation of acoustic waves in the implementation of measurement techniques. Urbanczyk offers a wide review of measurement techniques and devices based on surface acoustic wave (SAWs). Fellah et al. consider the propagation of transient acoustic waves in a homogeneous isotropic slab of porous material, having a rigid or an elastic frame. Taking into account the inertial, viscous and thermal losses of the medium, they presents a model of the direct and inverse scattering problem. This analysis suggests several characterization methods based on the inverse problem, using experimental data of reflected and/or transmitted waves. These methods were validated experimentally on samples of air saturated porous materials, and are useful for materials like bone tissue. Hosseini Fouladi et al. focus on Malaysian Natural Fibers exploited as sound absorbers. They give a comparative characterization of various types of fibers. Md Nor and co-workers focus instead on the measurement of acoustic velocities in reinforced concrete beams: such velocities are a required input in the analysis of non destructive tests. Di Fonzo et al. consider the mechanical characterization of thin films, and show how the results obtained by acoustic techniques can be usefully improved by the joint exploitation of a completely different technique, indentation.

The third section addresses the techniques developed for an accurate modelling of the behavior of acoustics based microdevices. Sveshnikov presents a highly efficient analytical model, able to describe any multilayer bulk acoustic wave (BAW) device, by a flexible one-dimensional modeling and a phenomenological consideration of dissipation. The model can treat structures with an arbitrary number and sequence of dielectric and metal layers, as well as multiple electrodes. To verify the validity of any simulation, an original integral method is proposed, based on fundamental principles like energy balance, the second law of thermodynamics, and reciprocity. Laroche & Ballandras summarize a development work that was pursued for over a decade, until the integration of different techniques, namely Finite Elements Analysis, Boundary Elements Method and Perfectly Matched Layer method. This wide modelling effort led to the selection of the most appropriate tool for each aspect of the simulation, and results in a modelling tool which can treat both bulk waves and surface waves microdevices. Guo & Chen present a new version of the analytical method of reverberation-ray matrix, devoted to the integrated analysis of film bulk acoustic wave resonators (FBARs) with an underlying Bragg Cell which decouples the resonator behavior from that of the substrate.

Chapters in the fourth section present design and development efforts which led to the fabrication of microdevices. Baron et al. report the development work which led to bulk acoustic resonators, in particular operating in the high overtone mode. The microfabrication technologies are addressed, and the behavior of the resonators is characterized. The exploitation of these resonators for the realization of devices of various types, like oscillators, sensors, wireless sensors is reviewed, including the detailed characterization of aspects like temperature compensation. Chatras and co-workers give a complete account of the whole process of design, prototype realization and testing of electronic devices based on bulk acoustic waves, including the characterization of the overall performance in terms of signal processing. The whole process shows how crucial the bulk acoustic wave based devices can be for the development of hardware meeting the requirements of last generation mobile radio (the UMTS standard). The importance of the analysis of the acoustic behavior for the optimization of the performance also emerges. Li et al., after a more general discussion of SAW based sensors, in particular of magnetic field sensors, discuss a recent development: the design, fabrication and characterization of an integrated SAW based passive sensor (a transponder) for magnetic field, which exploits a giant magnetoimpedance effect. Babkin discusses a peculiar method for the conversion between electromagnetic and acoustic energy, which is the basis of both the excitation and the detection of acoustic waves. Conversion by piezoelectric materials is widely adopted; Babkin discusses instead the conversion by magnetic means. Some devices are presented, including devices for the analysis of materials and of their defects. Mishra demonstrates the performance of in-fiber optical devices obtained coupling fibers of different thicknesses with a piezoelectric transducer driven by a RF generator. The acoustic energy is efficiently transferred to the fiber by a machined aluminum horn, and by the acousto-optic effects tunable filters are obtained, whose performance is characterized and optimized.

The fifth section addresses the integration of acoustic based devices into more complex engineered systems. El Hassan et al. consider FBARs and Solidly Mounted Resonators (SMRs), equipped with passive elements (capacitances and inductances), and demonstrate a tunable BAW-SMR filter which matches the requirements for mobile communications (the WLAN 802.11 b/g standard). Furthermore, a digitally tunable

BAW-SMR filter implementation is shown. Filipiak & Steczko consider SAW based vibration sensors, characterize their performance and show how they can be integrated into a warning system. Pirk and co-workers focus on a specific case: the payload of space flights. At rocket lift off, and during the subsequent transonic flight, severe acoustic loads are imposed to the payloads. Pirk and co-workers discuss techniques to measure such loads and to model them, aiming at the optimization of acoustic load mitigation measures.

The sixth section is devoted to the applications of acoustic waves into microfluidic devices. Luo et al. give an extensive overview of how acoustic waves can be exploited to perform various tasks, including manipulation of droplets and mixing of liquids.

On a completely different length scale, the seventh section is devoted to the acoustics of the oceans. De Sousa Costa & Bauzer Medeiros introduce the reader into the field, offering a synthetic and comparative overview of the modelling methods for acoustic propagation in the oceans, in which reflection phenomena at the surface and at the underlying interface with rocks or sediments have an important role. Hovem focuses on the ray tracing method. He shows the effects of the depth dependent, and season dependent, properties of water, which cause curved trajectories of individual rays. He also gives a detailed modelling of the reflection at the interface between the water and the underlying sediment layer.

As mentioned above, this book offers a representative sampling of the wide horizon of present research in acoustics. It includes, on one side, deep theoretical and numerical analyses of specific phenomena and devices, and, on the other side, more technically oriented contributions, which show the usefulness of acoustics based devices in building engineered systems. The book is therefore of interest for both the specialized reader, who can find deep insights into some specific topic, and for a more general readership, who, by exploring a wide variety of ideas and of approaches, can find interesting suggestions.

**Marco G. Beghi**
Politecnico di Milano,
Energy Department and NEMAS Center,
Milano,
Italy

Section 1

# Analyses and Models of Basic Phenomena

Chapter 1

# Electric Components of Acoustic Waves in the Vicinity of Nonpiezoactive Directions

V.I. Alshits, V.N. Lyubimov and A. Radowicz

Additional information is available at the end of the chapter

http://dx.doi.org/10.5772/56067

## 1. Introduction

The acoustic wave of displacements in piezoelectric media is usually accompanied by a quasistatic wave of the electric potential. This implies that, using acoustic waves, electric signals can be transmitted over a crystal at the velocity of sound. Such possibility opened the way to numerous applications of acoustic waves in electronic devices and even led to the formation of a special field of science called acoustoelectronics. The applied aspect provides an important stimulus for extensive investigations devoted to various features of acoustic fields in piezoelectric crystals (Royer & Dieulesaint, 2000). These investigations are also stimulated by basic interest in the study of new effects in media with electromechanical couplings (Lyubimov, 1968; Balakirev & Gilinskii, 1982; Lyamov, 1983). The acoustics of piezoelectric crystals is still an extensively developing field of solid state physics [see, e.g., the review article by Gulyaev (1998)], the more so that even purely basic investigations in this field frequently contain ideas for fruitful, however not immediately evident, applications.

It should also be noted that the anisotropy often influences the properties of piezoelectric crystals in a nontrivial way, and may sometimes lead to qualitatively new phenomena. In particular, it is very important from the practical standpoint to know the wave propagation directions **m** for which the electric field components possess maximum amplitudes (Alshits & Lyubimov, 1990) and, on the contrary, to reveal the nonpiezoactive directions (Royer & Dieulesaint, 2000; Lyamov, 1983) in which the electric signals are not transmitted. Taking into account that, irrespective of the anisotropy, the electric field in an acoustic wave is always longitudinal ($\mathbf{E} \parallel \mathbf{m}$) and the electric induction is always transverse ($\mathbf{D} \perp \mathbf{m}$), we have to distinguish (Lyamov, 1983) between the directions of longitudinal and transverse nonpiezoactivity in which $\mathbf{E} = 0$ and $\mathbf{D} = 0$, respectively. This paper presents the results of investigations aimed at a detailed analysis of the nonpiezoactivity of both types.

Another important aspect of this problem is related to directions **m**, in the vicinity of which the vector fields of displacements (**u**) and the acompanying electric components (**E**, **D**) exhibit singularities. According to the results obtained by Alshits, Sarychev & Shuvalov (1985), Alshits *et al* (1987), and Shuvalov (1998) this very situation takes place near the acoustic axes, where the orientational singularities in the degenerate branches of eigenwaves are observed for the **u** and **D** fields, and the amplitude singularities, for the **E** field. This paper deals with orientational singularities of another type, which occur in the vicinity of the transverse nonpiezoactivity directions in the vector fields **D**(**m**), i.e. around the points $\mathbf{m}_0$ on the unit sphere such that $\mathbf{D}(\mathbf{m}_0)=0$ (Alshits, Lyubimov & Radowicz, 2005 a, b).

Below we will formulate the equations determining special directions **m** for which either $\mathbf{E}_a(\mathbf{m}) = 0$ or $\mathbf{D}_a(\mathbf{m}) = 0$ for all three branches of the acoustic spectrum ($a = 1, 2, 3$). These directions have different dimensionalities: the typical solutions appear as lines of zero electric field ($\mathbf{E}_a = 0$) and points of zero induction ($\mathbf{D}_a = 0$) on the unit sphere $\mathbf{m}^2 = 1$. The equations obtained will be analyzed both in the general case and in application to various particular crystal symmetry classes. The two types on nonpiezoactivity are closely related to the crystal symmetry, but they can also exist in triclinic crystals possessing no elements of symmetry. The corresponding theorems of existence are proved.

The possible types of singularities in the vector field $\mathbf{D}_a(\mathbf{m})$ in the vicinity of the transverse nonpiezoactivity directions will be considered. In particular, it will be shown that, depending on the material moduli, the singularity in an isolated point $\mathbf{m}_0$ may be characterized by the Poincaré indices (topological charges) $n_{\mathrm{D}} = 0, \pm1, \pm2$. The general analytical expressions will be obtained for the $n_{\mathrm{D}}$ values in triclinic crystals with arbitrary anisotropy and specified for a large series of crystals belonging to particular crystal symmetry classes. Only the solutions corresponding to singularities with $n_{\mathrm{D}} = \pm1$ are topologically stable, while singularities of the other types either split or disappear upon an arbitrary triclinic perturbation of the material tensors. However, the sum of indices for any splitting must be equal to the initial index $n_{\mathrm{D}}$.

The chapter is mainly based on our papers (Alshits, Lyubimov & Radowicz, 2005 a, b).

## 2. Statement of the problem and general equations

In piezoelectric crystals, purely mechanical characteristics, the elastic displacement vector **u**(**r**, *t*), the distortion tensor $\hat{\beta}(\mathbf{r}, t)$, and the stress tensor $\hat{\sigma}(\mathbf{r}, t)$, are related to such electrical quantities as the potential $\phi(\mathbf{r}, t)$ and the electric field strength **E**(**r**, *t*), and induction **D**(**r**, *t*). The fields of $\hat{\beta}(\mathbf{r}, t)$ and **E**(**r**, *t*) can be expressed in terms of their own potentials as

$$\hat{\beta}(\mathbf{r},t) = \nabla\mathbf{u}(\mathbf{r},t), \qquad \mathbf{E}(\mathbf{r},t) = -\nabla\phi(\mathbf{r},t). \tag{1}$$

The interrelation of these characteristics is determined by the constitutive equations (Landau & Lifshitz, 1984):

$$\sigma_{ij} = c_{ijkl}\beta_{kl} - e_{kij}E_k, \qquad D_i = e_{ikl}\beta_{kl} + \varepsilon_{ik}E_k, \tag{2}$$

where $c_{ijkl} = \hat{c}$ is the tensor of elastic moduli, $e_{kij} = \hat{e}$ is the tensor of piezoelectric moduli, and $\varepsilon_{ik} = \hat{\varepsilon}$ is the permittivity tensor. In such a piezoelectric medium, the bulk acoustic wave with the phase velocity $v$ and the wave vector $\mathbf{k} = k\mathbf{m}$ must be a superposition of mechanical and electrical dynamic fields:

$$\{\mathbf{u},\ \varphi\} = \{\mathbf{u}_0,\ \varphi_0\}\exp\{ik(\mathbf{m}\cdot\mathbf{r} - vt)\}. \tag{3}$$

These fields obey the usual equations of motion (Landau & Lifshitz, 1984):

$$\mathrm{Div}\ \hat{\sigma} = \rho\ddot{\mathbf{u}}, \qquad \mathrm{Div}\mathbf{D} = 0, \tag{4}$$

where $\rho$ is the density of medium. Here, we use the well-known quasi-static approximation valid to within the terms proportional to the ratio $(v/c)^2 \sim 10^{-10}$ ($c$ is the velocity of light). Combining the above relations, we readily obtain a homogeneous equation for the polarization vector $\mathbf{u}_0$ (Landau & Lifshitz, 1984):

$$\hat{F}(v,\mathbf{m})\mathbf{u}_0 \equiv [\hat{F}^{(0)} + \mathbf{e}\otimes\mathbf{e}/\varepsilon]\mathbf{u}_0 = 0, \tag{5}$$

where

$$\hat{F}^{(0)} = \mathbf{m}c\mathbf{m} - \rho v^2\hat{I}, \qquad \mathbf{e} = \mathbf{m}\hat{e}\mathbf{m}, \qquad \varepsilon = \mathbf{m}\cdot\hat{\varepsilon}\mathbf{m}, \tag{6}$$

symbol $\otimes$ denotes the dyadic product, and $\hat{I}$ is the unit matrix. A necessary condition for the existence of nontrivial solutions of the homogeneous equation (5) is

$$\det\hat{F}(v,\mathbf{m}) = 0. \tag{7}$$

This is a cubic equation for the square of phase velocity $v$, which determines the three branches of the velocity of the bulk acoustic waves $v_\alpha(\mathbf{m})$ ($\alpha = 1, 2, 3$).

Orientations of the corresponding mutually orthogonal polarization vectors $\mathbf{u}_{0\alpha}(\mathbf{m})$ of the isonormal eigenwaves can be expressed in terms of the $\bar{\hat{F}}_\alpha$ matrix, which is adjoint to the matrix $\hat{F}_\alpha(\mathbf{m}) \equiv \hat{F}_\alpha(v_\alpha(\mathbf{m}),\mathbf{m})$ and is determined from the condition $\bar{\hat{F}}_\alpha\hat{F}_\alpha = \hat{I}\det\hat{F}_\alpha$. As can be readily checked, Eq. (5) for any vector $\mathbf{c}$ such that $\bar{\hat{F}}_\alpha\mathbf{c} \neq 0$ is satisfied for

$$\mathbf{u}_{0\alpha}(\mathbf{m}) \,||\, \bar{\hat{F}}_\alpha(\mathbf{m})\mathbf{c}. \tag{8}$$

It should be emphasized that the direct relation (8) between the polarization vector $\mathbf{u}_{0\alpha}(\mathbf{m})$ and the wave normal $\mathbf{m}$ will be widely used in the subsequent analysis.

Once the field of elastic displacements for a given wave branch $\mathbf{u}_a(\mathbf{r},t)$ is known, we can also determine the corresponding electric components (Landau & Lifshitz, 1984). For the subsequent analysis, these components are conveniently represented [by analogy with Eqs. (3)–(8)] in a coordinate-free form as

$$\phi_a = \mathbf{e}\cdot\mathbf{u}_a / \varepsilon, \qquad \mathbf{E}_a = -ik\phi_a\mathbf{m}, \qquad \mathbf{D}_a = ik\hat{N}\mathbf{u}_a, \tag{9}$$

$$\hat{N} = \hat{e}\mathbf{m} - (\hat{\varepsilon}\mathbf{m})\otimes\mathbf{m}\cdot\hat{e}\mathbf{m} / \mathbf{m}\cdot\hat{\varepsilon}\mathbf{m}. \tag{10}$$

Relations (9) together with condition (8) determine the functions $\mathbf{E}_a(\mathbf{m})$ and $\mathbf{D}_a(\mathbf{m})$ necessary for the subsequent analysis.

As can be readily seen, $\mathbf{m}\hat{N} \equiv 0$. This identity and the third relation in (9) clearly illustrate the well-known property (see Introduction) according to which the electric field $\mathbf{E}_a(\mathbf{m})$ is purely longitudinal, whereas the induction $\mathbf{D}_a(\mathbf{m})$ is purely transverse:

$$\mathbf{E}_a \parallel \mathbf{m}, \qquad \mathbf{D}_a \perp \mathbf{m}. \tag{11}$$

On the other hand, the same identity $\mathbf{m}\hat{N} = 0$ implies one useful property of the $\hat{N}$ matrix:

$$\det \hat{N} = 0, \tag{12}$$

which indicates that this matrix is planar and, hence, can be represented as a sum of two dyads.

## 3. Examples of transversely isotropic piezoelectrics

There are three groups of piezoelectrics which exhibit a transverse isotropy of their acoustic properties. They belong to the following classes of symmetry (Sirotin & Shaskolskaya, 1982):

$$\infty 2, 622, \tag{13}$$

$$\infty m, 6mm, \tag{14}$$

$$\infty, 6. \tag{15}$$

Owing to the transverse isotropy, the formulas presented below contain only the polar angle $\theta$ between the **m** vector and the $z$ axis coinciding with the principal axis of symmetry. Without loss of generality, we may proceed with the analysis upon selecting any cross section containing the main axis. Here, it is convenient to choose

$$\mathbf{m} = (m_1,\ 0,\ m_3) = (\sin\theta,\ 0,\ \cos\theta). \tag{16}$$

In these coordinates, the $\hat{F}^{(0)}$ matrix in Eq. (6) for all the six classes of symmetry (13)–(15) has the same quasi-diagonal form (Fedorov, 1968):

$$\hat{F}^{(0)} = \begin{pmatrix} c_{11}m_1^2 + c_{44}m_3^2 - \rho v^2 & 0 & dm_1m_3 \\ 0 & c_{66}m_1^2 + c_{44}m_3^2 - \rho v^2 & 0 \\ dm_1m_3 & 0 & c_{44}m_1^2 + c_{33}m_3^2 - \rho v^2 \end{pmatrix}, \quad (17)$$

where $d = c_{13} + c_{44}$. The $\hat{\varepsilon}\mathbf{m}$ vector and, hence, the $\varepsilon$ scalar in Eq. (6) are also the same for all symmetry classes (13)–(15) (Sirotin & Shaskolskaya, 1982):

$$\hat{\varepsilon}\mathbf{m} = (\varepsilon_1 m_1,\ 0,\ \varepsilon_3 m_3), \quad \varepsilon = \varepsilon_1 m_1^2 + \varepsilon_3 m_3^2. \quad (18)$$

However, the form of the electric vector **e** according to Eq. (6) for the transversely isotropic crystals of three types is different. For the piezoelectric media belonging to classes (13) and (14), the electric vectors are expressed as

$$\mathbf{e} = e_{14}(0,\ m_1m_3,\ 0), \quad (19)$$

$$\mathbf{e} = \{(e_{15} + e_{31})m_1m_3,\ 0,\ e_{15}m_1^2 + e_{33}m_3^2\}, \quad (20)$$

respectively. For a medium of the symmetry class (15), the electric vector is given by a sum of expressions (19) and (20). Thus, the structure of the $\hat{F}$ matrix in (5) for classes (13) and (14) is the same as in (17), but this conclusion is not valid for the $\hat{F}$ matrix of the symmetry classes (15), which contains no vanishing elements. In the same coordinates, the matrix for the piezoelectric media belonging to classes (13) and (14) has the following forms:

$$\hat{N} = e_{14}\begin{pmatrix} 0 & (\varepsilon_3/\varepsilon)m_3^2 & 0 \\ -m_3 & 0 & -m_1 \\ 0 & -(\varepsilon_3/\varepsilon)m_1m_3^2 & 0 \end{pmatrix}, \quad (21)$$

$$\hat{N} = \begin{pmatrix} e_{15}m_3 & 0 & e_{15}m_1 \\ 0 & e_{15}m_3 & 0 \\ e_{31}m_1 & 0 & e_{33}m_3 \end{pmatrix} - \varepsilon^{-1}\begin{pmatrix} (e_{15}+e_{31})\varepsilon_1 m_1^2 m_3 & 0 & (e_{15}m_1^2 + e_{33}m_3^2)\varepsilon_1 m_1 \\ 0 & 0 & 0 \\ (e_{15}+e_{31})\varepsilon_3 m_1 m_3^2 & 0 & (e_{15}m_1^2 + e_{33}m_3^2)\varepsilon_3 m_3 \end{pmatrix}, \quad (22)$$

respectively. For a medium of the symmetry classes (15), the matrix $\hat{N}$ is (by analogy with vector **e**) given by a sum of expressions (21) and (22). In classes (13) and (14) of higher symmetry, one of the eigenwave branches for any direction **m** is purely transverse:

$$\mathbf{u}_t \ || \ (0,1,0), \qquad \rho v_t^2 = c_{66}m_1^2 + c_{44}m_3^2 + (e_{14}^2/\varepsilon)m_1^2 m_3^2. \quad (23)$$

Such purely transverse waves of the $t$ mode polarized orthogonally to the propagation plane are frequently called SH waves. The other two branches are polarized in the $\{m_1, m_3\}$ plane:

$$\mathbf{u}_{l,t'} \,||\, (2dm_1m_3,\ \ 0,\ \ -\Delta_{14}^{-}m_1^2 + \Delta_{34}^{-}m_3^2 \pm R),$$
$$\rho v_{l,t'}^2 = (\Delta_{14}^{+}m_1^2 + \Delta_{34}^{+}m_3^2 \pm R)/2, \tag{24}$$

where

$$R = \sqrt{(\Delta_{14}^{-}m_1^2 - \Delta_{34}^{-}m_3^2)^2 + (2dm_1m_3)^2},$$
$$\Delta_{ij}^{\pm} = c_{ii} \pm c_{jj}. \tag{25}$$

The electric components of the above wave fields can be also determined for an arbitrary direction $\mathbf{m}$. For a medium of the symmetry class (13):

$$\phi_t = (e_{14}/\varepsilon)m_1m_3u_t,$$
$$\mathbf{D}_t \,||\, e_{14}(\varepsilon_3/\varepsilon)m_3^2(m_3,\ 0,\ -m_1)u_t,$$
$$\phi_{l,t'}(\mathbf{m}) \equiv 0,$$
$$\mathbf{D}_{l,t'} \,||\, -e_{14}(0,\ 1,\ 0)[m_1(\mathbf{u}_{l,t'})_3 + m_3(\mathbf{u}_{l,t'})_1]. \tag{26}$$

For the less simple symmetry class (14), we present only the result for the SH-branch (23):

$$\phi_t(\mathbf{m}) \equiv 0, \qquad \mathbf{D}_t \,||\, (0,\ e_{15}m_3,\ 0)u_t. \tag{27}$$

The structure of acoustic waves in media of the symmetry classes (15) is more complicated. In this case, even a purely transverse solution ($\mathbf{u}_t \,||\, y$) exists only in the $xy$ basis plane.

## 4. Lines of zero electric field on the unit sphere

According to the second relation in (9), the electric field amplitude distribution on the unit sphere of the wave propagation directions is described by the equation

$$\mathbf{E}_a(\mathbf{m}) = \text{const} \cdot \phi_a(\mathbf{m})\mathbf{m}, \tag{28}$$

which shows that zero values of $\mathbf{E}_a(\mathbf{m})$ coincide with those of the potential $\phi_a = \mathbf{e} \cdot \mathbf{u}_a / \varepsilon$. According to condition (8), these directions are determined by the equation,

$$\mathbf{e}(\mathbf{m}) \cdot \bar{\hat{F}}_a(\mathbf{m})\mathbf{c} = 0. \tag{29}$$

The acoustic waves (3) propagating in these directions contain no electrostatic components $\mathbf{E}_a$, as in a nonpiezoelectric medium. Even a nonzero induction field $D_i = e_{ijk}u_{k,j}$ in these directions does not influence the parameters of the displacement wave.

The scalar equation (29) poses only one limitation on the direction of the wave normal $\mathbf{m} \equiv \mathbf{m}(\theta,\varphi)$ as a function of two spherical angular coordinates. In other words, Eq. (29) determines a line (or several lines) of nonpiezoelectric directions (in which $\mathbf{E}_a = 0$) on the sphere $\mathbf{m}^2 = 1$. It should be noted that the condition of longitudinal nonpiezoactivity,

$$\mathbf{e}(\mathbf{m}) \perp \mathbf{u}_a(\mathbf{m}), \tag{30}$$

in some special cases can be satisfied even on the whole $\mathbf{m}^2 = 1$ sphere. This takes place, in particular, in the transversely isotropic crystals belonging to the symmetry classes (13) (for the $l$ and $t'$ modes (26)) and (14) (for the $t$ mode (27)). For all other crystals, including transversely isotropic crystals belonging to the symmetry classes (15), the geometric locus of the longitudinal nonpiezoactivity has the form of lines on the unit sphere $\mathbf{m}^2 = 1$. Such lines also exist in the piezoactive branches of the aforementioned high-symmetry media belonging to symmetry classes (13) and (14). For example, the zero-field lines $\mathbf{E}_a = 0$ in the $l$ and $t'$ branches of the media of classes (14) and (15) appear at the intersection of the $\mathbf{m}^2 = 1$ sphere with the cones of directions defined by the polar angles $\theta_l$ and $\theta_{t'}$ as

$$\tan^2\theta_l = -e_{33} / (2e_{15} + e_{31}), \quad \tan^2\theta_{t'} = (e_{15} + e_{31} - e_{33}) / e_{15}. \tag{31}$$

For simplicity, these expressions are written in an approximate form corresponding to the case of a weak electromechanical interaction and a small elastic anisotropy. Nevertheless, one can readily check that the exact condition for the existence of the aforementioned nonpiezoactivity cones is the positive determinacy of the right-hand parts of the approximate formulas (31).

It is possible to prove that the longitudinal nonpiezoactivity lines in fact exist practically in all (even triclinic) crystals. Let us consider a crystal with arbitrary anisotropy, which contains at least one acoustic axis of the general (conical) type. Here, it should be noted that no one real crystal without acoustic axes and no one triclinic crystal without conical axes are known so far. As was demonstrated by Alshits & Lothe (1979) and Holm (1992), the polarization fields of elastic displacements $\mathbf{u}_{0a}(\mathbf{m})$ for the bulk eigenwaves in such a crystal can be arranged on the $\mathbf{m}^2 = 1$ sphere so that one is even,

$$\mathbf{u}_{02}(-\mathbf{m}) = \mathbf{u}_{02}(\mathbf{m}), \tag{32}$$

and two are odd,

$$\mathbf{u}_{01}(-\mathbf{m}) = -\mathbf{u}_{01}(\mathbf{m}), \qquad \mathbf{u}_{03}(-\mathbf{m}) = -\mathbf{u}_{03}(\mathbf{m}). \tag{33}$$

The nondegenerate branch $\mathbf{u}_{03}(\mathbf{m})$ is always odd and continuous on the entire sphere of wave directions. As for the degenerate branches, $\mathbf{u}_{01}(\mathbf{m})$ and $\mathbf{u}_{02}(\mathbf{m})$, their evenness depends on the representation and can be changed simultaneously. These branches are continuous at all points of the sphere except for some open-ended lines on which the $\mathbf{u}_{01}(\mathbf{m})$ and $\mathbf{u}_{02}(\mathbf{m})$ functions change sign. Such "anti-sign" lines can be arbitrarily deformed on the unit sphere without changing the positions of end points (coinciding with the points of degeneracy). In fact, the representation is chosen by setting certain fixed positions of the anti-sign lines (coinciding for both degenerate branches).

One can readily check that the aforementioned properties of the fields of elastic displacements, which were established in (Alshits & Lothe, 1979; Holm, 1992) for purely elastic media, are also valid for piezoelectrics. Taking into account that, according to relations (33), the $\mathbf{u}_{03}(\mathbf{m})$ function is odd and the $\mathbf{e}(\mathbf{m})$ function is [by definition (6)] even, we may conclude that the potential

$$\phi_{03}(\mathbf{m}) = \mathbf{e}(\mathbf{m}) \cdot \mathbf{u}_{03}(\mathbf{m}) / \varepsilon$$

is an odd function

$$\phi_{03}(-\mathbf{m}) = -\phi_{03}(\mathbf{m}). \tag{34}$$

This result implies that, for any path connecting the opposite points $\mathbf{m}$ and $-\mathbf{m}$ on the unit sphere, there exists at least one point $\mathbf{m}_0$ such that $\phi_{03}(\mathbf{m}_0) = 0$ . In scanning the paths on the unit sphere, points $\mathbf{m}_0$ will apparently form a closed line representing a geometric locus of the directions of longitudinal nonpiezoactivity for the nondegenerate branch.

For the degenerate branches $\phi_{01}(\mathbf{m})$ and $\phi_{02}(\mathbf{m})$, the considerations should be somewhat modified, while being still generally analogous to those used in solving a similar problem (Alshits & Lothe, 1979) concerning the existence of the lines of solutions for exceptional bulk waves related to the same degenerate branches in semi-infinite elastic media. Not reproducing these considerations here, we only formulate the result: the longitudinal nonpiezoactivity lines exist in both degenerate branches and pass from one branch to another at the degeneratcy points. Thus, the following theorem of existence is valid:

*All three wave branches in an arbitrary crystal, which contains conical acoustic axes, must possess lines of longitudinal nonpiezoactivity directions on the unit sphere.*

It should be also noted that, when a wave propagates along an acoustic axis $\mathbf{m}_d$ of any type, the continuum of possible orientations of the wave polarization $\mathbf{u}$ in the plane of degeneracy always contains a vector orthogonal to the $\mathbf{e}(\mathbf{m}_d)$ direction. In view of the criterion (30), this ensures nonpiezoactivity of the corresponding wave. Therefore,

*acoustic axes must belong to the lines of longitudinal nonpiezoactivity.*

The elements of crystal symmetry can become an additional factor accounting for the phenomenon of nonpiezoactivity. According to (Royer & Dieulesaint, 2000), also

*symmetry axes determine the directions of longitudinal nonpiezoactivity for purely transverse modes,*

while

*a symmetry plane is the geometric locus of directions of longitudinal nonpiezoactivity for the related SH waves.*

One can add that

*the planes orthogonal to symmetry axes of even order are the geometric locus of directions of longitudinal nonpiezoactivity for in-plane polarized waves.*

Let us consider, for example, a monoclinic piezoelectric crystal belonging to one of the two possible symmetry classes: $m$ or 2. In the first case, the electric vector $\mathbf{e}$ of any wave propagating in a plane of symmetry $m$ must, obviously, lie in the same plane being, hence, orthogonal to the polarization vector $\mathbf{u}_{0t}$ of all SH waves of the $t$ branch. In the second case, the $\mathbf{e}$ vector for a wave normal occurring in the plane perpendicular to the dyad (2-fold) axis of symmetry must be parallel to this axis and, hence, orthogonal to polarization vectors (belonging to said plane) of the $l$ and $t'$ waves. Naturally, the latter property is valid for any other symmetry axis of even order. In monograph (Royer & Dieulesaint, 2000), this rule was formulated for planes orthogonal to the tetrad (4-fold) and hexad (6-fold) axes.

The above theorems are summarized in Table 1.

| Directions of nonpiezoactivity | *Directions of acoustic and symmetry axes* | *Any direction in symmetry planes* | *Any direction in planes orthogonal to symmetry axes of even order* |
|---|---|---|---|
| Wave branches | *Transverse waves* | *SH waves* | *In-plane polarized waves* |

**Table 1.** Directions of obligatory longitudinal nonpiezoactivity ( $\mathbf{E}_a = 0$) in crystals

In particular, the coordinate planes of the crystal system orthogonal to the tetrad and dyad axes in cubic piezoelectrics (symmetry classes $\bar{4}3m$ and 23) must be nonpiezoactive for the corresponding $l$ and $t'$ branches. At the same time, the diagonal symmetry planes {110} are nonpiezoactive for the corresponding $t$ (SH) waves. One can check that, in the vicinity of the coordinate axes, the potential amplitudes for these branches can be represented in spherical coordinates ($\theta$, $\varphi$) as (Fig. 1)

$$\phi_{l,t'} \propto \theta^2 \sin 2\varphi, \tag{35}$$

$$\phi_t \propto \theta^2 \cos 2\varphi. \tag{36}$$

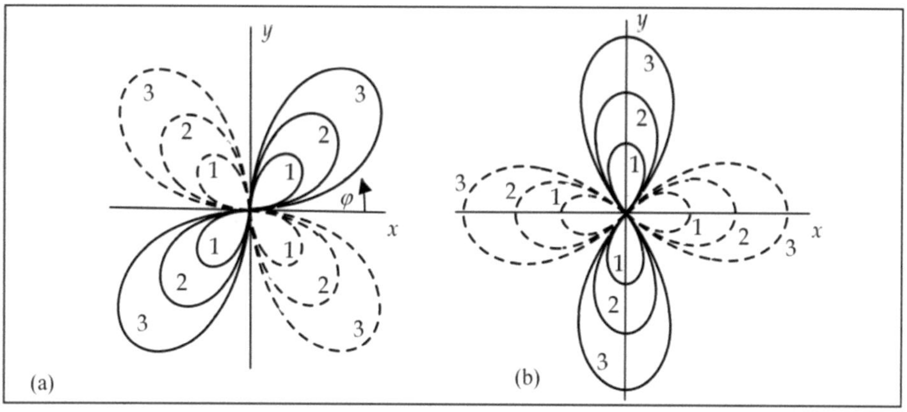


**Figure 1.** Polar diagrams of the electric potentials (a) $\phi_{l,t'}(\varphi)$ and (b) $\phi_t(\varphi)$ at $\theta = \text{const}$ in the vicinity of the (0, 0, 1) direction in a cubic piezoelectric crystal. Numbers 1, 2, and 3 refer to the angles $\theta_1 < \theta_2 < \theta_3$; solid and dashed lines relate to the potentials of different signs.

# 5. Zero-induction points on the unit sphere

## 5.1. General case of arbitrary anisotropy

Now let us consider the conditions determining the propagation directions $\mathbf{m}_0$ in which the electric induction vector $\mathbf{D}_a = ik\hat{N}\mathbf{u}_a$ defined in (9) vanishes. Taking into account identity (12) and the definition of the adjoint tensor

$$\hat{N}\bar{\hat{N}} = \hat{I}\det\hat{N}, \qquad (37)$$

one can readily check that $\mathbf{D}_a = 0$ for the directions $\mathbf{m}_0$ such that $\mathbf{u}_a \,||\, \bar{\hat{N}}\mathbf{d}$, where $\mathbf{d}$ is any vector obeying the condition $\bar{\hat{N}}\mathbf{d} \neq 0$. For these directions $\mathbf{m}_0$, according to condition (8), we also have

$$\bar{\hat{F}}_a(\mathbf{m})\mathbf{c} \,||\, \bar{\hat{N}}\mathbf{d} \qquad (38)$$

In the general case, this condition gives two equations with two unknowns $\theta$ and $\varphi$, which determine the positions of isolated points $\mathbf{m}_0(\theta,\varphi)$ such that $\mathbf{D}_a = 0$ on the unit sphere $\mathbf{m}^2 = 1$. There is the well-known Brouwer theorem in the topology, according to which

*any continuous transform on a sphere, not mapping any point by its antipode, has at least two stationary points.*

Now let us consider a distribution of vectors $\mathbf{D}_a(\mathbf{m})$ continuous everywhere on the unit sphere. The continuity of $\mathbf{D}_a(\mathbf{m})$ is ensured when the corresponding branch $\alpha$ is nondegenerate. According to Brouwer's theorem, this distribution of $\mathbf{D}_a$ vectors tangent to the sphere must have two stationary points for which $\mathbf{D}_a = 0$. On the other hand, relations (9) and (10) imply that this distribution also possesses an additional property: $\mathbf{D}_a(\mathbf{-m}) \,||\, \mathbf{D}_a(\mathbf{m})$. For this reason, the pair of points stipulated by Brouwer's theorem includes the inversion-equivalent stationary points $\mathbf{m}_0$ and $-\mathbf{m}_0$. Thus, the following theorem of existence of the transverse nonpiezoactivity directions is valid:

*In any crystal of unrestricted anisotropy each nondegenerate branch must contain at least one pair of inversion-equivalent zero-induction points* $\pm\boldsymbol{m}_0$ *such that* $\boldsymbol{D}_a(\pm\boldsymbol{m}_0) = 0$ *on the unit sphere.*

Therefore, the zero-induction points in a wave field $\mathbf{D}_\alpha(\mathbf{m})$ must exist even in triclinic crystals. Of course, the positions of such points in the general case (i.e., the solutions of Eq. (38) in the general form) cannot be found analytically. However, in some more symmetric crystals, nonpiezoactive directions $\mathbf{m}_0$ can be found without cumbersome computations.

## 5.2. Zero-induction points related to elements of the crystal symmetry

### *5.2.1. Longitudinal waves propagating along symmetry axes*

Let us consider a wave propagating along the direction $\mathbf{m}_0$, which coincides with a symmetry axis of any order except for dyad axes (e.g., this can be the 3, 4, $\bar{4}$, 6, or $\bar{6}$-fold axis). As is known (Fedorov, 1968), any symmetry axis (including a dyad axis) is a longitudinal normal. Evidently, the electric induction $\mathbf{D}_l(\mathbf{m}_0)$ accompanying the longitudinal wave $(\mathbf{u}_l \,||\, \mathbf{m}_0)$ must be zero, otherwise the $\mathbf{D}_l(\mathbf{m}_0)$ vector would possess two equivalent orientations, in contradiction with the single-valued third relation in (9):

$$\mathbf{D}_l = ik\hat{N}(\mathbf{m}_0)\mathbf{m}_0. \tag{39}$$

It should be noted that this argument does not work in the case of transverse branches. For the selected symmetry directions they are always degenerate, that is, possessing equal phase velocities $(v_t = v_{t'})$ and, hence, arbitrary orientations of $\mathbf{u}_{t,t'}$ and $\mathbf{D}_{t,t'}$ in the plane:

$$\{\mathbf{u}_{t,t'},\, \mathbf{D}_{t,t'}\} \perp \mathbf{m}_0. \tag{40}$$

The wave propagating along a dyad axis should be treated separately (albeit with the same result). In the general case, this direction is not an acoustic axis. On the other hand, the transverse isonormal vectors $\mathbf{D}_a \perp \mathbf{m}_0$ ($a = t, t', l$) are determined to within the sign (like $\mathbf{u}_a$) and, hence, their symmetry rotations due to the dyad axis cannot be considered as different solutions. So, one can readily check that, for a propagation direction along the dyad axis, the transverse branches $a = t, t'$ are again characterized by nonzero induction vectors. However, the longitudinal branch in the same direction always obeys the relation $\mathbf{D}_l(\mathbf{m}_0) = 0$. Indeed, combining Eqs. (10) and (39) for $\mathbf{m} = \mathbf{m}_0$, we obtain

$$\mathbf{D}_l / ik = \hat{N}(\mathbf{m}_0)\mathbf{m}_0 = (\hat{e}\mathbf{m}_0)\mathbf{m}_0 - \frac{(\hat{\varepsilon}\mathbf{m}_0)[(\mathbf{m}_0\hat{e}\mathbf{m}_0)\mathbf{m}_0]}{\mathbf{m}_0 \cdot \hat{\varepsilon}\mathbf{m}_0}. \tag{41}$$

Let us check that the right-hand part of this expression vanishes even for a monoclinic crystal of the class 2. Selecting the $z$ axis in (41) along the dyad axis $(2 \,||\, \mathbf{m}_0)$, we obtain

$$\mathbf{D}_l / ik = \left( e_{13} - \frac{\varepsilon_{13} e_{33}}{\varepsilon_{33}},\; e_{23} - \frac{\varepsilon_{23} e_{33}}{\varepsilon_{33}},\; 0 \right). \tag{42}$$

However, according to (Royer & Dieulesaint, 2000; Sirotin & Shaskolskaya, 1982), the off-diagonal components of $\hat{e}$ and $\hat{\varepsilon}$ tensors, entering into Eq. (42) for the symmetry class 2 in this coordinate system, are vanishing: $e_{13} = e_{23} = \varepsilon_{13} = \varepsilon_{23} = 0$ and, hence,

$$\mathbf{D}_l(\mathbf{m}_0 \,||\, 2) = 0. \tag{43}$$

Evidently, Eq. (43) is valid for all crystals of various classes possessing dyad axes. Thus, the following statement is valid:

*A longitudinal wave propagating along any axis of symmetry in a piezoelectric crystal is accompanied by an electric component with zero induction.*

For example, let us consider a piezoelectric crystal of the orthorhombic symmetry class 222. According to the above theorem, all three dyad axes in this crystal are the zero-induction directions $\mathbf{m}_0$ for the longitudinal modes. However, it can be shown that another four inversion-nonequivalent asymmetric directions $\mathbf{m}_0$ with zero induction ( $\mathbf{D}_l = 0$ ) may exist in a quasi-longitudinal branch of this crystal:

$$(\theta_0, \pm\varphi_0),\ (\theta_0, \pm\varphi_0 + \pi), \tag{44}$$

where the angles of spherical coordinate system are determined by approximate relations

$$\theta_0 = \operatorname{arccot}\sqrt{\frac{e_{36}\varepsilon_1\varepsilon_2}{(e_{14}\varepsilon_2 + e_{25}\varepsilon_1)\varepsilon_3}}, \qquad \varphi_0 = \arctan\sqrt{\frac{e_{25}\varepsilon_1}{e_{14}\varepsilon_2}}. \tag{45}$$

For simplicity, solutions (45) are written in the approximation of small piezoelectric moduli and weak elastic anisotropy. In this approximation, a necessary condition for the existence of the above series of zero induction points is that all the piezoelectric moduli entering into relations (45) must have the same sign (Fig. 2). It should be noted that cubic piezoelectric crystals (symmetry classes $\bar{4}3m$ and 23) are always described by Fig. 2b, since additional zero-induction directions (44), (45) always appear along the triad axes.

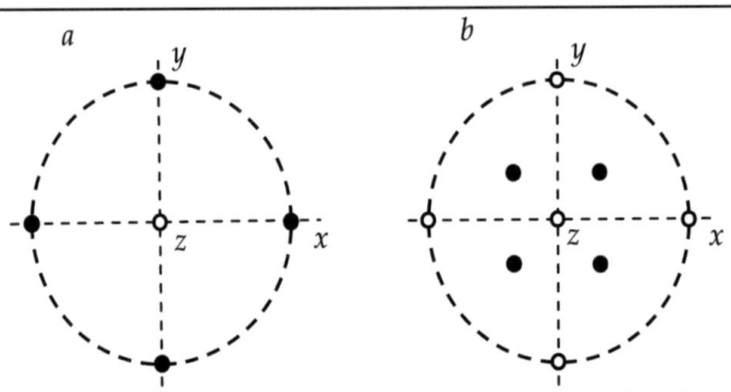


**Figure 2.** Diagrams of the directions of propagation of the transversely nonpiezoactive acoustic waves of quasi-longitudinal branch in crystals of the symmetry class 222. The stereographic projections are given for the cases when (a) the sign of the piezoelectric modulus $e_{36}$ is opposite to that of $e_{14}$ and/or $e_{25}$ and (b) all piezoelectric moduli have the same sign.

### *5.2.2. Transverse (SH) waves propagating in symmetry planes*

**Example 1: symmetry class *m*.** Let the *z* axis be perpendicular to the plane of symmetry of a monoclinic crystal ($z \perp m$) and consider the *t* branch of a wave propagating in this plane:

$$\mathbf{m} = (m_1,\ m_2,\ 0), \qquad \mathbf{u}_t \ || \ (0,\ 0,\ 1). \tag{46}$$

One can readily check that in such waves

$$\mathbf{D}_t \ || \ (0,\ 0,\ e_{35}m_1 + e_{34}m_2)u_t \ || \ \mathbf{u}_t. \tag{47}$$

Therefore, the symmetry plane always contains a single direction $\mathbf{m}_0$ corresponding to zero induction $\mathbf{D}_t$, which is determined by the azimuthal angle $\varphi_0$ (counted as in Fig. 1):

$$\mathbf{m}_0 = (m_{01},\ m_{02},\ 0), \qquad \tan\varphi_0 = m_{02} / m_{01} = -e_{35} / e_{34}. \tag{48}$$

The maximum amplitude of $\mathbf{D}_t$ in this plane corresponds to the direction

$$\mathbf{m}_{max} = (m_{02},\ -m_{01},\ 0), \tag{49}$$

which is perpendicular to $\mathbf{m}_0$.

This is the most general example. Thus, the following theorem is valid:

*In any symmetry plane there is always at least one direction for propagation of an SH wave with vanishing electric induction.*

The other examples below just specify orientations of the zero-induction direction in various symmetry classes.

**Example 2: symmetry class 3*m*.** In trigonal crystals, the situation with transverse nonpiezoactive directions for the $t$ waves in each of the three symmetry planes containing the triad axis is completely analogous to the above case of a monoclinic crystal. For example, in the $yz$ symmetry plane, relations (46)–(48) have to be replaced by

$$\mathbf{m} = (0,\ m_2,\ m_3), \qquad \mathbf{u}_t \ || \ (1,\ 0,\ 0), \tag{50}$$

$$\mathbf{D}_t \ || \ (-e_{22}m_2 + e_{15}m_3,\ 0,\ 0)u_t \ || \ \mathbf{u}_t, \tag{51}$$

$$\mathbf{m}_0 = (0,\ m_{02},\ m_{03}), \qquad \tan\theta_0 = \frac{m_{02}}{m_{03}} = \frac{e_{15}}{e_{22}} \tag{52}$$

where $\theta_0$ is the polar angle between $\mathbf{m}_0$ and the triad axis.

**Example 3: symmetry class *mm*2.** For a $t$ wave propagating in the $yz$ symmetry plane of an orthorhombic crystal, we have

$$\begin{gathered} \mathbf{m} = (0,\ m_2,\ m_3), \quad \mathbf{u}_t \ || \ (1,\ 0,\ 0), \\ \mathbf{D}_t \ || \ (e_{15}m_3,\ 0,\ 0)u_t \ || \ \mathbf{u}_t. \end{gathered} \tag{53}$$

Evidently, in this case

$$\mathbf{m}_0 = (0,\ 1,\ 0), \qquad \mathbf{m}_{\max} = (0,\ 0,\ 1). \tag{54}$$

Relations (53) and (54) are also valid for tetragonal crystals of the symmetry class $4mm$.

**Example 4: symmetry classes $\bar{4}2m$, $\bar{4}3m$, and 23.** For the $x$ axis parallel to the dyad axis $(x \parallel 2)$, transverse waves propagating in the diagonal plane (1, $\bar{1}$, 0) obey the relations

$$\mathbf{m} = (m_1,\ -m_1,\ m_3), \qquad \mathbf{u}_t \parallel (1,\ 1,\ 0). \tag{55}$$

These waves exhibit electric components with the amplitude of induction

$$\mathbf{D}_t \parallel e_{14} m_3 (1,\ 1,\ 0) u_t \parallel \mathbf{u}_t \tag{56}$$

and, hence, have the following special directions:

$$\mathbf{m}_0 = (1,\ -1,\ 0)/\sqrt{2}, \qquad \mathbf{m}_{\max} = (0,\ 0,\ 1). \tag{57}$$

The found above orientations of zero-induction direction $\mathbf{m}_0$ for a series of crystal classes are summarized in Table 2.

| Class of symmetry | $m$ | $3m$ | $mm2$, $4mm$ | $\bar{4}2m$, $\bar{4}3m$, 23 |
|---|---|---|---|---|
| Symmetry plane | $m \perp z$ | $m \parallel yz$ | $m \parallel yz$ | $m \parallel (1\bar{1}0)$ |
| Direction $\mathbf{m}_0$ | $\mathbf{m}_0 = (m_{01}, m_{02}, 0)$ $m_{02}/m_{01} = -e_{35}/e_{34}$ | $\mathbf{m}_0 = (0, m_{02}, m_{03})$ $m_{02}/m_{03} = e_{15}/e_{22}$ | $\mathbf{m}_0 = (0, 1, 0)$ | $\mathbf{m}_0 = (1, -1, 0)/\sqrt{2}$ |

**Table 2.** Propagation directions of transversely nonpiezoactive SH waves in the symmetry planes of crystals of various symmetry systems. Certainly, in Table 2 for crystals more symmetric than monoclinic ($m$) and containing other equivalent symmetry planes, the directions $\mathbf{m}_0$ of zero induction found are accordingly multiplied. For instance, in crystals of the orthorhombic ($mm2$) and tetragonal ($4mm$) classes there is also the symmetry plane $m \parallel xz$ where the corresponding transversely non-piezoactive direction is $\mathbf{m}_0 = (1, 0, 0)$.

In conclusion, let us consider the more exclusive case of hexagonal symmetry classes (14).

**Example 5: symmetry classes $6mm$ and $\infty m$ (14).** Any plane containing the principal symmetry axis in such a crystal is the plane of symmetry $m$. According to relations (27), the electric induction vector of $t$ waves propagating in such planes is orthogonal to $m$ and proportional to $m_3$. Therefore, $\mathbf{D}_t$ vanishes along the entire equator $m_3 = 0$:

$$\mathbf{D}_t(m_1,\ m_2,\ 0) = 0. \tag{58}$$

Note in passing that at the same equator for the same symmetry classes the other transverse branch ($t'$) polarized along the principal axis also forms a line of zero electric displacement:

$$\mathbf{D}_{t'}(m_1,\ m_2,\ 0) = 0. \tag{59}$$

# 6. Orientational singularities of the induction fields

The vector fields $\mathbf{D}_a(\mathbf{m})$, which are orthogonal to the wave normal **m**, may exhibit orientational singularity in the vicinity of directions of the two types: zero-induction points, where rotations are topologically allowed (Alshits, Lyubimov & Radowicz, 2005a, 2005b), and along acoustic axes, where inductions of degenerate branches, as a rule, do not vanish, but rotate together with the corresponding displacement vectors $\mathbf{u}_{0a}$ (Alshits *et al*, 1987).

Below we shall consider the both types of singularities concentrating our attention basically on the directions of transverse nonpiezoactivity.

## 6.1. General case of arbitrary anisotropy

### *6.1.1. Orientational singularities in the vicinities of zero-induction points*

As was mentioned above, the vector fields $\mathbf{D}_a(\mathbf{m})$ in the zero-induction points $\mathbf{m}_0$ may exibit rotations (Fig 3). Let us consider the $\mathbf{D}_a(\mathbf{m})$ function for $\mathbf{m} = \mathbf{m}_0 + \Delta\mathbf{m}$, where $\Delta\mathbf{m} \perp \mathbf{m}_0$ and $|\Delta\mathbf{m}| \ll 1$. Using condition (8) and the third relation in (9), we obtain

$$\mathbf{D}_a(\mathbf{m}) \;||\; \hat{N}(\mathbf{m})\bar{\hat{F}}_a(\mathbf{m})\mathbf{c}. \tag{60}$$

Taking into account that $\mathbf{D}_a(\mathbf{m}_0) = 0$, one has from (60) to a first approximation:

$$\mathbf{D}_a(\mathbf{m}) \;||\; \Delta\mathbf{m}\hat{Q}_a \equiv \Delta m_i \left\{ \frac{\partial}{\partial m_i} [\hat{N}(\mathbf{m})\bar{\hat{F}}_a(\mathbf{m})\mathbf{c}] \right\}_{\mathbf{m}=\mathbf{m}_0} . \tag{61}$$

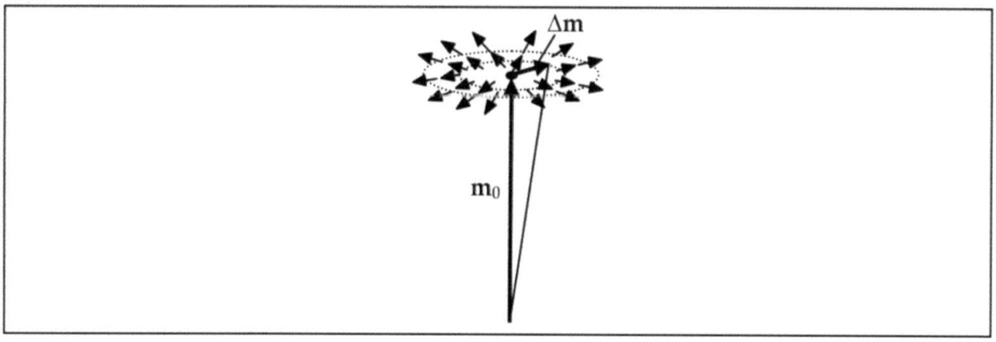


**Figure 3.** A singular vector distribution $\mathbf{D}_a(\mathbf{m})$ in the vicinity of a zero-induction point $\mathbf{m}_0$.

For the transverse $\mathbf{D}_a(\mathbf{m})$ field [see (11)], the asymmetric tensor entering into formula (61),

$$\hat{Q}_a = \nabla_{\mathbf{m}} \otimes \hat{N}(\mathbf{m})\bar{\hat{F}}_a(\mathbf{m})\mathbf{c}\,|_{\mathbf{m}=\mathbf{m}_0}, \tag{62}$$

must be planar, that is, its spectral expansion can be represented as a sum of two dyads:

$$\hat{Q}_a = \lambda_{a1}\tilde{\mathbf{e}}_{a1} \otimes \mathbf{e}_{a1} + \lambda_{a2}\tilde{\mathbf{e}}_{a2} \otimes \mathbf{e}_{a2}, \tag{63}$$

where $\lambda_{aj}$, $\tilde{\mathbf{e}}_{aj}$, and $\mathbf{e}_{aj}$ are the eigenvalues and eigenvectors (left and right) of the $\hat{Q}_a$ tensor ($\mathbf{e}_{a1}$ and $\mathbf{e}_{a2}$ must be orthogonal to $\mathbf{m}_0$). Note that the $\tilde{\mathbf{e}}_{aj}$ eigenvectors (in contrast to $\mathbf{e}_{aj}$) in the general case do not belong to a plane orthogonal to $\mathbf{m}_0$, but their components $\tilde{\mathbf{e}}_{aj}^{||}$ oriented along $\mathbf{m}_0$ are insignificant for our analysis.

Let us decompose each eigenvector into two components

$$\tilde{\mathbf{e}}_{aj} = \tilde{\mathbf{e}}_{aj}^{||} + \tilde{\mathbf{e}}_{aj}^{\perp}, \quad \tilde{\mathbf{e}}_{aj}^{||} \,||\, \mathbf{m}_0, \quad \tilde{\mathbf{e}}_{aj}^{\perp} = (\hat{I} - \mathbf{m}_0 \otimes \mathbf{m}_0)\tilde{\mathbf{e}}_{aj} \perp \mathbf{m}_0, \tag{64}$$

and form a more convenient matrix

$$\hat{Q}_a^{\perp} = (\hat{I} - \mathbf{m}_0 \otimes \mathbf{m}_0)\hat{Q}_a = \lambda_{a1}\tilde{\mathbf{e}}_{a1}^{\perp} \otimes \mathbf{e}_{a1} + \lambda_{a2}\tilde{\mathbf{e}}_{a2}^{\perp} \otimes \mathbf{e}_{a2}, \tag{65}$$

which will be used below instead of $\hat{Q}_a$:

$$\mathbf{D}_a \,||\, \Delta\mathbf{m}\hat{Q}_a^{\perp}. \tag{66}$$

Let the orientation angle $\Phi$ of the $\mathbf{D}_a(\mathbf{m})$ vector be measured from the $\mathbf{e}_{a1}$ direction, and the analogous angle $\varphi$ for $\Delta\mathbf{m}$ in the same plane, from the $\tilde{\mathbf{e}}_{a1}^{\perp}$ direction (Fig. 4). In these terms, we can write

$$\tan\Phi = \frac{\mathbf{D}_a \cdot \mathbf{e}_{a2}}{\mathbf{D}_a \cdot \mathbf{e}_{a1}} = \frac{\lambda_{a2}}{\lambda_{a1}}\frac{\Delta\mathbf{m} \cdot \tilde{\mathbf{e}}_{a2}^{\perp}}{\Delta\mathbf{m} \cdot \tilde{\mathbf{e}}_{a1}^{\perp}} = \frac{\lambda_{a2}}{\lambda_{a1}}\tan\varphi. \tag{67}$$

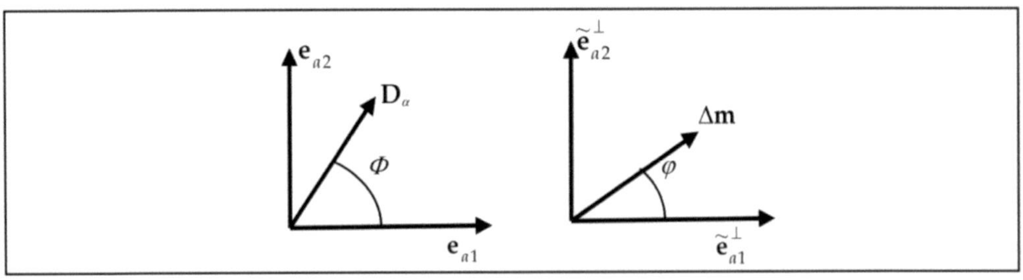


**Figure 4.** The angles of orientation of the $\mathbf{D}_a$ and $\Delta\mathbf{m}$ vectors in the plane orthogonal to $\mathbf{m}_0$.

Thus, the complete turn of $\Delta\mathbf{m}$ around $\mathbf{m}_0$ in the plane orthogonal to $\mathbf{m}_0$ implies the complete turn of $\mathbf{D}_a(\mathbf{m})$ in the same or in0 the opposite direction (depending on the sign of $\det\hat{Q}_a^{\perp} = \lambda_{a1}\lambda_{a2}$), which corresponds to the Poincaré index of the given singular point (Fig. 5)

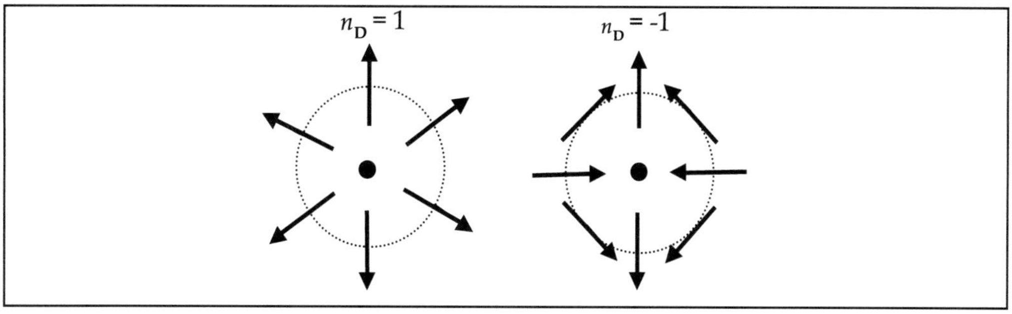


**Figure 5.** The two main types of singularities in the vicinity of zero points of the induction vector distribution in the normalized directed representation.

$$n_D = \operatorname{sgn}\det\hat{Q}_a^{\perp}. \tag{68}$$

The above considerations fail to be valid in particular cases, when one of the eigenvalues ($\lambda_{a1}$ or $\lambda_{a2}$) of matrices (63) and (65) vanishes. In such cases, $\det\hat{Q}_a^{\perp} = 0$, but formula (68) is not applicable. Indeed, let $\lambda_{a2} = 0$ at $\mathbf{m}_0$. Then,

$$\mathbf{D}_a \parallel \Delta\mathbf{m}\hat{Q}_a^{\perp} = \lambda_{a1}(\Delta\mathbf{m}\cdot\tilde{\mathbf{e}}_{a1}^{\perp})\mathbf{e}_{a1} \tag{69}$$

and a zero-induction line can pass via $\mathbf{m}_0$ in the direction of $\Delta\mathbf{m} \perp \tilde{\mathbf{e}}_{a1}^{\perp}$, but only provided that $\lambda_{a2} = 0$ is valid. In this situation, the very concept of the Poincaré index is inapplicable. However, if the vanishing of $\lambda_{a2}$ has a strictly local character and takes place only along $\mathbf{m}_0$, then we are dealing with a very special singularity analogous to a local-wedge degeneracy known in the theory of acoustic axes (Alshits, Sarychev & Shuvalov, 1985). It can be shown that a topological charge of the corresponding singularity in the $\mathbf{D}_a(\mathbf{m})$ field in this case can take one of three values: $n_D = 0, \pm1$. However, both situations (point and line) of this type with zero induction amplitude are very exclusive and never encountered in real (even symmetric) crystals. Below we will consider zero-induction lines of this kind in model crystals. However it will be demonstrated that the examples of the $\mathbf{D}_a = 0$ lines, existing in hexagonal crystals and described by Eqs. (58) and (59), belong to a different type.

On the other hand, the ordinary singular points (68) with indices $n_D = \pm1$ depicted in Fig. 5 are rather widely encountered in real crystals. For example, all directions $\mathbf{m}_0$ in Fig. 2b corresponding to orthorhombic (222) or cubic ($\bar{4}3m$ and 23) crystals are characterized by topological charges $n_D = \pm1$ (in Fig. 2, filled and empty circles correspond to +1 and –1, respectively). Figure 6 is a schematic diagram of the $\mathbf{D}_l(\mathbf{m})$ distribution in the central region of the circle in Fig. 2b.

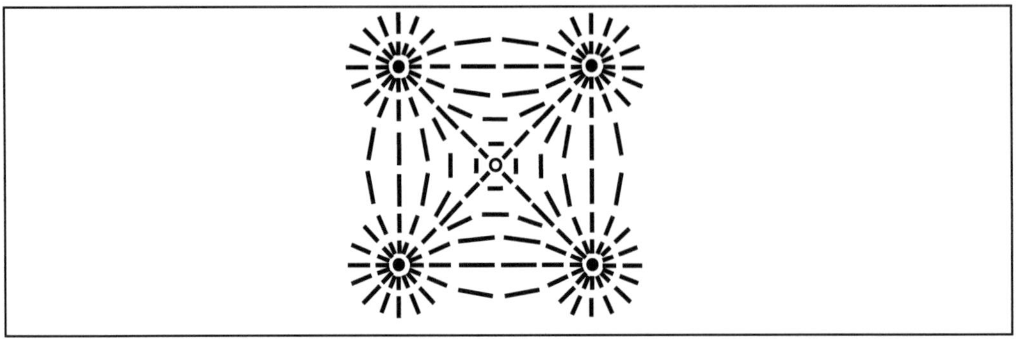

**Figure 6.** A schematic image of the $\mathbf{D}_l(\mathbf{m})$ vector field distribution over a group of five singular points in the central region of Fig. 2b in the representation of nondirected segments.

### 6.1.2. Orientational singularities in the $\mathbf{D}_a(\mathbf{m})$ fields around acoustic axes

Let us consider the vector polarization fields $\mathbf{u}_{01,2}(\mathbf{m})$ of degenerate branches in the vicinity of the direction $\mathbf{m}_d$ of the acoustic axis. In this region the considered vector distributions should be very close to the plane orthogonal to the unit polarization vector $\mathbf{u}_{03}(\mathbf{m}_d)$ of the non-degenerate isonormal eigenwave being singular at $\mathbf{m}_d$. Their rotations around the acoustic axis are equal to each other being described by the Poincaré index $n_\mathbf{u}$ which is determined by the type of the acoustic axis (Alshits *et al*, 1987). The appropriate fields of electric induction $\mathbf{D}_{1,2}(\mathbf{m})$ due to the coupling (9) $\mathbf{D}_a \,||\, \hat{N}\mathbf{u}_a$ have similar rotations characterized by the Poincaré index $n_\mathbf{D}$ which ordinarily may differ from $n_\mathbf{u}$ only by a sign. In order to find this sign we should take into consideration that the $\hat{N}$ matrix is degenerate ($\mathbf{m}\hat{N} = 0$). However, one can replace $\hat{N}$ by a matrix $\hat{N}'$ such that we have $\hat{N}'\mathbf{u} = \hat{N}\mathbf{u}$ for any $\mathbf{u} \perp \mathbf{u}_{03}$, but simultaneously $\det\hat{N}' \neq 0$. These conditions are satisfied by, for example, the matrix

$$\hat{N}' = \hat{N}(\mathbf{m}_d) + \mathbf{m}_d \otimes \mathbf{u}_{03}. \tag{70}$$

Indeed, in accordance with (Fedorov, 1968) we have $\det\hat{N}' = \mathbf{u}_{03} \cdot \hat{N}\mathbf{m}_d \neq 0$. Then following to (Alshits *et al*, 1987) one obtains

$$n_\mathbf{D} = n_\mathbf{u} \,\mathrm{sgn} \det\hat{N}'. \tag{71}$$

This equation is valid until $\hat{N}(\mathbf{m}_d) \neq 0$ which holds for any known acoustic axes except of the direction $\mathbf{m}_d \,||\, \bar{6}$ (see below).

It should be noted that, in contrast to the mutually orthogonal vectors $\mathbf{u}_{0a}(\mathbf{m})$ ($a = 1, 2, 3$), the three vectors $\mathbf{D}_a(\mathbf{m}) \perp \mathbf{m}$ are coplanar and generally unorthogonal in pairs. At the same time, it is clear that the vectors $\mathbf{D}_1(\mathbf{m})$ and $\mathbf{D}_2(\mathbf{m})$ are not collinear with any $\mathbf{m}$.

Consequently, these two vector fields are homotopic with each other. That is why they correspond to the same value of the index $n_{\rm D}$.

Now let us consider some examples of crystals belonging to particular symmetry systems.

## 6.2. Waves propagating along symmetry axes

### *6.2.1. Longitudinal waves along symmetry axes*

**Example 1.** For a longitudinal wave propagating along the $\mathbf{m}_0 \,||\, 2$ direction in a monoclinic crystal with dyad axis, we have

$$n_{\rm D} = \mathrm{sgn}\{(a_1 + b_1)(a_2 + b_2) - c_1 c_2\}. \tag{72}$$

where

$$a_1 = \frac{e_{14}c_{36}}{\Delta_{34}^-},\quad b_1 = \frac{e_{15}d_1}{\Delta_{35}^-} - \frac{e_{33}\varepsilon_1}{\varepsilon_3},\quad c_1 = \frac{e_{14}d_2}{\Delta_{34}^-} + \frac{e_{15}c_{36}}{\Delta_{35}^-}, \tag{73}$$

$$a_2 = \frac{e_{25}c_{36}}{\Delta_{35}^-},\quad b_2 = \frac{e_{24}d_2}{\Delta_{34}^-} - \frac{e_{33}\varepsilon_2}{\varepsilon_3},\quad c_2 = \frac{e_{25}d_1}{\Delta_{35}^-} + \frac{e_{24}c_{36}}{\Delta_{34}^-}, \tag{74}$$

$d_i = c_{i3} + c_{33}$. The $\hat{\varepsilon}$ tensor is assumed to be diagonal, which can be ensured by the appropriate choice of the $x$- and $y$ axes of the crystallographic coordinate system with the $z$ axis parallel to the dyad axis.

**Example 2.** For the $\mathbf{m}_0 \,||\, 2$ direction in an orthorhombic crystal belonging to the symmetry class *mm*2, we have

$$n_{\rm D} = \mathrm{sgn}(b_1 b_2). \tag{75}$$

**Example 3.** For the $\mathbf{m}_0 \,||\, 2 \,||\, z$ direction in an orthorhombic crystal belonging to the symmetry class 222, we have

$$n_{\rm D}^{(z)} = \mathrm{sgn}(a_1 d_2 \,/\, a_2 d_1). \tag{76}$$

Analogous formulas for the $n_{\rm D}^{(x)}$ and $n_{\rm D}^{(y)}$ are obtained from (76) by cyclic rearrangement of the indices. Note that, in the isotropic limit, we obtain

$$d_1 \,/\, \Delta_{35}^- = d_2 \,/\, \Delta_{34}^- \to 2, \tag{77}$$

and only a very large elastic anisotropy can change the signs of the ratios in formula (76). Therefore these signs for most orthorhombic crystals are determined only by the piezoelectric moduli: $n_{\rm D}^{(z)} = \mathrm{sgn}(e_{14}e_{25})$.

**Example 4.** As can be readily checked, for the principal symmetry axes in crystals of the symmetry classes 422, 622, ∞22, 4*mm*, 6*mm*, ∞*mm*, 4, 6, ∞, 32, 3*m*, and 3 we have

$$n_{\mathrm{D}} = 1, \tag{78}$$

and in crystals of the symmetry classes $\bar{4}2m$, $\bar{4}$, $\bar{4}3m$, and 23,

$$n_{\mathrm{D}} = -1. \tag{79}$$

### 6.2.2. *Degenerate transverse waves along symmetry axes*

In accordance with Eq. (71), the transverse waves propagating along directions near acoustic axes, which coincide with symmetry axes in piezoelectric crystals, are characterized by rotations of both polarization fields $\mathbf{u}_{01,2}(\mathbf{m})$ and accompanied induction fields $\mathbf{D}_{1,2}(\mathbf{m})$. The direct analysis for various types of symmetry axes gives the related Poincaré indices $n_{\mathbf{u}}$ and $n_{\mathrm{D}}$ shown in Table 3.

| N | ∞, 6 | | $\bar{6}$ | | 4 | | $\bar{4}$ and 2 ∈ 23 | | 3 | |
|---|---|---|---|---|---|---|---|---|---|---|
| branch | trans | long | trans | long | trans | long | trans | long | trans | long |
| $n_{\mathbf{u}}$ | 1 | 0 | 1 | 0 | ±1 | 0 | ±1 | 0 | -1/2 | 0 |
| $n_{\mathrm{D}}$ | 1 | 1 | -2 | -2 | $n_{\mathbf{u}}$ | 1 | $-n_{\mathbf{u}}$ | -1 | -1/2 | 1 |

**Table 3.** The Poincaré indices of vector fields of polarizations and inductions along acoustic axes coinciding with symmetry axes *N* for transverse and longitudinal wave branches. The sign of $n_{\mathbf{u}} = \pm 1$ when it is not universal may be found from the equations given in (Alshits, Sarychev & Shuvalov, 1985; Alshits & Shuvalov, 1987; Shuvalov, 1998). The indices $n_{\mathrm{D}}$ for the direction of a $\bar{6}$-fold symmetry axis are found in the next sub-section.

### 6.2.3. *The both types of induction singularities along a* $\bar{6}$*-fold symmetry axis*

An interesting configuration of the vector fields $\mathbf{D}_a(\mathbf{m})$ arises near an acoustic axis $\mathbf{m}_d \,||\, \bar{6}$. In this case we have an exclusive situation $\hat{N}(\mathbf{m}_d) = 0$. So, by Eq. $(9)_3$, along the direction $\mathbf{m}_d$ the induction components vanish, $\mathbf{D}_a(\mathbf{m}_d) = 0$, in all branches, both degenerate ($a = 1,2 \equiv t,t'$) and nondegenerate ($a = 3 \equiv l$). Accordingly, in the vicinity of the $\bar{6}$-fold symmetry axis these fields should be small. Let us consider the direction

$$\mathbf{m} = \mathbf{m}_d + \Delta\mathbf{m} \tag{80}$$

where

$$\begin{aligned} &\mathbf{m}_d = (0,\ 0,\ 1) \,||\, \bar{6}, \qquad \Delta\mathbf{m} = \mu\boldsymbol{\rho}(\varphi), \\ &\boldsymbol{\rho}(\varphi) = (\cos\varphi,\ \sin\varphi,\ 0), \quad 0 \le \mu << 1. \end{aligned} \tag{81}$$

In these terms, the vector fields $\mathbf{D}_a(\mathbf{m})$, in the main order, have the following form

$$\begin{aligned} \mathbf{D}_l &= \mu^2(d/\Delta_{34})\{e_{11}\boldsymbol{\rho}(-2\varphi) + e_{22}[\boldsymbol{\rho}(-2\varphi)\times\mathbf{m}_0]\}, \\ \mathbf{D}_t &= \mu\{e_{22}\boldsymbol{\rho}(-2\varphi) - e_{11}[\boldsymbol{\rho}(-2\varphi)\times\mathbf{m}_0]\}, \\ \mathbf{D}_{t'} &= \mu\{e_{11}\boldsymbol{\rho}(-2\varphi) + e_{22}[\boldsymbol{\rho}(-2\varphi)\times\mathbf{m}_0]\}. \end{aligned} \tag{82}$$

Here the notation is introduced: $d = c_{13} + c_{44}$, $\Delta_{34} = c_{33} - c_{44}$. We note that in the symmetry classes $\bar{6}m2$ ($m \perp X_1$) and $\bar{6}2m$ ($2 \parallel X_1$) one can put in (82), respectively, $e_{11} = 0$ and $e_{22} = 0$.

It is easily seen that the induction vectors $\mathbf{D}_t$ and $\mathbf{D}_{t'}$ of the degenerate branches are mutually orthogonal, while their absolute value is proportional to $|\Delta\mathbf{m}|$ and does not depend on the orientation $\varphi$ of the vector $\Delta\mathbf{m}$. The induction vector $\mathbf{D}_l$ of the non-degenerate quasi-longitudinal branch has much smaller length, $|\mathbf{D}_l| \propto |\Delta\mathbf{m}|^2$, which also does not depend on $\varphi$. The direction of $\mathbf{D}_l$ in the accepted main order coincides with that of $\mathbf{D}_{t'}$. Certainly, in the next approximation this coincidence disappears. It is clear from (82) that during the full rotation of the vector $\Delta\mathbf{m}$ around the axis $\mathbf{m}_d \parallel \bar{6}$ each of three vectors $\mathbf{D}_\alpha$ twice circumvents the same axis in opposite direction. This means that the point $\mathbf{m}_d \parallel \bar{6}$ in all three vector fields is characterized by the Poincaré index $n_D = -2$ (Table 3). The corresponding singular configuration for one of these vector fields is shown in Fig. 7.

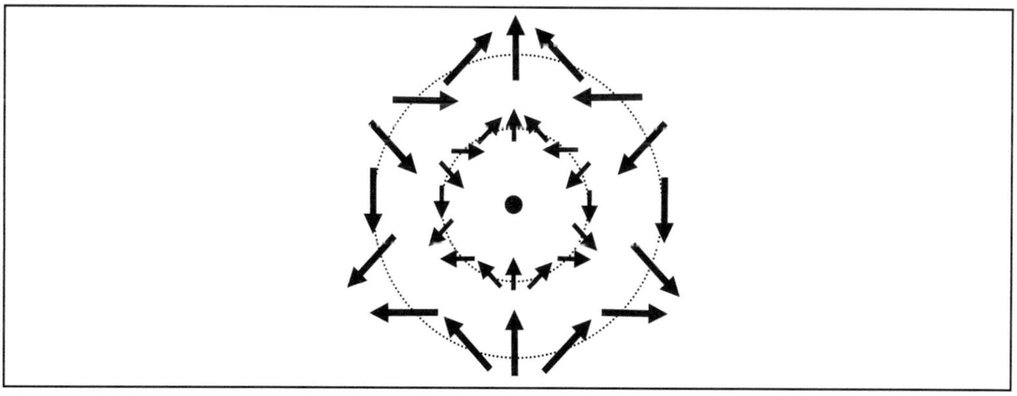

**Figure 7.** Vector induction field $\mathbf{D}_a$, $a = t, t'$ or $l$, near the acoustic axis $\mathbf{m}_d \parallel \bar{6}$ related to the Poincaré index $n_D = -2$ (top view of the plane orthogonal to $\mathbf{m}_d$; the central point corresponds to the direction of $\mathbf{m}_d$).

### 6.3. Transverse (SH) waves propagating in symmetry planes

The directions of transverse nonpiezoactivity in symmetry planes (Table 2) are also characterized by rotations $n_D$ in the induction vector fields of appropriate SH wave branches. We will not write lengthy expressions determining the choice between $n_D = \pm 1$

indices for the waves along $\mathbf{m}_0$ directions in monoclinic and trigonal crystals (see relations (49) and (52), respectively) and instead start our analysis from orthorhombic crystals.

**Example 1.** For the transverse acoustic waves (53) and (54) propagating in the vicinity of the zero-induction direction $\mathbf{m}_0 = (0, 1, 0)$ in an orthorhombic crystal belonging to the symmetry class $mm2$, the singular induction field is characterized by the Poincaré index

$$n_{\mathbf{D}} = \operatorname{sgn}\left[\frac{e_{32}(c_{12}+c_{66})/\Delta_{26}^{-} - e_{31}}{e_{15}}\right] \tag{83}$$

**Example 2.** For the same direction in a tetragonal crystal belonging to the symmetry class $4mm$, we have

$$n_{\mathbf{D}} = \operatorname{sgn}\left[\frac{e_{31}(c_{12}-c_{11}+2c_{66})}{e_{15}\Delta_{16}^{-}}\right] \tag{84}$$

**Example 3.** For the direction $\mathbf{m}_0 = (1, -1, 0)/\sqrt{2}$ in a crystal belonging to the other tetragonal symmetry class $\bar{4}2m$, we obtain

$$n_{\mathbf{D}} = -\operatorname{sgn}(e_{14}/e_{36}). \tag{85}$$

**Example 4.** For the same direction in a cubic crystals of the symmetry class $\bar{4}3m$ or 23 in the diagonal symmetry plane, as well as for the symmetry-equivalent direction $\mathbf{m}_0 = (-1, 1, 0)/\sqrt{2}$, we have (for any combinations of the moduli)

$$n_{\mathbf{D}} = -1 \tag{86}$$

The results presented in this subsection are summarized in Table 4.

| Classes of symmetry | $mm2$ | $4mm$ | $\bar{4}2m$ | $\bar{4}3m$, 23 |
|---|---|---|---|---|
| Symmetry plane | $m \parallel yz$ | | $(1\bar{1}0)$ | |
| Direction of propagation | $\mathbf{m}_0 = (0, 1, 0)$ | | $\mathbf{m}_0 = (1, -1, 0)/\sqrt{2}$ | |
| Poincare index $n$ | $\operatorname{sgn}\left(\frac{e_{32}}{e_{15}}\kappa_2 - \frac{e_{31}}{e_{15}}\right)$ | $\operatorname{sgn}\left(\frac{e_{31}}{e_{15}}(\kappa_1 - 1)\right)$ | $\operatorname{sgn}\left(-\frac{e_{14}}{e_{36}}\right)$ | −1 |

**Table 4.** Topological charges of **D** fields for SH acoustic waves propagating in symmetry planes of various crystals. In this table the notation is introduced:

$$\kappa_p = \frac{c_{12}+c_{66}}{c_{pp}-c_{66}} \tag{87}$$

### 6.4. Special types of singularities

The general analysis in Sec. 6.1 is exhaustive only provided that the $\hat{Q}_a$ tensor (62) is nonzero. As was shown above, usual systems have $\hat{Q}_a \neq 0$. However, in some very exclusive cases, this tensor may vanish in some special directions because of high symmetry or as a result of vanishing of certain combinations of the material tensor components. In such cases, the general expressions are very lengthy and we only present here some final results. For $\hat{Q}_a(\mathbf{m}_0) = 0$, the distribution of the induction vector field in the vicinity of $\mathbf{m}_0$ has four additional variants depicted in Fig. 8. The first three of them correspond to isolated singular points with the Poincaré indices $n_D = 0, \pm 2$ (Figs. 8a–8c), while the fourth variant corresponds to the existence of a $\mathbf{D} = 0$ line passing via the $\mathbf{m}_0$ point (Fig. 8d). This very situation is observed on the equator $m_3 = 0$ for the transverse tangentially polarized $t$ waves (58) in all transverse-isotropic media (13)–(15) and for the transverse $t'$ waves (59) polarized along the principal symmetry axis in the media of symmetry classes $6mm$ and $\infty m$. The only known alternative example of $\hat{Q}_a$ matrix vanishing is offered by a crystal with hexad axis $\bar{6}$. In this case, all three wave branches have along the direction $\bar{6}$ the identical singularities with $n_D = -2$ (Figs. 7 and 8c).

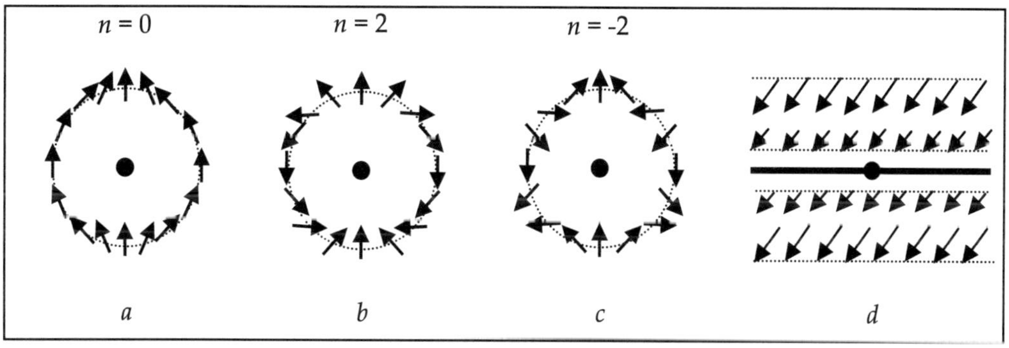


**Figure 8.** Four possible types of the $\mathbf{D}_a(\mathbf{m})$ vector field distribution around a zero-induction point $\mathbf{m}_0$, where $\mathbf{D}_a(\mathbf{m}_0) = \hat{Q}_a(\mathbf{m}_0) = 0$.

#### *6.4.1. A model crystal of the symmetry class mm2*

Let us assume that one piezoelectric modulus in the crystal under consideration is much smaller than the other moduli. In particular, we consider a conventional crystallographic coordinate system with the $x$ and $y$ axes perpendicular to the symmetry planes and the $z$ axis parallel to the dyad axis, in which

$$|e_{31}| << |e_{15}|, |e_{24}|, |e_{32}|, |e_{33}|. \tag{88}$$

One can readily check that, in a zero-order approximation with $e_{31} = 0$, a quasi-longitudinal nondegenerate wave branch along the $\mathbf{m}_0 = (1, 0, 0)$ direction features the above special situation, whereby simultaneously $\mathbf{D}_l(\mathbf{m}_0)=0$ and $\hat{Q}_l(\mathbf{m}_0)=0$. In this case, the $\mathbf{D}_l(\mathbf{m})$ vector field distribution in the $yz$ plane in the vicinity of $\mathbf{m}_0$ is described by the expression

$$\mathbf{D}_l \parallel \{0,\ g_2\sin 2\varphi,\ g_1+\varepsilon_1 e_{32}\gamma_2-(g_1-\varepsilon_1 e_{32}\gamma_2)\cos 2\varphi\} \tag{89}$$

where $\varphi$ is a polar angle of the $\Delta\mathbf{m} = \mathbf{m} - \mathbf{m}_0$ direction measured from the $y$ axis in the $yz$ plane and

$$\begin{gathered} g_1=\gamma_1(\varepsilon_1 e_{33}-\varepsilon_3 e_{15}), \\ g_2=(\gamma_1+\gamma_2)\varepsilon_1 e_{24}-\varepsilon_2 e_{15}, \\ \gamma_1=\bar{d}_5/\bar{\Delta}_{15},\ \gamma_2=d_6/\Delta_{16}, \\ \bar{d}_5=c_{13}+\bar{c}_{55},\ \bar{\Delta}_{15}=c_{11}-\bar{c}_{55}, \\ \bar{c}_{55}=c_{55}+e_{15}^2/\varepsilon_1,\ d_6=c_{12}+c_{66}. \end{gathered} \tag{90}$$

Expression (89) shows that, depending on the material constants, the $\mathbf{D}_l(\mathbf{m})$ field always corresponds to one of the possible variants depicted in Fig. 8. The Poincaré indices for the point singularities corresponding to Figs. 8a–8c are as follows:

$$n_l=\begin{cases} 0, & g_1 e_{32}\gamma_2>0, \\ 2\operatorname{sgn}[(g_1-\varepsilon_1 e_{32}\gamma_2)g_2], & g_1 e_{32}\gamma_2<0. \end{cases} \tag{91}$$

A condition for the existence of zero-induction lines in the $\mathbf{D}_l(\mathbf{m})$ field for Fig. 8d is

$$g_1 e_{32}\gamma_2=0 \quad \text{or} \quad g_2=0,\ \ g_1 e_{32}\gamma_2<0. \tag{92}$$

According to expression (89), a zero-induction line for $g_1=0$ passes via the vector $\mathbf{m}_0$ along the $z$ axis (Fig. 9a). For $e_{32}\gamma_2=0$, a similar zero-induction line is directed along the $y$ axis (Fig. 9b). If $g_1=0$ simultaneously with $e_{32}\gamma_2=0$, the two lines coexist (Fig. 9c).

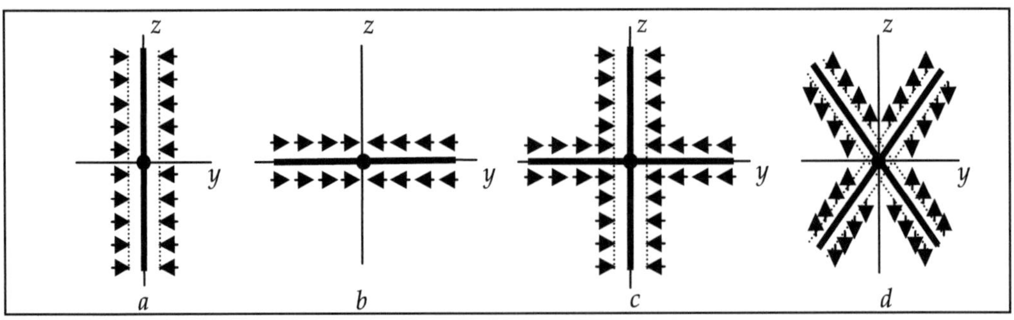


**Figure 9.** Four possible types of the $\mathbf{D}_l(\mathbf{m})=0$ lines in a model orthorhombic crystal.

Finally, when $g_2 = 0$ and $g_1 e_{32}\gamma_2 < 0$, the system features an oblique cross of zero-induction lines (Fig. 9d) with the mutual orientation determined by the equation

$$\cos 2\varphi = \frac{g_1 + \varepsilon_1 e_{32}\gamma_2}{g_1 - \varepsilon_1 e_{32}\gamma_2}. \quad (93)$$

### 6.4.2. *Behavior of point singularities in response to perturbations in the material moduli*

The point singularities of various types in vector fields $\mathbf{D}_a(\mathbf{m})$ behave differently in response to perturbations in the material moduli: they shift, split, or disappear. An analysis of this situation, analogous to that carried out by Alshits, Sarychev & Shuvalov (1985), showed that singularities with $n_D = \pm 1$ (Fig. 5) are topologically stable and can only be displaced by such perturbations. The singular points of other types (Figs. 8a–8c) are unstable and either split (in accordance with the law of topological charge conservation) or disappear (provided only that $n_D = 0$). The zero-induction lines (Fig. 8d and Fig. 9) are also unstable and disappear either completely or leaving a certain number of isolated zero points.

**Example 1.** The above general properties can be illustrated by a particular example using a model crystal of the symmetry class *mm*2 with a small modulus $e_{31}$ described above. It should be recalled that relations (89)–(93) were obtained in the zero-order approximation for $e_{31} = 0$. In the next order with respect to the small parameter $e_{31}$, the initial singularity along the direction $\mathbf{m}_0 = (1, 0, 0)$ exhibits splitting so as to form two or four singular points:

$$\mathbf{m}_0 + \delta\mathbf{m} = \begin{cases} (1, \ \pm\mu_2, \ 0), & \mu_2^2 = -e_{31} / e_{32}\gamma_2, \\ (1, \ 0, \ \pm\mu_3), & \mu_3^2 = -e_{31}\varepsilon_1 / g_1. \end{cases} \quad (94)$$

As is seen from Eq.(91) with $g_1 e_{32}\gamma_2 < 0$, the zero-order approximation along $\mathbf{m}_0$ corresponds to a singularity with $n_D = +2$ or $-2$. The introduction of a small $e_{31}$ modulus leads to a symmetric splitting of this singularity into a pair of zero-induction points with equal indices $n_D = +1$ or $-1$ along the *y* or *z* axis, depending on the sign of $e_{31} / g_1$ (Fig. 10a).

For $g_1 e_{32}\gamma_2 > 0$, when the initial topological charge in the zero-order approximation is zero, the perturbed pattern comprises either four singularities with a zero total index $n_D$ (for $e_{31} / g_1 < 1$) or none of them (which corresponds to the absence of zero-induction points in the vicinity of the given direction for $e_{31} / g_1 > 1$) as depicted in Fig. 10b.

**Example 2.** It should be noted that the discussed splitting of unstable singularities is by no means reduced to abstract mathematical games. Perturbations in the material moduli of real crystals are frequently caused by various external factors such as electric fields, mechanical stresses, or temperature fluctuations arising in the vicinity of phase transitions. For example, the phase transition from a crystal of the symmetry class $\bar{6}2m$, $\bar{6}m2$, or $\bar{6}$ to a trigonal crystal of the symmetry class 32, 3*m*, or 3, respectively, leads to replacement of the hexad axis by a triad axis. Simultaneously, in accordance with Table 3, the polarization singularity $n_{\mathbf{u}} = 1$ in the vector fields $\mathbf{u}_{0t}$ and $\mathbf{u}_{0t'}$ of degenerate branches along this direction radically changes into the point with the index $n_{\mathbf{u}} = -1/2$. And the induction vector fields $\mathbf{D}_t$ and $\mathbf{D}_{t'}$ of the same branches transform the singular pattern of the index $n_{\mathrm{D}} = -2$ into that with the index $n_{\mathrm{D}} = -1/2$. The requirement of the conservation of the topological charge (Poincaré index) is realized in the appearance of three additional conical acoustic axes with indices $n_{\mathbf{u}} = 1/2$ (Alshits, Sarychev & Shuvalov, 1985) and $n_{\mathrm{D}} = -1/2$ (Fig. 11a, b).

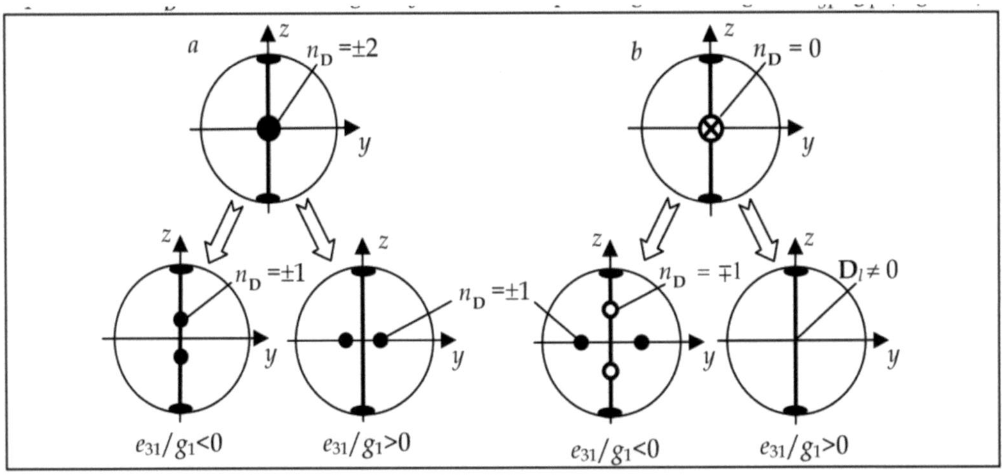


**Figure 10.** Diagrams illustrating the splitting of singular points with $n_{\mathrm{D}} = \pm 2$ (*a*) and $n_{\mathrm{D}} = 0$ (*b*) in a model crystal of the symmetry class *mm*2.

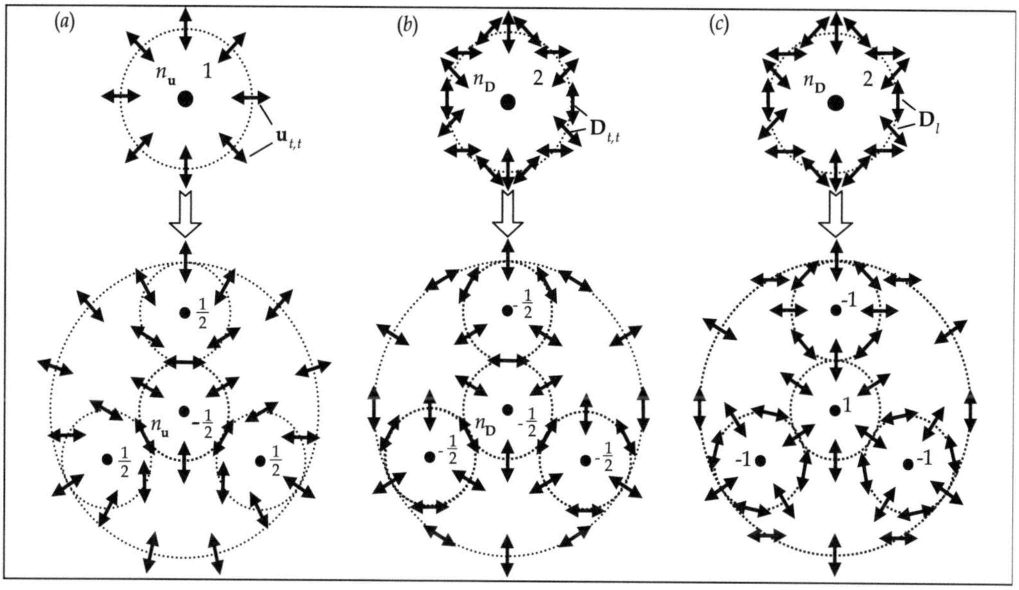


**Figure 11.** The three topological transformations in the vector fields $\mathbf{u}_{0t,t'}(\mathbf{m})$, $\mathbf{D}_{t,t'}(\mathbf{m})$ and $\mathbf{D}_l(\mathbf{m})$ (in the non-directed representation) after the phase transition $\overline{6} \rightarrow 3$.

In the nondegenerate quasi-longitudinal branch the index $n_D = -2$ is replaced after the transition by $n_D = 1$. In accordance with the same conservation law of the Poincaré index and with the final crystal symmetry, three additional zero-induction points $\mathbf{D}_l = 0$ with the indices $n_D = -1$ must be created along with the central zero-point (Fig. 11c). Thus the considered phase transition ($\overline{6} \rightarrow 3$) causes the three different topological transformations in the vector fields $\mathbf{u}_{0t,t'}(\mathbf{m})$, $\mathbf{D}_{t,t'}(\mathbf{m})$ and $\mathbf{D}_l(\mathbf{m})$ near the principal symmetry axis (Fig.11).

## 7. Conclusions

Two electric components, the electric field **E** and the electric induction **D**, accompanying a bulk acoustic wave which propagates in a piezoelectric medium, exhibit significantly different properties. The electric field is always purely longitudinal, whereas the electric induction vector is, in contrast, always purely transverse. On the unit sphere ($\mathbf{m}^2 = 1$) of wave propagation directions, the directions of zero electric field (**E** = 0) form lines, while the zero-induction directions (**D** = 0) are usually isolated and appear as singular points of the tangential vector field **D**(**m**) orientations. The nonpiezoactive directions of both types exist practically in all (even triclinic) crystals, although the presence of crystal symmetry elements is the additional factor determining the appearance of such directions.

The topological singularities of the $\mathbf{D}_a(\mathbf{m})$ vector fields in the vicinity of zero-induction points in most crystals are characterized by the Poincaré indices $n_D = \pm 1$, where the sign coincides with that of the determinant of the matrix (62) [see also Eq. (68)]. However, in some specific cases, this tensor may vanish ($\hat{Q}_a = 0$) in some special directions because of a high symmetry or as a result of vanishing of certain combinations of the material tensor components. In this case, the system has either an isolated zero-induction point $\mathbf{m}_0$ (and has the Poincaré indices $n_D = 0, \pm 2$) or a zero-induction line. Such special orientations are topologically unstable and, in response to any change in the anisotropy, either split into stable points with $n_D = \pm 1$ or disappear.

It is interesting to note that singularities of the induction vector field $\mathbf{D}_a(\mathbf{m})$ in the vicinity of the zero-induction points substantially differ from analogous singularities near the acoustic axes (see Eq. (71) and Table 3). According to (Alshits *et al*, 1987), stable singularities in the latter case are characterized by the Poincaré indices $n_D = \pm 1/2$, while the unstable ones have $n_D = 0, \pm 1$. The only exception to this rule is the acoustic axis along the hexad axis $\bar{6}$, for which the all branches, both degenerate $a = t, t'$ and non-degenerate $a = l$, are characterized by $\hat{Q}_a = 0$, $\mathbf{D}_a = 0$ and $n_D = -2$. However, as we have seen, in the latter case the transformation of the same singularity $n_D = -2$ due to the phase transition $\bar{6} \to 3$ has radically different topology [see Fig. 11 (*b*) and (*c*)].

## Author details

V.I. Alshits
*A.V. Shubnikov Institute of Crystallography, Russian Academy of Sciences, Moscow, Russia*
*Polish-Japanese Institute of Information Technology, Warsaw, Poland*

V.N. Lyubimov
*A.V. Shubnikov Institute of Crystallography, Russian Academy of Sciences, Moscow, Russia*

A. Radowicz
*Kielce University of Technology, Kielce, Poland*

## Acknowledgement

This study was performed within the framework of the Agreement on Cooperation between the Shubnikov Institute of Crystallography (Russia) and the Kielce University of Technology (Poland). V.I.A. and V.N.L. are grateful to the Kielce University of Technology for a hospitality and support.

## 8. References

Alshits, V.I. & Lothe, J. (1979). Elastic waves in triclinic crystals I, II, and III. *Kristallografiya*, Vol. 24, No. 4, 6 (Aug., Dec. 1979) 972-993, 1122-1130, ISSN 0023-4761 [*Sov. Phys. Crystallography*, Vol. 24, No. 4, 6 (1979) 387-398, 644-648, ISSN 1063-7745]

Alshits, V.I. & Lyubimov, V.N. (1990). Acoustic waves with extremal electro- (magneto-) mechanical coupling in piezocrystals. *Kristallografiya,* Vol. 35, No. 6 (Dec. 1990) 1325-1327, ISSN 0023-4761 [*Sov. Phys. Crystallography,* Vol. 35, No. 6 (1990) 780-782, ISSN 1063-7745]

Alshits, V.I.; Lyubimov, V.N. & Radowicz, A. (2005a). Special features of the electric components of acoustic waves in the vicinity of nonpiezoactive directions in crystals. *Zh. Eksp. Teor. Fiz.,* Vol. 128, No. 1 (Jan. 2005) 125-138, ISSN 0044-4510 [*JETP,* Vol. 101, No. 1 (2005) 107-119, ISSN 1063-7761]

Alshits, V.I.; Lyubimov, V.N. & Radowicz, A. (2005b). Non-piezoactivity in piezoelectrics: basic properties and topological features. *Arch. Appl. Mech.,* Vol. 74, No. 11-12 (Dec. 2005) 739-745, ISSN 0939-1533

Alshits, V.I.; Lyubimov, V.N.; Sarychev, A.V. & Shuvalov, A.L. (1987). Topological characteristics of singular points of the electric field accompanying sound propagation in piezoelectrics. *Zh. Eksp. Teor. Fiz.,* Vol. 93, No. 2 (8) (Aug. 1987) 723-732, ISSN 0044-4510 [*Sov. Phys. JETP,* Vol. 66, No. 2 (8) (1987) 408-413, ISSN 1063-7761]

Alshits, V.I.; Sarychev, A.V. & Shuvalov, A.L. (1985). Classification of degeneracies and analysis of their stability in the theory of elastic waves in crystals. *Zh. Eksp. Teor. Fiz.,* Vol. 89, No. 3(9) (Sept. 1985) 922-938, ISSN 0044-4510 [*Sov. Phys. JETP,* Vol. 62, No. 3 (1985) 531-539, ISSN 1063-7761]

Alshits, V.I. & Shuvalov, A.L. (1988). On acoustic axes in piezoelectric crystals. *Kristallografiya,* Vol. 33, No. 1 (Jan. 1988) 7-12, ISSN 0023-4761 [*Sov. Phys. Crystallography,* Vol. 33, No. 1 (1988) 1-4, ISSN 1063-7745]

Balakirev, M.K. & Gilinskii, I.A. (1982). *Waves in Piezoelectric Crystals,* Nauka, ISBN, Novosibirsk [in Russian]

Fedorov, F.I. (1968). *Theory of Elastic Waves in Crystals,* Plenum Press, ISBN, New York

Gulyaev, Yu.V. (1998). Review of shear surface acoustic waves in solids. *IEEE Trans. Ultrason. Ferroel. Freq. Control,* Vol. 45, No. 4 (April 1998) 935-938, ISSN 0885-3010

Holm, P. (1992). Generic elastic media. *Phys. Scr.,* Vol. T44 (1992) 122-127, ISSN 0031-8949

Landau, L.D. & Lifshitz, E.M. (1984). *Electrodynamics of Continuous Media,* Pergamon Press, ISBN 0080302750, New York

Lyamov, V.E. (1983). *Polarization Effects and Interaction Anisotropy of Acoustic Waves in Crystals,* Moscow State University, ISBN, Moscow [in Russian]

Lyubimov, V.N. (1969). Consideration of the piezoelectric effect in the theory of elastic waves for crystals of different symmetries. *Doklady AN SSSR,* Vol. 186, No. 5 (May 1969) 1055-1058, ISSN 0869-5652 [Sov. Phys. Doklady, Vol. 14, No. 5 (1969) 567-570, ISSN 1085-1992]

Royer, D. & Dieulesaint, E. (2000). Elastic *Waves in Solids*. Springer, ISBN 3-540-65932-3, Berlin

Shuvalov, A.L. (1998). Topological features of polarization fields of plane acoustic waves in anisotropic media. *Proc. R. Soc. Lond. A*, Vol. 454, (Nov. 1998) 2911-2947, ISSN 1471-2946

Sirotin Yu.I. & Shaskolskaya, M.P. (1979). *Fundamentals of Crystal Physics* (in Russian), Nauka, Moscow [(1982) translation into English, Mir, ISBN, Moscow]

Chapter 2

# Crack Detecting via Newton's Method

Kuo-Ming Lee

Additional information is available at the end of the chapter

http://dx.doi.org/10.5772/55876

## 1. Introduction

Crack detecting is an important issue in many areas of science and engineering. Often it is dangerous or it is only laborious to find the cracks directly, e.g., the detection of cracks within the nuclear power plants or of the cracks buried beneath the earth. One possible way out is to make use of the phenomenon of wave scattering. Sending an incident wave into the area in which we are interested, the information about the crack is hidden in the scattered wave that can be measured at some convenient place distance away. Thus, crack detection is equivalent to the extraction of the hidden information from the measured data. This is what we want to deal with here.

To understand the extraction of information, one has to realize the constitution of the data first. This leads to the concept of *inverse problems*. Following Keller [1], two problems are inverse to each other if the formulation of each of them requires all or partial knowledge of the other. Often, for historical reasons, one of the two problems has been studied extensively for some time, while the other is newer and not so well understood. In such cases, the former problem is called the direct problem, while the latter is called the inverse problem.

The direct problem in our case is the computation of the scattered wave from a prescribed crack. From the aspect of wave scattering, the space setting for our task is the unbounded domain outside the crack. This is certainly not a good place to begin with. We therefore employ the method of boundary integral equations which has the advantage of transforming a physical problem in the unbounded domain into one at the boundary which is just the crack itself. Thus, the boundary integral equation approach is advantageous both from a theoretical and a numerical point of view. On one hand, the boundary integral equations method allows an elegant and concise analysis of the unique solvability and the stability of the solution based on the powerful Riesz theory from functional analysis. On the other hand, this approach reduces the computational cost by decreasing the dimension. Having sucessfully solved the direct problem by the boundary integral equations method, it is natural for us to choose the same scheme to solve the inverse problem, i.e., the extraction of the information about the crack from the measured data. Later we will see that both for the direct and the inverse problems, the same equations are used almost all the time. This is beneficial both in the apprehension and in the computation.

## 2. Direct scattering problem

The scattering problem is mathematically modelled by an exterior boundary value problem governed by Helmholtz equation with some prescribed boundary conditions. The two dimensional problem arises when the obstacle is an infintely long cylinder. Thus, a planar crack can be seen as the intersction of a plane with an infintely long thin cylinder which runs in the direction normal to the plane. Mathematically, a planar crack can be given by a regular non-intersecting $C^3$- smooth open arc $\Gamma \subset \mathbb{R}^2$ which can be described as

$$\Gamma = \{z(s) : s \in [-1,1], z \in C^3[-1,1] \text{ and } |z'(s)| \neq 0, \forall s \in [-1,1]\}.$$

The two end points of the crack are denoted by $z^*_{-1}, z^*_1$ respectively. The left hand side and the right hand side of the crack are written by $\Gamma_+$ and $\Gamma_-$ respectively. The unit normal to $\Gamma_+$ is denoted by $\nu$. Further we set $\Gamma_0 := \Gamma \setminus \{z^*_{-1}, z^*_1\}$. The direct scattering problem that we are considering is as follows:

**Problem 1. (DP)**
*Given an incident planar wave $u^i(x,d) := e^{ik<x,d>}$ with a wave number $k > 0$ and a unit vector $d$ giving the direction of propagation, find a solution $u^s \in C^2(\mathbb{R}^2 \setminus \Gamma) \cap C(\mathbb{R}^2 \setminus \Gamma_0)$ to the Helmholtz equation*

$$\Delta u^s + k^2 u^s = 0, \quad \text{in } \mathbb{R}^2 \setminus \Gamma \tag{1}$$

*which satisfies the Neumann boundary conditions*

$$\frac{\partial u^s_\pm}{\partial \nu} = -\frac{\partial u^i}{\partial \nu} \qquad \text{on } \Gamma_0 \tag{2}$$

*on both sides of the crack and the Sommerfeld radiation condition*

$$\lim_{r\to\infty} \sqrt{r}\left(\frac{\partial u^s}{\partial r} - iku^s\right) = 0, \qquad r := |x| \tag{3}$$

*uniformly for all directions $\hat{x} := \frac{x}{|x|}$.*

In (2), the limits

$$\frac{\partial u^s_\pm(x)}{\partial \nu} := \lim_{h\to 0} < \nu(x), \mathrm{grad} u^s(x \pm h\nu(x)) >, \quad x \in \Gamma_0$$

are required to exist in the sense of locally uniform convergence.

Note that the boundary conditions (2) can be reformulated as homogeneous Neumann conditions for the total field $u := u^i + u^s$, i.e., $\frac{\partial u_\pm}{\partial \nu} = 0$. This means that the normal component of the velocity of the total wave vainishes on the crack, i.e., the crack is sound-hard. Using boundary integral equations, this direct problem can be solved via the layer approach, see for example [2] and [3]. Indeed, in terms of the fundamental solution to the Helmholtz equation in $\mathbb{R}^2$

$$\Phi(x,y) := \frac{\mathrm{i}}{4} H_0^{(1)}(k|x-y|), \qquad x \neq y$$

we have the following theorem which ensures the unique solvability of the direct scattering problem.

**Theorem 1.** *The direct Neumann problem 1 has a unique solution given by*

$$u^s(x) = \int_\Gamma \frac{\partial \Phi(x,y)}{\partial \nu(y)} \varphi(y) ds(y) \quad x \in \mathbb{R}^2 \setminus \Gamma. \tag{4}$$

*where $\varphi \in C^{1,\alpha,*}(\Gamma)$ is the (unique) solution to the following integral equation*

$$\frac{\partial}{\partial \nu(x)} \int_\Gamma \frac{\partial \Phi(x,y)}{\partial \nu(y)} \varphi(y) ds(y) = -\frac{\partial u^i(x)}{\partial \nu(x)} \tag{5}$$

*and for $0 < \alpha < 1$, the function space $C^{1,\alpha,*}(\Gamma)$ is defined by*

$$C^{1,\alpha,*}(\Gamma) := \left\{ \varphi | \, \varphi(z_{-1}^*) = \varphi(z_1^*) = 0, \frac{d\varphi(z(s))}{ds} = \frac{\tilde{\varphi}(\arccos s)}{\sqrt{1-s^2}}, \tilde{\varphi} \in C^{0,\alpha}[0,\pi] \right\}$$

Applying the Green's integral theorem, the uniqueness is ensured by the Rellich's lemma and the radiation condition. Using the potential ansatz, the solvability of the boundary value problem is then converted to the solvability of the induced boundary integral equation (5) which can be determined by the Riesz theory. For details we refer to [4].

At this place we want to point out that in the scattering problem, one is particularly interested in the far-field pattern $u_\infty$ of the scattered field $u^s$. The far-field pattern describes the behavior of the scattered wave at infinity

$$u^s(x) = \frac{e^{\mathrm{ik}|x|}}{\sqrt{|x|}} \left\{ u_\infty(\hat{x}) + O\left(\frac{1}{|x|}\right) \right\} \qquad |x| \to \infty$$

uniformly for all directions $\hat{x} \in \Omega := \{x \in \mathbb{R}^2 | |x| = 1\}$. The one-to-one correspondence between radiating waves and their far field patterns is established by the Rellich's lemma. In the case of a sound-hard crack, the far-field pattern is declared via

$$u_\infty(\hat{x}) = \rho \int_\Gamma < \nu(y), \hat{x} > \mathrm{e}^{-\mathrm{ik}<\hat{x},y>}{}'(y)ds(y). \tag{6}$$

with the density function $\varphi$ given by theorem 1 and the constant $\rho = \sqrt{\frac{k}{8\pi}} \mathrm{e}^{-\mathrm{i}\frac{ß}{4}}$.

For further treatment, we would like to transform our integral equation (5) into operator form. For this purpose, the following integral operators may be defined:

$$(S\varphi)(x) := \int_\Gamma \Phi(x,y)\varphi(y)ds(y)$$

$$(T_0\varphi)(x) := \frac{\partial}{\partial \nu(x)} \int_\Gamma \frac{\partial \Phi(x,y)}{\partial \nu(y)} \varphi(y) ds(y)$$

To reduce the hypersingularity of the operator $T_0$, we use the Maue's identity to split it into two milder parts (for a proof of this splitting, see for example Theorem 7.29 in [3])

$$T_0\varphi = \frac{\partial}{\partial \vartheta} S \frac{\partial \varphi}{\partial \vartheta} + k^2 < \nu, S\varphi\nu >,$$

where $\vartheta$ is the unit tangent vector. The integral equation (5) now becomes

$$\frac{\partial}{\partial \vartheta} S \frac{\partial \varphi}{\partial \vartheta} + k^2 < \nu, S\varphi\nu >= g \tag{7}$$

where we set $g = -\frac{\partial u^i}{\partial \nu}$.

## 3. Inverse scattering problem

After introducing the direct scattering problem in the last section, we consider the following inverse problem:

**Problem 2. (IP)**

*Determine the crack $\Gamma$ if the far-field pattern $u_\infty$ is known for one incident wave.*

About this inverse problem, the first thing to ask is the uniqueness, that is, the identifiability of the crack. Unfortunately, there exists no theoretical result for our setting of inverse problem. However, if all the far-field patterns from all possible incident directions are measured, the crack can be uniquely identified. We quote the following result from [5].

**Theorem 2.** *Assume that $\Gamma_1$ and $\Gamma_2$ are two sound-hard cracks with the property that for a fixed wave number $k > 0$, the far-field patterns $u_{1,\infty}$ and $u_{2,\infty}$ coincide for all incident directions $d$. Then we have $\Gamma_1 = \Gamma_2$.*

Our aim is to find and to reconstruct the crack from the measured far-field data resulted from just one single incident wave. Despite of the lack of a theoretical proof, our setting of inverse problem arises naturally from the viewpoint of the real applications. It is always desirable to find the crack with less effort.

For the detection of the crack, the only available information is the measured far-field data. Equation (6) relates the far-field pattern $u_\infty$ with the crack $\Gamma$ and is therefore suitable as the starting point for the reconstruction. For future reference, we rewrite it in the more compact form

$$F(\Gamma) = u_\infty \tag{8}$$

and call it the far-field equation as in the literature. The inverse problem is then equivalent to the solving of this equation. However, the sloving of (8) is not as straight forward as it appears. Mathematically this equation is not even solvable because of the nature of the operator $F$. Being a compact operator in infinite dimensional function space, $F$ cannot have a bounded inverse. This means that (8) cannot be solved in any reasonable normed space.

Therefore, certain regularization techniques must be introduced at this place. For the sake of completeness, we briefly discuss regularization for linear compact operators in the next section. We note that the operator $F$ will be differently interpreted for different methods in the following sectons. But the right hand side of (8) will remain the same.

## 4. Regularization

**Definition 3** (Regularization). *Assume that $X, Y$ are normed spaces. Let the operator $A : X \to Y$ be linear, bounded and injective. A family of bounded linear operators $R_\alpha : Y \to X, \alpha > 0$ is called a* regularization scheme *for*

$$A\varphi = f \tag{9}$$

*if it satisfies the following pointwise convergence*

$$\lim_{\alpha \to 0} R_\alpha A\varphi = \varphi, \text{ for all } \varphi \in X$$

*In this case, the parameter $\alpha$ is called the* regularization parameter.

From the above definition, it is easy to see that if the operator $A$ has a bounded inverse, then we can simply choose $R_\alpha = A^{-1}, \forall \alpha > 0$. In the case where the inverse of the injective operator $A$ is not bounded, for example when $A$ is compact and the dimension of $X$ is infinite, then difficulty arises since we are trying to approximate an unbounded operator with a sequence of bounded operators. The task of regularization is to find a stable approximate solution. The following theorem, which is proved in [3] (theorem 15.6), shows the limitations of the regularization.

**Theorem 3.** *Let $X$ and $Y$ be normed spaces with $\dim X = \infty$ and let $A : X \to Y$ be linear and compact. Then for a regularization scheme the operators $R_\alpha$ cannot be uniformly bounded with respect to $\alpha$, and the operators $R_\alpha A$ cannot be norm convergent as $\alpha \to 0$.*

Because of the lack of uniform convergence, the choice of the regularization parameter $\alpha$ is apparently crucial. To have a vivid impression of this, let's examine the aproximation error in a more pratical setting where measurement errors or noises are present. Let's denote the contaminated data by $f^\delta$ assuming an error level $\delta$, i.e., $\|f - f^\delta\| \leq \delta$. For $\alpha > 0$, we wirte the regularized approximation as

$$\varphi_\alpha^\delta := R_\alpha f^\delta$$

The total approximation error of the regularization scheme can be written as

$$\varphi_\alpha^\delta - \varphi = R_\alpha f^\delta - R_\alpha f + R_\alpha A\varphi - \varphi$$

From this, we have the following error estimate

$$\|\varphi_\alpha^\delta - \varphi\| \leq \delta \|R_\alpha\| + \|R_\alpha A\varphi - \varphi\| \tag{10}$$

The right-hand side of (10) reveals the difficulty of the approximation. As $\alpha$ tends to 0, the second term of the right-hand side becomes small because of the regularization. At the same time, the error $\delta$ will be amplified by the factor $\|R_\alpha\|$ which tends to infinity. Therefore, the total error is dominated by the first term and is much larger than the datat error. For larger $\alpha$, the first term on the right-hand side is smaller but the second term will be larger. Hence certain strategy for choosing $\alpha$ must be developed to balance the effects of the two factors. The discrepancy principle proposed by Morozov ([6], [7]) which based on the consideration that a problem cannot be solved more accurate than the data error, is a natural *a posteriori* strategy for determining the parameter $\alpha$. More precisely, one looks for an $\alpha > 0$ such that

$$\|AR_\alpha f^\delta - f^\delta\| = \gamma\delta$$

for some prescribed $\gamma \geq 1$. It can be shown that there always exists a smallest $\alpha > 0$ for which the approxiamte solution $f^\delta$ satisfies the above condition. (see [3]) Although the existence for a "best" $\alpha$ is theoretically assured, there is no obvious way to select it in practice. In most cases, the only way to choose a good parameter is by trial and error. For our purpose, it is sufficient to consider regularization in the Hilbert space setting. We denote by $A^*$ the adjoint operator of $A$. Without going into details, we adapt the following regularization scheme.

**Theorem 4** (Tikhonov Regularization). *Let x and Y be Hilbert spaces. The operator $A : X \to Y$ is assumed to be compact and linear. Then for every $\alpha > 0$, the operator*

$$\alpha I + A^*A : X \to X$$

*is bijective and has a bounded inverse. Furthermore, if the operator A is injective, then*

$$R_\alpha := (\alpha I + A^*A)^{-1} A^*, \qquad \alpha > 0$$

*describes a regularization scheme with $\|R_\alpha\| \leq \frac{1}{2\sqrt{\alpha}}$.*

Applying the Tikhonov Regularization scheme to find an approximate solution to (9), we actually solve the follwing equation

$$(\alpha I + A^*A)\varphi_\alpha = A^*f \tag{11}$$

for $\varphi_\alpha$.

## 5. Classical Newton's method

As noted earlier, (8) connects the far-field pattern with the crack and is therefore suitable for the inverse task of finding the crack. It is therefore natural to solve it for the unknown crack when one has the measured far-field data at hand. For this nonlinear problem, Newton's method provides a simple way to handle it. Utilizing the first Frechét derivative which existence is proved in [5], Newton's method approximates the nonlinear equation (8) with a linear one. This means that instead of solving (8), we endeavor to solve the linear equation

$$F'(\gamma)q = u_\infty - F(\gamma) \tag{12}$$

for an update $q$. We remark here that the derivative $F'$ is still a compact operator, this means that the above equation can't be solved directly. However, we can apply the regularization technique from the last section. Thus, the following equation has to be solved

$$(\alpha I + F'^*(\gamma)F'(\gamma))q = F'^*(\gamma)(u_\infty - F(\gamma)) \tag{13}$$

Numerically, starting from an initial guess $\gamma$ of the unknown crack, (13) finds an update $q$. The same equation is then solved iteratively with $\gamma$ replaced by $\gamma + q$ until some convergence criterion is met.

This scheme is conceptually very simple, yet the computation of the Fréchet derivative $F'$ is not straightforward. To calculate $F'$, one must first solve another direct problem since $F'$ is characterized as the solution of that problem (see [5]). Thus additional computational cost arises.

At this place, we want to discuss the uniqueness of the equation (12). As pointed out in [8], the Fréchet derivative $F'$ is not injective. The null space of this operator is given by

$$N(F') = \{q : \nu \cdot q = 0\}$$

This reflects the fact that different parametrizations of the arc leading to the same set of points which in turn give the same far-field pattern. We can avoid this ambiguity by limiting our solution space to the set of arcs representable as the graph of a function as suggested in [8]. Once this restriction is made, the operator $F'$ is then an injective linear compact operator. The same restriction will aslo be made in the next two sections.

## 6. Nonlinear integral equations method

Based on the reciprocity gap principle, Kress and Rundell [9] proposed a method using nonlinear integral equations for solving an inverse conducting problem avoiding the computation of a forward problem at each iteration step. The idea behind this approach is quite simple: in a doubly connected bounded domain to which the second Green's theorem applies, if one knows the Cauchy data of a harmonic function on the outer boundary and the type of the boundary condition on the interior boundary, then one can determine the Cauchy data on the interior boundary and also the interior boundary itself. This method is extended to other boundary conditions and cracks in [10] and to obstacle scattering in [11]. Based on a different system of two nonlinear integral equations, a method with far-field data is proposed in [12] without the need of an auxiliary curve around the obstacle.

To understand this method, let's first reformulate our equations for the direct problem. In terms of $\Gamma$ and $\varphi$, we can define the operators $B : C^3[-1,1] \times C^{1,\alpha,*}(\Gamma) \to C^{0,\alpha}(\Gamma)$ and $F : C^3[-1,1] \times C^{1,\alpha,*}(\Gamma) \to C^{0,\alpha}(\Omega)$ via

$$B(\Gamma,\varphi)(x) := \frac{\partial}{\partial\nu(x)} \int_\Gamma \frac{\partial\Phi(x,y)}{\partial\nu(y)} \varphi(y)ds(y), \quad x \in \Gamma \tag{14}$$

and

$$F(\Gamma,\varphi)(\hat{x}) := \rho \int_\Gamma < \nu(y), \hat{x} > e^{-ik<\hat{x},y>} \varphi(y)ds(y), \quad \hat{x} \in \Omega \tag{15}$$

respectively. Now we can consider the following system of operator equations

$$\begin{cases} B(\Gamma,\varphi) = -\frac{\partial u^i}{\partial \nu}|_\Gamma \\ F(\Gamma,\varphi) = u_\infty \end{cases} \tag{16}$$

Note that the operator $B$ is just the boundary operator which maps the unknowns to the Neumann boundary data. The operator $F$ calculates the far-field pattern. From the viewpoint of the direct problem, this system is just the same as the equations system (5) and (6) if $\Gamma$ is given in advance.

To solve the corresponding inverse problem, the idea of [12] is to use this same system (16). The reasoning is very simple. If $\Gamma$ solves the inverse scattering problem, then it follows directly from the solution theory of the direct problem that system (16) is satisfied. Conversely, if the pair $(\Gamma,\varphi)$ solves (16), the first equation of (16) ensures that the total field $u$ defined in theorem 1 satisfies the homogeneous Neumann boundary conditions on $\Gamma$. The second equation in (16) then ascertains the correct far field pattern for the scattered field $u^s$. From the uniqueness theorem it follows that $\Gamma$ is the solution of the inverse problem. We have thus the following main theorem (see [12]).

**Theorem 5.** *$\Gamma$ is the solution of the inverse problem if and only if $\Gamma,\varphi$ solve the system of nonlinear integral equations* (16).

We note that, mathematically, the system (16) illustrates the spirit of "inverse problems". Indeed, giving the type of the boundary conditions, (16) is just our system of equations for the direct problem which solves the far field pattern of the scattered field if one knows the crack. On the other hand, this very same system is used to find the crack from the knowledge of the far field pattern which is just the solution of the corresponding direct problem. This aspect also demonstrate one of the important features of our method. No new equations are created. What we use to solve the inverse problem is just what we have from the direct problem.

According to theorem 5, our task now is to solve the system (16). After parametrising the above system with the cosine substitution $t = \cos\tau$ (see [13]) incorporated to take care of the square root singularities of the solution $u^s$ at the crack tips, the system (16) can be written in the following form

$$A_0(\gamma,\psi') - A_1(\gamma,\psi) = a(\gamma) \tag{17a}$$

$$A_\infty(\gamma,\psi) = u_\infty \tag{17b}$$

where

$$\begin{aligned} A_0(\gamma,\psi') &:= \int_0^{2\pi} K_0(\tau,\sigma)\psi'(\sigma)\sin\tau d\sigma \\ A_1(\gamma,\psi) &:= \int_0^{2\pi} K_1(\tau,\sigma)\psi(\sigma)\sin\tau\sin\sigma d\sigma \\ A_\infty(\gamma,\psi) &:= \rho\int_0^{2\pi} K_\infty(\hat{x},\sigma)\psi(\sigma)\sin\sigma d\sigma \\ a(\gamma) &:= -2ik < n(\tau),d > u^i(\gamma(\tau),d)\sin\tau \end{aligned}$$

with the kernels

$$K_0(\tau,\sigma) := \frac{ik}{4} H_0^{(1)'}(k|\gamma(\tau)-\gamma(\sigma)|) \frac{<\gamma(\tau)-\gamma(\sigma),\gamma'(\tau)>}{|\gamma(\tau)-\gamma(\sigma)|}$$

$$K_1(\tau,\sigma) := \frac{ik^2}{4} <\gamma'(\tau),\gamma'(\sigma)> H_0^{(1)}(k|\gamma(\tau)-\gamma(\sigma)|)$$

$$K_\infty(\hat{x},\sigma) :=< n(\sigma),\hat{x} > e^{-ik<\gamma(\sigma),\hat{x}>}$$

and $\gamma(\tau) := (z \circ \cos)(\tau)$, $\psi(\tau) = \text{sign}(\pi - \tau)((\varphi \circ \cos)(\tau))$, $n = (\nu \circ \gamma) \cdot |\gamma'|$. Note that we make an odd extension of our system (16) to the closed interval $[0, 2\pi]$ which is not necessary but helpful in the theoretical treatment [4].

It is noted that all the operators are nonlinear with respect to $\gamma$. Since our numerical method will be based on the Newton's method, we need also the Fréchet derivatives of the operators w.r.t. the boundary $\gamma$. The Fréchet derivatives of the integral operators $A's$ are simply given by the Fréchet derivatives of their kernels (see [14]). For brevity, we set $\mathbf{a}^\perp = (a_2, -a_1)^t$ if $\mathbf{a} = (a_1, a_2)^t$. Also we write $\Delta\gamma = \gamma(\tau) - \gamma(\sigma)$ and $\Delta q = q(\tau) - q(\sigma)$ . Using the differentiation rules of the Hankel functions $H_0'(z) = -H_1(z)$ and $(zH_1(z))' = zH_0(z)$, we have

$$A_0'(\gamma,\psi';q) = \int_0^{2\pi} K_0'(\tau,\sigma,\gamma;q)\psi'(\sigma)\sin\tau d\sigma$$

$$A_1'(\gamma,\psi;q) = \int_0^{2\pi} K_1'(\tau,\sigma,\gamma;q)\psi(\sigma)\sin\tau\sin\sigma d\sigma$$

$$A_\infty'(\gamma,\psi;q) = \rho\int_0^{2\pi} K_\infty'(\gamma,\psi;q)\psi(\sigma)\sin\sigma d\sigma$$

$$a'(\gamma;q) = -2ik\left\{< q'(\tau)^\perp, d > +ik < q(\tau), d >< n(\tau), d >\right\} u^i(\gamma(\tau), d)\sin\tau$$

where

$$K_0'(\tau,\sigma,\gamma;q) = -\frac{ik}{4}\Big\{\frac{<\Delta\gamma,\Delta q><\Delta\gamma,\gamma'(\tau)>}{|\Delta\gamma|^2} kH_0^{(1)}(k|\Delta\gamma|) +$$

$$H_1^{(1)}(k|\Delta\gamma|)\Big(\frac{<\Delta q,\gamma'(\tau)> + <\Delta\gamma, q'(\tau)>}{|\Delta\gamma|} - 2\frac{<\Delta\gamma,\Delta q><\Delta\gamma,\gamma'(\tau)>}{|\Delta\gamma|^3}\Big)\Big\}$$

$$K_1'(\tau,\sigma,\gamma;q) = \frac{ik^2}{4}\Big\{(< q'(\tau),\gamma'(\sigma) > + <\gamma'(\tau), q'(\sigma) >) H_0^{(1)}(k|\Delta\gamma|)$$

$$-k <\gamma'(\tau),\gamma'(\sigma)> H_1^{(1)}(k|\Delta\gamma|)\frac{<\Delta\gamma,\Delta q>}{|\Delta\gamma|}\Big\}$$

$$K_\infty'(\gamma,\psi;q) = \left\{< q'(\sigma)^\perp, \hat{x} > -ik < q(\sigma),\hat{x} >< n(\sigma),\hat{x} >\right\} e^{-ik<\gamma(\sigma),\hat{x}>}\sin\tau$$

Therefore, to solve the parametrized system (17), Newton's method suggests the solving of the following system of linear approximations

$$A_0(\gamma,\psi') + A_0(\gamma,\chi') + A_0'(\gamma,\psi';q) - A_1(\gamma,\psi) - A_1(\gamma,\chi) - A_1'(\gamma,\psi;q)$$
$$= a(\gamma) + a'(\gamma;q) \tag{18a}$$
$$A_\infty(\gamma,\psi) + A_\infty(\gamma,\chi) + A_\infty'(\gamma,\psi;q) = u_\infty \tag{18b}$$

Numerically this system is solved iteratively: with a current approximation $(\psi,\gamma)$, the system (18) has to be solved for the pair $(\chi,q)$. The updated data are then given by $\psi+\chi$ for the density function of the integral operators and $\gamma+q$ for the unknown crack. For brevity, we rewrite the system (18) in the operator form:

$$\mathcal{A}(\chi,q) = \mathrm{Y} \tag{19}$$

Since this linear equation of the first kind is still ill-posed, we have to incorporate some regularization scheme. Instead of solving (19) directly, we solve the following regularized equation

$$\left(\begin{bmatrix} \alpha I & 0 \\ 0 & \beta I \end{bmatrix} + \mathcal{A}^*\mathcal{A}\right)\begin{bmatrix} \chi \\ q \end{bmatrix} = \mathcal{A}^*\mathrm{Y} \tag{20}$$

with two to be determined regularization parameters $\alpha$ and $\beta$.

The major advantage of this method as compared to the classical Newton's method is that the Fréchet derivatives of the integral operators can be easily computed. They are obtained simply by solving the above system which is nothing more than the Gaussian elimination.

## 7. Two-steps method

The idea of the nonlinear integral equations method is based on the observation that the inverse problem can be formulated in an equivalent two-by-two system (16) of nonlinear equations. This system is then treated as a coupled system of two variables $(\gamma,\psi)$. In the regularized version (20), two parameters $(\alpha,\beta)$ are therefore needed at the same time. Since these parameters are in general selected by trial and error, it is rather complicated. However, the system (16) can be solved in another way. Inspired from the solution theory of the direct problem, This system can be seen as the whole procedure for the calculation of the far-field pattern of the scattered field from a given crack resulted from an incident planar wave in two steps. Indeed, the first equation of (16) solves the scattered field from the crack and the second one calculates the far-field pattern from the solution of the first equation. This motivates our inverse scheme: the inverse solver should be *inverse* to the direct solver. There are several possibilities to solve (16) in two steps separately. One can solve the first equation first and the second one next, or the other way round. One has further choice concerning which quantity is to be solved at each individual step. Since our method is based on Newton's iteration method, an initial guess for $\gamma$ or for $\psi$ or for both is unavoidable. It is therefore natural to start with a initial guess for the crack and solve the first equation for $\psi$. This is beneficial since this is just the direct problem defined by problem 1. Even better is, theorem 1 ensures the unique solvability of the equation for any crack. Next we can solve the second equation for an update of the crack. Note that this step is nonlinear and ill-posed. With slightly abused notations, we have thus splitted the system (16) into the following two separate steps:

$$A_0(\gamma;\psi') - A_1(\gamma;\psi) = a(\gamma) \tag{21a}$$

$$A_\infty(\psi;\gamma) = u_\infty \tag{21b}$$

Note that this system is formally the same as system (17) except that the first variable is fixed in each single equation. That is, $\gamma$ is fixed in (21a) while $\psi$ is fixed in (21b). In the actual numerical computation, we have to solve a regularized version of (21b). We therefore solve

$$(\alpha I + A'^{*}_\infty(\psi;\gamma)A'_\infty(\psi;\gamma))q = A'^{*}_\infty(\psi;\gamma)(u_\infty - A_\infty(\psi;\gamma)) \tag{21b$'$}$$

instead of (21b).

The algorithm for the two-steps method can be summarized as follows:

1. Given an initial guess $\gamma_0$ for the unknown crack.
2. iterative steps, for $n = 0, 1, 2, \ldots$
   (a) Step 1: solve
   $$A_0(\gamma_n;\psi') - A_1(\gamma_n;\psi) = a(\gamma_n) \tag{22}$$
   for $\psi$.
   (b) Step 2: solve
   $$(\alpha I + A'^{*}_\infty(\psi;\gamma_n)A'_\infty(\psi;\gamma_n))q_n = A'^{*}_\infty(\psi;\gamma_n)(u_\infty - A_\infty(\psi;\gamma_n)) \tag{23}$$
   for the update $q_n$ of $\gamma_n$.
   (c) Update in each step
   $$\gamma_{n+1} = \gamma_n + q_n \tag{24}$$
3. Stopping criterion: Discrepancy principle

At this place we comment that from the numerical point of view this scheme is very attractive. As a variant of the nonlinear integral equations method, the Fréchet derivative of the integral operator is computed directly by solving the integral equation (23). Besides, the derivative is very simple as compared to those in the last section, since only the far-field operator which has a smooth kernel is to be differentiated. Numerically it can be easily solved via the rectangular rule, for example. Another advantage of this method is that it splits the problem into two smaller parts and thus makes the computation cheaper.

## 8. Numerical results

In this section we will demonstrate the applicability of the two-steps method via some examples. One can compare the results with those from the other methods in [5] and [12]. Using just one incident wave, we reconstruct the unknown crack from the measured far-field pattern at 16 points which are evenly distributed on the unit circle. For the direct problem, the forward solver is applied with 63 collocation points. To avoid committing an inverse crime, the number of collocation points used in the inverse problem is chosen to be different from that of the forward solver. We choose 31 collocation points in the inverse algorithm. In all our examples, we take the incident plane wave coming from the direction $d = \frac{1}{\sqrt{2}}(1,1)$.

For the reconstruction, we take the k-th Chebyshev monomials $T_k(x) = \cos(k\cos^{-1}x), k = 1,\ldots,m$ as our basis functions. The selection of the Chebyshev polynomials is based on the fact that they can take care of the square root singularity in their own right. The updates for the crack in each iterative step can be written in the form

$$q(x) = (\sum_{k=0}^{m-1} b_k^1 T_k(x), \sum_{k=0}^{m-1} b_k^2 T_k(x))$$

with to be determined coefficients $b_k^1, b_k^2, k = 0,\ldots,m-1$. We test our scheme for two different wave numbers, $k = 1,3$. The dimension of the test space is taken to be 5, i.e., $m = 5$. The starting curve, that is, the initial guess for the regularized Newton's method, is taken to be the straight line $y = 0$ in all examples. According to the discrepancy principle, the stopping criterion for the iterative scheme is given by the relative error

$$\frac{\|u_\infty - u_{\infty,n}\|_2}{\|u_\infty\|_2} \leq \epsilon$$

which is taken to be 0.001 in case of exact data and 0.03 in the case where 3% data error are present. In all our figures below, the dotted line (blue) represents the initial guess. We denote by the dashed line (red) the true solution and by the solid line (black) the reconstruction.

**Example 1.** For the first example, we take the arc

$$\Gamma = (x, 0.4x^3 + 0.2x^2 - 0.4x - 0.2), \quad x \in [-1,1]$$

which is a polynomial. The numerical results are given in the figures 1, 2 for $k = 1$ and $k = 3$, respectively.

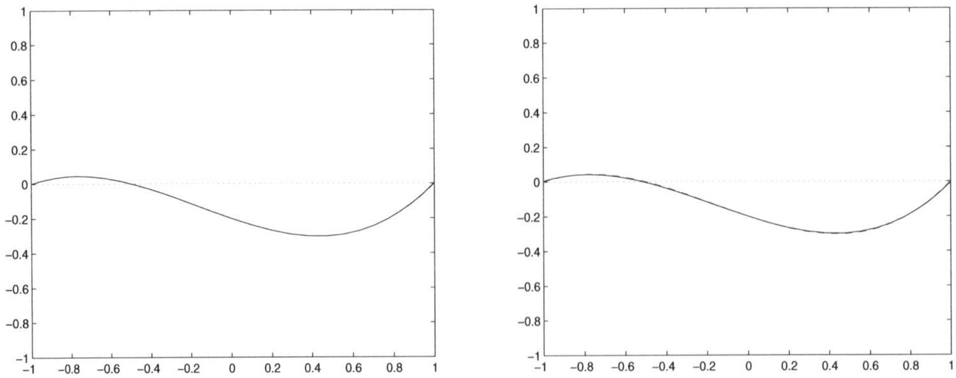


(a) exact data, $N = 11$

(b) 3% error, $N = 10$

**Figure 1.** $k = 1$

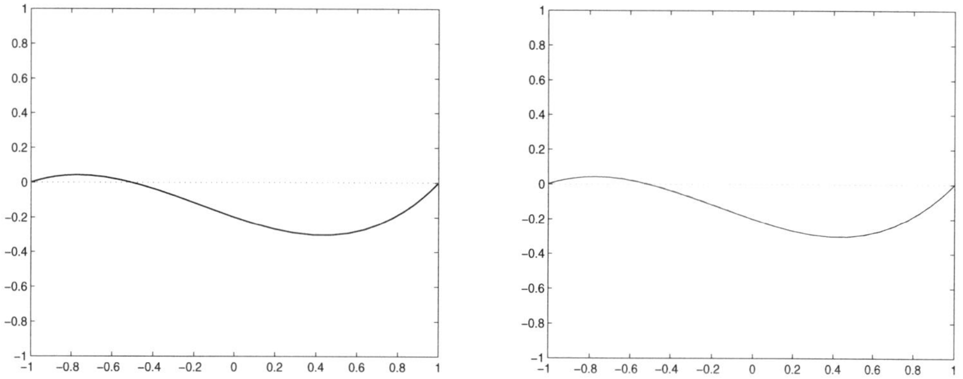


(a) exact data, $N = 9$

(b) 3% error, $N = 9$

**Figure 2.** $k = 3$

From the numerical results, we see that the reconstructions for exact data are very good. In the case where random errors are present, the reconstructions are not bad at all.

**Example 2.** To demonstrate the benefit of our method, we choose a curve which does not belong to our solution space. To this end, we take the arc

$$\Gamma = (x, 0.5\cos(\frac{\pi}{2}x) + 0.2\sin(\frac{\pi}{2}x) - 0.1\cos(\frac{3\pi}{2}x)), \quad x \in [-1,1]$$

.

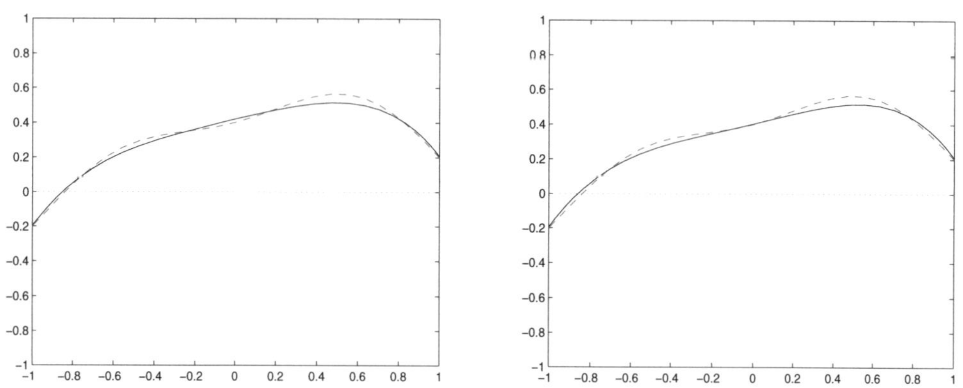


(a) exact data, $N = 6$

(b) 3% error, $N = 4$

**Figure 3.** $k = 1$

We see from the figures 3, 4 that the reconstructions are very good, even in the case of erroneous data.

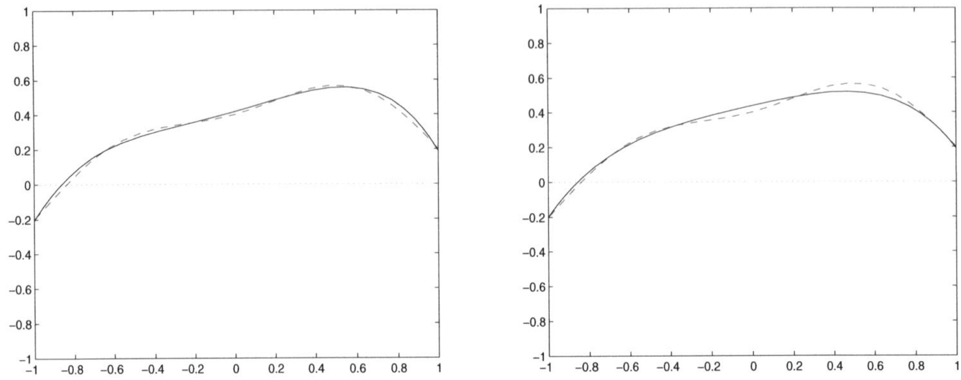

(a) exact data, $N = 9$ (b) 3% error, $N = 10$

**Figure 4.** $k = 3$

**Example 3.** To demonstrate the applicability of our method, we choose a curve which is not a graph of a function. To this end, we take the arc

$$\Gamma = (2\sin(\frac{3\pi}{8}(x+4/3)), -\sin(\frac{3\pi}{4}(x+4/3))), \quad x \in [-1,1]$$

.

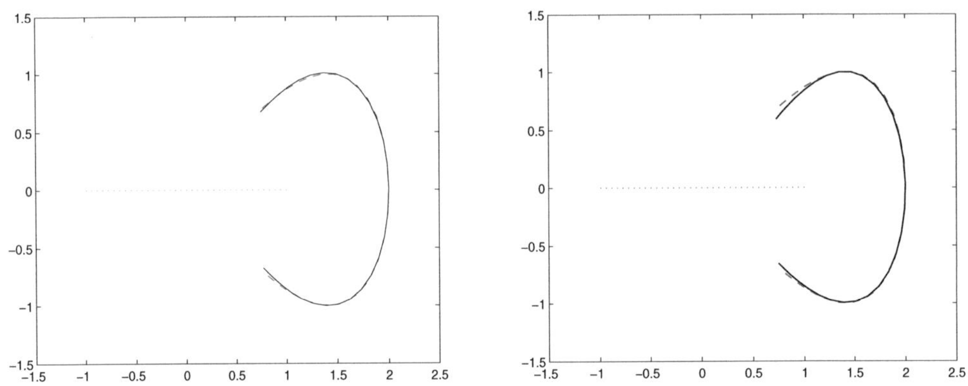

(a) exact data, $N = 35$ (b) 3% error, $N = 26$

**Figure 5.** $k = 1$

We see that in the case $k = 1$, the reconstructions are very good both with exact and erroneous data (Fig 5). In the case of more oscillations ($k = 3$, Fig 6), the result is quite good with respect to the bad initial guess for the iteration. We also note that although in this case, the Fréchet derivative of the far field operator is not injective, the actual reconstruction process is still running without any problem.

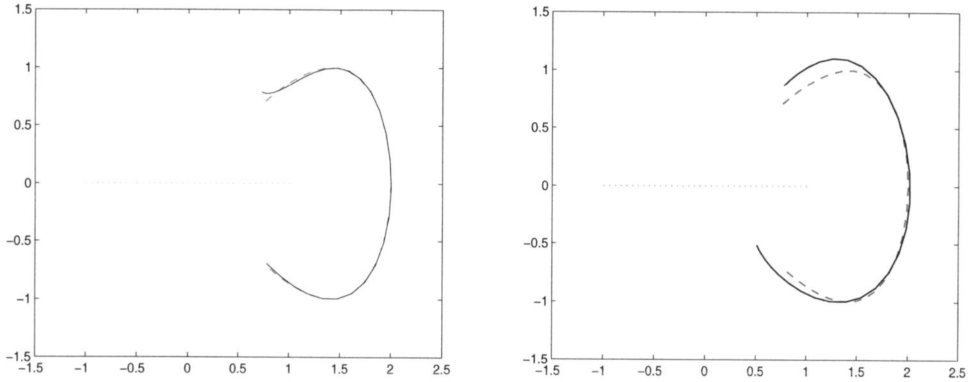


(a) exact data, $N = 26$

(b) 3% error, $N = 21$

**Figure 6.** $k = 3$

## 9. Conclusion

In this article, we try to detect a planar crack via iterative Newton's method with the knowledge of the measured far-field pattern. By the one to one correspondence of the scattered field and its far-field, the measurements can be taken anywhere outside the crack. The choice of boundary integral equations method reduces the dimension of the problem by one which makes the computation cheaper. The conceptual simplicity and its numerical accuracy make the Newton's method attractive. Besides, the two variants of the Newton's method (nonlinear integral equations method and two-steps method) on one hand simplifiy the calculation of the Fréchet derivatives and on the other hand help clearifying the idea of the inverse problem.

## Author details

Kuo-Ming Lee
*Department of Mathematics, National Cheng Kung University, Taiwan*

## Acknowledgements

This work is partially supported by the NSC grant NSC-100-2115-M-006-003-MY2.

## 10. References

[1] Keller J B 1976 Inverse problems. *Amer. Math. Monthly* 83 no. 2, 107-118.

[2] Colton D and Kress R 1998 *Inverse Acoustic and Electromagnetic Scattering Theory* 2nd edn (Berlin: Springer)

[3] Kress R 1999 *Linear Integral Equations* 2nd edn (Berlin: Springer)

[4] Mönch L 1996 On the numerical solution of the direct scattering problem for a sound-hard open arc *Comput. Appl. Math.* 71 343-356

[5] Mönch L 1997 On the inverse acoustic scattering problem by an open arc: the sound-hard case *Inverse Problems* 13 1379-1392

[6] Morozov V A 1966 On the solution of functional equations by the method of regularization *Soviet Math. Doklady* 7 414-417 (English translation)

[7] Morozov V A 1967 Choice of parameter for the solution of functional equations by the regularization method *Soviet Math. Doklady* 8 1000-1003 (English translation)

[8] Kress R 1995 Inverse scattering from an open arc *Math. Meth. Appl. Sci.* 18 267-293

[9] Kress R and Rundell W 2005 Nonlinear integral equations and the iterative solution for an inverse boundary value problem *Inverse Problems* 21 1207-1223

[10] Ivanyshyn O and Kress R 2006 Nonlinear integral equations for solving inverse boundary value problems for inclusions and cracks. *J. Integral Equations and Appl.* 18 13-38

[11] Ivanyshyn O and Kress R 2005 Nonlinear Integral Equations in Inverse Obstacle Scattering In *Proceedings of the 7th International Workshop on Mathematical Methods in Scattering Theory and Biomedical Engineering, Nymphaio, Greece*

[12] Lee K-M 2006 Inverse scattering via nonlinear integral equations for a Neumann crack *Inverse Problems* 22 1989-2000

[13] Yan Y and Sloan I H 1988 On integral equations of the first kind with logarithmic kernels *J. Integral Equations Appl.* 1 549-579

[14] Potthast R 1994 Fréchet differentiability of boundary integral operators in inverse acoustic scattering *Inverse Problems* 10 431-447

Chapter 3

# Neutron Diffraction on Acoustic Waves in a Perfect and Deformed Single Crystals

E. Raitman, V. Gavrilov and Ju. Ekmanis

Additional information is available at the end of the chapter

http://dx.doi.org/10.5772/55064

## 1. Introduction

The effect of crystal vibrations on the scattering of neutrons and X-rays in single crystals has already been studied for several decades. Crystals subjected to ultrasonic excitations have been investigated for some time by X-ray and neutron diffraction methods. The first X-ray diffraction experiments on oscillating crystals were performed in 1931 [1,2] stimulating large discussion to explain the observed increase in intensity of the Laue spots. Neutron experiments go back to the 60's [3-5]. At present time both neutrons and X-rays have become important tools in observing and understanding time-dependent matter-wave optics [6-8]. And vice versa the studying of the neutron and X-rays scattering in the time- and space-modulated with acoustic waves condensed matter represents a great interest. Applications from focusing effects [9], monochromators with tunable bandwidths [10-12], the characterization of static but tiny strain fields [13,14] have been discussed as well as fundamental questions about the formation of satellites [8] and its applications, inter-branch scattering, gradient crystal effects and the fundamental difference between neutron- and X-ray diffraction found their audience [15,16]. Theoreticians have tried to explain the variety of effects and have predicted even more challenging fields for experimentalists, like, for example, the formation of caustics in Laue diffraction [17]. The studies above all involve the spatial characteristics, but also attempts to investigate the temporal parameter have been made earlier in special cases of X-ray [18,19] and neutron diffraction [20]. Some work has been carried out with modulated ultrasonic waves [21-24] impregnating an artificial time structure to the carrier wave.

In the case of neutron scattering the effect of energy exchange between a neutron and an acoustic phonon is observed, which stems from the fact that the neutron velocity is comparable with that of ultrasound wave. In such exchange the amount of the energy transmitted is rather small - for an ultrasound wave with a frequency of 100 MHz it is approximately 400 neV. Studying the efficiency of energy exchange at the diffraction of thermal neutrons is therefore an involved task calling for special methods. The authors of work [25] observed very week extra peaks (angular satellites) in the proximity of the main reflection peak, which were caused by the energy exchange at the grazing diffraction of neutrons in a quartz single crystal on the grating of the surface acoustical wave. Impressive experiments have been performed to elucidate the concept of particle-wave dualism [26]. The authors observed the energy exchange of ~10 neV and the transition from quantum-mechanical to classical behavior.

In the chapter the neutron interactions with acoustic waves in solids observed by diffraction modern methods is discussed. Fundamental questions such as the investigation of the diffraction process itself as a response to the ultrasonic field as well as its applications to beam optics or materials research are of interest.

# 2. Diffraction on a perfect crystals Laue geometry

## 2.1. Theoretical background

The theoretical problem of ultrasonic influence on neutron and X-ray diffraction was under intensive investigation [8,9,16, 27,28]. Depending on the ultrasound acoustic waves (AW) frequency, a distinction can be made between two different physical mechanisms. At AW frequency $\nu_s < \nu_{res}$ , where $\nu_{res}$ is the AW frequency at neutron-acoustic resonance condition (see below) dispersion surface (DS) varies insignificantly and the problem is solved in terms of the usual perturbation theory. The correction to the eigen functions are of the order of $Hw$ ($H$ is the reciprocal lattice vector, $w$ the AW amplitude), and additions to the scattering intensity are of the order of $(Hw)^2$. Hence at sound frequencies $\nu_s < \nu_{res}$ there takes place nearly complete "pumping" of the elastic component of scattering into inelastic one. The Mossbauer diffraction spectra obtained on single Si crystal confirm the theoretical results [29,30]. The more interesting phenomena arise when the magnitude of the ultrasound wave vector $k_s$ is of order or greater of the gap $\Delta K_0$ between the branches of the DS (in the two-beam approximation). Interaction between modified Bloch states by means of high-frequency ultrasonic perturbation leads to new physical effects, such as the resonant ultrasonic suppression of the Borrmann transmission [31], the ultrasound induced *Pendellosung* beatings in diffraction intensity [14,32] and some another features. The effect of US (ultrasound) on neutron and X-rays diffraction in a perfect crystal is schematically shown in Fig. 1. The new energy gaps $k_s$ appear on DS which correspond to the additional regions of total reflection for the case of Bragg geometry. The US phonon's absorption (emission) by neutron mixes Bloch states [33] and displaces the dispersion surface of the neutron by the value of quasi-momentum $\delta q$:

$$\delta q = \nu_s / (v_n \cos\,\theta_B) \pm k_s. \tag{1}$$

The neutron-acoustic resonance frequency $\nu_{res}$ is determined by expression (2)

$$\nu_{res} \approx \left(v_n \cos\,\theta_B\right) / \tau \tag{2}$$

where $\theta_B$ and $v_n$ are the Bragg angle and the neutron velocity, and $\tau$ is the extinction length. Expression (1) is valid, if $H \perp k_s$ and this corresponds to the conditions of transversal AW. Taking into account the Debye–Waller factor, $v_{res}$ does not depend on the observed reflection [16]. For the case $v_s > v_{res}$ a linear increase in the diffraction intensity should be observed depending on the US wave's amplitude $w$ [14].

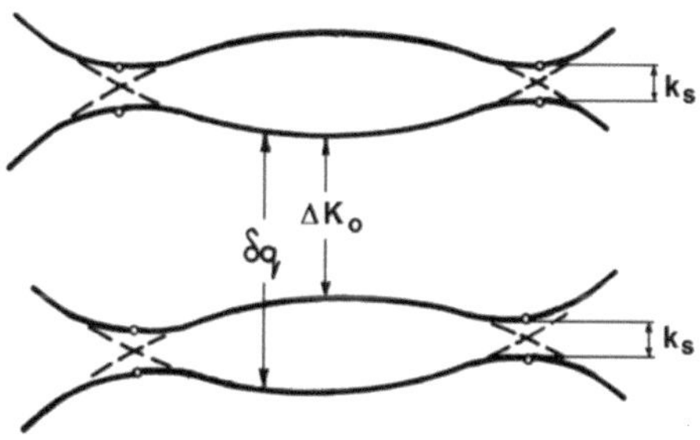


**Figure 1.** Schematic image of the dispersion surface modified by ultrasound in a perfect crystal (Laue geometry): $\Delta K_0$ is the gap between the sheets of dispersion surface determining the resonance frequency ($\tau = 2\pi/\Delta K_0$); and $k_s$ are the new US energy gaps determining the additional regions of Bragg reflection.

The change in the diffracted neutrons intensity distribution for the crystal by the AW can be interpreted quantitavely and qualitatively by the dynamical diffraction theory. Calculations taking into consideration the strains created by AW which affect the neutron wave field inside the crystal can be carried out using the Takagi–Taupin equations [34,35]:

$$i\beta\frac{\partial\Psi_0}{\partial t} = -i\frac{\partial\Psi_0}{\partial z} - i\tan\Theta_B\frac{\partial\Psi_0}{\partial x} + \frac{\Delta k_0}{2}\exp(iHu)\Psi_h \tag{3}$$

$$i\beta\frac{\partial\Psi_h}{\partial t} = -\frac{\partial\Psi_h}{dz} + i\tan\Theta_B\frac{\partial\Psi_h}{\partial x} + \frac{\Delta k_0}{2}\exp(-iHu)\Psi_0 \tag{4}$$

where

$$\beta = 1/(v_n\cos\Theta_B)$$

and

$$u = w\cos(\omega t)\cos(k_s z) \tag{5}$$

is the displacement of the nucleus for standing transverse waves excited between the two parallel surfaces of the sample with waves amplitude $w$, wave-length $\lambda_s$, wave vector $k_s = 2\pi/\lambda_s$. $\Psi_0$, $\Psi_h$ are the amplitudes of the incident and diffracted beams, respectively, $v_n$ is the velocity of the neutron wave and $\Delta K_0 = 2\pi/\tau$ where $\tau$ is the extinction length. The full solution of the Takagi–Taupin equations isn't possible now but some approximations can be done and this allows calculating diffraction intensities at the center of the Bormann fan in the quasi-classical approximation depending on $w$, $k_s$, $\lambda_n$ and $\tau$.

## 2.2. Experimental

The experimental layout for Laue geometry is shown in Fig. 2. As the sample, a Ge ((111) reflection single crystal (V= 52x22 x20 $mm^3$= YxLxT $mm^3$) was taken. A monochromatic and well-collimated neutron beam with a cross section of 1x10 $mm^2$ was directed to the sample. The sample quality was checked preliminarily and the FWHM of the rocking curve was 3.1″, close to the theoretical expectation. For the observation of the diffracted neutrons distribution two techniques were used. One of them using analyzing slit similar to the classic experiments [36-38] but with width 0.5 mm in difference of 0.1 mm slits in [36] to avoid "non-sound" intensities oscillations (Shull's fringes). At other method an analyzing slit was not used in general. Instead, the spatial distribution of the reflected neutron beam was measured with a position sensitive detector (PSD) After finding the optimal relation $\Theta$ - $2\Theta$ (adjustment) the sample and PSD were kept motionless and the reflected intensity distribution was directly measured in $2\Theta$ – coordinates then recalculated for current x-coordinates . This procedure is similar to the scanning with slit method, however, if PSD has good resolution enough, allows a much faster data acquisition.

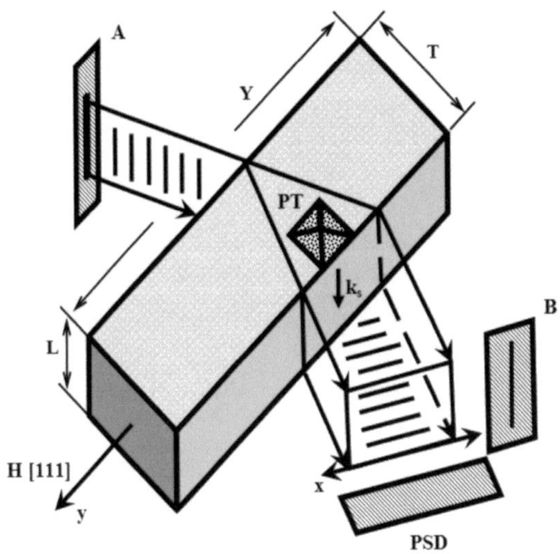


**Figure 2.** The layout of the experiment: A and B are forming and analyzing slits, respectively; PSD is position sensitive detector; PT is a piezo transducer; H is the vector of the reciprocal lattice; $k_s$ is AW wave number; y is direction of samples movements (slit B is not moving), x is a current coordinate in the base of Borrmann's fan and a direction of slit B movement. In this case, the sample is not moving.

## 2.3. Acoustic field in the sample

The transverse AW propagated perpendicular to the scattering vector ($k_s \perp H$, $u \| H$). Polarized AW with amplitude $w$ was parallel to the vector H. Piezotransducer was glued to the sample by salol. As shown below, the diffraction intensity is proportional to $|Hw|$, where $w$ is the AW amplitude, therefore its values could be used for estimation of the acoustic field inside the crystal. Owing to the diffractive divergence of acoustic waves this field is concentrated not only in the region around the piezotransducer but is distributed uniformly enough through the whole sample, especially at the levels of the weak and moderate excitation.

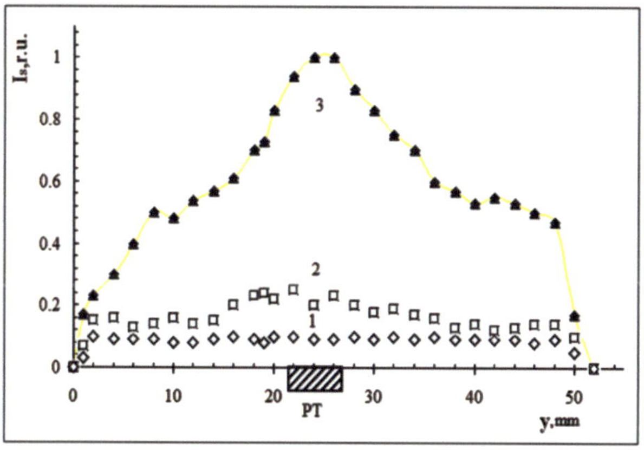


**Figure 3.** Diffraction intensity distribution in relative units at sample displacement along y-axis: 1 - in the absence of ultrasonic excitation; 2 - $V_G$=0.1V; 3- $V_G$=0.8V. The $I_s$ value is proportional to the AW amplitude.

The uniformity of acoustic field distribution in the region of Borrmann's fan is an important prerequisite to a sure observation of spatial oscillations of the diffraction intensity (see below). Assume that in the diffraction process a finite volume of the sample participates, in which the amplitude $w$ of AW oscillations is distributed in the interval $w \pm \Delta w$. This leads to a smearing of the pulse ($k_s \pm \Delta k_s$), therefore the phase difference on the crystal surface will be $\Delta\varphi = \Delta k_s T = \Delta k_0 H T(w \pm \Delta w)$. Since the intensity oscillations are approximately described by the function $I \sim cos^2(\Delta k_s T)$, the $n$-th intensity maximum corresponds to the condition $\Delta\varphi = \Delta k_0 H wT$. At a phase shift of $\Delta\varphi_1 = \pi/2$ the maximum of $I_s$ becomes its minimum. Therefore, to observe oscillations it is necessary that the phase shift $\Delta\varphi_1$ be less than $\pi/2$. From this follows the estimation $\Delta w/w < 1/2n$. At reasonable $w$ values and $\Delta w/w \approx 10\%$ this leads to the situation when in the experiment only few first oscillations could be observed (see below). From the data shown in Fig. 3 it is seen that these conditions are fulfilled well at low amplitudes of AW (curve 2, Fig. 3).

## 2.4. Acoustic waves velocities

The evidence of the standing AW existence in the single crystals is shown on the Figs. 4 and 5 (a-c). As a rule, theoretical estimations of the AW effect on the diffraction intensity were made for the case of coherent sound. As follows from Fig. 4a, in our experiment a wide enough frequency spectrum of excitations and AW cannot be considered as single mode. However, as was previously shown [28], the assumption of sound coherence is not always necessary for the correct interpretation of experimental results. When the diffraction intensity depends on frequency under scanning with a smaller step (Fig. 4b), a fine structure appears which is the evidence of the presence of an AW standing wave in a crystal. When the diffraction intensity depends on frequency under scanning with a smaller step (Fig. 4b), a fine structure appears which is the evidence of the presence of an AW standing wave in a crystal. For AW standing waves the relation

$$L = i\lambda_S/2 \tag{6}$$

should be fulfilled, where *i* is an integer of AW half-waves, $\lambda_s$ is the AW wavelength, and *L* is the distance of wave propagation (Fig.2). Knowing *L* and the distance between the maxima (minima) of the fine structure of the frequency dependence $\Delta v_s = \left[ v_{si} - v_{s(i \pm 1)} \right]$ (Fig. 4b), it is possible to determine the velocity of sound propagation in the[100] direction. Determining $\Delta v_S$=(0.08±0.01) MHz from Fig. 4b data we obtain $i$ = 517-524, which is testimony to a very high quality of the sample, and $v_{s[100]}$= (3.52±0.03)$10^5$ cm/s. This value is in very good agreement with the reference data 3.55.$10^5$ cm/s [36] for the velocity of a shear AW for Ge in the indicated direction.

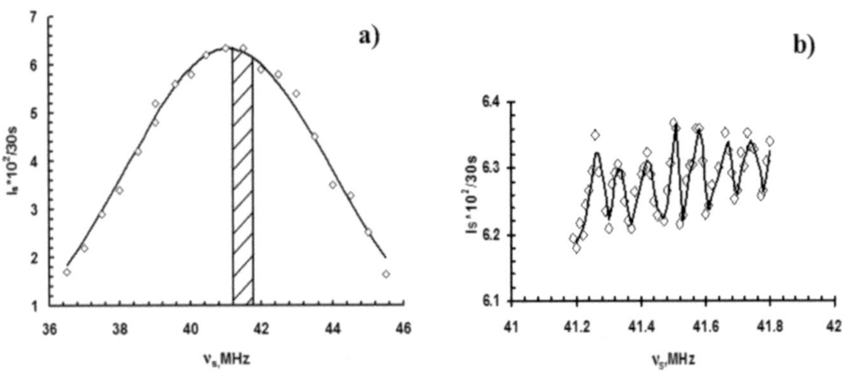


**Figure 4.** Dependence of the Laue diffraction's intensity $I_s$ at the Bormann's fan center on the ultrasound frequency ($\lambda_n$=0.410 nm, *S*=1 mm, *L*=22 mm): a) the frequency scanning step is 0.5 MHz; the solid curve is a result of data fitting by Gaussian with FWHM=6.34 MHz; the region in which the detailed frequency scanning was carried out is shaded (41.2-41.8 MHz); b) the same as in (4a) only for the narrow frequency range, shown in the Fig. 4a as a shaded. The frequency scanning step is ~0.01 MHz; the solid curve was obtained using the averaging on three points data (spline approximation).

The presence of frequency satellites shown in Fig. 5(*a–c*), which exist at equal distance from the main peak independently of the chosen frequency interval, is the one more evidence for excitation of standing waves in the crystal. And the sound velocity $v_s$ in the [111] direction parallel to the scattering plane can be determined. Defining $\Delta v_s$ from the dependences shown in Fig. 5*a–c*, where $\Delta v_s$ = = (1.475. ± 0.046) MHz one can see that the number of transverse half-waves *i*=10-12 for $v_s$= 14.84 MHz and $v_{s[111]}$ = 5.13 $10^5$ cm/s. This value is very close to the reference data of 5.09 $10^5$ cm/s [39]. This means that the values of US wave velocities in crystals can be determined with neutron and X-ray diffraction technique, and that this method is applicable for, e.g., determination of the sound velocity at phase transitions.

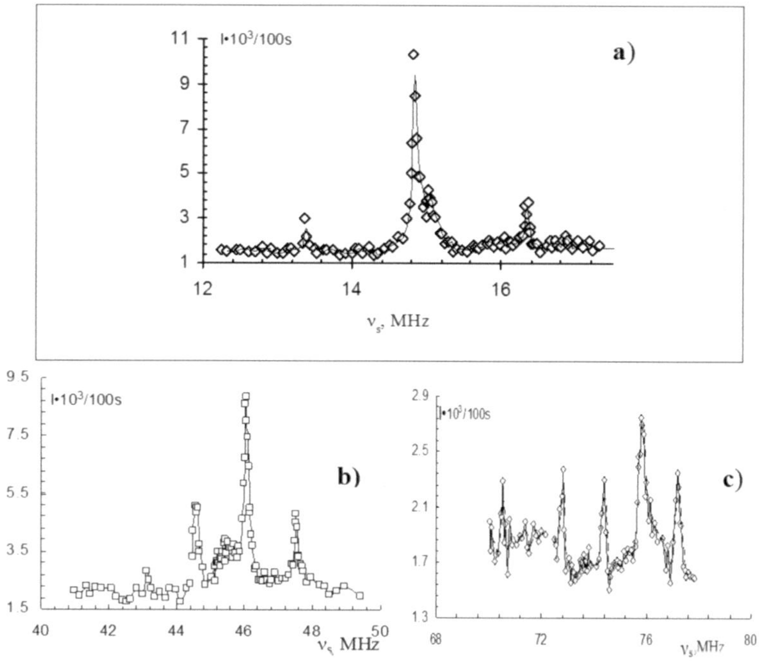


**Figure 5.** Standing AW observation in the single thin Si crystal (T=1.73 mm):a) for main transducer harmonic (15 MHz); b) the same for third harmonic; c) the same for fifth harmonic. Solid curves were obtained using the averaging on three points data (spline approximation).

## 3. Spatial intensity distribution of diffracted neutron beam

Owing to the interference of Bloch's waves the distribution of diffracted neutron beam intensity in a crystal arising on its outside surface has characteristic beats at the center and intensity is increasing toward the edges of the Bormann fan. These effects are described by the expression (7) [40]:

$$I_0(\Gamma) = c(1-\Gamma^2)^{-1/2}\cos^2(A(1-\Gamma^2)^{-1/2}) \tag{7}$$

where $\Gamma = x/(2T\ tg\ \Theta_B)$ is the deviation of a neutron wave from the atomic plane trace; $A=\pi T/\tau$; $c$ is a normalized constant, and $x$ is a current coordinate in the base of Bormann's fan. Expression (7) is valid for the diffraction of a plane monochromatic wave in the symmetrical Laue case. A more rigorous expression for the shape of the reflected beam distribution can be obtained with the help of spherical neutron waves [40,41]:

$$I_0(\Gamma) = \frac{A}{2} J_0^2(A(\sqrt{1-\Gamma^2}) \tag{8}$$

where $J_0$ is Bessel function of the zeroth order. According to Exprs. (7,8) the intensity of a diffracted beam oscillates on the exit surface of a crystal, with the oscillation period depending on parameter $\pi T/\tau$ and fast decreases outward from the center of profile ($\Gamma$=0).

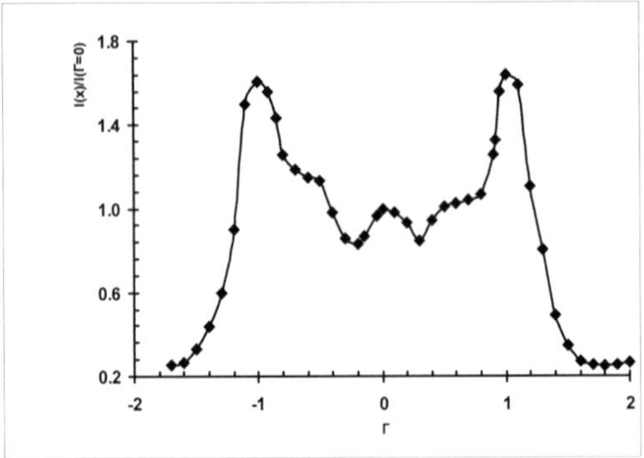


**Figure 6.** Normalized intensity distribution for Si single crystal (reflection reflex (111)), slit B width S≤ 0.2 mm. $T$=11 mm. Neutron wavelength $\lambda_n$ is equal 0.1 nm. The Shull's fringe is clearly seen at the center of profile [42].

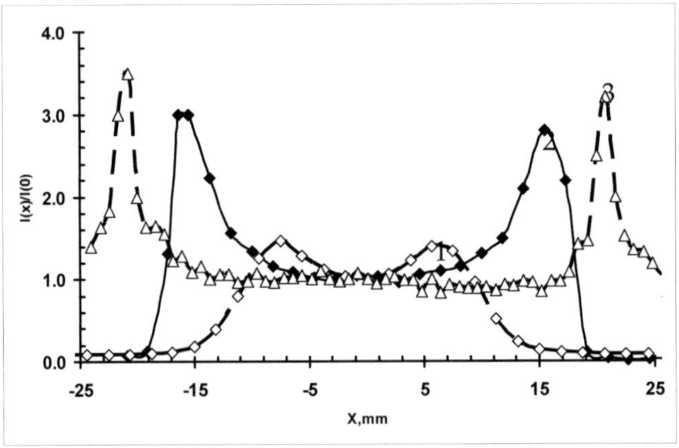


**Figure 7.** Normalized intensity distribution for reflection (111) Ge ($T$=20 mm) (Bormann fan) for different neutron wavelengths without sound: 1- $\lambda_n$= 0.471 nm; 2- $\lambda_n$ =0.410 nm; 3- $\lambda_n$=0.243 nm. Curve 2 was obtained using a slit with $S$= 1 mm. Curves 1 and 3 were obtained without analyzing slit (PSD only).

## 3.1. High-frequency sound effect on the spatial distribution of diffracted neutron intensity

The presence of new energy gaps $k_s$ (Fig. 1) on the dispersion surface under ultrasound should lead to the appearance of an additional structure in the intensity distribution of diffracted neutrons at the exit of a crystal. The new distribution, as is shown below, depends on the AW amplitude, $w$, while "sound" oscillations are superimposing on the initial ones. The size of the first soundless oscillation for the Ge (111) reflex according to formula (7) is $\Delta x_1=(T\ \tau)^{0.5} tg\Theta_B \approx 1.5$ mm (in the vicinity of Γ=0). The numerical analysis of experimental data based on the known theoretical expressions involves difficulties, since they have been obtained in the plane neutron wave approximation. Such approximation cannot be applied to description of the intensity distribution of diffracted neutrons on the Bormann fan's base when the whole dispersion surface is excited.

We have derived expressions for the diffraction intensity of a spherical neutron wave in an ideal crystal under ultrasound excitation. In this case the total diffraction intensity $I_t$ is composed of the elastic and inelastic components:

$$I_t = I_{el} + I_{in}$$
$$I_{el} = I_0 + \Delta I_{el} \tag{9}$$

where $\Delta I_{el}$ and $I_{in}$ are the elastic and inelastic addends, respectively; $I_0$ is determined from expression (8), while the elastic addend to the diffraction intensity is given by the expression (10):

$$\Delta I_{el} = \frac{2\Delta K_0^2 (Hw)^2 \sin^2(\delta q T/2) J_1^2\left(AHw(1-\Omega)^{0.5}\right)}{\left(\delta q^2 - \Delta K_0^2\right)(1-\Omega)} \tag{10}$$

where $J_1$ is Bessel function of the first order, $\delta q$ is the shift of the DS at absorption or emission of an ultrasound phonon. The main contribution to the intensity of inelastic (one-phonon) scattering is made by the term:

$$I_{in} = (Hw)^2 \left(1 + \frac{\delta q^2}{\left(\delta q^2 - \Delta K_0^2\right)}\right) J_0^2\left[AHw(1-\Omega)^{0.5}\right] \tag{11}$$

where

$$\Omega = \frac{\Gamma^2 \delta q^2}{\left(\delta q^2 - \Delta K_0^2\right)}, \quad A = \pi\, T/\tau \tag{12}$$

Formulas (10-12) describe the central part of a Bormann fan:

$$-\left(1 - \Delta K_0^2/\delta q^2\right)^{0.5} \le \Gamma \le \left(1 - \Delta K_0^2/\delta q^2\right)^{0.5} \tag{13}$$

besides, these are valid at $\Delta K_0\, Hw \ll (\delta q - \Delta K_0)$, which corresponds to the weak interaction of satellites with the main Bragg maximum. At $v_s \gg v_{res}$, ($\delta q \gg \Delta K_0$), i.e. under the conditions of our experiment, $I_{in}$ far exceeds $\Delta I_{el}$, which was previously confirmed by our experiments with measuring the spin-echo in a silicon single crystal under ultrasound pumping [43]. The spatial distribution of diffraction intensity is described by the expression:

$$I_s(\Gamma, Hw) = I_0 + I_{in} \tag{14}$$

where $I_{in}$ is taken in the form of equation (11), and $I_0$ — of equation (8). The general character of the spatial distribution of diffraction intensity is dictated by three parameters: $A$, $Hw$, and the $\delta q/\Delta K_0$ ratio. The choice of a thick ($T \approx 70\ \tau$) Ge crystal is not quite optimal, since the first AW oscillations will appear when the "acoustic" extinction length becomes equal $\tau_s=\tau\,|H\,w|^{-1}$. On the other hand, a thick crystal provides long enough Bormann fan, which allows for easy and detailed measurements of the spatial distribution of diffracted beam intensity (see Fig. 8)

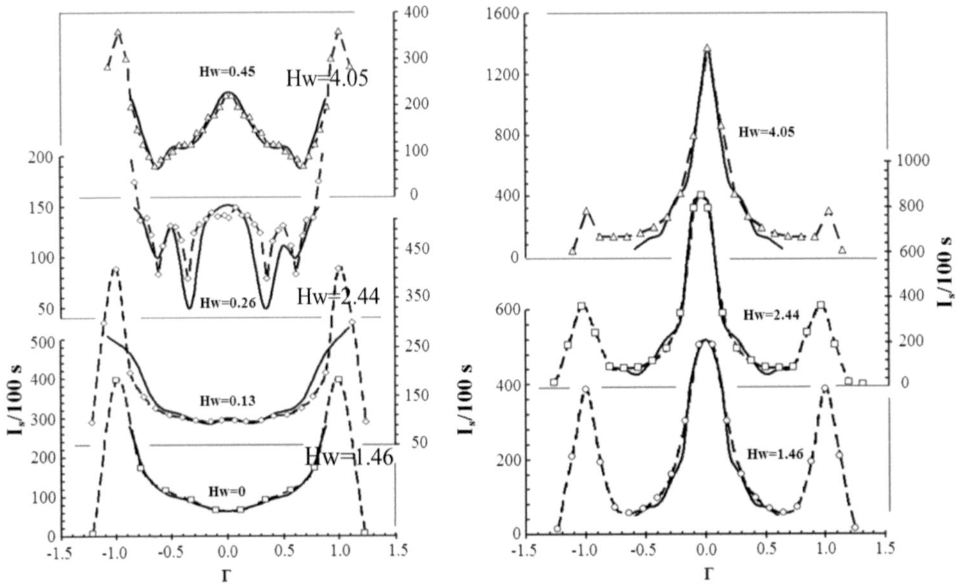


**Figure 8.** Spatial distribution of a diffracted beam intensity *vs* AW amplitude $w$ and $\Gamma$ value. Dashed curves and points – averaged experimental data, solid curves is a result of data fitting by Exprs. (11 and 8)

Already at a small amplitude of the acoustic wave $Hw$=0.26 ($w$=0.013 nm) the diffraction intensity in the center starts rising noticeably. The width of the first oscillation (Fig. 8) on the $x$-axis (close to $\Gamma$=0) at $Hw$=0 could be estimated from expression (8): $\Delta\,x_1=(2T\,\tau)^{1/2}tg\,\Theta_B$ and $\Delta\,x_1\approx$6 mm for the Ge (111) reflex. Since at the appearance of sound a new extinction length is $\tau_s=\tau\,|Hw|^{-1}$, at $Hw<<1$ the linear size of the first "sound" fringe will be increasing as $\Delta\,x_s=\Delta\,x_1\,|Hw|^{-0.5}$. As follows from Fig. 8, at $Hw$=0.26 the length of the first acoustic fringe is 11.4 mm, which coincides well with the calculated value of 12 mm. Besides, at least two next oscillations could be clearly discerned. As $Hw$ is further increasing, at the profile center a linear intensity rise is observed, while the diffraction intensity on the Bormann fan edges ($\Gamma=\pm 1$) remains practically unchanged. To compare the theoretical and experimental data a high-precision determination of the $A$ and $Hw$ parameters is needed, since the Bessel function is fast oscillating at large values of the argument. The calculated $A$ value is 2071, i.e. close to $A$=2060, at which "soundless" distribution of the diffraction intensity ($Hw$=0) is described most satisfactorily by expression (8), taking into account integration over the width of analyzing slits and correct averaging. The AW amplitude can be determined by changes in relative intensity variations at the center of the spatial profile ($\Gamma$=0):

$$\eta=\frac{I_s-I_0}{I_0}=\frac{I_{in}}{I_0}=\frac{(Hw)^2\left[1+\delta q^2/\left(\delta q^2-\Delta K_0^2\right)\right]J_0^2(HwA)}{J_0^2(A)} \qquad (15)$$

At large values of argument

$$J_0(z) \to \sqrt{2/\pi z}\cos(z-\pi/4), \tag{16}$$

and the relative contribution of the neutron inelastic scattering by AW of the lattice with a frequency above resonant is:

$$\eta\,(Hw) = 4Hw\frac{\cos^2(HwA)}{\cos^2(A)} \simeq 4Hw \tag{17}$$

The experimental data shown in Fig. 9 were obtained for one and the same sample but on different neutron diffractometers using different RF-generators and wide-band amplifiers. Besides, every time the piezotransducer was glued again; however, in all cases a linear dependence of $\eta$ on the generator voltage $V_G$ was observed , which makes it possible to find a calibration constant C and to determine the AW amplitude $w$ from the relationship: $\eta$=C $V_G = |Hw|$.

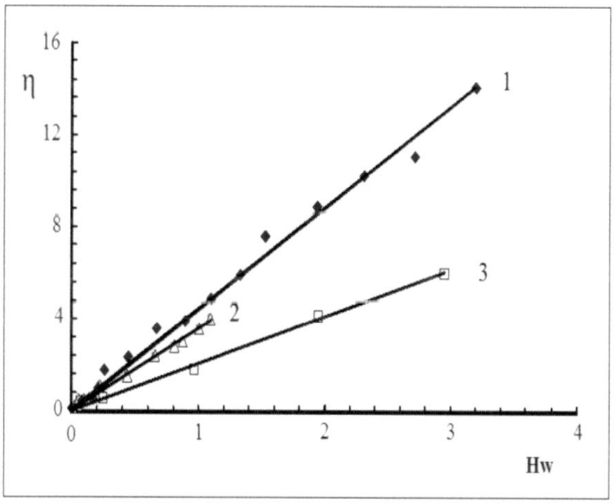


**Figure 9.** Diffraction intensity at the center of a Bormann fan (Γ=0) *vs.* acoustic wave amplitude for neutrons of different wavelength: 1) $\lambda_n$=0.41 nm; 2) $\lambda_n$= 0.471 nm; 3) $\lambda_n$=0.243 nm. Solid curves: fitting with eq.(18) taking into account $v_s \gg v_{res}$ $(\Delta K_0 \ll \delta q)$

The effect of saturation was observed in the $\eta$ dependence versus $V_G$ for large $Hw$. The heating of the organic salol gluing leads to the decrease of its viscosity and brings about violation of the proportionality between $V_G$ and $w$. This fact was taken into account, and the measurements were considered reliable up to $Hw \le 5$

Fig. 10 shows the spatial distribution of diffraction intensity $I_t$ depending on the acoustic wave amplitude for the neutrons of different wavelength. Independently on the neutron waves length, $I_t$ at the Bormann fan centers increases with $Hw$ growth.

Figure 11 demonstrates a sharp decrease in the spatial half-width of the central peaks in dependence on the AW amplitude. This result is seems to be unexpected, since in some earlier works where the possibility to create the ultrasound-driven/controlled monochromators was discussed [12,27,44] the gain in intensity with rising AW amplitude is always compensated by a loss in resolution (due to FWHM rise). In contrast, in our experiments the FWHM *decreases* 3-4 times at the Hw increased twice. To study this effect in more detail, additional investigations are needed – for example, measuring the FWHM of double-crystal rocking curves in dependence on the *Hw* at the center of Bormann's fan.

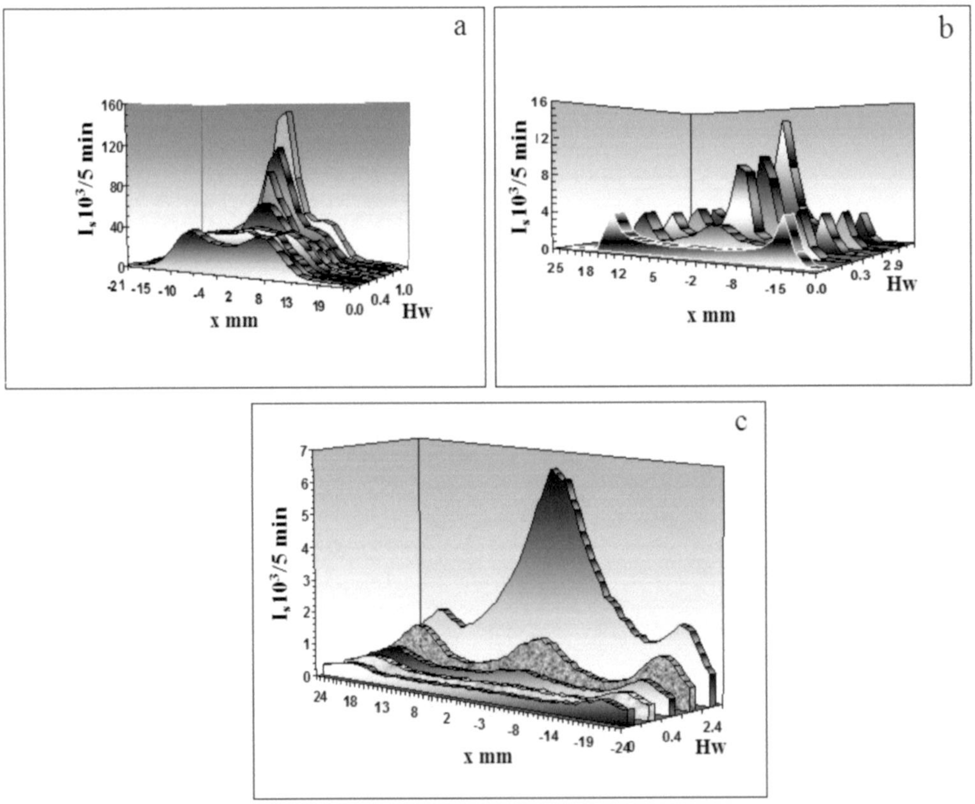


**Figure 10.** The neutron diffraction intensity as a function of acoustic wave amplitudes *Hw* and running coordinate *x* in the base of a Bormann fan for neutrons of different wavelength: a) $\lambda_n$=0.410 nm; b) $\lambda_n$= 0.243 nm; c) $\lambda_n$= 0.471 nm.

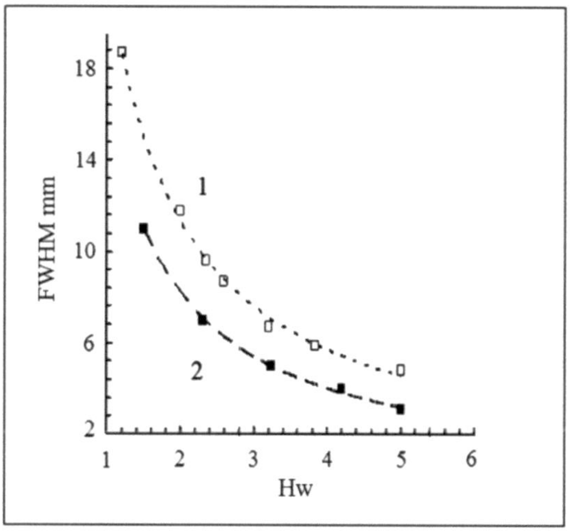


**Figure 11.** Full width at half maximum of the central peaks vs AW amplitude for neutron with different waves length: 1) $\lambda_n$=0.410 nm, 2) $\lambda_n$=0.243 nm

# 4. Diffraction - Deformed crystal

## 4.1. Theoretical background

In contrast to a perfect crystal where the US effect leads, as a rule, to an increase in the diffraction intensity, in a bent smoothly deformed sample a drastic decrease in the intensity $I_{ds}$ is observed already for very small AW amplitudes. In deformed crystals, which are of great practical interest, the effect of US on diffraction has been investigated much less. In [14, 45, 46] the results of theoretical and experimental investigations of neutrons and X-rays for the case of Laue diffraction in deformed silicon single crystals under high-frequency ($\nu_s >> \nu_{res}$) and in conditions of the neutron (X-ray) - acoustic resonance ($\nu_s \approx \nu_{res}$) are reported. The analysis has shown that in a deformed Si crystal ultrasound results in violation of the adiabatic conditions for the movement of tie points on the dispersion surfaces. Owing to this, a drastic decrease in the diffraction intensities was observed for low acoustic wave amplitudes. With the AW amplitudes increasing the diffraction intensity also increases, reaching the kinematical limit. In absolute values, the influence of the US is much stronger manifested in the diffraction on a deformed crystal than on a perfect one. A substantial role is played by multiphonon processes. The presence of static strains leads to the appearance of a new type of oscillations, which depend on the deformation gradient.

As distinguished from the perfect crystal case where for the whole crystal a unitary DS exists, in the Penning–Polder–Kato model [47,48] to each point of a bent crystal an own two-sheet dispersion surface corresponds, and the neutron is travelling adiabatically inside the crystal without transitions of the excited tie point between the effective DS sheets (Fig. 12)

The role of ultrasound in this model consists in the resonant suppression of adiabatic movement of the tie points. The one-phonon absorption corresponds to path 1–2–6–7–8 of the tie point along the dispersion surface. The neutron incident on the crystal excites points 1

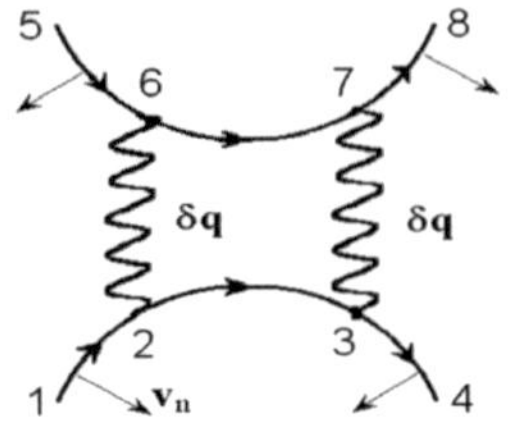


**Figure 12.** Dispersion surfaces of a crystal modified by ultrasound. Dynamics of travelling of tie points in a deformed crystal. Laue geometry. Directions of tie points and neutron group velocities are shown by arrows.

and 8 on the DS. In the absence of ultrasound, point 1 travels along path 1–2–3–4 and passes into the state corresponding to a diffracted wave. Point 8 makes no contribution to the diffraction. When US is switched on, the movement of point 8 is not disturbed. As concerns point 1, it can reach state 4 by two ways: a) 1–2–6–7–3–4 or b) 1–2–3–4. If the probability for the excitation point to remain on DS sheets at points 2, 3 6, and 7 is *P*, and the probability to pass onto another DS sheet owing to the US disturbance is *M* (*P*+*M* = 1), then the probability of the former process is equal to $M^2$, the probability of the latter process is equal to $P^2$.The change in the relative diffraction intensity will be

$$\eta=(I_{ds}-I_{d0})/I_{d0}=P^2+M^2-1=-2MP<0. \tag{18}$$

The probability of transition for the excitation point 1 in the case of Laue's diffraction is approximately described as

$$M=1-\exp[-\pi(Hw)^2/2B] \tag{19}$$

where *B* is the deformation gradient .

The maximum transition probability is 0.5, (for $v_s > v_{res}$) and, correspondingly, $\eta_{max}= 0.5$ according to expression (19). Interference between the trajectories of the two scattering process leads to the oscillating dependence on $B^{-1}$ of the maximum "dip" of the diffraction intensity. These oscillations modify the expression (19) as follows:

$$\eta=-2MP\left(1-\cos 2\varphi\right) \tag{20}$$

with phase factor ϕ [14]:

$$\varphi = (a/B)(a^2-1)^{0.5}; a=k_s/\Delta K_0 \tag{21}$$

For the Laue-symmetrical diffraction of neutrons and X-rays in a two-wave approximation when the propagation of neutron waves in a crystal with static and/or dynamic (US) deformations is described by the Takagi–Taupin equations (Expr. 3-5), where the displacement of nuclei, *u*, is introduced in forms [46]

$$\begin{aligned} Hu &= Hu_s + Hu_d \\ Hu_s &= 4Hw\cos[k_s(s_0+s_h)\cos\Theta_B]\cos(\omega_s t) \end{aligned} \tag{22}$$

$$Hu_d = 2as_0^2 + 4bs_0 s_h + 2cs_h^2$$
$$B = b[2/(\Delta k_0 \cos\Theta_B)]^2 \quad C = c[2/(\Delta k_0 \cos\Theta_B)]^2 \tag{23}$$

where *a, b,* and *c* are numerical constants describing inhomogeneous deformation; $u_s$ is the nuclear displacement in the acoustic field, and $u_d$ is the nuclei static displacement associated with the lattice deformation.

In the quasi-classical approximation ($B$ <<1 but $BT$ >>1), for $Hw$< 1 we will have analytical expressions (24-26) obtained by method of successive approximations for relative intensity variations depending on the relationship between $k_s$ and $k_0$. If the frequency of acoustic waves $\nu_s$ is higher than the resonance frequency $\nu_{res}$, ($\alpha > 1$) the second term in (26) describes the so-called "deformations oscillations" whose period depends on $B$ and $\alpha$ .

$$\eta = 1 - \frac{I_h(Hw,B)}{I(0,0)} = \frac{\pi[Hw\cos(\omega t)]^2}{B[8(1-\alpha)]^{1/2}} \times \exp\left(-\frac{2^{5/2}(1-\alpha)^{3/2}}{3B}\right) \quad \alpha < 1 \tag{24}$$

$$\eta = \frac{\left(Hw\cos(\omega t)\Gamma(1/3)\right)^2}{\pi B^{4/3}} \quad \alpha \gg 1 \tag{25}$$

$$\eta = \frac{\pi\left[Hw\cos(\omega t)\right]^2 \alpha}{B(\alpha^2-1)^{1/2}} \times \left[1 + \sin\left(\frac{(\alpha^2-1)\alpha}{2B} + \text{arcsinh}(\alpha^2-1)^{1/2}\right)\right] \quad \alpha > 1 \tag{26}$$

where $\alpha = k_s/\Delta k_0$.

The experimental verification of the qualitative validity of expressions (23-26) is presented on the Figs. 13-15

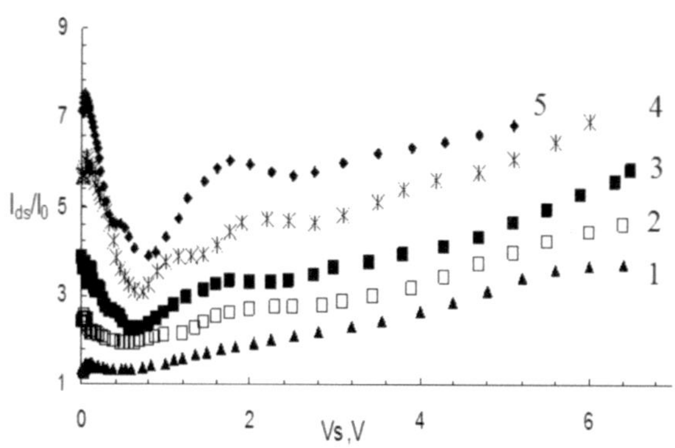


**Figure 13.** Variations in the relative intensity of diffraction in a deformed crystal *vs.* voltage on piezotransducer for $\alpha$ =3.13 ($\nu_s$ = 46.04 MHz) and different $B$: *1* – 0.023(▲); *2* – 0.052 (□); *3* – 0.083 (■); *4* – 0.124 (*); *5* –0.166 (♦); $I_0$ is the diffraction intensity in a perfect crystal without sound; $I_{ds}$ the same for a deformed one.

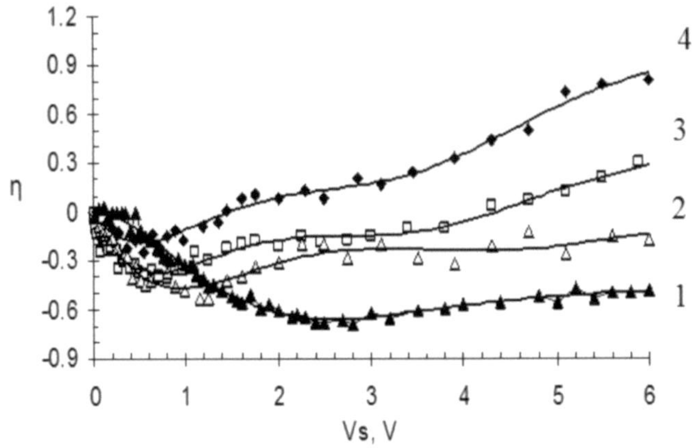


**Figure 14.** Relative diffraction intensity *vs.* piezo transducer voltage for different values of parameter $B$: 1) – 0.06, 2) – 0.11, 3) –0.23, 4) –0.89; $\nu_s$=14.84 MHz ($\alpha\approx 1$). The best fitting is obtained using series expansion of the Bessel functions $J_0\left[\frac{(HW)^2}{B^{4/3}}\right]$ (solid curves).

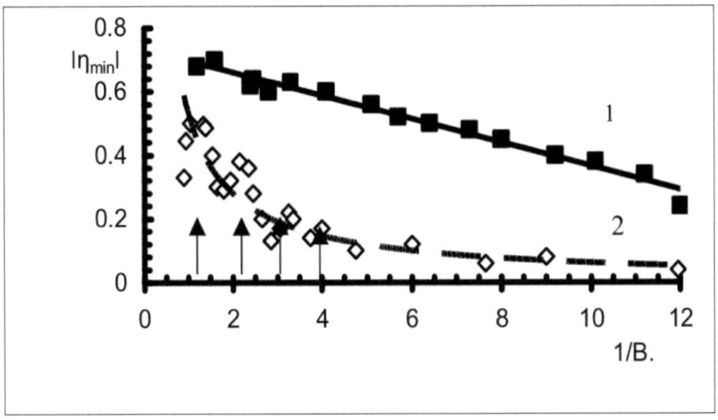


**Figure 15.** The module of "dip" maximum depth $\eta = |I_{ds}-I_{d0}|/I_{d0}$ *vs.* the inverse value of the deformation gradient $B$: 1(■) – for $\nu_s$ = 14.84 MHz (near to neutron acoustic resonance, $\nu_{res}$ = 14.71 MHz; $\alpha\approx 1$); 2(♦)– for $\nu_s$ = 46.04 MHz; $\alpha\approx 3$. Arrows show the maxima positions calculated by Eqs. 20 and 21

The solid (1) and dashed (2) curves in Fig. 15 were obtained from experimental data by the least squares method. The best fitting is at $|\eta_{min}| = 0.53\cdot(1/B)$ for curve (2) and $|\eta_{min}| = -0.04B^{-1} + 0.74$ for curve (1). From these it follows that at $B \sim 1$ the "dip" of the relative diffraction intensity $\eta$ reaches the maximum of ~0.5 for $\nu_s >> \nu_{res}$ and of 0.7 for $\nu_s \approx \nu_{res}$. The strong effect near the neutron-acoustic resonance ($\alpha \approx 1$) follows from Eq. 25 and by analogy with the resonance destruction of Bormann's effect [31] at the X-ray – acoustic resonance can be qualitatively explained by the fact that the transitions between DS branches in the case of their contact occur in a much greater volume of the pulse space.

## 4.2. Ultrasound effect on neutron bragg diffraction in a deformed silicon single crystal [23]

The scheme of the bending device, the neutron scattering geometry, and the US wave's propagation direction are shown in Fig. 16. The (220) and (440) reflections intensities were simultaneously measured at a standard single-crystal spectrometer by the time-of-flight technique [15]. The neutron wave lengths were 0.96 and 1.92 Å. The radius of curvature $R$ of the reflecting surface was determined by sag $h$ measured by a micrometer ($R= L^2/8\,h$), where $L$ is the sample length equal to 12 cm. A sag was measured from 0 to 40 μm and $R$ from 6 km to 51 m.

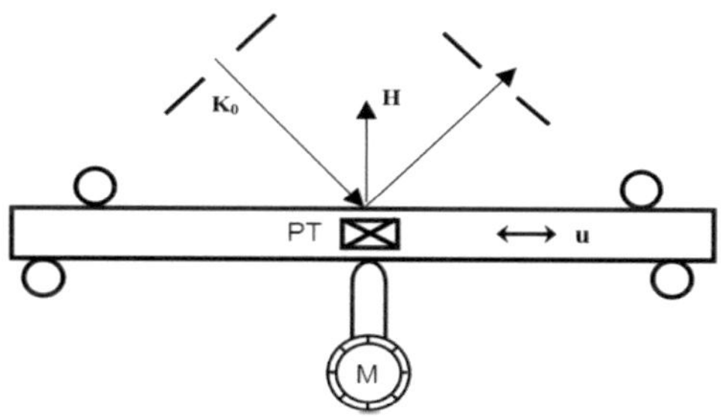


**Figure 16.** Scheme of experiment: H is the reciprocal lattice vector; $K_0$ is the wave vectors of neutron and ultrasound wave.$u=w\cos(\omega_s t)\cos(k_s x)$ is nuclei displacement in a US wave with amplitude $w$. PT is piezotransducer. M is micrometer. $R=L^2/8h$, where $L$=12 cm is a samples length, $h$ is the sag.

The transversal US wave ($k_s \perp K_0, H\|w$; $w\|H$), where $H$ is the reciprocal lattice vector, $k_s$ and $w$ are the US wave vector and amplitude, was excited in the [111] direction by a $LiNbO_3$ piezo transducer glued to the sample by salol. The fundamental piezoelectric transducer harmonic ($\nu_s$= 26.5 MHz) was used. Fig. 17 shows typical plots of the relative diffraction intensities $\eta=(I_s-I_0)/I_0$ vs the piezo transducer voltage for the two reflections (220) and (440) in the "perfect" crystal (without bending). It is seen that linear dependence is a good approximation.

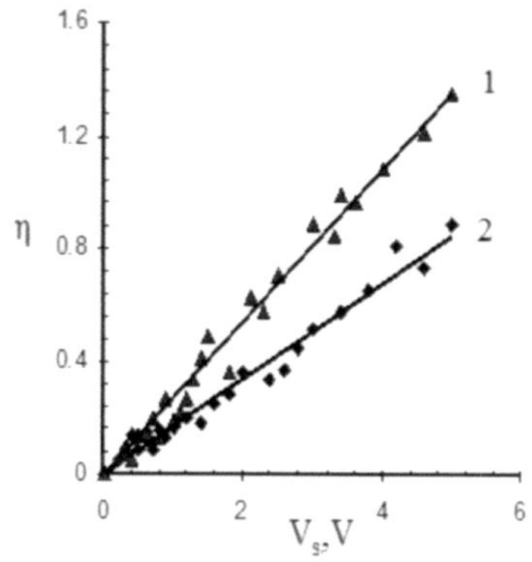


**Figure 17.** Relative changes of the diffraction intensities $\eta = (I_s - I_0)/I_0$ for perfect crystal for two reflections: 1-(▲) for (440), 2-(♦) for (220)

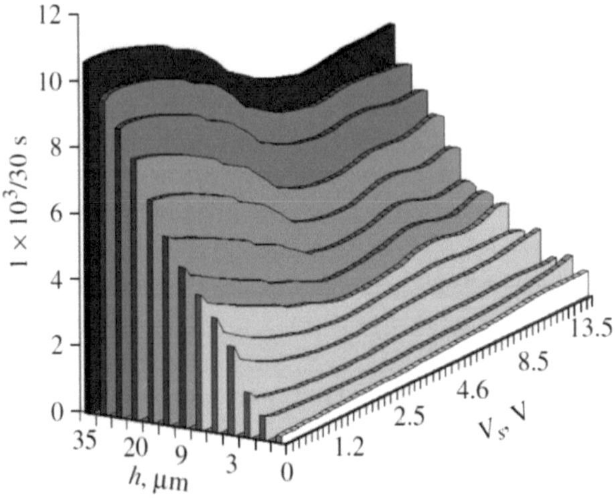


**Figure 18.** Common result of the simultaneous effect of ultrasound wave and bending of a crystal on the diffraction intensity of neutrons.

The total change in the intensity $I_{ds}$ with a simultaneous increase in $V_s$ and $h$ is shown in Fig. 18. The appearance of dips (minima) in these dependences is a characteristic phenomenon similar same as for Laue case. The changes in $I_{ds}$ (440 reflection) as a function of $V_s$ are shown in details in Figs. 19a and b for a bent crystal for different values of the sag $h$. Instead of the linear increase in $I_{ds}$ for $h = 0$ (line *1*), a dip in $I_{ds}$ reaches 30% for $h$ =0.3 μm ($R \approx 51$ m) at $V_s$ = 1.1 V ($w \approx 0.05$ Å) (curve 2). With a further decrease in the bending radius the dip depth reaches the maximal value of 52% (Fig. 19a, curve 3), flattening out (Fig. 19b, curves *3–10*). Curve *11* demonstrates the shift of the $I_{ds}$ minimum over $V_s$ with an increase in $h$.

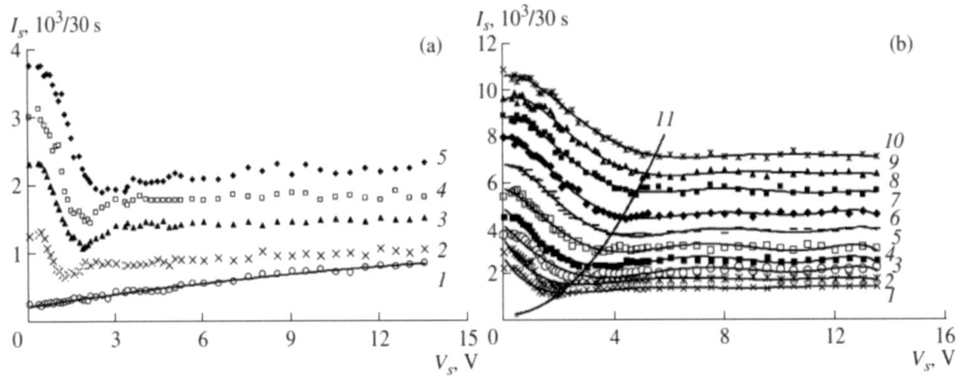


**Figure 19.** Diffraction intensity of the (440) reflection as a function of the US wave amplitude for different $h$, (a) (*1*) $h = 0$ (perfect crystal); (2) 1 μm; (*3*) 3 μm; (*4*) 5 μm; (*5*) 7 μm. (b) (*1*) 5 μm; (*2*) 7 μm; (*3*) 9 μm; (*4*) 12 μm; (*5*) 16 μm; (*6*) 20 μm; (*7*) 25 μm;(*8*) 30 μm; (*9*) 35 μm; (*10*) 40 μm; (*11*) displacement of the diffraction minimum along *Vs* as a function of *h*.

The simultaneous effect of both deformation types of a crystal lattice, static bending and dynamic atomic displacement in an US wave, on the neutron diffraction in the Laue geometry was taken into account in [14,46] where numerically (by the method of successive approximation) and, for separate cases, analytically by a modified Takagi–Taupin equation was solved. In [49], for this purpose Green's functions were used for Laue case. In the case of the Bragg geometry the mathematical task is substantially complicated because of specific boundary conditions and, as we know, has no analytical solutions up to now. The semi-phenomenological model explaining qualitatively the experimental results and allowing us to obtain quantitative characteristics in some cases is considered below in the framework of the dynamic scattering theory. The problem solution for the case of Laue diffraction in smoothly deformed crystals was described by us above and this model is also applied for the case of the Bragg geometry.

The role of US in a modified PPK model is the resonance suppression of adiabatic motion of tie points by analogy to Laue diffraction and schematically transition to the Bragg case can be done if to turn the Fig. 12 on 90° [23]. The relative change in the diffraction intensity $\eta$ can be again written in the form:

$$\eta=(I_{ds}-I_{d0})/I_{d0}=P^2+M^2-1=-2MP<0 \tag{27}$$

$$\eta = \left(I_{ds}-I_{d0}\right)/I_{d0} = P^2+M^2-1 = -F\left(B\right)MP < 0, \tag{28}$$

where $0 < F(B) < 2$ for the case $\nu_s > \nu_{res}$..

One can see that Expr. 27,28 coincides closely with Expr. 18,19 for the Laue diffraction.

Then we obtain $(Hw)_{min}$, the position of the $\eta$ minimum depending on the deformation gradient.

$$\left(Hw\right)_{min} = \left(ln\ 2B\right)^{0.5} \tag{29}$$

The changes of $\eta$ as a function of Hw are shown in Figs. 20a and 20b for comparison of the experimental data to the described model. Expressions (27-29) satisfactory describe the experiment only for the case of small *Hw* and small *B* (large bending radii) ($Hw < 0.3$ and $h <$ 7 μm). With an increase in both parameters these expressions do not correspond to the experimental data shown in Fig. 21b which are rather well described by the Bessel functions expanded in the alternating series.

Expressions (27-29) are valid for one-phonon processes of the energy exchange between the neutron and US phonon.

Using relation $Hw=cV$ and assuming $B = \alpha\, h$, Eq. (29) can be rewritten in the form

$$V_{smin} = \alpha_1 h^m \tag{30}$$

The graphs of the positions of minima $\eta$ as a function of $h$ are shown in Fig. 21 for the (220) and (440) reflections. The exponent $m$ is 0.49 for the (220) reflection (curve 2) and 0.46 for the

(440) reflection (curve *1*) and $\alpha_{1(440)}/\alpha_{1(220)} = 1.83$, which is rather close to the theoretical estimations of the exponent of 0.5 and ratio *H*440/*H*220 = 2. The value of the deformation gradient *B* is easily determined from here.

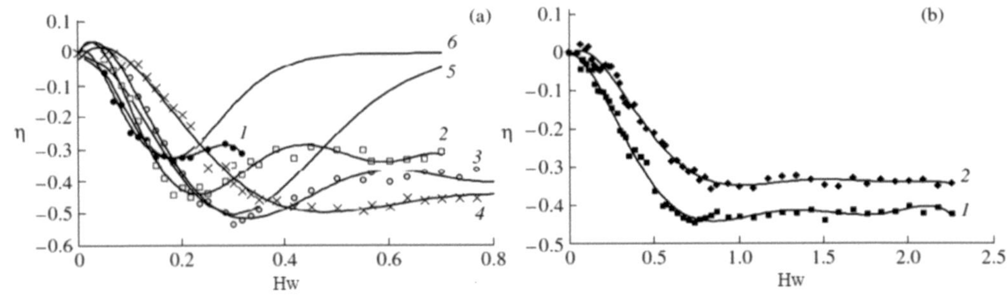


**Figure 20.** Relative changes in the diffraction intensity as a function of the US wave amplitude for different *h*: (a) (*1*) *h* = 0.4 μm; (2) 1 μm; (3) 3 μm; (4) 7 μm; curve *5* corresponds to Eq. (5) for *F*(*B*) = 2.0, *F*(*B*) = 1.3 for curve *6*. Solid curves *1–4* are fitting (Exprs.(27-29)) (b)1- *h* = 9 μm; 2- *h* =16 μm

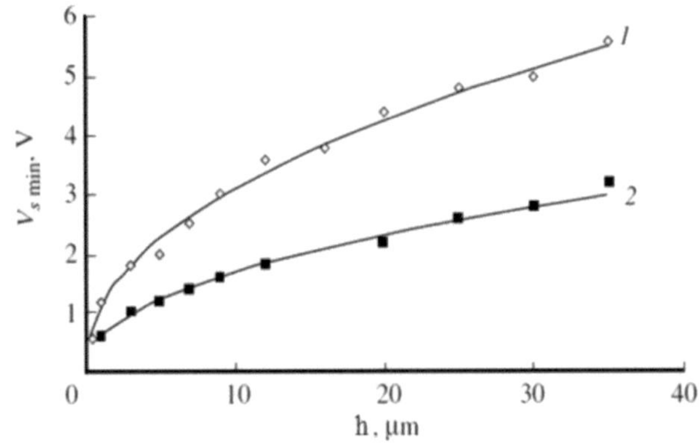


**Figure 21.** Shift of the minimum of $V_{s\,min}$ for relative changes in the diffraction intensity η with an increase in the value of bending deflection *h*: *1*: (*1*) for (220) reflection; (2) for (440) reflection.

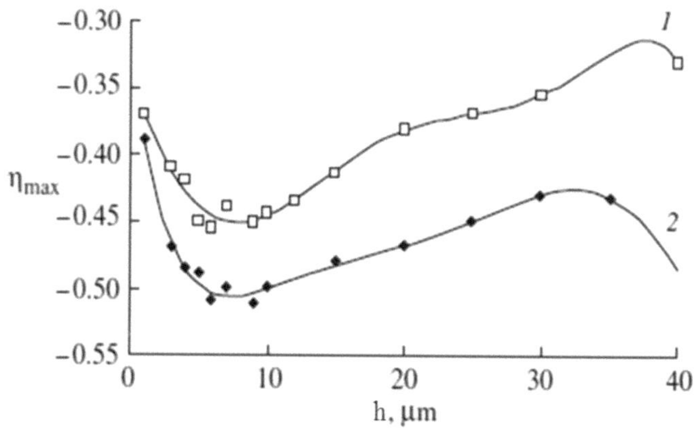


**Figure 22.** Dependence of the value of maximal dip of the relative diffraction intensity of neutrons $\eta_{max}$ on the value of bending of a crystal: (1) for (220) reflection; (2) for (440) reflection. Solid curves are fitting obtained by an expansion of the Bessel function in a Taylor series.

As it follows from Eq. (29), the maximal dip of the relative deformation's intensity $\eta_{max}$ shown in Fig. 22 is close to 0.5 for the (220) reflection and is reached for the bending radius $R$ =200–600 m. Thus, the PPK model modified for the case of ultrasound excitation satisfactorily describes some features of the neutron Bragg scattering from a bent silicon single crystal. The intensity dip of the Bragg diffraction is caused by the resonance transitions of the tie points between the sheets of dispersion surface related to the presence of ultrasound phonons. The position of the maxima depending on the investigated reflections and bending radius of a crystal is determined by the most probable single-phonon scattering of neutrons. With an increase in the US wave's amplitude and the value of crystal bending the role of less probable multi -phonon processes is enhanced, which results in the formation of a plateau in the curves of the voltage dependence of intensity on a piezoelectric transducer and bending radius.(Fig. 18).

## 5. Conclusion

For describing the diffraction in single perfect crystals the theory of dynamical neutron scattering has been employed. This approach considers the neutron wave pattern propagating in the periodic potential of an ideal lattice, and a variety of features is explained by this time, including an oscillating form for the distribution of the diffracted intensities within the Bormann fan in the case of Laue diffraction. One of the most important dynamical diffraction effects being the strongly limited intensity diffracted by an ideal crystal in Bragg's position. While, the interference effects disappear very quickly if the translation symmetry in a perfect crystal is violated due to any disturbance (static deformation strain, low-frequency sound excitations, etc.) and the Bragg-reflected intensity increases. In our work it is shown that even a very sharp rising diffraction intensity at the center of Bormann's fan ($I_s / I_0 \geq 10$) can be described quantitatively using the dynamic diffraction model for the neutrons propagating in a thick crystal as spherical wave.

The PPK model modified for the case of ultrasound excitation satisfactorily describes some features of the neutron Bragg scattering from a bent silicon single crystal. The intensity dip of the Bragg diffraction is caused by the resonance transitions of the imaging points between the sheets of dispersion surface related to the presence of ultrasound phonons. The position of the maxima depending on the investigated reflection and bending radius of a crystal is determined by the most probable single-phonon scattering of neutrons. With an increase in the US wave's amplitude and the value of crystal bending the role of less probable multiphonon processes is enhanced, which results in the formation of a plateau in the curves of the voltage dependence of intensity on a piezoelectric transducer. These effects described phenomenological in the given work are not taken into consideration by the simple PPK model.

## Author details

E. Raitman, V. Gavrilov and Ju. Ekmanis
*Institute of Physical Energetics, Riga, Latvia*

## Acknowledgement

The authors thank Prof. V.A. Somenkov (Kurchatov's Center, Moscow, Russia) for unique Ge sample used in this work and Dr. A.Hoser for help in the measurements. The research at HZB BENSC (Germany) was supported by EU programme contract NM13 II-1523.

## 6. References

[1] G. W. Fox, P. H. Carr, (1931) Phys. Rev, 37, 1622,
[2] W.J.Spencer, and G.T.Pearman, (1970), Adv.X-ray Analysis 13,307,
[3] T. F. Parkinson, M. W. Moyer, (1966)Nature 211, 400,
[4] J.W.Knowles, (1956), Acta Cryst. 9, 61,
[5] .D.Zippel, D.Kleinstuck, and G.E.R.Shulce, (1965)Phys.Lett.14, 174,
[6] A.V.Kozlov, and A.I.Frank (2005) , Phys.At.Nuclei 68,7,1104,
[7] J.Summhammer (1993)Phys.RevA47,556,
[8] M.Moshinsky (1952) Phys.Rev. 88,625,
[9] A. R. Mkrtchyan, M. A. Navasardian, R. G. Gabrielyan, and L. A. Kocharian, (1986) Solid St.Comm. 59(3), 147,
[10] R. Hock, T. Vogt, J. Kulda, Z. Mursic, H. Fuess, A. Magerl, (1993) Zeit. Phys. B 90, 143,
[11] A. Remhof, K.-D. Liß, A. Magerl, (1997) NIM A391, 485,

[12] J.P. Guigay, P. Mikula, R. Hock, R. Baruchel, and A. Waintal, (1990)Acta Cryst. A46, 821,
[13] E. Zolotoyabko, B. Sander, (1995) Acta Cryst., A51, 163.
[14] Iolin, E., Raitman, E., Kuvaldin, B., Gavrilov, and V. & Zolotoyabko, (1988) E. Zh. Eksp. Teor. Fiz. 94, 218,
[15] S. D. LeRoux, R. Colella, and R. Bray, (1975) Phys. Rev.Lett., 35(4), 230.
[16] E.Iolin,and I.Entin, (1983) Zh. Eksp. Teor. Fiz.,85,1692,
[17] V. L. Nosik, (1994 ) Cryst. Rep., 39(4),526,
[18] S. Kikuta, T. Takahashi, S. Nakatani, (1984) Jap. Journ. Appl. Phys., 23(4), L193,
[19] R. Tucoulou, I. A. Schelokov, D. V. Roshchupkin, M. Brunel, L. Ortega, and P. Cevallier, (1995) Opt. Commun. 118, 175-180,
[20] K.-D. Liss, A. Magerl, R. Hock, A. Remhof, B. Waibel, (1997) Europhys. Lett., 40(4), 369,
[21] E.M.Iolin, L.L.Rusevich,M.Vrana *et. al* (1996) Phys.Status Sol.(b) 195, 21,
[22] E.Iolin, E. Raitman,. *et. al.* (1986) Sov. Phys. JETP. 66, 218.,
[23] E.Raitman, V.Gavrilov *et. al.*, (2009) Surf. Invest. 6, 3,
[24] A. Blagov, M. Kovalchuk,.V .Kohn,( 2006) Crystallography 51,1,
[25] V.L.Aksenov, H.Fritzshe , V.Gavrilov , Yu.Nikitenko , V. Progliado,and E.Raitman, (2004) Latv. Journ. Phys. Techn. Sci., N 6, 1,
[26] Felber J., Gähler R., Rauch C., and Golub R. (1996) Phys. Rev.A53, 319,
[27] P.Micula, P.Lucas,,and J.Kulda, (1992) Acta.cryst. A48, 72 ,
[28] E.Iolin, L.Rusevich, M.Vrana,P.Mikula, and P.Lukas, (1996) phys.stat.sol.(b),195,21,
[29] V.N.Gavrilov, E.V.Zolotoyabko, and E.M.Iolin (1990)Hyperfine Intereactions 58, 2427,
[30] V.N.Gavrilov, E.V.Zolotoyabko, and E.M.Iolin, J.Phys. C21, 471, 1988
[31] I.Entin (1977) Sov.Phys.JETP Lett.50, 269,
[32] I.Entin, and I.Puchkova (1984) Sov. Phys.Solid State, 26,1995,
[33] R.Kohler, W.Mohling , and H.Peibst (1974) Phys.Stat. Solidi B61,173,
[34] S. Takagi, (1962) Acta Crystallogr. 15 1311,
[35] D. Taupin, (1961) Bull. Soc. Fr. Miner. Crist. 84 51,
[36] C.G.Shull J. Appl.Cryst. 6, 257, 1973
[37] A.Leilinder, C.G.Shull,J.Arthur, and M.A.Horn, (1983) Phys.Rev. A28,487,
[38] C.G.Shull (1968) Phys.Rev.Lett., 21,1585,
[39] H.J.mc Skimin (1950) J.Acoustic.Soc.of Am. 22, ,413,
[40] Yu.G. Abov, N.O. Elyutin, et al. (2002), Phys. At. Nucl. 65 (11), 1933 ,
[41] Kato, N. (1964) *J. Phys. Soc. Jpn.* 19,
[42] B.Kuvaldin, (1991), Private Comm.,Riga,
[43] E. Iolin, B. Farago, E. Raitman, F. Mezei.( 1998), Physica B 241-243, 1213,
[44] R.Hock,T.Vogt, J.Kulda , Z.Mursic, H.Fuess, and A.Magerl, 1993 Z.Phys.B90, 143,
[45] F.N .Chukhovskii,., V.L.Nosik, & E. M. Iolin, (1993) .JTEP (Sov.) 104, 2452,
[46] Iolin E., Raitman, E., Gavrilov, V,. & Rusevich, L.. (1995. *J. Phys. D.: Appl. Phys.*28, A281
[47] Penning, P.(1966 ) *Philips Res. Rep.*, suppl.,5, 1

[48] N.Kato and Y.Ando (1966)J.Phys.Soc. Jpn. 21,964,
[49] E.A.Raitman, V.N.Gavrilov *et. al.* (2008), Latvian J. Phys.Tech. Sci., 2, 13,

Chapter 4

# Enhanced Transmission of Acoustic Waves Through Subwavelength Holes in Hard Plates

Bo Hou and Weijia Wen

Additional information is available at the end of the chapter

http://dx.doi.org/10.5772/54196

## 1. Introduction

In 1944, Bethe found the transmittance of electromagnetic (EM) waves through a tiny hole in a perfectly conducting screen varies as being proportional to $(kr)^4$, where $k=2\pi/\lambda$, $\lambda$ is the wavelength and $r$ is the hole radius [1]. This result shows that a small hole has extremely low transmission or negligible cross section for EM waves of very long wavelength. For a hole of finite thickness, the transmittance is found to be reduced further, because no propagating mode exists inside the hole [2].

However, in 1998, Ebbesen *et al.* has observed enhanced transmission of light through either a lattice of subwavelength holes or a single hole surrounded by surface periodical patterns on thin metallic films, where the optical transmission can be much larger than the area fraction of the holes at specific frequencies [3-13]. The holes, once organized or decorated, have the transmission cross section larger than themselves' area, which is different radically from the theory by Bethe. Since then, the remarkable phenomenon has inspired a tremendous amount of attention and works on resonant transmissions of EM waves through various apertures on either metallic or dielectric structure [14-25]. Phenomenologically, various observed transmission resonances are associated with two geometrical factors: structural factor (SF) emerging globally from the lattice periodicity and aperture factor owned locally by the individual unit [26-29]. Structural-factor-related resonances typically have the transmission wavelength comparable to the lattice constant and are dependent strongly on the incidence angle. In sharp contrast, aperture-factor-related resonances have the wavelength determined mainly by the transversal/longitudinal dimensions of the aperture and are not sensitive to the incidence angle.

It is well known that acoustic and EM waves share a lot of wave phenomena, but they have something in difference. In nature, acoustic wave is a scalar longitudinal wave in inviscid fluids, while EM wave is a vector transverse wave. Consequently, a subwavelength hole has no cutoff for acoustic wave, but does for EM wave, which underlies the distinct transmissions of acoustic/EM waves through a hole in an ideally rigid/conducting screen. The acoustic transmission of a single hole approaches a constant, $8/\pi^2$, dislike the EM case, with decreasing the ratio $r/\lambda$ [30].

Transmission/diffraction by an acoustical grating is an old problem, and the previous investigations addressed some cases: one-dimensional (1D) periodic slits in a rigid screen [31,32], a single hole in a thick wall [33,34], and a 1D grating composed of parallel steel rods with finite grating thickness [35,36]. Here we studied the acoustic transmissions through two structures: (1) a two-dimensional array (square lattice) of subwavelength hole and (2) a single hole surrounded by the surface periodic grooves. It is found that the acoustic transmission phenomenon for the structured thin plates is analogous completely to the case of EM wave, except for the transmission phase. For the hole array in thick plates, the transmission peaks are related to the Fabry-Perot-like (FP-like) resonances inside the holes and can occur to the frequencies well below Wood's anomalies.

## 2. Ultrasonic measurements

In our experiments, the measurements of far field transmissions of acoustic waves in the ultrasonic frequency regime (0.2–2.0 MHz) were performed in a large water tank. Two immersion transducers were employed as ultrasonic generator and receiver, and the sample was placed at a rotation stage located between the two transducers at an appropriate distance. The sample could be rotated, so that the oblique incidences were measured. The ultrasonic pulse was incident upon the sample and the transmitted signal was collected by the receiver, collinear with the incident wave. Transmission magnitude, $T$, and transmission phase, $\phi$, of the sample were obtained by normalizing the Fourier transformed spectra of the signal through the sample, $|A_s(f)|\exp[j\phi_s(f)]$, with respect to the signal through the water background (without the sample in place), $|A_b(f)|\exp[j\phi_b(f)]$, where $f$ is the frequency and $j^2=-1$. Consequently, $T=|A_s(f)/A_b(f)|$, and $\phi=\phi_s(f)-\phi_b(f)-2\pi f t/c$ ($c$ =1490 m/s being the speed of acoustic wave in water, $t$ being the sample thickness).

In the context, the term "transmission" when referring to the spectrum means the amplitude ratio, $T$, of the transmitted and the incident waves. To figure out whether the transmission is enhanced, we need calculate the ratio of transmitted energy flow to the energy flow incident on all holes. Within a unit cell of the hole array, the ratio can be expressed by:

$$\frac{I_t \cdot a^2}{I_i \cdot \pi d^2/4} = \frac{|A_t/A_i|}{\pi d^2/\left(4a^2\right)} = \frac{T^2}{\xi} = \frac{\mathcal{T}}{\xi} \tag{1}$$

where $I_t$ ($A_t$) and $I_i$ ($A_i$) denote the intensities (amplitudes) of transmitted and incident acoustic waves, respectively, $a$ is the lattice constant, $d$ is the hole diameter, and $\xi=\pi d^2/\left(4a^2\right)$ is the area fraction of the holes. We call the squared transmission magnitude $T^2$ as transmittance $\mathcal{T}$ representing the acoustic intensity transmission. If $\mathcal{T}/\xi>1$, then the enhanced transmission is obtained.

Apart from measuring the transmission spectrum, we also implemented point-by-point scanning to detect the pressure field distribution in the transmission process. A pinducer (1.5 mm in diameter) replaced the receiving transducer and was located at a distance, $z$, from the rear surface of the sample to detect the pressure field distribution there. The pinducer was mounted on a two-dimensional translation stage. The scanning was done along the $x$-$y$ plane parallel to the sample surface, with a spatial step of 0.1mm×0.1mm.

## 3. Experimental results

### 3.1. Enhanced transmission of acoustic waves through hole array structure

First we measured the acoustic transmission of a hole array with the hole diameter $d$ = 0.5 mm, the lattice constant $a$ = 1.5 mm, and the plate thickness $t$ = 0.5 mm. Figure 1 shows the transmittance of the hole array at normal incidence, compared to the transmittance of a smooth brass plate with identical thickness. For the smooth brass plate, very low transmittance is seen because of the acoustic impedance mismatch ($\eta_{brass} / \eta_{water} \approx 25$). It is noticed that the transmittance rises at lower frequencies, which indicates a thin brass plate can not block acoustic waves of very long wavelength or very low frequency. This fact is different from the EM case where a sheet of metal as thin as skin depth works well. For the hole array, a pronounced peak is seen at 0.85 MHz and followed by a transmittance zero close to 1.0 MHz which is just Wood's anomaly $\lambda=a$. The peak has the transmittance (68%), much lager than the area fraction (8.7%) of holes occupation in the array structure, and shows an acoustic transmission enhancement through the hole array, similar to the EM case.

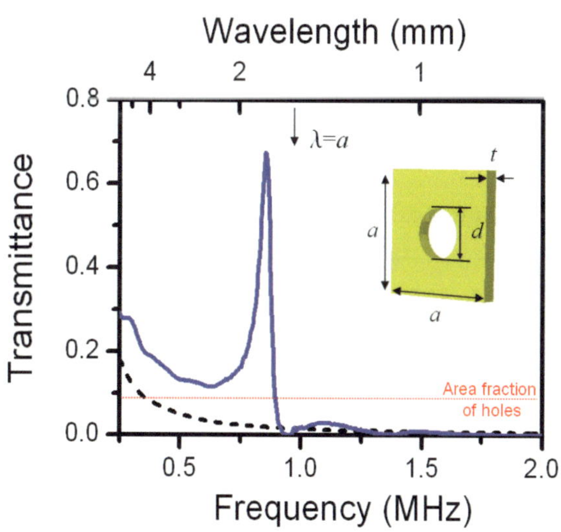


**Figure 1.** Transmittance of the acoustic waves at normal incidence through a hole array (solid line) and a smooth brass plate (dashed line) with the same thickness. The hole array has the parameters: the hole diameter $d$ = 0.5 mm, the lattice constant $a$ = 1.5 mm, and the plate thickness $t$ = 0.5 mm. A schematic picture of the unit cell is illustrated as the inset. The level dot line represents the area fraction of holes in the array. Reprinted with permission from J. Appl. Phys. 104, 014909 (2008). Copyright 2008 American Institute of Physics.

We also investigated the dependence of the transmission peak on the lattice constant. Figure 2 shows the normal transmissions of the hole arrays with identical lattice constant $a$ = 2.0 mm and different hole diameters. The transmission peak and two Wood's anomalies (pointed by arrows) are identified at ~0.75 MHz and ~1.1MHz. With the larger diameter holes ($d$ = 1.2 mm), the peak becomes more pronounced. Comparing with the array of $a$ = 1.5 mm in Figure 1, it is clear to show that the peaks and Wood's anomalies downshift to lower frequencies as the lattice constant increases. In Figure 2, we also plotted the measured transmission phase $\phi$ for the hole array of $d$ = 1.2, $a$ = 2.0, and $t$ = 0.5 mm, and found $\phi = -0.98\pi$ at the peak frequency. The approximate $-\pi$ phase change reveals the oscillations of the acoustic field on the front and rear surfaces of the plate are out-of-phase, which is distinct from the corresponding characteristic in the EM case. For EM wave transmitted through a hole array, the hole acts as barrier due to the transmission frequency much lower than the cutoff frequency of the hole, and the wave has to tunnel through the hole in a form of evanescent field. So the phase change of the EM wave across the holey film/plate assumes nearly zero [37]. This difference in transmission phase bares the distinct behaviors of a hole to acoustic and EM waves, again.

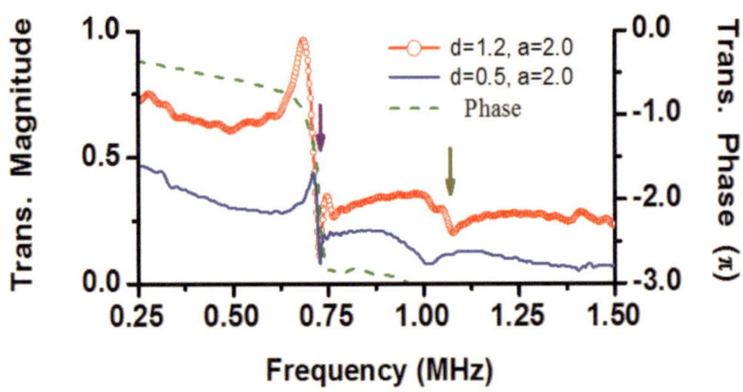


**Figure 2.** Acoustic transmissions at normal incidence through the hole arrays with the same lattice constant, 2.0 mm, and the same thickness, 0.5 mm, but different diameters as denoted. The measured transmission phase curve is for the hole array: $d$ = 1.2, $a$ = 2.0, and $t$ = 0.5 mm. The arrows indicate the Wood's anomalies, $\lambda = a$ and $\lambda = a/\sqrt{2}$ . Reprinted with permission from J. Appl. Phys. 104, 014909 (2008). Copyright 2008 American Institute of Physics.

Figure 3 shows the transmission spectra at oblique incidence measured with the incident angle $\theta$ varying from 0° to 25° for the hole array of $d$ = 1.2, $a$ = 2.0, and $t$ = 0.5 mm. The transmission map is plotted as a function of both the frequency and the incidence angle. The predicted variation of Wood's anomalies versus angle is plotted as solid lines and is superposed on the map. Derived from the conservation of momentum, the variation relation reads:

$$af^{(l,m)}/c = \left(l\sin\theta + \sqrt{l^2 + m^2\cos^2\theta}\right)/\cos^2\theta \quad (2)$$

for the Wood's anomaly frequency $f^{(l,m)}$ of order ($l$, $m$). It is seen from the map that the measured shifting of Wood's anomalies with the incidence angle agrees well with the solid lines. On the other hand, the peaks exhibit a strong angle-dependent behavior in the same way as Wood's anomalies.

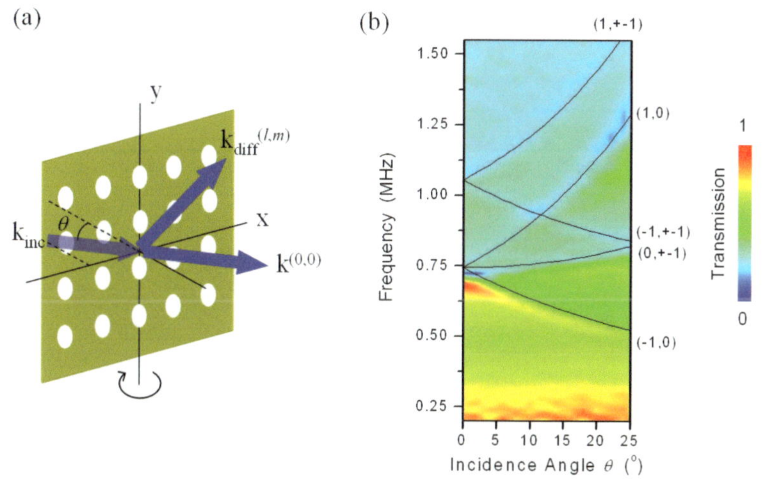


**Figure 3.** (a) The schematic picture of oblique incidence of acoustic waves. The wave vector $\mathbf{k}_{inc}$ represents the incident wave illuminating the hole array at the incidence angle $\theta$. The obliquity occurs along the [1,0] direction of the array. The wave vector $\mathbf{k}^{(0,0)}$ represents the (0,0)-order transmitted wave. The wave vector $\mathbf{k}_{diff}^{(l,m)}$ represents the $(l,m)$-order diffraction wave. (b) Acoustic transmission magnitude plotted as a function of the wave frequency and the incidence angle for the hole array, $d$ = 1.2, $a$ = 2.0, and $t$ = 0.5 mm. The solid lines superposed are the variation curves of Wood's anomalies with the incidence angle. Reprinted with permission from J. Appl. Phys. 104, 014909 (2008). Copyright 2008 American Institute of Physics.

In recent investigations, it is demonstrated that the SF resonance can be responsible for enhanced transmissions of EM waves through subwavelength hole arrays [26,27]. We have considered that the acoustic surface wave at the brass-water interface might play no role in the present transmission phenomenon, and shown that the SF resonance holds for acoustic waves by generalizing the proof of EM waves [38,39]. The SF resonance has some spectral features: the resonant wavelength is determined essentially by the lattice constant and is very sensitive to the incidence angle with accompanied by Wood's anomalies. Here, the experimental results for the enhanced acoustic transmission through the hole array in the 0.5 mm thick plate manifests the features of SF resonance.

When the plate thickness becomes larger, the situations begin to divide for two types of waves. For EM wave, the transmission peak will diminish after the metallic film/plate becomes thick enough, because the holes have the cutoff. In sharp contrast, there is no cutoff for acoustic waves to propagate through the holes. When the thickness is large enough, for instance $t$ = 2.3 mm, there can be multiple transmission peaks well below the Wood's anomaly, as shown in Figures 4(a) and 4(b). The measured spectra show the typical characteristics of FP resonance in terms of the phase values at the transmission maxima and minima. From Figure 4(c), the peaks are not sensitive to the incidence angle. In fact, these transmission peaks are caused by standing-wave-formed resonances of the acoustic wave establishing inside the hole channel. However, these resonances undergo a tuning, to some degree, by diffraction evanescent waves parasitical to a grating, and consequently deviate from the ordinary FP conditions while the plate thickness becomes comparable to the lattice constant, which will be further discussed later.

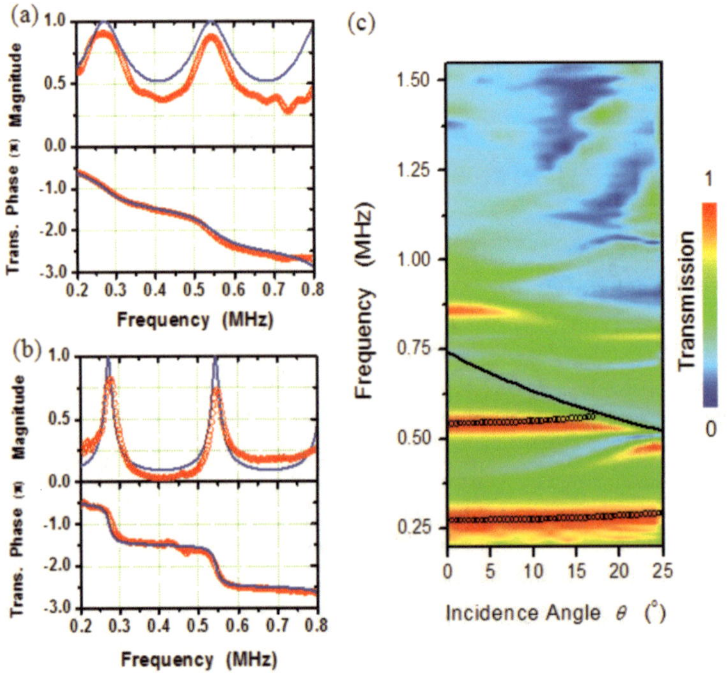


**Figure 4.** (a) Normal transmission of acoustic waves through the hole array (open circles), $d = 1.2$, $a = 2.0$, and $t = 2.3$ mm, immersed in water and through the effective fluid layer (solid line) immersed in water alike. Open circles are the measurement data, and the solid lines are the calculation based on the effective fluid model with $\overline{n} = 1.19$ and $\overline{\eta} = 0.28\eta_{water}$. (b) Normal transmission for the hole array, $d = 0.5$, $a = 2.0$, and $t = 2.3$ mm. Open circles are the measurement data, and the solid lines are the calculation based on the effective fluid model with $\overline{n} = 1.19$ and $\overline{\eta} = 0.05\eta_{water}$. (c) Acoustic transmission magnitude plotted as a function of the wave frequency and the incidence angle for the hole array $d = 1.2$, $a = 2.0$, and $t = 2.3$ mm. The obliquity occurs along the [1,0] direction of the array. The solid line superposed are the variation curve of Wood's anomaly (-1,0) with the incidence angle. The open circles superposed denote the variation of the transmission peaks, calculated from the FP resonance condition of the effective fluid model at oblique incidence. Reprinted with permission from J. Appl. Phys. 104, 014909 (2008). Copyright 2008 American Institute of Physics.

## 3.2. Enhanced transmission of acoustic waves through bull's eye structure

Soon after the discovery of extraordinary optical transmission through a metallic film with two-dimensional array of sub-wavelength holes, it was found that there can be enhanced and collimated transmission through a single sub-micron hole surrounded by finite periodic rings of indentations (denoted as bull's eye) [4]. We also examined the transmission of a bull's eye structure for acoustic waves. The bull's eye structure, shown in the inset of Figure

5, was fabricated by patterning both sides of a thin brass plate with concentric periodic grooves around a single cylindrical hole. The thickness of the brass plate is 1.6 mm, and the diameter of the central hole is 0.5 mm. The groove period is 2.0 mm, and there are a total of 15 grooves. The width and depth of each groove are 0.5 mm and 0.3 mm, respectively.

In Figure 5, we showed the measured transmittances as a function of frequency for both bull's eye structure and the reference sample (a smooth brass plate of the same thickness). It can be seen that there is a transmission peak at 0.71 MHz for bull's eye structure, while such peak is missing for the reference sample. In Figure 5, we also plotted the power transmittance calculated by using COMSOL MULTIPHYSICS, a commercial finite-element solver. It can be seen that the predicted peak position agrees well with the experimental data. However, the measured transmittance is much lower than that predicted and the precise reason for this disagreement is yet to be uncovered.

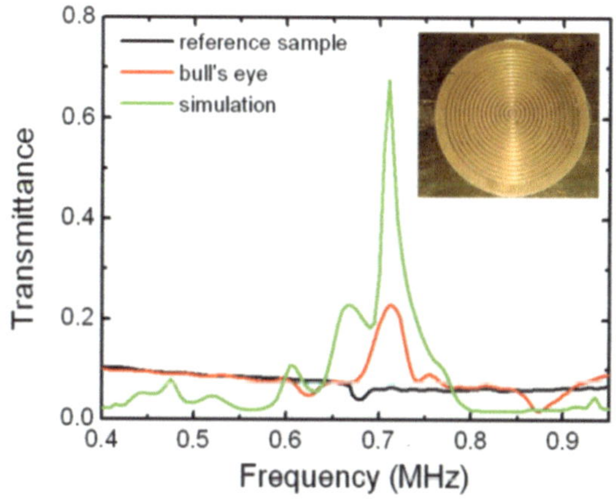


**Figure 5.** Measured transmittances for both the bull's eye structure and the reference sample, together with calculated power transmittance for bull's eye. Inset shows an image of the sample, fabricated by patterning both sides of a thin brass plate with concentric periodic grooves around a single cylindrical hole. The thickness of the brass plate is 1.6 mm, and the diameter of the central hole is 0.5 mm. The groove period is 2.0 mm, and the groove width and depth are 0.5 mm and 0.3 mm, respectively. Reprinted with permission from Appl. Phys. Lett. 92, 124106 (2008). Copyright 2008 American Institute of Physics.

For ultrasonic waves in water, wavelength corresponding to 0.71 MHz is 2.1 mm, which is slightly larger than the groove period of bull's eye, 2.0 mm. This close correspondence is a strong clue indicating that the enhanced transmittance is due to the diffraction effect. It has been shown that enhanced acoustic wave transmission through hole arrays in perfectly rigid thin plate, where there can be no surface wave, may be related (and understood via Babinet's principle) to "resonant" reflection by its complementary structure, i.e., planar arrays of perfectly rigid disks [26, 39]. In fact, both were found to be associated with the *divergence* in the scattering structure factor, owing to the coherent addition of the Bragg scattering amplitudes from the periodic array of holes or disks. As a result, a quasi surface mode with frequency close to the onset of the first diffraction order (wavelength $\lambda$ slightly larger than the lattice constant $a$) always exists. Such modes are denoted "structure-factor-induced surface modes," or SF resonances. Since diffraction is the ultimate mechanism for the SF resonances, we expect the same to also apply to bull's eye structure, which can be viewed as having 1D periodicity along the radial direction.

Besides the transmission enhancement, the collimation effect of the bull's eye structure is very striking [4,40]. As shown in Figure 6(a), the far-field acoustic wave on the transmission side is also in the form of a tight beam with a lateral dimension not exceeding the groove periodicity. The full width at half maximum (FWHM) divergence is ±2°. As analyzed above, it is the coherent scattering which leads to the emergence of a strongly collimated beam in the far-field region. In Figures 6(b) and 6(c) we also plotted the scanned results at a distance of about 15 wavelengths from the transmission side of the surface, for both the bull's eye structure and the reference sample. Compared with the reference sample, the collimation effect for bull's eye structure is very evident. In addition, it is found by simulation that both the intensity of the acoustic wave field around the central hole region, as well as the collimation effect, would increase with the number of concentric grooves. This is reasonable, since the coherent scattering effect becomes stronger if more concentric grooves are involved.

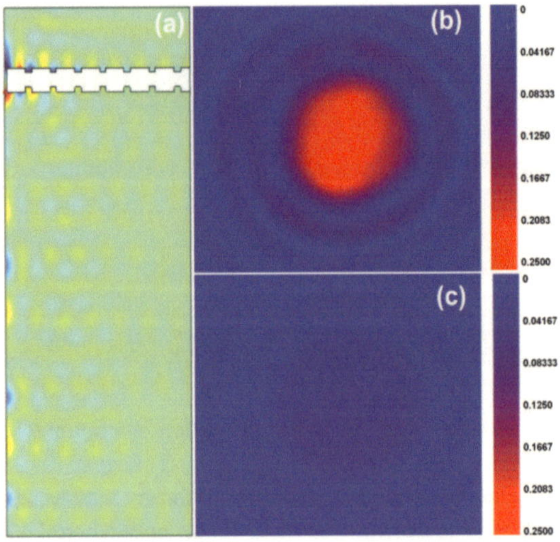


**Figure 6.** (a) Calculated far-field pressure amplitude distribution at 0.71 MHz in the axial symmetry coordinates. (b) Experimentally scanned far-field (~15 wavelengths from the transmission side of the plate surface) pressure magnitude distributions in an area of 40x40 mm², for the bull's eye structure at 0.71 MHz. (c) Same as (b), for the reference sample. Reprinted with permission from Appl. Phys. Lett. 92, 124106 (2008). Copyright 2008 American Institute of Physics.

## 4. Discussions

### 4.1. Fabry-Perot resonances tuned via diffraction evanescent waves

For the hole arrays, we measured the acoustic samples with various plate thicknesses ranging from 0.5 mm to 3.1 mm, as plotted in the inset of Figure 7. In theory, we employed the mode expansion method to calculate analytically the transmission [39]. We found that the observed transmission peaks are the manifestation of a type of resonance mode that has

FP and SF resonances as the two limits. The diffraction evanescent modes play an important role in interpolating between the two limits. To make explicit the role of diffraction evanescent waves, we retained the lowest cylindrical mode inside the holes and 5 lowest plane wave modes outside the holes, and obtained the resonant mode equation as

$$\tan(2t\omega / c) = \frac{-a^3 k_0 k_1^2}{\frac{\pi(1-\xi^2)}{16 J_1^2(2\pi\, r/a)} a^3 k_1^3 + \pi^2 \frac{2a}{t}}, \tag{3}$$

in which $k_0 = \omega / c = 2\pi f / c$, $k_1 = \sqrt{(2\pi / a)^2 - (\omega / c)^2}$ is the diffraction evanescent wavevector, and $J_1$ is the first order Bessel function. Equation (3) is instructive, since a vanishing right-hand side would directly yield the FP resonance condition $t = n\lambda/2$, $\lambda$ being the wavelength. A combination of hole and periodic diffraction evanescent wave effects constitute the correction to the usual FP condition in the form of a non-zero right hand side, implying that the FP resonance can be tuned by varying the periodicity and area fraction of holes. We denote such resonances the FPEV resonances.

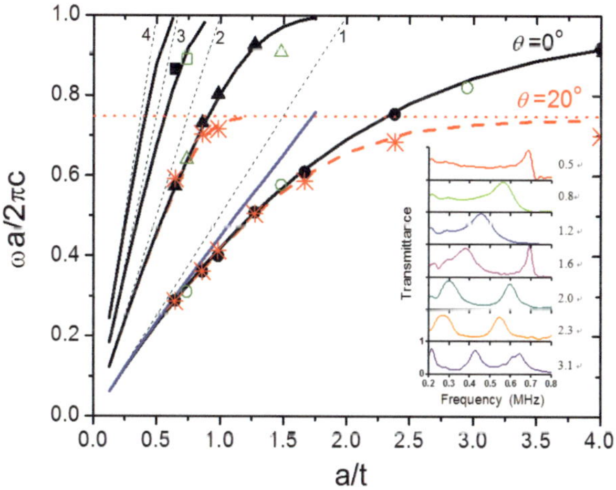


**Figure 7.** Solid symbols (measurement) and black solid lines (calculation) are the resonant transmission frequencies of normally incident acoustic wave through a square lattice of holes with $r = 0.3a$. Star symbols (measurement) and red dashed lines (calculation) denote the case of acoustic wave at 20° incidence angle along [0,1] direction, for which the Wood's anomaly frequency is the horizontal dot line. Black dashed lines delineate the FP condition with the order $n$ denoted. Blue line is determined by Equation (3), with a slope of ~0.42 (i.e. $t \approx 0.42\lambda$). Open symbols represent the measured transmission peaks of normally incident microwave on a metallic grating, 1D periodic slits, with area fraction 0.32. The inset shows the measured acoustic transmittances at normal incidence for various plate thicknesses, with $a = 2.0$ mm and $r = 0.6$ mm. The thickness (in mm) of each plate is given to the right of the spectrum. Reprinted with permission from Phys. Rev. B 76, 054303 (2007). Copyright 2007 by The American Physical Society.

In Figure 7 we show the measured and calculated FPEV resonance frequencies plotted as a function of inverse plate thickness. The FP condition is indicated by the black dashed straight lines, for $n$=1, 2, 3, 4, with slopes of 0.5, 1, 1.5 and 2, respectively. The FPEV frequencies are shown as solid black (normal incidence), and red dashed (20$^o$ oblique incidence) lines. Compared with the FP resonances, it is seen that the FPEV resonances always occur at lower frequencies, as though the effective plate thickness is greater than $t$. The prediction of Equation (3), for the $n$=1 FPEV resonance, is shown as the blue line. Except in the region of very small $a/t$ values, the blue line has a slope of 0.42. Thus the effect of the diffraction evanescent waves is to shift the resonance condition by ~16%, in the direction of smaller channel length. The difference between the prediction of Equation (3) and the black lines appears at the small $t$ limit, where the transmission peak frequency shows a clear dependence on the incidence angle. This is characteristic of the surface-wave-like mode induced by the SF resonance. In fact, these transmission peak frequencies all occur at close to the Wood's anomaly, as required by the SF resonance condition. Thus the lowest frequency FPEV resonance, which shows little or no dependence on the incidence angle, is smoothly converted to the structure-factor-induced surface mode in the thin-plate limit. The diffraction evanescent wave contributions are dominant at the intermediate values of $a/t$. To a lesser degree, similar behavior can be observed for the higher order FPEV's.

It is seen that as the ratio $a/t$ increases, the lowest order evanescent waves (Equation (3)) can no longer account for the resonant frequency trajectory. Also, in the large $a/t$ limit the curves also display pronounced incident angle dependence, in contrast to FP resonances which are nearly independent of the incidence angle. These are the signals for (1) the lateral scattering interaction is contributing much more to the resonant modes, hence the lowest order evanescent modes are no longer sufficient to account for such strong lateral interactions, and (2) with the increased lateral interaction, SF effect becomes more pronounced, implying incidence angle dependence. These spectral features also correspond to the different field distributions, as shown in Figure 8, where the surface field is localized on the holes for FP-like resonances and the interference pattern is seen in the region between the holes for SF resonances.

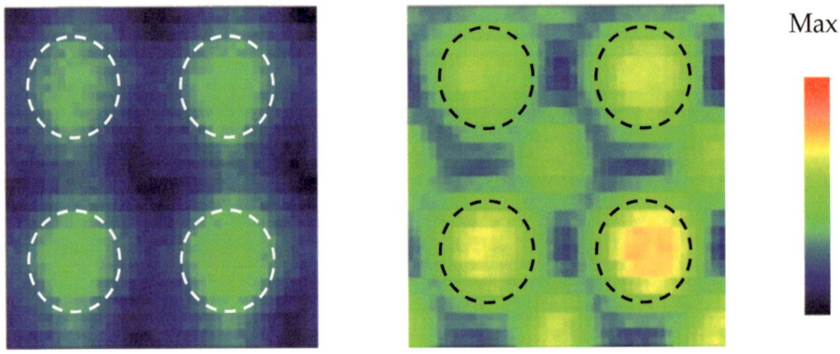


**Figure 8.** The measured distribution of pressure field at the rear surface of the hole array $a$ =2.0, $r$ =0.6, and $t$ =1.6 mm for two resonance frequencies 0.38MHz ($a/\lambda$=0.51) and 0.69MHz ($a/\lambda$=0.93). The holes are delineated by dashed lines.

It is interesting to note that the transmission of microwave through a metallic grating of 1D slits has the similar FPEV resonances for the incident polarization with E-field perpendicular to the silts, see open symbols in Figure 7 [37]. This is because the slits have no cutoff to the perpendicular polarization of EM waves, in the same physics as the holes to acoustic waves.

## 4.2. The effective fluid model for thick plates

For a very small *a*/*t* ratio, these resonance wavelengths are much larger than the lattice constant, allowing us to take a view of effective media. Here we employ a simple argument in the same fashion as the EM case [41] with the assumption of brass plate being rigid, and find that the hole array structure fabricated in a rigid plate and filled with a fluid (mass density $\rho_0$ and bulk modulus $\kappa$) may be viewed as an effective fluid with the same thickness, effective mass density $\bar{\rho}_0$ and bulk modulus $\bar{\kappa}$. It is known that the acoustic wave is characterized by the pressure field, $p$ , and velocity field, $\boldsymbol{u}$. Averaging the pressure field in the holes, we get the effective pressure field in the effective fluid $\bar{p} = \xi p$ . Requiring the acoustic energy flow across the surface to be the same for the hole array and the effective fluid, i.e. $p\boldsymbol{u} \cdot \pi d^2 / 4 = \bar{p}\bar{\boldsymbol{u}} \cdot a^2$ , we obtain the effective velocity $\bar{\boldsymbol{u}} = \boldsymbol{u}$ . Also the total acoustic energy for both systems are required to be the same, $\left(\frac{1}{2}\rho_0 \boldsymbol{u}^2 + \frac{1}{2}p^2 / \kappa\right) \cdot \pi d^2 / 4 \cdot t = \left(\frac{1}{2}\bar{\rho}_0 \bar{\boldsymbol{u}}^2 + \frac{1}{2}\bar{p}^2 / \bar{\kappa}\right) \cdot a^2 \cdot t$, which gives us the effective parameters $\bar{\rho}_0 = \xi \rho_0$ and $\bar{\kappa} = \xi \kappa$ . Thereafter, the acoustic speed and impedance of the effective fluid are $\bar{v} = v = \sqrt{\kappa / \rho_0}$ and $\bar{\eta} = \xi \eta = \xi \sqrt{\rho_0 \kappa}$ , where $v$ and $\eta$ are the acoustic speed and impedance of the filling fluid, respectively. The relations indicate the acoustic speed remains unchanged and the impedance is scaled by a factor of the area fraction of holes for the effective fluid.

The above argument is applicable under long wavelength limit ($a/\lambda \to 0$) where diffraction evanescent waves are negligible. For the sample with thickness comparable to lattice constant, the diffraction evanescent waves tune the FP resonances and the resultant transmission resonances can occur for channel length ~16% thinner than required by the FP resonances. Superficially, this diffractive effect is substitutable by a slowing of acoustic wave propagation inside the holes, i.e. $v = 0.84c$ . Based on the above argument, the effective fluid equivalent to a 2.3 mm thick sample has the acoustic speed $\bar{v} = v = 0.84c$, or the acoustic refractive index $\bar{n} = c / \bar{v} = 1.19$ , and the effective impedance $\bar{\eta} = \xi \eta = \xi \eta_{water}$ . With these two effective parameters at hand, we calculated the transmission spectra, both magnitude and phase, of the effective fluid layer at normal incidence according to the formula:

$$T\exp(j\phi) = \frac{1}{\cos(k_0 \bar{n} t) + \frac{j}{2}\left(\frac{\bar{\eta}}{\eta_{water}} + \frac{\eta_{water}}{\bar{\eta}}\right)\sin(k_0 \bar{n} t)} \tag{4}$$

where $k_0 = 2\pi f / c$ is the wavenumber of the incidence wave. The calculated results (solid lines) are shown in Figures 4(a) and 4(b) to compare with the experimental data (open circles) for two samples with identical thickness 2.3 mm, and identical lattice constant 2.0 mm, but different hole diameters 1.2 and 0.5 mm. Good agreement between the calculations and the experiments is seen at the frequencies below the Wood's anomaly (0.75 MHz), which verifies the applicability of the effective fluid model at normal incidence.

In Figure 4(c), we see two flat bands appear below 0.75 MHz, and they are the first and second FP resonances, see Figure 4(a) where the phase values indicate the order of FP resonances. The open circles superimposed are the variations of the transmission peaks of the effective fluid layer which are obtained from the FP resonance condition at oblique incidence, $\sin\left[k_0 \bar{n} t \sqrt{1-\left(\sin\theta / \bar{n}\right)^2}\right] = 0$ . The agreement between the calculated variations of the transmission peaks and the measured results indicates the applicability of the effective fluid model persists to a range of incidence angle. The discrepancy at $\theta$ >15° for the second FP transmission peak is due to the emerging of the (-1,0) diffraction order nearby, and the red flat band is seen to terminate upon crossing with Wood's anomaly (-1,0). Where the cross happens, there will be the strong coupling interaction between SF resonance of the array and FP-like resonance localized at each hole, which possibly gives rise to the 0.48 MHz band at $\theta$ =25°.

The effective fluid model allows us to use the holey or slotted hard plate to realize an acoustic medium, and provides some freedom to design acoustic materials, because some material parameters, difficult to be tuned, are related simply to the structural factors of the hole array or the slits. In Figure 9, we illustrated conceptually an acoustic prism made of such slotted hard plate. A detailed discussion is seen in reference [42].

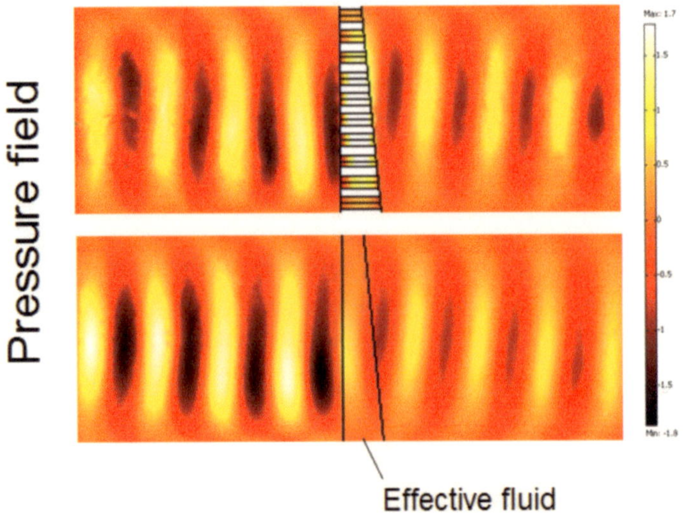


**Figure 9.** Simulation results for a structured hard plate (upper) with tapered thickness to function as the acoustic prism. The pressure field is compared with a fluid with the effective parameters (lower).

## 5. Conclusion

We investigated experimentally the acoustic transmission through subwavelength holes fabricated on brass plates at normal and oblique incidence within ultrasonic frequencies regime. The transmission phenomena for both hole array and bull's eye structure in thin brass plates, analogous to the observed enhanced transmission of EM waves through subwavelength hole arrays in a metallic film, exhibit the transmission enhancement because of the SF resonance. At the peak frequency, the transmission phase is nearly $-\pi$, indicating the out-of-phase oscillations of the acoustic field at two surfaces of the plate. For the hole array in thick brass plates, the transmission peaks of acoustic waves are related to the FP-like resonances inside the holes and therefore occur well below Wood's anomaly, since a hole has no cutoff frequency for acoustic propagation. By varying the plate thickness or channel length, one makes the transition from the FP resonance (thick plate limit) to the SF resonance (thin plate limit). Between the two limits there can be interesting deviation from FP resonance conditions, owing to the interaction of the diffraction evanescent waves. In the case of thick plates, the structure can be viewed as a new fluid with effective mass density and bulk modulus scaled, under long wavelength limit, by a factor of area fraction of the holes. The effective medium model describes well the transmission properties of the hole array within a range of incidence angle.

Our discussion assumed the approximation of hard plates and did not take acoustic surface waves into account. With acoustic surface waves being involved, transmissions of acoustic waves through structured plates have found far richer and more complicated physical phenomena in the past few years and will attract more attentions in the future [43-45]. Although this subject is an old problem, its new phenomena may appear from time to time and the underlying mechanism waits to be unlocked.

## Author details

Bo Hou
*School of Physical Science and Technology, Soochow University, Suzhou, Jiangsu, China*

Weijia Wen
*Department of Physics, Hong Kong University of Science and Technology, Clear Water Bay, Kowloon, Hong Kong, China*

## Acknowledgement

We would like to thank our collaborators in this project. B. Hou is supported by the National Natural Science Foundation of China (Grant No. 11104198) and a Project Funded by the Priority Academic Program Development of Jiangsu Higher Education Institutions (PAPD). W. Wen is supported by RGC Grant HKUST2/CRF/11G.

## 6. References


[1] Bethe H. A. (1944) Theory of diffraction by small holes, Phys. Rev. 66: 163.

[2] Roberts A. (1987) Electromagnetic theory of diffraction by a circular aperture in a thick, perfectly conducting screen, J. Opt. Soc. Am. A 4: 1970.

[3] Ebbesen T. W., Lezec H. J., Ghaemi H. F., Thio T., and Wolff P. A. (1998) Extraordinary optical transmission through subwavelength hole arrays, Nature 391: 667.

[4] Lezec H. J., Degiron A., Devaux E., Linke R. A., Martin-Moreno L., Garcia-Vindal F. J., and Ebbesen T. W. (2002) Beaming light from a subwavelength aperture, Science 297: 820.

[5] Ghaemi H. F., Thio T., Grupp D. E., Ebbesen T. W., and Lezec H. J. (1998) Surface plasmons enhance optical transmission through subwavelength holes, Phys. Rev. B 58: 6779.

[6] Grupp D. E., Lezec H. J., Ebbesen T. W., Pellerin K. M., and Thio T. (2000) Crucial role of metal surface in enhanced transmission through subwavelength apertures, Appl. Phys. Lett. 77: 1569.

[7] Krishnan A., Thio T., Kim T. J., Lezec H. J., Ebbesen T. W., Wolff P. A., Pendry J., Martin-Moreno L., and Garcia-Vidal F. J. (2001) Evanescently coupled resonance in surface plasmon enhanced transmission, Opt. Commun. 200: 1.

[8] Martín-Moreno L., García-Vidal F. J., Lezec H. J., Pellerin K. M., Thio T., Pendry J. B., and Ebbesen T. W. (2001) Theory of extraordinary optical transmission through subwavelength hole arrays, Phys. Rev. Lett. 86: 1114.

[9] Martín-Moreno L. and García-Vidal F. J. (2004) Optical transmission through circular hole arrays in optically thick metal films, Opt. Express 12: 3619.

[10] Barnes W. L., Murray W. A., Dintinger J., Devaux E., and Ebbesen T. W. (2004) Surface plasmon polaritons and their role in the enhanced transmissions of light through periodic arrays of subwavelength holes in a metal film, Phys. Rev. Lett. 92: 107401.

[11] Barnes W. L., Dereux A., and Ebbesen T. W. (2003) Surface plasmon subwavelength optics, Nature 424: 824.

[12] Genet C. and Ebbesen T. W. (2007) Light in tiny holes, Nature 445: 39.

[13] Garcia-Vidal F. J., Martin-Moreno L., Ebbesen T. W., and Kuipers L. (2010) Light passing through subwavelength apertures, Rev. Mod. Phys. 82: 729.

[14] Porto J. A., Garcia-Vidal F. J., and Pendry J. B. (1999) Transmission resonances on metallic gratings with very narrow slits, Phys. Rev. Lett. 83: 2845.

[15] Rosenberg A. and Bolden E. A. (2000) Optical filters consisting of metallic waveguide arrays, J. Opt. Soc. Am. A 17: 1461.

[16] Fan W., Zhang S., Minhas B., Malloy K. J., and Brueck S. R. (2005) Enhanced infrared transmission through subwavelength coaxial metallic arrays, Phys. Rev. Lett. 94: 033902.

[17] Bonod N., Enoch S., Li L., Popov E., and Neviere M. (2003) Resonant optical transmission through thin metallic films with and without holes, Opt. Express 11: 482.

[18] Baida F. I. and Van Labeke D. (2002) Light transmission by subwavelength annular aperture arrays in metallic films, Opt. Commun. 209: 17.

[19] Quemerais P., Barbara A., Le Perchec J., and Lopez-Rios T. (2005) Efficient excitation of cavity resonances of subwavelength metallic gratings, J. Appl. Phys. 97: 053507.
[20] Suckling J. R. and Sambles J. R. (2005) Remarkable zeroth-order resonant transmission of microwaves through a single subwavelength metal slit, Phys. Rev. Lett. 95: 187407.
[21] Matsui T., Agrawal A., Nahata A., and Vardeny Z. V. (2007) Transmission resonances through aperiodic arrays of subwavelength apertures, Nature 446: 517.
[22] Rivas J. G., Janke C., Bolivar P. H., and Kurz H. (2005) Transmission of THz radiation through InSb gratings of subwavelength apertures, Opt. Express 13: 847.
[23] Azad A. K., Zhao Y., and Zhang W., (2005) Transmission properties of terahertz pulses through an ultrathin subwavelength silicon hole array, Appl. Phys. Lett. 86: 141102.
[24] Wen W., Zhou L., Hou B., Chan C. T., and Sheng P. (2005) Resonant transmission of microwave through subwavelength fractal slits in a metallic plate, Phys. Rev. B 72: 153406.
[25] Ruan Z. and Qiu M. (2006) Enhanced transmission through periodic arrays of subwavelength holes: The role of localized waveguide resonances, Phys. Rev. Lett. 96: 233901.
[26] García de Abajo F. J., Gómez-Medina R., and Sáenz J. J. (2005) Full transmission through perfect-conductor subwavelength hole arrays, Phys. Rev. E 72: 016608.
[27] García de Abajo F. J. (2007) Light scattering by particle and hole arrays, Rev. Mod. Phys. 79: 1267.
[28] van der Molen K. L., Klein Koerkamp K. J., Enoch S., Segerink F. B., van Hulst N. F., and Kuiperts L. (2005) Role of shape and localized resonances in extraordinary transmission through periodic arrays subwavelength holes : experiment and theory, Phys. Rev. B 72: 045421.
[29] Cao H. and Nahata A. (2004) Influence of aperture shape on the transmission properties of a periodic array of subwavelength apertures, Opt. Express 12: 3664.
[30] Spence R. D. (1948) The diffraction of sound by circular disks and apertures, J. Acoust. Soc. Am. 20: 380.
[31] Lamb H., *Hydrodynamics*, 6th ed. (Dover, New York, 1991)
[32] Miles J. (2002) On resonant reflection by a plane grating, Wave Motion 35: 311.
[33] Wilson G. P. and Soroka W. W. (1965) Approximation to the diffraction of sound by a circular aperture in a rigid wall of finite thickness, J. Acoust. Soc. Am. 37: 286.
[34] Seo J. S., Eom H. J., and Lee H. S. (2000) Acoustic scattering from two circular apertures in a thick hard plane, J. Acoust. Soc. Am. 107: 2338.
[35] Zhang X. (2005) Acoustic resonant transmission through acoustic gratings with very narrow slits: Multiple-scattering numerical simulations, Phys. Rev. B 71: 241102(R).
[36] Lu M. H., *et al.* (2007) Extraordinary acoustic transmission through a 1D grating with very narrow apertures, Phys. Rev. Lett. 99: 174301.
[37] Hou B. and Wen W. (2008) Transmission resonances of electromagnetic wave through metallic gratings: phase and field characterizations, Opt. Express 16: 17098.
[38] Here, acoustic surface wave refers to a type of elastic interface wave, Stoneley wave, which can be excited at the interface between a solid material and a fluid, though the

surface-wave-like mode induced by the structure factor may be defined on structured surfaces made of ideally conducting/rigid material.

[39] Hou B., Mei J., Ke M., Wen W., Liu Z., Shi J., and Sheng P. (2007) Tuning Fabry-Perot resonances via diffraction evanescent waves, Phys. Rev. B 76: 054303.

[40] Christensen J., Fernandez-Dominguez A. I., de Leon-Perez F., Martin-Moreno L., and Garcia-Vidal F. J. (2007) Collimation of sound assisted by acoustic surface waves, Nature Physics 3: 851.

[41] Shen J. T., Catrysse P. B., and Fan S. (2005) Mechanism for designing metallic metamaterials with a high index of refraction, Phys. Rev. Lett. 94: 197401.

[42] Cai F., Liu F., He Z., and Liu Z. (2007) High refractive-index sonic material based on periodic subwavelength structure, Appl. Phys. Lett. 91: 203515.

[43] Estrada H., García de Abajo F. J., Candelas P., Uris A., Belmar F., and Meseguer F. (2009) Angle-dependent ultrasonic transmission through plates with subwavelength hole arrays, Phys. Rev. Lett. 102: 144301.

[44] Zhou Y., *et al.* (2010) Acoustic surface evanescent wave and its dominant contribution to extraordinary acoustic transmission and collimation of sound, Phys. Rev. Lett. 104: 164301.

[45] He Z., *et al.* (2010) Acoustic transmission enhancement through a periodically structured stiff plate without any opening, Phys. Rev. Lett. 105: 074301.

Section 2

# Measurement Techniques

Chapter 5

# The Application of Surface Acoustic Waves in Surface Semiconductor Investigations and Gas Sensors

Marian Urbańczyk and Tadeusz Pustelny

Additional information is available at the end of the chapter

http://dx.doi.org/10.5772/53717

## 1. Introduction

An ever-growing interest in the physical properties of the semiconductor surface results both from the influence of processes occurring at the surface, on the properties of semiconductors, and from the influence of the physical surface structure on the operation of semiconductor devices. Among the methods of investigations of semiconductor surfaces, there are no methods of investigating the dynamic properties of electrical carries in fast and very fast energetic surface states [1-5]. Up to now the existing methods allow only to investigate the surface sates with a carrier life-time $\tau$ of above $10^{-6}$s [6]. In the case of extrinsic semiconductors the surface states may, however, be considerably faster (the carrier life-time in surface traps is usually less than $10^{-8}$s). In such cases the existing methods of determining the parameters of fast surface states allow only to estimate these parameters, since the obtained results exhibit a considerable uncertainty. Therefore investigations of the kinetic properties of fast surface states are not popular and there are no new results concerning their determination.

For some years attention has been paid to the influence of the physical state of the near-surface region of a semiconductor on the results of investigations of the acoustoelectric effects in piezoelectric-semiconductor systems. Recently also attention has been paid to the possibility of applying Rayleigh's acoustic surface waves in investigations of various parameters of solid states [7-9].

Theoretical and experimental investigations of the application of acoustoelectric effects for the determination of carrier properties in the near surface region (e.g. the electrical surface potential, carrier concentration, electrical conductivities,...) are being run all the time. Problems of the influences of chemical and mechanical surface treatments realized in the first step of preparations of semiconductor plates (wafers) on kinetic and electron features of electrical carriers have not often been taken up. The quantitative data concerning the effective life-time $\tau$ and the velocity of carrier trapping $g$ are presented very seldom. Results of investigations have shown that acoustic methods based on the acoustoelectric effects with using surface acoustic waves can be applied to determine the parameters of fast and very fast surface states in semiconductors. An analysis of the methods shows that the accuracy of the obtained surface parameters are standard about some presents, which is a rather good accuracy in the determination of the parameters of fast surface states. By means of acoustic methods both element-semiconductors and compound semiconductors can be investigated (among others – semiconductors of the III-V group).

There are many methods of gas detection and a comprehensive review can be found in literature [10-17]. Since the early MOS field transistors and Schottky diodes sensors through the first SAW devices made by D'Amico [18] we have now many different fiber optic hydrogen sensors [19,20]. For example, hydrogen gas is used as a reducing agent and as a carrier gas in the process of manufacturing semiconductors and it has been increasingly noted as a clean source of energy or a fuel gas. A leak of hydrogen in large quantities should be avoided because if mixed with air in a ratio of 4.65 - 93.9 vol. %, hydrogen is explosive. Thus a fast and precise detection of hydrogen near and especially before the explosive concentration and at room temperature is still a great problem. In nowadays very important problem is connected with detection extremely low concentration of explosives material.

Surface acoustic wave gas sensors are especially attractive because of their remarkable sensitivities due to changes of the boundary conditions for propagating acoustic Rayleigh waves, introduced by the interaction of a thin active sensor layer with specific gas molecules. This unusual sensitivity results from the simple fact that most of the wave energy is concentrated near the crystal surface within one or two wavelengths. Consequently, the surface wave is in its first approximation highly sensitive to changes of the physical or chemical properties of the thin active sensor layer placed on the surface of the piezoelectric delay line. As long as the whole thickness of the sensor material is much smaller than the surface wave wavelength one can speak of a perturbation of the Rayleigh wave [21, 22]. Otherwise, we have to take into account other types of waves, such as Love waves, which can propagate in layered structures [23].

A very interesting feature of SAW sensors is the fact that the layered sensor structure on a piezoelectric substrate provides new possibilities of detecting gas in a SAW sensor system by using the acoustoelectric coupling between the surface wave and the free charges in a semiconductor sensor layer [24]. For selective detection gas particles must be used with specific chemical sensor layers placed on the piezoelectric acoustic line. Any change in the physical properties of this sensor layer results from changes of gas atmosphere, varies the conditions of acoustic wave propagation. Therefore, the propagation velocity and attenuation of the acoustic wave are changed, too. Mass and electrical effects, such as perturbations in mechanical and electrical boundary conditions, will be considered separately. Perturbations of the mechanical boundary conditions are important in a sensor with a dielectric sensor layer. Perturbations in electrical boundary conditions are important in a sensor with a conducting or semiconducting sensor layer. These two effects occur simultaneously in the interaction time of an active film with a specific gas and are additive because of their small values. For instance, in thin phtalocyanine films as organic semiconductors both of these exist, but the electric effect is much greater (several times, depending on the gas concentration) [25].

For sensor construction and optimization conditions of working it is important to get an analytic model of the SAW sensor. Here we provide theoretical information concerning the acoustoelectric effect in the piezoelectric acoustic delay line and semiconducting sensor layer configuration. This will be the starting point to construct a multilayer analytical model, because a sensor layer with absorbed gas molecules may be treated as the sum of partial layers with a different electrical conductivity. Inside the sensor layer Knudsen's model of gas diffusion was used [26-28]. The analysis summarizes the acoustoelectric theory, i.e. Ingebrigtsen's formula [23,29], the impedance transformation law and gas concentration profiles, and predicts the influence of a thin semiconductor sensor layer with Knudsen's gas diffusion model on the SAW wave velocity in a piezoelectric acoustic waveguide in steady-state and non-steady-state conditions [30]. Basing on these results the sensor structure can be optimized. Some numerical and experimental results will be shown in the part 2 of this chapter.

## 2. Surface acoustic wave for semiconductor surface investigations

The fast development of micro- and nanoelectronics is connected with the development of various sensor and measurement techniques for semiconductor characterizations, and requires searching of new research methods. It seems that, the acoustic methods provide prospective possibilities of their applications in the technology of micro- and nanoelectronic devices. The surface acoustic wave methods may prove to be a useful, non-destructive tool for researches of the semiconductor surface in high and very high frequency ranges. Generally, these SAW methods are based on the interaction between the surface acoustic waves of the Rayleigh type and the free carriers in a semiconductor. This interaction is realised in the layered structure: the piezoelectric waveguide and the semiconductor [31-35] (**Figure** 1).

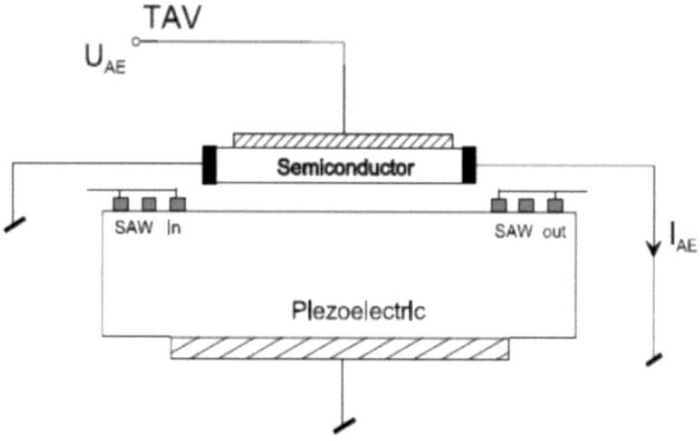


**Figure 1.** Piezoelectric-semiconductor structure for investigation of acoustoelectric interactions [35]

The electrical and electronic semiconductor surface properties may be determined by means of such parameters as: the surface potential, the carrier trapping velocity by fast and slow energetic surface states, the type of impurities of atoms and molecules, their concentration and location in the energy bandgap in the target material, as well as by the surface mobility of carriers and the lifetime of majority and minority carriers, the velocity of carriers trapping into surface states in semiconductor, and the effective live time of electrical carriers in fast surface states [36-39].

In the technology of electronic devices, silicon is a commonly applied material, but the III-V group semiconductors become more and more popular and important. This follows from the very interesting optical properties of these new materials. The coherent and non-coherent light sources were elaborated by using the electroluminescence effect observed in GaAs, GaP, InAs, InSb. Nowadays, special interest is devoted to the InAs semiconductor. The special electrical and electronic InAs properties arise mainly from the very high mobility of carriers, makeing it possible to construct electronic devices in very high frequency ranges. Applying the galvanomagnetic effect in InAs the hallotrons and gaussotrons were constructed [36]. This material is also often used in various sensors applications [6, 40].

Among the methods of semiconductor surface investigations, there are no methods which determine the surface parameters in high frequency ranges. The existing high frequency field effect method is used to deremine the kinetic parameters of near-surface regions, but its possibility of measurements is practically restricted to some MHz [1,5,41].

For this reason surface investigations of silicon and composed semiconductors in the high frequency range have not been performed. This gap can fill the methods based on SAWs.

## 2.1. Changes of SAW propagation conditions in results of acoustoelectric interaction

When the surface acoustic wave (SAW) propagates in a layered structore: piezoelectric – semiconductor, the electric field which accompanies this wave, penetrates the near- surface region of the semiconductor (**Figure** 1). The penetration depth of the electric field inside the semiconductor is of the order of the semiconductor extrinsic Debye length or the acoustic wavelength (whichever is shorter) [33, 42]. This electric field changes the free carrier concentration in the near-surface region of the semiconductor and causes a drift of the carriers [34,43]. Interaction of a surface acoustic wave (propagating on the border: piezoelectric crystal - semicondictor) with electrical carriers in the semiconductor may cause two types of effects. The first type forms effects observed in piezoelectric – changes of the velocity of SAW and changes of its attenuation [44]. The second type of effects are so called the acoustoelectric effects, observed in semiconductors [45,46].

Below the acoustic method of determining some parameters of fast surface states in semiconductors will be presented. This method is based on so-called interactions of the phonon-electron type [47] for determining both the effective carrier life-time $\tau$ influenced by the fast surface energetic states and the velocity $g$ of the carrier trapped by the surface states.

As mentioned above, among the methods of investigations of semiconductor surfaces, there are no methods of investigating the kinetic properties of electrical carriers in fast and very fast surface states [1]. The existing methods allow only to investigate the surface sates with a carrier life-time $\tau$ of above $10^{-6}$s [5,6]. In the case of extrinsic semiconductors the surface states may, however, be considerably faster (the carrier life-time in surface traps is usually less than $10^{-8}$s). The existing electrical and spectroscopic methods of determining the parameters of fast surface states allow only to estimate these parameters, since the obtained results exhibit a considerable uncertainty. For this reason, investigations of the kinetic properties of fast surface states are not popular and there are not any new results concerning their determinations.

Problems connected with the determination of the chemical and mechanical procedures of surface treatments in the first step of preparation of semiconductor plates (wafers) for technology on their electron kinetic properties have rarely been taken up. The elaborated method is based on the phenomena of surface acoustic wave SAW propagation in the system: piezoelectric waveguide-semiconductor. The electric field, which accompanies a surface wave in the piezoelectric waveguide, penetrates the semiconductor to a depth equal to Debye's screening length [32,33]. Thus, the interactions of the acoustic wave and electrical carriers are manifested in the semiconductor in the form of acoustoelectric effects [34] and in the piezoelectric waveguide - by an additional SAW attenuation (the so called - electron attenuation) and by changing the velocity of the acoustic wave [38] **Figure** 2.

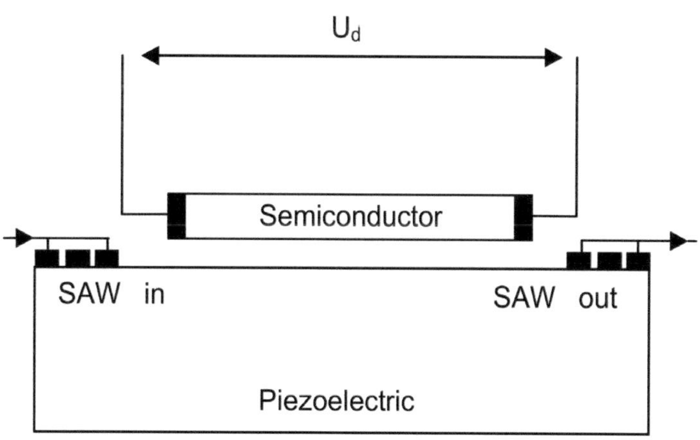


**Figure 2.** The idea of the piezoelectric waveguide-semiconductor system for investigations of the near-surface region in semiconductors [33]

Let us consider the surface wave which propagates on a piezoelectric waveguide in the system presented in **Figure** 2. The external electric drift field is applied to the semiconductor in the same direction. The coefficient of the electron attenuation of the surface wave $\alpha_e$ in such a system may be presented by the formula [33]:

$$\alpha_e = \frac{\eta \cdot \frac{\varepsilon_1}{\varepsilon_2} \cdot \frac{\omega}{\omega_c}(\gamma + a)\left[1 + \frac{\gamma + a}{A_D \cdot \frac{\omega}{\omega_c}\left[(\gamma+a)^2 + b^2\right]}\right]}{\left[1 + \frac{\omega}{\omega_c}\left(1 + \frac{\varepsilon_1}{\varepsilon_2}\right)b\right]^2 + \left[(\gamma + a)\left(1 + \frac{\varepsilon_1}{\varepsilon_2}\right)\frac{\omega}{\omega_c}b\right]^2} H \tag{1}$$

where:

$$a = \frac{g}{v_w} \cdot \frac{(\omega \cdot \tau)}{1 + (\omega \cdot \tau)^2} \tag{2}$$

$$\gamma = 1 - \frac{\mu_o \cdot E_d}{v_w} \tag{3}$$

$$A_D = \frac{1 + L_D^2 \cdot k^2}{L_D^2 \cdot k^2} \tag{4}$$

$$b = (\omega \cdot \tau) \cdot a \tag{5}$$

and $v_w$ is the velocity of SAW, $E_d$ the electric drift field, $\mu_o$ the mobility of electrons inside semiconductor, $L_D$ the Debye length, $k$ the wave number of SAW, $\omega$ the angular surface wave frequency, $\omega_c$ the so called "frequency of Maxwell's conductivity relaxation," $\eta$ the square of the electromechanical coupling coefficient, $\gamma$ the drift parameter, $\varepsilon_1$, $\varepsilon_2$ the permittivity of waveguide and semiconductor, respectively, $g$ the velocity of carriers trapping into surface states in semiconductor, $\tau$ – the effective live time of electrical carriers in fast surface states, $H$ – constant, the value of which depends on the elastic and piezoelectric properties of the waveguide and investigated semiconductor.

The process of electrical carriers trapped in the surface states in a semiconductor, under the influence of a surface wave which propagates in a piezoelectric crystal, was considered in detail in the paper [47]. **Figure** 3 presents changes of electron attenuation in the external electric field $E_d$ in the semiconductor.

The critical electrical drift field is a field at which the electron attenuation of the wave is equal to zero:

$$\alpha_{cr}(E_{dkr}) = 0 \tag{6}$$

In [47] it was shown that the equation for the critical drift field takes the following form:

$$E_{d_{kr}} = \frac{v_w}{\mu_0}\left[1 + \frac{\omega \cdot \tau}{1 + (\omega \cdot \tau)^2}\right] \tag{7}$$

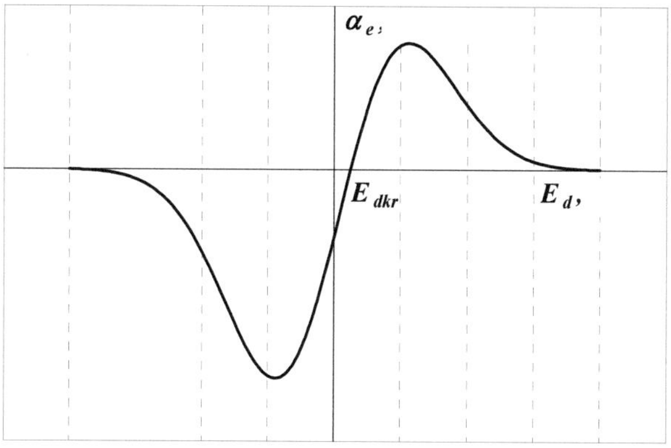


**Figure 3.** Electron attenuation $\alpha_e$ of a surface acoustic wave in the function of an external electric field $E_d$ [47]

Accordingly

$$E^o_{d_{kr}} = \frac{v_w}{\mu_o} \tag{8}$$

where $E^0_{d_{kr}}$ is the critical drift field for the theoretical case, when no surface states exist in the semiconductor surface. From equations (7) and (8) it follows that the relative change of the critical drift field in real structure, caused by surface states, is given by:

$$\frac{E_{d_{kr}} - E^o_{d_{kr}}}{E^o_{d_{kr}}} = \frac{\Delta E_{d_{kr}}}{E^o_{d_{kr}}} = \frac{g}{v_w} \cdot \frac{\omega \cdot \tau}{1 + (\omega \cdot \tau)^2} \tag{9}$$

This relation is presented in **Figure** 4.

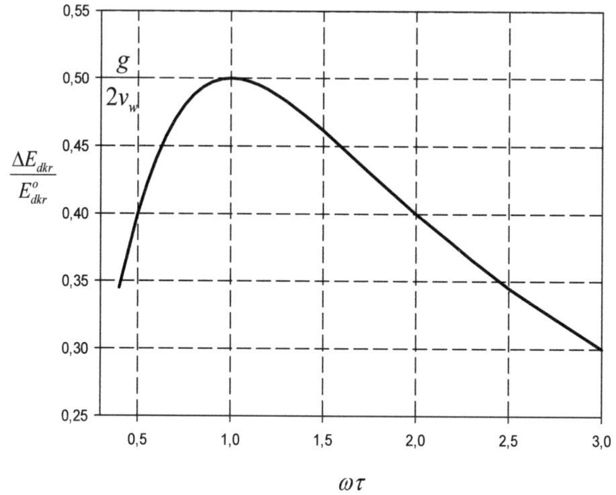


**Figure 4.** The characteristics $\frac{\Delta E_{dkr}}{E^o_{dkr}} = f(\omega\tau)$ [47]

The idea of assigning the parameters $\tau$ and $g$ of surface states consists in the determination of the electron attenuation coefficient as the drift field function for different frequencies of the surface waves. From the characteristics $\alpha_e = f(E_d)$ can be determined $E_{dkr}$ for each angular frequency $\omega$; the next one determines the characteristics $\Delta E_{dkr}/E^0_{dkr} = f(\omega)$ and the last one calculates the carrier life-time $\tau$ in the surface states as (**Figure** 4):

$$\tau = \frac{1}{\omega} \tag{10}$$

The velocity of trapping of the carriers $g$ in fast surface states is defined by the relation (Fig. 3):

$$g = 2v_w \left[ \frac{\Delta E_{dkr}}{E^0_{dkr}} \right]_{\max} \tag{11}$$

If, therefore the surface wave propagates in the structure of a piezoelectric waveguide-semiconductor, two essential parameters of surface states in a semiconductor can be determined by measurements of the velocity and attenuation of the SAW.

### 2.1.1. *Experimental set-up*

As piezoelectric waveguides the lithium niobate $LiNbO_3$ plates (propagation plane [Y] with the wave propagation direction [Z]) were used. The idea of the experimental set-up is presented in **Figure** 5. The waveguide system permitted to perform the investigations at the frequency range: 2 – 350 MHz, by using some $LiNbO_3$ plates with the SAW transducers, obtained photolithographically. The application of a monochromator in the set-up permits to illuminate the semiconductor surface and to realise investigations of semiconductor plates for various semiconductor photoconductivities.

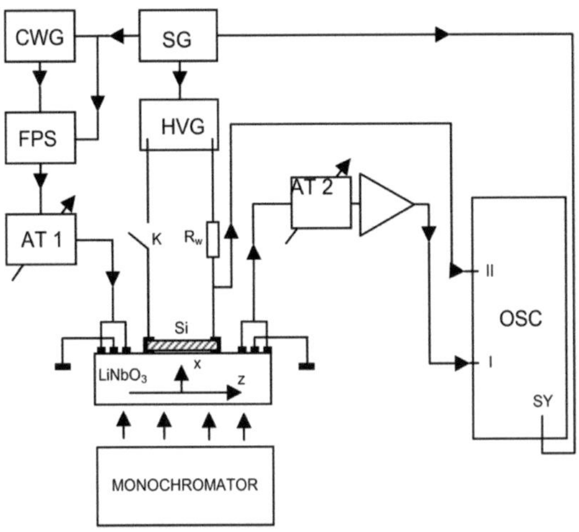


**Figure 5.** The scheme of the system for monitoring the electron attenuation coefficient $\alpha e$ as a function of the drift field Ed; (SG – synchronous generator, CWG – continuous wave generator, FPS – forming packet system, AT –attenuator, OSC – digital oscilloscope) [47]

The presented results concerning the plates of Si with the following parameters were obtained:

- Si[111], n-type, electrical conductivity $\sigma = 1{,}2[\Omega m]^{-1}$, volumetric mobility of the carrier $\mu_o = 0{,}13[m^2/V\cdot s]$, and the geometrical dimensions $10\times7\times0{,}05mm^3$.

The characteristics of $\alpha_e = f(E_d)$ for the real Si(111) surfaces after changing their electrical conductivity by means of optical excitations are presented in **Figure** 6. The measurements were performed for the following photoconductivities of the Si sample: $\sigma_1 = 1{,}2\ [\Omega m]^{-1}$; $\sigma_2 = 1{,}6\ [\Omega m]^{-1}$; $\sigma_3 = 1{,}9[\Omega m]^{-1}$.

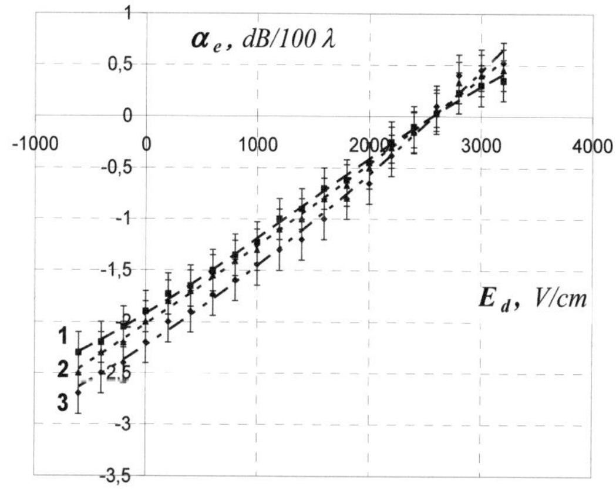


**Figure 6.** The experimental characteristics $\alpha_e = f(E_d)$ for the real surfaces of Si(111) concerning various photoconductivities: (1) $\sigma_1 = 1{,}2\ [\Omega m]^{-1}$; (2) $\sigma_2 = 1{,}6\ [\Omega m]^{-1}$; (3) $\sigma_3 - 1{,}9\ [\Omega m]^{-1}$ [17]

### *2.1.2. Results of investigations of fast surface state parameters in the Si single-crystal samples*

By means of the method presented above the parameters $\tau$ and $g$ for the Si single-crystal samples were determined whose surfaces were treated in various way.

**Figure** 7 presents the characteristics $\Delta E_{dkr} / E^o_{dkr} = f(\omega\tau)$ of the Si samples, but after their heating in vacuum at elevated temperature (~600K) as well as after heating at elevated temperature (~600K) but in atmosphere with saturated vapour. On the base of these investigations one can see that the parameters $\tau$ and $g$ in Si are essentially different.

The results of these investigations are presented in Table 1.

The approximate values of the fast surface states concentrations $N_t$ are quoted in Table 1. The procedure of $N_t$ determination is based on the well known relation [1]:

$$g = S_n \cdot V_T \cdot N_t \tag{12}$$

where: $V_T$ is the thermal velocity of electrons in semiconductors and $S_n$ is the effective cross-section for electron trapping.

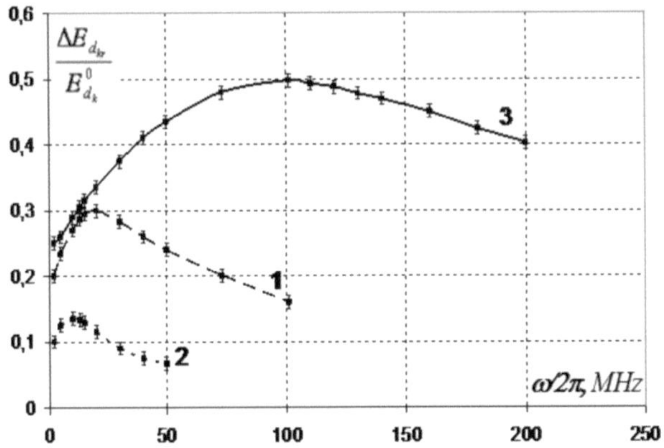


**Figure 7.** Relative changes of the critical field $\Delta E_{dkr} / E^{o}_{dkr} = f(\omega\tau)$ in Si(111): (1) in vacuum; (2) after heating in vacuum at ~600K; and (3) heating in steam at ~600K [47]

In the case of the Si single-crystals at room temperature the parameter $S_n$ is approximately equal to ~$3 \cdot 10^{-16}$cm$^2$ [1,6]. Therefore the surface concentration $N_t$ of the fast surface states can be estimated in tested semiconductors. The concentrations $N_t$ are of the order $10^{13}$cm$^{-2}$ and depend considerably on the type of surface treatments.

| Parameters of Si(111) | in vacuum | after heating in vacuum | after heating in water vapour |
|---|---|---|---|
| $\tau$, s | $8{,}0 \cdot 10^{-9}$ | $1{,}6 \cdot 10^{-8}$ | $1{,}8 \cdot 10^{-9}$ |
| $g$, m/s | 1000 | 450 | 3400 |
| $N_t$, cm$^{-2}$ | ~$3 \cdot 10^{13}$ | ~$1 \cdot 10^{13}$ | ~$8 \cdot 10^{13}$ |

**Table 1.** Parameters of the fast surface states in Si(111) [33]

### *2.1.3. Conclusion*

The results presented above concerning the interaction between the surface acoustic wave in piezoelectric crystal and the electrical carriers in semiconductor have shown that the acoustic method can be applied to determine the parameters $\tau$ and $g$ in investigations of the fast surface states in semiconductors.

The analysis of the method shows that the accuracy of the obtained results is better 5%, which is a rather good accuracy in the determination of the parameters of fast surface states. It can be pointed out that this method permits dynamic measurements of the surface state parameters over the frequency range up to several hundreds MHz (or even to some GHz), also for different, programmable changed photoconductivity of the tested semiconductors.

## 2.2. Acoustlectric effects in applications for semiconductor surface investigations

In a semiconductor, as a result of interaction between the electric carrier and the surface acoustic waves (SAW) the followning phenomena may be observed (**Figure** 1):

a. an electric current in the direction of surface wave propagation (i.e. longitudinal acoustoelectric effect LAE) [34],
b. a difference of electric potential between the semiconductor surface and its bulk (i.e. transverse acoustoelectric voltage TAV) [43].

In [26] the theoretical analysis of both acoustoelectric effects (LAE and TAV) was presented. These results formed the theoretical basis for new acoustic methods of determining the surface potential in semiconductors [46].

The experimental results of the surface potential as well as the lifetime of minority carrier investigations in some GaAs and GaP crystals performed by means of longitudinal and transverse acoustoelectric methods have been presented in [44,45].

The transverse acoustoelectric effect seems to be particularly important in semiconductor investigations. From its theoretical analysis [34, 44] it results that in the semiconductors in which one type of conductivity is predominant (n- or p-type), the sign of the acoustoelectric voltage depends on the type of electrical conductivity in the near-surface region. If the surface conductivity is negative (n-type), the transverse acoustoelectric voltage (TAV) has a positive value and if the surface conductivity is of the p-type, the value of TAV is negative. The results of investigations of the conductivity type in GaP and InP crystals have been presented in [46]. The measurement of the sign of TAV voltage is a fast and very easy method of determining the surface conductivity type.

The transverse acoustoelectric method of semiconductor surface investigations (based on the transverse acoustoelectric effect) is nondestructive. It does not require ohmic contacts on the investigated samples but, first of all, the method provides values of the surface parameters obtained at high frequencies.

Some samples of semiconductors of the III-V group were investigated by means of the TAV method. In [35] the investigations of semi-insulating GaAs at a low temperature and with a different surface wave power were reported.

Using the TAV methods we studied different semiconductor crystals, such as: GaAs [32], GaP [48], InP [45] crystals. To our knowledge, there are no papers dealing with these methods of surface investigations.

Further on, in this chapter the experimental results of the surface investigations of InAs (111) crystals after various surface treatments are quoted, as well as the results of the theoretical analysis of TAV concerning the surface electrical conductivity and surface potential.

### *2.2.1. Experimental*

The set-up for investigations of the surface semiconductor by means of the transverse acoustoelectric method is shown in **Figures** 8 and 9. The layered structure consists of the tested semiconductor is presented in **Figures** 8a and 8b [42,48].

The short (some μ-seconds) impulses of 134 MHz are applied to the input interdigital transducer to generate surface acoustic waves (SAWs) on the Y-cut, Z- propagating in LiNbO3 delay line. The semiconductor is placed on the surface of the delay line in an isolating distance bars in order to assure a non-acoustic contact between the semiconductor and the piezoelectric waveguide. The transverse acoustoelectric signal (TAV) across the semiconductor is detected by placing one Al plate on the back surface of the semiconductor (TAV electrode 2), and another one under the semiconductor sample placed on the acoustic wave guide (TAV electrode 1). The thickneses of the TAV electrode 1 and the isolating distance bars amounts to about 400 nm.

In order to provide the best contact between the investigated semiconductor surface and the TAV electrode 1, this electrode consists of narrow strips (**Figure** 8b). The width of a single metallic strip is ~ $2{\odot}10^{-2}$ mm and the distances between them are about 0.5 mm.

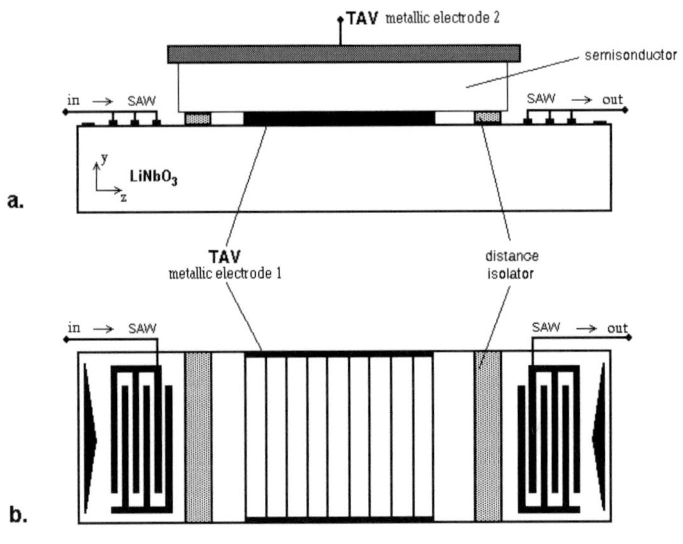


**Figure 8.** a,b. Schematic presentation of the layered structure: piezoelectric wave guide and investigated semiconductor [42]

The distance between the strips is much larger than the acoustic waveength ($\lambda \approx 2.5{\odot}10^{-2}$ mm), and for this reason the TAV electrode 1 is not essential for the SAW propagation and does not disturb it. The TAV electrode 1 in the form of a grating provides a good electric field distribution between both TAV electrodes. It relates to acoustoelectric voltage as well as to external voltage $U_d$.

In order to change the value of the surface potential in the investigated semiconductor, we applied to the semiconducter a d.c. voltage or the pulse voltage across the TAV electrodes. The set-up is shown in **Figure** 9.

For registration of the acoustoelectric voltage, in our set-up we used the digital scope (IWATSU DS-86359 ) and the computer scope converter( PFS-20).

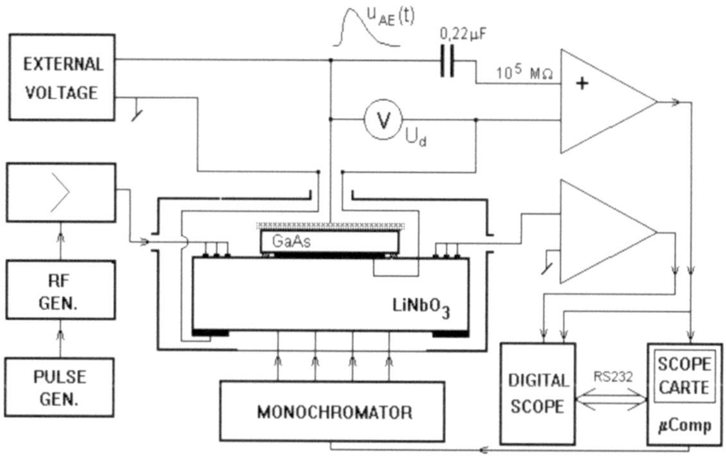


**Figure 9.** The experimental set-up for investigations of the semiconductor surface by means of theTAV method [42]

Using the higher harmonics of the acoustic transducer it is possible to investigate semiconductors above one GHz.

The interdigital transducers and the aluminium TAV electrode 1 are produced by standard evaporation and photolithography techniques.

The transverse acoustoelectric voltage method was used to investigate the InAs (111) crystal samples with the following bulk parameters:

- n-type electrical conductivity;
- carrier mobilities: $\mu_n = 33\,000$ cm$^2$V$^{-1}$s$^{-1}$; $\mu_p = 460$ cm$^2$V$^{-1}$s$^{-1}$permittivity: $\varepsilon = 11.5$;
- band gap : Eg = 0.44 eV,
- electron concentration: $N_d = 1.6 \cdot 10^{15}$ cm$^{-3}$,
- resistivity: $\rho = 2.5 \cdot 10^5$ Ωcm.

### 2.2.2. Results and discussion

In [49] we presented the results of our theoretical analysis of the acoustoelectric effects (longitudinal acoustoelectric current and transverse acoustoelectric voltage) in a layered piezoelectric - semiconductor structure. We have shown the theoretical relations between the acoustoelectric voltage and the semiconductor surface parameters such as the surface electrical conductivity and the surface potential. **Figure** 10 presents the transverse acoustoelectric voltage (TAV) as a function of surface electrical conductivity in the investigated InAs sample.

This function was calculated by means of our theoretical results presented in [49]. At some values of the p-type surface conductivity the minimum transverse acoustoelectric voltage is observed (about $\sigma \approx 10^{-4}$ 1/Ωcm). In the case of the p-type surface electrical conductivity the TAV amplitudes have negative values. Then, starting from $\sigma \approx 10^{-2}$ 1/Ωcm, the amplitude

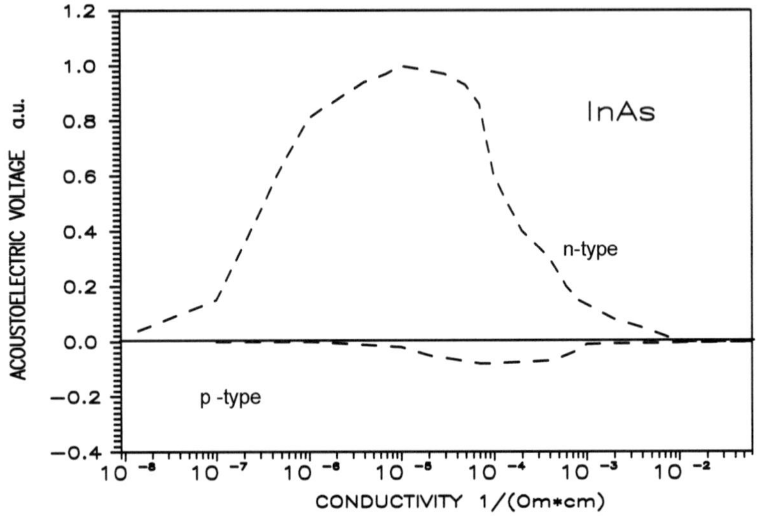


**Figure 10.** Transverse acoustoelectric voltage as a function of the electrical surface conductivity in InAs [49]

TAV is equal to zero up to $\sigma \approx 10^{-8}$ 1/Ωcm n-type surface conductivity. Next, the TAV starts to grow, it has positive values of TAV amplitude and reaches the maximum at $\sigma \approx 10^{-5}$ 1/Ωcm. For higher values of conductivity the TAV grows small and for $\sigma \approx 10^{-2}$ 1/Ωcm it is practically equal to zero. The qualitative explanation of the dependence of this kind is simple. For very high carrier concentrations, near the intrinsic region, the interactions between the electric field (accompanied the acoustic wave in piezoelectric waveguide) and carriers in the near-surface region are very weak. The TAV, as the results of these interactions are also very small. In the opposite case, if the carrier concentration in the semiconductor is high, the electric field of the wave is practically complet screened by these carriers. Then, the TAV amplitude is also small. One can see in **Figure** 10. that the transverse acoustoelectric voltage spectroscopy is a very sensitive method for high resistivity semiconductors. It is an ideal technique to study the InAs samples, because their electrical properties (high resistivity and high carrier mobility) are just in the high sensitivity TAV range.

As it was mentioned before, the electronic properties of the near-surface region may be changed by external electrical voltage perpendicular to the surface. It is a very simple method of influencing the surface semiconductor properties. In **Figure** 11 the experimental dependence of the amplitude of the transverse acoustoelectric voltage $U_{AE}$ upon the external voltage $U_d$ for InAs sample is presented.

**Figure** 12 shows the theoretical function of the amplitude of TAV versus the surface potential $U_s$ for InAs singlecrystals.

$U_s$ denotes the value of the surface potential in kT units: $U_s = e\Phi_s/kT$ and $\Phi_s$ denotes the surface potential in the semiconductor [6] (in Volt unites).

The function presented in **Figure** 12 was calculated basing on the theoretical relations quoted in [49]. In order to determine this relation, we had to know the following bulk parameters presented above: the mobility concentration of the carriers, the permittivity and the electrical conductivity for the investigated InAs crystals.

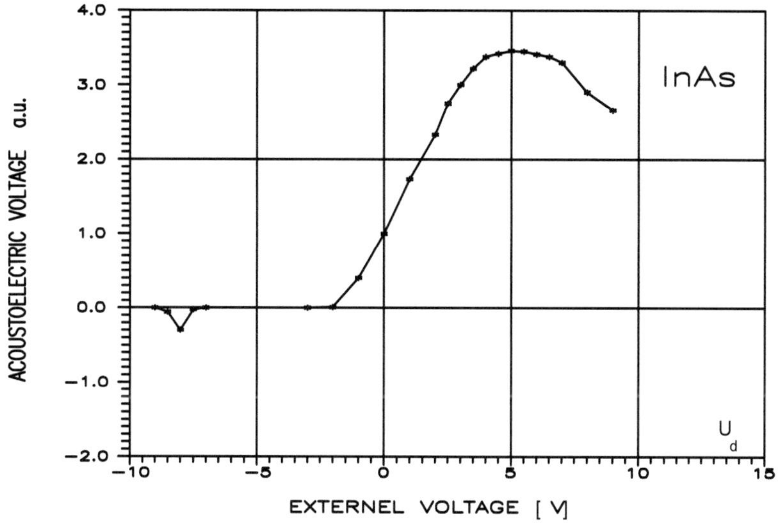


**Figure 11.** The experimental dependence of $U_{AE}$ on the external voltage $U_d$ for InAs surface

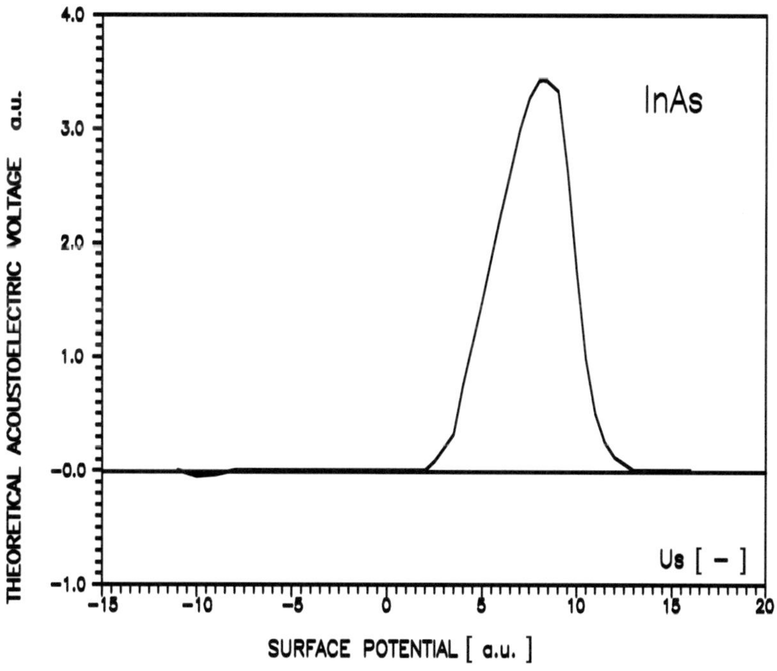


**Figure 12.** The theoretical dependence of $U_{AE}$ on the surface potential $U_S$ for InAs ($U_S$=e$\Phi_S$/kT) [49]

By using both functions: the experimental relation $U_{AE} = f(U_d)$ and the theoretical relation $U_{AE} = f(U_S)$, one may propose the method of determining the surface potential value in semiconductors (for zero external bias voltage). This method of determining the surface potential was discussed in detail in [45].

The theoretical dependence of the TAV amplitude upon the surface potential $U_S$ is calculated with an accuracy up to an arbitrary constant multiplier K [45]. (The constant K may be interpreted as a so called aperture constant). For K we may substitute the value for which the maximum in the theoretical relation $U_{AE} = f(U_S)$ and the maximum in the experimental relation have the same values. For the experimental characteristic $U_{AE} = f(U_d)$, the value $U_{AE} = 1$ corresponds to the case for which the external voltage $U_d$ is equal to zero. For the theoretical relation $U_{AE} = f(U_S)$, the value $U_{AE} = 1$ corresponds to the surface potential of the investigated semiconductor sample.

For the investigated InAs sample, the value of the surface potential $\Phi_S$ obtained by applying this method was equal to $\Phi_S = - 0.11 \pm 0.02$ V [49].

The transverse component of the electric field of the acoustic wave in the piezoelectric waveguide acting on electrical carriers in semiconductors changes its concentration in the near-surface region. The new steady state of the concentration is reached after the time period which depends on the carrier lifetime. The properties of the electrical surface semiconductor differ from the bulk ones. Therefore, the values of lifetime in the bulk and in the surface may be quite different, too.

The lifetime of the minority carriers may be determined from the shape of the transverse acoustoelectric signal. It was shown in [49] that the transverse acoustoelectric signal $u_{AE}(t)$ is described by the following mathematical formula:

$$u_{AE}(t) = U_{AE} \frac{t}{\tau_a - \tau_e} \left[ e^{-t/\tau_a} - e^{-t/\tau_e} \right] \tag{13}$$

where:

$U_{AE}$ - transverse acoustoelectric amplitude,

$\tau_a$ - time constant of the experimental set-up,

$\tau_e$ - effective life time of the minority carriers.

For the case, when the time constant of the experimental set up $\tau_a$ is much longer than the life time of minority carriers ($\tau_a >> \tau_e$), the value of the time constant of the acoustoelectric signal is practically equal to the life time of minority carriers $\tau_e$.

(The condition $\tau_a >> \tau_e$ is usually easy to be satisfied). The life time $\tau_e$ was determined making use of this method. For the InAs sample under investigation the life time of the minority carriers is equal to $\tau_e = 2.0 \cdot 10^{-4}$ s (before the surface treatment).

The experimental measurements of the TAV amplitude versus the surface acoustic wave power allow to estimate the intensiy of the influence of the surface processes on the total conductivity of the investigated samples. If the interactions between the electric carriers and surface traps are not intensive, the amplitude $U_{AE}$ versus the acoustic power of the surface acoustic wave is a linear function. If the interactions of the electric carriers with the surface tarps are intensive, relation $U_{AE} = f(P_{ak})$ is non-linear [42].

The very important advantage of the method based on the transverse acoustolectric effect is the possibility of checking the semionductor surface in the course of the alteration of the surface properties. We think that this fact will allow to control the changes of the surface parameters after various surface treatments due to the technological process.

In [44,45] we presented the results of the determination of the surface potential, the life time of the minority carriers, and the type of electrical conductivity obtained for Si, GaP and GaAs single-crystal samples, which vary in their bulk and surface properties.

The main purpose of these investigations was to explore the influence of the surface treatments for the values of the surface parameters. The samples were measured after mechanical and chemical surface treatments which are applied during the technology of electronic devices. We also tested the surface parameters after mechanical grinding with alumna powders of various granulations and after polishing with diamond paste. We also tested the surface parameters after the cleaning of the sample in acetone, benzene, methanol and after chemical etching in HF acid or in $3HNO_3 + 10H_2O_2$. After the surface treatments, the samples were rinsed in methanol and deionized water. We applied various combinations of simultaneous mechanical and chemical treatments.

**Figure** 13 presents the changes of life time of the minority carriers in the InAs sample after: a) grinding with alumna powder (grains No100) , b) diamond paste polishing, c) HF acid etching [49].

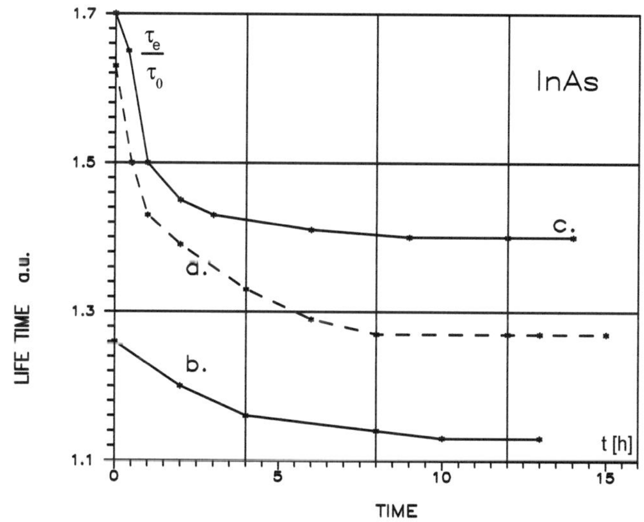


**Figure 13.** The life time of minority carriers in InAs after different surface treatment: a) alumna powder grinding, b) diamond paste polishing, c) HF acid etching [49]

In these studies we observed that the life time of the carriers changed very essentially after alumna powder grinding.

The changes of $\tau_e$ stopped after some hours, but with different steady state values than before the surface treatments.

The steady state value of the minority carriers life time after grinding with alumna powder is equal to $\tau_e = 2.5\odot10^{-4}$ s. In **Figure** 13, the time $\tau_0$ denotes the steady state of life time of the minority carriers before the surface treatments, equal to $\tau_o = 2.0\odot10^{-4}$ s.

The very long time of the transient state of the life time shows that in the surface processes not only fast surface states take part but also slowl surface states in the oxide layer.

### 2.2.3. *Conclusion*

It follows from the presented results that the transverse acoustoelectric method (TAV) is useful tool for studying the electrical and electron surface properties of semiconductors. On the example of IaAs (111) it was shown, that the TAV method allows to determine experimentally the surface potential and the effective live time of minority carriers, as well as the type of the surface conductivity in the near-surface segion. Indium arsenide is a relatively new semiconductor material. This material is used, among others, in the technology of luminescence and electroluminescense devices, as well as target material for laser diodes, hallotrons and gaussotrons.

Basing on the theory of acoustoelectric effects, one can present also the results of the theoretical analysis ofr TAV versus the surface potential. These theoretical results and the experimental relation $U_{AE}$ versus the external voltage $U_d$ allowed to determine the value of the surface potential in the investigated InAs sample by zero bias voltage.

In the case of the IaAs (111) sample the surface potential determined by means of our acoustic method was equal to $U_S = -0.11 \pm 0.01$ V.

We presented the theoretical results of the acoustoelectric voltage versus the electrical conductivity in the near-surface region. It follows from this result that TAV is a very strong function of the carriers concentration and only in the case of restricted concentration partitions the TAV amplitude differs from zero value. In the investigated InAs (111) sample these concentrations are in the range: $1.3 \odot 10^{15} \div 7.0 \odot 10^{16}$ cm$^{-3}$.

We proved that the values of $\tau_e$ depends on technology the type of technology of the surface preparation. The largest changes of $\tau_e$ in the InAs sample were observed in the case of mechanic alumna grinding (No 100) and for $HNO_3$ acid etching. The changes of $\tau_e$ after these surface treatments were of some tens percents. After cleaning and boiling in different alcoholes, benzene and acetone the changes of life times were less, only about some percents. The smallest changes of $\tau_e$ were observed after benzene cleaning (2-3 %). After diamond paste polishing the changes of $\tau_e$ amounted to about ten percents.

The Transverse Acoustoelectric Voltage Method is non destructive. Moreover, this method does not require ohmic contacts with the tested semiconductor samples. This method provides the possibility of determining the values of the parameters in high and very high frequency ranges.

The results presented above were obtained for surface acoustic wave frequency of 134 MHz. The surface semiconductors may be investigated by means of the SAW techniques up to some GHz. Results of the semiconductors surface investigations of III-V group in a very high frequency range have been presented in [48,49].

From the presented measurements it follows, that acoustic methods may be very useful complement for existing electrical, photospectroscopy and photoemission methods of semiconductor surface investigations.

## 3. SAW application in gas sensors

The SAW sensor set-up is based on frequency changes in a surface acoustic wave dual delay line system, which is nowadays well known – **Figure** 14.

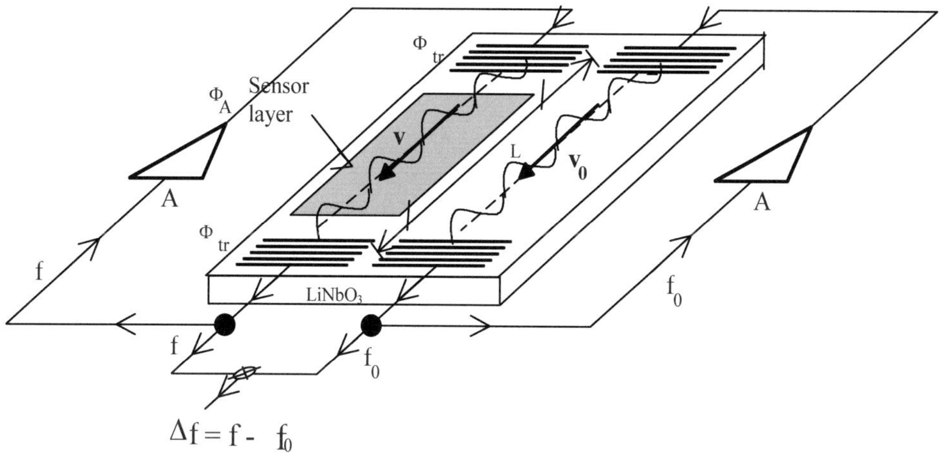


**Figure 14.** The SAW dual delay line system based on frequency changes. The abbreviations: $\Phi_A$ ,$\Phi_{tr}$ - phase shift introduced by amplifier and transducer respectively, A - amplifier; v, $v_0$ - velocity of the SAW in a measuring and reference paths [24]

On a piezoelectric substrate (usually $LiNbO_3$) two identical acoustic paths were formed using interdigital transducers [50]. Next an active multilayer sensor structure is formed in the measuring line by vacuum deposition processes. The second path serves as a reference and can compensate small variations of ambient temperature and pressure. Both delay lines are placed in the feedback loop of oscillator circuits and the response to the particular gas of the active multilayer is detected as a change in the differential frequency Δf, i.e. the difference between the two oscillator frequencies f and $f_0$.

Principally, any change in the physical properties of a thin active multilayer, due to its interaction with gas molecules on a piezoelectric surface can affect SAW propagation. However, from the practical point of view only the two following effects have a potential meaning. Namely, a change in the mass density of the multilayer, and a change in its electrical conductivity cause significant changes in the velocity and attenuation of the SAW and consequently changes in the frequency f and Δf .

At first, according to D'Amico [51], we used a palladium (Pd) layer as a sensitive material in experimental results. A palladium film absorbs easily hydrogen molecules and is a well known material for the detection of hydrogen. Hydrogen absorption and desorption cause changes in the density, elastic properties and conductivity of the Pd film. The process is completely reversible. The obtained result was unfortunately very weak – **Figure** 15. The changes of the frequency Δf do not exceed 50 Hz for 0.5 % and 1 % for hydrogen in nitrogen at 30 °C and this is almost the detection limit of the equipment. At higher temperatures there were no better results. When a 720 nm CuPc layer was used the sensor response amounted to about 600 Hz, but only at higher temperatures than70 °C. The result is shown in **Figure 16**. It is also a well- known fact, that phthalocyanine films must be thermally activated to detect gas. The investigated CuPc layer with a thickness of about 720 nm was obtained by means of the vacuum sublimation method, using a special aluminium mask.

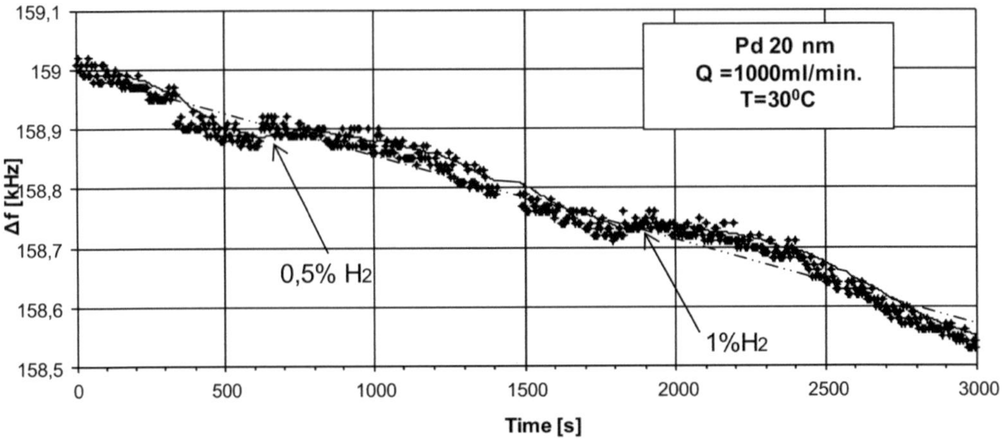


**Figure 15.** Changes of the differential frequency versus time at two concentrations of hydrogen for 20 nm of palladium at 30 °C [51]

The source temperature was about 600 °C and the thickness was measured by the interference method. Before the specific process of sublimation, CuPc powder (Aldrich) was initially out-gassed at 200 °C for 15 to 20 min in vacuum ($10^{-4}$ Torr). The CuPc vapour source was a crucible placed in a properly formed tungsten spiral. A copper-constantan thermocouple was used to control the temperature. In the case of multilayer structure the obtained results are very promising. We can observe recurrent changes of the differential frequency of the measurements system in the function of the medium hydrogen concentration in nitrogen at 30 °C and 50 °C of interaction – **Figure 16.**

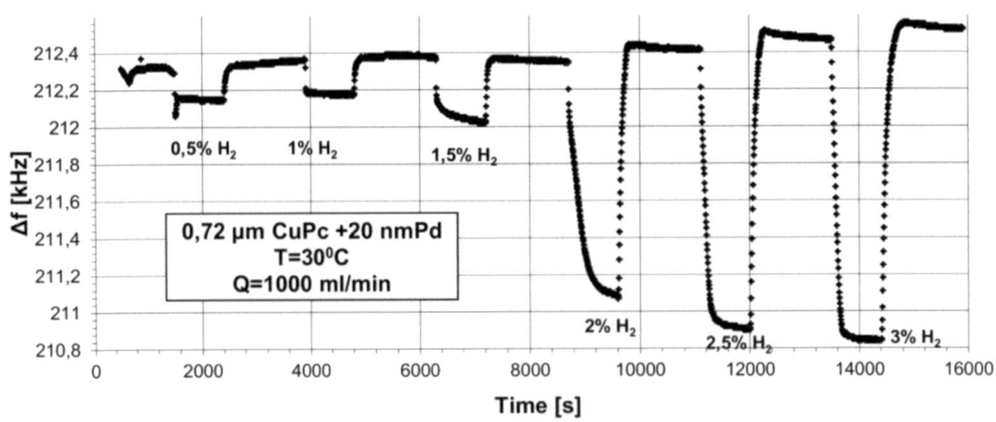


**Figure 16.** Changes of the differential frequency Δf versus time concerning a multilayered sensor structure at 30 °C with different concentrations of hydrogen in nitrogen [24]

Such a good work of the sensor may be explained as follows. As already mentioned above, it is well known, that the palladium film easily absorbs hydrogen molecules. However, when we use only a metal (palladium) layer on a piezoelectric substrate (which was $LiNbO_3$ Y-Z), the metal layer shortens the electric field associated with the surface wave. Consequently, the sensor can detect only the mass loading. When only a CuPc layer is used, the sensitivity of this compound is too weak to detect hydrogen molecules, besides, the conductivity of the layer at room temperature is too high and this sensor must be temperature-activated. Nowadays it is well established, that in the case of phthalocyanine compounds in a SAW system the electric effect is much greater than the mass loading. So that, to take full advantage of the high sensitivity offered by the SAW sensor, the thin film conductivity must be in some particular range [25,52,53]. The multilayer structure (CuPc 720 nm + 20 nm Pd) on a $LiNbO_3$ Y-Z substrate has probably its resultant electrical conductivity well fitted to the high sensitivity range of the SAW device and can detect hydrogen molecules even at room temperature.

We have also used hydrogen phtalocyanine covered with a thin palladium film. Interaction results with hydrogen gas are shown in **Figure** 17.

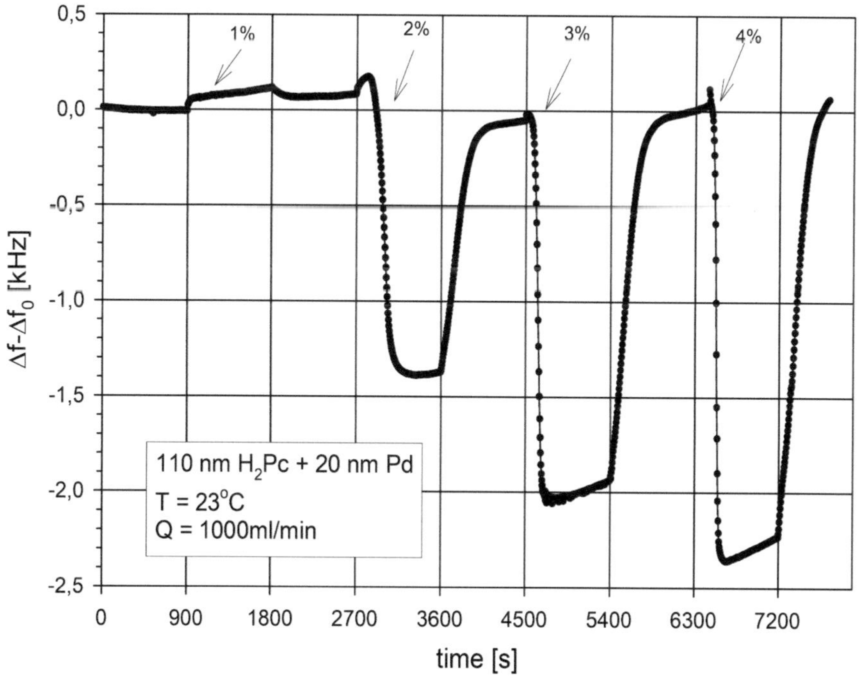


**Figure 17.** Changes of the differential frequency Δf versus time in a multilayered sensor structure at 23 $^{0}$C and at different concentrations of hydrogen in nitrogen [52,53]

## 3.1. Acoustoelectric effect in the piezoelectric – semiconductor layer

In the case of small disturbances, both mass and the electric load mentioned above may be considered separately, assuming that the total effect of a relative change of the wave vector

$\Delta k/k_0$ and the velocity of propagation $\Delta v/v_0$, is the sum of both these component disturbances [22].

$$\frac{\Delta k}{k_0} \approx \left(\frac{\Delta k}{k_0}\right)_m + \left(\frac{\Delta k}{k_0}\right)_\sigma \tag{14}$$

Making use of the theory of disturbances [54], the contribution of each of these effects can be determinates theoretically. From the physical point of view it seems to be essential that the influence of each effect (mass and electrical) should be considered separately from its reaction to the interaction of the layer with gas. In the case of an electric effect we may assume that the mass of the layer is $m$=0. The electric "load" results from the effect of the interaction of the electric potential associated with the surface wave with mobile charge carriers in the layer.

## 3.2. Electrical surface perturbations

In these problems the mechanical boundary conditions are unperturbed. If the electrical boundary conditions are perturbed only at the upper surface $y$=0, the perturbation formula is [23,54]:

$$(\Delta k)_\sigma \approx \frac{\omega\left[\varphi'(0)D_y^*(0) - \varphi^*(0)D_y'(0)\right]}{4P} \tag{15}$$

where the index $\sigma$ refers to the change of the wave number due to disturbances, electrical surface perturbations.

In these problems the mechanical boundary conditions ($T$=0) are unperturbed. Unperturbed electrical boundary conditions are usually neither a short-circuit [$\sigma(0) = 0$] nor an open-circuit [$D_y(0)$=0]. The best agreement between the perturbation theory and exact numerical calculations is obtained by using the so-called weak coupling approximation, in which the stress field $T$ is assumed to be unchanged by the perturbation.

We can define an impedance per unit area [30,55]:

$$Z_E(0) = \left(\frac{\varphi}{i\omega D_y}\right)_{y=0} \tag{16}$$

It will be assumed, that unperturbed problem provides free electrical boundary conditions at the substrate surface, that is, the region above the substrate (y<0 in **Figure** 18) is a vacuum and extends to y→ ∞. The space-charge potential satisfies the general form of Laplace. For this region Laplace's equation is reduced to [30]:

$$\nabla^2\Phi = 0 \tag{17}$$

The unperturbed potential function is therefore:

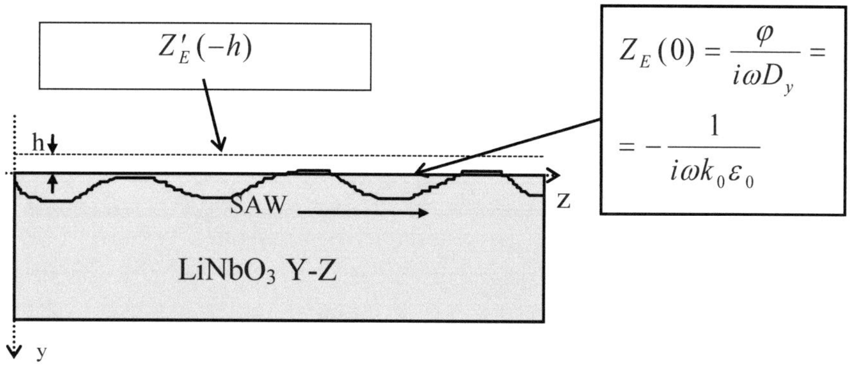


**Figure 18.** Electrical surface perturbations [30]

$$\Phi = \Phi_R(y)e^{-ikz} = e^{ky}e^{-ikz}, \; y < 0 \tag{18}$$

and the normal component of electrical displacement is:

$$D_y = -k\varepsilon_0 e^{ky}e^{-ikz}, \; y < 0 \tag{19}$$

Consequently, the unperturbed surface impedance is:

$$Z_E(0) = -\frac{1}{i\omega k_0 \varepsilon_0} \tag{20}$$

The perturbed boundary conditions will be specified in terms of the normalized surface impedance:

$$z'_E(0) = \frac{Z'_E(0)}{|Z_E(0)|} = -ik_0\varepsilon_0 \left( \frac{\varphi'}{D'_y} \right)_{y=0} \tag{21}$$

The potential of the perturbed fields $\varphi'(0)$ and the electrical displacement $D_y(0)$ are now related to the unperturbed fields:

$$\begin{aligned} \varphi'(0) &= \varphi(0) + A \\ D'_y(0) &= D_y(0) + k_0 \varepsilon_p^T A \end{aligned} \tag{22}$$

where

$$\varepsilon_p^T = \sqrt{\varepsilon_{yy}^T \varepsilon_{zz}^T - \left(\varepsilon_{yz}^T\right)^2}$$

and the approximation $k_0' = k_0$ has been made. The electrical displacements $D_y(0)$ and $D_y'(0)$ are expressed in terms of potential. At the end the electrical potential and induction may be expressed as:

$$\varphi'(0) = -iz'_E(0)\frac{\left(\varepsilon_0 + \varepsilon_p^T\right)}{\varepsilon_0 - i\varepsilon_p^T z'_E(0)}\varphi(0) \tag{23}$$

$$D'_y(0) = -\frac{k_0\varepsilon_0\left(\varepsilon_0 + \varepsilon_p^T\right)}{\varepsilon_0 - i\varepsilon_p^T z'_E(0)}\varphi(0) \tag{24}$$

At the end we obtain:

$$\frac{\Delta k}{k_0} = -\left(\frac{\Delta v}{v_0}\right)_{sc}\frac{1 + iz'_E(0)}{1 - i\frac{\varepsilon_p^T}{\varepsilon_0}z'_E(0)} \tag{25}$$

for the unperturbed boundary conditions (electrically free), and perturbation of the velocity due to an electrical short-circuit on the boundary (*sc*):

$$\left(\frac{\Delta v}{v_0}\right)_{sc} = -\omega\left(\varepsilon_0 + \varepsilon_p^T\right)\frac{\left|\varphi(0)\right|^2}{4P} \tag{26}$$

This is the Ingebrigtsen formula for electrical surface perturbations of SAW Rayleigh waves. It is also applicable to other types of piezoelectric surface waves when the appropriate normalized surface wave potential is used in $\left(\frac{\Delta v}{v_0}\right)_{sc}$. Since the unperturbed wave satisfies free electrical boundary conditions, then Δk→0 in (2.13) when:

$$z'_E(0) = \frac{Z'_E(0)}{\left|Z'_E(0)\right|} \to i \tag{27}$$

The influence of the effect between the electric potential associated with the acoustic wave and the carriers of the electric charge in this layer leads to a decrease of the velocity. This effect depends on the electromechanical coupling factor $K^2$. The Ingebrigtsen formula for electrical surface perturbations of SAW Rayleigh waves is reduced to the form [23,30]:

$$\left(\frac{\Delta k}{k_0}\right) = \frac{K^2}{2}\frac{1 + iz'_E(0)}{1 - i\frac{\varepsilon_p^T}{\varepsilon_0}z'_E(0)} \tag{28}$$

where

$$K^2 = 2\left(\frac{\Delta v}{v_0}\right)_{sc}$$

### 3.3. Infinitesimally thin semiconducting sensor layer

We consider first the RF interaction with an infinitesimally thin semiconductor, regarded to be thinner than the Debye length. For the width $w$ in the $x$ direction, the one-dimensional equation of motion is:

$$I_z = \rho_{01}\mu E_z - D\frac{d\rho_1}{dz} \tag{29}$$

where $I_z$ is the total RF current in the direction $z$, $\rho_1$ and $\rho_{01}$ are, respectively, the RF and dc charge density per unit length. The quantities $\mu$ and $D$ are, respectively, the carrier mobility and diffusion parameters. By combining the one-dimensional equation of continuity:

$$\frac{dI_z}{dz} + i\omega\rho_1 = 0 \tag{30}$$

with (2.17) and assuming that all RF quantities vary as $\exp[i(\omega t - kz)]$, it follows that the RF charge field is related by:

$$\rho_1\left(\omega - iDk^2\right) = k\rho_{01}\mu E_Z \tag{31}$$

Although thin, the semiconductor layer has a finite thickness $h$, width $w$ and the volume dc charge density is:

$$\rho_0 = \rho_{01}/wh$$

Following the conventions established by White [49] for the volume acoustic wave in configurations, where we define the relaxation frequency $\omega_c = \frac{\mu|\rho_0|}{\varepsilon}$ and the diffusion frequency $\omega_D = \frac{v_0^2}{D} = \frac{k_B T}{q}\mu$, where $k_B$ - Boltzman's constant, $T$ – temperature.

We then rewrite (31) in the form:

$$\rho_s = \frac{\left(\frac{\omega_c}{\omega}\right)\varepsilon kh}{1 - i\frac{\omega}{\omega_D}} E_z \tag{32}$$

where $\rho_s = \rho_1/w$

### 3.4. Surface impedance of single semiconducting sensor layer

**Figure** 19 presents the schematic diagram of the layered SAW gas sensor.

When $y < 0$ in **Figure** 19, we assume isotropic conditions. The space-charge field boundary conditions for single semiconductor layer in the plane $y=0$ are respectively: $\varphi(0^-) = \varphi(0^+)$ and $D_y(0^-) - D_y(0^+) = \varrho_s$, where $D_y(0^-) = \varepsilon_0 E_y(0^-)$. Since we assume that the field solution has an electrostatic form $\vec{E} = -\nabla\varphi$, we imply that $E_z^- = ik_0\varphi(y,z)$.

By applying Gauss' law and potential continuity in the plane $y=0$, the impedance $z_E'(0)$ can be expressed in the following form:

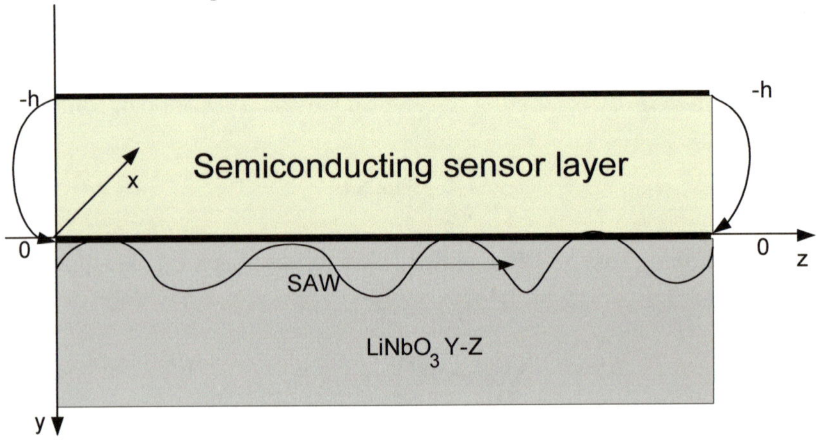


**Figure 19.** Schematic diagram of the layered SAW gas sensor [30]

$$z_E'(0) = \left( -i + \frac{\dfrac{\sigma_s}{\varepsilon_0 v_0}}{1 - i\dfrac{\omega}{\omega_D}} \right)^{-1} \tag{33}$$

where $\sigma_s = \sigma\, h$ is the surface conductivity of the layer.

In the case of the diffusion effects are small ($D$~0), the surface conductivity is independent of the frequency of the wave propagation and surface impedance $z_E'(0)$is only function of the surface conductivity $\sigma_s$

$$z_E'(0) = \left( -i + \frac{\sigma_s}{\varepsilon_0 v_0} \right)^{-1} \tag{34}$$

### 3.5. Dependence of the SAW velocity vs. surface layer conductivity

Imaginary and real parts in the equations indicate changes in the attenuation and velocity of an acoustic wave [22]:

$$\frac{\Delta\alpha}{k_0} = \mathrm{Im}\left\{\frac{\Delta k}{k_0}\right\} \qquad \frac{\Delta v}{v_0} = -\mathrm{Re}\left\{\frac{\Delta k}{k_0}\right\} \tag{35}$$

When the diffusion effect vanishes (*D*=0), the perturbation of velocity will be independent of the frequency of the propagated wave.

The reduction of the velocity of propagation is mainly caused by two phenomena [22,30]:

a. The mass load on the crystal surface, resulting this time from the phenomenon of adsorption of gas particles on the layer surface, the surface density may be disturbed due to the additional mass of absorbed gas particles and the elastic modulus of the isotropic layer and
b. The electric "load" resulting from the interaction of the electric potential associated with the surface wave with mobile charge carriers in the layer. This results of changes in the electric conductivity of the layer due to the interaction of this chemically active layer with gas.

The electric effect has one very interesting feature, namely it causes significant changes in the propagation velocity of the SAW only in some particular range, which depends only on the properties of the piezoelectric substrate (which was $LiNbO_3$, Y-Z); the metal layer shortens the electric field associated with the surface wave [22,30]. According to this theory we can prove that surface conductivity influences the detection features of SAW gas sensors. The initial value of the conductivity determines the point of sensor work. The value of acoustoelectric interaction in the sensor configurations fulfill an important role in the sensitivity of the sensor. **Figure** 20 presents the characteristics of the relative change of a surface acoustic wave and its attenuation in the function of conductivity. In the case of $LiNbO_3$ the cuts Y-Z as piezoelectric substrate the high sensitivity range of the sensor in the region:

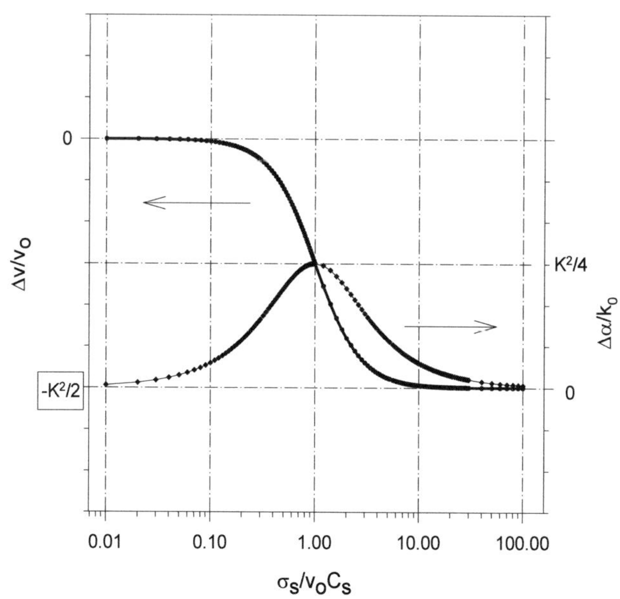


**Figure 20.** Velocity and attenuation of the SAW wave propagation versus the electrical conductivity of the sensor layer [22]

$$0.15 < \frac{\sigma_s}{v_0 C_s} < 5 \tag{36}$$

### 3.6. Analytical model of the SAW gas sensor

The concentration profile of the gas molecules in the sensor layer depends on the phenomena of gas diffusion in porous material. The mechanism of gas diffusion through a porous material depends on the size of the pores and the type of surface diffusion: Knudsen's diffusion and a molecular one. If the pores ranging from 1 to 100 nm in radius the Knudsen diffusion occurs. The Knudsen diffusion constant, $D_K$, depends on the molecular weight of the diffusing gas, $M$, the pore radius, $r$, temperature, $T$, and the universal gas constant, $R$, as follows [26,28]:

$$D_K = \frac{4r}{3}\sqrt{\frac{2RT}{\pi M}} \tag{37}$$

Two assumptions, i.e. Knudsen diffusion and first-order surface reaction, allow to formulate the well-known diffusion equation [26-28]:

$$\frac{\partial C_A}{\partial t} = D_K \frac{\partial^2 C_A}{\partial x^2} - kC_A \tag{38}$$

where $C_A$ – is the concentration of target gas, $t$ - time, $y$ distance from the bottom layer, counted from the piezoelectric substrate, $k$ is the rate constant.

The general solution of this equation in the steady–state condition $\frac{\partial C_A}{\partial t} = 0$ is:

$$C_A = C_1 \exp\left( y\sqrt{\frac{k}{D_K}} \right) + C_2 \exp\left( -y\sqrt{\frac{k}{D_K}} \right) \tag{39}$$

here $C_1$ and $C_2$ are integral constants.

For boundary conditions at the surface ($y$ =-$h$): $C_A = C_{A,s}$ and $\partial C/\partial y = 0$ the following equation can be obtained [26,28]:

$$C_A = C_{A,S} \frac{\cosh(|y|\sqrt{k/D_k})}{\cosh(|-h|\sqrt{k/D_k})} \tag{40}$$

$C_{A,s}$ is the target gas concentration outside the film at the surface $y = -h$. The concentration profile depends on the thickness of the sensor layer and the constants $k$ and $D_K$. The gas concentration inside the film is not constant. As in resistance sensors of the gas we may assume that the electrical conductance $\sigma(y)$ of the sheet exposed to the target gas is linear to the gas concentration ($C_A$) [26]: $\sigma(y) = \sigma_0(1 \pm a \cdot C_A)$, where $\sigma_0$ is the layer conductance in the air, $a$ is the sensitivity coefficient. In resistance sensors the electrical conductance of the whole film is given by integrating $\sigma(y)$ over the whole range of y ($y = -h$; -0). That treatment has been proposed by Sakai, Williams and Hilger [26,56]. We can't treat a semiconductor layer in a homogeneous way of thinking, because the profile of molecule gas concentration in a semiconducting sensor layer changes with the distance from the piezoelectric substrate and may influence the acoustoelectric interaction differently. To analyze such a sensor layer in SAW configuration we have to assume that the film is a uniform stack of infinitesimally thin sheets with a variable concentration of gas molecules and with a different electric conductance. Each sublayer is in another distance from the piezoelectric wave-guide (**Figure** 21).

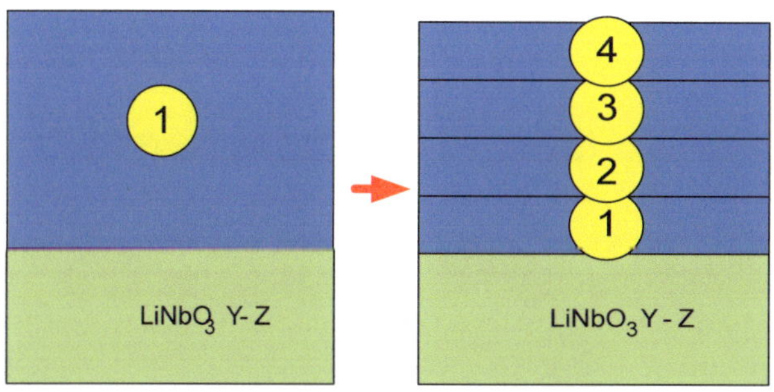


**Figure 21.** Model of a sensor layer in SAW system (n-sublayers) [57]

**Figure** 22 presents blocks whose common admittance is calculated basing on the law of impedance transformation and electronic equations. Then we applied the common impedance to the Ingebrigtsen formula.

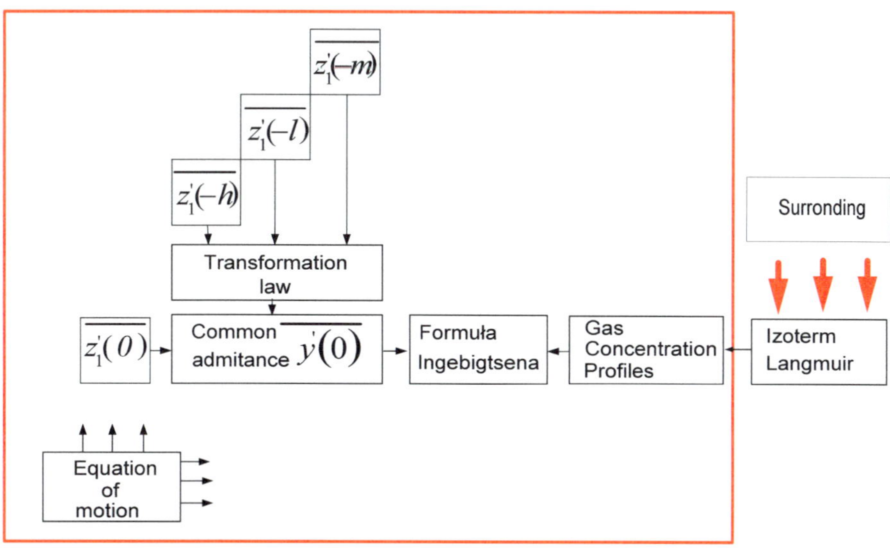


**Figure 22.** Diagram analysis of saw-type sensor [57]

The common admittance was applied in compliance with the Ingebrigtsen formula, which consisted in the defined impedance $z_E^{'}(0)$ on the surface at $y = 0$. The main problem was the calculation of the common normalized admittance which can be easy apply in the Ingebrigtsen formula. By using the impedance transformation law we counted the common admittance "which is seen" by an acoustoelectric wave on the plane $y = 0$.

We can see, that impedance transformation law takes the form [57]:

$$z_E^{'}(0) = \frac{i \cdot tgh\left(k_0 h\right) + z_E^{'}(-h)}{1 - i \cdot z_E^{'}(-h) \cdot tgh\left(k_0 h\right)} \tag{41}$$

The Ingebrigtsen formula for „$n$" sublayers is as follows [57]:

$$\frac{\Delta v}{v_0} = -\mathrm{Re}\left\{\frac{\Delta k}{k_0}\right\} = -\frac{K^2}{2}\frac{\left[\sigma_{T_2}(1 + aC_{A,y=0}) + \sum_{i=1}^{n-1}\sigma_{T_2}(y_i) f(y_i, \sigma_{T_2}(y_i))\right]^2}{\left[\sigma_{T_2}(1 + aC_{A,y=0}) + \sum_{i=1}^{n-1}\sigma_{T_2}(y_i) f(y_i, \sigma_{T_2}(y_i))\right]^2 + \left[1 + \sum_{i=1}^{n-1} g(y_i, \sigma_{T_2}(y_i))\right]^2 (v_0 C_S)^2}. \tag{42}$$

where: $C_S = \varepsilon_0 + \varepsilon_p^T$

$$f\left(y_i, \sigma\left(y_i\right)\right) = \frac{1 - \left[\mathrm{tgh}\left(ky_i\right)\right]^2}{\left[1 + \mathrm{tgh}(ky_i)\right]^2 + \left[\mathrm{tgh}(ky_i) \cdot \frac{\sigma\left(y_i\right)}{\varepsilon_0 v_0}\right]^2} \tag{43}$$

$$g\left(y_i, \sigma\left(y_i\right)\right) = \frac{\left[1 + \mathrm{tgh}(ky_i)\right]^2 + \mathrm{tgh}\left(ky_i\right) \cdot \left(\frac{\sigma\left(y_i\right)}{\varepsilon_0 v_0}\right)^2}{\left[1 + \mathrm{tgh}(ky_i)\right]^2 + \left[\mathrm{tgh}(ky_i) \cdot \frac{\sigma\left(y_i\right)}{\varepsilon_0 v_0}\right]^2} \tag{44}$$

and $\sigma_{T_2} = \sigma_{T_1} \exp\left(\frac{E_g}{2k_B} \cdot \frac{T_2 - T_1}{T_1 T_2}\right)$, $\sigma\left(y_i\right) = \sigma_0\left[1 + a \cdot C_A\left(y_i\right)\right]$, $\sigma_s = \sigma\left(0\right)$, $k_B$ – Boltzman constant, $E_g$ – band gap energy

### 3.7. Numerical results

According to the transformation law we calculated the common admittance on the surface y=0. The gas concentration profile was counted basing on the concentration of the profiles [58], which are different for gases, depending on Knudsen diffusion constants, temperature, molecular weight, the gas constants R and the thickness of the porous sensing materials with an assumed porous radius. A simple model of a structure consisting of porous sensing material is shown in **Figure** 19. Molecular hydrogen dissociated to atomic hydrogen on the outer catalytic surface of Pd. Adsorbed hydrogen atoms then act as electrical dipoles at the metal semiconductor interface and involve changes in the work function in the surface conductivity of the semiconductor material [53,58]. The target gas, diffusing in the sensing layer, is gradually consumed in the surface reaction. Under steady-state conditions the gas concentration inside the sensing layer decreases with the increasing diffusion depth, resulting in a gas concentration profile which depends on the rates of the surface reaction and diffusion reaction. Other gases like $NH_3$, $CO_2$, $NO_2$ can diffuse similarly as hydrogen into porous sensing material depending on its temperature, molecular weight and the physical parameters of the sensing layers.

Some numerical results concerning the gases mentioned above are presented in **Figures** 23, 24, 25. We assume the following values for the involved constant $\sigma_s = v_0 C_S = 1.6x10^{-8} [\Omega^{-1}]$, sensitivity coefficient a=1 [$ppm^{-1}$], thickness of the sensor layer 100 [nm], temperature 300 [K], pore radius 5 [nm]. As a sensor layer we use thin film of $WO_3$.

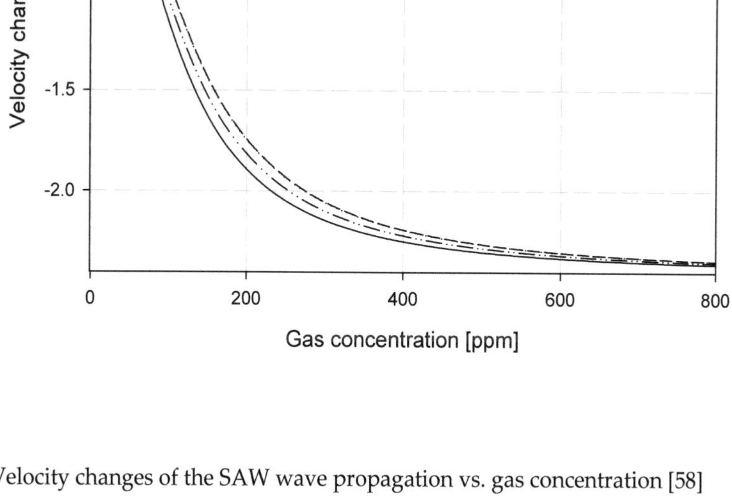


**Figure 23.** Velocity changes of the SAW wave propagation vs. gas concentration [58]

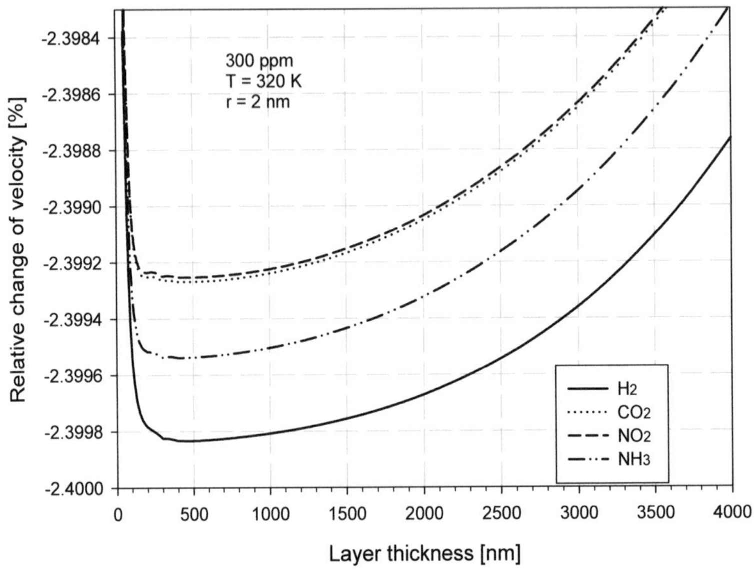


**Figure 24.** SAW wave propagation velocity changes vs. thickness of the sensor layer for 300 ppm $H_2$, $CO_2$, $NO_2$ or $NH_3$ [58]

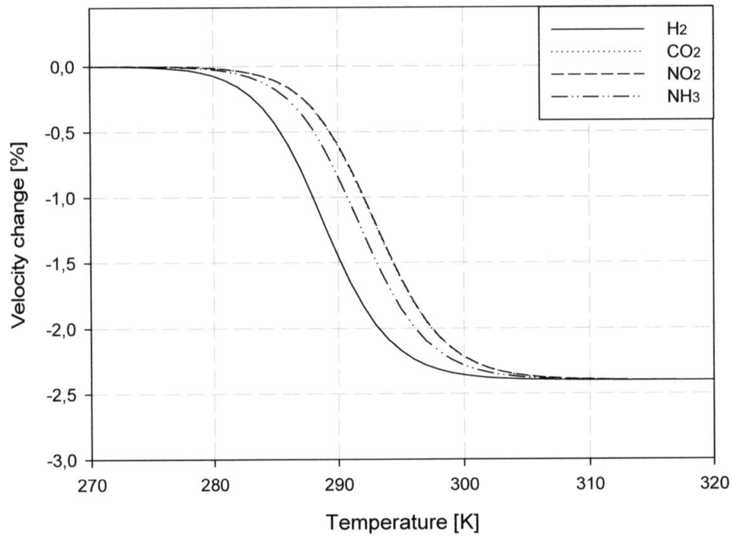


**Figure 25.** SAW wave propagation velocity changes vs. temperature of the sensor layer [58]

## 3.8. Conclusions

a. Experimental studies have shown high sensitivity dual sensor layer system especially: semiconducting polymer layer (CuPc) and palladium catalyst thin layer for hydrogen detection. It is also observed in the palladium phase transition at concentrations above 2% hydrogen in synthetic air.

b. The designed mathematical model of a SAW sensor is based on the electric effect which results from changes in the electric surface conductivity of the layers $\sigma_s(y)$, due to the interaction of this chemically active layer with gas.
c. In the analytical model we introduce sheet conductance which is assumed to be linear to the gas concentration.
d. We divided the porous semiconductor layer into sublayers. According to the transformation law we calculated the influence of the impedance above the piezoelectric substrate on the relative change of velocity versus the concentration, thickness, size of the pores and temperature.
e. We assumed the Knudsen diffusion in porous sensing materials in which pores in the range from 1 to 100 nm in radius do prevail.
f. The analytical model allows the optimization of sensor parameters of a sensor layer and the correct choice of sensor conditions

## Author details

Marian Urbańczyk and Tadeusz Pustelny
*Silesian Univ. of Technology, Faculty of Electrical Engineering, Department of Optoelectronics, Gliwice, Poland*

## Acknowledgement

This work is financed as a grant of the National Center of Research and Development PBR OR 00017912.

## 4. References

[1] Pieka A (2002) Physics of Surface Semiconductor, IUU, Kijev.
[2] Kittel C (2006) Introduction to solid state physics, John Wiley & Sons, Amsterdam.
[3] Ziman J.M (1996) Principles of the theory of solids, Cambridge Press.
[4] Calloway J (1997) Quantum theory of the solid state, Academic Press, Berlin.
[5] Deleven R. (1990) Introduction to applied state physics, Plenum, Roma.
[6] Seitz F (2008) Solid state physics, advances in research and applications, Academic Press, Berlin.
[7] Sliwinski A. (2001) Ultrasounds and their application, WNT, Warsaw.
[8] Gulayev Y (2000) On the nonlinear theory of ultrasound application in semiconductors, Phys. Usp. 48(8): 847-855.
[9] Das P (2006) Method for providing in-situ non-destructive monitoring of semiconductor surface, Molecular and Quantum Acoustics, vol. 26, No 11, 87-89
[10] Taguchi N (1971) Gas detecting device, US Patent, 3,631,436.
[11] Lundstrom I, Sodeberg D (1981/1982) Hydrogen sensitive MOS structures, part 2: characterization, Sensors and Actuators 2: 105-117.

[12] Mizsei J, Lantto V (1992) Air pollution monitoring with a semiconductor gas sensor array system, Sensors and Actuators B6:223-229.

[13] Comini E, Guidi V, Frigeri C, Ricco I, Sberveglieri G (2001) CO sensing properties of titanium and iron nanosized thin films, Sensors and Actuators B, vol.77:6-21.

[14] Bai Hua, Shi Gaoquan (2007) Gas sensors based on conducting polymers, Sensors vol.7:267-307.

[15] Adhikari B, Majumdar S (2004) Polymers in sensor applications, Prog. Polym. Scie. Vol.29:669-766.

[16] Harsanyi G (1995) Polymer films in sensor applications, Technomic, Budapest, Basel.

[17] Janata J, Huber R. (1985) Solid state chemical sensors, Academic Press, INC.

[18] D'Amico A, Verona E (1989), SAW sensors, Sensors and Actuators, 17:55-66.

[19] Maciak E, Opilski Z, Urbańczyk M (2005) Pd/$V_2O_5$ fiber optic hydrogen gas sensor, J.de Phys. IV France 129:137-141.

[20] Ignac-Nowicka J, Pustelny T, Opilski Z, Maciak, Jakubik W, Urbańczyk M (2003) Examination of thin films of phthalocyanines in plasmon system for application In $NO_2$ sensors, Opt. Eng. 42(10): 2978-2986.

[21] Thompson M, Stone D (1997) Surface-Launched Acoustic Wave Sensors. Chemical Sensing and Thin-Film characterization. Chemical analysis vol.144, J Willey and Sons.

[22] Jakubik W, Urbanczyk M (1997) The electrical and mass effect in gas sensors of the SAW type J. Tech. Phys. IV 38: 589-595.

[23] Auld B (1973) Acoustic Fields and Waves, vol. 1, 2 John Willey and Sons, New York.

[24] Jakubik W, Urbańczyk M, Kochowski S, Bodzenta J (2002) Bilayer structure for hydrogen detection in a surface acoustic wave sensor system, Sensors and Actuators B, 82:265-271.

[25] Jakubik W (2006), Investigations of bilayer sensor structure with copper phthalocyanine and palladium for hydrogen detection in SAW system, Journal de Physique IV, vol.137: .95-98.

[26] Sakai Go, Matsunaga Naoki, Shimanoe Engo, Yamazone Noboru (2001), Theory of gas-diffusion controlled sensitivity for thin film semiconductor gas sensor, Sensors and Actuators B 80:125-131.

[27] Gardner J W (1990), A non – linear diffusion – reaction model of electrical conduction in semiconductor gas sensors, Sensors and Actuators, B1:166-170.

[28] Pisarkiewicz T (2007), Mikrosensory gazów, AGH Kraków, Poland.

[29] Ingebrigtsen K A (1970), Linear and nonlinear attenuation of Acoustic Surface Waves in a Piezoelectric Coated with a semiconducting film, J.Appl.Phys. vol.41, No 2:454-459.

[30] Hejczyk T, Urbańczyk M, Jakubik W (2010) Semiconductor Sensor Layer in SAW Gas Sensors Configuration, Acta Physica Polonica A, vol. 118: 1152-1156.

[31] Adler R., Datta S (1988) Acoustoelectric Mobility Measurements on Films with Negligible Acoustic Loss, IEEE SU, 10: 139-144.

[32] Tabib-Azar M (1988) Characterization of Electrical Properties of Semi-Insulating GaAs Using Acousto-Electric Voltage Spectroscopy, Solid-State Electronics, v.31, 7:1197-1204.

[33] Pustelny T, Opilski A (1989) Investigations Concerning Fast Surface States of Semiconductors by Means of Acoustic Methods, Archives of Acoustics, 14, 3:253-260.

[34] Pustelny T, Kubik Z (1993) Acoustic Methods of the Determination of Surface Potential in GaAs, Molecular and Quantum Acoustic,. 14:112-120.

[35] Pustelny T, Kubik Z (1994) Investigation of Surface Potential of GaAs Surface by Means of Acoustics Effects, Archives Acoustics, 19, 2:271-280.

[36] Hoskins M.J. Morkov V. (1982) Charge transport by surface acoustic wave in GaAs, Appl. Phys. Lett., vol. 41:332-343.

[37] Barbe D.F. (1986) Charge-Coupled sefices, Topics Appl. Phys., vol. 38, Springer, Berlin, Heidelberg.

[38] Kino G.S. (1987) Acoustic Waves; Devices, Imaging and Analog Signal Processing, Prentice Hall. Englewood Cliffs, New York.

[39] Parker T.E., Monterss G.K (1988) precision surface acoustic wave (SAW) oscillators. IEEE Trans. on Ultrasonics, Ferroelectroics and Frequency Control, 35:342-364.

[40] Melen R, Buss D, (1999) Charge coupled Devices: Technology and Applications, Wiley & Sons, New York.

[41] Howes M.J, Morgan D.V (1991) Charge coupled devices and systems, Wiley & Sons, New York.

[42] Pustelny T, Pustelna B (2006) The acoustic method of investigating the electrical carrier mobilituy of the real GaP:Te (100) surface, Journal de Physique IV, 137:223-226.

[43] Gulajew J.W (1985) Issledovanije nieliniejnych akustoelektriczeskich effektow i akustoprowodimosti w sloistoj strukturie piezoelektrik- poluprowodnik, Fiz. Twiord. Tiela, 17, 12:3505-3513.

[44] Pustelny T, Pustelny B (2008) The acoustic method of the surface potential investigation in GaP:Te (110) real surface, European Physical Journal: Special Topics,Volume 154, Issue 1: 281-284.

[45] Pustelny T, Adamowicz B (1994) Transverse Acoustoelectric Effect and Surface Photovoltage Method in Surface Study of GaP and InP, J. Tech. Phys., 3:75-86.

[46] Pustelny B, Pustelny T (2009) Transverse acoustoelectric effect applying in surface study of Gap:Te(111), Acta Phys. Pol. A, 116(3):383-384.

[47] Pustelny T, Opilski A, Pustelny B, (2008) Determination of some kinetic parameters of fast surface states in silicon single crystals by means of surface acoustic wave method, Acta Phys. Pol. A, 114(6A):A183-A190.

[48] Pustelny T (1995) Acoustic method of investigating the surface potential in semiconductors. Study of the GaP:Te (110) real surface, Ultrasonics, 33(4):289-294.

[49] Pustelny T (1997) Surface Acoustic Wave Techniques for the Real-Surfaces Investigations of InP (111), InAs (111) Semiconductors, Acustica-acta acoustics, vol. 83 (3):482-488

[50] Matthews H (1997) Surface wave filters Design, construction and use, J Wiley and Sons, New York.

[51] D'Amico A, Palma A, Verona E (1982), Hydrogen sensor using a palladium coated surface acoustic wave delay-line, IEEE Ultrasonics Symposium: 308-311.

[52] Jakubik W, Urbańczyk M, Kochowski S, Nadolski M (2001) Surface acoustic wave sensor for hydrogen detection with a two-layered structure, Molecular and Quantum Acoustics, vol. 22: 123-131.

[53] Jakubik W (2007), Investigation of thin film structure of $WO_3$ and $WO_3$ with Pd for hydrogen detection in a surface acoustic wave sensor system, Thin Solid Films, 515: 8345-8350

[54] Tiersten H, Sinha B.K (1978) A perturbation analysis of the attenuation and dispersion of surface waves, J. Appl. Phys., vol. 49, No. 1.

[55] White R (1970), Surface elastic waves, Proc. IEEE, vol.58, No8:1238-1277.

[56] Williams D E, Hilger A, (1987), Solid state gas sensors, Bristol and Philadelphia.

[57] Hejczyk T, Urbańczyk M, Jakubik W (2010), Analytical Model of Semiconductor Sensor Layers in SAW Gas Sensors, Acta Physica Polonica A, vol 118:1148-1152.

[58] Hejczyk T, Urbańczyk M, Jakubik W (2010), Numerical Results of Modeling Semiconductor Sensor Layers in SAW Gas Sensors, Acta Physica Polonica A, vol 118:1157-1159.

Chapter 6

# Transient Acoustic Wave Propagation in Porous Media

Zine El Abiddine Fellah, Mohamed Fellah and Claude Depollier

Additional information is available at the end of the chapter

http://dx.doi.org/10.5772/55048

## 1. Introduction

A porous material is a two-phase medium consisting of a solid part (skeleton) and a fluid part (pore space). During the propagation of a sound wave in such a medium, interactions between these two phases of different nature take place, giving various physical properties that are unusual in classical media. The large contact area between solid and fluid, which is the main characteristic of porous media induces new phenomena of diffusion and transport in the fluid, in relation to micro-geometry of the pore space. Many applications are concerned with understanding the behavior of acoustic waves in such media. In geophysics, we are interested in the propagation of acoustic waves in porous rocks, for information on soil composition and their fluid content. Oil companies have greatly contributed to the study of acoustic properties of natural porous media. In medicine, the characterization of porous media such as trabecular bone, is useful for diagnosing osteoporosis, bone disease that is manifested by the deterioration of bone microarchitecture. Acoustic characterization of materials is often achieved by measuring the attenuation coefficient and phase velocity in the frequency domain [1] or by solving the direct and inverse problems directly in the time domain [2–4]. The attractive feature of a time domain based approach is that the analysis is naturally limited by the finite duration of ultrasonic pressures and is consequently the most appropriate approach for the transient signal. The objective of this chapter is to show the most recent theoretical and experimental methods developed by the authors for the acoustic characterization of porous materials. The direct and inverse scattering problems are solved in time domain using experimental reflected and transmitted signals. The physical parameters of the porous medium are recovered by solving the inverse problem at the asymptotic domain corresponding to the high frequency range (ultrasound), and at the viscous domain (low frequency range). Figures 1 and 2 give an example of porous materials commonly used in the characterization. The figure 1 shows a sample of air-saturated plastic foam used for sound absorption, and figure 2 shows a sample of cancellous bone used in the diagnosis of the disease of osteoporosis.

**Figure 1.** Air saturated Plastic foam

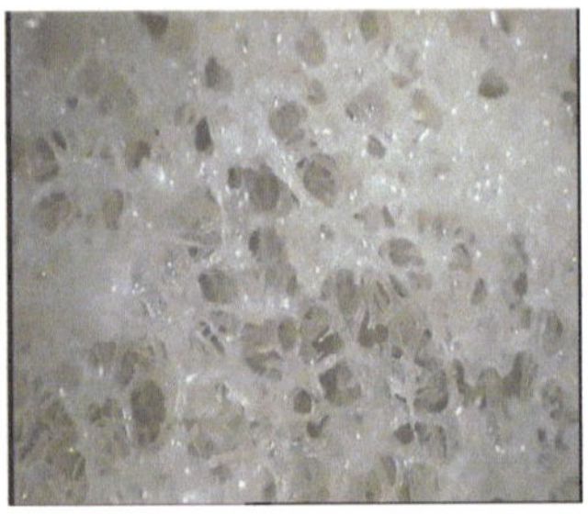

**Figure 2.** Human cancellous bone sample

## 2. Porous materials in rigid structure: Equivalent fluid model

### 2.1. Introduction

The air saturated porous materials are often used as a sound absorber in the fight against noise pollution. The polyurethane foam, felt or glass wools are three examples of common materials used in aerospace and automotive industries and in building construction. Generally, when a sound wave propagates through a porous material saturated with air, the structure remains stationary and non-deformable with respect to the acoustic excitation, this is due to the heaviness and stiffness of the skeleton of the structure relative in air. This is called porous rigid structure and, in this case, we use the equivalent fluid model where viscous and thermal effects are disconnected [5]. The fluid-structure interactions responsible for the sound attenuation (particularly important in porous media), are described by two dynamic susceptibilities, the dynamic tortuosity and compressibility. The dynamic tortuosity [6] decribes the viscous and inertial effects, the dynamic compressibility [7–9] represents the thermal effects. A prediction of the acoustic behavior of porous material requires the determination of both dynamic susceptibilities. The latter can depend only on the physical characteristics of the fluid and the geometry of the fluid domain. Sound propagation in the saturating fluid involves both transport and diffusion characteristics. The phenomenon of transport is the fact that a global movement fluid/solid is induced during the propagation. The phenomenon of diffusion is related to the expansion/compression of the fluid. They result in a change in temperature fluid/solid (the latter remaining at room temperature) driving a heat diffusion, not negligible if the fluid is a gas. Moreover, in the case of sound propagation, the frequency of movement is important. It is clear that the character of the movement is changed to different frequencies, the fluid probe different aspects of micro geometry. In principle one should expect that there are an infinite number of geometric parameters. But on the other hand, a practical viewpoint, the information provided by a finite set of parameters low and high frequencies may be sufficient to describe the dynamic susceptibilities. For a given frequency, one can define a boundary layer thickness, respectively viscous and thermal $\delta = (2\eta/\omega\rho_f)^{1/2}$, $\delta' = (2\kappa/\rho_f C_p\omega)^{1/2}$, where $\eta$ is the viscosity, $\rho_f$ the fluid density, $\kappa$ the coefficient of thermal conduction, $C_p$ the specific heat at constant pressure and $\omega$ the angular frequency. The quantities $\delta$ and $\delta'$ are respectively the penetration depths of isothermal viscous and rotational movement. The boundary layer is small if the penetration depth is small compared to a "radius" characteristic pore. It is then at high frequencies. The opposite situation corresponds to low frequencies. In the high frequencies, inertial effects are described by the tortuosity $\alpha_\infty$, the visco-thermal losses by viscous and thermal lengths $\Lambda$ et $\Lambda'$, respectively. At low frequencies, inertial effects are represented by the viscous tortuosity $\alpha_0$, and visco-thermal exchanges by the visco-thermal permeabilities $k_0$ and $k'_0$. Porosity $\phi$ is a parameter playing an important role both in high in low frequency.

## 2.2. Parameters describing the porous materials

There are several geometrical parameters that can be associated with a porous structure. Some depend only on the shape of the porous solid and do not vary when it expands or changes the scale of the solid. This is the case of the porosity and tortuosity. Others depend on the dimensions, in the case of viscous and thermal permeabilities that vary in shape as an equal surface, and viscous and thermal characteristic lengths that vary as lengths.

- **Porosity** $\phi$: The porosity of a porous material saturated with a fluid is defined by the ratio of the volume of fluid saturating the porous medium to the total volume of the sample. This quantity is dimensionless and can vary from 0 (non-porous solid material) to 1 (free fluid).
- **Tortuosity** $\alpha_\infty$: The tortuosity is a geometric parameter describing the appearance of tortuous pores, and inertial coupling between the fluid and the structure of the material in high-frequency regime. Tortuosity takes its lowest value ($\alpha_\infty = 1$) in the case of porous materials having a straight pores, and large values ($\alpha_\infty$ = 1.5, 2) for the most resistive material.
- **Inertial factor** $\alpha_0$ **(low frequency tortuosity)**: The inertial factor $\alpha_0$ corresponds to the low frequency approximation of the dynamic tortuosity given by Norris [10]; $\alpha_0 = \frac{<\mathbf{v}(\mathbf{r})^2>}{<\mathbf{v}(\mathbf{r})>^2}$, where $< \mathbf{v}(\mathbf{r}) >$ is the average velocity of the viscous fluid for direct current flow within an elementary volume, small compared to the relevant wavelength, but large compared to the individual grains/pores of the solid. Note that the tortuosity $\alpha_\infty$ (high frequency parameter) has a similar definition than $\alpha_0$, the difference is that for $\alpha_\infty$, $\mathbf{v}(\mathbf{r})$ corresponds to the velocity of the perfect incompressible fluid.
- **Thermal tortuosity** $\alpha'_0$: This low frequency parameter is the thermal counterpart of $\alpha_0$. Its definition is given by: $\alpha'_0 = \frac{<\theta_0^2>}{<\theta_0>^2}$, where $\theta_0$ is the scaled response excess temperature field in the low frequency range.
- **The specific flow resistivity** $\sigma$: When a porous material placed in a tube sealed to the walls, is traversed by a fluid, it is a pressure difference between the two free faces. This means that a proportional relationship exists between flow rate $Q_v$ and pressure drop $\Delta P$ given by the $\Delta P = RQ_v$, the coefficient R is the resistance to fluid flow and can therefore be defined as the ratio $\Delta P/Q_v$. For a homogeneous porous material, the resistance is proportional to

the length $d$ of the sample and inversely to his section $S$. Then can be defined, the resistivity as the specific resistance to the passage of a fluid. Denoted $\sigma$, the resistivity is defined by $\sigma = (S/d)R$.

- **Viscous permeability** $k_0$: Static viscous permeability is related to the flow resistivity by the relation $k_0 = \eta/\sigma$, where $\eta$ is the dynamic viscosity of the fluid. Viscous permeability is a geometrical parameter, homogeneous to a surface (it is expressed in $m^2$), independent of the nature of the saturating fluid and depending only on the internal geometry of the porous material. This parameter represents the effective pore section for the flow of fluid, it is connected to the low frequency behavior of the fluid exchange between viscous and saturating the porous structure.
- **Viscous characteristic length** $\Lambda$: The work of Johnson et al [6] in 1987 on the characterization of viscous effects at high frequencies has introduced the viscous characteristic length $\Lambda$ given by $\frac{2}{\Lambda} = \frac{\int_S u^2 dS}{\int_V u^2 dV}$ where $u$ is the speed of a microscopic incompressible perfect fluid, $S$ the area of the interface between solid and fluid phases and $V$ the volume of fluid. The length $\Lambda$ is a geometrical parameter. This definition applies to a smooth interface fluid/solid, and for a low boundary layer thickness to the radius of curvature characteristic of the interface. When the pore surface has singularities (peaks), this definition of the characteristic length is no longer valid. The parameter $\Lambda$ is an indicator of the size of the narrow neck of the pore, i.e. the privileged place of viscous exchanges.
- **Thermal characteristic length** $\Lambda'$: In 1991, Allard and Champoux [8], introduced by analogy with Johnson *et al* [6], a geometric parameter called thermal characteristic length given by $\frac{2}{\Lambda'} = \frac{\int_S dS}{\int_V dV}$. In other words $\Lambda'$ is twice the ratio of fluid volume on the total contact surface between the solid and fluid. The factor 2 is introduced so that in the case of cylindrical pores, $\Lambda'$ is the pore radius. The length $\Lambda'$ is an indicator of the size of large pores, privileged place of heat exchange.
- **Thermal permeability** $k_0'$: The study of viscous and thermal effects shows that there are similarities between these two effects. Starting from the equation of heat diffusion, Lafarge [9] introduced the equivalent of Darcy's law for the temperature excess, by introducing a new parameter: the thermal permeability $k_0'$, which is the inverse of the constant trapping.

## 2.3. Models of dynamic susceptibilities

There are several models describing the inertial viscous and thermal exchanges between the fluid and the structure. We present here the most elaborate.

### 2.3.1. *The model of Johnson et al. for the dynamic tortuosity*

Johnson et al [6] have proposed a simple model describing the function $\alpha(\omega)$, which is the dynamic tortuosity, when the porous structure is saturated with a Newtonian viscous fluid. This model can be applied to the case of our porous structures saturated with air. Two model parameters, the tortuosity $\alpha_\infty$ and the viscous characteristic length $\Lambda$ are involved in a high-frequency limit:

$$\alpha(\omega) = \alpha_\infty \left(1 - \frac{2}{\Lambda}\left(\frac{\eta}{j\omega\rho_f}\right)^{1/2}\right), \qquad \omega \longrightarrow \infty. \tag{1}$$

where $j^2 = -1$, $\eta$ is the dynamic viscosity of the fluid, $\omega$ the angular frequency, $\rho_f$ the density of the fluid. In the low frequency limit, the expression of the dynamic tortuosity in the frequency domain is given by:

$$\alpha(\omega) = -\frac{\eta\phi}{\rho_f k_0 j\omega}, \qquad \omega \longrightarrow 0, \tag{2}$$

where, $k_0$ is the viscous permeability given by: $k_0 = \frac{\eta}{\sigma}$ (with the dimension of a surface), $\sigma$ is the specific resistance passage of the fluid. The meaning of the quantities $\alpha_\infty$, $\Lambda$, $k_0$ have been given previously. Johnson et al [6] imply that these three geometric parameters provide sufficient information about the microgeometry. The proposed general expression for $\alpha(\omega)$ is then that of the simplest analytic function verifying these limits:

$$\alpha(\omega) = \alpha_\infty\left(1 - \frac{1}{jx}\sqrt{1 - \frac{M}{2}jx}\right) \quad \text{where} \quad x = \frac{\omega\alpha_\infty\rho_f}{\sigma\phi} \quad \text{and} \quad M = \frac{8k_0\alpha_\infty}{\phi\Lambda^2}. \tag{3}$$

*2.3.2. The equivalent model for the dynamic compressibility*

A model similar to the previous one has been set for the frequency dependence of the thermal exchanges between the fluid and the structure, because of the heat capacity of the solid part is significant. The result is a two-parameter model, $\Lambda'$ and $k_0'/\phi$, the quantities $\Lambda'$ and $k_0'/\phi$ for thermal effects playing respectively the role of $\Lambda$ and $k_0$ for viscous effects. In a high-frequency limit, Allard and Champoux [7, 8] showed the following behavior of $\beta(\omega)$:

$$\beta(\omega) = 1 - \frac{2(\gamma-1)}{\Lambda'}\left(\frac{\eta}{P_r\rho_f}\right)^{1/2}\left(\frac{1}{j\omega}\right)^{1/2}, \qquad \omega \longrightarrow \infty, \tag{4}$$

where $Pr$ is the number of Prandt and $\gamma$ the adiabatic constant. As before, this behavior implies that the thermal boundary layer thickness $\delta'$ becomes small compared to any characteristic radius of curvature of the interface. One can see that there is no equivalent of the tortuosity $\alpha_\infty$ for thermal effects, at very high frequency $\beta(\omega)$ tends to 1 as $\alpha(\omega)$ tends to the tortuosity $\alpha_\infty$.

In a low frequency limit, Lafarge [9] showed that:

$$\beta(\omega) = \gamma + \frac{(\gamma-1)\rho_f k_0' P_r}{\eta\phi j\omega}, \qquad \text{when} \qquad \omega \longrightarrow 0. \tag{5}$$

where $k_0'$, which has the same size (area) that the Darcy permeability of $k_0$, is a parameter analogous to the parameter $k_0$, but adapted to the thermal problem. The proposed general expression for $\beta(\omega)$ is then that of the simplest analytic function satisfying the limits ((4) and (5):

$$\beta(\omega) = \gamma - (\gamma-1)/\left[1 - \frac{1}{jx'}\sqrt{1 - \frac{M'}{2}jx'}\right] \quad \text{where} \quad x' = \frac{\omega\rho_f k_0' P_r}{\eta\phi} \quad \text{and} \quad M' = \frac{8k_0'}{\phi\Lambda'^2}.$$

the parameter $k_0'$ introduced by Lafarge called thermal permeability by analogy to the viscous permeability.

### 2.3.3. Extensions of models by pride and Lafarge

The development of Johnson et al in high frequency for the tortuosity dynamics stops after the term $\frac{1}{\sqrt{j\omega}}$, Pride *et al* [11] proposed a second correction term which is $\frac{1}{j\omega}$. The asymptotic expansion of the dynamic tortuosity now is written as follows:

$$\alpha(\omega) = \alpha_\infty \left(1 - \frac{2}{\Lambda}\left(\frac{\eta}{j\omega\rho_f}\right)^{1/2} - \frac{\sigma\phi(1-\wp)}{j\omega\rho_f\alpha_\infty}\right), \qquad \omega \longrightarrow \infty. \tag{6}$$

The Pride parameter $\wp$ [11] is a geometrical parameter, dimensionless, connected to a correction term $\alpha_0$ playing the role of inertia of fluid at low frequencies:

$$\wp = \frac{M}{4\left(\frac{\alpha_0}{\alpha_\infty} - 1\right)}, \qquad \text{where} \qquad M = \frac{8k_0\alpha_\infty}{\phi\Lambda^2}.$$

The expansion of low frequency dynamic tortuosity is given by:

$$\alpha(\omega) = -\frac{\eta\phi}{\rho_f k_0 j\omega} + \alpha_0, \qquad \omega \longrightarrow 0 \tag{7}$$

The new expression of the analytic function taking into account developments of (6) and (7) is given by:

$$\alpha(\omega) = \alpha_\infty \left(1 - \frac{1}{jx}\left(1 - \wp + \wp\sqrt{1 - \frac{M}{2\wp^2}jx}\right)\right).$$

The value of $\wp$ depends on the geometry of the pores, for example in the case of cylindrical pores of circular section, it is shown that $\wp = 3/4$. For rectangular sections, one obtains values close to $3/4$.

For thermal effects, Lafarge [9] suggested a similar expression for the dynamic thermal compressibility, also introducing a dimensionless parameter $\wp'$ and gives a correction to the Allard and Champoux model at high frequencies. The expression of the dynamic compressibility is then given by:

$$\beta(\omega) = 1 - (\gamma - 1)\left[\frac{2}{\Lambda'}\left(\frac{\eta}{P_r\rho_f}\right)^{1/2}\left(\frac{1}{j\omega}\right)^{1/2} + \left(\frac{\eta}{\rho_f P_r j\omega}\right)\cdot\left(\frac{(1-\wp')\phi}{k_0'} - \frac{4}{\Lambda'^2}\right)\right], \omega \longrightarrow \infty. \tag{8}$$

The parameter $\wp'$ is connected to $\alpha_0'$, thermal equivalent of $\alpha_0$, which describes the thermal inertia of fluid at low frequency:

$$\wp' = \frac{M'}{4(\alpha_0' - 1)}.$$

The general expression for the dynamic compressibility is then given by:

$$\beta(\omega) = \gamma - (\gamma - 1)/\left[1 - \frac{1}{jx'}\left(1 - \wp' + \wp'\sqrt{1 - \frac{M'}{2\wp'^2}jx'}\right)\right].$$

One can notice that the models of Pride and Lafarge are reduced to those of Johnson and Allard when $\wp = 1$ and $\wp' = 1$. We will see later the influence of these parameters on the propagation.

## 2.4. Temporal modeling of propagation: asymptotic regime

The use of transient signals in the acoustic characterization of porous media is widely used experimentally. These signals have a wide frequency content making some complicated frequency approaches. The frequency methods are very effective for monochromatic signals, however, for transient signals, the temporal approach [2, 12–15] is best suited for several reasons:

- Time domain analysis is naturally bounded (limited) by the finite duration of impulses.
- Time-modeling is often easier to develop because it is closer to the reality of experiments.
- In many situations, the introduction of the "time" parameter make easier the analysis of experimental results.
- For certain applications, it is fast because it avoids back and forth operations between time and frequency domains by FFT.
- It provides an elegant solution of the direct problem necessary for the resolution of the inverse problem.
- It is well suited for comparison between theory and experiment.

In return for these benefits, it is necessary to use a new mathematical formalism. In the frequency domain approach, the relevant quantities for the characterization of porous media are functions of frequency which can be interpreted as susceptibilities. While frequency methods lead to the frequency responses of the medium, in the temporal approach, one is interested in these impulse responses.

## 2.5. Concept of fractional derivative and propagation equation

In the asymptotic regime corresponding to high frequencies, the fluid-structure interactions are described by expressions (1) et (4) according to the Johnson-Allard model. Writing the equations in the time domain is equivalent to taking the inverse Fourier transform of (1) and (4). The temporal equivalent of $j\omega$ is $\partial/\partial t$, while the temporal equivalent of $\sqrt{j\omega}$ is a fraction0al derivative of order $1/2$. The definition of fractional derivative of order $\nu$ is given by [16]:

$$D^{\nu}[x(t)] = \frac{1}{\Gamma(-\nu)} \int_0^t (t-u)^{-\nu-1} x(u) du, \tag{9}$$

where $\Gamma$ is the Eulerian function of the second kind [17]. From the definition (9), the expressions of response factors $\alpha(\omega)$ and $\beta(\omega)$ are then given in the time domain by [2]:

$$\alpha(\omega) \xrightarrow{t} \alpha_{\infty}\delta(t) + \frac{1}{\rho_f}\chi_v(t),$$

$$\beta(\omega) \xrightarrow{t} \delta(t) + K_a \chi_{th}(t),$$

where $\delta(t)$ is the Dirac distribution and the operators, $K_a$ the modulus of the fluid, $\chi_v(t)$ and $\chi_{th}(t)$ are given by:

$$\chi_v(t) = \frac{2\rho_f\alpha_\infty}{\Lambda}\sqrt{\frac{\eta}{\pi\rho_f}}\,t^{-1/2},$$

$$\chi_{th}(t) = \frac{2(\gamma-1)}{K_a\Lambda'}\sqrt{\frac{\eta}{\pi P_r\rho_f}}\,t^{-1/2},$$

In this model, the time convolution of $t^{-1/2}$ with a function, is interpreted as an operator of fractional derivative. The acoustic fields satisfy the Euler equation and the constitutive equation [2] given by:

$$\rho_f\alpha_\infty\,\partial_t\mathbf{v}(\mathbf{r},t) + \int_0^t \chi_v(t-t')\,\partial_t\mathbf{v}(\mathbf{r},t')\,dt' = -\nabla p(\mathbf{r},t), \tag{10}$$

$$\frac{1}{K_a}\,\partial_t p(\mathbf{r},t) + \int_0^t \chi_{th}(t-t')\,\partial_t p(\mathbf{r},t')\,dt' = -\nabla.\mathbf{v}(\mathbf{r},t). \tag{11}$$

These constitutive relations in the time domain satisfy the principle of causality. In these equations, $p$ is the acoustic pressure, $\mathbf{v}$ the particle velocity. The term $\alpha_\infty\delta(t)$ reflects the instantaneous response of the porous medium and describes the inertial coupling between fluid and structure. For instantaneous responses, we mean that the time response is smaller than the characteristic time scale of the change in the acoustic field. The susceptibilities $\chi_v$ and $\chi_{th}$ are operators of memory that determine the dispersion of the medium.

We assume that the medium varies with the thickness $x$ only, and that the incident wave is plane and normal to the surface of the material. Sound pressure is denoted by $p(x,t)$. We assume that the pressure field is zero for moments earlier to 0. The wave equation for acoustic pressure field of a porous medium having a dispersive rigid structure is obtained from the constitutive equations (10, 11), and is of the form:

$$\partial_x^2 p(x,t) - \frac{1}{c_0^2}\left[\alpha_\infty\partial_t^2 p(x,t) + \left(\alpha_\infty K_a\chi_{th} + \frac{\chi_v}{\rho_f} + c_0^2\chi_{th} * \chi_v\right) * \partial_t^2 p(x,t)\right] = 0, \tag{12}$$

where $c_0 = (K_a/\rho_f)^{1/2}$ is the sound speed in the free fluid. The following notation is used for convolution integrals of two causal functions $f(x,t)$ and $g(x,t)$;

$$[f * g](x,t) = \int_0^t f(x,t-t')g(x,t')dt'.$$

The propagation equation (12) can be written as:

$$\frac{\partial^2 p(x,t)}{\partial x^2} - A\frac{\partial^2 p(x,t)}{\partial t^2} - B\int_0^t \frac{\partial^2 p(x,t)/\partial t'^2}{\sqrt{t-t'}}dt' - C\frac{\partial p(x,t)}{\partial t} = 0, \tag{13}$$

where the coefficients $A$, $B$ and $C$ are constant and given by:

$$A = \frac{\rho_f\alpha_\infty}{K_a}, \quad B = \frac{2\alpha_\infty}{K_a}\sqrt{\frac{\rho_f\eta}{\pi}}\left(\frac{1}{\Lambda} + \frac{\gamma-1}{\sqrt{Pr}\Lambda'}\right) \quad C = \frac{4\alpha_\infty(\gamma-1)\eta}{K_a\Lambda\Lambda'\sqrt{Pr}},$$

respectively, The coefficient $A$ gives the wave velocity $c = 1/\sqrt{\rho_f \alpha_\infty / K_a}$ in the air saturating the porous medium. $\alpha_\infty$ plays the role of the refractive index of the porous medium that changes the speed $c_0 = \sqrt{K_a/\rho_f}$ in free space to $c_0 = c/\sqrt{\alpha_\infty}$ in the porous medium. The other coefficients depend essentially on characteristic lengths $\Lambda$ and $\Lambda'$ and express the fluid-structure visco-thermal interactions. The constant $B$ describes the dispersion of the signal and $C$ the attenuation of the amplitude of the wave (without dispersion). The propagation equation (13) describes the evolution of the acoustic wave inside the porous material. The boundary conditions have not been introduced, the porous medium is assumed to be infinite. The coefficients of the equation of propagation do not depend on the porosity, in fact, porosity appears at the interfaces of the porous material via the relationship of continuity of acoustic flow (the flow rate is equal to the porosity multiplied by the speed). This parameter appears naturally at the reflection and transmission of an acoustic wave through a slice of porous material. The solution of the wave equation (13) is given by the Green function $G$ of the porous medium [3, 18], defined by:

$$p(x,t) = \int_0^t G(x, t-t')p(0,t)dt',$$

where $p(x,t)$ is the sound pressure in the porous medium and $p(0,t)$ the incident signal.

## 2.6. Solution of the propagation equation: Green function (Johnson-Allard model)

To solve the propagation equation and obtain the Green function of the medium, we solve the equation (13) using the method of Laplace transform [18], taking into account the following initial conditions of causality:

$$p(x,t)|_{t=0} = 0 \quad \text{and} \quad \frac{\partial p}{\partial t}|_{t=0} = 0. \tag{14}$$

Denote by $P(x,z)$ the Laplace transform of $p(x,t)$ defined by:

$$P(x,z) = \mathcal{L}\left[p(x,t)\right] = \int_0^\infty \exp(-zt)p(x,t)dt,$$

and the inverse Laplace transform by:

$$p(x,t) = \mathcal{L}^{-1}\left[P(x,z)\right].$$

Using the following relationship:

$$\mathcal{L}\left[\delta(t)\right] = 1, \qquad \mathcal{L}\left[H(t)\right] = \frac{1}{z} \quad \text{and} \quad \mathcal{L}\left[\frac{1}{\sqrt{t}}\right] = \sqrt{\frac{\pi}{z}},$$

where $H(t)$ is the Heaviside jump function, the Laplace transform of the wave equation (13) is:

$$\frac{\partial^2 P(x,z)}{\partial x^2} - z^2\left(\frac{1}{c^2}+\frac{C}{z}+B\sqrt{\frac{\pi}{z}}\right)P(x,z) = -\left(\frac{1}{c^2}+B\sqrt{\frac{\pi}{z}}\right)\left(zp(x,0)+\frac{\partial p(x,0)}{\partial t}\right) - Cp(x,0). \tag{15}$$

Taking into account the initial conditions (14), equation (15) simplifies to

$$\frac{\partial^2 P(x,z)}{\partial x^2} - z^2\left(\frac{1}{c^2} + \frac{C}{z} + B\sqrt{\frac{\pi}{z}}\right)P(x,z) = 0,$$

which is a second order differential equation with real coefficients constant. The general solution of equation is:

$$P(x,z) = e^{-\frac{x}{c}\sqrt{f(z)}}\varphi(z) + e^{\frac{x}{c}\sqrt{f(z)}}\psi(z),$$

where $\varphi(z)$ and $\psi(z)$ are two independent functions of $x$ and:

$$f(z) = z^2\left(1 + \frac{Cc^2}{z} + Bc^2\sqrt{\frac{\pi}{z}}\right) = z(z + b'\sqrt{z} + c'). \tag{16}$$

In equation (16), the constants $b'$ and $c'$ are positive and are given by:

$$b' = Bc^2\sqrt{\pi} \quad \text{and} \quad c' = Cc^2. \tag{17}$$

Retaining the finite solution at infinity, which corresponds to the physical solution of our problem, we have:

$$P(x,z) = e^{-\frac{x}{c}\sqrt{f(z)}}\varphi(z).$$

The solution of equation (13) is the inverse Laplace transform of $P(x,z)$. We obtain:

$$p(x,t) = \mathcal{L}^{-1}\left(e^{-\frac{x}{c}\sqrt{f(z)}}\varphi(z)\right) = \mathcal{L}^{-1}\left(e^{-\frac{x}{c}\sqrt{f(z)}}\right) * \mathcal{L}^{-1}\left(\varphi(z)\right).$$

The core problem is the calculation of the inverse Laplace transform of the term $\mathcal{L}^{-1}\left(e^{-\frac{x}{c}\sqrt{f(z)}}\right)$. This has been well studied in reference [18]. By letting:

$$\Delta^2 = b'^2 - 4c', \tag{18}$$

it is easy to check that $\Delta^2$ is always positive in the Johnson-Allard model. The expression of $f(z)$ (Eq. 16) can be written as

$$f(z) = \left(z + \frac{b'}{2}\sqrt{z}\right)^2 - \left(\frac{\Delta\sqrt{z}}{2}\right)^2. \tag{19}$$

When $\Delta = 0$, the solution of the propagation equation(13) is given [18] by:

$$p(x,t) = \begin{cases} 0, & \text{if} \quad 0 \le t \le x/c, \\ \frac{1}{4\sqrt{\pi}}\frac{b'x}{c}\int_{x/c}^{t}\frac{1}{(\tau - x/c)^{3/2}}\exp\left(-\frac{b'^2x^2}{16c^2(\tau - x/c)}\right)p(0, t-\tau)d\tau, & \text{if} \quad t > x/c, \end{cases} \tag{20}$$

where $p(0,t) = \mathcal{L}^{-1}\left(\varphi(z)\right)$. In this case the Green function is given [18, 19] by:

$$G(x,t) = \begin{cases} 0, & \text{if} \quad 0 \le t \le x/c, \\ \frac{1}{4\sqrt{\pi}}\frac{b'x}{c}\frac{1}{(t - x/c)^{3/2}}\exp\left(-\frac{b'^2x^2}{16c^2(t - x/c)}\right), & \text{if} \quad t > x/c, \end{cases} \tag{21}$$

When $\Delta^2 > 0$, the general solution of the propagation equation is given by:

$$p(x,t)=\begin{cases} 0, & \text{if} \quad 0 \le t \le x/c, \\ \frac{x}{c}\int_{x/c}^{t}\left(\frac{b'}{4\sqrt{\pi}}\frac{1}{(\tau-x/c)^{3/2}}\exp\left(-\frac{b'^2x^2}{16c^2(\tau-x/c)}\right)+\Delta\int_0^{\tau-x/c}h'(\xi)d\xi\right)p(0,t-\tau)d\tau, & t > x/c. \end{cases} \tag{22}$$

where

$$h'(\xi) = -\frac{1}{4\pi^{3/2}}\frac{1}{\sqrt{(\tau-\xi)^2 - x^2/c^2}}\frac{1}{\xi^{3/2}}\int_{-1}^{1}\exp\left(-\frac{\left(\mu\Delta\sqrt{(\tau-\xi)^2-x^2/c^2}+b'(\tau-\xi)\right)^2}{16\xi}\right)$$
$$\times\left(\frac{\left(\mu\Delta\sqrt{(\tau-\xi)^2-x^2/c^2}+b'(\tau-\xi)\right)^2}{8\xi}-1\right)\frac{\mu d\mu}{\sqrt{1-\mu^2}}.$$

In this case, the Green function [18] is given by:

$$G(x,t) = \begin{cases} 0, & \text{if} \quad 0 \le t \le x/c, \\ \frac{x}{c}\left(\frac{b'}{4\sqrt{\pi}}\frac{1}{(t-x/c)^{3/2}}\exp\left(-\frac{b'^2x^2}{16c^2(t-x/c)}\right)+\Delta\int_0^{t-x/c}h'(\xi)d\xi\right), & \text{if} \quad t > x/c. \end{cases} \tag{23}$$

## 2.7. Solution of the propagation equation: Green function (Pride-Lafarge model)

Extensions of Pride-Lafarge (6,8) are expressed in the time domain [3] using the definition of fractional derivative:

$$\tilde{\alpha}(t) = \alpha_\infty\left(\delta(t) + \frac{2}{\Lambda}\left(\frac{\eta}{\rho_f}\right)^{1/2}\frac{\partial^{-1/2}}{\partial t^{-1/2}} + \frac{\sigma\phi(1-\wp)}{\rho_f\alpha_\infty}\frac{\partial^{-1}}{\partial t^{-1}}\right),$$

$$\tilde{\beta}(t) = \delta(t) + (\gamma-1)\left(\frac{2}{\Lambda'}\left(\frac{\eta}{P_r\rho_f}\right)^{1/2}\frac{\partial^{-1/2}}{\partial t^{-1/2}} + \left(\frac{\eta}{\rho_f P_r}\right)\left[\frac{(1-\wp')\phi}{k_0'} - \frac{4}{\Lambda'^2}\right]\frac{\partial^{-1}}{\partial t^{-1}}\right),$$

$\tilde{\alpha}(t)$ and $\tilde{\beta}(t)$ are the operators of tortuosity and compressibility. In these equations, the operator $\partial^{-1}/\partial t^{-1}$ represents the time integral:

$$\frac{\partial^{-1}x(t)}{\partial t^{-1}} = \int_0^t x(t')dt',$$

and semi-operator $\partial^{-1/2}/\partial t^{-1/2}$ represents the fractional derivative. In this case, the basic equations are given by:

$$\rho_f\alpha_\infty\,\partial_t\mathbf{v}(\mathbf{r},t) + \int_0^t \chi_v'(t-t')\,\partial_t\mathbf{v}(\mathbf{r},t')\,dt' = -\nabla p(\mathbf{r},t), \tag{24}$$

$$\frac{1}{K_a}\,\partial_t p(\mathbf{r},t) + \int_0^t \chi_{th}'(t-t')\,\partial_t p(\mathbf{r},t')\,dt' = -\nabla.\mathbf{v}(\mathbf{r},t), \tag{25}$$

where the operators $\chi'_v(t)$ and $\chi'_{th}(t)$ are given by:

$$\chi'_v(t) = \frac{2\rho_f\alpha_\infty}{\Lambda}\sqrt{\frac{\eta}{\pi\rho_f}}\,t^{-1/2} + \sigma\phi(1-\wp)\frac{\partial^{-1}}{\partial t^{-1}},$$

$$\chi'_{th}(t) = \frac{2(\gamma-1)}{K_a\Lambda'}\sqrt{\frac{\eta}{\pi P_r\rho_f}}\,t^{-1/2} + \left(\frac{\eta(\gamma-1)}{K_a\rho_f P_r}\right)\left[\frac{(1-\wp')\phi}{k'_0} - \frac{4}{\Lambda'^2}\right]\frac{\partial^{-1}}{\partial t^{-1}},$$

In these expressions, the convolutions express the dispersive nature of the porous material. They take into account the memory effect, where the medium response to excitation of the wave is not instantaneous but takes some time.

The propagation equation can be easily obtained from (24, 25). The propagation equation obtained has exactly the same form as the equation (13), the only difference appears at the coefficient $C = \left(\frac{4\alpha_\infty(\gamma-1)\eta}{K_a\Lambda\Lambda'\sqrt{P_r}} + \frac{\sigma\phi(1-\wp)}{K_a} + \frac{\alpha_\infty(\gamma-1)\eta}{K_a k'_0 P_r}\left[\frac{(1-\wp')\phi}{k'_0} - \frac{4}{\Lambda'^2}\right]\right)$, However, we will see that its solution is different. Since the coefficient C exchange rate, the coefficients $c'$ and $\Delta$ given by equation ((17, 18) also change. With extensions of Pride-Lafarge, the coefficient $\Delta^2$ can be negative [3], it gives:

$$\Delta'^2 = -\Delta^2 \Longrightarrow \Delta'^2 = 4c' - b'^2. \tag{26}$$

In this case, $f(z)$ given by equation (16) can be written in the form:

$$f(z) = \left(z + \frac{b'}{2}\sqrt{z}\right)^2 + \left(\frac{\Delta'\sqrt{z}}{2}\right)^2. \tag{27}$$

Using expressions (26 and 27) and analytical calculations given in [3], we obtain the solution of the wave equation when $\Delta^2$ is negative:

$$p(x,t) = \begin{cases} 0, & \text{if } \ 0 \le t \le x/c, \\ \int_{x/c}^{t}\left[F_1(\tau) + \int_0^{\tau - x/c} H(\xi,\tau)d\xi + j\int_0^{\tau-x/c}\Xi(\xi,\tau)d\xi\right]p(0,t-\tau)d\tau & \text{if } \ t > x/c. \end{cases} \tag{28}$$

with

$$F_1(\tau) = \frac{1}{4\sqrt{\pi}}b'\frac{x}{c}\frac{1}{(\tau - x/c)^{3/2}}\exp\left(-\frac{b'^2x^2}{16c^2(\tau - x/c)}\right), \quad j^2 = -1.$$

The functions $H(\xi,\tau)$ and $\Xi(\xi,\tau)$ are given by the following equations where $\tau$ is replaced by $t$:

$$H(\xi,t) = \frac{\Delta'}{4\pi\sqrt{\pi}}\frac{x}{c}\frac{1}{\sqrt{(t-\xi)^2 - x^2/c^2}}\frac{1}{\xi^{3/2}} \times$$
$$\int_{-1}^{1}\left[\frac{\mathcal{AB}}{4\xi}\cos\left(\frac{\mathcal{AB}}{8\xi}\right) - \left(\frac{\mathcal{B}^2 - \mathcal{A}^2}{8\xi} - 1\right)\sin\left(\frac{\mathcal{AB}}{8\xi}\right)\right]\exp\left(-\frac{\mathcal{B}^2 - \mathcal{A}^2}{16\xi}\right)\frac{\mu d\mu}{\sqrt{1-\mu^2}}$$

and

$$\Xi(\xi,t) = -\frac{\Delta'}{4\pi\sqrt{\pi}}\frac{x}{c}\frac{1}{\sqrt{(t-\xi)^2 - x^2/c^2}}\frac{1}{\xi^{3/2}}$$
$$\int_{-1}^{1}\left[\left(\frac{\mathcal{B}^2 - \mathcal{A}^2}{8\xi} - 1\right)\cos\left(\frac{\mathcal{AB}}{8\xi}\right) + \frac{\mathcal{AB}}{4\xi}\sin\left(\frac{\mathcal{AB}}{8\xi}\right)\right]\exp\left(-\frac{\mathcal{B}^2 - \mathcal{A}^2}{16\xi}\right)\frac{\mu d\mu}{\sqrt{1-\mu^2}},$$

where

$$\mathcal{A} = \Delta' \mu \sqrt{(t-\xi)^2 - x^2/c^2} \quad \text{and} \quad \mathcal{B} = b'(t-\xi).$$

The Green function is given by [3]:

$$G(t,x) = \begin{cases} 0, & \text{if } 0 \le t \le x/c, \\ F_1(t) + \int_0^{t-x/c} H(\xi,t)d\xi + j\int_0^{t-x/c} \Xi(\xi,t)d\xi & \text{if } t > x/c. \end{cases} \tag{29}$$

## 2.8. The reflection and transmission operators

For a slice of porous material occupying the domain $0 \le x \le L$, the incident and scattered fields are connected by scattering operators (i.e. operators of reflection and transmission) of the material. These are operators [4] represented in integral form:

$$p^r(x,t) = \int_0^t \tilde{R}(\tau) p^i\left(t - \tau + \frac{x}{c_0}\right) d\tau, \tag{30}$$

$$p^t(x,t) = \int_0^t \tilde{T}(\tau) p^i\left(t - \tau - \frac{L}{c} - \frac{(x-L)}{c_0}\right) d\tau. \tag{31}$$

In equations (30) and (31) functions $\tilde{R}$ and $\tilde{T}$ represent the kernel of reflection and transmission operators, respectively. These are independently-owned operators of the incident field and depend only on the material properties. To express the operators of reflection and transmission, we assume [4] that the field of acoustic pressure and flow are continuous at the interfaces of the material:

$$\begin{array}{ll} p(0^+,t) = p(0^-,t), & p(L^-,t) = p(L^+,t) \\ v(0^-,t) = \phi v(0^+,t), & v(L^+,t) = \phi v(L^-,t) \end{array} \tag{32}$$

The expressions of $\tilde{R}$ and $\tilde{T}$ are given by (see Ref. [4])

$$\tilde{R}(t) = \left(\frac{-\phi + \sqrt{\alpha_\infty}}{\phi + \sqrt{\alpha_\infty}}\right) \sum_{n\ge 0} \left(\frac{\phi - \sqrt{\alpha_\infty}}{\phi + \sqrt{\alpha_\infty}}\right)^{2n} \left[G\left(t, 2n\frac{L}{c}\right) - G\left(t, (2n+2)\frac{L}{c}\right)\right],$$

$$\tilde{T}(t) = \frac{4\phi\sqrt{\alpha_\infty}}{(\sqrt{\alpha_\infty} + \phi)^2} \sum_{n\ge 0} \left(\frac{\phi - \sqrt{\alpha_\infty}}{\phi + \sqrt{\alpha_\infty}}\right)^{2n} G\left(t + \frac{L}{c_0}, (2n+1)\frac{L}{c}\right).$$

where $G$ is the Green function of the medium. These expressions take into account the multiple $n$ reflections within the porous material. Given the high attenuation of acoustic waves in air-saturated porous media, multiple reflections are negligible. For reflections at the interfaces $x = 0$ and $x = L$, the expressions of the operators of reflection and transmission are simplified (Ref. [4] ) as follows:

$$\tilde{R}(t) = \frac{\sqrt{\alpha_\infty} - \phi}{\sqrt{\alpha_\infty} + \phi}\delta(t) - \frac{4\phi\sqrt{\alpha_\infty}(\sqrt{\alpha_\infty} - \phi)}{(\sqrt{\alpha_\infty} + \phi)^3} G\left(t, \frac{2L}{c}\right), \tag{33}$$

$$\tilde{T}(t) = \frac{4\phi\sqrt{\alpha_\infty}}{(\phi + \sqrt{\alpha_\infty})^2} G\left(t + \frac{L}{c}, \frac{L}{c}\right). \tag{34}$$

The first term on the second member of the equation (33):
$((\sqrt{\alpha_\infty}-\phi)/(\sqrt{\alpha_\infty}+\phi))\,\delta(t)$ is equivalent to the instantaneous reflection response of the porous material. This term corresponds to the wave reflected by the first interface x = 0 . It depends only on the porosity and tortuosity of the material. The reflected wave to the first interface has the advantage not to be dispersive, but simply attenuated. This shows that it is possible to measure the porosity and tortuosity of the porous material by measuring just the first reflected wave.

The second term of equation (33): $-\frac{4\phi\sqrt{\alpha_\infty}(\sqrt{\alpha_\infty}-\phi)}{(\sqrt{\alpha_\infty}+\phi)^3}G\left(t,\frac{2L}{c}\right)$ corresponds to the reflection by the second interface $x = L$. This term depends on the Green function of the medium that describes the propagation and scattering of the acoustic wave having made a round trip in the slab of porous material. Green's function depends on the tortuosity, and viscous and thermal characteristic lengths $\Lambda$ and $\Lambda'$ material, but does not depend on the porosity. Experimentally, this second contribution to the debate can not be measured for low-resistive materials, because the acoustic signal is very attenuated.

Let us study the sensitivity [20] of porosity on the transmission operator $\tilde{T}(t)$. Taking the derivative of $\tilde{T}(t)$ versus the porosity $\phi$, we obtain:

$$\frac{\partial \tilde{T}}{\partial \phi}=\frac{4\sqrt{\alpha_\infty}(\sqrt{\alpha_\infty}-\phi)}{(\sqrt{\alpha_\infty}+\phi)^3}G\left(t+\frac{L}{c},\frac{L}{c}\right),$$

when $\phi \longrightarrow \sqrt{\alpha_\infty}$, the derivative $\frac{\partial \tilde{T}}{\partial \phi} \longrightarrow 0$. This shows that when the porous medium tends to a free fluid, or for weakly resistive porous material with low values of porosity and tortuosity (near 1), the sensitivity of the porosity of the transmitted wave tends to zero. More generally, for a wide range of air saturated porous materials, the term $\frac{\partial \tilde{T}}{\partial \phi}$ is very small. Finally, we conclude that the operator of transmission depends on all parameters, but the low sensitivity of the porosity, makes impossible the determination of this parameter from transmitted data.

## 2.9. Numerical simulations

Let us consider a sample of air saturated porous M1 having the following physical properties: thickness 0.8 cm, tortuosity $\alpha_\infty = 1.5$, viscous characteristic length $\Lambda = 25\ \mu$ m, thermal characteristic length $\Lambda = 75\mu$ m, specific resistance to fluid flow $\sigma$ = 200,000 Nm $^{-4}$ s, porosity $\phi = 0.82$ and thermal permeability $k_0' = 2.77\,10^{-10}$ m $^2$. A simulated transmitted signal was calculated from the equation (31). The input signal is given in figure 3 and its spectrum in Figure 4. Figure 5 shows a comparison between two signals, the first (solid line) corresponds to the real part of the solution (28) of Pride-Lafarge model when $\wp = \wp' = 0.7$ and the second (dotted) corresponds to the solution (22) model of Johnson-Allard ($\wp = \wp' = 1$). We choose the same values for $\wp$ and $\wp'$ to simplify the study. This specific value of 0.7 is only valid for porous materials with circular pores. In the general case, the values of $\wp$ and $\wp'$ may be different from 0.7. Note that it is possible to have positive values of $\Delta$ (Eq. 18) for other values of $\wp$ and $\wp'$. From Figure 5, we notice a significant change in the wave amplitude. By increasing $\wp$ and $\wp'$ from 0.7 to 1, the wave amplitude increases by 65% of its original value. This result can be predicted by the fact that when the values of $\wp$ and $\wp'$ increases, the coefficient $C$ decreases, and hence the wave amplitude decreases due to interactions; inertial, viscous and thermal between fluid and structure. This phenomenon is much more important for resistive porous materials. Figure 6 shows the same comparison as Figure 5 for another

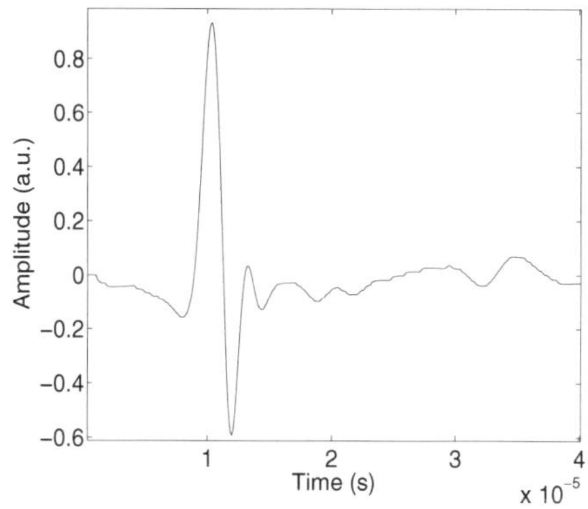


**Figure 3.** Incident signal

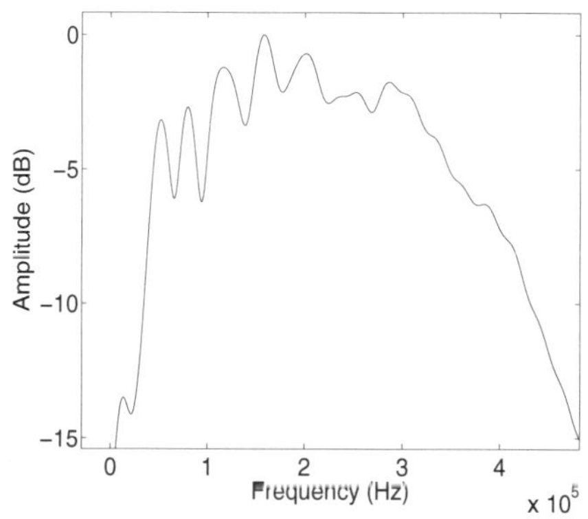


**Figure 4.** Spectrum of the incident signal

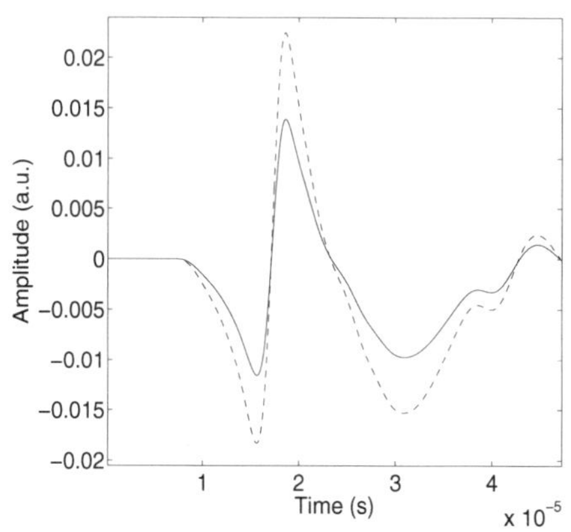


**Figure 5.** Comparison between simulated transmitted signal corresponding to the real part of the solution of Pride-Lafarge (Eq. 28) for $\wp = \wp' = 0.7$ (solid line) and simulated signal corresponding to the solution of Johnson-Allard (Eq. 22) for $\wp = \wp' = 1$ (dotted line) for the sample M1.

sample M2 less resistive, having the following parameters: thickness 0.8 cm, tortuosity $\alpha_\infty = 1.05$, viscous characteristic length $\Lambda = 300\ \mu$ m, thermal characteristic length $\Lambda'$ 900 = $\mu$ m,

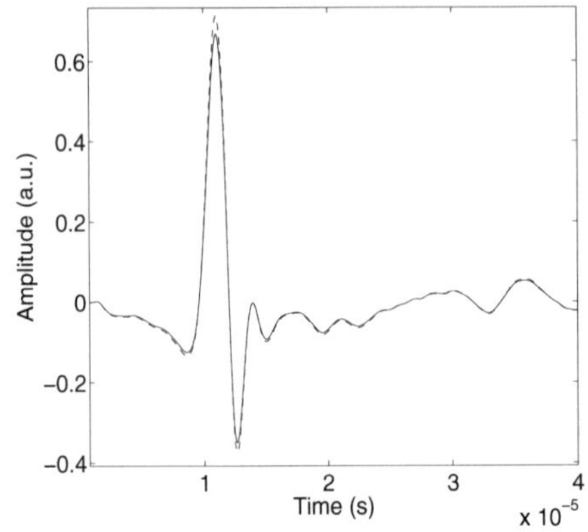


**Figure 6.** Comparison between simulated transmitted signal corresponding to the real part of the solution of Pride-Lafarge (Eq. 28) for $\wp = \wp' = 0.7$ (solid line), and the transmitted signal corresponding to the simulated Allard-Johnson solution (Eq. 22) for $\wp = \wp' = 1$ (dotted line) for the sample M2.

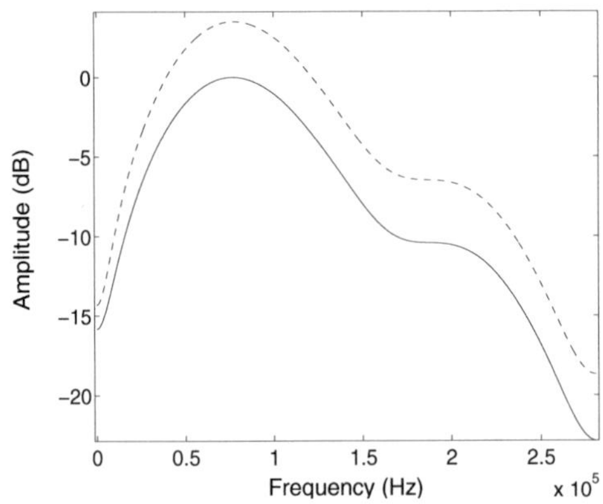


**Figure 7.** Spectrum of the transmitted signal simulated for $\wp = \wp' = 0.7$ Eq. (28) (solid line) and simulated transmitted signal spectrum for $\wp = \wp' = 1$ (Eq. 22) (poitillés).

specific resistance to the passage of fluid $\sigma$ =20000$Nm$−4 s, porosity $\phi = 0.96$ and thermal permeability $k_0' = 2.77\,10^{-9}$ m $^2$. In Figure 6, the influence of the parameters $\wp$ and $\wp'$ on the attenuation is smaller than that in Figure 5. We can conclude that the parameters $\wp$ and $\wp'$ play an important role in the acoustic attenuation, especially for resistive media.

It is possible to see from Figures 5 and 6, when the values of $\wp$ and $\wp'$ change from 0.7 to 1, the waveform changes only at the amplitude of the wave, but it is not dispersed. Figure 7 shows the spectra of two simulated signals given in Figure 5. From the spectra of two simulated signals, we can see they have the same bandwidth, which means that there is no dispersion. This last result shows that $\wp$ and $\wp'$ play an important role in attenuation the acoustic wave but not on its dispersion. Figure 8 shows the imaginary part of the solution (28) for $\wp = \wp' = 0.7$ (sample M1). The amplitude of the imaginary part of the solution is very small compared to the real part (figure 5), this is why only the real part of the solution corresponding to the physical solution is taken into account when a comparison with the experiment is performed. It is possible to write the Green function given by equation (29) as $G = G1 + iG2 =| G |e^{i\theta}$,

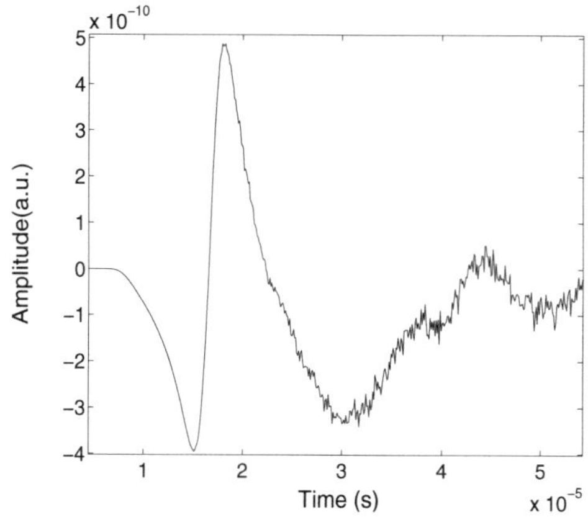


**Figure 8.** Simulated transmitted signal corresponding to the imaginary part of the solution (28) model of Pride-Lafarge.

$\tan\theta = G_2/G_1 \ll 1$. This leads to the conclusion that all components of the signal have the same phase (very small $\theta$, which is therefore a physical factor not essential.

## 2.10. Transmitted wave

The expression (34) shows that the transmitted wave depends on the porosity $\phi$, and parameters describing the high frequency acoustic propagation in the material (tortuosity $\alpha_\infty$ and viscous and thermal characteristic lengths $\Lambda$ and $\Lambda'$). The study of the sensitivity of the porosity (in the previous paragraph) showed that the effect of propagation described by the Green's function altogether mask the effect of interfaces: recall that the Green function $G$ is independent of porosity and the effect of the interface appears through the term $\frac{4\phi\sqrt{\alpha_\infty}}{(\phi+\sqrt{\alpha_\infty})^2}$ in the expression of the transmission operator (34). The low sensitivity of the porosity in transmission makes its determination impossible. Solving the inverse problem, using the transmitted waves allowed the determination of the tortuosity and viscous and thermal characteristic lengths. Note that the ratio of two lengths was set at 3 [20–23]. Currently the only method for the separate determination of the two lengths is based on the saturation of the porous by two different fluids [24]. The two characteristic lengths are acting the same way on the dispersion and attenuation of the transmitted signal. So it is mathematically impossible to trace two unknowns giving the same physical effect on the waveform. We limit ourselves in our work to set an arbitrary ratio between the two lengths. The tortuosity acts on the speed of the wavefront (delay of the transmitted signal with respect to the incident signal) and the attenuation of the amplitude of the waveform. While the characteristic lengths affect the dispersion and attenuation of the signal. A basic inverse problem associated with a slab of porous material may be stated as follows: from the measurements of the transmitted signals outside the slab, find the values of the parameters of the medium. As shown in the previous section, the solution of the direct problem is the system of two operators expressed as functions on $\alpha_\infty$, $\Lambda$ and $\Lambda'$. The inversion algorithm for finding the values of the parameters of the slab is based on a fitting procedure: find the values of the parameters $\alpha_\infty$, $\Lambda$ and $\Lambda'$ such that the transmitted and reflected signal describes the scattering problem in the best possible way (e.g., in the least-squares sense). Experiments were performed in air using a pairs of broadband Ultran transducers NCG200-D13 with a central frequency of 195 kHz and a bandwith of 6 dB extending from 140 kHz to 250 kHz. Pulses of 400 V are provided by a 5058PR Panametrics pulser/receiver. The received signals are filtered above 1 MHz to avoid high frequency noise. Electronic interference is eliminated by 1000 acquisition averages. The experimental setup is shown in Fig. 9. Consider a sample of plastic foam, of thicknesses

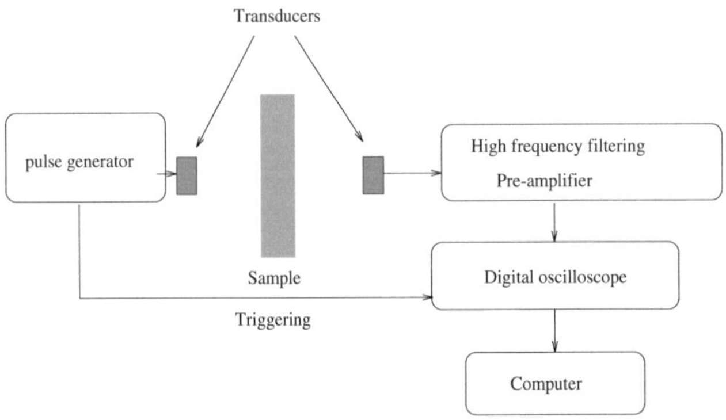


**Figure 9.** Experimental set-up of the ultrasonic measurements in transmitted mode.

$0.7 \pm 0.01cm$. The measured signal generated by the transducer (measured without the sample) is given in Fig. 10 as a dashed line. The measured transmitted signal (obtained with the sample inserted) is given in the same figure (Fig. 11) as a solid line. By solving the inverse problem using the experimental transmitted data, we find the following optimized values: $\alpha_\infty = 1.26$ and $\Lambda = 66\mu m$. In Figs. 11, we compare the experimental transmitted signal and simulated transmitted signal for the optimized values of tortuosity and viscous characteristic length. The difference between experiment and theory is slight, which leads us to conclude that the physical parameters are well-identified.

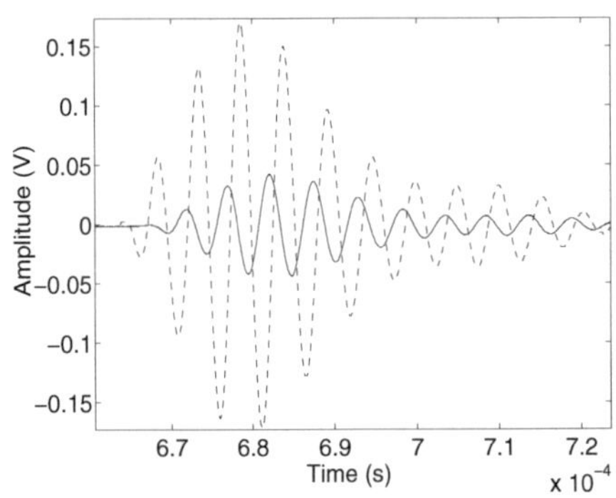


**Figure 10.** Experimental incident signal (dashed line) and transmitted signal (solid line) using the pair of transducers Ultran NCG200-D13 (140 kHz - 250 kHz).

## 2.11. Determination of transport parameters in air-saturated porous materials via reflected ultrasonic waves

A method for measuring transport parameters in porous materials simultaneously, using measurements of waves reflected at the first and second interface, has been proposed

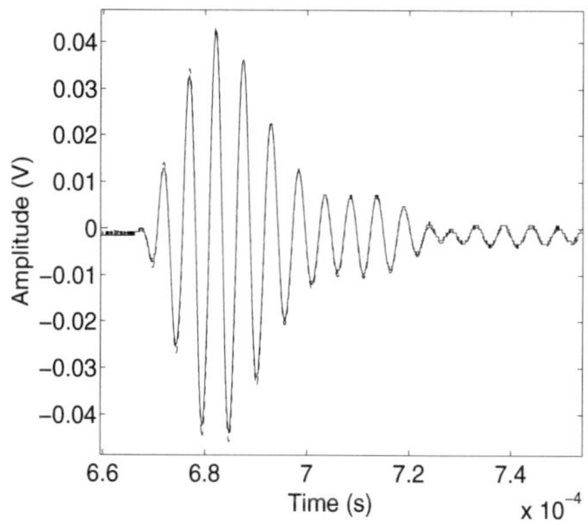


**Figure 11.** Comparison between experimental transmitted signal (solid line) and simulated transmitted signal (dashed line).

[25]. This method is based on a temporal model of direct and inverse scattering problems affecting the propagation of transient ultrasonic waves in a homogeneous isotropic slab of porous material with a rigid frame. Generally, porosity and tortuosity can be evaluated simultaneously by the wave reflected at the first interface at oblique incidence [26–29] but this is not possible at normal incidence [4, 30]. Viscous and thermal characteristic lengths can be estimated only by transmitted waves. Porosity cannot be determined from transmitted waves because of its weak sensitivity in this mode [4]. The advantage of the proposed method is that all the parameters can be determined at normal incidence (the ratio between viscous and thermal lengths is fixed as in classical methods [21–24] based on transmission measurement). The main principle of this method is the experimental detection of reflected contributions from the first and the second interface of the medium. The properties of these two contributions are used to estimate the four acoustical parameters needed for ultrasonic propagation in porous material with a rigid frame, by solving the inverse problem. Studying the sensitivity of each reflected wave parameter demonstrates the importance of each contribution) first and second interface! for the inversion. Numerical and experimental validation for weak resistive air-saturated industrial plastic foams is given to validate this method.

## 2.12. Reflected and transmitted waves

This section concerns the ultrasonic characterization of air-saturated porous materials by solving the inverse problem using experimental data. It is generally easy to solve the inverse problem via transmitted waves, obtaining optimized values of tortuosity, viscous and thermal characteristic lengths, but this is not the case for the porosity because of its weak sensitivity in the transmitted mode. The reflection mode is an alternative to the transmission mode, in that it gives a good estimation of porosity and tortuosity by processing the data relative to measurements of the wave reflected by the first interface. The viscous and thermal characteristic lengths cannot be determined via the first interface reflection. The wave reflected by the second interface can be experimentally detected only for the weakly resistive porous materials. In this case, the characteristic lengths can be estimated. But for common air-saturated porous materials, the second reflection is very damped and its experimental detection is difficult. We solve the inverse problem [31] numerically by the least-squares method, using both reflected and transmitted experimental data. We determine simultaneously all the physical parameters intervening in the propagation. The minimization between experiment and theory is made in the time domain. The inverse problem is well posed, and its solution is unique. As with the classic ultrasonic approach for characterizing porous material saturated with one gas, the characteristic lengths are estimated by assuming a given ratio between them. Tests are performed using industrial plastic foams. Experimental and numerical results, and prospects are discussed [31].

## 2.13. Ultrasonic characterization of air-saturated double-layered porous media

This section concerns a time-domain model of transient wave propagation in double-layered porous materials [32]. An analytical derivation of reflection and transmission scattering operators is given in the time domain. These scattering kernels are the medium's responses to an incident acoustic pulse. The expressions obtained take into account the multiple reflections occurring at the interfaces of the double-layered material. The double-layered porous media consist of two slabs of homogeneous isotropic porous materials with a rigid frame. Each porous slab is described by a temporal equivalent fluid model, in which the acoustic wave propagates only in the fluid saturating the material. In this model, the inertial effects are described by the tortuosity; the viscous and thermal losses of the medium are described by two susceptibility kernels which depend on the viscous and thermal characteristic lengths. the inverse problem is solved using experimental reflected signals at normal incidence. The sensitivity of porosity, tortuosity, and viscous characteristic length of each layer is studied showing their effect on the reflected interface waveforms. The inverse problem is solved numerically by the least-squares method. Five parameters are inverted: porosity and tortuosity of the two layers and the viscous characteristic length of the first layer. The minimization of the discrepancy between experimental and theoretical data is made in the time domain. The inverse problem is shown to be well posed and its solution to be unique. Experimental results for waves reflected by the interfaces of the double-layered porous material are given and compared with theoretical predictions [32].

## 2.14. Acoustic characterization of porous materials at low frequency range

In this frequency range, the viscous forces are important throughout the fluid, the cycle of compression/expansion in the porous medium is slow enough to allow heat exchange between fluid and structure. At the same time, the temperature of the structure is practically unchanged by the passage of the sound wave due to the large value of its specific heat conducting relation to that of the fluid: the structure acts as a thermostat. In this case the isothermal compression is directly applicable. This range corresponds to frequencies where the viscous skin depth $\delta = (2\eta/\omega\rho_f)^{1/2}$ is larger than the pore radius.

An acoustic method [33–37] using transmitted and reflected waves is proposed for measuring static viscous permeability $k_0$, flow resistivity $\sigma$, static thermal permeability $k'_0$ and the inertial factor $\alpha_0$ (low frequency tortuosity), of porous materials having a rigid frame at low frequencies. Flow resistivity of porous material is defined as the ratio between the pressure difference across a sample and the velocity of flow of air through that sample per unit cube. The static thermal permeability of porous material is a geometrical parameter equal to the inverse trapping constant of the solid frame. The inertial factor describes the fluid structure interactions in the low frequency range (1-3 kHz). These parameters are determined from the solution of the inverse problem. The propagation equation is given (along the $x-$axis) by:

$$\frac{\partial^2 p(x,t)}{\partial x^2} - \frac{1}{c^2}\frac{\partial^2 p(x,t)}{\partial t^2} - A\frac{\partial p(x,t)}{\partial t} + B\frac{\partial^3 p(x,t)}{\partial t^3} = 0, \tag{35}$$

$$\frac{1}{c^2} = \frac{\rho_f}{K_a}\left(\alpha_0\gamma - \frac{(\gamma-1)P_r k_0'}{k_0}\right), \quad A = \frac{\eta\phi}{\rho_f k_0},$$

$$B = -\frac{\alpha_0(\gamma-1)k_0' P_r \rho_f^2}{K_a\eta\phi} + \frac{\alpha_0'(\gamma-1){k_0'}^2 {P_r}^2 {\rho_f}^2}{\eta\phi k_0 K_a} - \frac{\gamma\alpha_\infty\rho_f^2\Lambda^2\left(\frac{\alpha_0}{\alpha_\infty}-1\right)^3}{4\eta K_a}$$

The solution of this equation gives the Green function [34] of the porous material:

$$G(x,t) = \frac{1}{\sqrt{\pi}}\frac{x}{c}\sqrt{b'}\left[\frac{\sqrt{\Delta}}{4\,b'}\frac{1}{t^{3/2}}\exp\left(-\frac{x^2\Delta}{16c^2\,b'\,t}\right) - \frac{2}{\pi\sqrt{\pi}}\int_{\frac{x}{c}\sqrt{b'}}^{\infty}\frac{F(\zeta,t)}{\sqrt{\zeta^2-\frac{x^2}{c^2}b'}}d\zeta\right],$$

$$F(\zeta,t) = \int_0^\infty \exp(-s\,t)\sqrt{s}(s+1/2b')\aleph(\zeta,s)ds,$$

$$\aleph(\zeta,s) = \int_{-1}^{1}\cos\left[\frac{\sqrt{\Delta}}{2b'}\zeta\sqrt{s} - y\sqrt{s}(s+1/2b')\sqrt{\zeta^2-\frac{x^2}{c^2}b'}\right]\frac{y\,dy}{\sqrt{1-y^2}},$$

$$b' = Bc^2, c' = Ac^2, \Delta = 1+4b'c'.$$

The incident $p^i(t)$ and transmitted $p^t(t)$ fields are related in time domain by the transmission scattering operator 31. Its expression is given by the relation given in Ref. [34]:

$$\tilde{T}(t) = \int_0^t D(\tau)G(t-\tau+L/c, L/c)d\tau, -* \tag{36}$$

$$D(t) = \frac{2}{\varpi\sqrt{\pi}}\frac{1}{t^{3/2}}\int_0^\infty\left(\frac{u^2}{2t}-1\right)u\exp\left(-\frac{u}{\varpi}-\frac{u^2}{4t}\right)du, \qquad \varpi = \sqrt{\frac{\rho_f k_0\gamma\pi}{\eta}}.$$

In the very low frequency range, the propagation equation 35 is reduced to a diffusive equation

$$\frac{\partial^2 p(x,t)}{\partial x^2} - d\frac{\partial p(x,t)}{\partial t} = 0, \tag{37}$$

where $d = \frac{\sigma\phi\gamma}{K_a}$ the diffusion constant. The solution of this diffusive equation gives the Green function of the porous material [33, 35, 36]:

$$G(x,t) = \frac{x\sqrt{d}}{2\sqrt{\pi}}\frac{1}{t^{3/2}}\exp\left(-\frac{x^2\,d}{4t}\right). \tag{38}$$

The transmission scattering operator is $\tilde{T}(t) = D(t) * G(L,t)$, where * denotes the time convolution operation. The operator $D(t)$ is given by

$$D(t) = -\frac{8}{B^2}\left(1+\frac{t}{B^2}\right)\exp\left(\frac{t}{B^2}\right)\text{Erfc}\left(\frac{\sqrt{t}}{B}\right) + \frac{4}{B\sqrt{\pi t}}\left(2\frac{t}{B^2}+1\right), \qquad B = \frac{1}{\eta}\sqrt{\phi^3\gamma\sigma^3\rho_f},$$

where Erfc is the complementary error function and $\rho_f$ is the fluid density.

The inverse problem is to find the parameters $k'$ and $\alpha_0$ [37] which minimize numerically the discrepancy function $U(k', \alpha_0) = \sum_{i=1}^{i=N} (p^t_{exp}(x, t_i) - p^t(x, t_i))^2$, wherein $p^t_{exp}(x, t_i)_{i=1,2,...n}$ is the discrete set of values of the experimental transmitted signal and $p^t(x, t_i)_{i=1,2,...n}$ the discrete set of values of the simulated transmitted signal predicted from Eq. (31). The inverse problem is solved numerically by the least-square method. For its iterative solution, we used the simplex search method (Nedler Mead) [38] which does not require numerical or analytic gradients. Experiments are performed in a guide [33–37] (pipe), having a diameter of 5 cm. The experimental set up is given in Fig. 12. For measuring the static thermal permeability and the inertial factor, a pipe of 3 m long, since the frequencies used in the experiment are between 1kHz and 4 kHz. However, for measuring the viscous permeability (and flow resistivity) [33, 35, 36], a pipe of 50 m long must be used (50Hz-1kHz) to avoid the reflections at its end. In this case, it is not important to keep the pipe straight, it can be rolled in order to save space without perturbations on experimental signals (the cut-off frequency of the tube $f_c \sim$ 4kHz). A sound source Driver unit "Brand" constituted by loudspeaker Realistic 40-9000 is used. Bursts are provided by synthesized function generator Standford Research Systems model DS345-30MHz. The signals are amplified and filtered using model SR 650-Dual channel filter, Standford Research Systems. The signals (incident and transmitted) are measured using the same microphone (Bruel&Kjaer, 4190) in the same position in the tube. The incident signal is measured without porous sample, however, the transmitted signal is measured with the porous sample. Consider a cylindrical sample of plastic foam M of diameter 5 cm, porosity $\phi = 0.85$ and thickness 4.15 cm. The viscous permeability $k_0$ of the porous sample is measured by solving the inverse problem in time domain at viscous domain (very low frequency range)[33, 35, 36]. The obtained value of $k_0$ is $(0.44 \pm 0.02) \times 10^{-9}\ m^2$. The static thermal permeability is measured using classic Kundt tube (continuous frequency) [39] (continuous frequency) obtaining the value of $k'_0 = (1.54 \pm 0.05) \times 10^{-9}$, with a ratio of 3.5 between $k'_0$ and $k_0$. The tortuosity $\alpha_\infty$ and the viscous characteristic length $\Lambda$ are measured in the high frequency range [4] obtaining the values: $\alpha_\infty = 1.2$ and $\Lambda = 65\mu m$. The value of the thermal tortuosity $\alpha'_0$ is approximated by the relation $\alpha'_0 \simeq \alpha_0/\alpha_\infty$. Fig. 13 shows the experimental incident signal (solid line) generated by the loudspeaker in the frequency bandwidth (2.5 - 3.5) kHz, and the experimental transmitted signal (dashed line). After solving the inverse problem numerically for the thermal permeability $k'_0$ and the inertial factor $\alpha_0$ , we find the following optimized values: $k'_0 = (1.06 \pm 0.2) \times 10^{-9} m^2$ and $\alpha_0 = 2.15 \pm 0.15$. We present in Fig. 14 the variation of the minimization function $U$ with the thermal permeability $k'_0$ and tortuosity $\alpha_0$. The obtained ratio of 2.4 between the inverted $k'_0$ and $k_0$ is in adequacy with the ratio given in literature [9, 39], which is found generally to be between 2 and 4 for plastic foams. In Fig. 15, we show a comparison between an experimental transmitted signal and simulated transmitted signal for the optimized values of thermal permeability and tortuosity. The difference between the two curves is slight, which leads us to conclude that the optimized values of the thermal permeability and inertial factor are correct. This study has been carried out, in the frequency bandwidths (1.5 - 2.5) kHz and has also given good results.

## 3. Modified Biot theory

When the structure of a porous material is not rigid, the equivalent fluid model, on which the previous sections are based, no longer applies because the waves propagate in both the

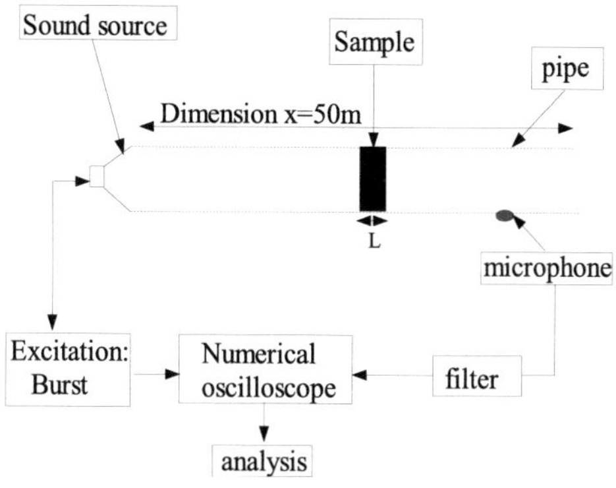


**Figure 12.** Experimental setup of acoustic measurements.

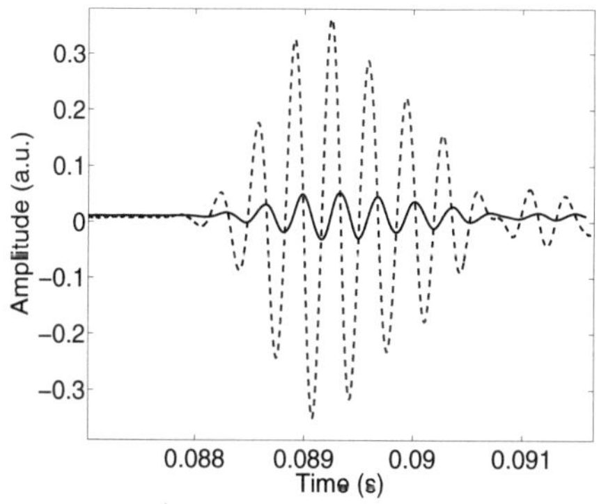


**Figure 13.** Experimental incident signal (solid line) and experimental transmitted signal (dashed line).

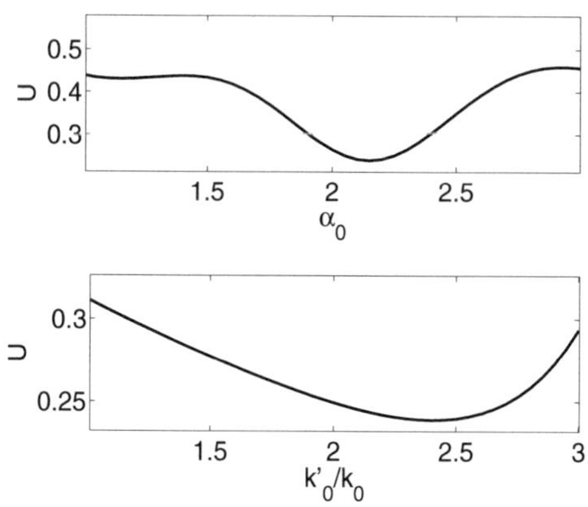


**Figure 14.** Minima of the inverse problem.

skeleton and the saturating fluid. The fluid-structure interactions then play an essential role in the propagation. The study of theses effects has been largely developed by Biot [40, 41] for applications in the domain of oil exploration since 1950. This semi-phenomenological theory provides a rigorous description of the propagation of acoustical waves in porous

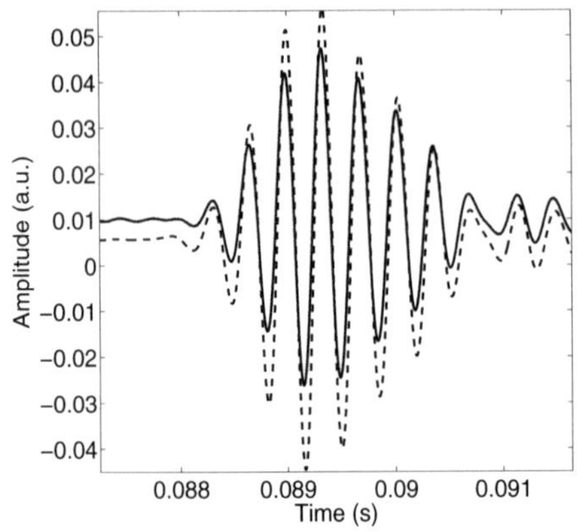


**Figure 15.** Comparison between the experimental transmitted signal (solid line) and the simulated transmitted signals (dashed line) using the reconstructed values of $k'_0$ and $\alpha_0$.

media saturated by a compressible viscous fluid. Such diphasic materials are supposed to be elastic and homogeneous. Biot's theory was initially introduced for petroleum prospecting and research. Due to its very general and rather fundamental character, it has been applied in various fields of acoustics such as geophysics, underwater acoustics, seismology, ultrasonic characterization of bones, etc. Cancelous bone is a porous material consisting of a matrix of solid trabeculae filled with soft bone marrow. The interaction between ultrasound and bone is highly complex. Modeling ultrasonic propagation through trabecular tissue has been considered using porous media theories, such as Biot's theory. This theory derives the equations of motion for each phase (i.e. the solid frame and the fluid) based on energy considerations which include the inertial, potential and viscous coupling between the two phases. For an isotropic porous medium, three different bulk modes are predicted, i.e. two compression waves and one shear wave. One compressional wave, the so-called wave of the first type or fast longitudinal wave, and the transverse wave are similar to the two bulk waves observed in an anisotropic linear elastic solid. The other longitudinal wave, called a wave of the second kind, or slow wave, is a highly damped and very dispersive mode. It is diffusive at low frequencies and propagative at high frequencies. In this section, the direct and inverse scattering problems for the ultrasonic propagation in porous material having an elastic frame are solved [42–46]. An experimental application on human cancelous bone is discussed [42, 43].

### 3.1. Model

The equations of motion of the frame and fluid are given by the Euler equations applied to the Lagrangian density. Here $\vec{u}$ and $\vec{U}$ are the displacements of the solid and fluid phases. The equations of motion are:

$$
\begin{aligned}
\widetilde{\rho}_{11}(t) * \frac{\partial^2 \vec{u}}{\partial t^2} + \widetilde{\rho}_{12}(t) * \frac{\partial^2 \vec{U}}{\partial t^2} &= P\vec{\nabla}.(\vec{\nabla}.\vec{u}) + Q\vec{\nabla}(\vec{\nabla}.\vec{U}) - N\vec{\nabla} \wedge (\vec{\nabla} \wedge \vec{u}), \\
\widetilde{\rho}_{12}(t) * \frac{\partial^2 \vec{u}}{\partial t^2} + \widetilde{\rho}_{22}(t) * \frac{\partial^2 \vec{U}}{\partial t^2} &= Q\vec{\nabla}(\vec{\nabla}.\vec{u}) + R\vec{\nabla}(\vec{\nabla}.\vec{U}).
\end{aligned} \tag{39}
$$

wherein $P$, $Q$ and $R$ are generalized elastic constants which are related, via gedanken experiments, to other, measurable quantities, namely $\phi$ (porosity), $K_f$ (bulk modulus of the pore fluid), $K_s$ (bulk modulus of the elastic solid) and $K_b$ (bulk modulus of the porous skeletal frame). $N$ is the shear modulus of the composite as well as that of the skeletal frame. The equations which explicitly relate $P$, $Q$, and $R$ to $\phi$, $K_f$, $K_s$, $K_b$, and $N$ are given by

$$P = \frac{(1-\phi)(1-\phi-\frac{K_b}{K_s})K_s + \phi\frac{K_s}{K_f}K_b}{1-\phi-\frac{K_b}{K_s}+\phi\frac{K_s}{K_f}} + \frac{4}{3}N, \quad Q = \frac{(1-\phi-\frac{K_b}{K_s})\phi K_s}{1-\phi-\frac{K_b}{K_s}+\phi\frac{K_s}{K_f}}, \quad R = \frac{\phi^2 K_s}{1-\phi-\frac{K_b}{K_s}+\phi\frac{K_s}{K_f}}.$$

The Young modulus and the Poisson ratio of the solid $E_s$, $\nu_s$ and of the skeletal frame $E_b$, $\nu_b$ depend on the generalized elastic constant $P$, $Q$ and $R$ via the relations:

$$K_s = \frac{E_s}{3(1-2\nu_s)}, \qquad K_b = \frac{E_b}{3(1-2\nu_b)}, \qquad N = \frac{E_b}{2(1+\nu_b)}. \tag{40}$$

The temporal operators $\widetilde{\rho}_{11}(t)$, $\widetilde{\rho}_{12}(t)$ et $\widetilde{\rho}_{22}(t)$ are the operators of mass coupling between fluid and structure and are given in the high frequency range by [47]:

$$\widetilde{\rho}_{11}(t) = \rho_{11} + \frac{2\phi\alpha_\infty}{\Lambda}\left(\frac{\rho_f \eta}{\pi t}\right)^{1/2}, \quad \widetilde{\rho}_{12}(t) = \rho_{12} - \frac{2\phi\alpha_\infty}{\Lambda}\left(\frac{\rho_f \eta}{\pi t}\right)^{1/2},$$
$$\widetilde{\rho}_{22}(t) = \rho_{22} + \frac{2\phi\alpha_\infty}{\Lambda}\left(\frac{\rho_f \eta}{\pi t}\right)^{1/2}.$$

Biot coefficients $\rho_{mn}$, are the "mass factors" and are connected to the densities of solid ($\rho_s$) and fluid ($\rho_f$) by $\rho_{11} + \rho_{12} = (1-\phi)\rho_s$ and $\rho_{12} + \rho_{22} = \phi\rho_f$, $\rho_{12} = -\phi\rho_f(\alpha_\infty - 1)$.

### 3.1.1. Longitudinal waves

As in the case of an elastic solid, the wave equations of dilatational and rotational waves can be obtained using scalar and vector displacement potentials, respectively. Two scalar potentials for the frame and the fluid $\Phi_s$ and $\Phi_f$

$$\vec{u} = \vec{\nabla}\Phi_s, \qquad \vec{U} = \vec{\nabla}\Phi_f.$$

In this case the equations 39 become for a propagation along the $x$ axis:

$$\begin{pmatrix} \rho_{11} & \rho_{12} \\ \rho_{12} & \rho_{22} \end{pmatrix} \frac{\partial^2}{\partial t^2} \begin{pmatrix} \Phi_s \\ \Phi_f \end{pmatrix} + \frac{2\phi\rho_f\alpha_\infty}{\Lambda}\sqrt{\frac{\eta}{\rho_f}} \begin{pmatrix} 1 & -1 \\ -1 & 1 \end{pmatrix} \frac{\partial^{3/2}}{\partial t^{3/2}} \begin{pmatrix} \Phi_s \\ \Phi_f \end{pmatrix} = \begin{pmatrix} P & Q \\ Q & R \end{pmatrix} \frac{\partial^2}{\partial x^2} \begin{pmatrix} \Phi_s \\ \Phi_f \end{pmatrix}, \tag{41}$$

The result of solving this system of equations is the existence of two distinct longitudinal modes called slow mode and fast mode. The system (41) can be expressed on the basis of slow and fast waves $\Phi_1$ and $\Phi_2$ respectively by:

$$\frac{\partial^2}{\partial x^2} \begin{pmatrix} \Phi_1 \\ \Phi_2 \end{pmatrix} = \begin{pmatrix} \widetilde{\lambda}_1(t) & 0 \\ 0 & \widetilde{\lambda}_2(t) \end{pmatrix} \begin{pmatrix} \Phi_1 \\ \Phi_2 \end{pmatrix},$$

where $\widetilde{\lambda}_1(t)$ and $\widetilde{\lambda}_2(t)$ are operators of time corresponding to the eigenvalues of the system (41). Their expressions are given by:

$$\widetilde{\lambda}_1(t) = \frac{1}{2}\left(\tau_1 - \sqrt{\tau_1^2 - 4\tau_3}\right)\frac{\partial^2}{\partial t^2} + \frac{1}{2}\left(\tau_2 - \frac{\tau_1\tau_2 - 2\tau_4}{\sqrt{\tau_1^2 - 4\tau_3}}\right)\frac{\partial^{3/2}}{\partial t^{3/2}}$$
$$- \frac{1}{4}\left(\frac{\tau_2^2}{\sqrt{\tau_1^2 - 4\tau_3}} - \frac{(\tau_1\tau_2 - 2\tau_4)^2}{2(\tau_1^2 - 4\tau_3)^{3/2}}\right)\frac{\partial}{\partial t},$$

$$\widetilde{\lambda}_2(t) = \frac{1}{2}\left(\tau_1 + \sqrt{\tau_1^2 - 4\tau_3}\right)\frac{\partial^2}{\partial t^2} + \frac{1}{2}\left(\tau_2 + \frac{\tau_1\tau_2 - 2\tau_4}{\sqrt{\tau_1^2 - 4\tau_3}}\right)\frac{\partial^{3/2}}{\partial t^{3/2}}$$
$$+ \frac{1}{4}\left(\frac{\tau_2^2}{\sqrt{\tau_1^2 - 4\tau_3}} - \frac{(\tau_1\tau_2 - 2\tau_4)^2}{2(\tau_1^2 - 4\tau_3)^{3/2}}\right)\frac{\partial}{\partial t}$$

with:

$$\tau_1 = R'\rho_{11} + P'\rho_{22} - 2Q'\rho_{12}, \quad \tau_2 = A\left(P' + R' + 2Q'\right), \quad \tau_3 = \left(P'R' - Q'^2\right)\left(\rho_{11}\rho_{22} - \rho_{12}^2\right),$$
$$\text{and} \quad \tau_4 = A\left(P'R' - Q'^2\right)\left(\rho_{11} + \rho_{22} - 2\rho_{12}\right).$$

Coefficients $R'$, $P'$ and $Q'$ are given by: $R' = \frac{R}{PR-Q^2}$, $Q' = \frac{Q}{PR-Q^2}$, and $P' = \frac{P}{PR-Q^2}$. The system of equations (41) shows that the slow and fast waves obey the same propagation equation developed in the framework of the equivalent fluid model (equation 13). The eigenvectors $(1, \widetilde{\Im}_1(t))$ and $(1, \widetilde{\Im}_2(t))$ associated with the eigenvalues $\widetilde{\lambda}_1(t)$ et $\widetilde{\lambda}_2(t)$ link the potential solid and fluid $\Phi_s$ and $\Phi_f$, respectively to slow and fast waves $\Phi_1$ and $\Phi_2$ by the following relations:

$$\begin{pmatrix}\Phi_s\\ \Phi_f\end{pmatrix} = \begin{pmatrix}1 & 1\\ \widetilde{\Im}_1(t) & \widetilde{\Im}_2(t)\end{pmatrix}\begin{pmatrix}\Phi_1\\ \Phi_2\end{pmatrix}, \tag{42}$$

the analytical expressions of temporal operators $\widetilde{\Im}_1(t)$ and $\widetilde{\Im}_2(t)$ are given by:

$$\widetilde{\Im}_1(t) = \frac{\tau_1 - 2\tau_5 - \sqrt{\tau_1^2 - 4\tau_3}}{2\tau_7} + \frac{t^{-1/2}}{4\tau_7^2\sqrt{\pi}} \times$$
$$\left[\left(\tau_2 - 2\tau_6 - \frac{\tau_1\tau_2 - 2\tau_4}{\sqrt{\tau_1^2 - 4\tau_3}}\right)2\tau_7 + \left(\tau_1 - 2\tau_5 - \sqrt{\tau_1^2 - 4\tau_3}\right)2\tau_6\right],$$

$$\widetilde{\Im}_2(t) = \frac{\tau_1 - 2\tau_5 + \sqrt{\tau_1^2 - 4\tau_3}}{2\tau_7} + \frac{t^{-1/2}}{4\tau_7^2\sqrt{\pi}} \times$$
$$\left[\left(\tau_2 - 2\tau_6 + \frac{\tau_1\tau_2 - 2\tau_4}{\sqrt{\tau_1^2 - 4\tau_3}}\right)2\tau_7 + \left(\tau_1 - 2\tau_5 + \sqrt{\tau_1^2 - 4\tau_3}\right)2\tau_6\right],$$

where

$$\tau_5 = \left(R'\rho_{11} - Q'\rho_{12}\right) \quad \tau_6 = A\left(R' + Q'\right), \quad \tau_7 = \left(R'\rho_{12} - Q'\rho_{22}\right).$$

For a slab of cancellous bone occupying the region $0 \leq x \leq L$, the incident $p^i(t)$ and transmitted $p^t(t)$ fields are related in the time domain by the transmission scattering operator $\tilde{T}$ given by Eq. 31. In the frequency domain, the expression of the transmission coefficient $\mathcal{T}(\omega)$, which is the Fourier transform of $\tilde{T}$ is given by [42]

$$\mathcal{T}(\omega) = \frac{j\omega 2\rho_f c_0 F_4(\omega)}{\left(j\omega\rho_f c_0 F_4(\omega)\right)^2 - \left(j\omega F_3(\omega) - 1\right)^2}, \tag{43}$$

where

$$F_i(\omega) = \left(1 + \phi\left(\Im_i(\omega) - 1\right)\right)\sqrt{\lambda_i(\omega)}\frac{\Psi_i(\omega)}{\sinh\left(l\sqrt{\lambda_i(\omega)}\right)}\frac{2}{\Psi(\omega)}, \qquad i = 1, 2.$$

$$F_3(\omega) = \rho_f c_0\left(F_1(\omega)\cosh\left(l\sqrt{\lambda_1(\omega)}\right) + F_2(\omega)\cosh\left(l\sqrt{\lambda_2(\omega)}\right)\right), \quad F_4(\omega) = F_1(\omega) + F_2(\omega).$$

The functions $\lambda_1(\omega)$, $\lambda_2(\omega)$, $\Im_1(\omega)$ and $\Im_2(\omega)$ are the Fourier transform of $\widetilde{\lambda}_1(t)$, $\widetilde{\lambda}_2(t)$, $\widetilde{\Im}_1(t)$ and $\widetilde{\Im}_2(t)$, respectively. The coefficients $\Psi_1(\omega)$, $\Psi_2(\omega)$ and $\Psi(\omega)$ are given by

$$\Psi_1(\omega) = \phi Z_2(\omega) - (1-\phi)Z_4(\omega), \qquad \Psi_2(\omega) = (1-\phi)Z_3(\omega) - \phi Z_1(\omega),$$
$$\Psi(\omega) = 2\left(Z_1(\omega)Z_4(\omega) - Z_2(\omega)Z_3(\omega)\right),$$

and the coefficients $Z_1(\omega)$, $Z_2(\omega)$, $Z_3(\omega)$ and $Z_4(\omega)$ by

$$Z_1(\omega) = (P + Q\Im_1(\omega))\lambda_1(\omega), \quad Z_2(\omega) = (P + Q\Im_2(\omega))\lambda_2(\omega),$$
$$Z_3(\omega) = (Q + R\Im_1(\omega))\lambda_1(\omega), \quad Z_4(\omega) = (Q + R\Im_2(\omega))\lambda_2(\omega).$$

## 3.2. Ultrasonic characterization of human cancellous bone using the Biot theory: Inverse problem

Osteoporosis is a degenerative bone disease associated with biochemical and hormonal changes in the ageing body. These changes perturb the equilibrium between bone apposition and bone removal, resulting in a net decrease in bone mass. This leads to a modification of the structure (porosity, trabecular thickness, connectivity etc.) and, to a lesser extent, the composition (mineral density) of the bone. These changes result in a decrease of the mechanical strength of bone and in an increase of the risk of fracture. Osteoporosis mainly affects the trabecular bone (located at the hip, vertebrae or heels for instance). Early clinical detection of this pathological condition is very important to insure proper treatment. The primary method currently used for clinical bone assessment is based on x-ray absorptiometry, and measures total bone mass at a particular anatomic site. Because other factors, such as architecture, also appear to have a role in determining an individual's risk of fracture, ultrasound is an alternative to X-rays that has generated much attention. In addition to their potential for conveying the architectural aspects of bone, ultrasonic techniques also may have advantages in view of their use as a nonionising radiation and inherently lower costs, compared with x-ray densitometric methods. Although ultrasonic methods appear promising for noninvasive bone assessment, they have not yet fulfilled their potential. Unfortunately, a poor understanding of the ultrasound interaction with bone has become one of the obstacles preventing it from being a fully developed diagnostic technique. Despite extensive research on the empirical relationship between ultrasound and the bulk properties of bone, the mechanism of how ultrasound physically interacts with bone is still unclear. Since trabecular bone is an inhomogeneous porous medium, the interaction between ultrasound and bone is a highly complex phenomenon. Modelling ultrasonic propagation through trabecular tissue has been considered using porous media theories, such as Biot's theory. As seen in the previous section, within the framework of the modified Biot theory, the propagation of ultrasonic waves in a slab of cancellous bone is conditioned by many parameters: porosity $\phi$, tortuosity $\alpha_\infty$, viscous characteristic length $\Lambda$, fluid viscosity $\eta$, Young's modulus of the elastic solid $E_s$, Young's modulus of porous skeletal frame $E_b$, Poisson's ratio of the elastic solid $\nu_s$, Poisson's ratio of the porous skeletal frame $\nu_b$, the solid density $\rho_s$, the bulk modulus of the saturating fluid $K_f$ and the fluid density $\rho_f$. It is therefore important to develop new experimental methods and efficient tools for their estimation. The basic inverse problem associated with the slab of cancellous bone may be stated as follows: from measurements of the signal transmitted outside the slab, find the values of the medium's parameters. Solving the inverse problem for all the Biot parameters using only the transmitted experimental data is difficult, if not impossible. To achieve this task, requires more experimental data for obtaining a unique solution. For this reason, in this contribution we limit the inversion to the five parameters: $E_b$, $\nu_b$, $\phi$, $\alpha_\infty$ and $\Lambda$. In our previous paper [42], we studied the sensitivity of transmitted waveforms to variations of $\phi$, $\alpha_\infty$ and $\Lambda$. The sensitivity of $E_b$ and $\nu_b$ has been examined in Ref. [43]

As an application of this model, some numerical simulations are compared with experimental results. Experiments are performed in water using two broadband Panametrics A 303S plane piezoelectric transducers with a central frequency of 1 MHz in water, and diameter of 1cm. 400 V Pulses are provided by a 5058PR Panametrics pulser/receiver. Electronic interference is removed by averaging 1000 acquisitions. The experimental setup is shown in Fig. 9. The parallel-faced samples were machined from femoral heads and femoral necks of human cancellous bone. The liquid in the pore space (blood and marrow) is removed from the bone sample and substituted by water. The size of the ultrasound beam is very small compared to the size of the specimens. The emitting transducer insonifies the sample at normal incidence with a short (in time domain) pulse. When the pulse hits the front surface of the sample, a part is reflected, a part is transmitted as a fast wave, and a part is transmitted as a slow wave. When any of these components, travelling at different speeds, hit the second surface, a similar effect takes place: a part is transmitted into the fluid, and a part is reflected as a fast or slow wave. The experimental transmitted waveforms are travelling through the cancellous bone in the same direction as the trabecular alignment (x direction). The fluid characteristics are: bulk modulus $K_f = 2.28$ GPa, density $\rho_f = 1000\text{Kgm}^{-3}$, viscosity $\eta = 10^{-3} Kg.m.s^{-1}$. Consider a sample of human cancellous bone M1 (femoral neck) of thickness 11.2 mm and solid density $\rho_s = 1990\text{Kgm}^{-3}$. The Young's modulus $E_s = 13GPa$ and Poisson ratio $\nu_s = 0.3$ of the solid bone are taken from the literature [48]. Figure 9 shows the experimental incident signal. The inverse problem is solved by minimizing the function $U(\phi, \alpha_\infty, \Lambda, E_b, \nu_s)$ given by:

$$U(\phi, \alpha_\infty, \Lambda, E_b, \nu_b) = \sum_{i=1}^{i=n} (p^t_{exp}(x, t_i) - p^t(x, t_i))^2,$$

wherein $p^t_{exp}(x, t_i)_{i=1,2,...n}$ is the discrete set of values of the experimental transmitted signal and $p^t(x, t_i)_{i=1,2,...n}$ the discrete set of values of the simulated transmitted signal. A large variation range is applied of each estimating parameter value in solving the inverse problem. The variation range of the parameters is ; $\alpha_\infty \in [1, 2]$, $\Lambda \in [1,\ 200]\mu m$, $\phi \in [0.5, 0.99]$, $\nu \in [0.1, 0.5]$ and $E_b \in [0.5, 5]$GPa. The variations of the cost function with the physical parameters present one clear minimum corresponding to the mathematical solution of the inverse problem. This shows that the inverse problem is well posed mathematically, and that the solution is unique. The minima, corresponding to the solution of the inverse problem, are clearly observed for each parameter. After solving the inverse problem, we find the following optimized values: $\phi = 0.64$, $\alpha_\infty = 1.018$, $\Lambda = 10,44\mu m$, $\nu_b = 0.28$ and $E_b = 4.49$ GPa. Using these values, we present in Figs. 16-19 the variations in the discrepancy function U with respect to two values of the inverted parameters. To show clearly the solution of the

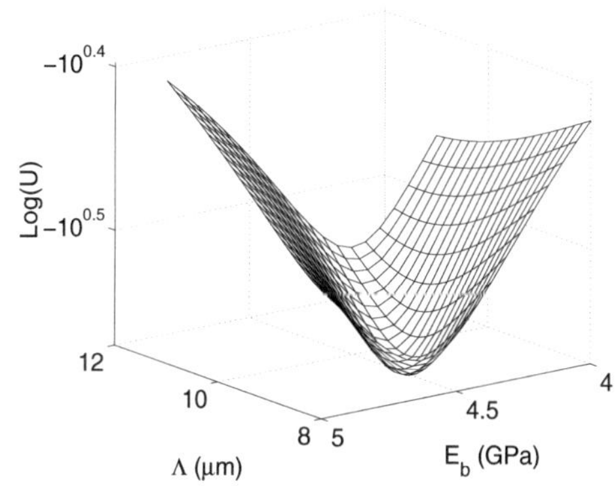


**Figure 16.** Variation of the minimization function $U$ with the viscous characteristic length $\Lambda$ and the Young modulus of the skeletal frame $E_b$.

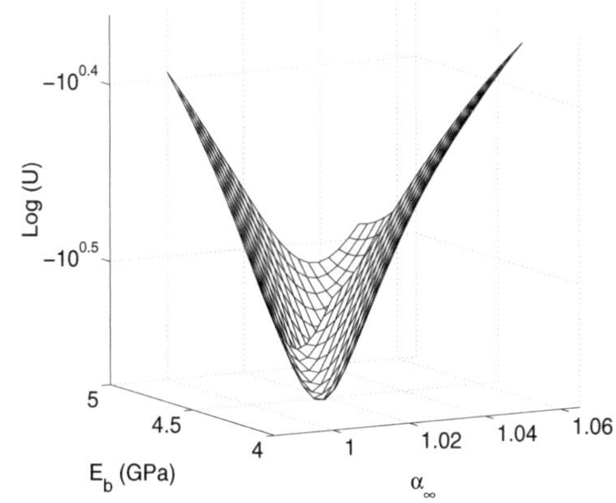


**Figure 17.** Variation of the minimization function $U$ with the the Young modulus of the skeletal frame $E_b$ and the tortuosity $\alpha_\infty$.

inverse problem, the variation of U in Figs. 16-19 is given only around the minima values of the inverted parameters. In Fig. 20, a comparison is made between the experimental transmitted signal and the simulated transmitted signals using the reconstructed values of

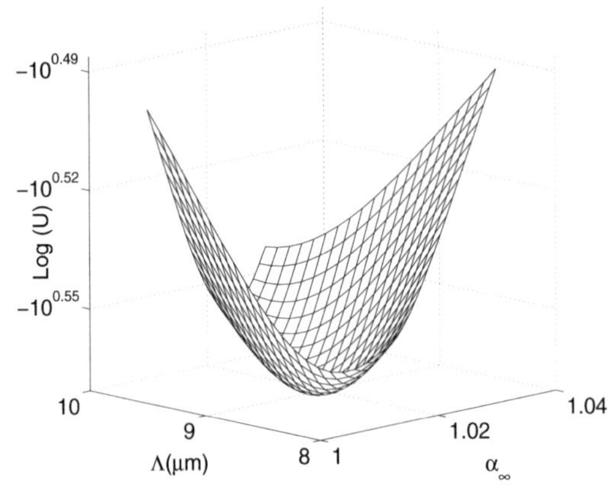


**Figure 18.** Variation of the minimization function $U$ with the viscous characteristic length $\Lambda$ and the tortuosity $\alpha_\infty$.

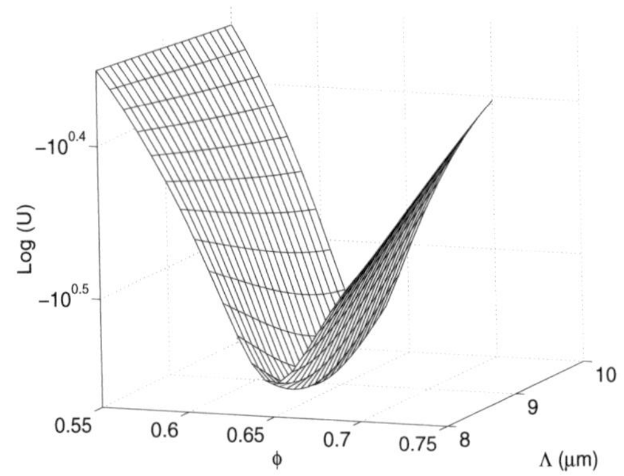


**Figure 19.** Variation of the minimization function $U$ with the porosity $\phi$ and the viscous characteristic length $\Lambda$.

$\alpha_\infty$, $\phi$, $\Lambda$, $\nu_b$ and $E_b$. The difference between the two curves is small, which leads us to conclude that the optimized values of the physical parameters are correct. The fast and slow waves predicted by the Biot theory are easily detected in the transmitted signal. The slow wave seems to be less attenuated than the fast wave. In the other applications, the slow wave is generally more-attenuated and dispersive than the fast wave. We usually observe the opposite phenomena for cancellous bone samples; this can be explained by the different orders of magnitude of the physical parameters (high porosity, low tortuosity...etc). Let us now solve the inverse problem for sample M2 (femoral neck) of thickness 12 mm. Using another sample of cancellous bone (femoral head) M2 of thickness 10.2 mm. The results after solving the inverse problem are: $\phi = 0.72$, $\alpha_\infty = 1.1$, $\Lambda = 14.97\mu m$, $\nu_b = 0.22$ and $E_b = 3.1GPa$. In Fig. 21, we compare the experimental transmitted signal to the transmitted simulated signal using reconstruct values of the physical parameters. The correlation between the two curves is excellent. The simulated transmitted signals obtained using optimized values (Figs. 20 and

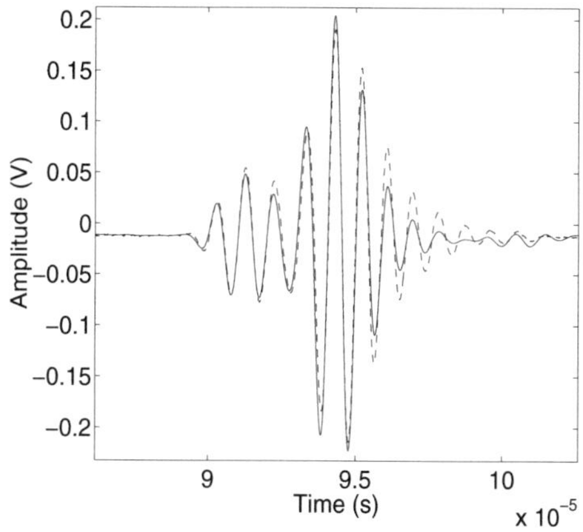


**Figure 20.** Comparison between the experimental transmitted signal (solid line) and the simulated transmitted signals (dashed line) using the reconstructed values of $\alpha_\infty$, $\phi$, $\Lambda$, $\nu_b$ and $E_b$ (sample M1).

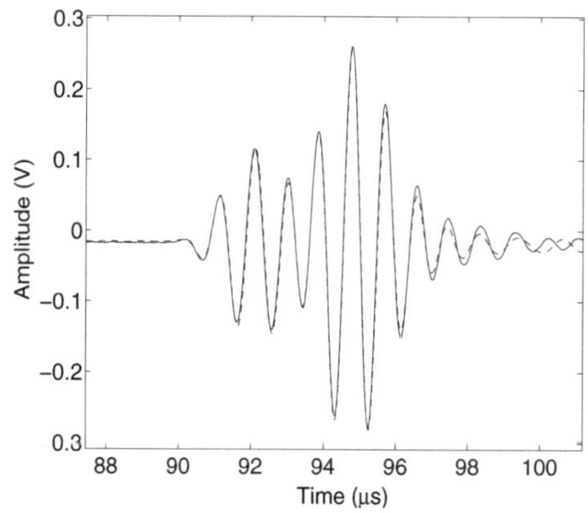


**Figure 21.** Comparison between the experimental transmitted signal (solid line) and the simulated transmitted signals (dashed line) using the reconstructed values of $\alpha_\infty$, $\phi$, $\Lambda$, $\nu_b$ and $E_b$ (sample M2).

21) reproduce correctly the experimental transmitted signals. This leads us to conclude that this method is well adapted for the characterization of cancellous bone.

## Author details

Zine El Abiddine Fellah
*LMA, CNRS, UPR 7051, Aix-Marseille Univ, Centrale Marseille, F-13402 Marseille Cedex 20, France*

Mohamed Fellah
*Laboratoire de Physique Théorique, Faculté de Physique, USTHB, BP 32 El Alia, Bab-Ezzouar, 16211, Alger- Algeria*

Claude Depollier
*Universite du Maine UMR CNRS 6613 Laboratoire d'Acoustique de l'Universite du Maine UFR STS Avenue O. Messiaen 72085 Le Mans CEDEX 09 France*

*LUNAM Universite du Maine. UMR CNRS 6613 Laboratoire d'Acoustique de l'Universite du MaineUFR STS Avenue O. Messiaen 72085 Le Mans CEDEX 09 France*

## 4. References

[1] Allard JF (1993), Propagation of Sound in Porous Media : Modeling Sound Absorbing Materials, Chapman and Hall, London.

[2] Fellah ZEA, Depollier C (2000) Transient acoustic wave propagation in rigid porous media: A time domain approach, J. Acoust. Soc. Am. 107: (2), 683-688.

[3] Fellah ZEA, Depollier C, Fellah M, Lauriks W et Chpaleon JY (2005), Influence of dynamic tortuosity and compressibility on the propagation of transient waves in porous media, Wave Motion 41 (2): 145-161.

[4] Fellah ZEA, Fellah M, Lauriks W, Depollier C (2003), Direct inverse scattering of transient acoustic waves by a slab of rigid porous material., J. Acoust. Soc. Am. 113: (1) 61-72.

[5] Zwikker C, Kosten CW (1949) Sound Absorbing Materials, Elsevier , New York .

[6] Johnson D.L, Koplik J, Dashen R (1987) Theory of dynamic permeability and tortuosity in fluid-saturated porous media. J. Fluid. Mech. 176: 379-402.

[7] Allard JF, Champoux Y (1992), New empirical equations for sound propagation in rigid frame fibrous materials . Acoust. Soc. Am. 91: p. 3346.

[8] Champoux Y, Allard JF (1191), Dynamic tortuosity and bulk modulus in air-saturated porous media J. Appl. Phys. 70: p. 1975.

[9] Lafarge D, Lemarnier P, Allard JF, Tarnow V (1996), Dynamic compressibility of air in porous structures at audible frequencies Journ. Acoust. Soc. Am, 102: 4.

[10] Norris AN (1986), On the viscodynamic operator in Biot's equations of poroelasticity, J. Wave Mat. Interact. 1: 365-380.

[11] Pride SR, Morgan FD, Gangi AF (1993), Drag forces of porous media acoustics. Phys. Rev. B, 47: p. 4964.

[12] Szabo TL (1994), Time domain wave equations for lossy media obeying a frequency power law, J. Acoust. Soc. Am. 96: p. 491.

[13] Szabo TL (1995), Causal theories and data for acoustic attenuation obeying a frequency power law, Journal of the Acoustical Society of America 97: p. 14.

[14] Norton V, Novarini JC (2003), Including dispersion and attenuation directly in time domain for wave propagation in isotropic media, J. Acoust. Soc. Am. 113: p. 3024.

[15] Chen W, Holm S (2003), Modified Szabo's wave equation models for lossy media obeying frequency power law, J. Acoust. Soc. Am. 113: p. 3024.

[16] Samko SG, Kilbas AA, Marichev OI (1993), Fractional Integrals and Derivatives Theory and Applications, Gordon and Breach Publishers, Amsterdam.

[17] Abramowitz M, Stegun TA (1968), Handbook of mathematical functions, National bureau of standards, U.S. Governement Printing Office, Washington, DC. Chap. 6, p. 358.

[18] Fellah ZEA, Fellah M, Lauriks W, Depollier C, Angel Y, Chapelon JY (2003), Solution in time domain of ultrasonic propagation equation in porous material, Wave motion, 38: p. 151-163.

[19] Fellah ZEA, Depollier C, Fellah M (2001), Direct and inverse scattering problem in porous material having a rigid frame by fractional calculus based method, J. Sound. Vib. 244: (2), p. 359-366.
[20] Fellah ZEA, Ogam E, Wirgin A, Fellah M, Depollier C et Lauriks W, Ultrasonic characterization of porous materials: Inverse problem, J. Sound and Vib. 203:746-759.
[21] Leclaire P, Kelders L, Lauriks W, Glorieux C, Thoen J (1996), Determination of the viscous characteristic length in air-filled porous materials by ultrasonic attenuation measurements, J. Acoust. Soc. Am. 99: p. 1944.
[22] Ayrault C, Moussatov A, Castagnède B, Lafarge D, (1999) Ultrasonic characterization of plastic foams via measurements with static pressure variations, Applied Physics Letters 74: 3224.
[23] Moussatov A, Ayrault C, Castagnède B (2001), Porous material characterization ultrasonic method for estimation of tortuosity and characteristic length using a barometric chamber, Ultrasonic 39: 195.
[24] Leclaire P, Kelders L, Lauriks W, Brown NR, Melon M, Castagnède B (1996), Determination of viscous and thermal characteristic lengths of plastic foams by ultrasonic measurements in helium and air, J. Appl. Phys. 80:, p. 2009.
[25] Fellah ZEA, Depollier C, Berger S, Lauriks W, Trompette P, Chapelon JY (2003), Detemination of transport parameters in air saturated porous materials via ultrasonic reflected waves, J. Acoust. Soc. Am. 113: (5), 2561-2569.
[26] Fellah ZEA, Berger S, Lauriks W, Depollier C, Chapelon JY (2003), Inverse problem in air-saturated porous media via reflected waves, Rev. Sci. Inst. 74: (5), p. 2871.
[27] Fellah ZEA, Berger S, Lauriks W, Depollier C, Aristégui C, Chapelon JY (2003), Measuring the porosity and the tortuosity of porous materials via reflected waves at oblique incidence, J. Acoust. Soc. Am. 113: (5), p. 2424.
[28] Fellah ZEA, Berger S, Lauriks W, Depollier C, Trompette P, Chapelon JY (2003), Ultrasonic measuring of the porosity and tortuosity of air- saturated random packings of beads, J. Appl. Phys. 93: p. 9352.
[29] Fellah ZEA, Mitri FG, Depollier D, Berger S, Lauriks W Chapelon JY (2003), Characterization of porous materials having a rigid frame via reflected waves, J. Appl. Phys. 94: p. 7914.
[30] Fellah ZEA, Berger S, Lauriks W, Depollier C, Fellah M (2003), Measurement of the porosity of porous materials having a rigid frame via reflected waves : A time domain analysis with fractional derivatives, J. App. Phys, 93: p. 296.
[31] Fellah ZEA, Mitri FG, Fellah M, Ogam E, Depollier C (2007), Ultrasonic characterization of porous absorbing materials: Inverse problem, J. Sound. Vib. 302: 746-759.
[32] Fellah ZEA, Wirgin A, Fellah M, Sebaa N, Depollier C, Lauriks W (2005), A time-domain model of transient acoustic wave propagation in double-layered porous media, J. Acoust. Soc. Am. 118: 661-670.
[33] Fellah ZEA, Fellah M, Sebaa N, Lauriks W, Depollier C (2006) , Measuring flow resistivity of porous materials at low frequencies range via acoustic transmitted waves, J. Acoust. Soc. Am. 119: 1926-1928.
[34] Fellah ZEA, Fellah M, Mitri FG, Sebaa N, Lauriks W, Depollier C (2007), Transient acoustic wave propagation in air-saturated porous media at low frequencies, J. Appl. Phys. 102: 084906.
[35] Sebaa N, Fellah ZEA, Fellah M, Lauriks W, Depollier C (2005), Measuring flow resistivity of porous material via acoustic reflected waves, J. Appl. Phys. 98: 084901.

[36] Fellah ZEA, Fellah M, Mitri FG, Sebaa N, Depollier C, Lauriks W, Measuring permeability of porous materials at low frequency range via acoustic transmitted waves, Rev. Sci. Instrum. 78: 114902-10.

[37] Sadouki M, Fellah M, Fellah ZEA, Ogam E, Sebaa N, Mitri FG, Depollier C, Measuring static thermal permeability and inertial factor of rigid porous materials, J. Acoust. Soc. Am. 130: 2627-2630.

[38] Lagarias JC, Reeds JA, Wright MH, Wright PE (1998), Convergence properties of the Nedler-Mead Simplex Method in low dimensions SIAM J. Optim. 9: 112-147.

[39] Henry M, Allard JF (1997), Acoustical measurement of the trapping constant of foams with open cells, C. R. Acad. Sci. Paris 325: 331-338.

[40] Biot MA (1956), The theory of propagation of elastic waves in fluid-saturated porous solid. I. Low frequency range, J. Acoust. Soc. Am. 28: 168.

[41] Biot MA (1956), The theory of propagation of elastic waves in fluid-saturated porous solid. I. Higher frequency range, J. Acoust. Soc. Am. 28: 179.

[42] Fellah ZEA, Chapelon JY, Berger S, Lauriks W and Depollier C (2004), Ultrasonic wave propagation in human cancellous bone: Application of Biot theory, J. Acoust. Soc. Am. 116: (1), 61-73.

[43] Sebaa N, Fellah ZEA, Fellah M, Ogam E, Wirgin A, Mitri FG, Depollier C, Lauriks W (2006), Ultrasonic characterization of human cancellous bone using the Biot theory: Inverse problem, J. Acoust. Soc. Amer. 120: 1816-1824.

[44] Ogam E, Fellah ZEA, Sebaa N, Groby JP (2011), Non-ambiguous recovery of Biot poroelastic parameters of cellular panels using ultrasonic waves. J. Sound and Vibr. 330(6):1074-1090.

[45] Ogam E, Fellah ZEA (2011), The direct and inverse problems of an airsaturated poroelastic cylinder submitted to acoustic radiation. AipAdvances, 1(3):032174.

[46] Ogam E, Depollier C, Fellah ZEA (2010), The direct and inverse problems of an air-saturated porous cylinder submitted to acoustic radiation. Rev. Sci. Inst. 81(094902):9.

[47] Sebaa N, Fellah ZEA, Lauriks W, Depollier C (2006), Application of fractional calculus to ultrasonic wave propagation in human cancellous bone, Signal Processing 10: 2668-2677, DOI : 10.1016/j.sigpro.2006.02.015.

[48] Ashman RB, Rho JY (1988), Elastic modulus of trabecular bone material, J. Biomech. 21: 177-181.

Chapter 7

# Utilizing Malaysian Natural Fibers as Sound Absorber

Mohammad Hosseini Fouladi, Mohamed H. Nassir, Masomeh Ghassem, Marwan Shamel, Sim Yeng Peng, Sin Yi Wen, Pang Zong Xin and Mohd Jailani Mohd Nor

Additional information is available at the end of the chapter

http://dx.doi.org/10.5772/53197

## 1. Introduction

Previous researchers [1, 2, 3, 4] successfully produced sound absorption panels using agricultural wastes. These panels play important role both in noise absorption and heat insulation preserving the comfort of indoor living spaces. Yang et al. [4] produced rice straw-wood particle composite boards. They found that the sound absorption coefficient of these boards is higher than other wood-based materials in the 500-8000 Hz frequency range. Reason was the low specific gravity of composite boards having high amount of porosity compared to wood-based materials [4]. Another study by Davern [5] aimed at producing airspace layers and examined the influence of porosity on the acoustic properties of materials. He found that the porosity of the perforated plate and the density of the porous material would significantly affect the acoustic impedance and sound absorption coefficient of the panel, in which case, the frequency band near the resonance frequency achieved high acoustic absorption. In addition to Davern's study, Lee and Chen [6] reported that the acoustic absorption of multi-layer materials is better with a perforated plate backed with airspaces. Other usages of natural fibres are in reduction of sound propagation in automotive interior spaces, or to improve the control of outdoor noise propagation [7, 8]. Recent studies show that researchers are focusing on coir fibre and oil palm fibre in replacing synthetic-based fibres for sound absorption applications simply due to their abundance in tropical countries such as Malaysia.

Acoustics, as science, deals with sound creation and transmission through materials. Sound and light waves share same vibrational system as these waves creating pressure during propagation due to the nature of fluctuation in the material. However, unlike electromagnetic waves, sound cannot travel through a vacuum. Consequently, sound wave has become important in many applications in materials science, medicine, dentistry, oceanology, marine navigation, communications, petroleum and mineral prospecting, industrial processes, music and voice synthesis and animal bioacoustics. Sound in the audio frequency range of approximately 20 Hz–20 kHz can be heard by humans. Excessive levels of sound can cause permanent hearing loss while continuous exposure could be physiologically and psychologically deleterious to one's well-being. This study, amongst many others, is very important in ergonomics since it is closely related to the sense of hearing; the crucial in our daily routine whether we are in schools, factories, offices and theaters. These places are in high demand for materials that are able to reduce noise level at various frequency ranges. When sound propagates in an enclosed room, the frequencies and decibel levels are subjected to reflections which lead to echoes. An acoustic absorption panel can be used to prevent echoes and reduce the intensity of the sound that is heard outside the room. Acoustic absorption panels are normally filled up with porous layers of materials capable of controlling reverberation and background noise [9]. The energy of incident sound is transformed into heat and vibration of fibers, and eventually dissipated [10].

Acoustic absorption panel is placed on ceilings and walls to improve the comprehensibility of speech in the room. Commercial acoustic panels are made from synthetic fibers that may be hazardous to health and environment. Current trend is to replace them with natural fibers that are cheap, environmental friendly and free of health risks. Nowadays, the concept of green technology has been incorporated in many fields in the industry. Companies are constantly researching for alternatives to further improve their choice of materials that are more environmental friendly. These waste materials should have reasonable absorption performance compared to synthetic fibers. Studies are going on the sound absorption of coir (Cocos nucifera) and palm oil (Elaeis Guinnesis) fibers at various thicknesses and frequency bands [11]. On the other hand, there are still other local agricultural resources that their acoustical properties are yet to be determined like sugar cane (Saccharum), corn (Zea mays) and grass (Axonopus compressus) fibers. Hence, the focal point of this project is to study the sound absorption coefficient of four different fibers; coir, corn, sugar cane and dry grass with different panel thicknesses.

## 2. Methodology

Coir, corn and sugarcane fibers were obtained from the wet market whereas grass was readily attained from a field and then dried out. Corn and Coir fibers were purchased in loose forms (see Figure 1 for coir and corn fiber). Samples were compressed separately into disks of diameters of 100 and 28 mm by fitting into molds of cylindrical shapes (Figures 2(a) coir fiber, (b) corn fiber) according to impedance tube diameter of 100 and 28 mm for low and high frequency measurements, respectively. In order to better observe the fiber structures, a magnified picture was taken for each type, see Figure 3(a) for coir fiber, (b) for corn fiber. Figure 4 shows the molds that were used to prepare samples in disk forms. Molds were manufactured using steel tubes while the punchers were made of wood. Samples were compressed using hydraulic compressor machine with an average pressure of 45 kg/cm$^2$. The absorption coefficient of fibers was evaluated using impedance tube apparatus as shown in Figure 5. For comparison of results, two different thicknesses of each sample were tested.

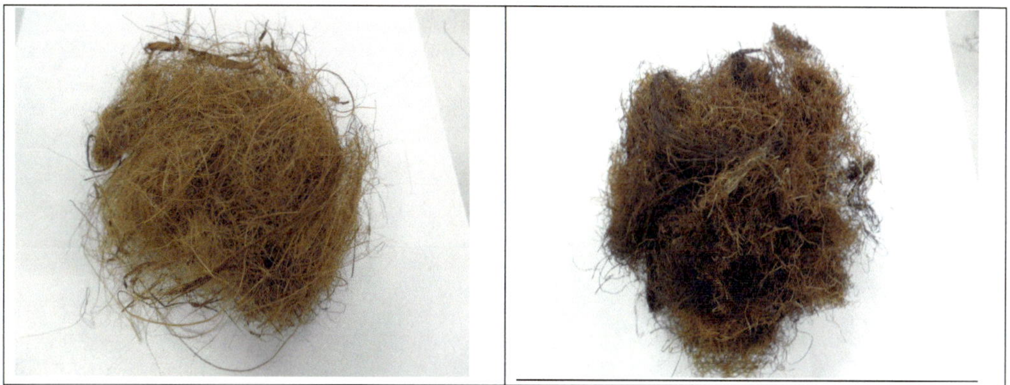

(a) Loose Coir Fiber | (b) Loose Corn Fiber

**Figure 1.** Loose forms (a) Coir Fiber (b) Corn Fiber

(a) Coir Fiber in Disk Form | (b) Corn Fiber in Disk Form

**Figure 2.** Fibers in disk forms (a) Coir fiber (b)Corn fiber

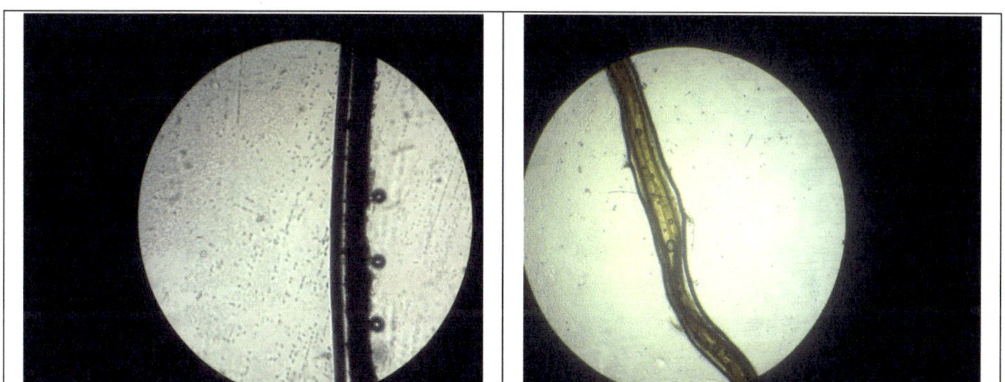

(a) Coir fiber under microscope | (b) Corn fiber under microscope

**Figure 3.** Fibers under microscope (a) Coir fiber (b) Corn fiber

**Figure 4.** Steel molds and wooden punchers were used to fabricate the samples.

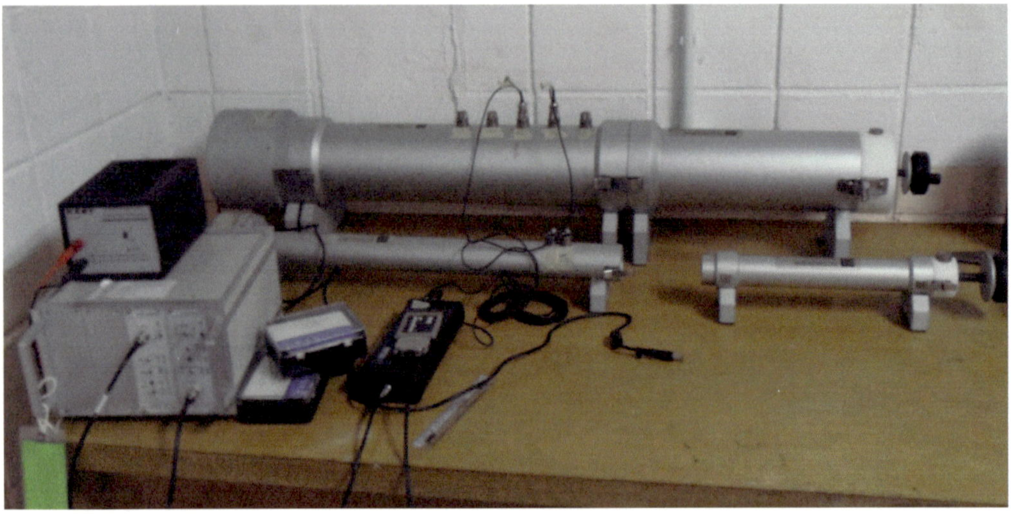

**Figure 5.** Measurement apparatuses consisting of impedance tubes, noise generator and 01dB analyzer.

Porosity is the ratio of the volume of openings to the total volume of the sample and affects the absorption efficiency of each sample. Mass, thickness and fiber diameter of each sample determines the porosity and can be calculated using Eq. (1):

$$\varphi = 1 - \frac{\rho_{bulk}}{\rho_{fiber}} \tag{1}$$

where $\rho_{bulk}$ and $\rho_{fiber}$ are the bulk density of sample and fiber density, respectively. The quantity $\rho_{bulk}$ is the ratio of mass and volume of each sample as a disk, whereas $\rho_{fiber}$ depends on physical properties of the fiber itself.

The impedance tube (Figure 5) creates sound by a generator mounted at one end, in the form of broadband stationary random waves. The generated sound travels to hit the sample which is attached at the other end. The sound pressure is measured at two fixed points using two ¼" microphones. An analyzer calculates the complex transfer function in order to obtain the sound absorption coefficient of material. The sensitivity of microphones was calibrated utilizing calibrator type GRAS-42AB at 114 dB levels and 1 kHz. The frequency span of experiment was 100-4500 Hz with 3 Hz resolution and it took approximately 10 seconds for the instrument to achieve the absorption spectrum. Based on Delany-Bazley model, the acoustic absorption coefficient, $\alpha$ can be defined as [12]:

$$\alpha = \frac{4R_r/\ \rho_0 C_0}{(R_r/\rho_0 C_0\ + 1)^2 + (X_r/\rho_0 C_0)^2} \tag{2}$$

where, $C_0$ is speed of sound,$\rho_0$ density of air, $R_r$ and $X_r$ real and imaginary components of surface acoustic impedance.

## 3. Results & discussion

First step was to obtain the physical characteristics of fibers. Masses of samples were measured by a precise electronic balance. Length and diameter were measured by caliper and used to calculate each fiber's volume assuming that they have perfect cylindrical shape. The last column in Table 1 is the ratio of mass and volume as the fiber density ($\rho_{fiber}$). A number of 15 samples from each batch were selected and results averaged. The quantity $\rho_{bulk}$ was calculated based on mass and volume of a disk-shape sample and together with $\rho_{fiber}$ were put in Eq. (1) to calculate the porosity of the sample. They are presented in Table 2 and 3 for 1 and 2 cm thicknesses, respectively.

Figure 6 and 7 show that acoustic absorption of fiber was improved as thickness of sample increased. The reason is observable from Table 2 and 3; in which, the porosities of fibers are decreased as the corresponding thicknesses increased. Basically more amounts of fibers were used to fabricate the 2 cm thickness samples and they have higher bulk density compared to 1 cm samples. According to Eq. (1), the increase in bulk density reduces the porosity of sample; i.e. less perforation exists which improve the absorption coefficient of sample and moves the resonance peaks to lower frequencies. Therefore resonance peaks of sugar cane and corn fiber samples were shifted from 3800 and 3200 Hz in Figure 6 to 1000 and 2800 Hz in Figure 7, respectively. Coir and grass do not show significant absorption behavior in Figure 6 but it is very much improved for higher thickness in Figure 7 as both having resonance peaks around 2000 Hz.

Fig. 7 is chosen to compare the absorption coefficient of fibres as the patterns are more significant in this plot. Coir and grass absorbed more than 70% of incident sound at frequencies higher than 1300 Hz whereas corn presented similar behaviour above 1800 Hz. For sugar cane absorption is generally lower than 70% except for a small region around 1000 Hz that the structural resonance happens because the diameter of sugar cane fibre is around 400 μm which is larger than the rest. A sample made of fibre with larger diameter does not show significant elastic behaviour; also have lower flow resistivity because there are more hallow spaces when putting fibres together. Reduction in flow resistivity decreases the absorption coefficient of sample.

| Sample No. | Mass (mg) | Length (mm) | Diameter (μm) | Volume ($mm^3$) | Density ($kg/m^3$) |
|---|---|---|---|---|---|
| 1 | 0.8 | 151 | 450 | 5.44 | 147.17 |
| 2 | 3 | 180 | 350 | 5.04 | 595.24 |
| 3 | 3.4 | 190 | 380 | 5.78 | 588.64 |
| 4 | 3.2 | 185 | 390 | 5.77 | 554.40 |
| 5 | 2.8 | 159 | 370 | 4.71 | 594.93 |
| 6 | 3.2 | 162 | 350 | 4.54 | 705.47 |
| 7 | 3.3 | 191 | 370 | 5.59 | 590.08 |
| 8 | 3.5 | 195 | 400 | 6.24 | 560.90 |
| 9 | 3.6 | 202 | 370 | 5.98 | 602.09 |
| 10 | 2.9 | 183 | 350 | 5.09 | 569.22 |
| 11 | 3.1 | 191 | 350 | 5.35 | 579.66 |
| 12 | 3.8 | 217 | 350 | 6.08 | 625.41 |
| 13 | 2.5 | 181 | 330 | 4.78 | 523.19 |
| 14 | 1.3 | 175 | 290 | 4.03 | 322.42 |
| 15 | 1.8 | 170 | 260 | 3.59 | 501.34 |
| Average Diameter (μm): | | | 357 | | |
| Average Density ($kg/m^3$): | | | 537 | | |

**Table 1.** Physical characteristics of corn fiber

| Type of Fiber | Porosity (%) Sample Diameter | |
|---|---|---|
| | 100 mm | 28 mm |
| Coir | 89.38 | 93.41 |
| Corn | 97.83 | 97.50 |
| Sugar Cane | 96.32 | 95.20 |
| Grass | 96.93 | 96.86 |

**Table 2.** Porosities of samples with thickness of 1 cm.

| Type of Fiber | Porosity (%) Sample Diameter | |
|---|---|---|
| | 100 mm | 28 mm |
| Coir | 88.67 | 89.68 |
| Corn | 96.12 | 96.24 |
| Sugar Cane | 95.91 | 94.71 |
| Grass | 94.22 | 96.07 |

**Table 3.** Porosities of samples with thickness of 2 cm.

Comparing the synthetic materials in Table 4 with natural fibres in Table 5 proves the possibility of implementing these natural materials in acoustic panels. Apart from fibre glass board; carpet, plywood and drapery generally have lower absorption coefficient than the selected natural fibres throughout the frequency spectrum which the difference is even more significant above 1000 Hz.

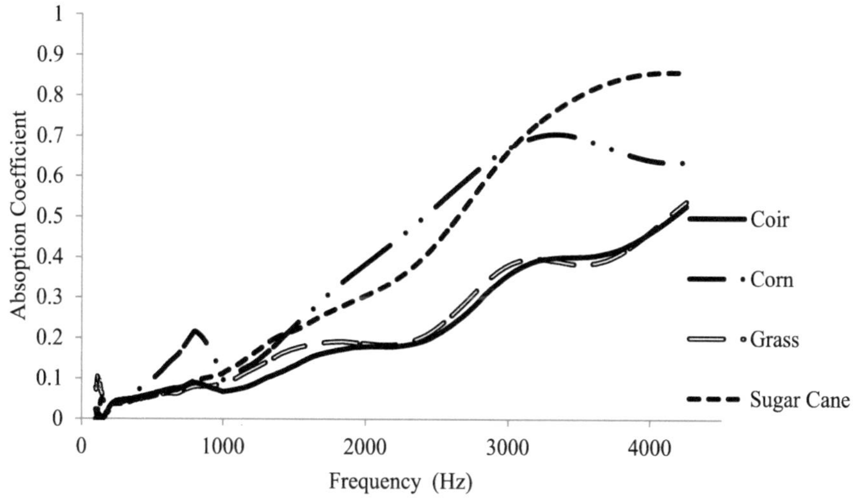


**Figure 6.** Acoustic absorption coefficient of 1 cm sample for the four types of fibers.

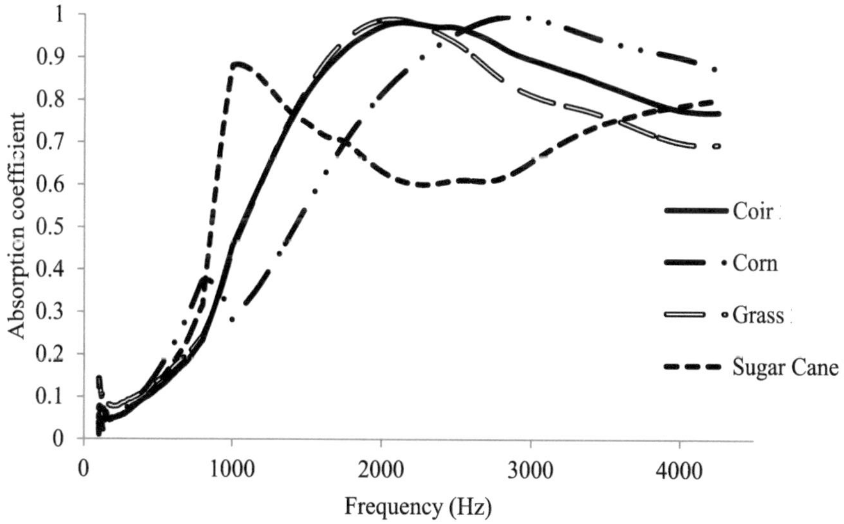


**Figure 7.** Acoustic absorption coefficient of 2 cm sample for the four types of fibers.

| Synthetic Materials | Frequency (Hz) | | | | | |
|---|---|---|---|---|---|---|
| | 125 | 250 | 500 | 1000 | 2000 | 4000 |
| Carpet | 0.01 | 0.02 | 0.06 | 0.15 | 0.25 | 0.45 |
| Plywood (19 mm) | 0.2 | 0.18 | 0.15 | 0.12 | 0.1 | 0.1 |
| Drapery (340 g/m²) | 0.04 | 0.05 | 0.11 | 0.18 | 0.3 | 0.35 |
| Fiberglass board (1" thickness) | 0.06 | 0.2 | 0.65 | 0.9 | 0.95 | 0.98 |

**Table 4.** Absorption coefficients of some common building and acoustic panel materials in octave bands [13].

| Natural fibres | Frequency (Hz) | | | | | |
|---|---|---|---|---|---|---|
| | 125 | 250 | 500 | 1000 | 2000 | 4000 |
| Coir | 0.04 | 0.06 | 0.12 | 0.46 | 0.97 | 0.78 |
| Corn | 0.04 | 0.06 | 0.16 | 0.28 | 0.81 | 0.90 |
| Grass | 0.11 | 0.08 | 0.14 | 0.45 | 0.98 | 0.70 |
| Sugarcane | 0.07 | 0.05 | 0.13 | 0.88 | 0.63 | 0.78 |

**Table 5.** Absorption coefficients of natural fibres of thickness 20 mm obtained from tests in octave bands.

Fibreglass shows better results in comparison with the abovementioned natural fibers. Unfortunately, there have been serious health issues related to fiberglass that may occur immediately or within a few days of exposure such as skin irritation and redness, eye, nose and throat irritation. Not only that, breathing in fiberglass dust may result in coughing, bronchitis, shortness of breath, and even permanent lung disease if exposure is too excessive [14]. Therefore fiberglass is not a good option despite its good acoustical absorption characteristics. Natural fibers not only exhibited good absorption behavior but also play an important role in design for ergonomics. They maintain a comfortable environment by reducing noise level and health risks at the same time.

There are possible measurement errors due to deformation and expansion of samples in impedance tube. The reason is absence of binder that enhances surface strength of the sample and maintains fibers together. This study was aimed at solely investigating the properties of pure natural fibers and any use of additives was prohibited.

## 4. Mechanism for the process

In Physics, resonance is defined as "the increase in amplitude of oscillation of an electric or mechanical system exposed to a periodic force whose frequency is equal or very close to the natural undamped frequency of the system"[15]. All structures have natural frequencies and when a similar vibrational mode exists, resonance occurs. Guitars and electronic devices are designed to resonate with external mode in order to detect that specific mode of frequency. When resonance occurs, the corresponding systems are able to store and easily transfer energy at resonance frequency modes. As the energy transfer between these modes, it is expected that the system losses some energy which entirely depends on damping factor. If the damping factor is small, the resonance frequency is approximately equal to the natural frequency of the system. Resonance phenomena have become the corner stone of so many applications in medicine, science and technology such as nuclear and electron spin resonance.

The study of the resonance phenomena in fibers is essential in improving their acoustical properties. Design of sound absorbers depend on the resonance frequency of the fiber or, in more complicated system, a perforated fiber panel. One of the factors that considerably change the acoustic impedance and absorption coefficient of the acoustic absorber is the porosity. Baranek and Ver [16] presented a compact expression for acoustic impedance of perforated plates. The expression indicated that the influence factor include thickness of the sample, hole radius, hole pitch, and porosity of the perforated plates and air contained in the holes. For porous material, the complex wave propagation constant and characteristic impedance could be expressed in terms of the flow resistivity, wave number, air density, and sound frequency. Sound absorption characteristic of porous material is not so much a function of type material but airflow resistivity and how well material construction can be executed to achieve desirable properties for sound absorbers [17].

Research is not only focusing on measuring the resonance frequency, but also its prediction. Within the frame of this approach, mathematical models were presented showing how the acoustical properties of fibrous materials are related to their characteristic impedance. The Delany-Bazley and Miki [17] models are well known as conventional prediction models. The methodology of preparing these models and subsequent mathematical formulas are not so different from each other when it comes to the scientific basis but they are different only in the values of coefficients and degrees in the formulas. The acoustic panel exhibits a shift of the absorption coefficient peak to lower frequency range when the thickness of the samples is increased. This mechanism depends on the physical properties of the fibers such as diameter and density.

## 5. Conclusions and recommendations

Current trend in green technology is to replace synthetic materials with natural alternatives that have similar functionalities. This study aimed at enlightening the acoustic behavior of natural fibers seeking possibility of implementing them as absorbers in acoustic absorption panel. The selected fibers were coir, corn, sugar cane and grass that are vastly available in tropical countries such as Malaysia. They were compressed into cylindrical samples prior to be tested in impedance tube. Physical characteristics of samples were measured and used to calculate their porosities. By using the same compression ratio, samples with larger thickness possessed lower porosity value. Impedance tube measurements revealed that samples with 1.0 cm thickness had generally absorption coefficient below 70% in the frequency span of 3000 Hz. Increasing the thickness to 2.0 cm reduced their porosity and improved the absorption coefficient. For this thickness; coir, corn and grass absorbed more than 70% of the incident sound at $f$ >1300 Hz. Sugar cane had same amount of absorption only for a narrow band around 1000 Hz and was not considered as a good absorber. Comparing the absorption coefficient of coir, corn and grass (20 cm thickness) in various octave bands with common building and acoustic panel materials showed that they are outstanding alternatives. Fiberglass was an exception; it had best absorption coefficient among the all but also known to have risks to health and environment and is advised to be replaced in the future.

## Author details

Mohammad Hosseini Fouladi, Mohamed H. Nassir, Marwan Shamel,
Sim Yeng Peng, Sin Yi Wen and Pang Zong Xin
*School of Engineering, Taylor's University, Subang Jaya, Selangor, Malaysia*

Masomeh Ghassem
*Faculty of Science and Technology, Universiti Kebangsaan, Malaysia, Bangi, Selangor, Malaysia*

Mohd Jailani Mohd Nor
*Department of Mechanical and Materials Engineering, Faculty of Engineering, Universiti Kebangsaan Malaysia, Bangi, Selangor, Malaysia*

## Acknowledgement

Authors would like to acknowledge the Ministry of Higher Education of Malaysia (MOHE) for supporting this research through the FRGS grant FRGS/2/2010/SOE/06.

## 6. References

[1] Khedari, J., S. Charoenvai and J. Hirunlabh, Building and Environment 38, 435 (2003).
[2] Karl, K. H., Katrin, K. *Ergonomics: How To Design For Ease And Efficiency*, 3rd Ed. Prentice Hall: Upper Saddle River, New Jersey, (2001).
[3] Wambua, P., J., Ivens, and I. Verpoest, Compos Sci Technol 63, 1259 (2003).
[4] Yang, H-S., D-J., Kim, and H-J. Kim, Bioresource Technology 86, 117 (2003).
[5] Davern, W.A., Appl Acous 10, 85 (1977).
[6] F. –C. Lee, and W.-H. Chen. Acoustic Transmission Analysis of Multi Layer Absorbers. Journal of Sound and Vibration, 248(4):621-634 (2001).
[7] Puglia, D., J., Biagiotti, and J.M., Kenny, Journal of Natural Fibers 1(3), 23 (2004).
[8] Wang C. N., and J.H., Torng, Appl Acoust 62, 447 (2001).
[9] Australian Government. Sound Absorption Panel. Available at: http://jobaccess.gov.au/Advice/ProductOrSolutionOne/Pages/Sound_absorption_panels.aspx [Accessed: 22 May 2011]
[10] Acoustical Surfaces Inc., Sound absorption, Available at: http://www.acousticalsurfaces.com/acoustic_IOI/101_7.htm [Accessed: 22 May 2011].
[11] Zulkifli, R., Mohd Nor, M.J., Ismail, A.R., Nuawi, M.Z., Abdullah, S., Tahir, M.F.M., Abd. Rahman, M.N. Comparison of Acoustic Properties between Coir Fiber and Oil Palm Fiber. *European Journal of Scientific Research* 33 (1), 144-152 (2009).
[12] Hosseini Fouladi, M., Ayub, Md., Mohd Nor, M.J. Analysis of Coir Fiber Acoustic Characteristic. *Applied Acoustics* 72 (1), 35-42 (2011).
[13] Absorption Coefficient Chart, Available at: http://www.sae.edu/reference_material/pages/Coefficient%20Chart.htm [Accessed 14 April 2011].
[14] Aerias Air Quality Sciences. Fiber Glass And The Indoor Environment, Available at: http://www.aerias.org/DesktopModules/ArticleDetail.aspx?articleId=38&spaceid=1&subid=4 [Accessed: 27 March 2011].
[15] http://www.thefreedictionary.com/resonance [Accessed 25 April 2012].
[16] Baranek and Ver www.sdcounty.ca.gov/dplu/.../docs/.../EIR_Section_2.11__Noise.pdf [Acsses 20 April, 2012]
[17] Takeshi Komatsu. Improvement of the Delany-Bazley and Miki models for fibrous sound-absorbing materials. Acoust. Sci & Tech 29 (2) 121-129 (2008).

Chapter 8

# An Investigation of an Acoustic Wave Velocity in a Reinforced Concrete Beam from Out-of Plane and in Plane Sources

Noorsuhada Md Nor, Norazura Muhamad Bunnori, Azmi Ibrahim,
Hamidah Mohd Saman, Shahiron Shahidan and Soffian Noor Mat Saliah

Additional information is available at the end of the chapter

http://dx.doi.org/10.5772/54448

## 1. Introduction

Acoustic Emission (AE) is a phenomenon of a transient stress waves resulting from a sudden release of elastic energy, caused by mechanical deformations, initiation and propagation of microcracks, dislocation movement, and other irreversible changes in material [1]. AE sensors are used to detect the acoustic waves at the surface of a structure, produced by AE events either on the surface or in the bulk of the material [2].

There are two types of signals in the AE system; namely burst and continuous signals. Burst signal is a separate type of signal of a very short duration (in the range of a few microseconds (μs) to a few miliseconds (ms)) and it is a broad frequency domain spectrum. Meanwhile the continuous signal is emitted close to each other or the burst is very high rate. The continuous signal is also occurred very close and sometimes overlaps. When the AE signal or output is transmitted in a structure, an array is identified. The output is always represented in a waveform which has information on a source location. A key to compute a source location is by determination of the wave velocity of the wave propagation. If incorrect wave velocity was used either owing to poor assumption or triggering of the system, it would affect to the determination of the source location [3]. Thus accurate wave velocity is important for the determination of the source location prior to any AE test.

The AE wave velocity can be determined by estimating the time of arrival (TOA) of the wave propagating in the structure; normally based on the threshold level [3] or frequency [4]. Typically, the propagation of wave in concrete can be categorized into three different types; namely dilatation wave (compression waves, longitudinal wave or P-wave for primary), the distortion wave (shear waves or S-wave for secondary) and the Rayleigh wave (surface wave or R-wave) [5–7]. Longitudinal wave is known as L – wave or P – wave; which the wave travels in the material and Rayleigh wave (R-wave) travels along the surface of the specimen. The initiation time of the P-wave is when the first TOA of the elastic wave reaches to each sensor. If the onset of the S-wave is detected, this information can be used either in combination with P-wave or be considered as P-wave onset [8]. Determining S–wave is challenging because the distance between sensor and receiver is only few wavelengths. Another problem is because the onset of the S–wave is hidden in the P–wave [8]. Generally, an S-wave is normal to the direction of the wave propagation and a P-wave, parallel. Determination of onset time can be done visually or automatically. This depends on the onset definition itself and also by picking algorithm. In relation with the wave velocity, the P-wave travels at a higher velocity than S-wave [9]. In concrete, the S-wave velocity is approximately 40% less than of the P-wave velocity and R-wave velocity is 92 % of the S-wave velocity and 56 % of the P-wave velocity [10]. However, only the P-wave would be considered in this study based on the TOA.

In order to determine the wave velocity, AE signals are produced by pencil lead fracture (PLF), known as Hsu-Nielsen technique. In this technique, PLF is the monopoles which is normally applied to the outside of the test sample, and the real AE signal is nearly dipoles in which the AE sources that originated from the points are buried inside the sample [11]. The previous PLF was used to generate simulated acoustic emission signals in an aluminum plate at different angles; 0, 30, 60 and 90 degree with respect to the plane of the plate [12]. Next, it is suggested that the plate wave analysis be used to determine the source orientation of acoustic emission sources. A few researchers have used the same method in the heterogeneous material as well as reinforced concrete (RC) [3, 13]. To verify the wave velocity propagation in the RC beam, corrected signal was used by Muhamad Bunnori [3], by normalizing each signal for peak value of amplitude up to 10 volts. The same method was also utilized for this analysis.

In this study, various threshold levels were performed to determine the TOA and the wave velocity in the RC beams with shorter source-to-sensor distance. Very little attention has been given on the shorter distance between sensor and sensor to source distance. An AE source was specified and focused on out-of plane and in-plane of the arrangement of sensor faces. Three main objectives are addressed. Firstly to study the relationship between wave velocity and sensor distance, secondly to investigate the relationship between wave velocity and threshold level and lastly to observe the relationship between arrival time and sensor distance.

## 2. Experimental procedure

### 2.1. Preparation of beam specimen

The test of wave velocity was carried out on a reinforced concrete (RC) beam with dimension of 150 mm x 150 mm x 750 mm; designed in accordance with British Standard [14] for grade C40. The concrete was made up from cement, water, fine aggregate and coarse aggregate with proportion of 1: 0.43: 2.16: 2.60, respectively. Then, 1 % of water reducer agent (Rheobuild 1100) of cement weight was added in the concrete mix to improve the workability of the fresh concrete. The maximum coarse aggregate of 20 mm was used.

The beams were designed as a singly RC beam with two high yield steels of 16 mm to strengthen the tension part and two mild steels of 8 mm as hanger bars. The bars were bent at both end to form a standard hook of 60 mm at tension part and 30 mm at compression part. In the stirrups, 12 diameter of mild steel with spacing of 100 mm centre to centre was used. In the preparation of each RC beam, the reinforcement was submerged in the standard mould of 150 mm x 150 mm x 750 mm with cover of 20 mm, before the concrete mix was cast. All beams were demoulded after ±24 hours and submerged in the water curing tank for 28 days. In order to ensure the design fulfilled the strength, the cubes were prepared and tested under compression test for 7 and 28 days. The compressive strength for 7 days was found to be 32.35 N/mm$^2$ and 44.65 N/mm$^2$at age of 28 days.

## 2.2. Acoustic emission system

Acoustic emission (AE) was monitored using a MicroSAMOS (μSAMOS) supplied by Physical Acoustic Corporation (PAC). The system consists of integral preamplifier acoustic emission sensors (transducers) R6I (40 – 100 kHz); a notebook acoustic emission system board (8 (channels) x 16 (hubs) bit acoustic emission channels, low peak and high peak filters, 2 MHz bandwidth, auto sensor test (AST), time definition display (TDD), digital signal processor (DSP) and waveform module; personal computer memory card international association (PCMCIA) interface card; cables; internal and external parametric cable set; a notebook personal computer with full suite of AEWin Software; universal serial bus (USB) license key; and magnetic clamps. The acquisition parameters in the AEWin software were summarized in Table 1.

| Parameter | Value |
|---|---|
| Hit definition time (HDT) | 2000μs |
| Peak definition time (PDT) | 1000μs |
| Hit lockout time (HLT) | 500μs |
| Preamplifier (R6I) | 40dB |
| Bandpass data acquisition filter | 400kHz |
| Sample rate | 100ksps |
| Analog filter (lower) | 1kHz |
| Pre-trigger | 250,000 |

**Table 1.** AE test parameters

## 2.3. AE sensor installation and sensitivity checking

Prior to the wave velocity test, three pairs of thin plates were prepared to attach the sensors S1, S2 and S3 with the distance of 200 mm and 100 mm between pair of plates. Spacing between two plates is 45 mm. The beam surface was polished smoothly using course sand paper to ensure good connectivity between sensors and beam surface. The plates were then fixed to the beam surface using epoxy and hardener.

Three sensors were coupled on the beam surfaces at the selected point using a good couplant such as a thin layer of high performance grease. The thinnest practical layer of couplant is usually the best [15]. The magnetic clamps were used to safeguard the sensors; which sensors held in position in magnetic clamps; then the magnetic clamps properly coupled to the steel plates.

With sensors in good mounting condition, calibration (sensitivity) checking was carried out to recognize the sensitivity of the sensors. In this experiment, a magnetic pencil with a Nielson shoe (Teflon shoe) was used to break a 0.5 mm 2 H lead to generate a simulated acoustic wave against the surface of the beam. The handling method of the magnetic pencil has been presented in ASTM E976 [16]. All sensors would significantly coupled if the wave was generated by at least three or more replicates of pencil lead fracture (PLF). This might produce high amplitude of 99 dB or the sensitivity within ± 3 dB in different [3, 17]. Based on the research done by others, it was found that the amplitude recorded by each sensor should not be permitted to vary more than 4 dB from the average of all sensors [18]. This technique was used to ensure that the sensor and the beam were in a good contact to provide an adequate result throughout the test. If these criteria are not met, the sensors on the beam surface would be remounted and sensitivity check was carried out until the amplitude fulfilled the requirement.

## 2.4. Wave velocity test

In this experimental work, three integral preamplifier R6I sensors with 55 kHz resonant frequency were used. Acoustic emission sensors are transducers that convert the mechanical waves into electrical signal [19], where the information about the existence and location of possible damage or stress released sources can be obtained. The specification and feature of the sensor is shown in Table 2.

| Specification | Value |
|---|---|
| Dimension (Ø x height) mm | 29 x 40 |
| Weight (gm) | 98 |
| Operating temperature (0°c) | -45 to +85 |
| Shock limit (g) | 500 |
| Peak sensitivity (V/m/s) | 120 |
| Directionality (dB) | +1.5 |

**Table 2.** Specification and feature of the R6I sensor

A schematic diagram of the test set up for beam size of 150 mm x 150 mm x 750 mm and 100 mm x 100 mm x 500 mm in a linear structure is shown in Figure 1 and Figure 2, respectively. The AE sources were identified on the out-of plane and in plane of the sensors arrangement or source parallel to the sensor face. Three sensors were mounted in a symmetrical arrangement on the RC beam using a thin layer of high vacuum grease as a couplant with the same spacing of 200 mm centre to centre for beam 150 mm x 150 mm x 750 mm and 100 mm centre to centre for beam 100 mm x 100 mm x 500 mm.

In this experiment, four different pre-set threshold levels were chosen namely 40 dB, 45 dB, 65 dB and 70 dB. At each threshold level, 15 replicates of PLF were applied to generate a simulated AE source at the same spot or the same AE source location. When the lead is pressed against the RC beam surface, the applied force produces a local deformation and the stress where the lead touched is suddenly being released. It is important to handle the pencil properly while breaking the lead against the testing specimen to get significant value. Improper handling of the pencil would give imprecise value; which will affect the TOA. The proper breaking of the lead creates a very short-duration, localized impulse that is quite similar to a natural acoustic emission source such as crack [19]. The lead was extended to 0.1 inches and the pencil was slanted down $30^0$ to the plane of the beam surface, as suggested by other researchers [18, 20]. The out-of plane AE source was located at the centre of the cross section of the end beam. As for in-plane AE source, it was located at 100 mm from Sensor 1 as illustrated in Figures 1 and 2.

In the AE hardware, for in-plane source, Sensor 1 was set as individual and the rest were synchronized. Meanwhile, for out of plane source, Sensor 3 was set as individual and the rest were synchronized. The same method applied for beam size of 100 mm x 100 mm x 500 mm.

In acoustic emission, waveform parameters normally used are frequency and amplitude [22]. Amplitude of AE signal parameter was used in identification of wave velocity. It is defined as the magnitude of the peak voltage of the largest excursion attained by the signal waveform from single emission event [1]. Amplitude is reported in decibels (dB) to measure signal size and typical AE signal is represented as a voltage versus time curve. Voltage is converted to dB using the following equation:

$$A = 20 \log (V/V_{ref}) \tag{1}$$

Where: $A$ is an amplitude (dB), $V$ is voltage of peak excursion and $V_{ref}$ is the reference voltage. Generally the dB scale runs from 0 to 100 [21]; which depend on the threshold set.

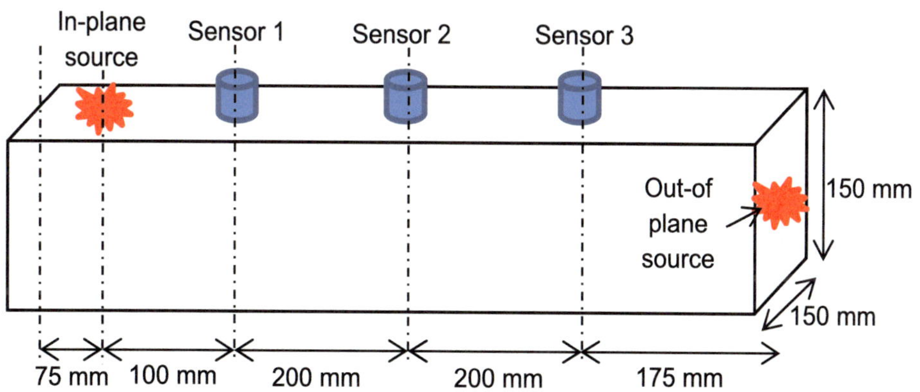


**Figure 1.** Schematic diagram of sensors location on the RC beam 150 mm x 150 mm x 750 mm

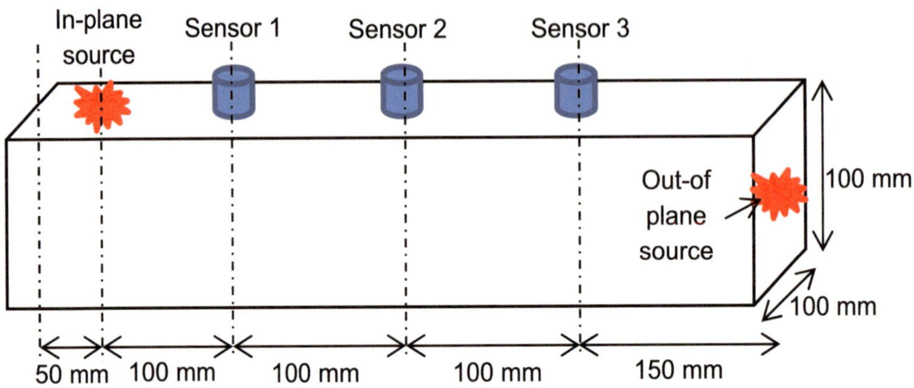


**Figure 2.** Schematic diagram of sensors location on the RC beam 100 mm x 100 mm x 500 mm

In a relationship between threshold level and amplitude, preamplifier was considered; where equation (2) was used to calculate amplitude:

$$\text{Threshold} = \text{Amplitude} - \text{Preamplifier} \tag{2}$$

The rudimentary determination for location calculation is based on time-distance relationship implied by the velocity of the sound wave [4, 23]; where the absolute arrival time, $t$, of a hit in an event combine with the velocity, $v$ ($v$ is P-wave velocity [19]), of the sound wave to yield the distance, $d$, from sensor to the source as represented by

$$d = v\,t \tag{3}$$

Generally the distance between the two sensors depends on the geometry of the sample. Determination of TOA (the exact time the event originated) was made by recording many data hits for each AE event. $t_1$ represents the time of arrival (or arrival of longitudinal wave, P-wave) at Channel 1. Meanwhile $t_2$ is arrival of P-wave at Channel 2. The arrival time difference between the arrivals of the signal at the two sensors can be written as

$$\Delta T = (t_1 - t_2) \tag{4}$$

At selected source locations (either out-of plane or in-plane), 10 replicates of PLF were applied from the lower threshold level of 40 dB to the higher 70 dB. Distance for beam 150 mm x 150 mm x 750 mm from sensors 1-2 and sensors 1-3 is 200 mm and 400 mm, respectively. Meanwhile, for beam 100 mm x 100 mm x 500 mm, the distance from sensors 1 to 2 is 100 mm and sensors 1 to 3 is 200 mm.

The relationship between wave velocity and threshold level, and TOA and sensor distance were correlated. Both correlations were then verified by corrected AE signal. Corrected AE signal was used for a particular threshold level, the same temporal position in each waveform which involves the normalizing of each signal up to a peak value of 10 volts. In this research, the corrected signal was used for the evaluation of TOA for each threshold level. The corrected signal can be identified by normalizing the peak signal of a hit to 10 volts followed by other signal in a hit. Then, the TOA for each threshold was then determined.

# 3. Results and discussion

## 3.1. Sensitivity checking

Sensitivity checking was used to ensure the sensor and the beam was in a good contact to provide an adequate results. The sensors must have a higher sensitivity [24] in order to get good contact between sensors surface and beam surfaces. In this experimental in order to check the sensitivity, at least ten numbers of 99 dB were considered for each sensor. The responses of the transducer to PLFs are shown in Figure 3. Sensor 1 represents a good sensitivity since almost of all the PLFs produced amplitude of 99 dB. However, Sensor 2 needs several trials to ensure the sensors and the beam surface have a good contact. These scenarios that occurred in Sensors 1 and 2 are related with the signal waveform produced from the PLF. It is typically affected by the source characteristics, the path taken from the source to the sensors, the characteristic of the sensor and the measurement system [25]. The waveform for Hit 1 and Hit 12 are represented in Figure 4.

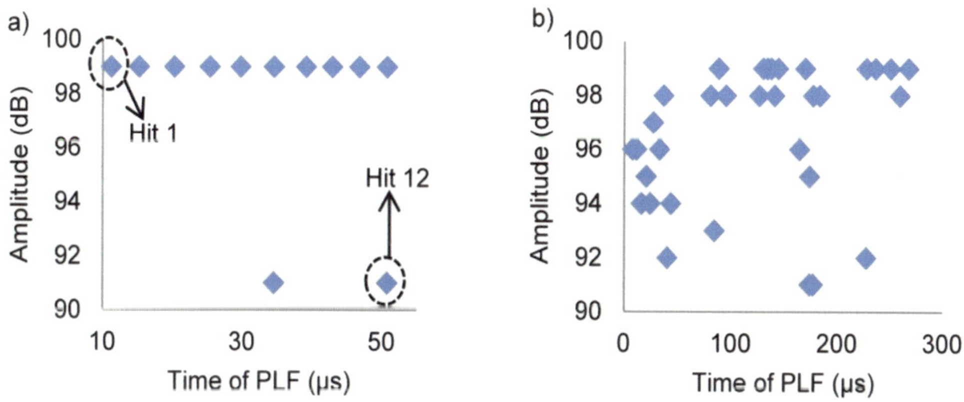


**Figure 3.** Response to PLF on reinforced concrete beam for a) Sensor 1 and b) Sensor 2

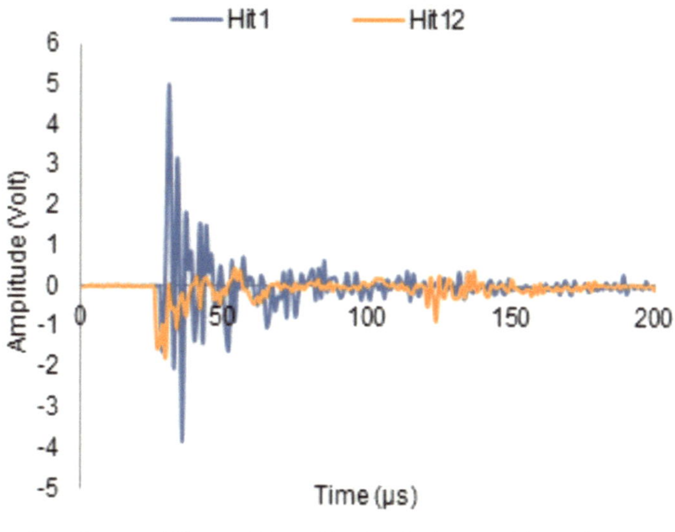


**Figure 4.** AE waveform for two different AE amplitudes during sensitivity checking

Generally, each PLF produces one signal of AE hit or more and in various form or shape of waves. Figure 4 shows different shapes of wave for two hits of PLF from Sensor 1. Hit 1 is a response of PLF that produces amplitude 99 dB and generates high peak wave. Hit 12 is for amplitude 91 dB response from the bounce back of the lead during PLF process and produced lower amplitude. Hit 1 generates high peak wave rather than Hit 12. Figure 3a indicated that Hit 12 occurred at the same time of PLF with other amplitudes. However, for sensitivity checking this will be ignored. The peak signal amplitude can be related to the intensity of the source in the material [21, 25]. Pullin et al. [26] stated that the good sensitivity between sensor and beam surface is enough when the amplitude signal exceeded 95 dB. Hence, sensitivity checking would be better when a higher amplitude is being produced during replicates of PLFs.

### 3.2. Identification of Time of Arrival (TOA)

Typical AE waveform parameter received by Sensor 1 type R6I from a 150 x 150 x 750 mm RC beam during PLF process, represented in amplitude (v) against time (μs) is shown in Figure 5. As lead breaks (in-plane or out-of-plane) on the surface of the beam, a wave propagates through a solid medium; it carries certain amount of energy. The energy can be consumed by scattering during propagation. The scattering effect principally relies on the defects such as micro-cracks inside the material. In wave velocity determination only several AE parameters would be considered such as time and amplitude. AE amplitude gives the information about the time at which AE signals take place [27]. Wave velocity is one of the methods to be considered for source location. Figure 6 presents the wave of the signal for threshold 40 dB. The first wave crosses the threshold level is known as P-wave.

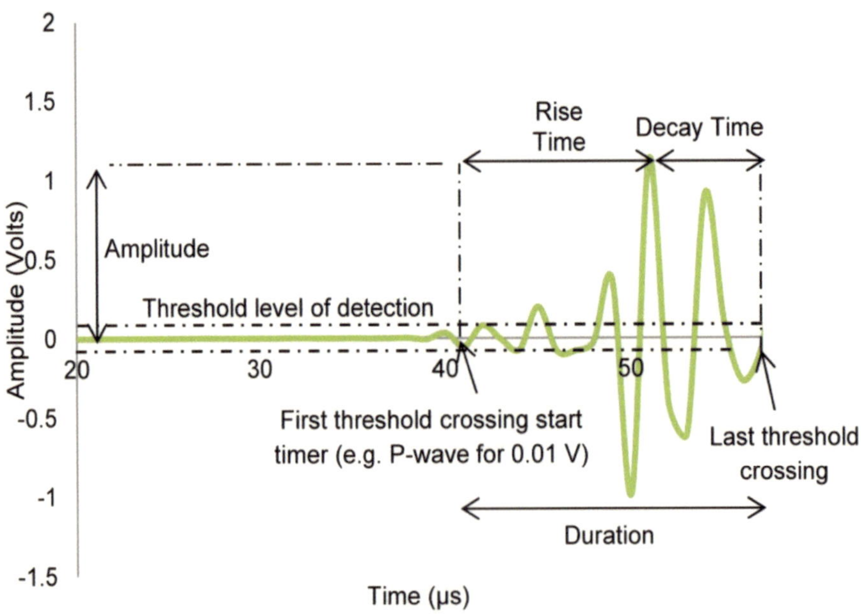


**Figure 5.** Simple waveform parameter

In this experiment, Sensor 1 was set as individual due to close to the AE source and the rest were considered as synchronize. For instance, the threshold was set at 40 dB prior to test. The Sensor 1 would be the first received the waves emitted by PLF at AE source followed by other sensors as depicted in Figure 6. In this figure, the TOA or longitudinal wave (P-wave) for Sensors 1, 2 and 3 were noted as $t_1$, $t_2$ and $t_3$ respectively. The AE signal for first P-wave, $t_1$ was examined out to 29 μs after the source initiation time followed by P-wave, $t_2$ for sensor 2 about 34 μs and P-wave for $t_3$ is 43 μs. In the test, gain 0 dB and preamplifier 40 dB were used with the sample interval of 10μs. In this case, the thresholds were assumed to be as lower as 40 dB (0.01V) to 70 dB (0.32V). Referring to eq. 1 and eq. 2, amplitude (represented in dB) can be calculated with the $V_{ref}$ equals 1μv. Then, volt for each threshold levels can be determined. Volt for each threshold can be used as a guideline determine the TOA or P-wave at each sensor from AE source.

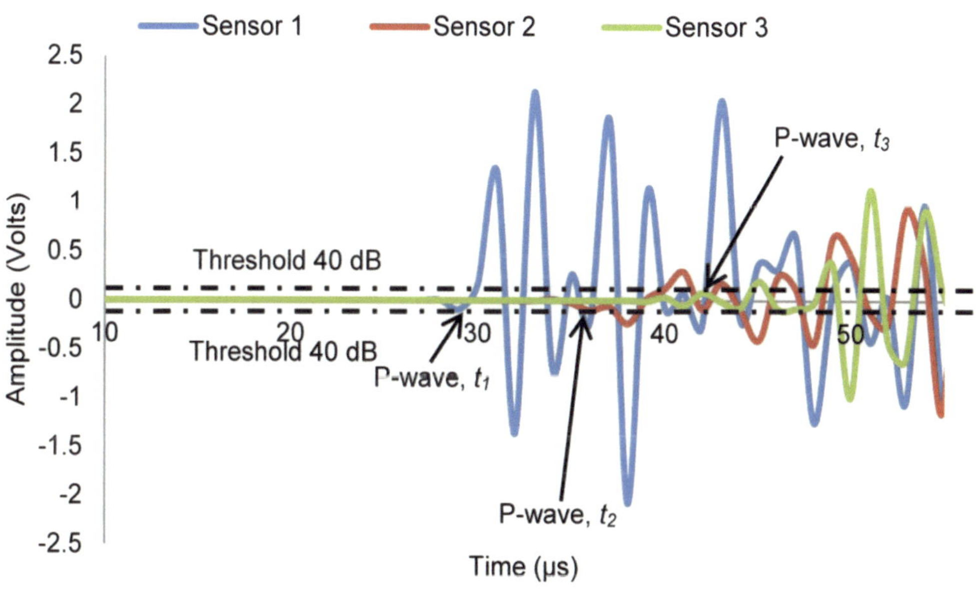


**Figure 6.** TOA or P-wave for threshold 40 dB at time 35.53 s for each sensor

## 3.3. Relationship between wave velocity and sensor distance for RC beam

The first method in the determination of wave velocity in 150 mm x 150 mm x 750 mm and 100 mm x 100 mm x 500 mm RC beam is based on the changes of pre-set threshold level in AE hardware or threshold crossing technique. Four different pre-set threshold levels were chosen namely 40 dB (0.01V), 45 dB (0.02V), 65 dB (0.2V) and 70 dB (0.32V). For each pre-set threshold level, 10 replicates of PLF were carried out in-plane and out-of-plane sources. TOA to each sensor for each PLF was analysed and followed by other replicates of PLFs. Figure 7a shows the relationship between the average wave velocity and sensor distance for out-of-plane source at each threshold level. Figure 7b presents the relationship of average wave velocity and sensor distance for in-plane source. Both figures show the wave velocities are apparently depends on the sensor distance and threshold levels.

In determination of wave velocity, sensors 1, 2 and 3 was set of 0 m, 0.2 m and 0.4 m, respectively. Figure 7 illustrates that, for RC beams 150 mm x 150 mm x 750 mm with sensor distances 0.2 m, at lower threshold level (40 dB) and higher threshold level (70 dB) for out-of-plane source, the velocity is in the range of 7889 m/s to 4559 m/s and 2409 m/s to 1654 m/s, respectively. Meanwhile in in-plane source, velocities for lower threshold and higher threshold are in the range of 3086 m/s to 2608 m/s and 2619 m/s to 2192 m/s, respectively. It is found that the wave velocity for in-plane source has lower velocity than out-of-plane source. It might be due to the in-plane source location is parallel to the sensor surface and the wave propagates take longer time to reach the sensors. Hence, the source location in linear measurement of sensors for in-plane would prolong the TOA of the wave to reach the sensor and the velocity is reduced. Out-of-plane source is longitudinal to the sensor surface and hence the wave signal propagates inside the RC beam take shorter time to reach the sensors. Thus, it reduces the TOA of the wave to each sensor.

In wave velocity calculation for beam 100 mm x 100 mm x 500 mm, sensors 1, 2 and 3 was set as 0, 0.1 m and 0.2 m, respectively. Figure 8 presents the wave velocity in 100 mm x 100 mm x 500 mm with sensor distances of 0.1 m for out-of-plane. The figure shows that the lower threshold level is in the range of 3430 m/s to 2810 m/s and the higher threshold level is 2765 m/s to 1166 m/s. Velocity propagates in in-plane source is in the range of 5030 m/s to 3150 m/s for threshold 40 dB and 2751 m/s to 2603 m/s for threshold 70 dB. In this case, apparently the velocity in the in-plane source has higher velocity than out-of-plane source. Logically, the pattern of the wave velocity occurred in the both size of beams with different spacing of sensors should be synchronized. However, this phenomenon indicates that the wave velocity in the RC beam or heterogeneous material cannot be predicted.

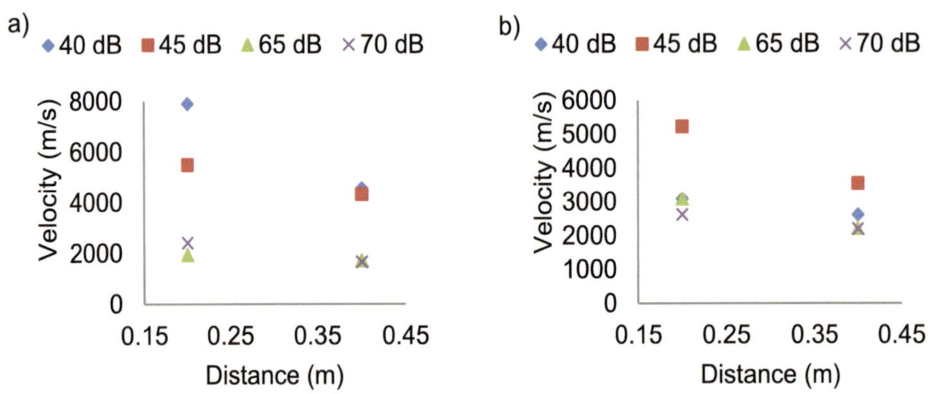


**Figure 7.** Wave velocity in 150 mm x 150mm x 750 mm RC beam for a) out-of-plane source and b) in-plane source

The results of the study were also compared to other research done in heterogeneous material as well as RC structure of the wave velocity of RC beam 100 mm x 150 mm x 2000 mm with the sensor distance of 250 mm, the value was in the range of 4000 m/s to 4500 m/s [3]. For large spectrum with frequency between 50 to 600 kHz, the wave velocity is 2350 m/s [28]. The wave velocity in ceramic matrix composite was 3200 m/s with the threshold level of 48 dB; however, the initial velocity was 10000 m/s [29]. If resonant sensor of 300 kHz was used, the wave velocity is 4000 m/s [30]. For carbon fibre reinforced polymer (CFRP) the wave velocity equal to 1000 m/s [13].

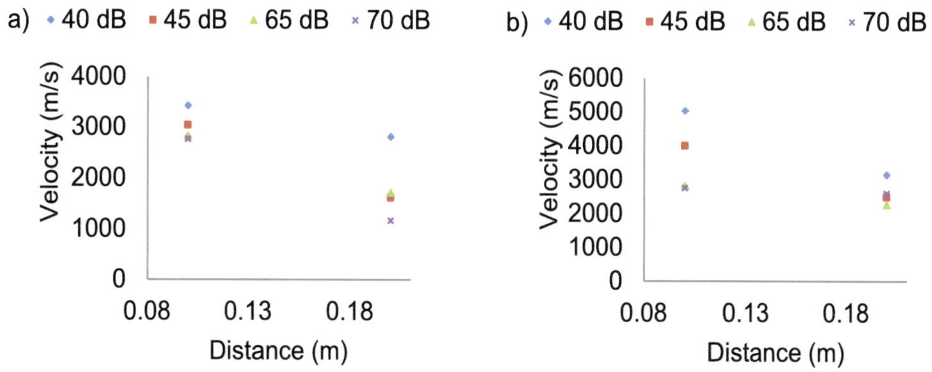


**Figure 8.** Wave velocity in 100 mm x 100 mm x 500 mm RC beam for a) out-of-plane source and b) in-plane source

### 3.4. Relationship between wave velocity and threshold level

Figures 9a and 9b display the velocity propagates in the 150 mm x 150 mm x 750 mm RC beam as threshold levels increased for out-of-plane and in-plane sources. As mentioned earlier, apparently the wave velocities depend on the threshold levels. Thus, the statistical coefficient can be used to enhance the prediction by identification of determination coefficient, $R^2$. For out-of-plane source, for sensors distance 0.2 m (Sensor 3-2), the $R^2$ is 91.3 % and for distance 0.4 m (Sensor 3-1) produced higher $R^2$ of 97.8 %. Thus, both distances have strong correlation between velocities and also the threshold level, where, the wave velocity in the RC beam is influenced by the threshold level crossing. The higher the threshold level, the lower of the wave velocity would be.

For in-plane source, distance between Sensors 1 to 2 and 1 to 3 is 0.2 m and 0.4 m, respectively. It is found that for in-plane source, both distances of sensors have weak correlation of 30.3 % for Sensors 1-2 and 49 % for Sensors 1-3. It is due to the wave velocities for 40 dB are lower than for threshold 45 dB with the approximate different in the range of 2100 m/s to 940 m/s. However, after threshold of 45 dB, the wave velocity seems has good correlation, which the threshold increased as the wave velocity decreased.

Figure 10 shows the relationship between wave velocity propagation and threshold level for 100 mm x 100 mm x 500 mm RC beam. Wave velocity for closet distance in Figure 10 shows a good relationship between velocity and threshold levels with the correlation of 83 % and 92 % for out-of-plane and in-plane source, respectively. However the longer distance represents a weak relationship with the correlation of 59 % and 40 % for out-of-plane and in-plane source, respectively. Overall the prediction of the wave velocity in short distance is influenced by threshold levels has strong correlation. However, for longer distance, the relationship is seemingly unpredictable.

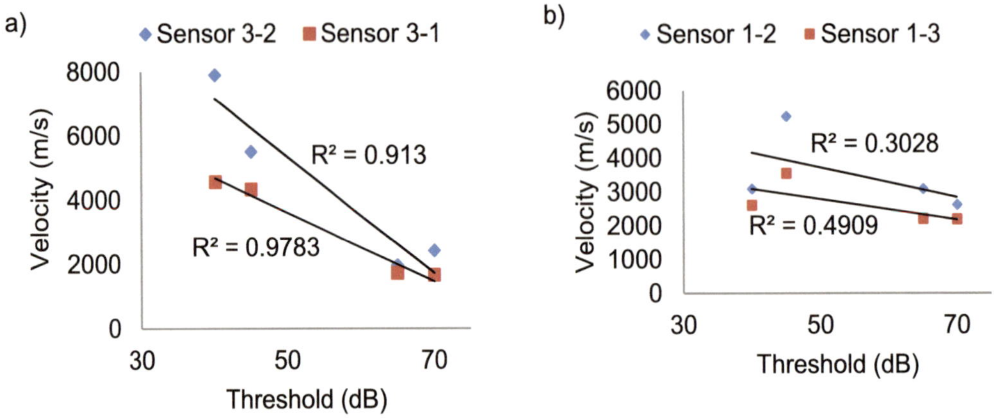


**Figure 9.** Relationship between velocity and threshold level on RC beam 150 mm x 150 mm x 750 mm for a) out-of-plane source and b) in-plane source

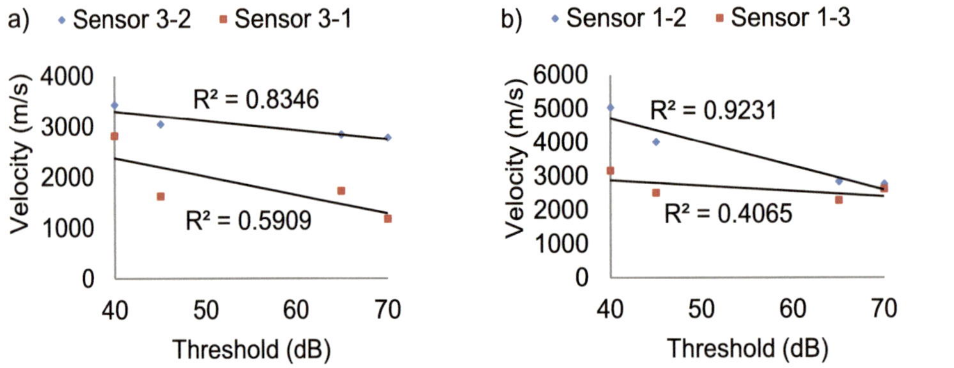


**Figure 10.** Relationship between velocity and threshold level on RC beam 100 mm x 100 mm x 500 mm for a) out-of-plane source and b) in-plane source

It can be concluded that the relationship of the velocity and threshold level is easy to predict for shorter distance, where the velocity decreases when the threshold level increases. It apparently indicates a good correlation for shorter distance. However, for longer distance the relationship is unpredictable and has weak relationship. This can be related with the material applied, where concrete always has imperfection of the composition such as voids.

### 3.5. Relationship between arrival time and sensor distance

Figure 11 shows the average of arrival time (μs) against sensor distance for original signal and corrected signal for out-of-plane source at threshold 40 dB, 45 dB, 65 dB and 70 dB on 150 mm x 150 mm x 750 mm RC beam. In the graphs, Cor stands for corrected signal and Ori stands for original signal. The corrected signal has been used to justify the correlation between TOA and distance. Sensor 3 (distance zero) close to the AE source, it receives the first TOA emits by PLF and follows by the other sensors. Sensor 1 would receive the last signal with the higher TOA. In some situation, the farthest sensors from the AE source would not receive any signal. This phenomenon occurs because wave attenuation in the RC beam cannot be avoided. Wave attenuation is loss of AE energy as waves travel through in a material [25]. It is also the loss of amplitude with distance as the wave travel through the beam [7]. This is also affected of the time taken path of the signal wave from the AE source to the sensors. The farthest sensor would receive the lower signal wave and hence take longer TOA. For instance, Original 40 dB at fist sensor is 29 μs and the farthest sensor is 40 μs. Thus, TOA takes longer time to reach sensor 1. Other factor causes the waveform amplitude loss such as intrinsic mechanisms and imperfections. It is because the intrinsic mechanism (thermal effects) and imperfections (void and misoriented grains) reduce the amplitude of the wave by using scattering and reflections [4]. Otherwise, RC beam is a heterogeneous material; many obstacles such as aggregates and bars affected to the movement of waves and yet delayed the TOA to the longest distance from the AE source. If the attenuation is too large or size of the structure is too big, the location of the hit cannot be defined. Thus, the attenuation depends on the types of material through which the waves are progressing and the source of the waves [7].

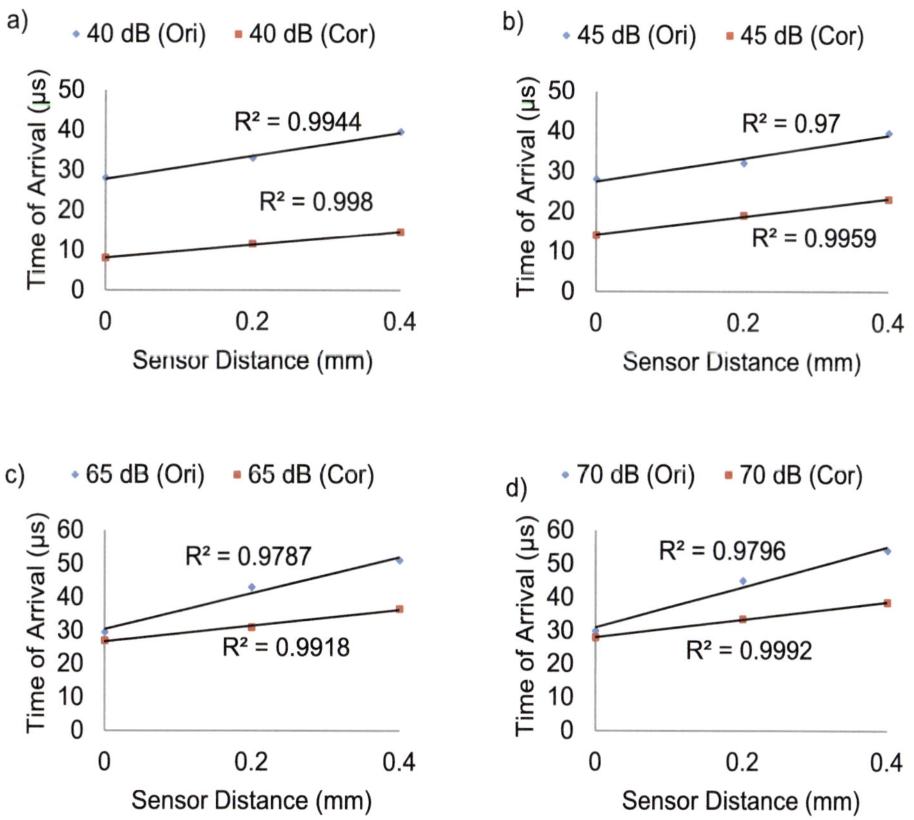


**Figure 11.** Regression analysis of arrival time against sensor distance for 150 mm x 150 mm x 750 mm RC beam (out-of-plane source) at a) 40 dB, b) 45 dB, c) 65 dB and d) 70 dB

All figures show very good correlation between TOA and sensor distance for out-of-plane source and there is insignificant difference between original signal and corrected signal. All threshold levels have strong correlation with the determination of coefficient more than 97 % between TOA and sensor distance.

Figures 12 a), b), c) and d) represent the regression analysis of arrival time against sensor distance for in-plane source at threshold 40 dB, 45 dB, 65 dB and 70 dB, respectively. Similar to the one represented in Figure 11, it shows that all threshold levels have good correlation between time of arrival (TOA) and sensor distance with all the coefficient of determinations more than 96 %. However, threshold level of 40 dB for corrected signal analysis apparently has a perfect fit of linear correlation by representing the coefficient equal to 1. The value indicates that the TOA is absolutely influenced by the sensor distance. Similar to out-of plane source, the relationship also indicates no significant difference between original signal and corrected signal.

Figure 13 represents the regression analysis of TOA against sensor distance for 100 mm x 100 mm x 500 mm in-plane source at all threshold levels. Similarly, it shows that it has good correlation with the regression above 80 % with no significant difference between original signal and corrected signal.

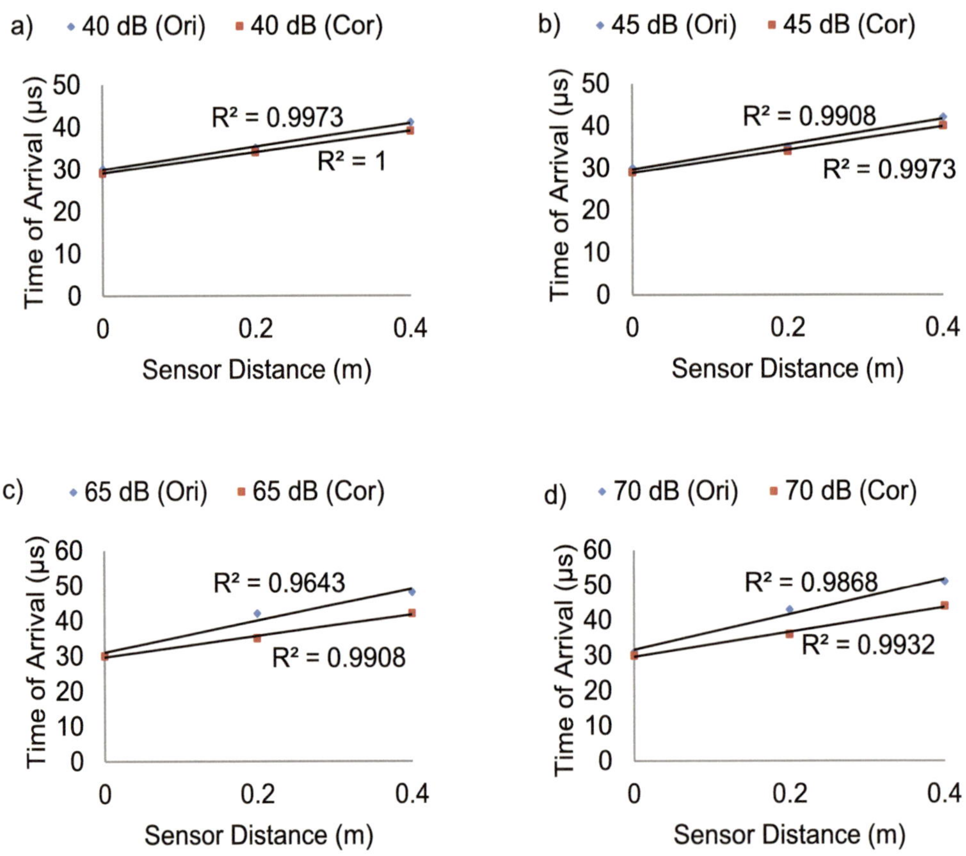


**Figure 12.** Regression analysis of arrival time against sensor distance for 150 mm x 150 mm x 750 mm RC beam (in-plane source) at a) 40 dB, b) 45 dB, c) 65 dB and d) 70 dB

Overall, the TOA of the wave propagation has good agreement with distance of sensors, where the longer the distance, the higher the time taken to reach the sensor. It can be seen in Figures 8 and 9, the slope of the best straight line drawn through the points are linear. However, other researchers found the relationship between TOA and sensor distance is not linear [3]. Since, the TOA is a good indication of wave velocity travels in the beam, hence the wave velocity decreases as the TOA increases.

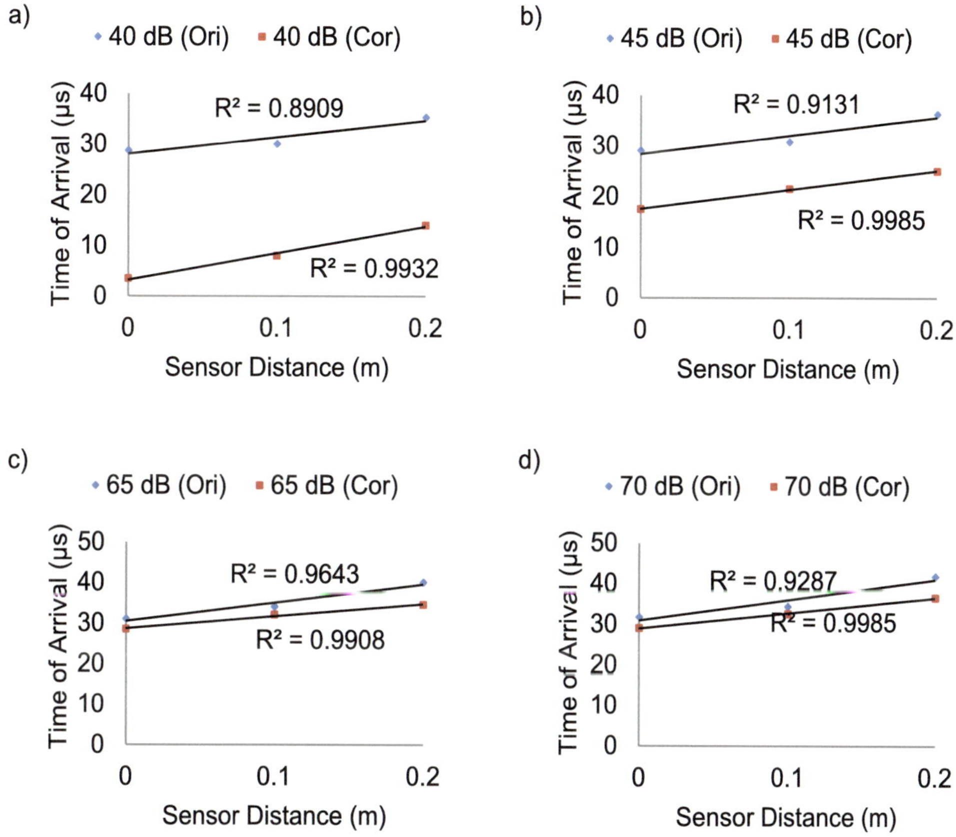


**Figure 13.** Regression analysis of arrival time against sensor distance for 100 mm x 100 mm x 500 mm (in-plane source) at a) 40 dB, b) 45 dB, c) 65 dB and d) 70 dB

## 4. Conclusions

Three sensors were used to determine the TOA and the wave velocity in RC beams. Sensor 1 was set as individual since it is close to the AE source and the first to receive the waves (P-Wave) that were emitted by PLF in the solid medium of concrete.

It can be concluded that AE wave velocities were found dependent on threshold levels and distance of sensor. It is found that the higher the threshold levels, the lower the wave velocity. This is because the threshold level prolonging arrival times of the waves and reducing velocities. Wave velocities for 150 mm x 150 mm x 750 mm were calculated to be in the range of 7889 m/s to 1654 m/s for out-of-plane source and 5229 m/s to 2192 m/s for in-plane source. Wave velocities for 100 mm x 100 mm x 500 mm were computed to be in the range of 3430 m/s to 1166 m/s for out-of-plane source and 5030 m/s to 2603 m/s for in-plane source. Wave velocities decreases with the increasing distance between sensors and threshold levels. It was enhanced by the relationship of wave velocities and sensor distance for each threshold level.

The relationship between wave velocity and threshold level were identified for each sensor distance and then the coefficient of correlation, $R^2$ was determined. It is found that $R^2$ for beam 100 mm x 100 mm x 500 mm at out-of-plane source has good correlation than in-plane source. Beam 150 mm x 150 mm x 750 mm has weak correlation for longer sensor distance, but it has good correlation for shorter distance.

The relationship of TOA and sensor distance was identified. Its correlation was presented for each original signal and corrected signal. In total, both original signal and corrected signal indicated strong correlation between TOA and sensor distance with domain regression of 90 %.

## Author details

Noorsuhada Md Nor and Soffian Noor Mat Saliah
*Faculty of Civil Engineering, Universiti Teknologi MARA (UiTM), Permatang Pauh, Pulau Pinang, Malaysia*

Norazura Muhamad Bunnori and Shahiron Shahidan
*School of Civil Engineering, Engineering Campus, Universiti Sains Malaysia (USM), Seri Ampangan, Seberang Perai Selatan, Nibong Tebal, Pulau Pinang, Malaysia*

Azmi Ibrahim and Hamidah Mohd Saman
*Faculty of Civil Engineering, Universiti Teknologi MARA (UiTM), Shah Alam, Selangor, Malaysia*

## Acknowledgement

The authors gratefully acknowledge the support provided by the Universiti Teknologi MARA (for financial support), Universiti Sains Malaysia (for the AE equipment) and Ministry of Higher Education, Malaysia (MOHE). A special thank you to all technicians in the Laboratory of Heavy Structure, Faculty of Civil Engineering, Universiti Teknologi MARA and Universiti Sains Malaysia.

## 5. References

[1] ASTM E 1316-06 (2006) Standard Terminology for Non-Destructive Examinations. ASTM International, 5.

[2] Theobald P, Dar F (2006) AE Sensor Calibration for Out-of-Plane and In-plane Displacement Sensitivity. Advanced Materials Research. 13–14: 91–98.

[3] Muhamad Bunnori N, Pullin R, Holford KM, Lark RJ (2006) A Practical Investigation into Acoustic Wave Propagation in Concrete Structures. Advanced Materials Research. 13–14: 205–212.

[4] Ervin BL (2007) Monitoring Corrosion of Rebar Embedded in Mortar using Guided Ultrasonics Waves. PhD Thesis, University of Illinois, Urbana-Champaign.

[5] Liu Z (2007) Evaluation of Reinforced Concrete Beams using Cyclic Load Test, Acoustic Emission and Acousto-Ultrasonics. PhD Thesis, Auburn University, Auburn Alabama.

[6] Lovejoy S (2006) Development of Acoustic Emissions Testing Procedures Applicable to Conventionally Reinforced Concrete Deck Girder Bridges Subjected to Diagnoal Cracking. PhD Thesis, Oregon State University.

[7] ARCHES (2009) Sustainable Development, Global Change and Ecosystems Sustainable Surface Transport. ARCHES.

[8] Kurz JH, Köppel S, Linzer LM, Schechinger B, Grosse CU (2008) Source Localization. In: Grosse CU, Ohtsu M, editors. Acoustic Emission Testing. Springer. pp. 101–147.

[9] Grosse CU, Linzer LM (2008) Signal-Based AE Analysis. In:Grosse CU, Ohtsu M, editors. Acoustic Emission Testing. Springer. pp. 53–99.

[10] Zein, Abd-Elbaky AS (2005) Nondestructive Evaluation of Reinforced Concrete T-Shaped Beams by Frequency Spectrum Analysis of Impact-Echo Waveforms. PhD Thesis, University of South Carolina.

[11] Hamstad MA (2006) Contrasts between the Acoustic Emission Signals Generated by Monopole versus Dipole Sources. Advanced Materials Research, 13–14: 61–68.

[12] Gorman MR, Prosser WH (1991) AE Source Orientation by Plate Wave Analysis. Journal of Acoustic Emission. 9: 283-288.

[13] Degala S, Rizzo P, Ramanathan K, Harries KA (2009) Acoustic Emission Monitoring of CFRP Reinforced Concrete Slabs. Construction and Building Materials. 23: 2016–2026.

[14] British Standard BS8110 Part 1 (1997) Structural use of Concrete: Part 1. Code of Practice for Design and Construction. British Standard Institution.

[15] ASTM E650-97 (2007) Standard Guide for Mounting Piezoelectric Acoustic Emission Sensors. ASTM International, Pennsylvania United State.

[16] ASTM E976-10 (2010) Standard Guide for Determining the Reproducibility of Acoustic Emission Sensor Response. ASTM International. 5.

[17] Md Nor N, Muhamad Bunnori N, Ibrahim A, Shahidan S, Mat Saliah SN (2011) An Investigation on Acoustic Wave Velocity of Reinforced Concrete Beam In-Plane Source. 7th International Colloquium on Signal Processing and its Application. IEEE. 19–22.

[18] Chotickai P (2001) Acoustic Emission Monitoring of Prestressed Bridge Girders with Premature Concrete Deterioration. Thesis of Master Science in Engineering, The University of Texas, Austin.

[19] Physical Acoustic Corporation (2005) Samos AE System; Use's manual. Princeton Junction, NJ.

[20] ASNT (1996) Nondestructive testing handbook 2nd edition. Columbus, OH: American Society for Nondestructive Testing.

[21] Xu J (2008) Nondestructive Evaluation of Prestressed Concrete Structures by means of Acoustic Emissions Monitoring. PhD Thesis, Auburn University, Auburn, Alabama.

[22] Kalyanasundaram P, Mukhopadhyay CK, Subba Rao SV (2007) Practical Acoustic Emission. Indian Society for Non-Destructive Testing – National Certification Board Series.

[23] Baron JA, Yip SP (1987) Acoustic Emission Source Location. In: Miller RK, McIntire P, editors. Nondestructive Testing Handbook, 2nd edition Vol 5. American Society for Nondestructive Testing.pp.135-154.

[24] Qin L, Huang S, Cheng X, Lu Y, Li Z (2009) The Application of 1-3 Cement Based Piezoelectri Transducers in Active and Passive Health Monitoring for Concrete Structures. Smart Materials and Structures. 18: 1-8.

[25] Spanner JC, Brown A, Robert Hay D, Mustafa V, Pollock A (1987) Fundamentals of Acoustic Emission Testing. In: Miller RK, McIntire P, editors. Nondestructive Testing Handbook, 2nd edition Vol 5. American Society for Nondestructive Testing.pp.11-42.

[26] Pullin R, Theobald P, Holford KM, Evans SL (2006). Experimental Validation of Dispersion Curves in Plates for Acoustic Emission. Advanced Materials Research. 13-14: 53-60.

[27] Ni QQ, Jinen E (1997) Fracture Behaviour and Acoustic Emission in Bending Tests on Single-Fiber Composites. Engineering Fracture Mechanics. 56: 779-796.

[28] Marrec LL, Chekroun M, Villain G, Abraham O, Rakotomanana L (2009) Analytical Study of the Sensitivity of Rayleigh Wave for the Non-Destructive Evaluation of Cover-Concrete. Non-Destructive Testing in Civil Engineering.

[29] Momon S, Moevus M, Godin N, R'Mili M, Reynaud P, Fantozzi G, Fayolle G (2010) Acoustic Emission and Lifetime Prediction During Static Fatigue Tests on Ceramic Matrix-Composite at High Temperature under Air. Composites: Part A. 41: 913-918.

[30] Roca LOY (1997) Acoustic Emission Examination of High Strength Prestressed Concrete Girders. Master of Science in Engineering. The University of Texas, Austin.

Chapter 9

# Combination of Acoustic Methods and the Indentation Technique for the Measurement of Film Properties

Fabio Di Fonzo, Francisco García Ferré, Dario Gastaldi, Pasquale Vena and Marco G. Beghi

Additional information is available at the end of the chapter

http://dx.doi.org/10.5772/56815

## 1. Introduction

New techniques are continuously being developed to produce films and thin films, which have a variety of properties and are exploited by an ever increasing number of technologies. The properties of films typically depend on the preparation process, and can be significantly different from those of the material in bulk form, due to the different micro- and nano-structures obtained by the various production processes. Although new techniques are also introduced to measure the film properties, the characterization of thin layers remains an open issue. A precise knowledge of the mechanical properties is crucial in several cases. Namely, whenever the films have structural functions, as it happens in several micro electro-mechanical systems (MEMS), and when the operation of a device is based on acoustic waves, as it happens in various types of micro-devices, for either sensing or signal processing purposes. But since the mechanical properties depend on the microstructural properties, they are also of more general interest, because they can be a good indicator also for properties which are not strictly mechanical, but for instance depend on the degree of compactness of the film.

A full mechanical characterization includes the determination of both the elastic properties, which characterize the reversible deformations, and the properties which characterize the irreversible behaviour. In most cases the elastic behaviour can be completely characterized by the elastic moduli, or equivalently by the components of the elastic tensor. It is well known that also in the simplest case, the homogeneous isotropic continuum, the elastic stiffness cannot be characterized by a single parameter, but needs two independent parameters. These can be taken as, for instance, the bulk modulus $B$, which characterizes the stiffness with respect to volume changes, and the shear modulus $G$, which characterizes the stiffness with respect to changes of shape, or can be taken (as it happens more frequently) as Young modulus $E$ and Poisson's ratio $\nu$. In the case of anisotropic solids the number of independent parameters increases. The inelastic behaviour is typically more complex, and only some overall parameters, like the yield stress or the hardness, is usually measured.

A wide variety of methods has been developed to perform the mechanical characterization of bulk specimens. Among them, a specific class exploits vibrations of acoustic nature as a probe of the material behaviour. These methods are non destructive, and involve only elastic strains; therefore, they are intrinsically unable to give indications about any inelastic behaviour. On the other hand, due to the complete absence of inelastic strains, the relationship between the raw measurement results and the stiffness parameters can be more straightforward, and less subjected to uncertainties or to spurious effects, possibly allowing better accuracies.

The mechanical characterization of supported films typically requires specific methods. The most widespread technique is indentation, for which a specific standard exists: instrumented indentation, ISO 14577. Indentation induces both elastic and inelastic strains, and supplies significant information about irreversible deformation, namely the hardness value, but the extraction of the information concerning the elastic behaviour is non trivial. The most frequently adopted analysis technique (Oliver & Pharr, 1992; Bhushan & Li, 2003) gives access, once the compliance of the indenter is taken into account, to a single parameter, usually referred to as "indentation modulus". If a reasonable assumption about the value of Poisson's ratio is available, a value of Young modulus can be derived, which obviously depends on the reliability of the adopted assumption. In the case of films, since the nano- and micro-structure can be different from that of bulk samples, a well grounded assumption about the value of Poisson's ratio might be not available. It is also well known that, when supported films are measured, care must be exercised to avoid the influence of the substrate properties.

Methods which exploit vibrations of acoustic character have been developed also for supported films. Acoustic properties depend on stiffness and inertia; therefore, as it happens for bulky samples, acoustic methods require a value of mass density, independently measured. However in acoustic methods the intrinsic absence of inelastic strains makes the derivation of the stiffness parameters less subjected to spurious effects, and less dependent on specific modelling assumptions.

Among the techniques based on acoustic excitations, the so called laser ultrasonics techniques rely on impulsive, therefore broadband, excitation, requiring an analysis of the response in the frequency domain. Excitation is performed by a laser pulse, and the response is measured by piezoelectric or optical means. Quantitative acoustic microscopy relies instead on monochromatic excitation by a piezoelectric transducer, which typically is also exploited as sensor. In the detection of vibrational excitations, substantial advantages are offered by light, a contact-less and inertia-less probe; such advantages are particularly relevant in the measurement of films and small structures. They are exploited by Brillouin spectroscopy, which relies on Brillouin scattering: the inelastic scattering of light by acoustic excitations. Brillouin spectrometry, like Raman spectrometry, is not performed by specifically exciting vibrations, but relies on thermal excitation (anti-Stokes events) and on the excitation produced by the scattering process itself (Stokes events). In both cases excitation has a small amplitude, but it has the broadest band, allowing access to the GHz and multi GHz band, accessible by piezo-electric transducers only with specific transducer structures.

For all these methods, the outcome is the measurement of the propagation velocity of one or more acoustic modes. If sufficient information is gathered, because several acoustic modes are detected and/or their velocities are measured for a sufficiently wide interval of frequencies, a full elastic characterization can be achieved by purely vibrational means, if an independent value of the mass density is available (Comins et al., 2000; Zhang et al., 2001; Djemia et al., 2004; Beghi et al., 2011).

However, a complete elastic characterization by only acoustic means is not always achievable, as it happens when only one acoustic mode is measurable. The results of acoustic methods and of indentation can therefore be combined, with the purpose to obtain a complete elastic characterization, not achievable by each of the techniques alone. This can be particularly useful in the case of new materials or of films of unconventional structures, for which a reliable assumption about the value of Poisson's ratio, needed by indentation, is not available. And even when a full elastic characterization by acoustic techniques is achievable, the combination with indentation offers a useful cross-check among techniques based on completely different principles and also supplies the hardness value, intrinsically out of reach of the acoustic techniques.

This chapter is devoted to this combination of indentation with acoustic techniques, namely quantitative acoustic microscopy (Bamber et al., 2001; Goodman & Derby, 2011) and Brillouin spectroscopy (Garcia Ferré et al, 2013).

## 2. Acoustic modes in elastic solids

Some basic elements of the acoustics of homogeneous and of layered solids are briefly recalled here for sake of completeness. For an elastic continuum undergoing deformation the mechanical state can be represented by the displacement vector field $\mathbf{u}(\mathbf{r},t)$, where $\mathbf{r}=(x_1,x_2,x_3)$ is the position vector and $t$ is time. The local state of the solid being represented by the strain and stress tensors, if the behaviour is linear elastic the stiffness is represented, for any type of anisotropy, by their proportionality: the tensor of the elastic constants $C_{ijmn}$, which can be conveniently cast in the form of the matrix of the elastic constants $C_{ij}$; inertia is represented by the mass density $\varrho$. In a homogeneous linear elastic continuum with no body forces the equations of motion for the displacement vector field are homogeneous, and have the form (Auld, 1990; Every, 2001; Kundu, 2012)

$$\rho \frac{\partial^2 u_i}{\partial t^2} = \sum_{j,m,n} C_{ijmn} \frac{\partial^2 u_m}{\partial x_j \partial x_n}, \quad i = 1,2,3. \tag{1}$$

These equations admit solutions in the form of travelling plane acoustic waves, or modes (Every, 2001; Kundu, 2012):

$$\mathbf{u} = \mathbf{e}\Re\left\{\tilde{A}\exp\left[i\left(\mathbf{k}\cdot\mathbf{r} - \omega t\right)\right]\right\}, \tag{2}$$

where $\mathbf{k}$ is the wavevector, $\omega = 2\pi f$ the circular frequency, $f$ the frequency, $\tilde{A}$ an arbitrary complex amplitude, and $\mathbf{e}$ the polarization vector, which is normalized. The elastic continuum description, underlying Eqs. (1) and (2), is appropriate whenever the wavelength $\lambda = 2\pi / |\mathbf{k}|$ is much larger than the interatomic distances. The three translational degrees of freedom of each infinitesimal volume element correspond, for each wavevector $\mathbf{k}$, to three independent modes, having different polarization vectors $\mathbf{e}_1$, $\mathbf{e}_2$, and $\mathbf{e}_3$, and frequencies. In general the phase velocity $v = \omega / |\mathbf{k}| = \lambda f$ depends on the directions of both $\mathbf{k}$ and $\mathbf{e}$. In an infinite homogeneous medium, waves of the form of Eq. (2) exist for any frequency $f$ compatible with the mentioned lower limit for wavelength.

In the simplest case, the isotropic continuum, which we will mainly consider, the matrix of the elastic constants is fully determined by only two independent quantities; the only non null matrix elements are $C_{11} = C_{22} = C_{33}$, $C_{44} = C_{55} = C_{66}$, $C_{12} = C_{13} = C_{23} = C_{11} - 2C_{44}$. In this case the shear modulus $G$ coincides with $C_{44}$, while Young modulus $E$, Poisson's ratio $\nu$ and bulk modulus $B$ are respectively given by (Every, 2001; Kundu, 2012)

$$E = \frac{C_{44}\left(3C_{12} + 2C_{44}\right)}{C_{12} + C_{44}} = \frac{C_{44}\left(3C_{11} - 4C_{44}\right)}{C_{11} - C_{44}} = \frac{9GB}{3B + G} \tag{3}$$

$$\nu = \frac{C_{12}}{C_{11} + C_{12}} = \frac{C_{11} - 2C_{44}}{2\left(C_{11} - C_{44}\right)} = \frac{E}{2G} - 1 = \frac{3B - 2G}{2\left(3B + G\right)} = \frac{3B - E}{6B} \tag{4}$$

$$G = C_{44} = \frac{E}{2\left(\nu + 1\right)} = \frac{3BE}{9B - E} \tag{5}$$

$$B = \frac{C_{11} + 2C_{12}}{3} = C_{11} - \frac{4}{3}C_{44} = \frac{E}{3\left(1 - 2\nu\right)} = \frac{GE}{3\left(3G - E\right)} \tag{6}$$

In an isotropic continuum the phase velocities do not depend on the direction of $\mathbf{k}$, but only on the relative orientation of $\mathbf{e}$ with respect to $\mathbf{k}$; one of the three modes is longitudinal ($\mathbf{e}_1 \,||\, \mathbf{k}$) and has velocity $v_l$, the other two are transversal ($\mathbf{e}_2 \perp \mathbf{k}$, $\mathbf{e}_3 \perp \mathbf{k}$), are independent ($\mathbf{e}_2 \perp \mathbf{e}_3$) and degenerate: they have the same velocity $v_t$ (Auld, 1990; Kundu, 2012). The two velocities are

$$v_l = \sqrt{\frac{C_{11}}{\rho}} = \sqrt{\frac{E}{\rho}} \sqrt{\frac{(1-\nu)}{(1+\nu)(1-2\nu)}} \qquad \text{(a)}$$

$$\text{and} \quad v_l = \sqrt{\frac{C_{44}}{\rho}} = \sqrt{\frac{E}{\rho}} \sqrt{\frac{1}{2(1+\nu)}} \qquad \text{(b)} \tag{7}$$

In non isotropic continua more than two independent quantities are needed to determine the matrix of the elastic constants, and the phase velocities, beside depending on the direction of **k**, have a more complex dependence on the $C_{ij}$ values. They however remain of the form $v = \sqrt{C/\rho}$, where $C$ is the appropriate combination of elastic constants and, possibly, direction cosines of **k**.

In a homogeneous semi-infinite solid the free surface breaks the translational symmetry in the direction perpendicular to the surface, causing new phenomena: the reflection of bulk waves, and the existence of surface acoustic waves (SAWs). Namely, at a stress free surface, Eqs. (1) admit, for an isotropic continuum, a further solution: the Rayleigh wave, the paradigm of SAW. Such waves have peculiar characters (Farnell & Adler, 1972): a displacement field confined in the neighborhood of the surface, with the amplitude which declines with depth, a wavevector $\mathbf{k}_{\|}$ parallel to the surface, and a velocity lower than that of any bulk wave, such that the surface wave cannot couple to bulk waves, and does not lose its energy irradiating it towards the bulk. Pseudo-surface acoustic waves can also exist, which violate this last condition. The velocity $v_R$ of the Rayleigh wave cannot be given in closed form; a good approximation is (Farnell & Adler, 1972)

$$v_R \cong v_t \frac{0.862 + 1.14\nu}{1+\nu} = \sqrt{\frac{C_{44}}{\rho}} \frac{0.862 + 1.14\nu}{1+\nu} = \sqrt{\frac{E}{\rho}} \sqrt{\frac{1}{2(\nu+1)}} \frac{0.862 + 1.14\nu}{1+\nu} \tag{8}$$

The continuum model of a homogeneous solid does not contain any intrinsic length scale. Accordingly, all the solutions for this model are non dispersive, meaning that the velocities (Eqs. (7) and (8)) are independent from wavelength (or frequency).

More complex modes occur in non homogeneous media. Layered media, like a supported film, are a particularly relevant case, in which new types of acoustic modes can occur (Farnell & Adler, 1972); namely, modes confined around an interface (Stoneley waves) and modes which are essentially guided by one layer (Sezawa waves). All these modes have a wavevector $\mathbf{k}_{\|}$ parallel to the surface. In this case the medium does have an intrinsic length scale, given by the layer thicknesses, and the modes become dispersive: their velocities depend on wavelength, or, more precisely, on the wavelength to thickness ratio(s). When the wavelength is comparable to the layer thicknesses, the displacement fields of the acoustic modes extend over several layers: the modes are modes of the whole structure. Their velocities therefore depend on the properties of at least two adjacent layers, and can be computed only numerically, as non trivial functions of the properties of the substrate and the layer(s), and of the wavelength to thickness ratio. The dispersion relations $\omega\left(\left|\mathbf{k}_{\parallel}\right|\right)$ or $v(f)$ can thus be obtained.

In particular, in the design and analysis of measurement methods for a supported film, two basic cases can be distinguished: 'high' and 'low' thicknesses. Thickness is 'high' or low' in comparison with the acoustic wavelength involved in the measurement; therefore a same thickness can be 'low' in an experiment which exploits relatively low frequencies, and 'high' in an experiment performed at higher frequencies, i.e. smaller wavelengths.

The thickness is 'high' when it is significantly larger than the involved acoustic wavelength. In this case the film behaves acoustically as a semi-infinite medium: it supports acoustic waves of bulk type, both longitudinal and transverse, as if it was infinite, and, at the free surface, supports the Rayleigh surface wave, as if it was semi-infinite. In this case all these acoustic modes are non dispersive, their properties depend on the material properties of the film alone, and in the analysis of experimental results the thickness value is irrelevant. It turns out, experimentally, that this regime is achieved for thicknesses of the order of a few wavelengths, without requiring thicknesses which exceed the wavelengths by orders of magnitude.

Thicknesses are 'low' when they are comparable to or smaller than the acoustic wavelength. The bulk waves cannot be fully developed in the film, the Rayleigh wave becomes a so called 'modified Rayleigh wave', whose displacement field penetrates in the substrate, and Sezawa waves can be supported, whose displacement fields also penetrate in the substrate. Therefore the properties of all these acoustic modes depend on the material properties of the film, but also on the film thickness, whose value becomes a crucial parameter, and on the properties of the substrate. In particular, the velocities of both the modified Rayleigh wave and the Sezawa waves depend on the wavelength to thickness ratio: these modes are dispersive. This must be taken into account in the analysis of experimental results.

A gradual transition from the 'low' to 'high' thickness behaviour occurs at intermediate values. For instance, when the film thickness is very small the Rayleigh wave has a displacement field which mainly extends in the substrate, and a velocity which approaches the Rayleigh wave velocity of the substrate. For increasing thickness the fraction of the acoustic energy which is in the substrate decreases, and the velocity gradually approaches the Rayleigh wave velocity of the film material; the latter is reached when the displacement field amplitude becomes negligible before reaching the substrate.

## 3. Measurement techniques

Some observations are recalled here concerning two methods to measure the acoustic velocities, namely quantitative acoustic microscopy (AM) and Brillouin spectroscopy (BS), and micro-indentation. It should be recalled that, unlike polymers which exhibit a frequency dependent mechanical response, the viscous effects in ceramic-like materials and in metals is generally negligible, such that the high and low frequency properties are essentially indistinguishable. It can also be remembered that the acoustic methods measure the adiabatic stiffness, which does not coincide with the isothermal one, measured in monotonic tests (if strain rate is not too high); however, in elastic solids the difference between adiabatic and isothermal moduli seldom exceeds the measurement uncertainties (Every, 2001). It is also worth remembering that in most metals and ceramics the acoustic velocities are of the order of a few km/s = μm×GHz.

## 3.1. Acoustic microscopy

Acoustic microscopy (Zinin, 2001; Zinin et al., 2012) is one of the techniques which adopt harmonic excitation. It exploits a piezoelectric actuator coupled to sapphire acoustic lens, which operates also as a transducer, in a pulse-echo operation mode. The acoustic lens is mechanically coupled to the sample by a liquid drop, often distilled water. Acoustic microscopy can be operated with imaging purposes; in the quantitative acoustic microscopy version (Zinin, 2001) it allows to measure the acoustic properties of the sample.

Experiments are performed by first adjusting the lens to sample distance $z$ to give the maximum intensity of the reflected signal: at this distance the lens is said to be at focus. The distance is then decreased (the lens is defocused), measuring the intensity of the reflected signal. Since the direct outcome of the transducer is a voltage $V$, the so called $V(z)$ curve is obtained. Such a curve has characteristic oscillations, of period $\Delta z$, arising from the interference between two acoustic paths. One is the direct back reflection at the liquid/sample interface, due to the mismatch of acoustic impedances, the second involves a refraction into the solid sample: acoustic wave are excited in the solid, which propagate irradiating part of their energy towards the coupling liquid, back to the lens. In homogeneous solids, and in "high thickness" layers, the refracted bulk acoustic waves propagate towards the depth of the solid, and the energy irradiated back into the fluid mainly comes from the Rayleigh wave, propagating along the specimen surface. In specimens with thin coatings also other acoustic modes can play a significant role.

When the Rayleigh wave has the dominant role, the measurement of the period $\Delta z$ allows to derive its velocity $v_R$, since

$$\Delta z = \frac{v_0}{2f\left(1-\cos\theta_R\right)} \quad \text{with} \quad \sin\theta_R = \frac{v_0}{v_R} \tag{9}$$

where $v_0$ is the acoustic velocity in the fluid and $f$ is the acoustic frequency. Piezoelectric transducers coupled to acoustic lenses typically operate in the frequency range from tens of MHz up to 1 GHz, resulting in acoustic wavelengths from a few to many micrometres. Therefore only films having a thickness of several micrometres can be in the 'high thickness' regime mentioned above.

## 3.2. Brillouin spectroscopy

Brillouin scattering is the inelastic scattering of light by vibrational excitations of acoustic nature, or by long wavelength acoustic phonons; the scattering mechanisms are discussed elsewhere. In BS measurements a focused laser beam, of angular frequency $\Omega_i$, wavelength $\lambda_0$ and wavevector $\mathbf{q}_i$, impinges on the sample, and the light scattered with wavevector $\mathbf{q}_s$ is collected and spectrally analyzed. The strongest feature in the spectrum is due to the elastically scattered light, at frequency $\Omega_i$, but also Stokes/anti-Stokes doublets can be present, due to inelastic scattering by thermally excited vibrations (Sandercock, 1982; Comins, 2001; Grimsditch, 2001; Every, 2002; Beghi et al., 2012). The wavevectors $\mathbf{q}_i$ and $\mathbf{q}_s$ select the probed wavevectors $\mathbf{k} = \mathbf{q}_s - \mathbf{q}_i$ and $\mathbf{k}_{\|} = (\mathbf{q}_s - \mathbf{q}_i)_{\|}$. Inelastic scattering by a bulk wave (see Eq. (2)) or a SAW of such wavevectors gives one or more doublets, at frequencies $\Omega_s = \Omega_i \pm \omega$. Detection of a doublet thus allows to measure the velocity of an acoustic mode, either $v = \omega / |\mathbf{k}|$ or $v = \omega / |\mathbf{k}_{\|}|$. In the evaluation of $\mathbf{k}$ the effects of refraction must be taken into account, as recalled below.

BS thus measures the velocities of bulk and surface acoustic waves in a fully optical, and therefore contact-less, way. Obviously, bulk waves are detectable only in sufficiently transparent media. The probed acoustic wavelengths $\lambda = 2\pi / |\mathbf{k}|$ and $\lambda = 2\pi / |\mathbf{k}_{\|}|$ are determined by the optical wavelength and the scattering geometry: with visible light they are typically sub-micrometric, resulting in acoustic frequencies in the GHz to tens of GHz range. Wavelengths in this range are three orders of magnitude larger than interatomic distances, making the continuum description (Eq. (1)) fully appropriate.

Although BS can be performed in various geometries (Grimsditch, 2001; Beghi et al., 2011; 2012), the most frequent one is backscattering ($\mathbf{q}_s = -\mathbf{q}_i$), which maximizes $|\mathbf{k}|$. This geometry selects, for scattering by bulk waves in isotropic media, the specific probed wavevector $\mathbf{k} = \pm 2\mathbf{q}_i n$, where $n$ is the refractive index, while, for scattering by SAWs, it can explore a range of wavevectors of modulus $|\mathbf{k}_{\|}| = \pm 2|\mathbf{q}_i|\sin\theta$ which depend on the incidence angle $\theta$ (the angle between the incident beam and the surface normal) but do not depend on $n$. Therefore the analysis of experimental results for bulk scattering requires an independent value of the refractive index. Since the probed acoustic wavelengths are typically sub-micrometric, films having a thickness of a couple of micrometres already are in the 'high thickness' regime mentioned above.

### 3.3. Nanoindentation

During nano-indentation tests, an indenter is pressed onto the surface of the material that has to be characterized. The tip of the indenter, that can have different shapes, is made of a hard material, usually diamond. During the test, the load $P$ and the tip penetration $h$ are continuously recorded during a loading-unloading cycle, building up the *P-h* curve, whose analysis allows the computation of indentation modulus and indentation hardness, attributable to the elastic and post-elastic response of the material, respectively.

The key parameter when analyzing indentation is the reduced modulus $E_r$. Generally $E_r$ depends on the elastic properties of the material as well as on the shape of the indenter and, therefore, the shape of the contact area. With the assumption of axisymmetric indenters and homogeneous, elastic, isotropic materials, $E_r$ can be computed with the simple relation:

$$E_r = \frac{E}{1-\nu^2} \tag{10}$$

Where $E$ and $\nu$ are Young modulus and Poisson's ratio of the material, respectively.

From the theory of contact mechanics between an isotropic elastic half space and an elastic sphere (Hertz, 1881), the relationship between the contact force $P$ and the relative displacement $h$ between the bodies dependes on the elastic properties of both bodies, through one single parameter E* according to the relationship:

$$P = \frac{4}{3} E^* \sqrt{R} h^{3/2} \tag{11}$$

in which $R$ is the indenter radius and $E^*$, also referred to as indentation modulus, is such that:

$$\frac{1}{E^*} = \frac{1}{E_r} + \frac{1-\nu_i^2}{E_i} \tag{12}$$

where $E_i$ and $\nu_i$ are the Young modulus and the Poisson's ratio of the indenter. In case of diamond, it is $E_i = 1141\ \text{GPa}$ and $\nu_i = 0.07$ .

The above solution, holding for a spherical indenter, has been extended to conical indenters by Sneddon (Sneddon, 1965). By using simple geometrical arguments on the geometry of the tip, he obtained the following force-displacement relationship:

$$P = \frac{\pi a}{2} E^* a \cot(\alpha) = \frac{2}{\pi} E^* \tan(\alpha) h^2 \tag{13}$$

in which $\alpha$ is the opening half angle of the conical tip and $a$ is the contact radius.

A simple relationship between the elastic parameter $E^*$ and the the unloading contact stiffness $S=dP/dh$ is:

$$E^* = \frac{S}{2}\sqrt{\frac{\pi}{A}} \tag{14}$$

in which $A$ is the projected contact area having radius $a$. The above solution holds for an elastic contact between a perfect conical surface and the elastic isotropic semi-infinite solid.

Oliver and Pharr (Oliver and Pharr 1992) introduced a generalized method to determine the reduced modulus $E_r$ for pyramidal indenters and non linear materials with irreversible response. When elastic-plastic response is obtained, the unloading branch of the *P-h* curve will be lower with respect to the loading branch, and will exhibit non zero displacement at zero load, denoting a residual inprint on the material surface. The method of Oliver and

Pharr is based on the basic hypothesis that the initial portion of the unloading *P-h* curve is essentialy driven by elastic response of the material. Thus the contact stiffness *S* calculated at the top portion of the unloading curve can be used in a generalized version of equation (14) to estimate the indentation modulus:

$$E^* = \frac{S}{2\beta}\sqrt{\frac{\pi}{A}} \tag{15}$$

in wich an adjusting empirical parameter $\beta$ has been introduced with the purpose to account for specific pyramidal shapes. For the three sided pyramid (Berkovich tip) $\beta$= 1.034 is usually assumed.

The contact area *A* is the projected area of the material in contact with the indenter surface and it is a function of $h_c$ which is the height of the contact surface in the vertical direction (the indentation direction). The function $A(h_c)$ is generally calibrated for low depth indentations by running experiments on reference materials (typically fused silica); calibrated functions $A(h_c)$ accounts for the blunt geometry of real used tips and therefore it must be re-calibrated periodically as the number of idents made with the tip increases.

The Oliver and Pharr method provides a simple relationship for the experimental determination of the contact depth $h_c$

$$h_c = h_{max} - \frac{2(\pi - 2)}{\pi}\frac{P_{\max}}{S} \tag{16}$$

which is used to calculate the contact area and the reduced modulus $E_r$ using (15) and (12).

The ability of the nanoindentation technique to probe small amount of material by shallow penetration depths enables one to probe the mechanical properties of thin coatings without removing the film from the substrate.

The main difficulty in nanoindentation of thin films is to ensure that the properties of the substrate are not affecting the characterization of the coatings. To achieve this aim, a simple rule to restrict the maximum penetration depth to no more than 1/10 of the coating thickness is usually applied.

However, for measurement of the elastic properties, the influence of the substrate compliance is unavoidable. Despite this intrinsic difficulties, methods to determine film thickness out of nanoindentation on film/substrate systems are available. As an example, Doener and Nix (Doener and Nix 1986) determined a relationship between an effective reduced modulus (which is strictly related to $E^*$) and the properties of film, substrate and indenter, by using finite element simulations (Fischer-Cripps, 2011):

$$\frac{1}{E_{eff}} = \frac{1-\nu_f^2}{E_f}\left(1-e^{-\alpha t/\sqrt{A}}\right) + \frac{1-\nu_s^2}{E_s}\left(e^{-\alpha t/\sqrt{A}}\right) + \frac{1-\nu_i^2}{E_i} \tag{17}$$

in which the subscripts $f$ and $s$ denote film and substrate, respectively. In (17) $\alpha$ is a parameter which governs the decay of the effect of the substrate for increasing penetration depths of indentations and $t/\sqrt{A}$ is the geometric scaling parameter. For ideal conical indentation, $A$ is proportional to $h^2$, so the scaling parameter is proportional to $t/h$.

For deep indentations, i.e. for small $t/h$, equation (17) provides the properties of the substrate; whereas, for shallow indentation (i.e. at the limit $t/h \to 0$ )equation (17) provides the properties of the film.

## 4. Data analysis

Data analysis is first discussed for the case of an isotropic semi-infinite medium. This analysis applies to the 'high thickness' films as well, either supported or free standing; with AM this regime can be achieved only by thicknesses of many micrometres, while with BS it is already achieved by thicknesses of a couple of micrometres.

The assumption of a semi-infinite medium, or of a film of 'high thickness', implies that the velocities of all the acoustic modes do not depend on the wavevector (Farnell & Adler, 1972). The crude outcome of measurements is a set of $N_E$ values for the reduced modulus $\{E_{r,m}\}$ $m = 1,2, \ldots ,N_E$, and a set of $N_R$ values for the Rayleigh velocity $\{v_{R,i}\}$, $i = 1,2, \ldots ,N_R$; if the bulk acoustic wave is measurable, a third set of $N_l$ values for the bulk longitudinal velocity $\{v_{l,j}\}$, $j = 1,2, \ldots ,N_l$, is also available. For each value an uncertainty can be evaluated, like $\sigma_{v_{R,i}}$ for $v_{R,i}$; these uncertainties can be statistically evaluated *a posteriori*, and therefore be the same for all the values belonging to the same set, or can be evaluated for each single value, if a sufficiently detailed analysis of the measurement process is available. With AM only the $v_R$ values are accessible, the bulk waves being not measurable; with BS the bulk waves are measurable only for sufficiently transparent media (silicon, for example), and in this case also the bulk transversal wave can be measurable, beside the bulk longitudinal one, giving a fourth set of $N_t$ values for the bulk transversal velocity $\{v_{t,j}\}$. This does not always happen, because the spectral doublet due to scattering by the transversal wave is typically weaker than that due to the longitudinal wave. From the set of $N_E$ values for the reduced modulus $\{E_{r,m}\}$ an average value $\overline{E}_r$ can be derived, with its uncertainty $\sigma_{\overline{E}}$, and the same can be said for $\overline{v}_R$, $\overline{v}_l$ and $\overline{v}_t$, when they are measurable.

The assumption of an isotropic medium implies that the stiffness is fully identified by two independent parameters: therefore it is intrinsically a two-dimensional quantity, which can be represented by a point in a two dimensional 'stiffness space'. The 'stiffness space', in turn, can be represented by the $(E,\nu)$ couple, i.e. the $(E,\nu)$ plane, or, equivalently, by other planes, like the $(B,G)$ plane or the $(C_{11}, C_{44})$ plane, or the plane defined by the two Lamé constants. Any of these planes can be mapped into any other (see Eqs. (3)-(6)); the representation by the $(E,\nu)$ plane is first considered here. For any point of this plane the value $E_r(E,\nu)$ is immediately computed, and, if the mass density is known, the value $v_R(E,\nu)$ is also computed, as well as $v_l(E,\nu)$ and $v_t(E,\nu)$.

### 4.1. Graphical method

The condition $E_r(E,\nu)=\overline{E}_r$ identifies a line in the 'stiffness space', and the condition $\overline{E}_r-\sigma_{\overline{E}}\le E_r(E,\nu)\le \overline{E}_r+\sigma_{\overline{E}}$ identifies a confidence band, or a 'stripe'. The same happens for $\overline{v}_R$, $\overline{v}_l$ and $\overline{v}_t$, when available. A first analysis procedure is graphical, and basically consists of drawing the available curves, or bands, in the stiffness space. If two of them are available, typically $\overline{E}_r$ and $\overline{v}_R$ from AM (Bamber et al., 2001; Goodman & Derby, 2011), their intersection fully identifies the stiffness, and allows a semi-quantitative estimate of its uncertainty. If three or more lines are available, as it can happen with BS from sufficiently transparent samples (Beghi et al., 2011; Garcia Ferré et al, 2013), all their intersections typically do not coincide exactly, but the amplitude of the region which contains them allows a qualitative estimation of the consistency of the different measurement techniques. In particular, a confidence band which lies beside the intersection of the others is a hint to a possible systematic error in one of the measurement techniques.

Validations of the data analysis procedure have been performed exploiting fused silica samples supplied for calibration of the indentation equipment (Bamber et al., 2001; Garcia Ferré et al, 2013). In the derivation of stiffness values from the measured acoustic velocity the widely accepted value of 2200 kg/m³ was adopted for the mass density of fused silica. Fig. 1 shows a recent result obtained by BS, which allows to measure also the doublets due

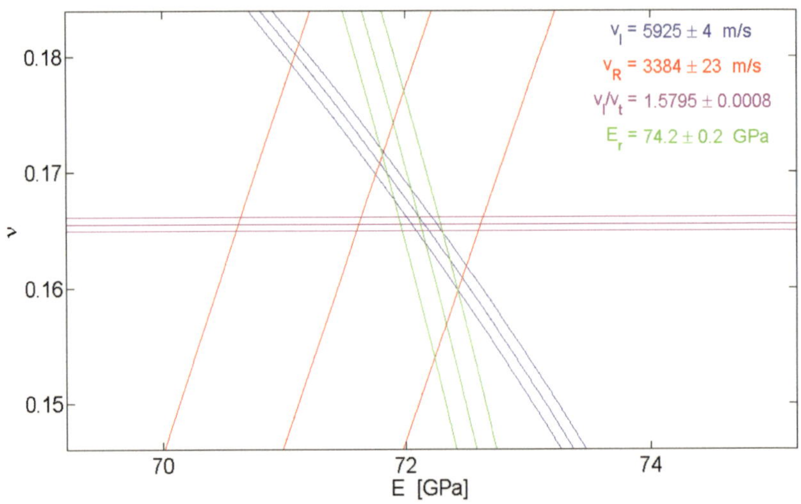


**Figure 1.** (adapted from Garcia Ferré et al. 2013). Representation in the $(E,\nu)$ plane of the measurements on a bulk silica platelet of nominal properties $E=72$ GPa and $\nu=0.17$. Results from indentation (reduced modulus $E_r$, green lines) and from BS (bulk longitudinal wave velocity $v_l$, blue lines; Rayleigh wave velocity $v_R$, red lines; ratio of bulk longitudinal to bulk transversal wave velocities $v_l/v_t$, violet lines). For each quantity: central values and uncertainty bands. The conversion from raw spectral data to $v_l$ is performed exploiting the measured value of refractive index $n_{si}=1{,}464$; the conversion from velocities $v_l$ and $v_R$ to the elastic moduli is performed by the widely accepted value of 2200 kg/m³. The ratio $v_l/v_t$ is directly obtained from raw spectral data, and is a function of ν which can be plotted without needing any of the above data.

to bulk waves (Garcia Ferré et al, 2013), In the derivation of the corresponding velocity the value of the refractive index $n$ is needed. The latter was measured by an ellipsometer, obtaining $n_{si} = 1{,}464 \pm 0.003$. An analogous outcome was obtained by AM (Bamber et al., 2001), measuring only the Rayleigh velocity but not needing the value of the refractive index.

## 4.2. Least squares estimation method

A more detailed statistical analysis can be performed by a least squares minimization procedure. Following standard non linear estimation theory (Seber & Wild, 2003), a least square estimator $S(E, \nu)$ is built as the weighted sum over each single result:

$$S(E,\nu)=\sum_m\left(\frac{E_r(E,\nu)-E_{r,m}}{\sigma_{E_{r,m}}}\right)^2+\sum_i\left(\frac{v_R(E,\nu)-v_{R,i}}{\sigma_{v_{R,i}}}\right)^2+\sum_j\left(\frac{v_l(E,\nu)-v_{l,j}}{\sigma_{v_{l,j}}}\right)^2, \tag{18}$$

the weights being determined by the uncertainties. Obviously, the sum over the $v_l$ values is present only if the measurement of the bulk longitudinal wave is available. If the bulk transversal wave is also measurable, a further summation over the $v_t$ values is also present in Eq. (18). The minimum $S_{\min}$ of $S(E,\nu)$ identifies the most probable value of the $(E,\nu)$ couple, and the isolevel curves of the normalized estimator

$$S'(E,\nu)=\frac{S(E,\nu)-S_{\min}}{S_{\min}} \tag{19}$$

identify the confidence region at any predetermined confidence level (Lefeuvre et al., 1999; Beghi et al., 2001, 2011, 2013). The value of $S'(E, \nu)$ that identifies a confidence region is determined by the confidence level (typically the 68%, 90% and 95% confidence levels are considered) and by the available number of measurements (velocities of acoustic modes and indentation data) (Seber & Wild, 2003).

The estimation of stiffness is performed computing the normalized estimator $S'(E, \nu)$ at the nodes of a discrete mesh, which is refined until discretization effects become negligible, and then identifying the minimum of $S'(E,\nu)$, as well as the confidence regions. It is often found that the precise position of the minimum, which by definition falls is in region in which the gradient of the estimator is small, is relatively more sensitive to numerical noise. The confidence regions typically have a regular shape; their boundaries, which fall in regions in which the gradient of the estimator is significant, turn out to be a more robust result.

The robustness of this outcome was exploited to define the following procedure to pick the final results of the evaluation: for each parameter, the minimum and the maximum of the values falling within the confidence region are taken. Their average value is taken as the estimate of the parameter, and the semi-amplitude of the interval is taken as the estimate of its uncertainty. Graphically, this means taking, instead of the point at which the estimator is at its minimum and of the precise shape of the confidence region, the rectangle, with sides parallel to the axes, circumscribed to the confidence region, and its centre (see Fig. 2a). This gives a simple and clearly defined algorithm to identify the measured value and its uncertainty, for each of the parameters, namely for $E$ and $\nu$ in the representation considered until here. However, it must be remembered that these values are only the one-dimensional projections of a quantity, the stiffness, which is intrinsically two-dimensional. A statistically more rigorous procedure to take into account its two-dimensional nature would require considering it as a bivariate normal distribution, and implementing an algorithm to estimate its whole covariance matrix, which includes the expectation values for each parameter and all the correlations. This implementation is left to future developments.

The two-dimensional nature of stiffness must however be taken into account when seeking the values of the other parameters (such as $G$, $B$ …). Once $E$ and $\nu$ are identified with their uncertainties $\sigma_E$ and $\sigma_\nu$, as indicated above, the functional dependences on $E$ and $\nu$, like Eqs. (5) and (6) can be exploited to derive the values of other parameters, like $G$, $B$, $C_{11}$ and $C_{44}$. However, care must be exercised in the evaluation of their uncertainties. The standard form of the error propagation formula holds for parameters which are uncorrelated. Making this assumption to estimate the uncertainties of, e.g., $G$ and $B$, from $\sigma_E$ and $\sigma_\nu$, would treat $E$ and $\nu$ as uncorrelated, disregarding the information about their correlation which is intrinsically given by the shape of the confidence regions. This might lead to a substantial overestimation of the uncertainties of all the other parameters.

The most rigorous way to estimate the values and the uncertainties of all the parameters would require, as mentioned above, the estimation of the whole covariance matrix. A simpler procedure, which however fully takes into account the two-dimensional nature of stiffness, replicates the 'circumscribed rectangle' algorithm indicated above. Considering for instance $G$ and $B$, it can be implemented in two different, but equivalent, ways. The first one substitutes the functional dependences of $E$ and $\nu$ on $G$ and $B$ (Eqs. (5) and (6)) into the definition of the estimator $S(E,\nu)$ (Eq. (18)) obtaining its expression as $S(B,G)$, and then repeats the whole procedure outlined above: computation of the normalized estimator at the nodes of a rectangular mesh in the $(B,G)$ plane, determination of the confidence regions in this plane, identification of the parameters and of their uncertainties by the 'circumscribed rectangle' algorithm. In a completely equivalent way, once the confidence regions are determined in the $(E,\nu)$ plane, their frontiers can be mapped into the $(B,G)$ plane by the functional dependences of $G$ and $B$ on $E$ and $\nu$, and the parameters and their uncertainties are then identified by the same algorithm. The two implementations only differ by the discretization errors, because a mesh which is rectangular and regular in the $(E,\nu)$ plane is neither rectangular nor regular in the $(B,G)$ plane, and vice versa.

### 4.3. Anisotropic and non homogeneous cases

Until now, the data analysis procedure was discussed under the assumptions of isotropic and homogeneous medium, remembering that 'homogeneous' also includes layers in the 'high thickness' regime, in which the presence and the properties of the substrate are essentially irrelevant. The extension of the procedure beyond these assumptions is briefly indicated here.

In the case of an anisotropic, but still homogeneous, medium, the number of independent parameters needed to fully identify the tensor of the elastic constants is larger than two, meaning that stiffness is not a two-dimensional quantity, but a multi-dimensional one. Considering here for example the cubic symmetry, the independent parameters are three, which can be taken as $C_{11}$, $C_{12}$ and $C_{44}$; for the hexagonal symmetry the number of independent parameters is five.

From the acoustic point of view, in a homogeneous anisotropic medium the acoustic waves can be not exactly longitudinal and transverse, and, more importantly, their velocity depends on the direction of the wavevector **k**, although not on its modulus. Therefore, any acoustic velocity, like e.g. $v_R$, is a function of not only $E$ and $\nu$, but of more variables: $v_R = v_R\left(C_{11}, C_{12}, C_{44}, \mathbf{k}/|\mathbf{k}|\right)$. In the same way, each experimental value, like $v_{R,i}$ in Eq.(18), must be associated to the propagation direction $\left(\mathbf{k}/|\mathbf{k}|\right)_i$ along which it was measured. From the indentation point of view, anisotropy means that each measurement depends on the direction of the normal to the surface on which indentation is performed, and, if the indenter is not axisymmetric, it might also depend on the angular position of the indenter around its axis.

It is thus evident that the anisotropic case requires a data analysis which is substantially more complex, and, perhaps even more importantly, it requires measurements performed on surfaces of various orientations. This typically requires crystals with different cuts, which, in the case of films, might be not achievable. In the case of supported films, and remembering that we are now dealing with the 'high thickness' regime, it seems that the only type of anisotropy which can be characterized in detail is the transverse isotropy, meaning a film which is isotropic in its plane and has different properties in the perpendicular direction.

Remaining instead in the isotropic case, but relaxing the assumption of homogeneous medium, a couple of cases seem to be tractable in some detail. The first one is that of a homogeneous substrate with a homogeneous supported film in the 'low thickness' regime, in which the substrate becomes relevant. In this case 'non homogeneous' means a sharp discontinuity between two homogeneous media. The second case is that of a mild gradient of properties, distributed over a superficial layer.

In a homogeneous supported film, beside the Rayleigh wave of velocity $v_R$, other surface waves can exist (Sezawa waves). For all of them, for given substrate properties $\left(E,\nu\right)_{subs}$ and film thickness $h$, the velocity depends on the film properties and on the modulus $k_{\|} = |\mathbf{k}_{\|}|$ of the wavevector: $v_R = v_R\left(E, \nu, k_{\|} \, h, \left(E,\nu\right)_{subs}\right)$. By the same token, each measured value, like $v_{R,i}$ in Eq.(18), must be associated to the modulus $k_{\|i}$ at which it was measured, obtaining the dispersion relations $v\left(k_{\|}\right)$. By AM the dispersion relation is obtained by measurements at different frequencies, while with BS it is obtained by measurements at different incidence angles.

From the indentation point of view, the critical point for a supported film in the 'low thickness' regime is the avoidance of the influence of the substrate. With BS the 'low thickness' regime occurs at thicknesses of the order of one micrometre or less, for which the avoidance of substrate influences is a crucial point, but it can be dealt with using sharp tips with accurate diamond area function calibrations (Barone et al. 2010) and shallow penetration depths. With AM, as discussed above, the 'low thickness' regime can occur at thicknesses of several micrometres, for which indentation can be easily performed avoiding the substrate influences. Therefore, for films of this thickness, the characterization by AM can be substantially improved by the combination with indentation.

The second case, that of a mild gradient of properties distributed over a finite depth, can be treated in an approximate way considering the average properties over the explored depth. In this case it is crucial that the indentation depth be comparable to the exploited acoustic wavelength(s) (Goodman & Derby, 2011). Although the two techniques do not average in the same way the properties at different depths, the above condition ensures that averaging is performed over the same depth interval. This procedure approximates the actual superficial gradient by a homogeneous equivalent film of average properties, over a perfectly homogeneous substrate of slightly different properties.

## 5. Experimental results

The combination of AM and indentation, after successful validation by a fused silica sample, was exploited to characterize TiN/NbN multilayers deposited by closed field magnetron sputtering (Bamber et al., 2001), assuming for the mass density of the multilayer the same value of TiN, 5210 kg/m$^3$. Since the thicknesses of the single layers are much smaller than the acoustic wavelengths at which they are probed, the whole multilayer coating acoustically behaves as an equivalent homogeneous medium. In this case, the same authors state that the results were not conclusive, and they present a detailed analysis, focusing on the interpretation of the indentation results: they show that a main source of uncertainty lies in the analysis of indentation results from a multilayer.

Other two possible sources of uncertainty can be considered. Firstly, since the thickness of the whole multilayer is smaller than the acoustic wavelength, the layer is not in the 'high thickness regime', meaning that also for the acoustic behaviour the influence of the substrate is not negligible, and that the speed of the Rayleigh wave depends on frequency. The authors mention that by performing AM measurements at various frequencies the dispersion relation can be measured, but they do not give details on how they exploited this possibility. Secondly, by its same geometry a multilayer is not isotropic. It has in-plane isotropy, but different properties in the normal direction. This type of anisotropy corresponds to hexagonal symmetry (transverse isotropy), which requires five independent parameters to fully specify the elastic tensor. An isotropic model is the best approximation currently available with the experimental information, but it obviously introduces approximations.

A similar investigation (Goodman & Derby, 2011) was aimed at comparing the properties of the air side and the tin side of float glass, which is produced by floating molten glass on molten tin. The mass density values were taken from previous studies on the properties of soda-lime glass, and on the influence of the tin content. In this case, despite the need of considering a possible gradient of properties in the superficial layer, due to the occurrence of microdefects and to diffusion of tin, higher precision results were obtained, allowing to detect a measurable difference between the two sides. Due to the possible gradient of properties close to the surface, it was ensured that the depths probed by indentation and by AM be strictly comparable.

In a recent investigation (Garcia Ferré et al, 2013), the combination of indentation and BS was exploited to achieve a complete mechanical characterization of $Al_2O_3$ coatings deposited by pulsed laser deposition (PLD). Coatings were deposited on silicon wafer and on stainless steel substrates, ablating a 99.99% pure polycrystalline alumina target, and obtaining ultra-smooth samples (RMS <0,1 nm). SEM observations show a uniform, compact and fully dense microstructure; Raman spectroscopy and X-ray diffraction supply evidence of the amorphous structure of the coatings. Due to the absence of a crystalline structure and of a structure at mesoscopic level, like e.g. a columnar structure, the coating properties are assumed to be isotropic. The thickness of the measured coatings were between 4 and 8 μm. This means that by BS the layer is acoustically in the 'high thickness' regime, and that indentation can be performed with penetration depths, ranging from 300 to 500 nm, below one tenth of the layer thickness, thus avoiding the effect of the substrate.

Nanoindentation measurements were carried out by a nanoindenter (Micro-Materials, Ltd., Wrexham, UK) equipped with a diamond Berkovich tip. The reduced Young's modulus and hardness were assessed from the indentation curves following the Oliver and Pharr approach discussed above. BS measurements were performed in the backscattering geometry, with incidence angles from 30° to 70°, using an Ar+ laser at the wavelength of 514,5 nm, and analyzing the scattered light by a tandem multipass Fabry-Perot interferometer of the Sandercock type.

Since the coatings are transparent, BS can measure bulk acoustic modes, but only the longitudinal one could be reliably measured. In the analysis of the corresponding data the value of the refractive index is needed: ellipsometric measurements supplied $n_a = 1.647 \pm 0.003$. The analysis of acoustic data also requires the value of the mass density. For $Al_2O_3$ data are available data for α–alumina, i.e. sapphire: mass density $\rho_s$ between 3970 and 3980 kg/m³, with index $n_s$, at a wavelength of 514,5 nm, between 1,770 and 1,773 (Malitson, 1962; Defranzo & Pazol, 1993). The significant difference between $n_a$ and $n_s$, for the same nominal composition, indicates an appreciably different structure, such that the above value of $\rho_s$ cannot be considered a sufficiently accurate estimate for the amorphous alumina: the coating mass density $\rho_a$ was therefore derived from $\rho_s$, $n_s$ and $n_a$ by the Clausius-Mossotti formula, also called Lorentz-Lorenz formula when written in terms of the refractive indexes (Beghi et al., 2011)

$$\rho_a = \rho_s \frac{n_a^2 - 1}{n_a^2 + 2} \frac{n_s^2 + 2}{n_s^2 - 1} \tag{20}$$

obtaining $\rho_a = 3471 \pm 17$ kg/m$^3$.

This combination of ellipsometry, BS and nanoindentation was called 'EBN procedure' (Garcia Ferré et al., 2013). Its validation was achieved by a bulk silica sample supplied by the producer of the indenter for calibration purposes, of nominal properties $E = 72$ GPa and $\nu = 0{,}17$ (see Fig. 1). On this sample BS measurements could detect, beside the Rayleigh wave, both the bulk longitudinal and the bulk transversal waves. When both types of bulk waves are measurable in the same spectrum, the ratio of their velocities $v_l/v_t$ can be directly obtained from the raw spectral data, irrespective of their calibration and of the values of the mass density and refractive index, and without being affected by possible imperfections of the scattering geometry. The ratio of velocities is thus measured with better precision than the single velocities.

For the silica sample the results from BS and from indentation are shown in Fig.1: the confidence bands resulting from the various measurements all intersect in a small region, showing a high degree of consistency among measurement techniques of completely different nature.

For the alumina coatings indentation measurements gave for the reduced modulus $E_r = 212.3 \pm 5.4$ GPa, and for hardness $H = 10{,}32 \pm 1{,}03$ GPa. BS measurements could reliably measure the Rayleigh wave and the bulk longitudinal waves, of velocities $v_R = 4328 \pm 42$ m/s and $v_l = 8655 \pm 36$ m/s. The conversion from raw spectral data to $v_l$ is performed exploiting the measured value of refractive index $n_a = 1.647$; the conversion from velocities $v_l$ and $v_R$ to the elastic moduli is performed by the value of $\rho_a = 3471$ kg/m$^3$ obtained, as indicated above, by the Lorentz-Lorenz formula. The procedures outlined above are performed in the $(E,\nu)$ plane, as well as in the $(B,G)$ and the $(C_{11},C_{44})$ planes, as shown in Fig. 2. The stiffness is thus characterized. The results obtained on the basis of the 95% confidence regions are summarized in Table 1. A detailed analysis of the uncertainties of statistical and of systematic nature was performed (Garcia Ferré et al., 2013). Basically, the errors in the measurement of the reduced modulus and the spectral frequencies are of statistical nature, and lead to a broadening of the confidence regions. An uncertainty in the value of mass density or of the refractive index acts instead like a calibration error, leading to a shift of the confidence regions, without appreciable broadening.

The results of Table 1 show that the stiffness of these PLD deposited alumina coating is significantly different from that of most ceramic coatings: namely, Young modulus is lower and Poisson's ratio is higher, resulting in a stiffness which is close to that of steel. This finding could be rationalized in the light of TEM analysis, which shows that the coatings consist of a homogeneous dispersion of ultra-fine (2-5 nm) $Al_2O_3$ nanoparticles in an amorphous alumina matrix (Garcia Ferré et al., 2013). This metal-like stiffness is interesting, because at the substrate/coating interface a stress concentration occurs, due to the discontinuity of mechanical properties. The stress concentration can promote delamination, therefore a smaller stiffness discontinuity, reducing the stress concentration, can favour the durability of adhesion.

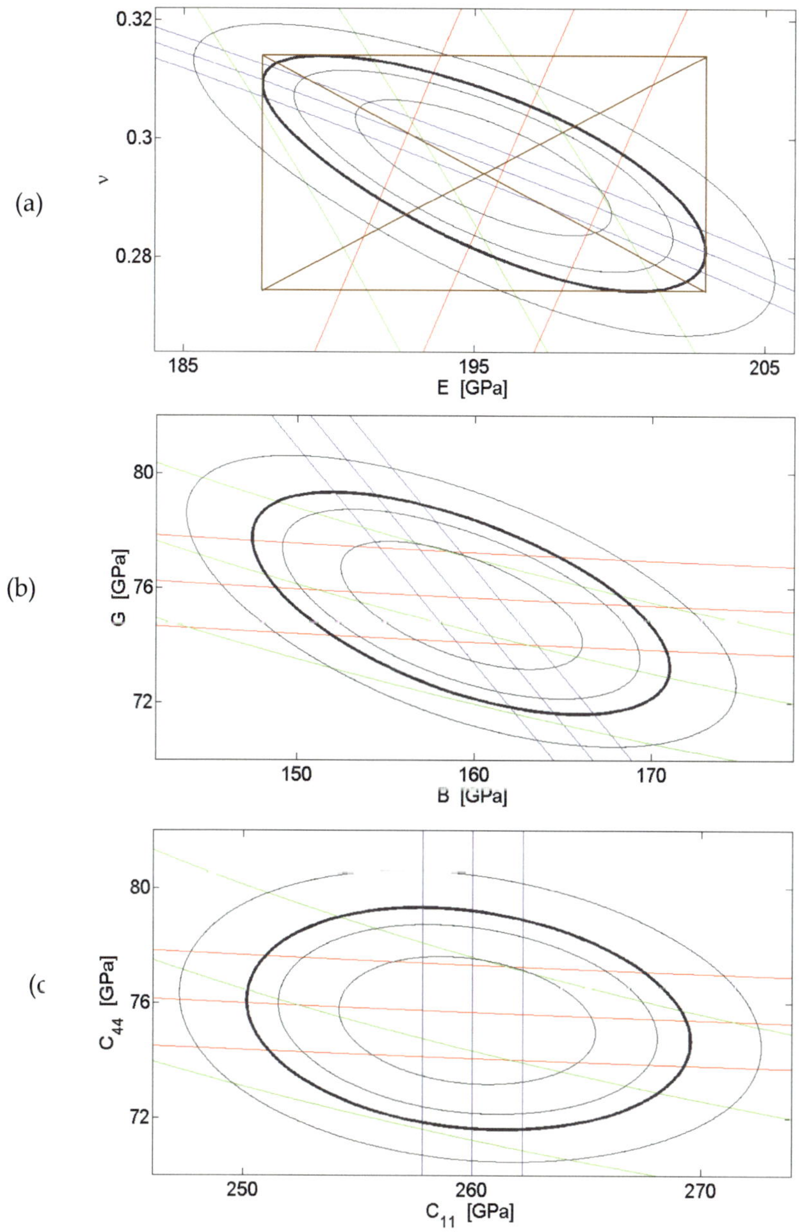


**Figure 2.** (2a is adapted from Garcia Ferré et al. 2013). Stiffness of the alumina coating, from results of indentation (reduced modulus $E_r$, green lines) and BS (bulk longitudinal wave velocity $v_l$, blue lines; Rayleigh wave velocity $v_R$, red lines). For each quantity: central values and uncertainty bands. The same stiffness data are represented in the $(E,\nu)$ plane (Fig. 2a), the $(B,G)$ plane (Fig. 2b) and the $(C_{11}, C_{44})$ plane (Fig. 2c). Thin black lines: the 68%, 90% and 99% confidence regions; thick black line: the 95% confidence region. In Fig. 2a the 'circumscribed rectangle algorithm' (see text) is also shown.

| | |
|---|---|
| $E$ (GPa) | $195.3 \pm 7.6$ |
| $\nu$ | $0.294 \pm 0.020$ |
| $G = C_{44}$ (GPa) | $75.5 \pm 3.8$ |
| $B$ (GPa) | $159.2 \pm 11.8$ |
| $C_{11}$ (GPa) | $259.8 \pm 9.7$ |

**Table 1.** (adapted from Garcia Ferré et al. 2013). Stiffness parameters for an alumina coating, obtained by the combination of ellipsometry, BS and indentation (see Fig. 2).

## 6. Conclusions

Nanoindentation is among the most widespread techniques to perform the mechanical characterization of materials. It is applicable to bulk materials, and it is the standard characterization route for supported films whose thickness is not too low. Beside the value of hardness, instrumented indentation can supply a measurement of stiffness, by the careful analysis of a deformation process which involves both elastic and inelastic strains. However, stiffness is a two-dimensional quantity for isotropic materials, and has a higher number of dimensions in the anisotropic cases. Indentation supplies a one-dimensional estimate, and can achieve a full estimate only by some kind of assumption, typically about the value of Poisson's ratio. The techniques which exploit propagating acoustic waves involve exclusively elastic strains: they therefore offer the most direct access to the elastic properties, and potentially their most accurate measurements. However, if only one acoustic mode can be detected, as it typically happens with AM and with BS of metallic samples, also acoustic techniques only supply a one-dimensional estimate. The combination of indentation and an acoustic method then allows to obtain the full two-dimensional measurement of stiffness, avoiding any assumption. With sufficiently transparent samples BS can detect further acoustic modes, supplying additional information and allowing to reduce the uncertainties. The consistency of results obtained by the two completely different methods could be checked by reference samples. The combined procedure, which is able to achieve the complete elastic characterization without needing assumptions, is particularly useful in the case of materials with novel properties, either for the novelty of the material or for the novelty of the production process. In these cases any assumption is not sufficiently grounded, and might lead to misleading results; the combined procedure then achieves a characterization of stiffness whose accuracy and reliability would not be reachable by each of the techniques alone. Furthermore, the presence of indentation simultaneously provides the measurement of hardness, intrinsically out of the reach of acoustic methods.

## Author details

Marco G. Beghi
*Politecnico di Milano, Energy Department and NEMAS Center, Milano, Italy*

Fabio Di Fonzo
*Center for Nano Science and Technology @PoliMi, Istituto Italiano di Tecnologia, Milano, Italy*

Francisco García Ferré
*Politecnico di Milano, Energy Department and NEMAS Center, Milano, Italy*
*Center for Nano Science and Technology @PoliMi, Istituto Italiano di Tecnologia, Milano, Italy*

Dario Gastaldi and Pasquale Vena
*Politecnico di Milano, Department of Chemistry, Materials and Chemical Engineering – Giulio Natta, Milano, Italy*
*IRCCS Istituto Ortopedico Galeazzi, Milano, Italy*

## 7. References

Auld, B. A. (1990). *Acoustic fields and Waves in Solids,* Robert E. Krieger Publishing Company, Malabar, Florida

Bamber, M.J., Cooke, K.E., Mann, A.B. & Derby B. (2001). Accurate determination of Young's modulus and Poisson's ratio of thin films by a combination of acoustic microscopy and nanoindentation, *Thin Solid Films,* Vol. 398-399, pp. 299-305.

Barone, A.C., Salerno, M., Patra, N., Gastaldi, D., Bertarelli, E., Carnelli, D., Vena, P., (2010). Calibration issues for nanoindentation experiments: direct atomic force microscopy measurements and indirect methods. Micros. Res. Tech. Vol. 73, pp. 996-1004.

Beghi, M. G., Bottani, C. E. & Pastorelli, R. (2001). High accuracy measurement of elastic constants of thin films by surface Brillouin scattering. In *Mechanical properties of structural films,* C. Muhlstein, C. & Brown, S.B. (Eds.), pp. 109-126. ASTM STP 1413, American Society for Testing and Materials, Conshohoken, PA

Beghi, M.G., Every, A.G., Prakapenka, V. & Zinin, P.V. (2012). Measurement of the Elastic Properties of Solids by Brillouin Spectroscopy, in *Ultrasonic and Electromagnetic NDE for Structure and Material Characterization: Engineering and Biomedical Applications,* T. Kundu (Ed.), chapter 10, CRC Press, Boca Raton, FL.

Beghi, M. G., Di Fonzo, F., Pietralunga, S., Ubaldi, C. & Bottani, C. E. (2011). Precision and accuracy in film stiffness measurement by Brillouin spectroscopy, *Review of Scientific Instruments,* Vol.102, Paper No. 053107

Bhushan, B. & Li, X.D. (2003). Nanomechanical characterisation of solid surfaces and thin films, *Int. Mater. Rev.*, Vol. 48, pp. 125-164.

Comins, J. D., Every, A. G., Stoddart, P. R., Zhang, X., Crowhurst, J. C. & Hearne, G. R. (2000). Surface Brillouin scattering of opaque solids and thin supported films. *Ultrasonics,* Vol.38, pp. 450-458.

Doerner, M.F., Nix, W.D. (1986). A method of interpreting the data from depth-sensing indentation instruments, *J. Mater. Res.* Vol 1, n. 4, pp. 601–609.

Every, A.G. (2001). The Elastic Properties of Solids: Static and Dynamic Principles, In: *Handbook of Elastic Properties of Solids, Liquids, and Gases,* M. Levy, H. Bass, R. Stern & V. Keppens (Eds.), Volume I: *Dynamic Methods for Measuring the Elastic Properties of Solids,* pp. 3-36, Academic Press, New York

Farnell, G.W. & Adler, E.L. (1972). Elastic wave propagation in thin layers. In: *Physical Acoustics,* W.P. Mason & R.N. Thurston (Eds.), Vol. 9, pp. 35-127, Academic, New York.

Fischer-Cripps A.C. (2011), *Nanoindentation*, Springer (third edition), Chap. 8.

Garcìa Ferré, F., Bertarelli, E., Chiodoni, A., Carnelli, D., Gastaldi, D., Vena, P., Beghi, M.G. & Di Fonzo, F. (2013). The mechanical properties of a nanocrystalline $Al_2O_3$/a-$Al_2O_3$ composite coating measured by nanoindentation and Brillouin spectroscopy, *Acta Materialia*, Vol. 61, pp. 2662-2670.

Goodman, O. & Derby, B. (2011). The mechanical properties of float glass surfaces measured by nanoindentation and acoustic microscopy, *Acta Materialia,* Vol. 59, pp. 1790-1799.

Hertz H. (1881). On the contact of elastic solids, *J. Reine Angew. Math.* Vol. 92, pp. 156–171; translated and reprinted in English in Hertz's Miscellaneous Papers, Macmillan & Co., London, 1896, Ch. 5.

Kundu, T. (2012). Mechanics of elastic waves and ultrasonics non-destructive evaluation, in *Ultrasonic and Electromagnetic NDE for Structure and Material Characterization: Engineering and Biomedical Applications,* T. Kundu (Ed.), chapter 1, CRC Press, Boca Raton, FL.

Oliver, W.C. & Pharr, G.M. (1992). An improved technique for determining hardness and elastic modulus using load and displacement sensing indentation experiments, *J Mater Res*, Vol. 7, pp. 1564-1583.

Seber, G.A.F. & Wild, C.J. (2003). *Nonlinear regression,* Wiley Interscience, Hoboken, NJ.

Sneddon, I. N. (1965). The relation between load and penetration in the axisymmetric Boussinesq problem for a punch of arbitrary profile. *Int. J. Engng. Sci.,* 3:47–57.

Zhang, X., Stoddart, P.R., Comins, J.D. & Every, A. G. (2001). High-temperature elastic properties of a nickel-based superalloy studied by surface Brillouin scattering, *Journal of Physics: Condensed. Matter*, Vol.13, pp. 2281-2294

Zinin, P.V (2001). Quantitative acoustic microscopy of solids, in *Handbook of Elastic Properties of Solids, Liquids, and Gases,* M. Levy, H. E. Bass and R. R. Stern & V. Keppens (Eds.), Volume I: *Dynamic Methods for Measuring the Elastic Properties of Solids,* pp. 187-226, Academic Press, New York

Zinin, P.V, Arnold, W., Weise, W., & Berezina, S. (2012). Theory and Applications of Scanning Acoustic Microscopy and Scanning Near-Field Acoustic Imaging, in *Ultrasonic and Electromagnetic NDE for Structure and Material Characterization: Engineering and Biomedical Applications,* T. Kundu (Ed.), chapter 11, CRC Press, Boca Raton, FL.

Section 3

# Modeling Methods for Microdevices

Chapter 10

# On Universal Modeling of the Bulk Acoustic Wave Devices

Boris Sveshnikov, Sergey Nikitov and Sergey Suchkov

Additional information is available at the end of the chapter

http://dx.doi.org/10.5772/53191

## 1. Introduction

Delay lines, as well as thin film acoustic resonators (FBAR) and solidly mounted resonators (SMR) of bulk acoustic waves (BAW), are widely used in the modern wireless communication networks due to their compactness and ability to operate in the ultra-high and super-high frequencies.

The most rigorous description of the characteristics of arbitrary BAW devices may be achieved by using the known three dimensional finite element method (e.g., Giraud et al, 2007). However, this approach demands enormous computational resources and a long computation time during each design of BAW resonators based on multilayer topologies. So, it is difficult to optimize the system construction promptly, with respect to the peculiarities of each particular specification.

In addition, the results obtained by means of FEM analysis do not have a clear physical sense. Consequently, every time it is not evident which factors of either material or constructive nature influence dominantly on the degradation of the real system performances, and one has to make rather expensive experiments to answer the necessary questions in the course of optimization of a concrete design. Thus, an evident interest appears to develop approximate but flexible analytical approach, allowing fast preliminary investigation and synthesis of any BAW system[1].

We would like to restrict our theoretical investigation here to the one-dimensional (1D) analysis: the model described below admits changes of the system properties along the longitudinal coordinate only. This approach doesn't merely simplify our analysis: it has also some objective advantages in the sense of its practical use due to the following.

[1] 3D-FEM approach may be aplyied at the final stage of needed design, if necessary.

In general, BAW devices represent a set of layers made of various materials. Where parallel facets that are perpendicular not only to the "working" X-direction, but also to Y & Z axes exist (Fig.1a), the synchronous resonant excitations of spurious lateral modes associated with the simultaneous excitations of transverse bending wave motion in plates with side edges parallel to one another become inevitable. The main shortcoming of one-dimensional models is their inability to take into account only those spurious lateral modes.

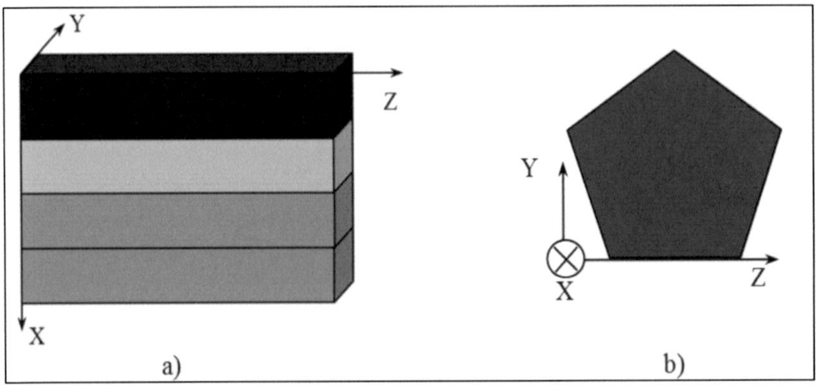


**Figure 1.** Schematic of layered BAW system with parallel planes being perpendicular to X,Y, & Z axes, where the desired bulk acoustic waves propagate along X-direction (a), and the preferred configuration of the "top-side" electrode (Bradley et al., 2002) in an actual BAW device (b).

However, a usage of irregular polygon electrode, in which no two sides are parallel to one in the directions being perpendicular to Y- & Z- axes (Fig.1b), allows considerable suppression of parasitic movements in similar systems (Bradley et al., 2002). Thus, we must only introduce - on the phenomenological level - the proper imaginary addends to the wave-numbers in every layer, taking into account the losses caused by non-synchronous excitation and transformation to heat of parasitic modes (in addition to propagation losses of acoustic waves because of the material viscosity).

Therefore, a one-dimensional model, which allows involving additional losses into analysis, is even more suitable for studying real BAW devices, than, say, two-dimensional models, which demand much more complex routine but cannot be used in design of actual devices with polygonal electrodes.

It is noteworthy that one-dimensional simulation of BAW device is a preferable and correct approach if the direction of the wave propagation in a crystal coincides with the axis of its symmetry. This is so at least in the case of the widely used orientations of ZnO, AlN, W, Mo, $SiO_2$, Ti, Al, $Al_2O_3$, Si, etc.

Despite numerous publications on research of BAW devices even in a simplified one-dimensional case (e.g., Hashimoto, K. (Ed.), 2009), to date no-one has provided reliable fast calculations of many real systems with complex structures. For example, an equivalent circuit network analysis (Ballato et al., 1974) becomes progressively difficult with increased number of electrodes. On the other hand, the most general modelling approach, based on the direct solution of the relevant equations for electro-acoustic fields (Novotny, H., et al., 1991), cannot describe adequately many different configurations of practical importance (e.g., a real SMR). Moreover, Novotny's model is based on a cumbersome cascading routine, where every layer is being represented by a (8×8) transfer matrix. Accordingly, it is difficult to arrange enough complex electrical circuitry in that manner while minimizing at the same time inevitable mistakes during preparation of corresponding software tools.

The novel one-dimensional theory is a more universally suitable designing tool, since it is both much clearer and simpler in use than predecessors. At the same time an effective self-checking algorithm based on satisfaction of three fundamental conditions (energy balance, the second law of thermodynamics and reciprocity) is proposed and utilized in the application software.

Applying the newly developed approach, one would be capable of analyzing - while remaining in the frame of the same modelling principles - any system with an arbitrary number and sequence of dielectric and metal layers under arbitrary inter-electrode connections[2]. Multiple electrodes may compose the multilayer transducers, forming either those based one-port and two-port networks, or tunable re-radiators, loaded by variable admittance. The last variant may be used in order to control electrically the frequency responses of various modern devices, based on usage of bulk acoustic waves.

## 2. Viscous losses in BAW devices

One of the most important aspects affecting the quality of real BAW devices, is the energy losses which emerge during propagation of acoustic waves in crystals. The main cause of the wave attenuation in this case is a viscosity of elastic medium, i.e., the friction arising due to the mechanical movements of the material particles with respect to the neighboring environment. A reliable quantitative estimation of propagation losses is possible only on the basis of experiments.

Many similar experiments have been carried out earlier by different groups of researchers (e.g., Gulyaev, Yu., & Mansfeld, G., 2004). Their results show that the logarithmic decrement characterizing propagation loss per BAW's wavelength depends on frequency almost linearly and can be expressed as $\zeta(f)=\zeta_1 f$. Table 1 shows, for example, a set of $\zeta_1$- values, related to some materials, commonly used in modern BAW resonators.

| Mat. | Al | AlN | ZnO | $SiO_2$ | W | Ti | Mo | Si | $Al_2O_3$ | Diam | LTO | LNO |
|---|---|---|---|---|---|---|---|---|---|---|---|---|
| $\zeta_1 \cdot 10^3$ (1/GHz) | 4.13 | 0.58 | 1.04 | 0.15 | 1.74 | 4.9 | 1.16 | 0.86 | 0.025 | 0.45 | 0.0058 | 0.014 |

**Table 1.** Values of logarithmic decrement (1/GHz), evaluating attenuation per wavelength of BAW in some popular materials.

[2] The infinite conductivity of electrodes is assumed below, when free charges may concentrate only in the infinitely thin skins at metal's borders.

## 3. Simulation principles

In order to calculate the characteristics of multilayer system, we should consider within each layer the motion equation and constitutive relations (Auld, 1973), reducing them to the one-dimensional case in application to the selected bulk wave mode - either longitudinal or the shear one (Kino, 1987).

Let us denote the normal component of the elastic stress tensor as $T$ (*Pa*), while $u$ = elastic displacement (*m*), $\rho$= mass density ($kg/m^3$), $D$ = electric displacement ($C/m^2$) in the presence of electric field with intensity $E$ (*V/m*). Besides, $c$, $\beta$, $\varepsilon$, and $\eta$ mean, respectively, the corresponding components of tensors, characterizing elastic stiffness (*Pa*), piezoelectric stress ($C/m^2$), relative permittivity (*F/m*), and viscosity ($s{\cdot}Pa$).

Then, a well known system of the motion equation (1) and constitutive relations (2-3) should be written for each layer (Auld, 1973):

$$\frac{\partial T}{\partial x} = \varrho \cdot \frac{\partial^2 u}{\partial t^2} \tag{1}$$

$$T = c \cdot \frac{\partial u}{\partial x} + \eta \cdot \frac{\partial^2 u}{\partial x \partial t} - \beta \cdot E \tag{2}$$

$$D = \varepsilon \cdot E + \beta \cdot \frac{\partial u}{\partial x} \tag{3}$$

At this point a quasi-static approximation of Maxwell's equations holds everywhere, except neighbor edges of the electrodes with different polarity:

$$\frac{\partial D_j}{\partial x} = 0 \tag{4}$$

It is convenient to convert (2-3) to the following relations for the electric intensity and elastic stress:

$$E = \frac{1}{\varepsilon} \cdot \left( D - \beta \cdot \frac{\partial u}{\partial x} \right) \tag{5}$$

$$T = \tilde{c} \cdot \frac{du}{dx} - \frac{\beta}{\varepsilon} \cdot D \tag{6}$$

The following denotations are used here:

$q = \omega \cdot \sqrt{\frac{\rho}{\tilde{c}}}$, $\tilde{c} = c(1+k^2) + i\omega\eta = c \cdot \left(1+k^2\right) \cdot \left(1 + i \cdot \frac{\zeta}{\pi}\right)$, $k^2 = \frac{\beta^2}{\varepsilon \cdot c}$, while the coefficient

$\zeta = \pi \frac{\omega\eta}{c \cdot (1+k^2)}$ is a logarithmic decrement, characterizing total distributed dissipation per wavelength of bulk acoustic waves in the chosen medium.

NB: $v=\sqrt{\frac{c\cdot\left(1+k^2\right)}{\rho}}$ is a BAW velocity in the lossless case.

Assuming, as usual, a harmonic solution and omitting the time oscillating factor $e^{i\omega t}$, one can simply obtain from Eqs. (1,4,6) the wave equation for dissipative medium:

$$\frac{\partial^2 u}{\partial x^2}+q^2\cdot u=0 \tag{7}$$

Therefore, it is possible to find a solution in every $j$-th layer, i.e. within a spatial interval $x_{j-1}\leq x\leq x_j$ ($j$ = 1,2…$N$), as a superposition of two counterpropagating waves:

$$u_j(x)=U_j\cdot e^{-iq_j\cdot\left(x-x_{j-1}\right)}+U_{j+N}\cdot e^{iq_j\cdot\left(x-x_{j-1}\right)} \tag{8}$$

In analogy with the previously developed approach to study arbitrary BAW devices , we are using below the sequential "end-to-end" subscripting of wave amplitudes: index of the backward wave comes to hand from an index of the forward wave within the same layer simply by adding $N$-figure, where $N$ is a number of the system layers, including electrodes and substrate (Sveshnikov, 2009). Amplitudes $U_m$ ($m$= 1,2... 2$N$) have to be found when satisfying all the boundary conditions at the layer interfaces $x_j$:

$$u(x_j-0)=u(x_j+0) \tag{9}$$

$$T(x_j-0)=T(x_j+0) \tag{10}$$

$$D(x_j+0)-D(x_j-0)=\sigma(x_j) \tag{11}$$

where $\sigma(x_j)$ means the surface charge density on $j$-th interface[3].

Before giving the general solution of the stated problem, we'd like to describe, first of all, solutions of a few typical tasks to facilitate understanding of the present modelling logics.

### 3.1. P-matrix of a circuitry containing a single piezoelectric layer

Let us describe a simple bulk acoustic wave transducer (BT), formed by a piezoelectric layer with thickness '$d$', placed between perfectly conductive metal electrodes, infinitely extended along the acoustic channel (Fig.2). This structure may be used either as a transducer directly (S is in position "1"), or as a tunable reflector loaded by variable admittance $Yo$ (S ⇔ "2").

As a three-port network containing one electrical and two acoustic ports it may be characterized by means of the usual $P$-matrix. Its sense is explained by the following relations, appropriate when switch S on Fig.2 is in the 1st position:

---

[3] $\sigma_j$ - values differ from zero only at the edges of neighbor electrodes with different polarity.

$$\begin{pmatrix} b_1 \\ b_2 \\ I \end{pmatrix} = \begin{pmatrix} P_{11} & P_{12} & P_{13} \\ P_{21} & P_{22} & P_{23} \\ P_{31} & P_{32} & P_{33} \end{pmatrix} \cdot \begin{pmatrix} a_1 \\ a_2 \\ V \end{pmatrix} \tag{12}$$

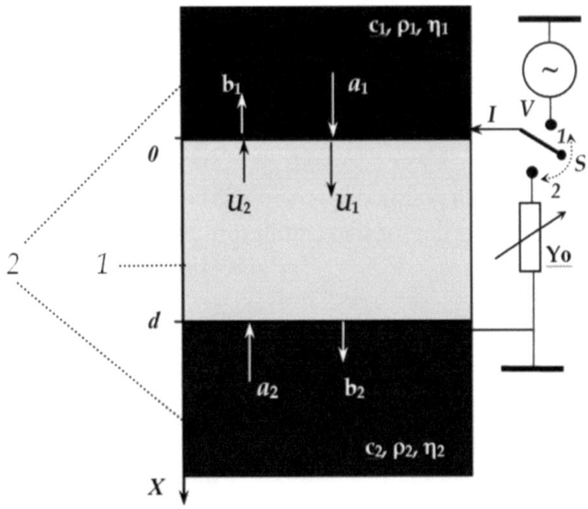


**Figure 2.** A piezoelectric layer (1), placed between two semi-infinite electrodes (2), made, in general, from different materials (we assume below that they are perfectly conductive).

One can find all needed terms of this *P*-matrix in two steps.

### *3.1.1. "One-layer" transducer of bulk acoustic waves*

Assuming that there no beams launching upon a BT from outside ( $a_1 = a_2 = 0$ , satisfaction of the boundary conditions (9-11) at the cross-sections *x*=0 & *x*=*d* gives us a set of linear equations for five unknowns:

$$U_1 + U_2 = b_1 \tag{13}$$

$$-i \cdot \tilde{c}q\left(U_1 - U_2\right) - \frac{\beta}{\varepsilon} D = i\tilde{c}_1 q_1 \cdot b_1 \tag{14}$$

$$U_1 \cdot e^{-2i\psi} + U_2 \cdot e^{2i\psi} = b_2 \tag{15}$$

$$-i \cdot \tilde{c}q\left(U_1 \cdot e^{-2i\cdot\psi} - U_2 \cdot e^{2i\cdot\psi}\right) - \frac{\beta}{\varepsilon} D = -i\tilde{c}_2 q_2 \cdot b_2 \tag{16}$$

$$D = \frac{\varepsilon}{d} V - i\beta \cdot q \cdot \frac{\sin(\psi)}{\psi} \cdot \left(U_1 \cdot e^{-i\cdot\psi} - U_2 \cdot e^{i\cdot\psi}\right) \tag{17}$$

Here $D = D(0+) = D(d-0)$ and $\psi = q \cdot d/2$ .

Note that equality (17) is obtained by integrating (5) over a piezoelectric layer:

$$\int_0^d E(x)\cdot dx = V \tag{18}$$

When substituting (13) & (15) into (14) & (16), one can get a couple of equations allowing us to express the amplitudes $U_{1,2}$ through applied voltage $V$:

$$\begin{cases} \left(1+\frac{Z_1}{Z}-K_e^2\cdot\frac{\sin(\psi)}{\psi}\cdot e^{-i\cdot\psi}\right)\cdot U_1-\left(1-\frac{Z_1}{Z}-K_e^2\cdot\frac{\sin(\psi)}{\psi}\cdot e^{i\cdot\psi}\right)\cdot U_2=\frac{i\beta\cdot V}{\tilde{c}\cdot q\cdot d} \\ \left(1-\frac{Z_2}{Z}-K_e^2\cdot\frac{\sin(\psi)}{\psi}\cdot e^{i\cdot\psi}\right)\cdot e^{-2i\cdot\psi}\cdot U_1-\left(1+\frac{Z_2}{Z}-K_e^2\cdot\frac{\sin(\psi)}{\psi}\cdot e^{-i\cdot\psi}\right)\cdot e^{2i\cdot\psi}\cdot U_2=\frac{i\beta\cdot V}{\tilde{c}\cdot q\cdot d} \end{cases} \tag{19}$$

$$Z_j=\sqrt{\rho_j\cdot\tilde{c}_j}=\frac{\tilde{c}_j q_j}{\omega} \tag{20}$$

Parameter $Z_j$ above has a sense of acoustic impedance (Pa·s/m) of $j$-th layer (subscript is omitted for a middle film), being complex valued quantity in the presence of dissipation, and $K_e^2=\frac{k^2}{1+k^2}$ is an effective piezoelectric coupling constant (Kino, 1987).

By substituting solutions $U_{1,2}$ of (19) to (13) & (15), we find the terms $P_{23,13}$ of the considered $P$-matrix, characterizing amplitudes of the "forward" and "backward" acoustic beams, radiated into the neighboring semi-infinite acoustic media.

$$P_{23,13}=\frac{\pm\frac{\beta}{\tilde{c}\cdot q\cdot d}\cdot\tan(2\psi)\left(\frac{Z_{1,2}}{Z}+i\cdot\tan(\psi)\right)}{\frac{Z_1+Z_2}{Z}\left(1-K_e^2\cdot\frac{\tan(2\psi)}{2\psi}\right)+i\cdot\tan(2\psi)\left(1+\frac{Z_1Z_2}{Z^2}-K_e^2\cdot\frac{\tan(\psi)}{\psi}\right)} \tag{21}$$

On the other hand, with the help of (17) one can calculate an admittance of considered "one-layer" transducer as $P_{33}=i\cdot\omega\cdot D(0<x<d)\cdot S/V$:

$$P_{33}=\frac{i\cdot\omega\cdot C_0\cdot\left\{\frac{Z_1+Z_2}{Z}+i\cdot\tan(2\psi)\cdot\left(1+\frac{Z_1Z_2}{Z^2}\right)\right\}}{\frac{Z_1+Z_2}{Z}\left(1-K_e^2\cdot\frac{\tan(2\psi)}{2\psi}\right)+i\cdot\tan(2\psi)\left(1+\frac{Z_1Z_2}{Z^2}-K_e^2\cdot\frac{\tan(\psi)}{\psi}\right)}, \tag{22}$$

where $C_0=\frac{\varepsilon\cdot S}{d}$ is the static capacitance of a BT, if an area of its electrodes equals $S$. In the lossless case ($\zeta$=0) conductance of similar transducer $Ga$ = Re($P_{33}$) is expressed as follows:

$$Ga = \frac{\omega \cdot C_0 \cdot K_e^2 \cdot \frac{Z_1+Z_2}{Z} \cdot \frac{(\tan(2\psi))^2}{2\psi} \cdot \left\{ \frac{Z_1 Z_2}{Z^2} + (\tan(\psi))^2 \right\}}{\left\{ \frac{Z_1+Z_2}{Z}\left(1 - K_e^2 \cdot \frac{\tan(2\psi)}{2\psi}\right)\right\}^2 + \left\{ \tan(2\psi)\left(1 + \frac{Z_1 Z_2}{Z^2} - K_e^2 \cdot \frac{\tan(\psi)}{\psi}\right)\right\}^2} \quad (23)$$

In particular, at frequency $f = f_0$, for which $\psi = \pi/2$, i.e., when the film thickness equals half of a wavelength within the piezoelectric layer ($d = \lambda_0 / 2 = v / 2$), one can obtain:

$$Ga = \omega \cdot C_0 \cdot \frac{K_e^2 \cdot \frac{4}{\pi} \cdot \frac{Z}{Z_1+Z_2}}{1 + \left(K_e^2 \cdot \frac{4}{\pi} \cdot \frac{Z}{Z_1+Z_2}\right)^2} \quad (24)$$

This value is maximized ($Ga = \omega C_0/2$) under the evident relation between the acoustic impedances of neighboring media: $Z_1 + Z_2 = \frac{4}{\pi} \cdot Z \cdot K_e^2$.

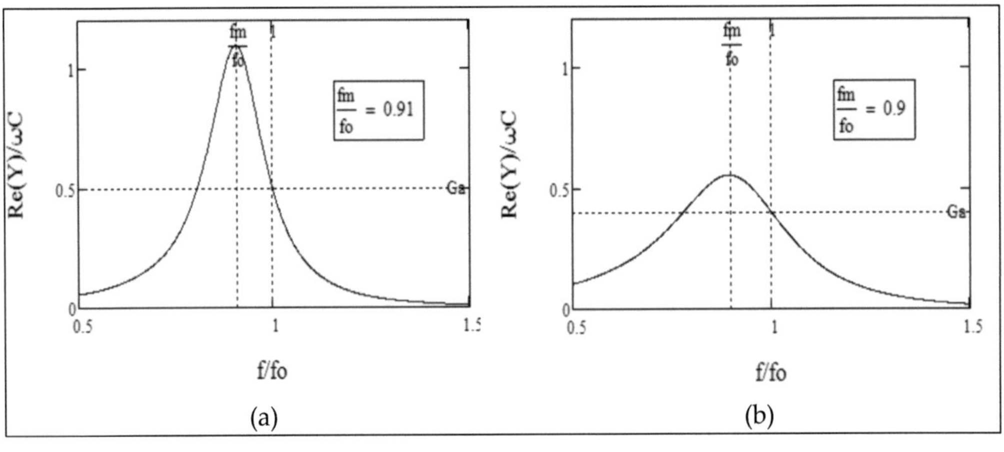


**Figure 3.** Frequency dependence of normalized conductance, related to a simple transducer with the AlN film under $d=\lambda_0/2$, when $Z_1 = Z_2 = \frac{2}{\pi} \cdot K_e^2 \cdot Z$ (a), and $Z_1 = Z_2 = \frac{4}{\pi} \cdot K_e^2 \cdot Z$ (b).

However, this is not the maximal conductance meaning over a whole spectrum (look, e.g., at Fig.3). As one can see, in order to maximize the BT's conductance at the desired frequency, one should make a piezoelectric film thinner than $\lambda_0/2$. Besides, by varying $Z_{1,2}$ -values one can change both magnitude and working bandwidth of the main BT's characteristic.

Anyway, an optimization of the relation between the material parameters and thickness of piezoelectric film is needed even in this simple case. In general, a lot of input parameters (including thickness of metal electrodes) should be involved in optimization routines to improve performances of any real system. With this aim we have to apply the more rigorous modeling tools, being developed below (in Section 4).

*3.1.1.1. Isolated BAW resonators with infinitely thin electrodes*

One can shorten Eq. (22) notably in two particular cases, appropriate for the ideal transducer with infinitely thin electrodes, when:

1. Both edges of the piezoelectric layer to be free of external stress ($Z_{1,2} = 0$) (Kino, 1987):

$$P_{33} = Y_{OO} = i \cdot \omega \cdot C_0 \cdot \left(1 - K_e^2 \cdot \frac{\tan(\psi)}{\psi}\right)^{-1}; \tag{25}$$

2. One edge of the BT is free of stresses, while another border is a rigidly clamped surface ($Z_1 = 0$; $Z_2 \to \infty$):

$$P_{33} = Y_{OC} = i \cdot \omega \cdot C_0 \cdot \left(1 - K_e^2 \cdot \frac{\tan(2\psi)}{2\psi}\right)^{1} \tag{26}$$

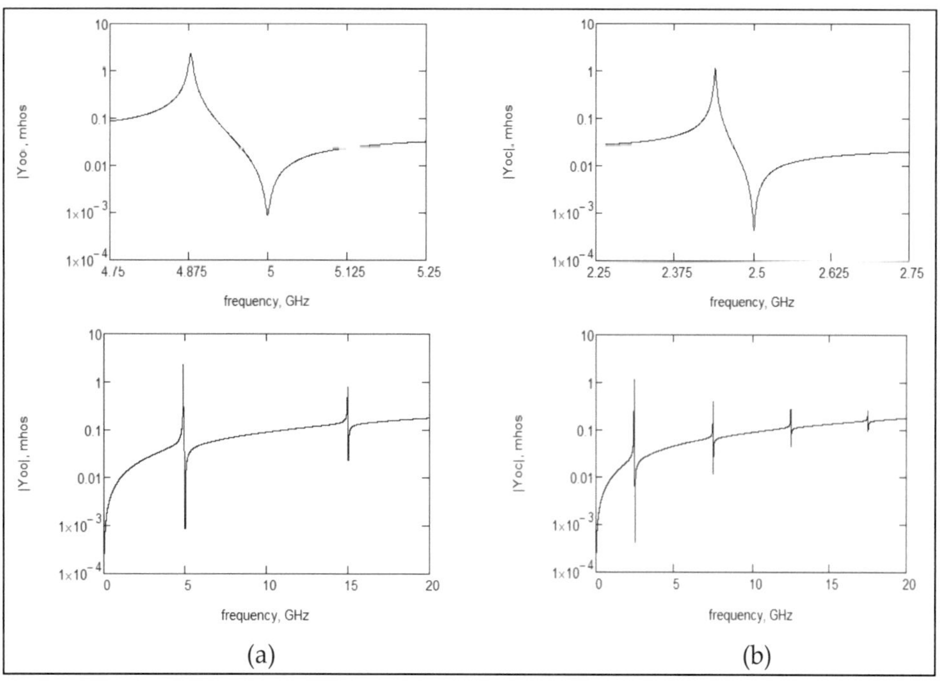


**Figure 4.** Admittance magnitudes of the considered idealized embodiments - $|Yoo(f)|$ (a) & $|Yoc(f)|$ (b) - related to BT, formed by a single aluminum nitride (AlN) film with thickness d=1.093 microns, calculated in a narrow (top) and wide (bottom) frequency ranges of analysis.

The parallel resonances for the considered transducers exist under $\psi = \pi \cdot (2n+1)/2$, when *Yoo*=0, and if $\psi = \pi \cdot (2n+1)/4$, when *Yoc*=0 ($n = 0,1,2...$).

Figures 4 illustrate functions $|Yoo(f)|$ and $|Yoc(f)|$ in logarithmic scale both for a narrow (top) and wide (bottom) frequency bandwidths of analysis, taking into account the propagation losses in AlN film, found from the Table 1 ($\zeta \approx 2.88 \cdot 10^{-3}$ under $f$=5 GHz). It should be noted that the frequency interval between the harmonic resonances in the 2nd case is twice smaller than in the first embodiment. However, the fractional ratio for $n$-th couple of resonant ($|Y(f_r)|$=max) and anti-resonant ($|Y(f_a)|$=min) frequencies satisfies the same relation for both constructions:

$$K_e^2 \cdot \tan\left(\frac{f_n^{(r)}}{f_n^{(a)}} \cdot \frac{\pi}{2}(2n+1)\right) = \frac{f_n^{(r)}}{f_n^{(a)}} \cdot \frac{\pi}{2}(2n+1) \tag{27}$$

Besides, figures 5(a,b) demonstrate the corresponding conductance frequency responses, which may be non-zero values ($\mathrm{Re}(Y_{oo,oc}) \neq 0$) only due to dissipation: a single reason of the power consumption within those systems, isolated hypothetically from the neighboring environment, is the transformation of acoustic energy to heat.

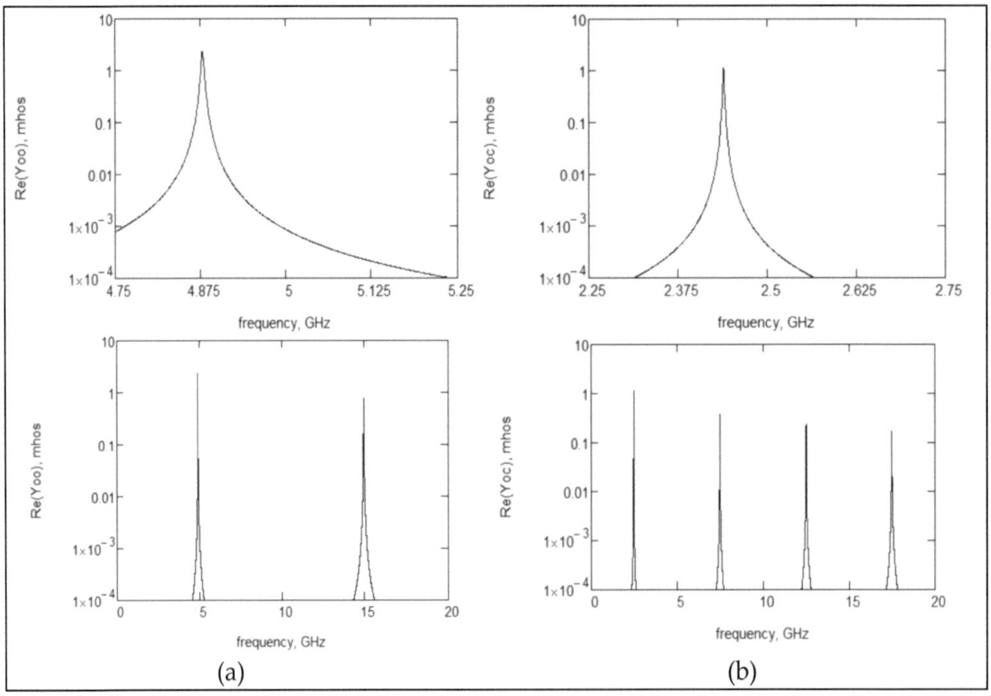


**Figure 5.** The conductance of considered idealized BTs, formed by a single AlN layer with thickness equaled to d=1.093 microns and placed between infinitely thin electrodes. Calculations are made in a narrow (top) and wide (bottom) frequency ranges.

### 3.1.2. *Tunable BAW reflectors containing a single piezoelectric layer*

In the recent years a demand in the electrically tunable BAW devices has revived as a result of the ability of the thin film bulk acoustic wave resonators to enable development of advanced reconfigurable/adaptable microwave circuits. In particular, the focus of the attention remained up till now on tunability which is provided when external variable bias voltage is applied to FBAR. This voltage influences the BAW velocity in the layers forming the system and therefore on its resonant frequencies (Vorobiev & Gevorgian, 2010; Defaÿ, E. et al., 2011).

Another possibility exists to achieve the same goal on a basis of the tunable reflectors with variable electrical load, as a part of modern FBARs or SMRs.

When switch *S* on Fig.2 is turned to the 2nd position, one can consider a transducer with electric load *Yo*, as the tunable reflector of bulk acoustic waves. Aiming to describe its operation more in detail, one should find a solution of the wave equation (8) when using a different, as compared to (13-17), set of boundary conditions.

Assuming that there is no voltage, applied to BT from an external source while looking at Fig.2, one can could conclude the following:

$$U_1 + U_2 = a_1 + b_1 \tag{28}$$

$$-i \cdot \tilde{c} \cdot q(U_1 - U_2) - \frac{\beta}{\varepsilon} D = -i\tilde{c}_1 \cdot q_1 \cdot (a_1 - b_1) \tag{29}$$

$$U_1 \cdot e^{-2i\psi} + U_2 \cdot e^{2i\psi} = a_2 + b_2 \tag{30}$$

$$-i \cdot \tilde{c} \cdot q\left(U_1 \cdot e^{-2i\cdot\psi} - U_2 \cdot e^{2i\cdot\psi}\right) - \frac{\beta}{\varepsilon} D = i\tilde{c}_2 \cdot q_2 \cdot (a_2 - b_2) \tag{31}$$

$$D = \frac{\varepsilon}{d} V_r - i\beta \cdot q \cdot \frac{\sin(\psi)}{\psi} \cdot \left(U_1 \cdot e^{-i\cdot\psi} - U_2 \cdot e^{i\cdot\psi}\right) \tag{32}$$

If $V_r$=0 (this is so under $|Yo| \to \infty$), then, by substituting (32) into (29) & (31), we find a system of two coupled equations for $U_{1,2}(a_1, a_2)$:

$$\begin{cases} \left(1 + \frac{Z_1}{Z} - K_e^2 \cdot \frac{\sin(\psi)}{\psi} \cdot e^{-i\cdot\psi}\right) \cdot U_1 - \left(1 - \frac{Z_1}{Z} - K_e^2 \cdot \frac{\sin(\psi)}{\psi} \cdot e^{i\cdot\psi}\right) \cdot U_2 = a_1 \cdot \frac{2Z_1}{Z} \\ \left(1 - \frac{Z_2}{Z} - K_e^2 \cdot \frac{\sin(\psi)}{\psi} \cdot e^{i\cdot\psi}\right) \cdot U_1 - \left(1 + \frac{Z_2}{Z} - K_e^2 \cdot \frac{\sin(\psi)}{\psi} \cdot e^{-i\cdot\psi}\right) \cdot U_2 = a_2 \cdot \frac{2Z_2}{Z} \end{cases} \tag{33}$$

By solving it and setting alternately two combinations of a couple $\{a_1, a_2\}$ ({1, 0} or {0,1}), which define the fields incident on a BT, one can find all the remaining terms of *P*-matrix, mentioned above:

$$P_{11} = \frac{\left(\frac{Z_1 - Z_2}{Z}\right)\left(1 - K_e^2 \cdot \frac{\tan(2\psi)}{2\psi}\right) - i \cdot \tan(2\psi) \cdot \left(1 - \frac{Z_2 Z_1}{Z^2} - K_e^2 \frac{\tan(\psi)}{\psi}\right)}{\frac{Z_1 + Z_2}{Z}\left(1 - K_e^2 \cdot \frac{\tan(2\psi)}{2\psi}\right) + i \cdot \tan(2\psi) \cdot \left(1 + \frac{Z_1 Z_2}{Z^2} - K_e^2 \cdot \frac{\tan(\psi)}{\psi}\right)} \tag{34}$$

$$P_{22} = \frac{\left(\frac{Z_2 - Z_1}{Z}\right)\left(1 - K_e^2 \cdot \frac{\tan(2\psi)}{2\psi}\right) - i \cdot \tan(2\psi) \cdot \left(1 - \frac{Z_2 Z_1}{Z^2} - K_e^2 \frac{\tan(\psi)}{\psi}\right)}{\frac{Z_1 + Z_2}{Z}\left(1 - K_e^2 \cdot \frac{\tan(2\psi)}{2\psi}\right) + i \cdot \tan(2\psi) \cdot \left(1 + \frac{Z_1 Z_2}{Z^2} - K_e^2 \cdot \frac{\tan(\psi)}{\psi}\right)} \tag{35}$$

$$P_{21} = \frac{\frac{2Z_1}{Z}\left(\frac{1}{\cos(2\psi)} - K_e^2 \frac{\tan(2\psi)}{2\psi}\right)}{\frac{Z_1 + Z_2}{Z}\left(1 - K_e^2 \cdot \frac{\tan(2\psi)}{2\psi}\right) + i \cdot \tan(2\psi) \cdot \left(1 + \frac{Z_1 Z_2}{Z^2} - K_e^2 \cdot \frac{\tan(\psi)}{\psi}\right)} \tag{36}$$

$$P_{12} = \frac{\frac{2Z_2}{Z}\left(\frac{1}{\cos(2\psi)} - K_e^2 \frac{\tan(2\psi)}{2\psi}\right)}{\frac{Z_1 + Z_2}{Z}\left(1 - K_e^2 \cdot \frac{\tan(2\psi)}{2\psi}\right) + i \cdot \tan(2\psi) \cdot \left(1 + \frac{Z_1 Z_2}{Z^2} - K_e^2 \cdot \frac{\tan(\psi)}{\psi}\right)} \tag{37}$$

$$P_{31} = i\omega \cdot D(a_1 = 1; a_2 = 0) = \frac{\omega \cdot S \cdot \frac{2\beta}{d} \cdot \frac{Z_1}{Z} \cdot \tan(2\psi)\left(\frac{Z_2}{Z} + i \cdot \tan(\psi)\right)}{\frac{Z_1 + Z_2}{Z}\left(1 - K_e^2 \cdot \frac{\tan(2\psi)}{2\psi}\right) + i \cdot \tan(2\psi)\left(1 + \frac{Z_1 Z_2}{Z^2} - K_e^2 \cdot \frac{\tan(\psi)}{\psi}\right)} \tag{38}$$

$$P_{32} = i\omega \cdot D(a_1 = 0; a_2 = 1) = \frac{-\omega \cdot S \cdot \frac{2\beta}{d} \cdot \frac{Z_2}{Z} \cdot \tan(2\psi)\left(\frac{Z_1}{Z} + i \cdot \tan(\psi)\right)}{\frac{Z_1 + Z_2}{Z}\left(1 - K_e^2 \cdot \frac{\tan(2\psi)}{2\psi}\right) + i \cdot \tan(2\psi) \cdot \left(1 + \frac{Z_1 Z_2}{Z^2} - K_e^2 \cdot \frac{\tan(\psi)}{\psi}\right)} \tag{39}$$

Otherwise (if $Y_0^{-1} \neq 0$), one can express a voltage between the electrodes with the help of the Ohm's law for external electrical circuitry:

$$i\omega \cdot D \cdot S + V_r \cdot Y_0 = 0 \tag{40}$$

As a result, Eq. (32) may be simply transformed to the relation (41):

$$D = -i\beta \cdot q \cdot K_Y \cdot \frac{\sin(\psi)}{\psi} \cdot \left(U_1 \cdot e^{-i\cdot\psi} - U_2 \cdot e^{i\cdot\psi}\right), \tag{41}$$

where $K_Y = \dfrac{Y_o}{Y_o + i\omega C_0}$ is the variable parameter, characterizing an electrical load.

Omitting the intermediate calculations, we present here just the resulting expressions for the modified $\tilde{P}_{11,22}$ & $\tilde{P}_{21,12}$ terms, of the total *P*-matrix, describing the tunable scattering (reflection and transmission) of bulk acoustic waves by the electrically loaded „one layer" tunable reflector (TR):

$$\tilde{P}_{11}(K_Y) = \frac{\left(\frac{Z_1 - Z_2}{Z}\right)\left(1 - K_e^2 \cdot K_Y \cdot \frac{\tan(2\psi)}{2\psi}\right) - i \cdot \tan(2\psi)\left(1 - \frac{Z_2 Z_1}{Z^2} - K_e^2 \cdot K_Y \cdot \frac{\tan(\psi)}{\psi}\right)}{\frac{Z_1 + Z_2}{Z}\left(1 - K_e^2 \cdot K_Y \cdot \frac{\tan(2\psi)}{2\psi}\right) + i \cdot \tan(2\psi)\left(1 + \frac{Z_1 Z_2}{Z^2} - K_e^2 \cdot K_Y \cdot \frac{\tan(\psi)}{\psi}\right)} \tag{42}$$

$$\tilde{P}_{22}(K_Y) = \frac{\left(\frac{Z_2 - Z_1}{Z}\right)\left(1 - K_e^2 \cdot K_Y \cdot \frac{\tan(2\psi)}{2\psi}\right) - i \cdot \tan(2\psi)\left(1 - \frac{Z_2 Z_1}{Z^2} - K_e^2 \cdot K_Y \cdot \frac{\tan(\psi)}{\psi}\right)}{\frac{Z_1 + Z_2}{Z}\left(1 - K_e^2 \cdot K_Y \cdot \frac{\tan(2\psi)}{2\psi}\right) + i \cdot \tan(2\psi)\left(1 + \frac{Z_1 Z_2}{Z^2} - K_e^2 \cdot K_Y \cdot \frac{\tan(\psi)}{\psi}\right)} \tag{43}$$

$$\tilde{P}_{21}(K_Y) = \frac{\frac{2Z_1}{Z}\left(\frac{1}{\cos(2\psi)} - K_e^2 \cdot K_Y \frac{\tan(2\psi)}{2\psi}\right)}{\frac{Z_1 + Z_2}{Z}\left(1 - K_e^2 \cdot K_Y \cdot \frac{\tan(2\psi)}{2\psi}\right) + i \cdot \tan(2\psi)\left(1 + \frac{Z_1 Z_2}{Z^2} - K_e^2 \cdot K_Y \cdot \frac{\tan(\psi)}{\psi}\right)} \tag{44}$$

$$\tilde{P}_{12}(K_Y) = \frac{\frac{2Z_2}{Z}\left(\frac{1}{\cos(2\psi)} - K_e^2 \cdot K_Y \frac{\tan(2\psi)}{2\psi}\right)}{\frac{Z_1 + Z_2}{Z}\left(1 - K_e^2 \cdot K_Y \cdot \frac{\tan(2\psi)}{2\psi}\right) + i \cdot \tan(2\psi)\left(1 + \frac{Z_1 Z_2}{Z^2} - K_e^2 \cdot K_Y \cdot \frac{\tan(\psi)}{\psi}\right)} \tag{45}$$

Assuming that reactance is used as the electrical load (either capacitor $\tilde{C}$ or inductor $\tilde{L}$), one can take into consideration also the finite resistance ($R_e$) of TR's electrodes, placing - formally – in parallel to TR a shunting conductance $G_S \approx \omega_0^2 \cdot C_0^2 \cdot R_e = \omega_0 \cdot C_0 / Q_0$:

$$Y_0 = \begin{cases} i \cdot \omega \cdot \tilde{C}\left(1 - \frac{i}{Q_C}\right) + \frac{\omega_0 \cdot C_0}{Q_0} \Leftrightarrow \textit{for capacitive load;} \\ \frac{1}{i \cdot \omega \cdot \tilde{L}}\left(1 + \frac{i}{Q_L}\right) + \frac{\omega_0 \cdot C_0}{Q_0} \Leftrightarrow \textit{for inductive load,} \end{cases} \tag{46}$$

where $Q_C$ - & $Q_L$- values mean the load's Q-factor (being different, generally, for capacitors and inductors).

It is convenient to characterize $Y_0$-value by the dimensionless parameter [4]:

$$\delta = -\frac{\mathrm{Im}(Y_0) + \omega_0 \cdot C_0}{\omega_0 \cdot C_0} \tag{47}$$

Thus, the „tuning" parameter $K_Y$, introduced above, may be represented in the following form:

$$K_Y = \frac{y_0}{y_0 + f/f_0}, \quad \text{where} \tag{48}$$

$$y_0 = \begin{cases} -(1+\delta)\frac{f}{f_0}\cdot\left(1-\frac{i}{Q_C}\right)-\frac{i}{Q_0}, & if \quad \delta \le -1\,(for\ capacitive\ load) \\ -(1+\delta)\cdot\frac{f_0}{f}\cdot\left(1+\frac{i}{Q_L}\cdot\frac{f_0}{f}\right)-\frac{i}{Q_0}, & otherwise\,(for\ inductive\ load) \end{cases} \tag{49}$$

Here $Q_C$ & $Q_L$ mean the Q-factors of capacitive and inductive loads, respectively, while $f_0 = \omega_0/2\pi$ is that frequency, at which external inductance compensates the BT's static capacitance ($\delta = 0$). Figures (6-8) show some examples illustrating the scattering features of TR in the considered simplified variant when using aluminum nitride (AlN) piezoelectric film, supporting the longitudinal bulk wave mode.

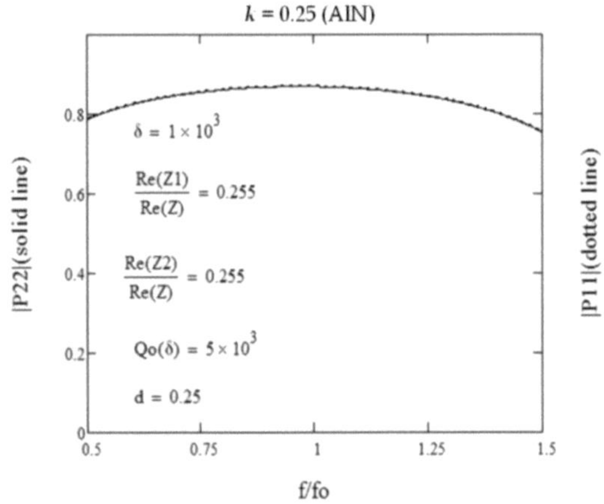


**Figure 6.** Reflectivity of the short-circuited BT ($\delta \to \infty \Leftrightarrow K_Y=1$) with the quarter-wavelength thickness of AlN film ($d = \lambda_0/4 = v/2f_0$).

Contrary to the short-circuited BT (Fig.6), the reflectivity may practically vanish under properly found (for a chosen frequency) load's reactance, if the load's Q-factor is enough high (Fig.7b).

[4] $\delta = -1$ & $\delta = 0$ mean the open- & short-circuited modes, correspondingly.

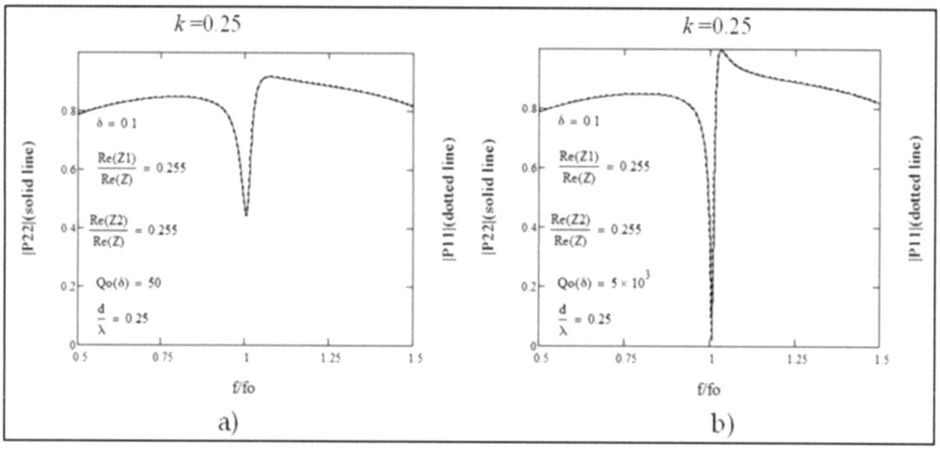


**Figure 7.** Antireflecting effect under different values of the load'd Q-factor.

Physical sense of the mentioned tunability is quite evident: a current, induced because of inverse piezoelectricity by incident waves on electrodes, provides "secondary" voltage over the load, generating - in turn - the waves in the acoustic channel (which propagate, generally, in both opposite directions).

The re-radiated waves interfere with the reflected and transmitted ones, which appear in the inhomogeneous elastic channel under electrical shorting ($Yo\to\infty$). The load' change results, surely, in the amplitude and phase variation of the reflected and passed through BT waves.

The antireflecting properties of a similar structure may appear when a wave, regenerated backward, comes into antiphase with the „elastic" reflections, cancelling the resulting backward wave almost at all. The finite Q-factor of a load results in a several degradation of the antireflecting effect (Fig.7b).

*3.1.2.1. Energy balance and the second law of thermodynamics as checking points*

There are two conditions here, which allow us to verify the modeling adequacy bearing in mind that the power flow, carried by acoustic wave with amplitude $U$ in the non-piezoelectric medium, equals $i\omega\cdot U^{*}\cdot T=\omega\cdot\mathrm{Re}(c\cdot q)\cdot|U|^{2}=\omega^{2}\cdot\mathrm{Re}(Z)\cdot|U|^{2}$ (Auld, 1973).

First of all, in the absence of the beams coming from the outside ($a_1=a_2=0$), the total power of acoustic waves, radiated by a transducer under applied voltage $V$, is equal to

$$P_T=\left|V^2\right|\cdot\omega^2\cdot S\cdot\left\{\mathrm{Re}\left(Z_2\right)\cdot\left|P_{23}\right|^2+\mathrm{Re}\left(Z_1\right)\cdot\left|P_{13}\right|^2\right\} \tag{50}$$

Accordingly, we should always check whether the energy balance condition is satisfied requiring the following obligatory relation between $P_T$ -value and the transducer conductance, indicating the total power consumed from electrical source:

$$P_T\leq\mathrm{Re}\left(P_{33}\right)\cdot\left|V^2\right| \tag{51}$$

Equality in (51) holds exactly only in the lossless case: in general, a part of acoustic power transforms to heat within BT's body and can't be radiated outside it.

On the other hand, in perfect accordance with the 2nd law of thermodynamics, the total part of acoustic power, passed through the reflector, cannot depend on the propagation direction of the incident wave (Fig.8a)[5]. This condition should always be satisfied, even in the presence of dissipation in asymmetrical case (if $Z_1 \neq Z_2$), when a layer may have different reflectivity with respect to the waves launching upon it from opposite directions (Fig.8b):

$$\left|\tilde{P}_{21}\right|^2 \cdot \mathrm{Re}(Z_2)/\mathrm{Re}(Z_1) = \left|\tilde{P}_{12}\right|^2 \cdot \mathrm{Re}(Z_1)/\mathrm{Re}(Z_2), \text{ that is} \quad \left|\tilde{P}_{21}\right| \cdot \mathrm{Re}(Z_2) = \left|\tilde{P}_{12}\right| \cdot \mathrm{Re}(Z_1) \tag{52}$$

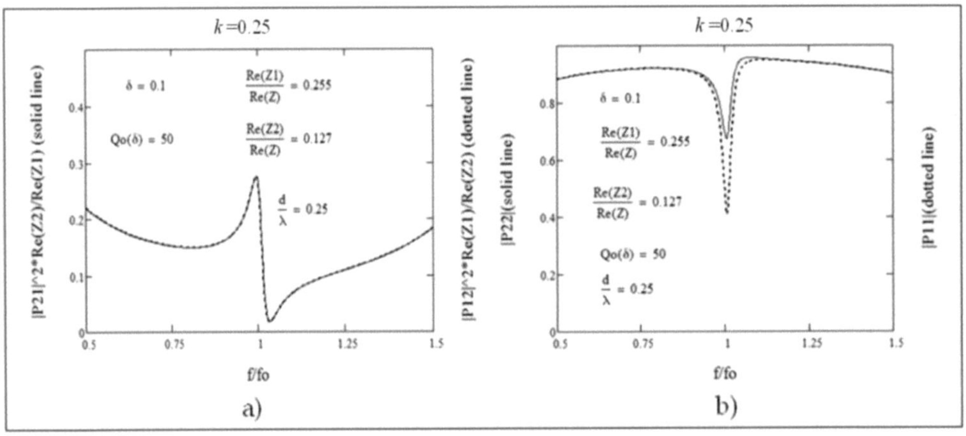


**Figure 8.** The peculiarity of the BAW scattering by asymmetrical tunable reflector ($Z_1 \neq Z_2$).

As calculations show, Eqs. (21-22) and (42-45) agree absolutely with the conditions (51-52).

*3.1.2.2. Tunability of the "frontal" BAW reflectors*

The tuning possibilities of acoustoetectric BAW transducers as reflectors have been investigated first long ago (Gristchenko, 1975). It was proved that one could have control over the reflected and transmitted acoustic power by means of variable reactance, connected to BT's electrodes.

Note that a physically analogous phenomenon, concerning the tunable scattering of surface acoustic waves (SAWs), has been widely investigated even earlier than tunability of BAW devices. Smith et al. (1969) first have presented an analysis of the interdigital transducer (IDT) basing on the equivalent circuit model in the absence of distributed feedback (DFB) caused by SAW reflections from electrodes as the periodic inhomogeneities. The corresponding expression for the regenerative reflection coefficient of IDT (with the total

[5] Otherwise, a dissipative half-space from a one side of reflector should be heated as compared with the neighbor half-space, disturbing the condition of thermodynamic equilibrium.

admittance $Y$) was obtained. It can be expressed, within the constant phase multiplier, in the following simple form:

$$R = \frac{\text{Re}(Y)}{Y + Y_0} \tag{53}$$

Later Sandler and Sveshnikov (1981), basing on the Coupling-of-Modes (COM) analysis, have developed the more general model for IDT, taking into account distributed feedback also. As they have shown theoretically and, in co-author with Paskhin, experimentally, if a $Q$-factor of a reactance is high enough, then the reflection coefficient of the interdigital reflector may be electrically varied practically from zero to unity for any efficacy of Bragg's reflections. This circumstance was used to manufacture the tunable SAW resonators, as well as to suppress electrically the triple-transit signals in ordinary bandpass SAW filters (Paskhin et al., 1981).

Then, the phase-shifting features of a reflector, manufactured as a single phase unidirectional transducer (SPUDT) with variable electrical load, has been discovered (Sveshnikov & Filinov, 1988). It was revealed that the phase value of the SPUDT reflection coefficient can be varied electrically over an interval [0; 2π] at the stopband frequency, if the SAW beam is launched upon a SPUDT from the direction of its predominant radiation. It means that by changing variable reactance $Y_0$ one can fluently change resonant frequency of the resonator, containing similar reflector, over a frequency interval $\Delta f$ between its neighboring resonant frequencies.

Due to the physical propinquity of SPUDT with the frontal BT (which has a unidirectional nature in principle), it became clear that the same phenomenon has to appear for the frontal BAW transducer, placed on the crystal surface and used as the „one-side " tunable mirror. This effect was well founded further both by simplified analytical model (Sveshnikov, 1995), and by numerical calculations made for multimode BAW resonators (Kucheryavaya, et al., 1995).

Indeed, assuming that there is the air above the boundary x=0 on Fig.2 (when a shift S there is in the second position) and neglecting a dissipation in the middle layer, a surface $x=0$ should be considered as the perfect elastic mirror: $\left|\frac{U_2}{U_1}\right|$ practically equals unity[6].

Owing to system's linearity, the total reflection coefficient of the loaded BT ($R = b_2/a_2$), being a superposition of the elastic and regenerative terms, within the inessential phase constant may be represented as follows:

$$\begin{gathered} R = 1 + \frac{A \cdot \text{Re}(Y)}{Y + Y_0} = 1 + \frac{A}{1 + \alpha + i \cdot \theta}, \text{ where} \\ \alpha = \frac{\text{Re}(Y_0)}{\text{Re}(Y)}; \quad \theta = \frac{\text{Im}(Y + Y_0)}{\text{Re}(Y)}, \end{gathered} \tag{54}$$

[6] Because, for example, $Z_{air}/Z_{AlN} \sim 10^{-5}$.

while $A$ is unknown constant coefficient. As it doesn't depend on $Y_0$-value, one can find it assuming for the nonce a load to be the lossless reactance (α=0). In this case, due to the energy balance, an equality $|R|=1$ holds too. The last relation may be satisfied for arbitrary $\theta$ -value only under A = - 2. Thus, we get the following simple formula, being valid even in the presence of Ohm loss in electrical load (α≠0), neglecting only the BAW attenuation in piezoelectric film (Sveshnikov, 1995):

$$R = \frac{1-\alpha-i\theta}{1+\alpha+i\theta} = |R|\cdot e^{i\Phi_R}, \text{where} \quad \Phi_R = -a\tan\left(\frac{\theta}{1-\alpha}\right) - a\tan\left(\frac{\theta}{1+\alpha}\right) \tag{55}$$

As a consequence of the reciprocity principle, under the electrical matching (α=1 and $\theta = 0$) the mentioned reflectivity should disappear (R=0), because of the perfect unidirectionality of this frontal BT (all the incident power is absorbed in a load).

Taking into account (21) & (54), by numerical calculations made under condition $Z_1=0$ one can always ascertain that Eq.(43), obtained rigorously, absolutely coincides with (55) in the absence of layer's viscosity[7]. For example, if $\psi = \pi/2$, then $\tan(2\psi)\cdot\tan(\psi) = -2$, and

$$\tilde{P}_{22} = \frac{1 - i\cdot\dfrac{4}{\pi}\cdot\dfrac{Z}{Z_2}K_e^2\cdot K_Y}{1 + i\cdot\dfrac{4}{\pi}\cdot\dfrac{Z}{Z_2}K_e^2\cdot K_Y} \tag{56}$$

It is clear from (55) also that under condition of enough small Ohm losses (α < 1), if the reactive load is widely varied, from a very large capacitance ($\theta \to \infty$) to a very small inductance ($\theta \to -\infty$), then the reflection phase is changed all around a circle: $-\pi < \Phi_R < \pi$. Nonetheless, under the finite Q - factor of a load, $|R|$-value may decrease notably during the tuning process. Figures 9a & 9b illustrate this effect in two cases: a) under real Q-factors of electrical circuitry, and b) when these Q-factors assumed to be ten times larger ($f_0$ = 2GHz).

The larger ratio $G=\mathrm{Re}(P_{33})/\omega C_0$ between the transducer conductance and its „static" susceptance, the smaller the mentioned falling of TR's reflectivity. As it is clear from (56) one can increase G-value by decreasing the ratio $|Z_2/Z|$. However, on this way we have not enough variety of the impedance combinations, when using real materials.

Another technological possibility must be analyzed also to improve the functional features of tunable BAW reflectors. It concerns manufacturing of the multi-layered BAW transducer, containing a number ($N_p$) piezoelectric layers placed between electrodes with alternating polarity - similarly to the interdigital transducer (IDT) of surface acoustic waves. It was clear that G-value in this case may rise almost linearly with increasing $N_p$. On the other hand, the wave propagation within the extended (in the longitudinal direction) reflector will bring us

[7] This fact is another confirmation of the simulation validity.

additional losses too. So, it is necessary to create a proper model, to be able optimize the parameters of actual BAW devices in the presence of unavoidable dissipation.

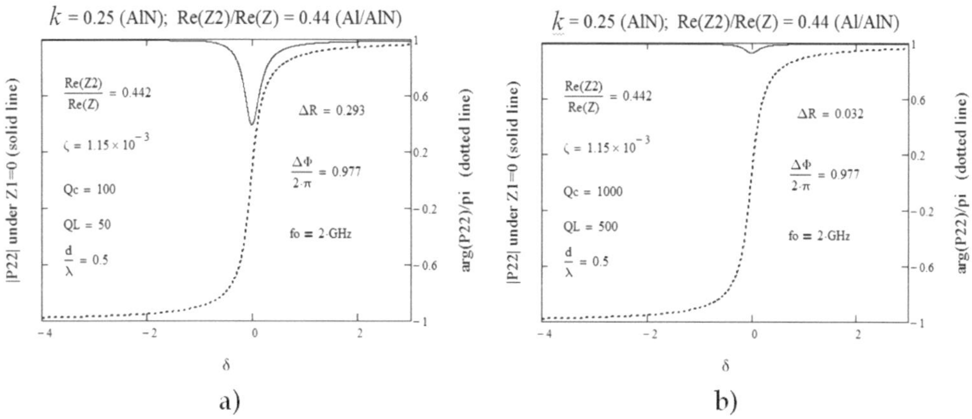


**Figure 9.** Tunability of the frontal BAW reflector by varying a load's reactance: an interval $\Delta\Phi$ of variation of the reflection coefficient phase $\Phi = \arg(P22(f_0))$ is close to $2\pi$ when $\delta \in [-4;+3]$.

## 4. P-matrix of BAW transducer with multiple electrodes

Suppose, there is a sequence of piezoelectric layers, placed between metal electrodes with alternating polarity, which form a multi-layer BAW transducer (MBT), in analogy with IDT (Fig.10).

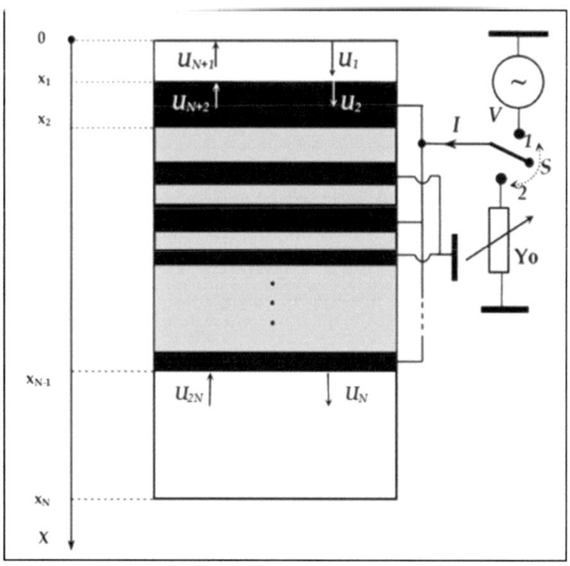


**Figure 10.** Schematic representation of multi-layer transducer of bulk acoustic waves, which becomes tunable reflector, when being loaded by variable reactance.

The *P*-matrix of an BT, containing a number of arbitrary layers in the amount of $N_t = 2N_p + 1$ (including electrodes), depends on materials of the neighboring media of acoustic track, where a transducer should be placed. So, when finding the BT's *P*-matrix we have to involve into consideration also a pair of the frontier non-piezoelectric layers, made from the arbitrary materials. The total number of layers, to be taken into account at this point, equals $N=N_t+2 = 2N_p+3$.

Denoting $d_j = x_j - x_{j-1}$ and following the above-mentioned numeration of acoustic waves, appeared inside a system either because of applied voltage ($V\neq 0$), or due to external beams under short-circuiting condition ($V=0$), we consider the boundary conditions (9-10) at the interfaces $x_1, x_2... x_{N-1}$, bearing in mind (in analogy with (17)) that within *j*-th film

$$D_j = \frac{\varepsilon_j}{d_j}\delta V_j - i\beta_j \cdot q_j \cdot \frac{\sin(\psi_j)}{\psi_j} \cdot \left(U_j \cdot e^{-i\cdot\psi_j} - U_{j+N} \cdot e^{i\cdot\psi_j}\right), \tag{57}$$

where $\delta V_j = V_{j-1} - V_j$ mean the voltage drop upon *j*-th layer.

Besides, there are two boundary conditions at the borders $x = x_0 = 0$ & $x = x_N$. One can characterize them by two parameters $B_{1,2}$, which may accept three meanings: 1) $B_{1,2} = 1$, relating to free edges ($T(x_{0,N}) = 0$); 2) $B_{1,2} = -1$, relating to the rigidly clamped borders ($u(x_{0,N}) = 0$), and 3) $B_{1,2} = 0$, imitating the perfect matching of transducer with neighboring acoustic channels (when there are no waves, coming to transducer from outside):

$$\begin{cases} U_1 = B_1 \cdot U_{N+1} \\ U_{2N} = B_2 \cdot e^{-i\cdot 2q_N \cdot d_N} \cdot U_N \end{cases} \tag{58}$$

So, one can get a linear system of 2*N* coupled equations for amplitudes $U_m$ ($m = 1...2N$), which may be written in the following matrix form for two vectors, characterizing the spatial distribution of acoustic fields ($\vec{U}^{(V)}$ & $\vec{U}^{(a)}$), induced within a device either by applied voltage, or by external beams incident on a short circuited transducer from outside:

$$\vec{U}^{(V,a)} = \hat{M} \cdot \vec{U}^{(V,a)} + \vec{H}^{(V,a)} \tag{59}$$

At this time the vectors $\vec{H}^{(V)}$ & $\vec{H}^{(a)}$ define in (59) the sources of the BAW excitation, appeared both because of the applied voltage *V* (when $a_1=a_2=0$) and due to external acoustic fields with unit amplitudes, launching on BT at the cross-sections $x=0$ ($a_1=1, a_2=0$) or $x=x_N$ ($a_1=0, a_2=1$):

$$H_m^{(V)} = \frac{i}{\omega} \cdot \begin{cases} 0, \ if\ m=1,\ or\ m=2N; \\ \dfrac{\dfrac{\beta_m}{d_m}\delta V_m - \dfrac{\beta_{m-1}}{d_{m-1}}\delta V_{m-1}}{Z_{m-1}\left(1-\Omega_{m-1}\cdot ex_{m-1}^{-1}\right)+Z_m\left(1-\Omega_m \cdot ex_m^{-1}\right)},\ if\ 1<m\le N; \\ \dfrac{ex_{m-N}^{-2}\cdot\left(\dfrac{\beta_{m+1-N}}{d_{m+1-N}}\delta V_{m+1-N} - \dfrac{\beta_{m-N}}{d_{m-N}}\delta V_{m-N}\right)}{Z_{m+1-N}\left(1-\Omega_{m+1-N}\cdot ex_{m+1-N}^{-1}\right)+Z_{m-N}\left(1-\Omega_{m-N}\cdot ex_{m-N}^{-1}\right)},\ if\ \ N<m<2N \end{cases}$$

$$H_m^{(a)} = a_1 \cdot \delta(m,1) + a_2 \cdot \delta(m,2N)$$

where $ex_j = \exp(i \cdot \psi_j)$; $\Omega_j = \frac{k_j^2}{1+k_j^2} \cdot \frac{\sin(\psi_j)}{\psi_j}$; $a_{1,2}$ =1 or 0, and $\delta(m, n)$ is the Kronecker symbol ($m$, $n$ = 1…2$N$).

The „local " scattering matrix in (59) $\hat{M} = \begin{pmatrix} \hat{M}11 & \hat{M}12 \\ \hat{M}21 & \hat{M}22 \end{pmatrix}$, characterizing the interaction of counter-propagating waves at all interfaces, is formed by four ($N \times N$) sub-matrices:

$$\hat{M}11 = \begin{pmatrix} 0 & 0 & \cdots & 0 & 0 & 0 \\ t_1^{(+)} \cdot ex_1^{-2} & 0 & \cdots & 0 & 0 & 0 \\ 0 & t_2^{(+)} \cdot ex_2^{-2} & \cdots & 0 & \vdots & 0 \\ 0 & 0 & \ddots & \vdots & 0 & \vdots \\ \vdots & \vdots & \vdots & t_{N-2}^{(+)} \cdot ex_{N-2}^{-2} & 0 & 0 \\ 0 & 0 & \cdots & 0 & t_{N-1}^{(+)} \cdot ex_{N-1}^{-2} & 0 \end{pmatrix};$$

$$\hat{M}22 = \begin{pmatrix} 0 & t_1^{(-)} \cdot ex_1^{-2} & 0 & \cdots & 0 & 0 \\ 0 & 0 & t_2^{(-)} \cdot ex_2^{-2} & \cdots & 0 & 0 \\ 0 & 0 & 0 & \cdots & \vdots & 0 \\ 0 & 0 & \vdots & \ddots & t_{N-2}^{(-)} \cdot ex_{N-2}^{-2} & \vdots \\ \vdots & \vdots & 0 & \cdots & 0 & t_{N-1}^{(-)} \cdot ex_{N-1}^{-2} \\ 0 & 0 & 0 & \cdots & 0 & 0 \end{pmatrix};$$

$$\hat{M}12 = \begin{pmatrix} B_1 & 0 & 0 & \cdots & 0 & 0 \\ 0 & r_1^{(+)} & 0 & \cdots & 0 & 0 \\ 0 & 0 & r_2^{(+)} & \cdots & \vdots & 0 \\ 0 & 0 & 0 & \ddots & 0 & \vdots \\ \vdots & \vdots & 0 & \cdots & r_{N-2}^{(+)} & 0 \\ 0 & 0 & 0 & \cdots & 0 & r_{N-1}^{(+)} \end{pmatrix};$$

$$\hat{M}21 = \begin{pmatrix} r_1^{(-)} ex_1^{-4} & 0 & 0 & \cdots & 0 & 0 \\ 0 & r_2^{(-)} ex_2^{-4} & 0 & \cdots & 0 & 0 \\ 0 & 0 & r_3^{(-)} ex_3^{-4} & \cdots & \vdots & 0 \\ 0 & 0 & 0 & \ddots & 0 & \vdots \\ \vdots & \vdots & 0 & \cdots & r_{N-1}^{(-)} ex_{N-1}^{-4} & 0 \\ 0 & 0 & 0 & \cdots & 0 & B_2 \cdot ex_N^{-4} \end{pmatrix}$$

Here $t_j^{(+)}$ & $t_j^{(-)}$ are the transmission coefficients of BAW beams through $j$-th interface for waves, propagating under $\delta V_j$=0 forward (in „+x" direction) & backward (in „-x" direction), respectively; $r_j^{(+)}$ & $r_j^{(-)}$ have a sense of reflection coefficients, describing (under $\delta V_j$=0) reflection at *j*-th interface of waves, propagating after reflection in „+x" & „-x" directions, correspondingly:

$$t_j^{(+)} = \frac{2\cdot\left(1-\Omega_j\cdot\cos(\psi_j)\right)\cdot Z_j}{Z_{j+1}\left(1-\Omega_{j+1}\cdot ex_{j+1}^{-1}\right)+Z_j\left(1-\Omega_j\cdot ex_j^{-1}\right)};$$
$$t_j^{(-)} = \frac{2\cdot\left(1-\Omega_{j+1}\cdot\cos(\psi_{j+1})\right)\cdot Z_{j+1}}{Z_{j+1}\left(1-\Omega_{j+1}\cdot ex_{j+1}^{-1}\right)+Z_j\left(1-\Omega_j\cdot ex_j^{-1}\right)}; \tag{60}$$

$$r_j^{(+)} = \frac{Z_{j+1}\left(1-\Omega_{j+1}\cdot ex_{j+1}\right)-Z_j\left(1-\Omega_j\cdot ex_j^{-1}\right)}{Z_{j+1}\left(1-\Omega_{j+1}\cdot ex_{j+1}^{-1}\right)+Z_j\left(1-\Omega_j\cdot ex_j^{-1}\right)};$$
$$r_j^{(-)} = \frac{Z_j\left(1-\Omega_j\cdot ex_j\right)-Z_j\left(1-\Omega_{j+1}\cdot ex_{j+1}^{-1}\right)}{Z_{j+1}\left(1-\Omega_{j+1}\cdot ex_{j+1}^{-1}\right)+Z_j\left(1-\Omega_j\cdot ex_j^{-1}\right)}, \tag{61}$$

where $Z_j$ means complex acoustic impedance of *j*-th layer (see (19a)).

The following evident equality solves (59) when introducing the (2*N*×2*N*) unit matrix $\hat{I}$:

$$\vec{U}^{(V,a)} = (\hat{I}-\hat{M})^{-1}\cdot\vec{H}^{(V,a)} \tag{62}$$

In accordance with (62) the spatial distribution of electric displacement $D_j$, expressed using (57) through the spatial distribution $U_j$ & $U_{j+N}$, may be also represented as a superposition of two terms induced either by the voltage or by the external acoustic beams: $D_j = D_j^{(V)} + D_j^{(a)}$.

Thus, when using (62), one could simply obtain all terms of the desired *P*-matrix related to arbitrary multi-layer transducer:

$$P_{11} = ex_1^2\cdot U_{N+1}^{(a)}\Big|_{a_1=1;\,a_2=0};\quad P_{12} = U_{N+1}^{(a)}\Big|_{a_1=0;\,a_2=1};\quad P_{13} = ex_1^2\cdot U_{N+1}^{(V)}/V \tag{63}$$

$$P_{21} = U_N^{(a)}\Big|_{a_1=1;\,a_2=0};\quad P_{22} = U_N^{(a)}\Big|_{a_1=0;\,a_2=1};\quad P_{23} = U_N^{(V)}/V \tag{64}$$

$$P_{31} = \frac{i\omega\cdot S}{V}\cdot\sum_j D_j^{(a)}\Big|_{a_1=1;\,a_2=0};\quad P_{32} = \frac{i\omega\cdot S}{V}\cdot\sum_j D_j^{(a)}\Big|_{a_1=0;\,a_2=1};\quad P_{33} = \frac{i\omega\cdot S}{V}\cdot\sum_j D_j^{(V)}\cdot\delta V_j \tag{65}$$

Note, the equalities (63-65) concern the same *P*-parameters as (20-21) & (34-39), derived above for the „one-layer" BT. In order to obtain the scattering parameters, characterizing tunable multi-layer reflector, loaded by variable electrical admittance $Y_0$, it is convenient to apply, using (63-65), the well-known relations (see, e.g., Hashimoto, 2009):

$$\tilde{P}_{11,22} = P_{11,22} - \frac{P_{31,32}\cdot P_{13,23}}{P_{33}+Y_0};\quad \tilde{P}_{21,12} = P_{21,12} - \frac{P_{32,31}\cdot P_{23,13}}{P_{33}+Y_0} \tag{66}$$

Naturally, the found relations (63-66) for generalized *P*-matrix meet both of the fundamental requirements (51-52), confirming the simulation validity.

Figs.11 illustrate some frequency responses, representing conductance of the multi-layered frontal transducer ($B_1$=1), which contains AlN films with thickness $d_p = v_p/(2f_0)$, placed between infinitely thin, though infinitely conductive, *Al* electrodes. The curves, found under the assumption that the "bottom" layer of transducer is perfectly matched with the adjacent acoustic channel ($B_2$=0), was normalized by the product of *N*p-value on a conductance, calculated at frequency $f_0$ for the "one layer" BT[8].

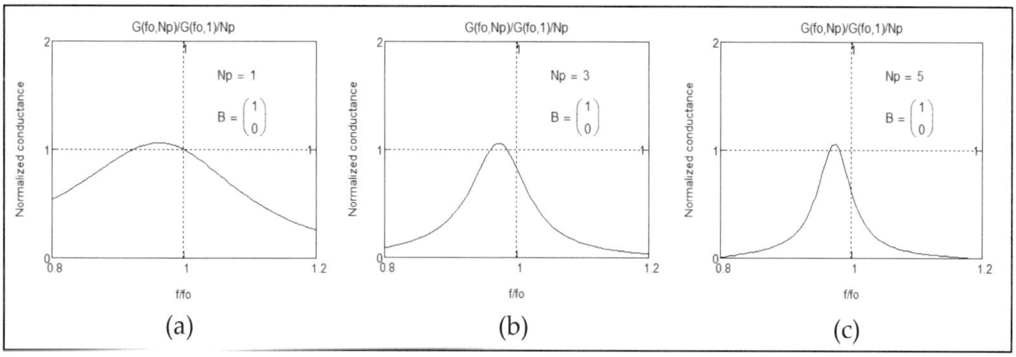


**Figure 11.** The normalized conductance of BTs, containing different numbers (*Np*) of AlN layers with thickness $d = v/(2f_0)$, placed between infinitely thin, though perfectly conductive *Al* electrodes, under a different number of piezoelectric layers: *Np*=1 (a), 3 (b) and 5 (c).

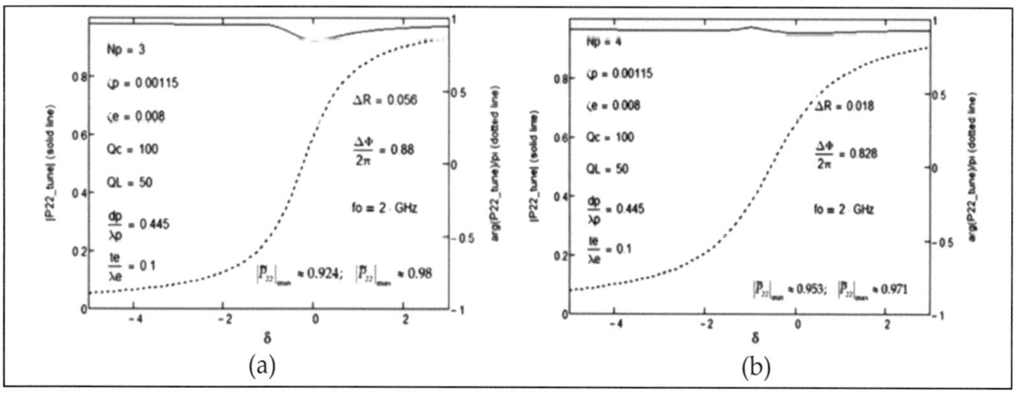


**Figure 12.** Tuning of multi-layered frontal TR, containing three (b) & four (c) piezoelectric films with thickness $d_p = 0.445 \cdot \lambda_p$, when $\delta \in [-4;+3]$ & $f = f_0 = v_p/\lambda_p$ = 2GHz (electric resistance of *Al* electrodes with thickness $t_e = 0.1 \cdot \lambda_e$ assumed to be negligible yet).

As it is clear from Figs.12 one can minimize the undesired variation of TR's reflectivity, occurring during the tuning process, by using multilayer structures (compare with Fig.9a). However, this is done at the expense of certain reduction of the reflection coefficient magnitude in consequence of the viscous damping in TR with increased longitudinal size.

[8] If *Np*=1, these results numerically coincide with the analytical ones, found above in Section 3.1.2.

We don't pay attention here to a separate task, concerning optimization of the system parameters, as its solution depends on concrete specifications and technological capabilities of the BAW device manufacturing. However, the developed model is just the well suited instrument facilitating a solution of the multi-parametric optimization problem, bearing in mind the finite thickness of electrodes as well.

## 5. Tunable SMR

As an example of universal modeling, we'd like to demonstrate, how one can calculate characteristics of solidly mounted resonators (SMR), containing, in particular, the above mentioned multi-layered tunable reflector.

Let us consider, for instance, a device, consisting of two parts: 1) frontal TR, placed in the domain $x < 0$ including *Np* piezoelectric films placed between *Al* electrodes, and 2) "two-port" domain, with input & output transducers (containing $NP_1$ & $NP_2$ piezoelectric films), mounted on the bottom substrate by using an intermediate acoustic Bragg reflector (BR). BR consists of alternating high and low acoustic impedance layers (e.g., Mo and $SiO_2$), in amount *Nr*, manufactured in order to select only one acoustic resonance in this potentially multi-resonant system.

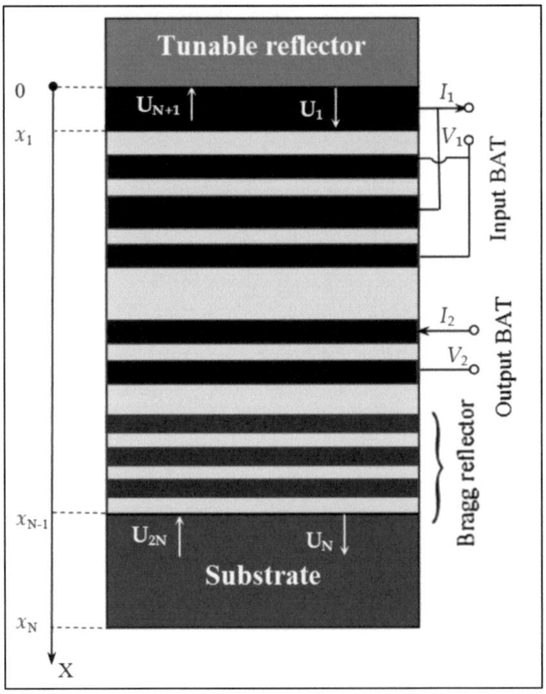


**Figure 13.** Schematic image of two-port solidly mounted BAW device.

The wave amplitudes $U1_{1...2N}$ & $U2_{1...2N}$ within domain $0 \le x \le x_N$ on Fig.13 may be found in analogy with the previous analysis, brought in "Sect. 3.3", when voltage is applied either to input or output transducers ($V_1$=1 & $V_2$=0 $\Rightarrow$ $U_m = U1_m$; $V_1$=0 & $V_2$=1 $\Rightarrow$ $U_m = U2_m$).

A couple of changes should be, yet, involved into the modeling, when forming the general scattering matrix $\hat{\mathrm{M}}$ of SMR in whole (besides, surely, the novel spatial distribution of materials and thickness of layers there). First, $\tilde{P}_{22}$-parameter of the "top side" tunable reflector, found under condition $B_2$=0, must be used as the 1st term in the sub-matrix $\hat{\mathrm{M}}12$:

$$\hat{\mathrm{M}}12_{11} = \tilde{P}_{22}(Y_0)\cdot BC_1 \tag{67}$$

Secondly, $\hat{\mathrm{M}}21_{NN} = BC_2 \cdot ex_N^{-4}$. Here $BC_{1,2}$ are the components of the boundary condition vector, relating to the domain $x \in [0, x_N]$ on Fig.13 ($BC_2$ =-1 means, e.g., the rigidly clumped bottom side of a substrate).

Then, using (57), one has to determine afresh the corresponding distribution of electrical displacements within the transducers for the voltages applied either to input or output MBTs, finding all the needed $Y$- parameters of arbitrary four-terminal network:

$$I_{1,2} = Y_{11,22}\cdot V_{1,2} + Y_{12,21}\cdot V_{2,1} \tag{68}$$

Accordingly to the energy balance condition at every frequency point of analysis and for arbitrary architecture of a device the normalized acoustic power hypothetically going outside a system (when $BC_1 \cdot BC_2 = 0$) under external voltage, applied either to input ($P_T$=$Pa_1$) or output ($P_T$=$Pa_2$) ports, in the lossless case absolutely coincides with the input/output conductance values, and becomes smaller them in the presence of dissipation (a part of energy, supplied by source, is transformed to a heat):

$$\mathrm{Re}(Y_{11,22}(f))\cdot\left|V_{1,2}\right|^2 \geq Pa_{1,2}(f)\ \text{, where} \tag{68}$$

$$\begin{cases} Pa_1 = \omega^2\cdot S\cdot\left|V_1^2\right|\cdot\left\{\mathrm{Re}(Z_N)\cdot\left|U1_N\right|^2\cdot\left(1-\left|BC_2\right|\right)+\mathrm{Re}(Z_1)\cdot\left|U1_{N+1}\right|^2\cdot\left(1-\left|BC_1\right|\right)\right\},\ if\ V_2=0 \\ Pa_2 = \omega^2\cdot S\cdot\left|V_2^2\right|\cdot\left\{\mathrm{Re}(Z_N)\cdot\left|U2_N\right|^2\cdot\left(1-\left|BC_2\right|\right)+\mathrm{Re}(Z_1)\cdot\left|U2_{N+1}\right|^2\cdot\left(1-\left|BC_1\right|\right)\right\},\ if\ V_1=0 \end{cases} \tag{69}$$

Note, as well, that one more checking rule must be applied here too. It is coupled with the reciprocity principle, being valid for any four-terminal network, based on acoustic waves:

$$Y_{21} = Y_{12} \tag{70}$$

The present model always passes through this test also - under the arbitrary combination of the topological and material parameters of a system.

Figures 14(a,b) illustrate how an input impedance of the one-port SMR ($Z_e = 1/|Y_{11}|$) may be influenced by the variable reactance for some combinations of SMR's input data. Two variants (with more realistic thicknesses of electrodes) have been considered here, when the TRs contain a) one (with thickness $dp_{TR} \approx 2.49$ μm), and b) three (with thickness $dp_{TR} \approx 2.57$ μm) AlN layers. At this point BT contains in both cases a one piezoelectric layer with thickness $dp_{BT} = \lambda_{AlN}/2 \approx 2.73$ μm. Thickness of electrodes (with area $S = 1$ mm$^2$) assumed to be different within TR ($te_{TR}$ ≈ 0.16 μm) and BT ($te_{BT}$ ≈ 1.58 μm). The Bragg reflector, containing *Nr* =10 "quarter-wavelength" layers, made from $SiO_2$ & Mo films, separates a transducer from the bottom substrate (Si) which has a thickness about 1 mm.

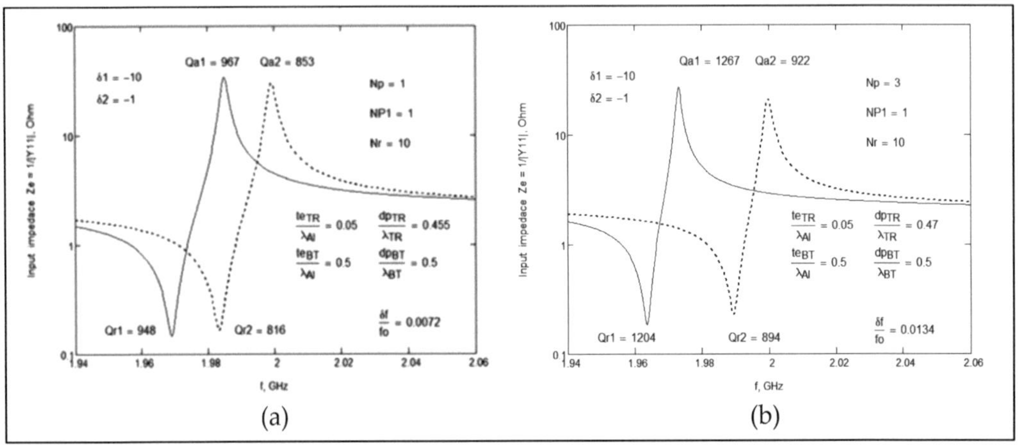


**Figure 14.** Input impedance of SMR with the rigidly clamped bottom surface of a substrate (Si). Tunable reflector is loaded only by variable capacitor, allowing changes of its capacitance From $\tilde{C} = 9 \cdot C_0$ (solid lines) to $\tilde{C} = 0$ (dotted lines), where $C_0$ is a static capacitance of TR. Calculations are made in two cases, when TR contains one (a) and three (b) AlN layers.

As one can see, the SMR's Q-factors at both resonant ($Q_r$) and anti-resonant ($Q_a$) frequencies increase when utilizing a multi-layer tunable reflector. Even these (not optimized yet) variant shows that under *Np*=3 a fractional interval of the frequency tuning reaches a rather large quantity ($\delta f/f_0 \approx$ 1.34 %), to be almost twice better than for SMR with a "one layer" TR. At this time only a capacitive reactance (varicap with Q-factor equaled to 100) is assumed to be used as a load, in order to prevent increasing of Ohm losses, which rise usually when using an inductor in the external electrical circuitry.

## 6. Conclusion

A highly efficient self-consistent analytical model, allowing us to describe an arbitrary BAW device, has been developed. Comprehensive solution of several typical tasks is given with the clear physical argumentation. Flexible one-dimensional modelling is based on a direct solution of the motion and constitutive acoustic equations, taking into account the relevant boundary conditions. Any kind of dissipation may be involved into consideration at the phenomenological level.

Using the proposed approach one may analyze and synthesize, while remaining within the frame of the same investigation manner, a structure with an arbitrary number and sequence of dielectric and metal layers. Multiple electrodes may compose the multilayer transducers forming those based one- and two-port networks.

A method to control over the resonant frequency of solidly mounted BAW resonators, based on the usage of multi-layered tunable reflectors, is investigated in detail. It improves substantially the SMR tunability and may be applied either with the aim to correct for frequency errors, caused by technological thickness variations of layers, or in order to compensate the temperature drifts of the device characteristics using variable electrical load.

An original and very useful integral method is proposed to verify the validity of the simulation, when basing on three fundamental principles, namely: energy balance, the second law of thermodynamics, and reciprocity. The presented checking algorithm, on the one hand, gives us assurance in the modeling correctness. On the other hand, it simplifies considerably the search for mistakes during preparation of the corresponding software tools needed to optimize the device parameters in the shortest time.

## Author details

Boris Sveshnikov
*Lebedev Research Center in Physics, Moscow, Russia*

Sergey Nikitov
*Institute of Radio engineering and Electronics of RAS, Moscow, Russia*

Sergey Suchkov
*State University, Saratov, Russia*

## Acknowledgements

This work was supported by the Russian Government Grant # 11.G34.31.0030.

The authors thank Ms. Tatyana Sveshnikova for helping with grammar editing of the text.

## 7. References

Giraud, S.; Bila, S.; Aubourg, M. & Cros, D. (2007). Bulk acoustic wave resonators 3D simulation, *2007 Joint with the 21st European Frequency and Time Forum*, pp. 1147-1151, IEEE International Digital Object Identifier: 10.1109/FREQ.2007.4319258

Bradley, P.; Ruby, R.; Barfknecht, A.; Geefay, F.; Han, C.; Gan, G.; Oshmyansky, Y. & Larson, J. (2002). A 5 mm x 5 mm x 1.37 mm Hermetic FBAR Du-plexer for PCS Handsets with Wafer-Scale Packaging, *IEEE Ultrasonics Symposium Proceedings*, 2002, pp. 931-934, ISSN : 1051-0117

Ruby R., Review and Comparison of Bulk Acoustic Wave FBAR, SMR Technology, *IEEE Ultrasonics Symposium Proceedings*, 2007, pp. 1029-1040

Fattinger, G. (2008), BAW Resonator Design Considerations - An Overview, *IEEE Ultrasonics Symposium Proceedings*, 2008, pp. 762-767

Hashimoto, K. (Ed.). (2009). *RF Bulk Acoustic Wave Filters for Communications*, ARTECH HOUSE, ISBN-13: 978-1-59693-321-7, Norwood, MA 02062

Ballato, A.; Bertoni, H. & Tamir, T. (1974). Systematic design of stacked-crystal filters by microwave network methods, *IEEE Trans. Microwave Theory Tech.*, MTT-22, pp. 14-25, 1974

Novotny, H. & Benes, E. (1991). Layered piezoelectric resonators with an arbitrary number of electrodes (general one-dimensional treatment), *Journ. Acoust. Soc. Am.*, Vol. 90, Sept. 1991, pp. 1238-1245

Auld, B. (1973). *Acoustic Fields and Waves in Solids, Vol. 1.*, New York: John Wiley & Sons, Inc., 1973

Kino, G. (1987). *Acoustic waves: devices, imaging, and analog signal processing*, New Jersey: Prentice-Hall, 1987

Sveshnikov, B. (2009). Universal Modeling of the Bulk Acoustic Wave Devices, *Proceedings of the EFFT-IFCS*, 2009, pp. 466-469

Gulyaev Yu. & Mansfeld G. (2004). Resonators and filters for microwave bulk acoustic wave devices - current status and trends, *Uspekhi sovremennoi radioelectroniki*, Moscow, 2004, # 5-6, pp. 13-28 (in Russian).

Vorobiev, A . & Gevorgian, S. (2010). Tunable thin film bulk acoustic wave resonators with improved Q factor, *Appl. Phys. Lett.*, Vol. 96, no. 21, art. no. 212904, 2010.

Gristchenko, E. (1975). Acoustic analog of the electro-optical gate, *Akust. Zh.*, Vol.21, no. 5, pp. 827-828 (in Russian)

Smith, W.; Gerard, R.; Collins, J.; Reeder, T. & Shaw, H. (1969). Analysis of inter- digital surface wave transducers by use of an equivalent circuit model, *IEEE Trans. On MTT*, Vol. MTT-17, pp. 856-864, 1969

Sandler, M. & Sveshnikov, B. (1981). An investigation of interdigital transducers of surface acoustic waves, taking into account the finite mass of the electrodes, *Radio Engng. and Electron. Phys.*, vol. 26, No. 9, pp. 9-17, 1981.

Paskhin, V.; Sandler, M. & Sveshnikov, B. (1981). A method to suppress the triple-transit signals in SAW filters, *Zk Tekh. Fiz.*, Vol. 51, no. 12, pp. 2595-2597, (in Russian)

Sveshnikov, B. & Filinov, V. (1988). Tunable SAW phase-shifting reflector using unidirectional IDT, *Sov. Tech. Phys. Lett.*, v. 14, no. 8, pp. 658-660, 1988

Sveshnikov, B. (1995). Tunable Phase-Shifting Reflectors of Surface and Bulk Acoustic Waves, *Ultrasonics World Congress Proceedings*, Berlin, 1995, pp. 387-390.

Kucheryavaya, E.; Mansfeld, G.; Sveshnikov, B. & Freik, A. (1995). Frequency control of composite volume acoustic-wave resonators, *Acoustical physics*, 41(2), 1995, pp. 302-304

Defaÿ, E.; Hassine, N.; Emery, P.; Parat, G.; Abergel, J. & and Devos, A. (2011). Tunability of aluminum nitride acoustic resonators: A phenomenological approach, *IEEE Transactions on UFFC*, Vol. 58, no. 12, December 2011, pp. 2516-2520

Chapter 11

# Progress in Theoretical and Numerical Tools Devoted to Understanding of Acoustic Devices Behavior

T. Laroche and S. Ballandras

Additional information is available at the end of the chapter

http://dx.doi.org/10.5772/56164

## 1. Introduction

Since several decades, the acoustic devices improvement depends largely on the theoretical and Numerical tools. In Acoustic domain, the simulations provide results in high agreement with the measurements. The numerical methods mainly used are either heuristic or algorithmic. The heuristic ones give results rapidly but not necessary optimized e.g. the Mixed Matrix formulation [1–3]. The second kind of numerical methods (algorithmic ones) requires advanced mathematical development and also complex numerical algorithms e.g. Finite Element Analysis (FEA)/Boundary Element Method (BEM) [4]. FEA also provides the data structures for the heuristic methods such as P-Matrix method. In this chapter, we go over the second kind of numerical methods i.e. the exact ones (FEA/BEM). First, the basis equations are detailed as well as the current level of the numerical tools (Periodic FEA/BEM [5]). Next, FEA improvements are demonstrated. Indeed, new research fields in acoustic need to consider extra configurations which cannot be treated by the periodic way e.g. in dual-mode SAW filters [6]. So, a non-periodic model was developed and implemented. The theoretical approach is identical to the one used in periodic case. The strategy of digitizing is also the same i.e. So only the inhomogeneous part of the studied configuration must be discretized. By contrast, the surrounding space is considered by using boundary conditions. Indeed, an acoustic devices is not flying in a free space but if one does not take care of the boundary that is exactly what it is simulate. The edges of the mesh used in a FEA act as a perfect mirror. So, without boundary conditions the component is in the vacuum. This remark is valid both for the non-periodic and periodic. To avoid any problems with the bottom and top limits of the mesh, one must first develop a BEM to simulate the radiation in a multilayered substrate or surrounding medium. Here, BEM is based on Green functions. The inverse Green problem is solved for the periodic case [7] while the direct one is used for the non-periodic systems [8]. In a second step the artificial reflexions on the side edges of the meshed grid must be eliminated. In periodic configuration (e.g. SAW transducers), this is considered by using periodic conditions on the edges [9]. In non-periodic case, a Perfectly Matched Layer (PML) method must be used and specially for spatially finite SAW resonators. This method avoids artificial reflexions on the edge side of the meshed.

This work presents first the basis of the FEA with a brief state of the art in simulation with the main known results until now both for the non-periodic and periodic cases. Next, the last improvements applied to these numerical investigations are introduced. The boundary conditions are mainly developed to address new configurations such as the spatially finite SAW devices. The radiation problem (BEM) is first shown and secondly the PML. The last part is dedicated to the new results obtained by using the latter improvements.

## 2. Analytical recipes and state of the art

### 2.1. Fundamental equations

The Finite Element model allows to simulate the behavior of acoustic devices against many parameters such as coated medium, electrodes apodization... This method was first initiated by Tiersten [10]. The principle consists in the equilibrium of the potential and kinetic energy in the volume with the electrical and mechanical excitation applied on the edges. Thus we obtained the variational formulation from the equilibrium point of the Lagrangian functional [11] :

$$\begin{aligned}&\iiint_{\Omega}\left(\frac{\partial\delta u_i}{\partial x_j}C_{ijkl}\frac{\partial u_l}{\partial x_k}+\frac{\partial\delta u_i}{\partial x_j}e_{kij}\frac{\partial\phi}{\partial x_k}+\frac{\partial\delta\phi}{\partial x_i}e_{ijk}\frac{\partial u_j}{\partial x_k}-\frac{\partial\delta\phi}{\partial x_j}\varepsilon_{jk}\frac{\partial\phi}{\partial x_k}-\rho\omega^2u_i\delta u_i\right)dV=\\&\iiint_{\Omega}F_i\delta u_idV+\iint_{\Gamma}\delta u_iT_{ij}n_jdS+\iint_{\Gamma}\delta\phi D_jn_jdS\end{aligned}\tag{1}$$

$u_i$ and $\phi$ are respectively the displacement and the potential unknowns, $C_{ijkl}$ the elastic constants, $e_{ijk}$ the piezoelectric ones, $\rho$ the density and $\varepsilon_{ij}$ the dielectric coefficients. $F_i$, $T_{ij}$ and $D_i$ are respectively the forces the stress and the displacement vector. $\Omega$ and $\Gamma$ are respectively the studied domain and its frontier with the outer space (See Fig. 1). The solution of such an equation is not trivial. The scheme of FEA allows to find a global solution from an exact one computed at local points. The global solution is obtained by polynomial interpolation in finite elements. The sum of each elements give the $\Omega$ domain. The variational equation (1) is given for the FEA scheme:

$$\begin{aligned}&\sum_{e=1}^{E}\iiint_{\Omega^{(e)}}\left(\frac{\partial\delta u_i^{(e)}}{\partial x_j}C_{ijkl}^{(e)}\frac{\partial u_l^{(e)}}{\partial x_k}+\frac{\partial\delta u_i^{(e)}}{\partial x_j}e_{kij}^{(e)}\frac{\partial\phi^{(e)}}{\partial x_k}+\frac{\partial\delta\phi^{(e)}}{\partial x_i}e_{ijk}^{(e)}\frac{\partial u_j^{(e)}}{\partial x_k}\right.\\&\left.-\frac{\partial\delta\phi^{(e)}}{\partial x_j}\varepsilon_{jk}^{(e)}\frac{\partial\phi^{(e)}}{\partial x_k}-\rho\omega^2u_i^{(e)}\delta u_i^{(e)}\right)dV=\\&\iiint_{\Omega^{(e)}}F_i^{(e)}\delta u_i^{(e)}dV+\iint_{\Gamma^{(e)}}\delta u_i^{(e)}T_{ij}^{(e)}n_j^{(e)}dS+\iint_{\Gamma^{(e)}}\delta\phi^{(e)}D_j^{(e)}n_j^{(e)}dS\end{aligned}\tag{2}$$

where $e$ defines the $e^{th}$ element and $E$ the total number of elements. In equation (2), all the unknowns ($\delta u$, $u$...) should be written using polynomial interpolation for each element [12]. Each quantities $\Delta^{(e)}(x_i)$ is written according to the following interpolation in one element:

$$\Delta^{(e)}(x_i)=\sum_{n=1}^{N_e}\Delta^{(e,n)}P^{(e,n)}(x_i),\tag{3}$$

where $x_i$ is one space direction, $N_e$ the number of nodes in the $e^{th}$ element, $\Delta^{(e,n)}$ the value of the quantity $\Delta$ at the nth node of the eth element and $P^{(e,n)}(x_i)$ is the Lagrangian interpolation polynomial for the same node. For the sake of clarity, only the elastic part is written below according to the FEA formulation for a dimensional device (for instance along $x_1$ in Fig. 1) with all the mechanical contributions:

$$\sum_{e=1}^{E}\sum_{n=1}^{N_e}\sum_{\mu=1}^{N_e}\left(\iiint_{\Omega^{(e)}}\frac{\partial P^{(e,n)}(x_1)}{\partial x_1}C_{i11l}^{(e)}\frac{\partial P^{(e,\mu)}(x_1)}{\partial x_1}dV\right. \\ \left. -\rho\omega^2\iiint_{\Omega^{(e)}}P^{(e,n)}(x_1)P^{(e,\mu)}(x_1)dV\right)u_i^{(e,\mu)}\delta u_i^{(e,n)}=0. \tag{4}$$

In the nodal expression (4), the right hand is zero. The boundary is actually not considered in this trivial model.

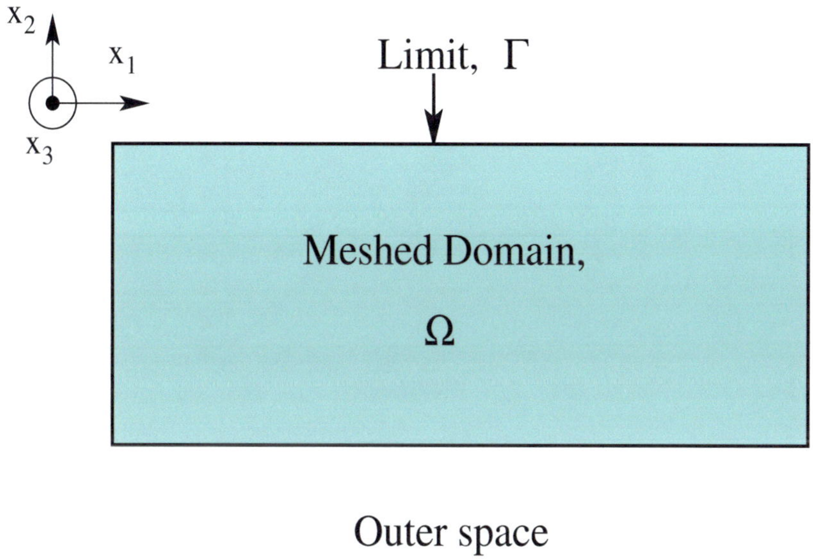


**Figure 1.** Splitting of space for a FEA. $\Omega$ is the inhomogeneous space under simulation. This is the meshed domain in which the FEA is applied. $\Gamma$ is the boundary of $\Omega$ with the remaining space. The latest is either simulated using boundary conditions or considered as vacuum.

The right hands of equations (1) and (2) show several boundary conditions. They are represented by the integration on the limit $\Gamma$. This domain is also discretized in all its elements. The boundary conditions can be applied on each element considering a condition on the stress as well as on the displacement or the potential. These parts are for example the beginning of the radiating conditions. We define below this condition and how consider it in the variational equation.

## 2.2. State of the art

The FEA is known since forty years in piezoelectric problem [13]. However, the finite dimension of the meshed space involves the problem of the spurious reflexions on the edges. Indeed, by default the edges act as perfect mirrors. The device is seldom in the vacuum.

Moreover, to consider a real case, the device must be supported by a substrate. So, the boundary conditions have to be set by considering all these constraints.

First, we were interested in the periodic acoustic devices e.g. Surface Acoustic Waves devices (SAW). So, the side edges constraints were easily avoided. Indeed, periodic condition was set on them [14]. to consider realistic case, the radiating medium must be modeled. The mix of FEA and BEM in periodic configuration was so implemented [15]. A lot of results with high accuracy are obtained from this numerical method. One of these results is depicted in figure 2 and 3. These results come from Ref. [15]. This is the comparison between experimental results

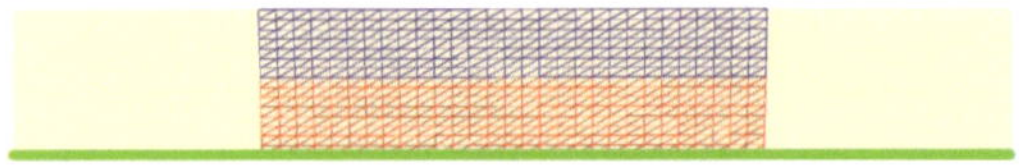

**Figure 2.** Mesh of the considered prominent electrode grating device Ref. [16] red part of the mesh: quartz; blue part: Al; and green section: interface between FEA and BEM.

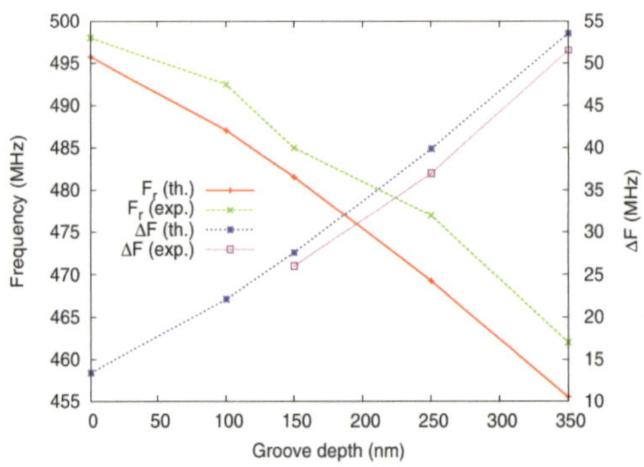


**Figure 3.** Comparison between theoretical and experimental resonance frequency and stop bandwidth for prominent electrode grating based resonators Ref [16].

(Ref. [16]) and numerical simulations combining both FEA and BEM for Surface Transverse Wave problem under massive electrodes. In that case, the grating's period was 5 m with a metalization ratio of 0.5. The strip height atop the resulting prominent quartz ridge was 300 nm with a groove depth ranging from 0 to 350 nm. FEA/BEM approach. The typical mesh for prominent electrode gratings is plotted in Fig. 2. Both the resonance frequency and frequency stop-band width were measured. The comparison between FEA results and experiments are reported in figure 3. It highlights a good prediction of the behavior of the device.

3D periodic problems have also be considered. We depict here a very original work on a two direction a 2D periodic transducer consisting of square dots or pads exhibiting a double excitation potential alternation (along x1 and x2, here the axes defining the surface) [17]. In other words, a two directions SAW resonator is investigated in this previous paper.We particularly focused on one excitation configuration consisting in potential alternation along both period d1 and d2 as illustrated in figure 4. It is however clear that the actual implementation of such an excitation configuration is quite difficult to achieve, as we can not imagine the fabrication of metal bridges for accessing all the square dots of the surface. We then imagine the use of a dielectric spacer allowing for preventing the presence of electrical charge at the surface between the dots but allowing for the connection we are looking for. This is graphically illustrated in figure 5, showing an example of mesh. This kind of mesh has been

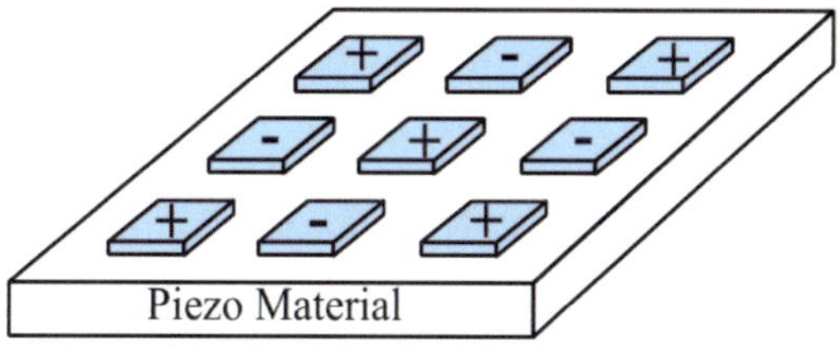


**Figure 4.** A 2D excitation structure based on a double potential alternation to promote the excitation of surface waves exhibiting complex polarization (the 2D equivalent of the usual interdigital transducer)

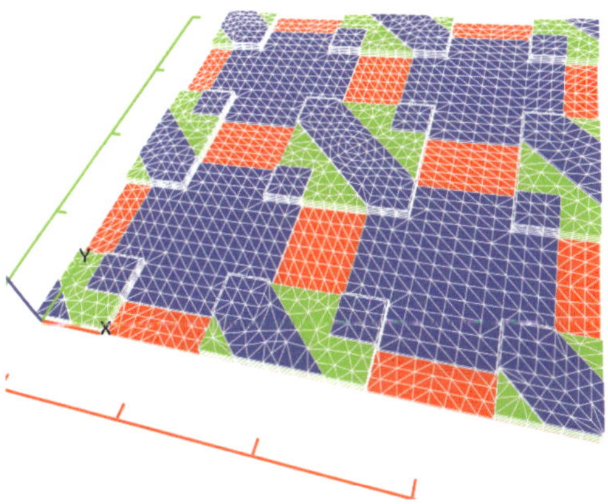


**Figure 5.** An example of simplified 3D mesh used to simulate the 2D periodic transducer. The period of the grating is set to 25Âţm. The green square dots localize the dielectric âĂŞ SiO2 âĂŞ spacers and the blue strips and squares show the deposition of the Al electrodes above the structure. The red background is the piezoelectric material.

used for the computation of the harmonic admittance of an infinite 2D periodic transducer according to the excitation conditions of figure 4. We then simply apply opposite potentials to the adjacent strips and set the two excitation parameters to an integer value (typically one). We particularly focused our interest on LiNbO3 and more specifically the (YXl)/128Âř cut because of its advantageous electromechanical coupling. The numerical and experimental results are drawn in figures 6. At a glance, the correspondence of the measured admittance with the harmonic conductance reported in is rather convincing about the fact that the device operates as theoretically announced. When focusing on the low frequency contributions, one can see that they are composed of two principal wide band contributions modulated by resonance peaks, expected to correspond to wave reflection on the surrounding devices in experimental measurements. Whatever the origin of this effect, this measurement proofs that the device actually can generate waves with clear electrical contribution on the device admittance and that quite sharp resonance can be expected, although the operation frequency remains rather low (near 100 MHz), corresponding rather fairly to theoretical predictions. To better understand the operation of the transducer, we have plotted the deformed shape of the surface for the two guided modes in figures 7. It turns out that the propagation takes place in both case along the median trace of the dielectric dots. In that case, the wavelength is then equal to 35.5$\mu$m, yielding the observed contributions near 100 MHz. In this configuration, the

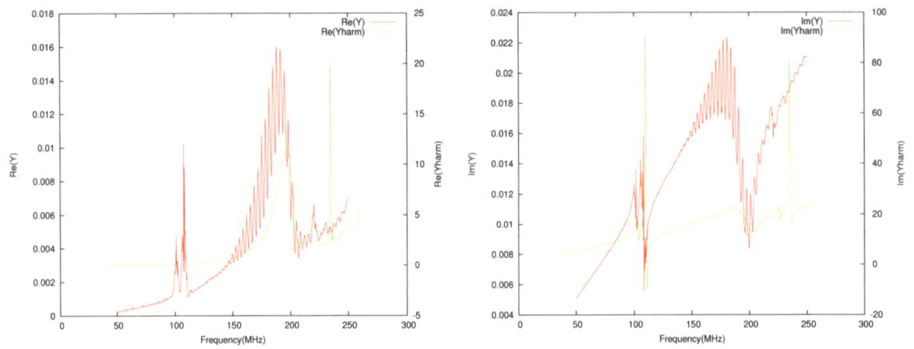


(a) Real part (b) Imaginary part

**Figure 6.** Harmonic admittance of the infinite 2D periodic grating of figure 5 focusing on the guided modes. In red experimental measurements and in green simulation.

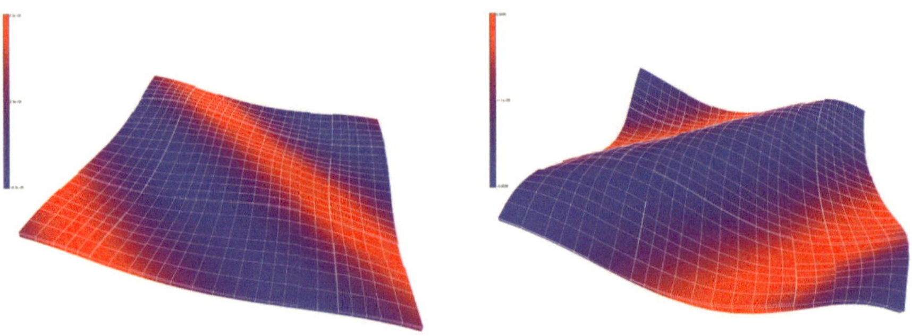

(a) Beginning of the stop-band (b) End of the stopband

**Figure 7.** Deformed mesh for both modes emphasized in figure 6 (isovalues correspond to displacement along x1)

transducer does not really allow for exciting complex propagation polarization but behaves more like a natural partially directionnal transducer (a small resonance at the beginning of the stop band but a large one at its end).

The improvements of experimental approaches lead to take into account new effects such as influence of SAW devices aperture. Indeed, up to now, the length of electrode is considered as infinite compare to the period of the grating in SAW resonators. So, the influence of the input buses and diffraction at the end of electrodes is neglected. However, this is a drastic assumption. Thus, to quantify such an influence, PML method is necessary to simulate the lateral leaky mode outside of the resonator. PML was already implemented in FEA [18, 19]. Nevertheless, a global approach to include PML in FEA is demonstrated below.

There are also more and more acoustic devices with finite lateral dimension such as BAW resonators or acoustic filter to avoid reflections at the end of SAW devices (Bragg mirrors). In some cases, it is possible to define a large periodicity in periodic FEA to model a non periodic structure and thus avoid the interaction between the neighbors (as in specific BAW resonators simulation). However, the computational time cost is very high. Moreover, in most cases, it is not possible to use periodic FEA with high width. For example, the simulation of non periodic SAW resonators cannot be achieved in this way. So, the non-periodic FEA is necessary as well as the related boundary conditions i.e. the BEM [20].

Below, the main advancements of the FEA scheme applied to acoustic resonators with finite lateral dimension are depicted (such as dual mode filters or realistic periodic SAW resonators with effective aperture). The PML method as well as the BEM are thus demonstrated separately even if they are used together to simulate embedded finite resonator on radiating medium for instance.

## 3. FEA improvements

### 3.1. Boundary Element method in non-periodic cases

A radiating surface in an non periodic acoustic problem can be included to the FEA formalism in the right hand side of the general variational equation (1).

The right hand of equation (1) limited to the stress part stands for the radiation part. It can thus be treated by using Green's function based relation [21],

$$T_{ij}n_j = Gi_{ijk}(\omega, S)n_j * u_k, \tag{5}$$

where $*$ denotes the convolution between the displacements $u$ and $Gi$, the inverse Green tensor relating the stress $T$ to the displacement. Knowing the Green tensor, one can just insert equation (5) in equation (1) to solve the problem with radiation boundaries conditions without any restriction. We developed a first approach of this method for isotropic radiation medium in further work [8]. However, we can't apply straight this method for anisotropic medium. Indeed, the spectral Green function $\hat{Gi}(\omega, s)$ (where $s$ is the slowness) is not holomorphe and so its inverse Fourier Transform (iFT) can't be computed directly for the general case (anisotropic medium). So to avoid this problem we use the reciprocal form of equation (5),

$$u_k = G_{ijk}(\omega, S) * T_{ij}n_j, \tag{6}$$

in which the direct green function $G$ relates the displacements to the stress [21]. This function is defined from its iFT in the slowness space,

$$G_{ijk}(\omega, x_1) = \frac{\omega}{2\pi}\int_{-\infty}^{+\infty} \hat{G}_{ijk}(\omega, s_1)\exp(j\omega s_1 x_1)ds_1, \tag{7}$$

where $\hat{}$ defines the Fourier Transform. $s_1$ is the slowness along the direction $x_1$. To avoid the problem of loading or time calculation, we use the canonical Green function $\hat{H}(s_1)$ [22] which is frequency independent. This function is related to the Green function as following

$$\hat{G}_{ijk}(\omega, s_1) = \frac{\hat{H}_{ijk}(s_1)}{\omega}. \tag{8}$$

Thus we are able to compute the canonical function before the FEA once and for all assuming the radiation medium is semi-infinite [23]. The actual Green function is obtained by including the equation (8) in equation (7),

$$G_{i2k}(\omega, x_1) = \frac{1}{2\pi}\int_{-\infty}^{+\infty} \hat{H}_{i2k}(s_1)\exp(j\omega s_1 x_1)ds_1. \tag{9}$$

(Here the notation takes into account the propagation and radiation assumptions of this work i.e., $j = 2$). However, the spectral canonical Green function $\hat{H}(s_1)$ gives rise to some singularities. So, we must to divide up this function into three parts [24]

$$\hat{H}_{i2k}(s_1) = \hat{H}_{i2k}^{(res)}(s_1) + \hat{H}_{i2k}^{\infty}(s_1) + \hat{H}_{i2k}^{(0)}(s_1) + \hat{H}_{i2k}^{R}, \tag{10}$$

where $\hat{H}_{i2k}^{(0)}(s_1)$ is the contribution for the slowness $s_1 = 0$, $\hat{H}_{i2k}^{\infty}(s_1)$ is the asymptotic one and $\hat{H}_{i2k}^{R}$ stands for the acoustic poles contribution. Thus, we can define $\hat{H}_{i2k}^{(res)}(s_1)$ as the residual Green function part without singularities. The computation in the spectral domain as well as in the spatial one always takes into account separately these four parts. The residual part can be numerically computed whereas the others need to be considered analytically [24].

Once, the Green function numerically and/or analytically defined, one can express the variational form of equation (6),

$$\begin{aligned}&\int_{x_1=-\infty}^{x_1=+\infty} \delta u_k^* u_k(x_1) dx_1 = \\ &\int_{x_1=-\infty}^{x_1=+\infty} \delta u_i^* \int_{x_1'=-\infty}^{x_1'=+\infty} G_{i2k}(\omega, x_1 - x_1') T_{i2}(x_1') dx_1' dx_1,\end{aligned} \tag{11}$$

in order to develop the Finite Element scheme for the radiation conditions,

$$\begin{aligned}&\sum_{e=1}^{N_e} \sum_{m=1}^{\eta_e} \delta u_k{}^{*(em)} \sum_{\mu=1}^{\eta_e} u_k^{(e\mu)} \int_{\Gamma_e} P^{(em)}(x_1^e) P^{(e\mu)}(x_1^e) dx_1 = \\ &\sum_{e=1}^{N_e} \sum_{m=1}^{\eta_e} \delta u_k{}^{*(em)} \sum_{\epsilon=1}^{N_e} \sum_{\mu=1}^{\eta_e} T_{i2}^{(\epsilon\mu)} \int_{\Gamma_e} P^{(em)}(x_1^e) \\ &\int_{\Gamma_\epsilon} P^{(\epsilon\mu)}(x_1^\epsilon) G_{i2k}(\omega, (x_1^e - x_1^\epsilon)) dx_1^\epsilon dx_1^e,\end{aligned} \tag{12}$$

where $N_e$ is the number of elements on $\Gamma_2$ and $\eta_e$ the number of nodes by element. $P^{(em)}$ are the FEA interpolation polynomials (first or second degree). The integrations from $-\infty$ to $+\infty$ are bound to the $\Gamma_2$ length. Indeed, anywhere else that the activated domain the Green function is null. Note also that the nodal displacement, variational unknown and stress are outside the convolution and the integrals on the elements.

So to match the FEA algorithm, we must write equation (12) in a nodal matrix relation,

$$< \delta u > (\Psi) \{u\} =< \delta u > (G) \{T\}, \tag{13}$$

where $(\Psi)$ is the nodal matrix relating the nodal vectors of the variational unknown and displacements. In the same way, $(G)$ is the nodal Green matrix relating the nodal vectors of the variational unknown and stress. Equation (13) is right whatever the variational unknown. So, on can write the reciprocal relation relating the stress to the displacement,

$$\{T\} = (G)^{-1} (\Psi) \{u\}, \tag{14}$$

where $(G)^{-1}$ is the inverse of the nodal Green matrix defined in equation (12,13). The general variational equation (1) must be also written in a matrix formulation (Only the elastic part is considered without losses of generality)

$$< \delta u > \left[K - M\omega^2\right] \{u\} =< \delta u > (\Upsilon) \{T\} . \tag{15}$$

Thus the final FEA/BEM system is obtained by including equation (14) in equation (15),

$$\left[K - M\omega^2 - (\Upsilon)(G)^{-1}(\Upsilon)\right] \{u\} = 0. \tag{16}$$

the global FEA/BEM system is similar to the ideal case with an additional correction which doesn't change the scheme to solve the problem.

## 3.2. Perfectly Matched Layer method

The PML method was first developped in electromagnetism [25] and well adapted to the Finite Difference in Time Domain (FDTD) method [26]. More recently, some works demonstrated the implementation of this approach for acoustic simulations based on FEA [18] [19]. The basic idea consists in rigorously simulating an exponential decrease of the acoustic field along at least one space direction. To clear the approach, let us consider the following incident plane wave

$$u = Aexp^{-j(k_x x - \omega)} = Aexp^{-j\omega(s_x x - t)} \tag{17}$$

We then consider that in the absorbing area, one can apply a geometrical transform in the complex plane to introduce the exponential decay. Since it must not modify the propagation phase, this transform can be written

$$\tilde{x} = x - jf(x) \tag{18}$$

where $f(x)$ growths from the origin of the absorbing area to its end along a defined rate. However, since this transform must be efficient for any frequency (we represent the problem in the spectral domain), it is wise to define this function as follow :

$$f(x_i) = \frac{1}{\omega} \int_0^{x_i} d(x)\mathrm{dx}, \tag{19}$$

$$d(x_i) = d_{max} \left(1 - \frac{(abs(x_i) - x_p)^2}{(x_a - x_p)^2}\right)^n, \tag{20}$$

which allows for an easy definition of the transform Jacobian linking the considered coordinate systems. This reads

$$\frac{1}{\partial \tilde{x}} = \frac{j\omega}{j\omega + d(x)} \frac{1}{\partial x} \rightarrow \frac{1}{\partial x} = \left(1 + \frac{d(x)}{j\omega}\right) \frac{1}{\partial \tilde{x}} = \alpha_x \frac{1}{\partial \tilde{x}} \tag{21}$$

Replacing $x$ by $\tilde{x}$ in (17) provides the wanted exponential decay if $f(x)$ unconditionally growths, imposing $d(x)$ even and positive to fulfil the absorbing condition for any $x$ (we assume the problem centred around $x = 0$). Since the absorbing function $d(x)$ is not frequency dependent, its efficiency should be constant along $\omega$. Conformably to Zheng and Huang [27], we develop a formulation based on the usual piezoelectricity equations, yielding significant modifications of the elastic, piezoelectric and dielectric constants to account for the absorption.

We now rewrite the elasticity equations in the absorbing region turning $x$ to $\tilde{x}$, using then (5) to express the result in the initial coordinates. As in [27], the absorbing effect is assumed along the three space directions for the sake of generality. The equilibrium equation then reads

$$-\rho\omega^2 u_i = \frac{\partial T_{ij}}{\partial \tilde{x}_j} = \frac{1}{\alpha_j}\frac{\partial T_{ij}}{\partial x_j} \tag{22}$$

where $\alpha_j$ is characterized by its specific function $d_j(x_j)$. $T_{ij}$ and $u_i$ respectively represent the dynamic stresses and displacements, and $\varrho$ is the mass density. We introduce a non symmetrical stress tensor, expressed in the transformed axis

$$\tilde{T}_{ij} = \frac{\alpha_1\alpha_2\alpha_3}{\alpha_j} C_{ijkl}\frac{\partial u_l}{\partial \tilde{x}_k} = \frac{\alpha_1\alpha_2\alpha_3}{\alpha_j\alpha_k} C_{ijkl}\frac{\partial u_l}{\partial x_k} = \tilde{C}_{ijkl}\frac{\partial u_l}{\partial x_k} \tag{23}$$

where $\tilde{C}_{ijkl}$ is the transformed elastic constant tensor relative to the absorption area. We multiply (22) by $\alpha_1\alpha_2\alpha_3$, thus yielding Newton relation for PMLs in the real coordinates

$$-\tilde{\rho}\omega^2 u_i = \frac{\partial \tilde{T}_{ij}}{\partial x_j} \tag{24}$$

where $\tilde{\rho} = \rho\alpha_1\alpha_2\alpha_3$ is the mass density relative to the transformed domain. Since the obtained form of the equilibrium equation complies with the classical expression for usual solids, it is liable to exploit the standard FEA formulation for PML as well, accounting for the frequency dependence of the transformed physical tensors. These developments of course can be extended to piezoelectricity by rewriting Poisson's equation and taking into account the piezoelectric coupling in the stress definition as follows

$$\begin{aligned}\tilde{T}_{ij} = \tfrac{\alpha_1\alpha_2\alpha_3}{\alpha_j}\left(C_{ijkl}\tfrac{\partial u_l}{\partial \tilde{x}_k} + e_{kij}\tfrac{\partial \varphi}{\partial \tilde{x}_k}\right) = \\ C_{ijkl}\tfrac{\partial u_l}{\partial x_k} + \tilde{e}_{kij}\tfrac{\partial \varphi}{\partial x_k}\end{aligned} \tag{25}$$

Poisson's equation expressed in the transformed system of axes reads

$$\frac{\partial D_i}{\partial \tilde{x}_i} = 0 \Rightarrow \frac{1}{\alpha_i}\frac{\partial D_i}{\partial x_i} = 0 \tag{26}$$

with $D_i$ the electrical displacement vector. To provide an homogeneous formulation, we proceed as for the stress definition (23), multiplying the electrical displacement by $\alpha_1\alpha_2\alpha_3$, yielding

$$\tilde{D}_i = \frac{\alpha_1\alpha_2\alpha_3}{\alpha_i} D_i \tag{27}$$

As for the propagation equation (22), the Poisson's condition is written accounting for these changes as

$$\frac{\partial \tilde{D}_i}{\partial x_i} = 0 \tag{28}$$

Conformably to the stress tensor transformation, we introduce modified piezoelectric and dielectric constants defined as follows

$$\tilde{D}_i = \frac{\alpha_1\alpha_2\alpha_3}{\alpha_i}\left(e_{ikl}\frac{\partial u_l}{\partial \tilde{x}_k} - \varepsilon_{ik}\frac{\partial \varphi}{\partial \tilde{x}_k}\right) = \tilde{e}_{ikl}\frac{\partial u_l}{\partial x_k} - \tilde{\varepsilon}_{ik}\frac{\partial \varphi}{\partial x_k} \quad (29)$$

We now are able to establish a FEA formulation exploiting these developments without fundamental changes of the existing code. Thus, equation (1) becomes

$$\iiint_{\Omega_{PML}}\left(\frac{\partial \delta u_i}{\partial x_j}\tilde{C}_{ijkl}\frac{\partial u_l}{\partial x_k} + \frac{\partial \delta u_i}{\partial x_j}\tilde{e}_{kij}\frac{\partial \phi}{\partial x_k} + \frac{\partial \delta \phi}{\partial x_i}\tilde{e}_{ijk}\frac{\partial u_j}{\partial x_k} - \frac{\partial \delta \phi}{\partial x_j}\tilde{\varepsilon}_{jk}\frac{\partial \phi}{\partial x_k} - \rho\omega^2 u_i \delta u_i\right) dV = \iint_{\Gamma_{PML}}\left(\delta u_i \tilde{T}_{ij} n_j + \delta\phi \tilde{D}_j n_j\right) dS \quad (30)$$

where $\Omega_{PML}$ and $\Gamma_{PML}$ are respectively the PML domain and its boundary.

We must note here that all the parameters of the absorbing polynomial (Eqs. (19) and (20)) are empirically defined. Indeed, the polynomial $d(x)$ must increase from zero to the maximum value $dmax$ by a progressive slope. The value of $d(x)$ must be zero at the beginning of the PML to verify the impedance match. Moreover, the derivative of $d(x)$ must be null too to avoid any singularities. However, the absorbing polynomial must increase as quick as possible to minimize the number of PML finite elements with respect to the better smoothy shape. Therefore one can choose the degree of polynomial $d(x)$ equal to 3. The depth of the PML domain $(x_p - x_a)$ is also defined according the absorbed waves. So, we defined that this length must be approximately equal to one wavelength.

## 4. Results

In this section, we first depict results from the Boundary Element Method based on the direct Green function. Next, the 2D and 3D results of the Perfectly Matched Layer method are exhibited. At last, we depict how powerful the mixed approached is on a 2D case.

### 4.1. A Green's function as boundary condition

#### *4.1.1. Validation by comparison with proofed periodic FEA scheme*

First, the non periodic numerical scheme developed here is compared to an existing periodic one that we developed in previous works [15]. In order to make this comparison, we choose the configuration depicted in figure 8. It seems to be odd to test a non periodic code with a periodic one. However, in the configuration of figure 8, the most of waves provided by the transducer propagate into the direction $x_2$. Of course, there are some components along the $x_1$ due to the coupling with the radiation surface or the finite width of the transducer. So, we choose a period $p = 1.2$mm which makes insignificant the coupling with nearest neighbor in the periodic case at $\Gamma_3$. This choice also allows us to neglect the reflections at the en of the mesh for the non periodic case on $\Gamma_3$. The results of the periodic computation have been

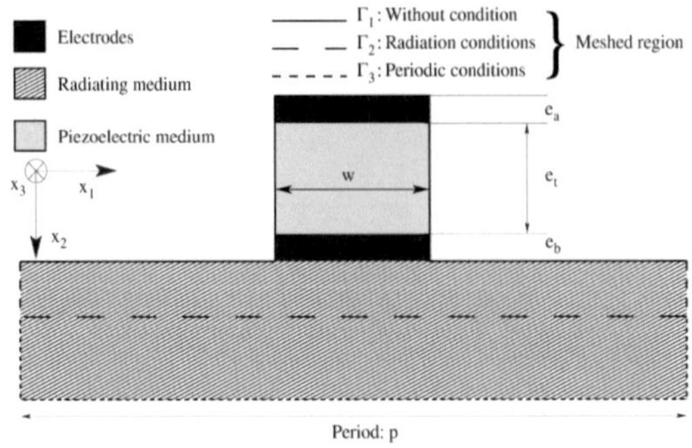


**Figure 8.** Chosen geometry for the validation by comparison with periodic FEA code. This configuration defines a single transducer providing mainly bulk wave.

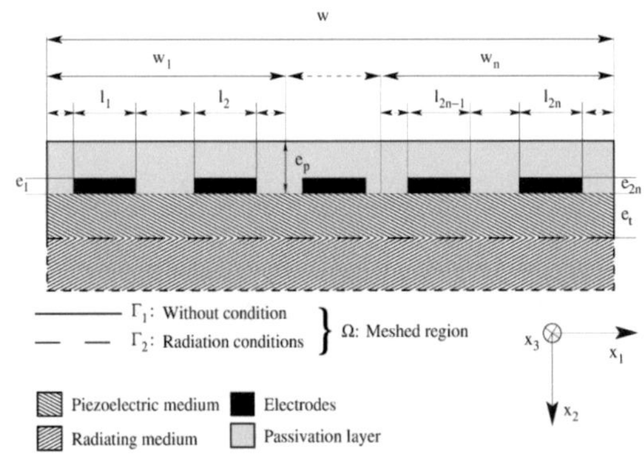


**Figure 9.** General geometry of resonator studied addressed by the investigated FEA-BEM. This configuration defines a non periodic interdigital transducers grating devoted to SAW applications

tested to be unchanged from the chosen period $p = 1.2\text{mm}$. The thickness of electrodes is set to zero and so we don't consider the influence of the mass loading and material. However, it's not a restriction due to the numerical method and a study of the influence of the electrode material could be achieved. The width of the piezoelectric transducer is set to $w = 100\mu\text{m}$, its thickness is $e_t = 4\mu\text{m}$. we also choose the same material for the radiation medium and the piezoelectric one. The constants used for computation are these of quartz with a YXl/36 cut. The surrounding medium is the vacuum. In figure 10, we show this comparison between the periodic and the non periodic code. In this computation, we consider all the degrees of freedom i.e., all modes can be observed (the in plane one as well as the shear one). The figure 10(a) depicts the conductance whereas figure 10(b) draws the susceptance both computed for the above configuration. The good agreement between these two different methods is the first proof of validation for the non periodic scheme presented in this work.

### *4.1.2. Validation by comparison with an other non periodic scheme*

The second test presented here is the comparison with a non periodic method based on the integral equations method [24]. The configuration is the one depicted in figure 9. It addresses a SAW devices of finite width equal to $w = \sum_{i=1}^{n} w_i$. $w_i$ is the width of the pattern and $n$ is the number of electrodes pairs. The pattern $k$ is defined by a pair of metallic electrodes. The widths and the thicknesses of each electrodes for the considered pattern are respectively denoted by $l_{2n-1}, e_{2n-1}$ and $l_{2k}, e_{2n}$. The metal ratio is constant and set to 0.5. The width of the

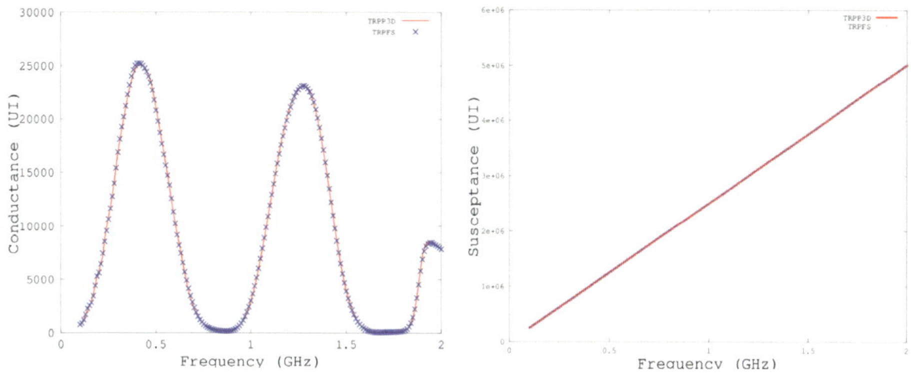


(a) Conductance ($\mu S$)

(b) Susceptance ($\mu S$)

**Figure 10.** Comparison between the results obtained with a proofed periodic computation (TRPP3D) and the method presented in this work (TRPFS). All the degrees of freedom are taken into account and the FEA interpolation is quadratic. All elements of the mesh are identical (triangle, 2 elements per micron in $x_1$ and 10 elements in $x_2$).

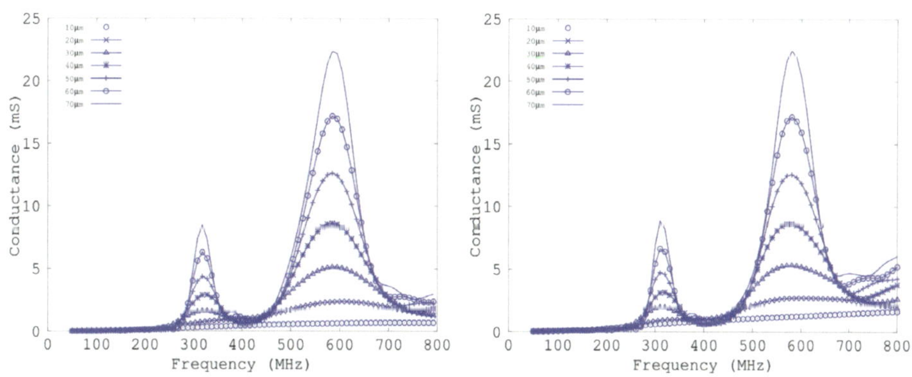


(a) Integral equations method

(b) FEA/BEM

**Figure 11.** Comparison between the results obtained with an integral equations based method and the FEA/BEM method presented in this paper. All the degrees of freedom are taken into account and the FEA interpolation is quadratic. The mesh is the same as previously.

piezoelectric medium is $w$ and its thickness is $e_t = 200nm$. The piezoelectric material constants used for computation are still these of quartz with a YXl/36 cut. The radiation medium is also the same as the previous test with periodic method. As previously, the surrounding medium is still the vacuum. Here, all the electrodes widths are equal to 2.5$\mu$m and they are without thickness i.e., there is no mass loading. So the total width of the SAW device is $w = 10n\mu$m. We increase the number of pairs of electrodes from 1 to 7. Thus, $w$ varies from 10$\mu$m to 70$\mu$m. The excitation is the excitation is an symmetric alternate potential for each pattern i.e. $+U/-U$ where $U = 0.5V$.

The results of this comparison are drawn in figure 11. The figure 11(a) depicts the results for the integral equations method. Those of the FEA/BEM methods are in figure 11(b). Once

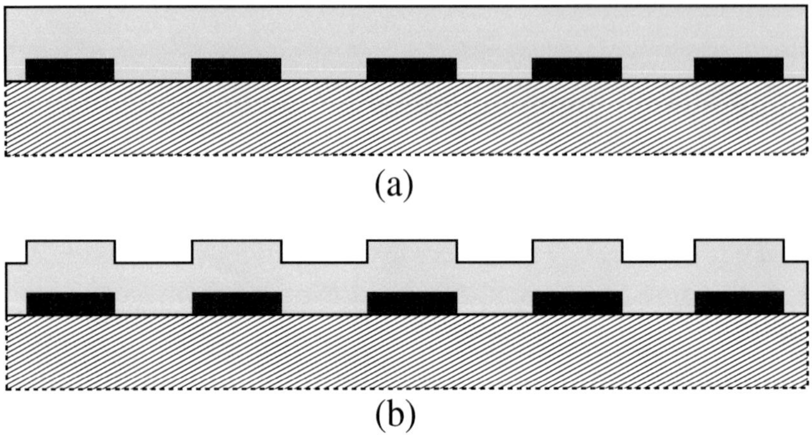


**Figure 12.** Geometrical configuration for a passivated electrodes grating with (a) non conformal and (b) conformal deposition. The electrodes thickness is either null or equal to 200$nm$. The other parameter still unchanged with the configuration depicted in figure 9.

again, the good agreement between the two method allows us to conclude that the FEA/BEM method is a good tools for the understanding the behavior of such a SAW devices.

### 4.2. Application to a wide non periodic passivated SAW devices

The method presented in this work is devoted to address non periodic passivated SAW devices among others. So, we give here some results about this configuration. We consider the same configuration as in the case depicted in figure 9. We only vary the thickness of electrodes. First we let them without thickness and look at the influence of the passivation layer and next we consider a electrodes thickness equal to 200$nm$. We then study again the device behavior against the passivation configuration i.e., without passivation, with conformal passivation and with non conformal passivation (See figure 12). The material constants of the passivation layer are to be the fused silicon ones. These results are depicted in figure 13. We can see that the global effect of then passivation layer red-shift the resonance frequencies for the non massive electrodes as well as for the massive one. Indeed, this layer acts as a mass loading effect. So, the resonance frequency must decreases with the mass over the electrodes. Moreover, this conclusion is corroborated by the evolution between the cyan and magenta curves of figure 13. The frequency is slightly lower for the conformal passivation which leads to a higher weight over electrodes. We also point out that the conductance level increases when we put the passivation layer. This comment can be explained by considering the velocity of the mode in the passivation layer is higher than in the guide. So the guidance is better for the case with passivation layer. These results are in good agreement with previous results obtained on periodic structures [15].

### 4.3. Perfectly Matched Layer Method: an other approach

The efficiency of the PML implemented in FEA is depicted in three parts. First, a 2D-case is investigated showing the absorbing due to the PML domain and the effects of the effects of the finite lateral size on the behavior of a SAW resonator. Next, the same study is repeated but in a 3D configuration in order to validate the general PML approach. At last, a realistic SAW

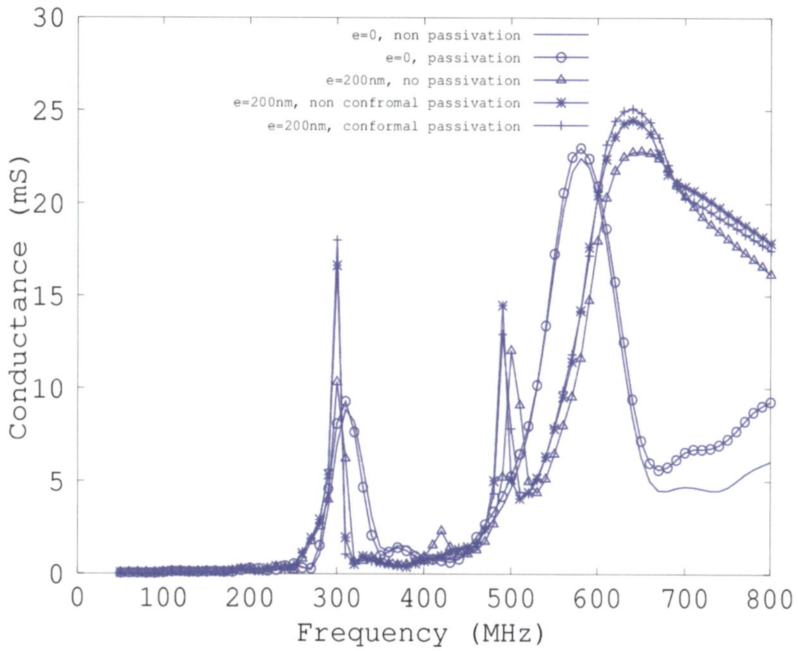


**Figure 13.** Conductance in (mS). Influence of the passivation layer on the acoustic device behavior (figure 9). All the degrees of freedom are taken into account and the FEA interpolation is quadratic. The same FEA mesh is used.

problem is addressed by considering the aperture of the resonator and absorbing the lateral leaky modes.

First, A 2D SAW resonator problem is addressed. The geometrical configuration is depicted in figure 14. A piezoelectric medium (quartz YXl/36) is driven by 4 pairs of electrodes. The electrodes are non massive and alternatively activated with $V = 1V$ and $V = 0V$. The period is $10\mu m$. The depth of PML on both sides is $35\mu m$. The eight of the mesh is $75\mu m$. The absorbing parameters are set to $dmax = 10^6$ and $n = 3$. The result of this simulation is shown in figure 15. We depicted the vibrations for the x-displacements in the XY plane. The vibrations in both the right and left PML domains are strongly reduced as and as they enter into. The decreasing factor is around $10^{-5}$. We also hardly observed the phenomena of diffraction due to the finite lateral size of the resonator. Indeed, weak lobes appears at the both lateral end of the grating and give rise to bulk wave and so losses in the medium.

Next, we repeat the same simulation as the one depicted in figure 14 but for a 3D geometry. The configuration is drawn in figure 16. The piezoelectric is once more quartz YXl/36. The same set of non massive electrodes is still powered in the same way. The absorbing conditions are also the same.

The x-displacement obtained by FEA/PML is shown in the perspective figure 17. It is clearly demonstrated that the vibration have the same absorption as in the 2D-case even if the absorbing factor is slightly worse. One more time, the losses in the medium can also be observed at the end of the resonator. We notice that there is no relection at the end of the mesh for both side edges and bottom boundaries.

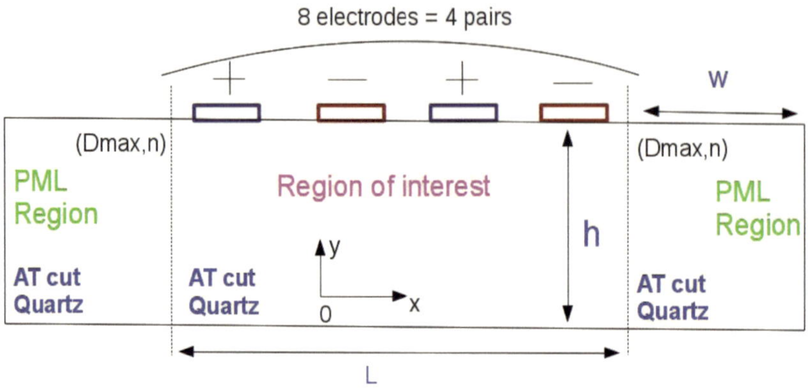


**Figure 14.** Characteristic configuration for finite SAW resonators considering the mixed PML/FEA approach. Eight non massive electrodes are activated alternatively ($V_1 = 1V$, $V_2 = 0V$). The width of electrodes is equal to $2.5\mu m$ and the period of the resonator is set to $10\mu m$. The piezoelectric medium is quartz YXl/36. Its thickness vary from zero. On the left and right parts of the scheme, two PML domains are set. No boundary conditions are defined neither at the top nor at the bottom. Th depth of the piezoelectric $h$ is chosen to avoid any interaction with the penetrating bulk wave at the bottom interface. $dmax = 10^6$ and $n = 3$.

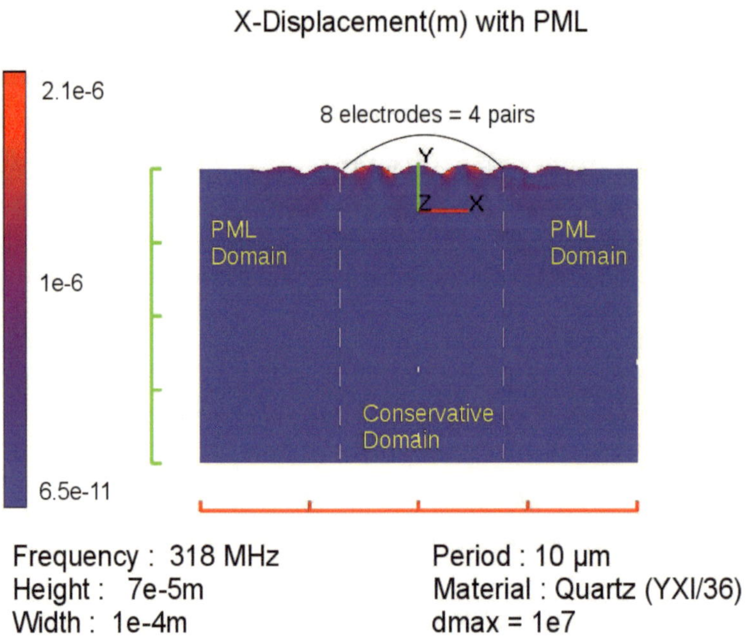


**Figure 15.** The vibrations for x-displacement in the XY plane for the 2-D problem depicted in figure 14. The piezoelectric medium is quartz YXl/36 activated by eight electrodes alternatively powered by $V = 1V$ and $V = 0V$. The PML domains are delimited by the white dashed line at the let and right sides. The frequency is $318MHz$. $h = 7e - 5m$, $dmax = 10^6$ and $n = 3$.

The last configuration highlight a new point to design SAW resonator. Indeed, up to now, the devices modelings were most often considered as 2D systems infinitely periodic in the direction of propagation and infinite in the perpendicular one. We just demonstrate that it

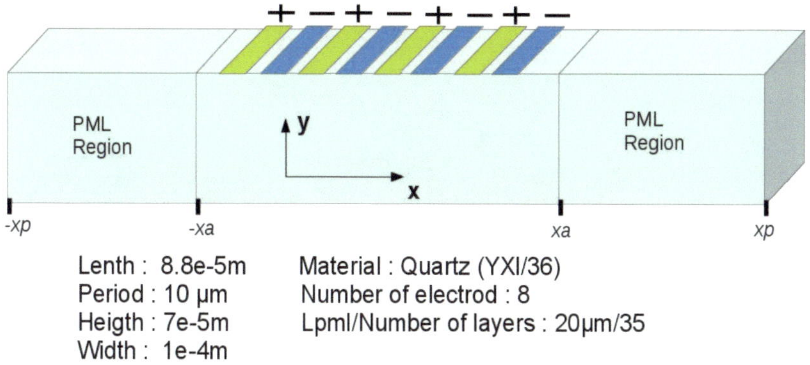


**Figure 16.** Same configuration as in figure 14 but in 3D-case. The z-direction is periodically infinite. So the electrodes are infinitly long in the z-direction. Eight non massive electrodes are activated alternatively ($V_1 = 1V, V_2 = 0V$).$h = 7e - 5m, dmax = 10^6$ and $n = 3$.

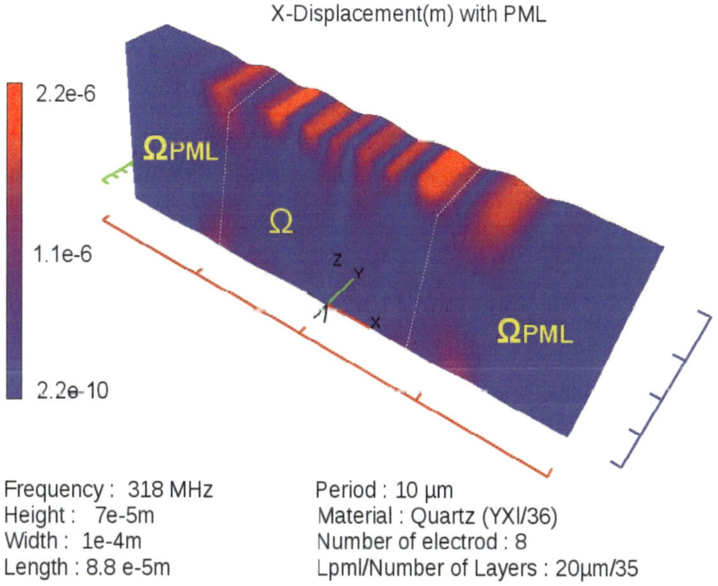


**Figure 17.** The vibrations for x-displacement for the 3-D problem depicted in figure 16. The piezoelectric medium is quartz YXl/36 activated by eight electrodes alternatively excited by $V = 1V$ and $V = 0V$. The PML domains are delimited by the white dashed line at the let and right sides. The frequency is $318MHz$. $h = 7e - 5m, dmax = 10^6$ and $n = 3$.

is now possible to take into account the effects due to finite dimension along the direction of propagation. In this part, we depict the way to address the problem of real aperture of a SAW resonator. In other words, we consider a finite dimension in the perpendicular direction of the propagation. In this study, the number of electrodes is infinite. The geometrical configuration is depicted in figure 18. The materials properties, excitation and dimensions of the grating are still the same as previously in figure 16. We now consider a length of the electrodes equal to $54\mu m$ for a period in the direction of propagation equal to $10\mu m$. The buses on the both right and left gratings are infinite along the propagation and $20\mu m$ wide. We also assume that the piezoelectric medium continue towards the infinity on the both sides of the resonator. The PML method allows this assumption.

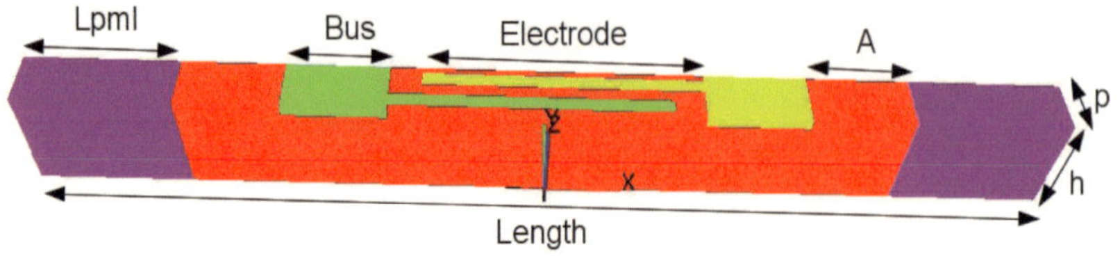


**Figure 18.** Configuration of an infinitely periodic SAW resonator in the propagation direction but with finite lateral dimension. The z-direction is infinitely periodic and the non massive electrodes are alternatively excited with $V_1 = 1V$ and $V_2 = 0V$. The piezoelectric medium is quartz YXl/36. The length of electrodes is $54\mu m$ while the period is $10\mu m$. $h = 30\mu m$, $dmax = 1e - 6$ and $n = 3$.

Once again, we show the vibration for the x-displacement in the perspective figure 19. The factor of absorption is still very high even if we notice a slight decrease. The $\Omega$ domain stands for the physical space in which the SAW is generated. We observe the Rayleigh wave in the middle of $\Omega$. On each side of this vibration, the presence of th buses is denoted by two maxima of displacement. This displacements give rise to a lateral mode which is reflected on the side edges if there is no activated PML. But, in figure 19, all PML are turned on. So, the lateral modes can be detected at the very end of the $\Omega$ area, just before the PML domains. However, due to the presence of the PML, this mode do not reflect on the side edges and moreover its amplitude decreases at the time it progress in the PML from the beginning to the end where it almost vanishes by then.

These three results show the efficiency of the combining of PML and FEA to simulate the effects due to the consideration of the real length or width of a SAW resonator. Thus, using this kind of method, we are able to simulate realistic effects in SAW. This method can also be applied to other kind of resonator.

### 4.4. The mixed BEM/PML approach

In the last part of the results exhibiting, the complete mixed BEM/PML approach is highlighted. This is done considering the general SAW configuration in figure 9 and adding PML area. This configuration is depicted in figure 20. A finite SAW resonator is characterized considering both the PML and the BEM as boundary conditions. The piezoelectric medium is a YXl/36 quartz. Its thickness can increase from zero to a non null value. It's excited by two non massive electrodes. The width of each electrodes is equal to $2.5\mu m$ and the length of the resonator is $10\mu m$. The excitation is symmetric and $V_1 = -V_2 = 0.5V$. The number of finite elements by length unit is equal to $1.6e6$ $elements/m$ along $x$ direction and $6e5$ $elements/m$ along $y$. The PML domains have the same FEA properties except for the absorbing conditions. The absorbing coefficient $dmax$ is equal to $1e - 6$ and the order of the polynomial is set to $n = 3$. The BEM stands for the bottom radiation on $\Gamma_\beta$ and simulates the propagation to the infinite in the same medium as the piezoelectric one. It's important to note

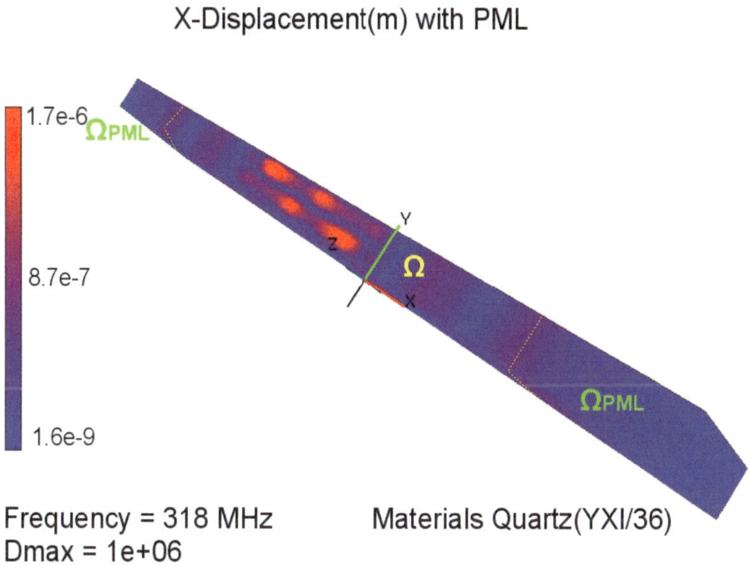


**Figure 19.** The vibrations for x-displacement for the 3-D problem depicted in figure 18. The piezoelectric medium is quartz YXl/36 activated by eight electrodes alternatively excited by $V = 1V$ and $V = 0V$. The PML domains are delimited by the white dashed line at the let and right sides. The frequency is $318MHz$. $h = 7e-5m$, $dmax = 10^6$ and $n = 3$.

that the BEM introduces an infinite slab in the $x$ direction under the piezoelectric medium. So if the thickness of the piezoelectric is zero, the two electrodes are placed on an infinite layer of YXl/36 quartz.

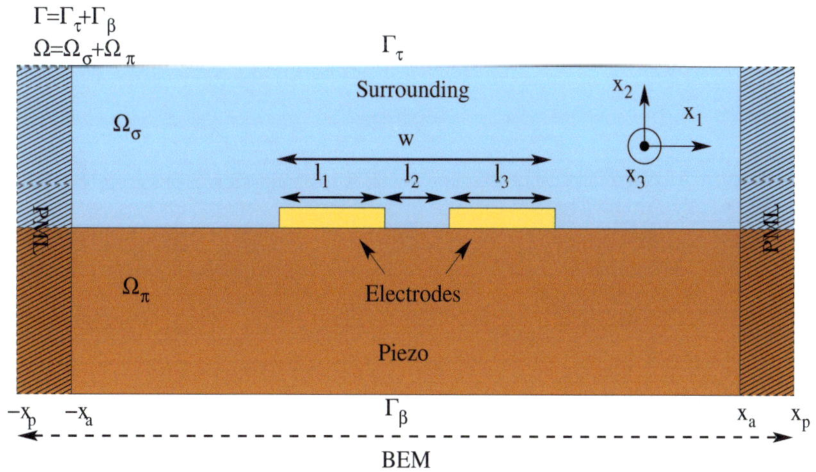


**Figure 20.** Characteristic configuration for finite SAW resonators considering the mixed BEM/PML approach. Two non massive electrodes are activated ($V_1 = 0.5V$, $V_2 = -0.5V$). The width of electrodes is equal to $2.5\mu m$ and the period of the resonator is set to $10\mu m$. The piezoelectric medium is quartz YXl/36. Its thickness vary from zero. Only the radiation on $\Gamma_\beta$ is active and the radiating medium is the same as the piezoelectric one. The surrounding medium is vacuum so there is no radiation on $\Gamma_\alpha$. On the left and right parts of the scheme, two PML domain are added.

We first focused on the effects of the thickness of the piezoelectric medium on the response of the resonator to an electric excitation. The excitation and the geometric configuration are depicted in figure 20. This response is exhibited in figure 21. We depict the logarithm of the conductance expressed in micro Siemens against the frequency for different thicknesses of the piezoelectric medium. The depth of the piezoelectric medium increases from zero for the red solid line to 2 microns for the red stars. It can be noted that when the thickness decreases the conductance converges to the result of the infinite slab, i.e. when the thickness is zero. This evolution can be explained by the diffraction at the end of the resonator when the depth is non zero. So even is the behavior is nearly the same whatever the thickness of the piezoelectric, the influence of this parameter is actually important.

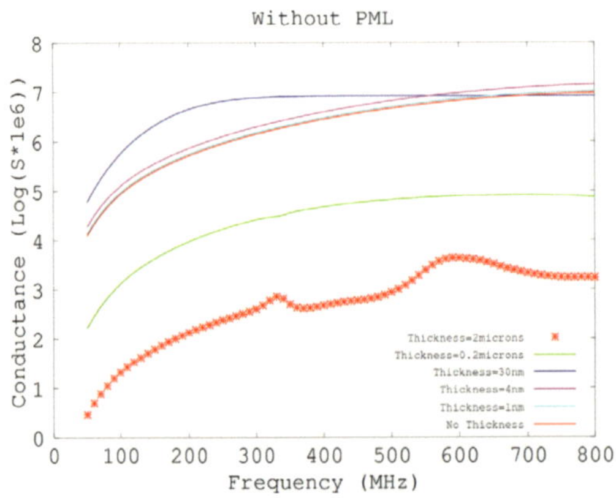


**Figure 21.** The logarithm of the conductance in $\mu S$ against the frequency for different thicknesses of the piezoelectric medium when the PML is switch off. The radiation in the substrate is simulated with BEM. The width of the transducer is finite and equal to $10\mu m$ and the PML domains vanish. The electrodes are non massive. The number of finite elements by length unit is equal to $1.6e6$ $elements/m$ along $x$ direction and $6e5$ $elements/m$ along $y$.

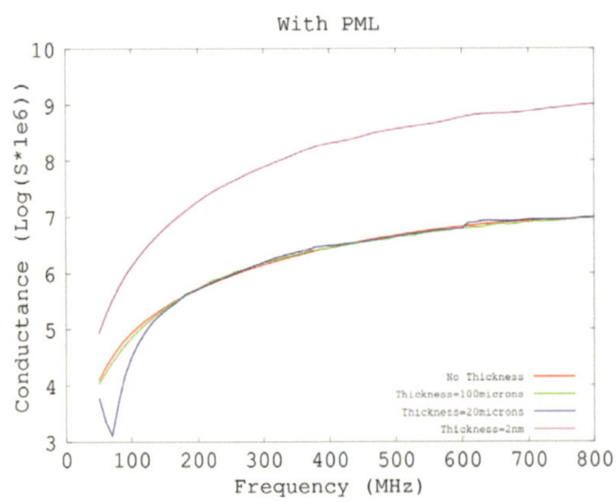


**Figure 22.** The logarithm of the conductance in $\mu S$ against the frequency for different thickness of the piezoelectric medium when the PML is switch on. Except for PML domains, the configuration is the same as in the figure 21. The width of PML Domains is equal to $30\mu m$ for the both right and the left part. The number of finite elements by length unit is the same as outside PML domains. The absorbing coefficient is set to $dmax = 1e6$ and the polynomial order is equal to $n = 3$.

So to highlight the PML efficiency, we compare the results of the thickness equal to zero in figure 21 to the non-zero ones when the PML are activated. The same configuration as in figure 21 is used for the non PML area, we only add two PML on the right and the left parts. The depth of the PML domains is equal to $x_p - x_a = 3\mu m$. The absorbing coefficients are defined above. Again, the variation of the conductance against the frequency is depicted for different thickness of piezoelectric when PML is switch on in figure 22. We can notice again that the conductance converges to the result for a zero thickness when the piezoelectric depth increases. That means there is a non zero interaction between PML and BEM giving rise to undesirable effects for low thickness. So, It's again important to notice that the mixed of the two boundary conditions PML and BEM must be used in a drastic validity domain. One must consider a sufficient thickness of piezoelectric medium or used a zero thickness without PML. The last way can be used if no surrounding medium is taken into account. However, in the configuration of figures 12, the PML must be used for an infinite passivation layer and we must check if the simulation is in the validity domain.

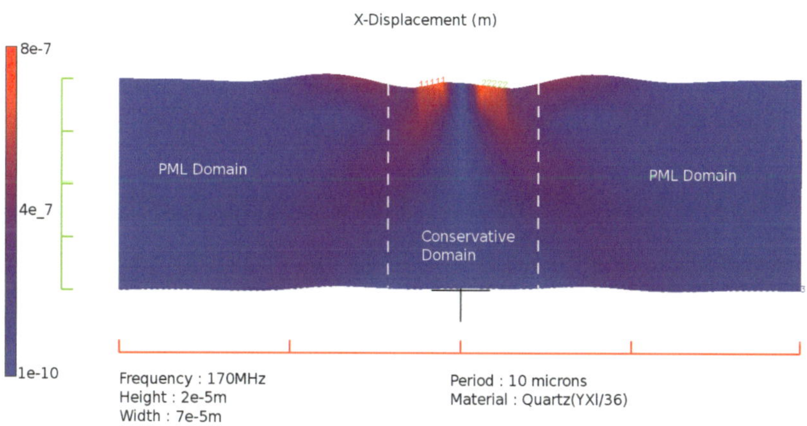


**Figure 23.** The map of x-displacement in the XY plane for the fixed frequency $170MHz$. The configuration is the one depicted in figure 20. The height of the piezoelectric medium is set to $2e5m$. The electrodes are denoted by the references 1 and 2. The PML domain are delimited by the white dashed line. The BEM is set at the bottom of the meshed domain also in PML area.

The last part of the results on mixed PML/BEM approach shows the effects of the boundary conditions on the $x$-displacement in the sagittal plane (XY). The distortion is depicted in figure 23. This is due to the electric excitation of the piezoelectric medium located in references 1 and 2. This result is obtained for a fixed frequency equal to 170MHz for the previous geometrical configuration of figure 22 for a $20\mu m$ thickness. The PML areas are denoted by the white dashed line and the conservative part (or the physical one) is under the electrodes. First, we can notice that the $x$-displacement is absorbed in the PML domains with a $1e4$ factor. Indeed, we really pay attention to the validity domain and thus the PML actually acts as a propagation towards the infinity in the same material that initially. Secondly, we can also note the presence of lobes on the left and the right parts of the resonator due to its finite lateral dimension. We already signified the lateral absorbed ones in the PML. However, there are two other which propagate in the $y$-direction. They are not only absorbed by PML too but also not reflected at the BEM interface indicating thus that the Green radiation conditions work well too. These results show the possibility of combining these two boundary conditions.

## 5. Conclusions

In this paper, we first demonstrated the theoretical and numerical basis of FEA. We next illustrated the state of the art in simulation based on FEA for 2D and 3D acoustic devices (especially SAW resonators). These examples introduced how the last research in acoustic research needs further improvements of the numerical tools. For instance, the wide and finite SAW resonators cannot be well numerically understood without a new theoretical approach. We drew these numerical improvement based on a finite FEA combining with two different kind of boundary conditions i.e. the PML method and BEM. At last, we show the efficiency of these approach for several example and more precisely we also demonstrated that we can use simultaneously the two boundary conditions to allow the simulation of realistic resonators e.g. we can consider the numerical aperture of an acoustic device.

So we introduced here a new complete numerical based on FEA combining with PML and BEM. We thus are capable to simulate a wide spectrum of resonators. We can simulate periodic as well as non periodic devices. Moreover, new parameters can be considered e.g. aperture or number of electrodes (for SAW resonators). In other worlds, we have now a strong and efficient numerical tool allowing the simulation of nearly all the configurations for SAW and BAW resonators.

## Author details

T. Laroche and S. Ballandras
*FEMTO-ST institute, Time & Frequency departement, France*

## 6. References

[1] Tobolka, G.: Mixed matrix representation of SAW transducers, IEEE Transactions on Sonics and Ultrasonics 26(6), pp426-428, 1979

[2] Kubat, F., Ruile, W., Rosler, U., Reindl, L.M.: A numerical method for calculating the dynamic stress in SAW devices, Microelectronic Engineering 82(3-4, SI), 670-674, DEC 2005, European Workshop on Materials for Advanced Metalization, Dresden, Germany, 06-09, 2005

[3] Hermelin, D., Daniaux, W., Ballandras, S., Belgacem, B.: Fabrication of Surface Acoustic Wave wireless pressure sensor, 2009 Joint meeting of the European frequency and time forum and the IEEE International Frequency Control Symposium, VOLS 1 and 2, pp96-99, 2009,

[4] Ballandras, S., Wilm, M., Edoa, P.F., Soufyane, A., Laude, V., Steichen, W., Lardat, R.: Finite-element analysis of periodic piezoelectric transducers, Journal of Applied Physics 93(1), pp702-711, JAN 1 2003

[5] Wilm, M., Ballandras, S., Reinhardt, A., Laude, V., Lardat, R., Armati, R., Daniau, W., Lanteri, F., Gelly, J.F., Burat, O.: 3-D mixed finite-element/boundary-element model for the simulation of periodic ultrasound transducers radiating in layered media, 2003 IEEE Ultrasonics Symposium Proceedings, VOLS 1 and 2, 1654-1657

[6] Wang, W.B., Zhang, X.D., Shui, Y.G., Wu, H.D., Zhang, D., Plessky, V.P.: Minimizing the bulk-wave scattering loss in dual-mode SAW devices, IEEE Transactions on Ultrasonics Ferroelectrics and Frequency Control 53(1), 193-198, JAN 2006

[7] Ballandras, S., Laude, V., Pastureaud, T., Wilm, M., Daniau, W., Reinhardt, A., Steichen, W., Lardat, R.: A FEA/BEM approach to simulate complex electrode structures devoted to guided elastic wave periodic transducers, 2002 IEEE Ultrasonics Symposium Proceedings, VOLS 1 and 2, 321-324, Eds: Yuhas, DE and Schneider, SC, 2002, IEEE International Ultrasonic Symposium, MUNICH, GERMANY, OCT 08-11, 2002

[8] Wilm, M., Ballandras, S., Laude, V.: Non periodic acoustic devices radiating in semi-infinite solids simulated by a combination of finite element analysis and a boundary element method, 2004 IEEE Ultrasonics Symposium, Vols 1-3, 1620-1623, Eds: Yuhas, MP, 2004, IEEE Ultrasonics Symposium, Montreal, CANADA, AUG 23-27, 2004

[9] Pastureaud, T., Daniau, W., Laude, V., Wilm, M., Malecamp, Y., Ballandras, S.: Characterization and prediction of transverse plate resonators built using mixed strip and groove gratings, 2001 IEEE Ultrasonics Symposium Proceedings, VOLS 1 and 2, 93-96, Eds: Yuhas, DE and Schneider, SC, 2001, IEEE International Ultrasonic Symposium, ATLANTA, GA, OCT 07-10, 2001

[10] Tiersten, H.F., Hamilton's principle for linear piezoelectric media, Proc. of the IEEE, Vol. 55, pp. 1523- 1524, 1967.

[11] Courant, R., Hilbert, D.: Methods of Mathematical Physics, Interscience Publishers, Inc., 1953

[12] Zienkiewicz, O. C., Taylor, R. L., Zhu, J. Z.: The finite element method: its basis and fundamentals, Butterworth-Heinemann, 2005

[13] Eernisse, E.P., Variational method for electro-elastic vibration analysis, IEEE Transactions on Sonics and Ultrasonics, SU-14, pp. 153-160, 1967.

[14] Clatot, S., Laude, V., Reinhardt, A., Wilm, M., Daniau, W., Ballandras, S., Lardat, R., Solal, M.: Sensitivity of interface acoustic waves to the nature of the interface, 2003 IEEE Ultrasonics Symposium Proceedings, VOLS 1 and 2, 2126-2129, Eds: Yuhas, DE and Schneider, SC, 2003, IEEE International Ultrasonics Symposium, Honolulu, HI, OCT 05-08, 2003

[15] Ballandras, S., Lardat, R., Wilm, M., Pastureaud, Th., Reinhardt, A., Champavert, N., Steichen, W., Daniau, W.; Laude, V., Armati, R., Martin, G., A mixed finite element/boundary element approach to simulate complex guided elastic wave periodic transducers, Journal of Applied Physics, pp014911, 105, 2009.

[16] Daniau, W., Ballandras, B., Briot, J.B., Proceeding of the IEEE International Frequency Control Symposium, pp. 800-806, 1997

[17] Daniau, W., Baron, T., Garcia, J., Laroche, T., Ballandras, S., A 2D Transducer Structure for the Excitation of Surface Acoustic Wave, EFTF 2010, 24th European Frequency and Time Forum, April 2010

[18] Mayer, M., Zaglmayr, S., Wagner, K.,Schöberl, J., Perfectly Matched Layer Finite Element Simulation of Parasitic Acoustic Wave Radiation in Microacoustic Devices, 2007 IEEE Ultrasonics Symposium Proceedings

[19] Tajic, A., Volatier, A., Aigner, R., Solal, M., Simulation of Solidly Mounted BAW Resonators using FEM combined with BEM and/or PML, 2010 IEEE International Ultrasonics Symposium Proceedings

[20] Laroche, T., Ballandras, S., Daniau, W., Garcia, J., Dbich, K., Mayer, M., Perois, X., Wagner, K., Mixed finite element analysis/boundary element method based on canonical Green's function to address non periodic acoustic devices., International Ultrasonics Symposium, 2011

[21] Ventura, P.,Hodé, J.M., Solal, M., Desbois, J. Ribbe, J., Numerical methods for SAW propagation characterization, Proc. of the IEEE Ultrasonics Symposium, pp175-186, 1998.
[22] Hodé, J.M., Desbois, J., Original basic properties of the Green's functions of a semi-infinite piezoelectric substrate, Proc. of the IEEE Ultrasonics Symposium, pp131-136, 1999.
[23] Boyer, L., Étude des phénomènes de réflexion/réfraction dóndes planes acoustiques dans les milieux piézoélectriques, Thèse de lÚniversité de Paris VII an Acoustique Physique, Paris, 1994
[24] Ribbe, J., On the coupling of integral equations and finite element/Fourier modes for the simulation of piezoelectric surface wave component, Phd Thesis, École Polytechnique, 2002.
[25] Berenger, J.P., Three-dimensional perfectly matched layer for the absorption of electromagnetic waves, J. Comp. Phy. 127, 363–79, 1996
[26] Laroche, T., Baida, F.I., Van Labeke, D., Three -dimensional finite-difference time domain study of enhanced second-harmonic generation at the end of a apertureless scanning near-field optical microscope metal tip, josab 22, 1045–1051, 2005
[27] Zheng, Y., Huang, X., Anisotropic Perfectly Matched Layers for Elastic Waves in Cartesian and Curvilinear Coordinates, 2002 MIT Earth Resources Laboratory Industry Consortium Report

Chapter 12

# Precise Analysis and Design of Multi-Layered Acoustic Wave Devices with Bragg Cell

Yongqiang Guo and Weiqiu Chen

Additional information is available at the end of the chapter

http://dx.doi.org/10.5772/54799

## 1. Introduction

The bulk acoustic wave (BAW) devices first emerged in 1920s and the surface acoustic wave (SAW) devices first appeared in 1960s (Royer & Dieulesaint, 2000). Since invented, these acoustic wave devices have been improved greatly in their performance and applications, along with significantly extended working parameters and application areas (Royer & Dieulesaint, 2000; Hashimoto, 2000). Nevertheless, in the last two decades, even more rigorous demands such as high operational frequency, high sensitivity, high reliability, multiple functionality, broad environment applicability, low attenuation and low cost, arise from the consumer, commercial and military applications. These demands challenge the conventional acoustic wave devices in which single crystalline piezoelectric materials are used as the wave medium. Therefore, the scheme of innovative acoustic wave devices utilizing piezoelectric multi-layered (stratified) structures was presented to cater for these demands. Fortunately, the successes of thin film deposition, etching and lithography technologies lead to the availability of piezoelectric multi-layered structures (Benetti et al., 2005). Recently, high-performance acoustic wave devices with multi-layered structures have been contrived and successfully fabricated (Kirsch et al., 2006; Benetti et al., 2008; Nakanishi et al., 2008; Brizoual et al., 2008). To further reduce the acoustic loss and enhance the quality factor of the multi-layered acoustic wave devices, Bragg Cell composed of many thin periodic alternate high- and low-impedance sublayers can be inserted between the propagation layer and the substrate. Efforts have been made on the fabrication of integrated piezoelectric multi-layered materials with Bragg Cell (Yoon & Park, 2000) and on the realization of superior multi-layered acoustic wave devices with Bragg Cell (Chung et al., 2008), especially aiming at the film bulk acoustic resonators (FBAR).

To ensure the well and stable performance of multi-layered acoustic wave devices, clear understanding of their operation status, especially the acoustic wave propagation behavior, is indispensable in the design process. Therefore, accurate and reliable modeling methods are necessary. By far, three sorts of matrix methods, including the analytical methods based on continuous (distributed-parameter) models, numerical methods based on discrete models and analytical-numerical mixed methods, have been presented for analyzing multi-layered acoustic wave devices. Analytical matrix methods, such as the transfer matrix method (TMM) (Lowe, 1995; Adler, 2000), the effective permittivity matrix method (Wu & Chen, 2002), the scattering matrix method (Pastureaud et al., 2002), and the recursive asymptotic stiffness matrix method (Wang and Rokhlin, 2002), usually give accurate results with low computational cost. However, some of these analytical methods are numerically instable. One reason is that both exponentially growing and decaying terms with respect to frequency and thickness are incorporated in a same matrix, and the other is that matrix inversion is involved in the formulation. For example, TMM ceases to be effective for cases of high frequency-thickness products. Tan (2007) compared most analytical methods in their mathematical algorithm, computational efficiency and numerical stability. Very recently, Guo et al. (Guo, 2008; Guo & Chen, 2008a, 2010; Guo et al., 2009) have presented a new version of the analytical method of reverberation-ray matrix (MRRM) formerly proposed by Pao et al. (Su et al., 2002; Pao et al, 2007), based on three-dimensional elasticity/piezoelectricity (Ding & Chen, 2001), state-space formalism (Stroh, 1962) and plane wave expansion for the analysis of free waves in multi-layered anisotropic structures. The new formulation of MRRM deals with the exponentially growing and decaying terms separately and refrains from matrix inversion. It is a promising analytical matrix method, which bearing unconditionally numerical stability, for accurately modeling the multi-layered acoustic wave devices (Guo and Chen, 2008b). Numerical methods, including the finite difference method (FDM), the finite element method (FEM), the boundary element method (BEM) and the hybrid method of BEM/FEM (Makkonen, 2005), are powerful for modeling multi-layered acoustic wave devices with complex geometries and boundaries. However, they are less accurate and efficient, especially for high frequency analysis. The reason is that the wave media should be modeled by tremendous elements of small size to ensure computational convergence. Analytical-numerical mixed methods, such as the finite element method/boundary integral formulation (FEM/BIF) (Ballandras et al., 2004) and the finite element method/spectral domain analysis (FEM/SDA) (Hashimoto et al., 2009; Naumenko, 2010), are usually powerful for modeling both the small-sized accessories and the large-dimensioned wave media with high accuracy. They seem to be promising as long as the uniformity of their formulation is improved (Hashimoto et al., 2009; Naumenko, 2010). Although some of these matrix methods are extendable to modeling the multi-layered acoustic wave devices with Bragg Cell, there are few investigations focused on this subject. Few studies have been reported on the effects of a Bragg Cell on wave propagation characteristics of multi-layered acoustic wave devices either. To the authors' knowledge, all existing references aimed at Bragg Cell in solidly mounted resonators. Zhang et al. (2006, 2008) and Marechal et al. (2008) studied both the resonant transmission in Bragg Cell and acoustic wave propagation in multi-layered bulk acoustic devices with Bragg Cell. Tajic et al., (2010) presented FEM combined with BEM and/or PML to simulate the solidly mounted BAW resonators with Bragg Cell. The formation mechanisms of the frequency bands in Bragg Cell are still an untouched topic. It should be pointed out that the Bragg Cell, as a kind of reliable wave guiding and isolating structure, is potential for utilizing in multi-layered acoustic wave devices working with various acoustic modes including Rayleigh modes, Love modes, Lamb modes, SH modes and bulk longitudinal/transversal modes, so as to improve their performances (Yoon & Park, 2000; Chung et al., 2008). Moreover, for acoustic wave devices working with a specific acoustic mode, other spurious modes inevitably exist. Therefore, for appropriately designing the multi-layered acoustic wave devices with Bragg Cell, modeling methods should be established by considering various wave modes and based on an integrated model, which reckoning on the propagation media, electrodes, Bragg Cell, support layer and substrate. In addition, for appropriately designing the Bragg Cell to improve the performance of multi-layered acoustic wave devices, the features and the mechanisms of frequency bands in the Bragg Cell should be studied. The influence of inserted Bragg Cell on acoustic wave propagation in the working layer should also be clearly revealed.

In this chapter, the wave behavior in the Bragg Cell and the design rules of a Bragg Cell are studied by taking SH wave mode as illustration and by using the Method of Reverberation-Ray Matrix (MRRM). The MRRM is also proposed for accurate analysis and design of multi-layered acoustic wave devices with Bragg Cell, based on an integrated model involving the effects of electrodes, Bragg Cell, support layer and substrate on the working media. Firstly, the MRRM is extended to the analysis of SH wave dispersion characteristics of a ternary Bragg Cell, whose unit cell consisting of three isotropic layers. Based on the resultant closed-form dispersion equations, the formation mechanisms of the SH wave frequency bands are revealed. The design rules of the Bragg Cell according to specific isolation requirements of SH waves are summarized. Secondly, the integrated model, which incorporates the effects of electrodes, Bragg Cell, support layer and substrate on the working piezoelectric media by modeling them as individual non-piezoelectric or piezoelectric layers, is proposed for accurately analyzing acoustic wave propagation in multilayered acoustic wave devices. The formulation of MRRM for the integrated multi-layered structures based on the state space formalism is derived, by which the propagation characteristics of waves can be investigated. In view of the achieved dispersion characteristics, the operating status of various acoustic wave devices can be decided. Thirdly, numerical examples are given to validate the proposed MRRM, to show the features and the formation of SH-wave bands in the Bragg Cell and to indicate the resonant characteristics of multi-layered acoustic wave devices. Finally, conclusions are drawn concerning the SH wave behavior in the Bragg Cell, the advantages of the integrated model and MRRM, and the resonant characteristics of multi-layered acoustic wave devices.

## 2. The features and formation of SH-wave bands in the Bragg Cell

Consider an infinite periodic layered structure with each unit cell containing three isotropic elastic layers. A unit cell is depicted in Fig. 1, which can completely determine the band features of the infinite periodic layered structure by invoking the Floquet-Bloch principle (Mead, 1996). The surfaces and interfaces of the unit cell are denoted by numerals 1 to 4 from top to bottom, and the layers are represented by numerals 1 to 3 from top to bottom. Due to the isotropy of the layers, the in-plane wave motion is decoupled from the out-of-plane one. We limit our discussion to the out-of-plane (transverse) wave motion, i.e. only the SH type mode is present.

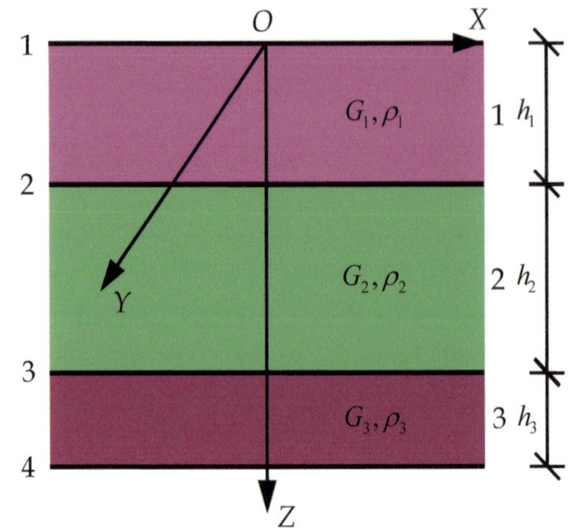


**Figure 1.** The schematic of the unit cell of a periodic ternary layered structure and its description in the global coordinates

## 2.1. SH wave dispersion characteristics of the Bragg Cell

Within the framework of the method of reverberation-ray matrix (MRRM) (Su et al., 2002; Pao et al., 2007; Guo, 2008), constituent layers of the unit cell are individually described in the corresponding local dual coordinates. Fig. 2 depicts the local dual coordinates of a typical layer $j$ ($j=1,2,3$) with its top and bottom surfaces denoted respectively as $J\,(=j)$ and $K\,(=j+1)$, and the SH wave amplitudes along the thickness in the typical layer $j$ as the wavenumber along $X$ is $k$ for all of the constituent layers. Meanwhile, superscripts $JK$ or $KJ$ will be attached to physical variables of the typical layer $j$, which is also called as $JK$ or $KJ$ according to the related coordinates $(x^{JK}, y^{JK}, z^{JK})$ or $(x^{KJ}, y^{KJ}, z^{KJ})$, to indicate the layer and its pertaining coordinate system. The displacement and stress are deemed to be positive as they are along the positive direction of the pertaining coordinates.

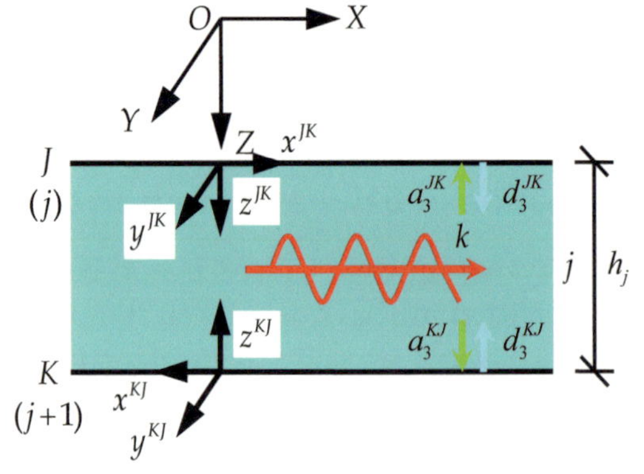


**Figure 2.** Description of a typical constituent layer $j$ of the unit cell in local dual coordinates

According to the elastodynamics of linear isotropic media (Eringen & Suhubi, 1975), the plane wave solutions to the out-of-plane displacement $v$ and shear stress $\tau_{zy}$ in any constituent layer $JK$ (or $j$) in its pertaining coordinates $(x^{JK}, y^{JK}, z^{JK})$ can be expressed as follows (with superscripts $JK$ omitted)

$$v(x,z,t)=\hat{v}(z)\mathrm{e}^{\mathrm{i}(\omega t-kx)}=(a_3\,\mathrm{e}^{\mathrm{i}\gamma_s z}+d_3\,\mathrm{e}^{-\mathrm{i}\gamma_s z})\mathrm{e}^{\mathrm{i}(\omega t-kx)}, \tag{1}$$

$$\tau_{zy}(x,z,t)=\hat{\tau}_{zy}(z)\mathrm{e}^{\mathrm{i}(\omega t-kx)}=\zeta_3(a_3\,\mathrm{e}^{\mathrm{i}\gamma_s z}-d_3\,\mathrm{e}^{-\mathrm{i}\gamma_s z})\mathrm{e}^{\mathrm{i}(\omega t-kx)}, \tag{2}$$

where $\hat{v}(z)$ and $\hat{\tau}_{zy}(z)$ signify the corresponding quantities in the frequency-wavenumber ($\omega-k$) domain (i.e. the transformed quantities), $\mathrm{i}=\sqrt{-1}$ is the imaginary unit, $\gamma_s$ ($=+\sqrt{k_s^2-k^2}$ or $=-\sqrt{k_s^2-k^2}$) and $k_s=\omega/c_s$ are respectively the $z$-direction shear wavenumber and the total shear wavenumber with $\omega$ being the circular frequency and $c_s=\sqrt{G/\rho}$ the shear wave velocity, $\zeta_3=\mathrm{i}\gamma_s G$ is the shear stress coefficient, constants $G$ and $\rho$ are respectively the shear modulus and the mass density. It is clearly seen that the terms with $a_3$ and $d_3$ at the right-hand sides of Eqs. (1) and (2) signify backward and forward traveling waves along the thickness coordinate, i.e. the arriving and departing waves relative to the surface $J$, with $a_3$ and $d_3$ being the corresponding undetermined wave amplitudes.

First, we consider the spectral equations within the layers. The transformed displacement $\hat{v}$ (stress $\hat{\tau}_{zy}$) at an arbitrary plane $z^{JK}$ of any layer $j$ expressed in one coordinate system $(x^{JK}, y^{JK}, z^{JK})$ should be compatible with that expressed in the other coordinate system $(x^{KJ}, y^{KJ}, z^{KJ})$, due to the uniqueness of the physical essence. Referring to the sign convention of displacement (stress), we have

$$\hat{v}^{JK}(z^{JK})=\hat{v}^{KJ}(h^{JK}-z^{JK}), \tag{3}$$

$$\hat{\tau}_{zy}^{JK}(z^{JK})=-\hat{\tau}_{zy}^{KJ}(h^{JK}-z^{JK}). \tag{4}$$

Substituting Eq. (1) into Eq. (3) and Eq. (2) into Eq. (4), and noticing the functions $\mathrm{e}^{\mathrm{i}\gamma_s z}$ and $\mathrm{e}^{-\mathrm{i}\gamma_s z}$ are nonzero for finite $\gamma_s$ and $z$, one obtains the phase relation of layer $j$ ($JK$ or $KJ$)

$$a_3^{KJ}=\mathrm{e}^{-\mathrm{i}\gamma_s^{JK}h^{JK}}\,d_3^{JK}=P_3^{KJ}(k,\omega)d_3^{JK} \tag{5}$$

where $\gamma_s^{JK}=\gamma_s^{KJ}=\gamma_{sj}$, $h^{JK}=h^{KJ}=h_j$ and $P_3^{JK}=P_3^{KJ}=P_{3j}$ are the wavenumber along $z$ direction, the thickness and the phase coefficient of layer $j$ ($JK$ or $KJ$), respectively. It is noted that the thickness wavenumber $\gamma_{sj}$ can always be chosen to satisfy $\mathrm{Re}[-\mathrm{i}\gamma_{sj}h_j]\le 0$, so that no exponentially growing function is included in the phase relation.

Second, we consider the spectral equations at the interfaces between adjacent layers. The compatibility of displacements and equilibrium of stresses at the interfaces 2 and 3 are expressed as

$$\hat{\upsilon}^{21}(0)=\hat{\upsilon}^{23}(0),\ \ \hat{\tau}_{zy}^{21}(0)+\hat{\tau}_{zy}^{23}(0)=0,\ \ \hat{\upsilon}^{32}(0)=\hat{\upsilon}^{34}(0),\ \ \hat{\tau}_{zy}^{32}(0)+\hat{\tau}_{zy}^{34}(0)=0 \tag{6}$$

Substituting Eqs. (1) and (2) into Eq. (6), one obtains the scattering relations at interfaces 2 and 3

$$\begin{aligned} a_3^{21}+d_3^{21}-a_3^{23}-d_3^{23}=0,\ \ \varsigma_3^{21}a_3^{21}-\varsigma_3^{21}d_3^{21}+\varsigma_3^{23}a_3^{23}-\varsigma_3^{23}d_3^{23}=0 \\ a_3^{32}+d_3^{32}-a_3^{34}-d_3^{34}=0,\ \ \varsigma_3^{32}a_3^{32}-\varsigma_3^{32}d_3^{32}+\varsigma_3^{34}a_3^{34}-\varsigma_3^{34}d_3^{34}=0 \end{aligned} \tag{7}$$

Third, we consider the spectral equations at the top and bottom surfaces. The Floquet-Bloch principle of periodic structures (Brillouin, 1953; Mead, 1996) requires that the displacement (stress) of bottom layer at the bottom surface 4 should relate to that of top layer at the top surface 1 by

$$\hat{\upsilon}^{43}(0)=\mathrm{e}^{\mathrm{i}qh}\,\hat{\upsilon}^{12}(0),\ \ \hat{\tau}_{zy}^{43}(0)=-\mathrm{e}^{\mathrm{i}qh}\,\hat{\tau}_{zy}^{12}(0) \tag{8}$$

where $h=\sum_{j=1}^{3}h_j$ is the thickness of the unit cell, $q$ is the wavenumber of the characteristic waves in the periodic ternary layered media. The real part $q_R h$ and the imaginary part $q_I h$ of dimensionless wavenumber $qh$ denote the phase constant and the attenuation constant of the characteristic wave, respectively (Mead, 1996). Substitution of Eqs. (1) and (2) into Eq. (8) gives the scattering relations at surfaces 1 and 4

$$\mathrm{e}^{\mathrm{i}qh}\,a_3^{12}+\mathrm{e}^{\mathrm{i}qh}\,d_3^{12}-a_3^{43}-d_3^{43}=0,\ \ \mathrm{e}^{\mathrm{i}qh}\,\varsigma_3^{12}a_3^{12}-\mathrm{e}^{\mathrm{i}qh}\,\varsigma_3^{12}d_3^{12}+\varsigma_3^{43}a_3^{43}-\varsigma_3^{43}d_3^{43}=0 \tag{9}$$

Finally, introducing the phase relations of all layers as given in Eq. (5) to the scattering relations of all interfaces and surfaces as given in Eqs. (7) and (9), we obtain the system equations with all the departing wave amplitudes as basic unknown quantities

$$\begin{bmatrix} P_3^{21} & 1 & -1 & -P_3^{23} & 0 & 0 \\ \varsigma_3^{21}P_3^{21} & -\varsigma_3^{21} & -\varsigma_3^{23} & \varsigma_3^{23}P_3^{23} & 0 & 0 \\ 0 & 0 & P_3^{32} & 1 & -1 & -P_3^{34} \\ 0 & 0 & \varsigma_3^{32}P_3^{32} & -\varsigma_3^{32} & -\varsigma_3^{34} & \varsigma_3^{34}P_3^{34} \\ \mathrm{e}^{\mathrm{i}qh} & \mathrm{e}^{\mathrm{i}qh}\,P_3^{12} & 0 & 0 & -P_3^{43} & -1 \\ -\varsigma_3^{12}\,\mathrm{e}^{\mathrm{i}qh} & \varsigma_3^{12}\,\mathrm{e}^{\mathrm{i}qh}\,P_3^{12} & 0 & 0 & \varsigma_3^{43}P_3^{43} & -\varsigma_3^{43} \end{bmatrix} \begin{Bmatrix} d_3^{12} \\ d_3^{21} \\ d_3^{23} \\ d_3^{32} \\ d_3^{34} \\ d_3^{43} \end{Bmatrix} = \begin{Bmatrix} 0 \\ 0 \\ 0 \\ 0 \\ 0 \\ 0 \end{Bmatrix}$$

or

$$\mathbf{R}_d(k,\omega,q)\mathbf{d}=\mathbf{0} \tag{10}$$

where $\varsigma_3^{JK}=\varsigma_3^{KJ}=\varsigma_{3j}$ ($J=K-1=j=1,2,3$) is the shear stress coefficient of layer $j$, $\mathbf{R}_d$ is the system matrix, and $\mathbf{d}$ is the global departing wave vector.

The dispersion equation governing the characteristic SH waves in periodic ternary layered media is obtained by vanishing of the determinant of system matrix

$$\det[\mathbf{R}_d(k,\omega,q)] = \mathbf{0}. \tag{11}$$

Further expansion of the determinant in Eq. (11) gives the closed-form dispersion relation of characteristic SH waves in periodic ternary layered media as follows

$$\begin{aligned}
&8\zeta_{31}\zeta_{32}\zeta_{33}\,\mathrm{e}^{-\mathrm{i}\gamma_{s1}h_1}\,\mathrm{e}^{-\mathrm{i}\gamma_{s2}h_2}\,\mathrm{e}^{-\mathrm{i}\gamma_{s3}h_3}(1+\mathrm{e}^{2\mathrm{i}qh})\\
&-2\zeta_{31}\zeta_{32}\zeta_{33}\,\mathrm{e}^{\mathrm{i}qh}(1+\mathrm{e}^{-2\mathrm{i}\gamma_{s1}h_1})(1+\mathrm{e}^{-2\mathrm{i}\gamma_{s2}h_2})(1+\mathrm{e}^{-2\mathrm{i}\gamma_{s3}h_3})\\
&-\left[(\zeta_{31})^2+(\zeta_{32})^2\right]\zeta_{33}\,\mathrm{e}^{\mathrm{i}qh}(1-\mathrm{e}^{-2\mathrm{i}\gamma_{s1}h_1})(1-\mathrm{e}^{-2\mathrm{i}\gamma_{s2}h_2})(1+\mathrm{e}^{-2\mathrm{i}\gamma_{s3}h_3})\\
&-\left[(\zeta_{32})^2+(\zeta_{33})^2\right]\zeta_{31}\,\mathrm{e}^{\mathrm{i}qh}(1-\mathrm{e}^{-2\mathrm{i}\gamma_{s2}h_2})(1-\mathrm{e}^{-2\mathrm{i}\gamma_{s3}h_3})(1+\mathrm{e}^{-2\mathrm{i}\gamma_{s1}h_1})\\
&-\left[(\zeta_{33})^2+(\zeta_{31})^2\right]\zeta_{32}\,\mathrm{e}^{\mathrm{i}qh}(1-\mathrm{e}^{-2\mathrm{i}\gamma_{s3}h_3})(1-\mathrm{e}^{-2\mathrm{i}\gamma_{s1}h_1})(1+\mathrm{e}^{-2\mathrm{i}\gamma_{s2}h_2}) = 0.
\end{aligned} \tag{12}$$

Dispersion equation (12) assures unconditionally numerical stability because $\mathrm{Re}[-\mathrm{i}\gamma_{sj}h_j] \le 0$ is already guaranteed by the properly established phase relation and $\mathrm{Re}[\mathrm{i}qh] \le 0$ can also be guaranteed by properly specifying $q$ in the solving process, since $+q$ and $-q$ signifying wavenumbers in opposite direction should give the same propagation characteristics. Due to $\mathrm{e}^{-\mathrm{i}\gamma_{sj}h_j} \ne 0$ and $\mathrm{e}^{\mathrm{i}qh} \ne 0$, Equation (12) can be simplified, by virtue of relations between trigonometric functions and exponential functions, to

$$\begin{aligned}
&2\zeta_{31}\zeta_{32}\zeta_{33}\left[\cos(qh)-\cos(\gamma_{s1}h_1)\cos(\gamma_{s2}h_2)\cos(\gamma_{s3}h_3)\right]=\\
&-\left[(\zeta_{31})^2+(\zeta_{32})^2\right]\zeta_{33}\sin(\gamma_{s1}h_1)\sin(\gamma_{s2}h_2)\cos(\gamma_{s3}h_3)\\
&-\left[(\zeta_{32})^2+(\zeta_{33})^2\right]\zeta_{31}\sin(\gamma_{s2}h_2)\sin(\gamma_{s3}h_3)\cos(\gamma_{s1}h_1)\\
&-\left[(\zeta_{33})^2+(\zeta_{31})^2\right]\zeta_{32}\sin(\gamma_{s3}h_3)\sin(\gamma_{s1}h_1)\cos(\gamma_{s2}h_2).
\end{aligned} \tag{13}$$

Define

$$Z_{SHj} = \zeta_{3j}/\mathrm{i}\omega = \gamma_{sj}G_j/\omega = \sqrt{\rho_j G_j - k^2 G_j^2/\omega^2} \tag{14}$$

as the characteristic impedance of SH wave in layer $j$, which is dependent on not only the shear modulus $G_j$ and mass density $\rho_j$, but also the frequency $\omega$ and wavenumber $k$. Therefore, the characteristic impedance of SH wave in an isotropic layer is not a constant and can be imaginary below the cutoff frequency of SH wave. It is very different from the characteristic impedance of bulk shear wave $Z_{Tj} = \sqrt{\rho_j G_j}$ in an isotropic medium $j$, which is a real-valued constant. But $Z_{SHj}$ equals to $Z_{Tj}$ as $k=0$. If $Z_{SH1}Z_{SH2}Z_{SH3}=0$, i.e. $(\omega \pm kc_{s1})(\omega \pm kc_{s2})(\omega \pm kc_{s3}) = 0$, the dispersion equation (13) is automatically satisfied regardless of $qh$. This case merely gives the cutoff frequency of SH wave mode $\omega_c = |k|c_s$

within one of the constituent layers, and we shall not discuss it any further. Otherwise for $Z_{SH1}Z_{SH2}Z_{SH3} \neq 0$, the dispersion relation is further simplified to

$$\cos(qh) = \cos(\gamma_{s1}h_1)\cos(\gamma_{s2}h_2)\cos(\gamma_{s3}h_3) - \frac{1}{2}(F_{1/2} + F_{2/1})\sin(\gamma_{s1}h_1)\sin(\gamma_{s2}h_2)\cos(\gamma_{s3}h_3)$$
$$-\frac{1}{2}(F_{2/3} + F_{3/2})\sin(\gamma_{s2}h_2)\sin(\gamma_{s3}h_3)\cos(\gamma_{s1}h_1) - \frac{1}{2}(F_{3/1} + F_{1/3})\sin(\gamma_{s3}h_3)\sin(\gamma_{s1}h_1)\cos(\gamma_{s2}h_2), \quad (15)$$

where

$$F_{j'/j''} = \frac{\zeta_{3j'}}{\zeta_{3j''}} = \frac{\mathrm{i}\gamma_{sj'}G_{j'}}{\mathrm{i}\gamma_{sj''}G_{j''}} = \frac{\gamma_{sj'}G_{j'}/\omega}{\gamma_{sj''}G_{j''}/\omega} = \frac{Z_{SHj'}}{Z_{SHj''}} \quad (j', j'' = 1,2,3, j' \neq j'') \quad (16)$$

signifies the contrast of characteristic impedances of SH waves in layer $j'$ and layer $j''$.

## 2.2. Formation mechanisms of SH-wave bands in the Bragg Cell

Based on Eq. (15), in which $Z_{SH1}Z_{SH2}Z_{SH3} \neq 0$ is implied, in what follows we will discuss the formation mechanisms of frequency bands of SH waves in periodic ternary layered media, according to the following two cases for the characteristic impedances of SH waves in the three constituent layers of the unit cell.

1. $Z_{SH1} = Z_{SH2} = Z_{SH3}$

Owing to $F_{j'/j''} = F_{j''/j'} = 1$ ($j', j'' = 1,2,3$, $j' \neq j''$), the dispersion relation (15) is reduced to

$$\cos(qh) = \cos(\gamma_{s1}h_1 + \gamma_{s2}h_2 + \gamma_{s3}h_3) = \cos(\gamma_{se}h)$$
$$= \cos(\frac{\omega}{c_{SH1}}h_1 + \frac{\omega}{c_{SH2}}h_2 + \frac{\omega}{c_{SH3}}h_3) = \cos[\omega(T_{SH1} + T_{SH2} + T_{SH3})] = \cos(\omega T_{SH}), \quad (17)$$

where $\gamma_{se} = \sum_{j=1}^{3}(\gamma_{sj}h_j)/h$ is the equivalent wavenumber of SH wave in the unit cell, $T_{SHj} = h_j / c_{SHj}$ is the parameter reflecting the characteristic time as SH wave traverses the thickness of constituent *j*, but may be imaginary number below the cutoff frequency $\omega_c$, $T_{SH} = \sum_{j=1}^{3} T_{SHj}$ is the parameter reflecting the characteristic time as SH wave traverses the thickness of the unit cell. Equation (17) has the solution

$$qh = \pm\gamma_{se}h + 2\frac{m}{n}\pi = \pm\omega T_{SH} + 2\frac{m}{n}\pi, \quad (18)$$

where $m$ and $n$ are arbitrary integers corresponding to positive and negative signs, respectively . Equation (18) indicates that when all the three constituent layers have the same characteristic impedance of SH wave, there is no bandgap above the maximum cutoff frequency $\omega_{c\max} = \max(\omega_{c1}, \omega_{c2}, \omega_{c3})$. The dispersion spectra are completely determined by the fundamental dispersion curve of the equivalent SH wave in the unit cell due to the zone folding effect (Brillouin, 1953) with the characteristic time of the unit cell being the essential parameter. In other words, the contrast of characteristic impedances determines whether the band gaps exist or not above $\omega_{c\max}$, and the characteristic time of the unit cell decides the dispersion spectra of the periodic layered media as no bandgap exists above $\omega_{c\max}$.

2. $Z_{SHj} \neq Z_{SHj'}, Z_{SHj} \neq Z_{SHj''}$ ($Z_{SHj'} = or \neq Z_{SHj''}$, $j, j', j'' = 1$ or 2 or 3, $j \neq j' \neq j''$)

If $Z_{SHj'} = Z_{SHj''}$, then the dispersion relation (15) is reduced to

$$\begin{aligned}\cos(qh) &= \cos(\gamma_{sj}h_j)\cos(\gamma_{sj'}h_{j'} + \gamma_{sj''}h_{j''}) - \frac{1}{2}(F_{j/j'} + F_{j'/j})\sin(\gamma_{sj}h_j)\sin(\gamma_{sj'}h_{j'} + \gamma_{sj''}h_{j''}) \\ &= \cos(\gamma_{sj}h_j)\cos(\gamma_{si}h_i) - \frac{1}{2}(F_{j/i} + F_{i/j})\sin(\gamma_{sj}h_j)\sin(\gamma_{si}h_i),\end{aligned} \tag{19}$$

which is the dispersion relation for periodic binary layered media already obtained (Shen & Cao, 2000; Wang et al., 2004). $\gamma_{si} = (\gamma_{sj'}h_{j'} + \gamma_{sj''}h_{j''})/h_i$ and $h_i = h_{j'} + h_{j''}$ are the wavenumber along the $z$-direction and the thickness of the new equivalent layer $i$ composed of the two constituent layers with identical characteristic impedance of SH wave. If $Z_{SHj'} \neq Z_{SHj''}$, then the dispersion relation should be the general form of Eq. (15).

In any circumstances, as $\omega > \omega_{c\max}$ the characteristic impedances and the wavenumbers along the $z$-direction of SH waves in the constituent layers will be positive real number. Thus, we have

$$F_{j/\Gamma} + F_{\Gamma/j} = 1 - \varepsilon_{j/\Gamma} + \frac{1}{1-\varepsilon_{j/\Gamma}} = \frac{1}{1-\varepsilon_{\Gamma/j}} + 1 - \varepsilon_{\Gamma/j} = 1 - \varepsilon_{j\Gamma} + \frac{1}{1-\varepsilon_{j\Gamma}} = 2 + \sum_{b=2}^{+\infty}\varepsilon_{j\Gamma}^{b} \quad (\Gamma = j', j'') \tag{20}$$

where $\varepsilon_{j/\Gamma}$, $\varepsilon_{\Gamma/j}$ and $\varepsilon_{j\Gamma}$ are real numbers, $0 < \varepsilon_{j/\Gamma} < 1$ ($\varepsilon_{\Gamma/j} > 1$) when $F_{j/\Gamma} < 1$, $0 < \varepsilon_{\Gamma/j} < 1$ ($\varepsilon_{j/\Gamma} > 1$) when $F_{\Gamma/j} < 1$, $\varepsilon_{j\Gamma} = \min(\varepsilon_{j/\Gamma}, \varepsilon_{\Gamma/j}) \in (0,1)$. Similarly, we have

$$F_{j'/j''} + F_{j''/j'} = 1 - \varepsilon_{j'j''} + \frac{1}{1-\varepsilon_{j'j''}} = 2 + \sum_{b=2}^{+\infty}\varepsilon_{j'j''}^{b}, \tag{21}$$

where $\varepsilon_{j'j''} = 0$ as $Z_{SHj'} = Z_{SHj''}$, $0 < \varepsilon_{j'j''} < 1$ as $Z_{SHj'} \neq Z_{SHj''}$. Therefore, the dispersion equations (15) and (19) can be rewritten uniformly as

$$
\begin{aligned}
\cos(qh) &= \cos(\gamma_{se}h) - \frac{1}{2}\sum_{b=2}^{+\infty} \varepsilon_{jj'}^{b} \sin(\gamma_{sj}h_j)\sin(\gamma_{sj'}h_{j'})\cos(\gamma_{sj''}h_{j''}) \\
&- \frac{1}{2}\sum_{b=2}^{+\infty} \varepsilon_{jj''}^{b} \sin(\gamma_{sj}h_j)\sin(\gamma_{sj''}h_{j''})\cos(\gamma_{sj'}h_{j'}) - \frac{1}{2}\sum_{b=2}^{+\infty} \varepsilon_{j'j''}^{b} \sin(\gamma_{sj'}h_{j'})\sin(\gamma_{sj''}h_{j''})\cos(\gamma_{sj}h_j) \\
&= \cos(\gamma_{se}h) - \frac{\varepsilon_{jj'}^2}{2(1-\varepsilon_{jj'})}\sin(\gamma_{sj}h_j)\sin(\gamma_{sj'}h_{j'})\cos(\gamma_{sj''}h_{j''}) \\
&- \frac{\varepsilon_{jj''}^2}{2(1-\varepsilon_{jj''})}\sin(\gamma_{sj}h_j)\sin(\gamma_{sj''}h_{j''})\cos(\gamma_{sj'}h_{j'}) - \frac{\varepsilon_{j'j''}^2}{2(1-\varepsilon_{j'j''})}\sin(\gamma_{sj'}h_{j'})\sin(\gamma_{sj''}h_{j''})\cos(\gamma_{sj}h_j) \\
&= \cos(\omega T_{SH}) - \frac{\varepsilon_{jj'}^2}{2(1-\varepsilon_{jj'})}\sin(\omega T_{SHj})\sin(\omega T_{SHj'})\cos(\omega T_{SHj''}) \\
&- \frac{\varepsilon_{jj''}^2}{2(1-\varepsilon_{jj''})}\sin(\omega T_{SHj})\sin(\omega T_{SHj''})\cos(\omega T_{SHj'}) - \frac{\varepsilon_{j'j''}^2}{2(1-\varepsilon_{j'j''})}\sin(\omega T_{SHj'})\sin(\omega T_{SHj''})\cos(\omega T_{SHj}),
\end{aligned}
\tag{22}
$$

which indicates that the band structures of the periodic ternary layered media are not only determined by the fundamental dispersion curve of the equivalent SH wave according to zone folding effect, but also influenced by three disturbance terms with disturbing functions $\sin(\gamma_{sj'}h_{j'})\sin(\gamma_{sj''}h_{j''})\cos(\gamma_{sj}h_j)$ (or $\sin(\omega T_{SHj'})\sin(\omega T_{SHj''})\cos(\omega T_{SHj})$, $j,j',j''=1,2,3$, $j \neq j' \neq j''$) and disturbing amplitudes $\varepsilon_{j'j''}^2/[2(1-\varepsilon_{j'j''})]$. The value of the right-hand side of Eq. (22) determines the demarcation of frequency bands: $\pm 1$ gives the dividing lines of pass-bands and stop-bands; $0$ gives the central frequencies of pass-bands; those between $-1$ and $1$ give the pass-bands; and all other values give the stop-bands. The characteristic time of the unit cell, the characteristic times of constituent layers and the contrasts of characteristic impedances of SH waves in the constituent layers are the essential parameters for the band structure formation, which determine the shape of the dispersion curves of the equivalent SH wave (the pre-disturbed baselines), the shapes of the disturbing functions, and the amplitudes of the disturbance terms, respectively. When the disturbing functions satisfy $\sin(\gamma_{sj'}h_{j'})\sin(\gamma_{sj''}h_{j''})\cos(\gamma_{sj}h_j)=0$ ($\sin(\gamma_{sj'}h_{j'})\sin(\gamma_{sj''}h_{j''})\cos(\gamma_{sj}h_j)=\pm 1$), the band structures coincide exactly with (deviate most from) the fundamental and derivative dispersion curves of the equivalent SH wave in the unit cell. The corresponding points on the dispersion curves are called as the coincident (separating) points. The frequency equation can be simplified to ($j,j',j''=1,2,3$, $j \neq j' \neq j''$)

$$
\begin{aligned}
&\gamma_{sj'}h_{j'} = g_{j'}\pi,\ \gamma_{sj''}h_{j''} = g_{j''}\pi\ \left(g_{j'} \text{ and } g_{j''} \text{ are integers}\right) \text{or} \\
&\gamma_{sj}h_j = (2g_j+1)\pi/2,\ \gamma_{sj'}h_{j'} = (2g_{j'}+1)\pi/2,\ \gamma_{sj''}h_{j''} = (2g_{j''}+1)\pi/2
\end{aligned}
\tag{23}
$$

$$
\gamma_{sj}h_j = g_j\pi,\ \gamma_{sj'}h_{j'} = (2g_{j'}+1)\pi/2,\ \gamma_{sj''}h_{j''} = (2g_{j''}+1)\pi/2, \tag{24}
$$

for coincident and separating points, respectively.

In physics, at any interface $J$ of the unit cell there are one incident wave $w_{iJ}$ and one reflected wave $w_{rj}$ arising from the next interface except that there is no reflection at the interface where the two constituent layers with identical characteristic impedances connected. $\gamma_{sj}h_j$, $\gamma_{sj'}h_{j'}$ and $\gamma_{sj''}h_{j''}$ denote the phase changes as SH wave passes through layer $j$, $j'$ and $j''$, respectively. Thus, the former formula in Eq. (23) corresponds to the constructive interference condition of the incident wave and reflected wave at two interfaces, and the latter formula in Eq. (23) corresponds to the destructive interference condition at three interfaces. Equation (24) corresponds to destructive interference condition of incident wave and reflected wave at two interfaces and constructive interference condition at one interface. Therefore, it is concluded that the frequency bands are formed physically as a result of interference phenomenon as waves transmit and reflect in the constituent layers of a periodic ternary layered media. The specified combination of exact constructive and destructive interferences of the incident and reflected waves at some interfaces makes the equivalent SH wave travel through the unit cell without any change of its dispersion characteristic or be completely prohibited to travel. The specified combination of near constructive and destructive interferences of the incident and reflected waves at some interfaces makes the equivalent SH wave be capable of going through the unit cell with a change of its dispersion characteristic or be attenuated. The exact constructive and destructive interferences specified by Eq. (23) and Eq. (24) are only possible for special periodic ternary layered media with the characteristic times of constituent layers satisfying

$$T_{SHj'} : T_{SHj''} = g_{j'} : g_{j''} \quad \text{or} \quad T_{SHj} : T_{SHj'} : T_{SHj''} = (2g_j + 1) : (2g_{j'} + 1) : (2g_{j''} + 1) \tag{25}$$

$$T_{SHj} : T_{SHj'} : T_{SHj''} = g_j : (2g_{j'} + 1) : (2g_{j''} + 1) \tag{26}$$

However, the near constructive and destructive interferences specified by Eq. (23) and Eq. (24) can occur in general periodic ternary layered media.

In summary, the occurrence of some specified combination of exact or near constructive and destructive interference phenomena in the unit cell makes the equivalent SH wave travel through the unit cell and gives birth to the pass-bands, whereas the occurrence of other specified combination of exact or near destructive interference makes the equivalent SH wave unable to pass through the unit cell and brings about the stop-bands. Although the above discussion on the formation mechanisms of SH-wave bands is based on the periodic ternary layered structure, it is actually extendable to SH wave in general periodic layered media.

## 2.3. Design rules to the Bragg Cell concerning with SH wave bands

The discussion of formation mechanisms of SH wave bands in the layered Bragg Cell indicates that the contrasts of characteristic impedances of the constituent layers, the characteristic time of the unit cell and the characteristic times of the constituent layers are three kinds of essential parameters, which influence the band properties. First, the contrasts of characteristic impedances decide whether the stop-bands other than that due to SH wave cutoff property exist or not. When the characteristic impedances of all the constituent layers are identical, any SH waves above the maximum cutoff frequency can propagate in the periodic layer without attenuation and no stop-bands other than that due to the SH wave cutoff property exist. In other cases, stop-bands exist above the maximum cutoff frequency, and the contrasts of characteristic impedances decide the widths of the frequency bands. The characteristic time of the unit cell decides the slopes of the dispersion curves of equivalent SH waves, thus it definitely specifies the number of pass-bands/stop-bands in a given frequency range. The characteristic times of the constituent layers mainly decide the mid-frequencies of the frequency bands. It should be pointed out that the mass densities and shear moduli of constituent layers affect all the three kinds of essential parameters, while the thicknesses of the constituent layers only influence the characteristic times of the unit cell and of the constituent layers. These rules can be used for the design of layered Bragg Cells according to the SH-wave bands requirements.

## 2.4. Numerical examples

In this section, the above proposed MRRM for dispersion characteristic analysis and the mechanisms for band structure formation of SH waves in periodic ternary layered media are validated by considering a periodic ternary layered structure with the unit cell consisting of one Pb layer in the middle and two epoxy layers at the up and down sides. The thickness of the Pb layer is 10mm and that of the epoxy layers is 5mm. The material parameters of Pb and epoxy including the Young's modulus, shear modulus and mass density are $E_{Pb} = 40.8187$ GPa, $E_{epoxy} = 4.35005$ GPa, $G_{Pb} = 14.9$ GPa, $G_{epoxy} = 1.59$ GPa, $\rho_{Pb} = 11600$ kg/m$^3$ and $\rho_{epoxy} = 1180$ kg/m$^3$. The band structures of SH waves in this periodic ternary layered medium are calculated by the formulation presented in Section 2.1. For the convenience of presentation, the results are represented by the dimensionless wavenumbers $kh/\pi$ and $qh/\pi$ with $h = 0.02$ m being the total thickness of the unit cell, and the engineering frequency $f = \omega/2\pi$.

We first consider the property of SH-wave band structures in the exemplified periodic ternary layered medium. Figure 3 gives the band structures of SH waves below 140 kHz represented as various forms of graphs. Figure 3(a) depicts the phase constant surfaces in the pass-bands as the dimensionless wavenumbers $kh/\pi$ and $qh/\pi$ are in the range of $[-2,2]$. Figures 3(b) to 3(d) describe both the phase constant spectra in pass-bands (i.e. the relation between $f$ and $q_R h/\pi$) and the attenuation constant spectra in stop-bands (i.e. the relation between $f$ and $q_I h/\pi$) as the dimensionless wavenumber $kh/\pi$ is $1.0$, $0.5$ and $0.0$, respectively. Figure 3(e) plots the relation between $f$ and $|k|h/\pi$ when $qh/\pi$ is specified as $0.0+2I_1$, $0.5+2I_2$ and $1.0+2I_3$ with $I_1$, $I_2$ and $I_3$ being arbitrary integers. To validate our obtained results, in Fig. (3d) the phase constant spectra as $kh/\pi = 0$ are compared to the corresponding results calculated by Wang et al. (2004), and in Fig. (3e) the spectra of $f$ versus $|k|h/\pi$ as $qh/\pi = 0+2I_1$ and $qh/\pi = 1+2I_3$ are compared to their counterparts calculated by Wang et al. (2004).

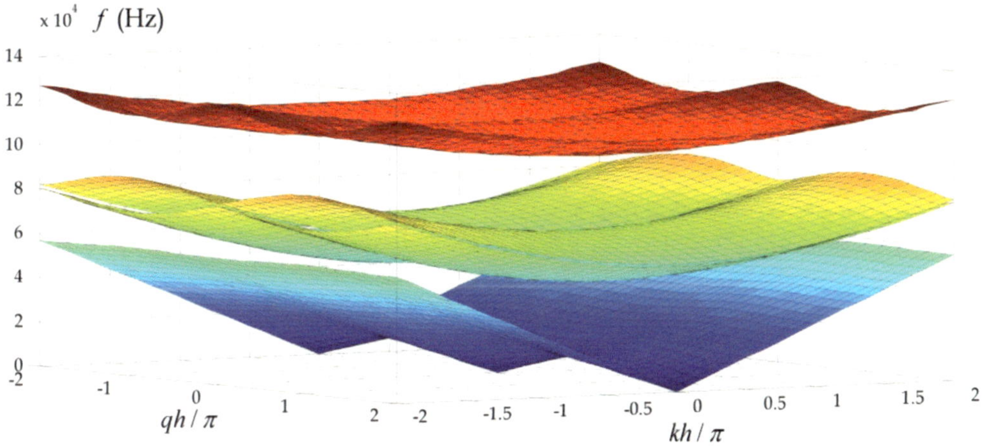


(a) The phase constant surfaces in the pass-bands

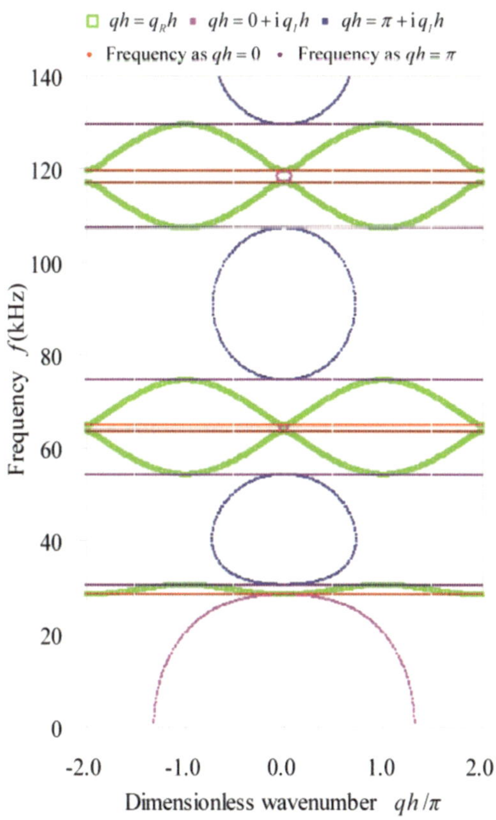


(b) The band structures as $kh/\pi$=1.0

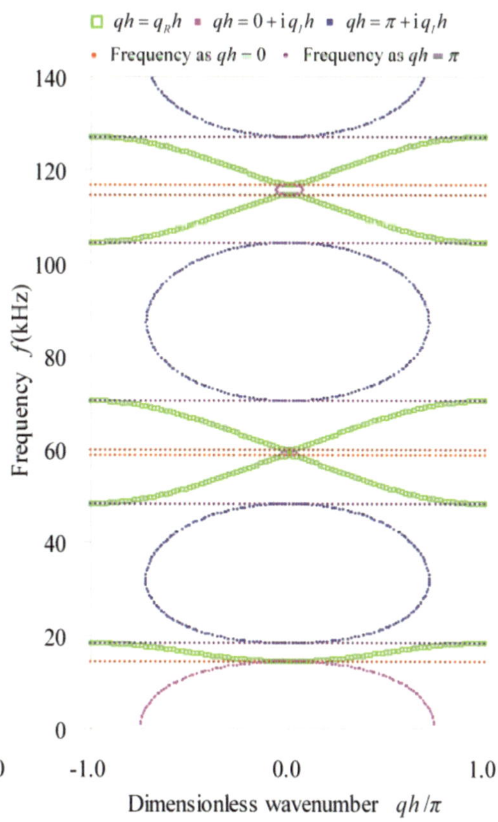


(c) The band structures as $kh/\pi$=0.5

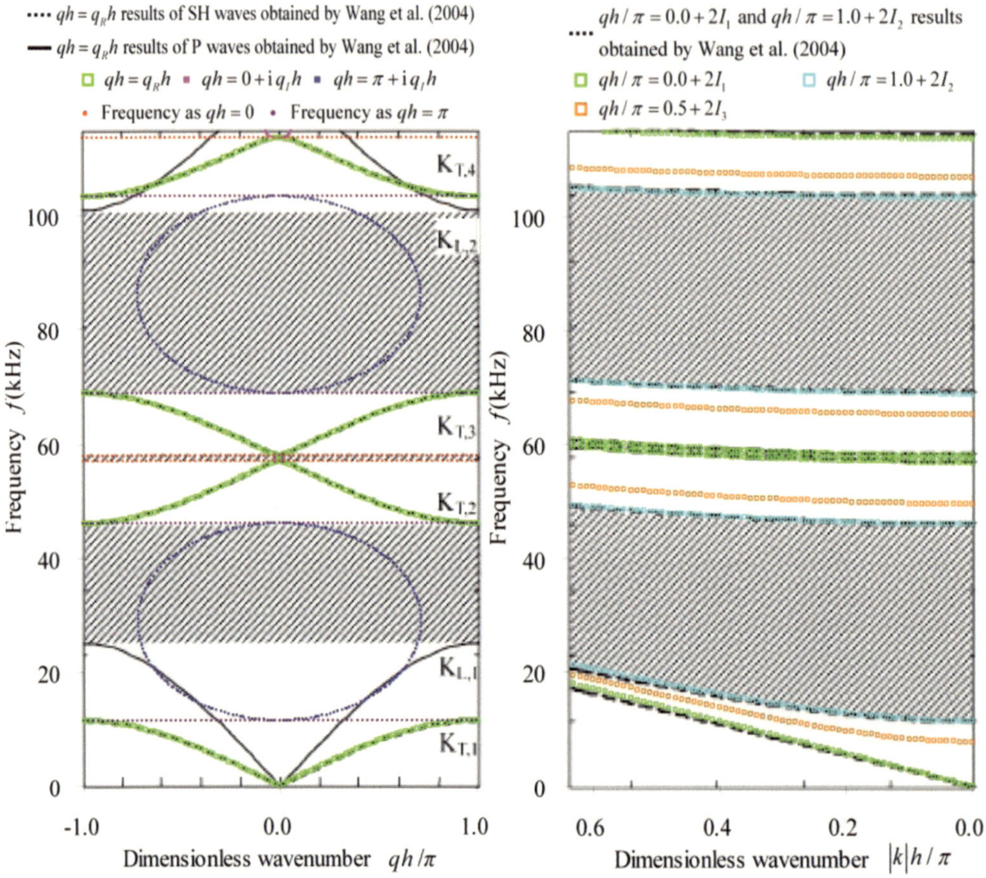


(d) The band structures as $kh/\pi$=0.0

(e) The relation between $f$ and $kh/\pi$ when $qh/\pi$ are specified as 0.0+2$I_1$, 0.5+2$I_2$ and 1.0+2$I_3$

**Figure 3.** The band structures of SH waves below 140 kHz in the periodic ternary layered medium consisting of one Pb layer and two epoxy layers

It is seen from Fig. 3(a) that the phase constant surfaces in the pass-bands are symmetrical with respect to the vertical plane $kh/\pi = 0$, which indicates that the SH waves along the positive and the negative $X$ directions have identical propagation properties. This is due to the symmetry of structural configuration and material parameters of the exemplified periodic ternary layered medium with respect to $YOZ$ plane. Likewise, the phase constant surfaces in Fig. 3(a) and the attenuation constant spectra in Figs. 3(b) to 3(d) are also symmetry with respect to $qh/\pi = 0$, which indicates that the upward and downward characteristic SH waves have identical band structures. In addition, the phase constant surfaces in Fig. 3(a) are periodical with respect to $qh/\pi$ as the minimum positive period being $qh/\pi = 2$, which indicates the periodicity of propagating characteristic SH waves along the thickness of the unit cell and reflects the zone folding effect of periodic structures.

Figs. 3(b) to 3(d) signify that for any $kh/\pi$, with the increasing of frequency the phase constant spectra and the attenuation constant spectra of characteristic SH waves in periodic ternary layered media occur alternately, i.e. the pass-bands and the stop-bands occur alternately. However, the attenuation spectra (the stop band) will advent first for any $kh/\pi$ except for $kh/\pi = 0$. It should be noted that as $kh/\pi \neq 0$, the first stop band is formed due to the cutoff property of the SH waves in the constituent layers. For any $kh/\pi$, the attenuation constant spectra in form of closed loops with phase 0 and phase $\pi$ appear alternately. It indicates from Figs. 3(b) to 3(e) that for any $kh/\pi$, the frequencies as $qh/\pi = 0$ and those as $qh/\pi = 1$, which are respectively identical to those as $qh/\pi = 0 + 2I_1$ and those as $qh/\pi = 1 + 2I_3$, are the demarcations between the pass-bands and the stop-bands. The ranges between two adjacent bounding frequencies, with one corresponding to $qh/\pi = 0 + 2I_1$ and the other corresponding to $qh/\pi = 1 + 2I_3$ are pass-bands, whereas the ranges between two adjacent bounding frequencies with both corresponding to $qh/\pi = 0 + 2I_1$ or $qh/\pi = 1 + 2I_3$ are stop-bands. Fig. 3(e) shows the $f$ - $|k|h/\pi$ spectra as $qh/\pi$ takes any other real value lie in the pass-bands between the $f$ - $|k|h/\pi$ spectra as $qh/\pi = 0$ and those as $qh/\pi = 1$. With the increasing of $|k|h/\pi$, the demarcating frequencies of the corresponding frequency-bands rise. The first demarcation frequency increases most obviously with the rise of $|k|h/\pi$.

In Fig. 3(d), the comparison between the phase constant spectra obtained by our proposed method and those calculated by Wang et al. (2004) indicates good agreement. In Fig. 3(e), the comparison between the $f$ - $|k|h/\pi$ spectra obtained by our method and those calculated by Wang et al. (2004) also manifests close coincidence. Furthermore, the $f$ - $|k|h/\pi$ spectra corresponding to $qh/\pi = 0 + 2I_1$ and $qh/\pi = 1 + 2I_3$ in our results are explicitly separated, which clearly denotes the pass-bands and stop-bands. All these validate the accuracy and excellence of the proposed MRRM for dispersion characteristic analysis.

Let us now consider the formation of SH-wave band structures in the exemplified periodic ternary layered structure. We plot in Figs. (4a) and (4b) the phase and the attenuation constant spectra of characteristic SH waves together with the dispersion curves of the equivalent SH waves in the exemplified periodic ternary layered structure, as $kh/\pi = 0.0$ and $kh/\pi = 0.5$, respectively, for illustrating the close relation between the dispersion curves of the equivalent SH waves and the band structures of characteristic SH waves. The fundamental real-part and the imaginary-part dispersion curves of the equivalent SH waves are obtained directly from the definition $\gamma_{se}h = \sum_{j=1}^{3} h_j \sqrt{\omega^2 / c_{sj}^2 - k^2}$, while the derivative real-part dispersion curve of the equivalent SH waves are attained by virtue of the zone folding effect.

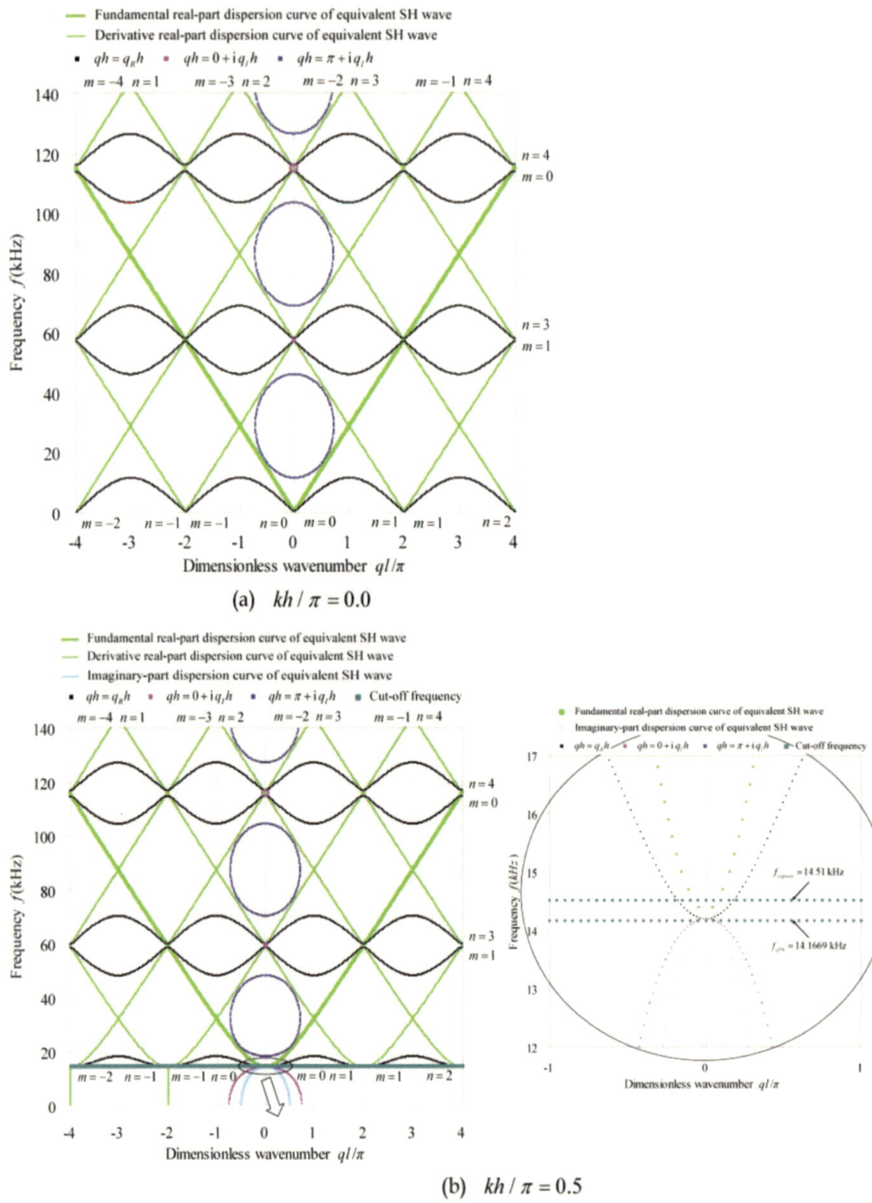


**Figure 4.** The band structures of characteristic SH waves and the dispersion curves of equivalent SH waves below 140 kHz in the periodic ternary layered medium consisting of one Pb layer and two epoxy layers

It is clearly seen from Figs. (4a) and (4b) that above the maximum cutoff frequency $\omega_{c\max}$ ( $\omega_{c\max} = 0 = \omega_{c\min}$ while $kh/\pi = 0.0$ ), the dispersion curves of the equivalent SH waves have only the real part. In this case the fundamental and derivative dispersion curves of equivalent SH waves serve as the baselines during the band-structure formation of characteristic SH waves. In the forming process of the band structures, the dispersion curves first separate with respect to frequency at the intersections $\gamma_{se}h + 2m\pi = -\gamma_{se}h + 2n\pi = I\pi$ ($I$ is an arbitrary integer) representing the boundaries of the Brillouin zone to give the stop-bands, in which the attenuation constant spectra with phase 0 and phase $\pi$ corresponding to even numbers and odd numbers of $I$, respectively. The intermediate segments of dispersion curves between the adjacent intersections are disturbed to form the phase constant spectra in the pass-bands. Figure (4b) also shows below the minimum cutoff frequency $\omega_{c\min}$ as $kh/\pi \neq 0.0$, the dispersion curves of the equivalent SH waves have only the imaginary part. In this case the dispersion curve serve as the baseline during the formation of the attenuation constant spectrum in the first stop-band, which is emerged due to the cutoff property of the SH waves in the constituent layers.

It should be emphasized that although the above example is a simple periodic ternary layered structure, the obtained property and formation of SH wave band structures are in fact also applicable to SH waves in all periodic layered isotropic media.

## 3. Analysis of acoustic waves in integrated multi-layered structures

Various multi-layered acoustic wave devices with Bragg Cell can be modeled by the multi-layered structures of infinite lateral extent depicted in Fig. 5, which including both non-piezoelectric layers and piezoelectric layers. Usually, the electrodes, support layers and substrate consist of elastic (non-piezoelectric) layers. The propagation media consist of

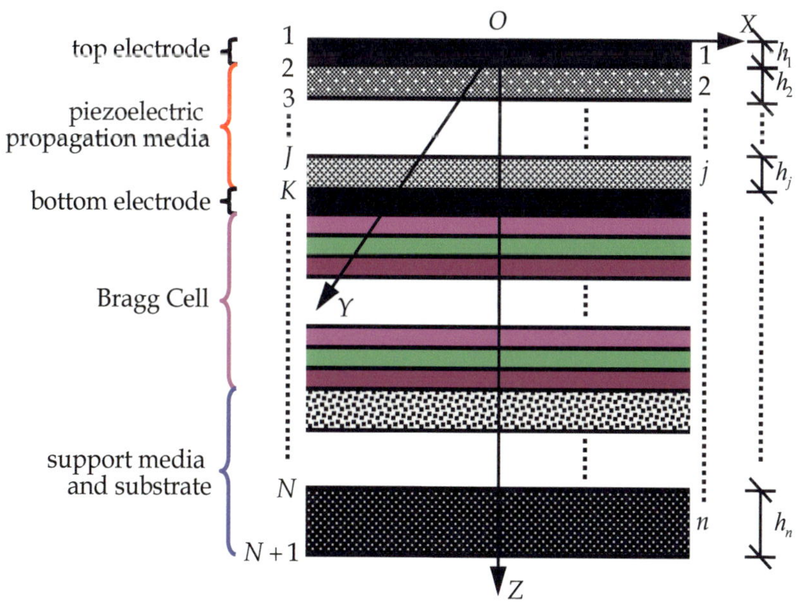


**Figure 5.** The schematic of multi-layered structures consisting of $n$ layers for modeling the multi-layered acoustic wave devices

piezoelectric single-layer or multi-layers. The Bragg Cell can currently be made of alternate elastic layers such as W and $SiO_2$ or alternate elastic and piezoelectric layers such as $SiO_2$

and AlN (Lakin, 2005), and may in the future be made of alternate piezoelectric layers. Assume in the multi-layered model, each one of the $n$ layers is homogeneous and the adjacent two layers are perfectly connected. To establish a general formulation for the analysis of various multi-layered acoustic wave devices with Bragg Cell, each layer in the multi-layered model is assumed as arbitrarily anisotropic. From up to down, the layers are denoted in order by numbers 1 to $n$, and the top surface, interfaces and bottom surface in turn are denoted by numbers 1 to $N+1$ ($N=n$). Thus, the upper and lower bounding faces of an arbitrary layer $j$ ($j=1,2,\cdots,n$) are denoted by $J$ ($J=j$) and $K$ ($K=j+1$), respectively, and the layer $j$ will also be referred to as $JK$ or $KJ$. Moreover, a global coordinate system $(X,Y,Z)$ with its origin located on the top surface and the $Z$-axis along the thickness direction, as shown in Fig. 5, is utilized to describe the integrated multi-layered structure.

### 3.1. Modeling of the non-piezoelectric layers (electrode, Bragg Cell, support layer and substrate)

Based on the three-dimensional linear elasticity (Stroh, 1962), the equations governing the dynamic state of a homogeneous, arbitrarily anisotropic elastic medium in absence of body forces can be written as

$$\sigma_{ij} = c_{ijkl}(u_{k,l} + u_{l,k})/2, \quad \sigma_{ij,j} = \rho \ddot{u}_i \tag{27}$$

where the comma in the subscripts and the dot above the variables imply spatial and time derivatives, $\sigma_{ij}$ and $u_i$ are respectively the stress and the displacement tensors, $c_{ijkl}$ denotes the elastic constant tensor having at most 21 independent components, and $\rho$ is the material density.

In the case of layer configuration, the state space formalism (Tarn, 2002a) can be adopted to describe mathematically the dynamic state of the medium. Referring to the global coordinate system $(X,Y,Z)$ in Fig. 5, we divide the stresses into two groups: the first consists of the components on the plane of $Z=\text{constant}$, and the second consists of the remaining components. The combination of the displacement vector $\mathbf{v}_u=[u,v,w]^{\mathrm{T}}$ and the vector of first group stresses $\mathbf{v}_\sigma=[\tau_{zx},\tau_{zy},\sigma_z]^{\mathrm{T}}$ gives the state vector $\mathbf{v}=[(\mathbf{v}_u)^{\mathrm{T}},(\mathbf{v}_\sigma)^{\mathrm{T}}]^{\mathrm{T}}$.

### 3.2. Modeling of the piezoelectric layers (propagation media and Bragg Cell)

According to the three-dimensional linear theory of piezoelectricity (Ding & Chen, 2001), the dynamic governing equations for the arbitrarily anisotropic piezoelectric medium in absence of both body forces and free charges are

$$\begin{cases} \sigma_{ij} = c_{ijkl}(u_{k,l} + u_{l,k})/2 + e_{kij}\varphi_{,k} \\ D_i = e_{ikl}(u_{k,l} + u_{l,k})/2 - \beta_{ik}\varphi_{,k} \end{cases}, \begin{cases} \sigma_{ij,j} = \rho \ddot{u}_i \\ D_{i,i} = 0 \end{cases} \tag{28}$$

where $D_i$ and $\varphi$ are respectively the electric displacement and the electric potential tensors, $e_{kij}$ and $\beta_{ik}$ are the piezoelectric and the permittivity constant tensors having at most 18 and 6 independent components, respectively, and all the remaining symbols have the same meanings as the corresponding ones in Eq. (27). It is seen from Eq. (28) that the coupling between the mechanical and electrical fields is considered.

Similar to the arbitrarily anisotropic elastic layer, an arbitrarily anisotropic piezoelectric layer can also be described mathematically by the state space formalism (Tarn, 2002b). In view of the global coordinate system $(X,Y,Z)$ in Fig. 5, the state vector is also represented as $\mathbf{v}=[(\mathbf{v}_u)^{\mathrm{T}},(\mathbf{v}_\sigma)^{\mathrm{T}}]^{\mathrm{T}}$, but with $\mathbf{v}_u=[u,v,w,\varphi]^{\mathrm{T}}$ being the generalized displacement vector and $\mathbf{v}_\sigma=[\tau_{zx},\tau_{zy},\sigma_z,D_z]^{\mathrm{T}}$ the generalized stress vector of the first group.

### 3.3. The state equations and solutions of non-piezoelectric and piezoelectric layers

By virtue of the triple Fourier transform pairs as follows

$$\begin{aligned}\hat{f}(k_x,k_y;z;\omega)&=\int_{-\infty}^{+\infty}\mathrm{e}^{-\mathrm{i}\omega t}\,\mathrm{d}t\int_{-\infty}^{+\infty}\int_{-\infty}^{+\infty}f(x,y,z,t)\mathrm{e}^{(\mathrm{i}k_x x+\mathrm{i}k_y y)}\,\mathrm{d}x\,\mathrm{d}y,\\ f(x,y,z,t)&=\int_{-\infty}^{+\infty}\frac{1}{2\pi}\mathrm{e}^{\mathrm{i}\omega t}\,\mathrm{d}\omega\int_{-\infty}^{+\infty}\int_{-\infty}^{+\infty}\left(\frac{1}{2\pi}\right)^2\hat{f}(k_x,k_y;z;\omega)\mathrm{e}^{-(\mathrm{i}k_x x+\mathrm{i}k_y y)}\,\mathrm{d}k_x\,\mathrm{d}k_y\end{aligned}\tag{29}$$

the dynamic governing equations (27) and (28) in the time-space domain can be transformed into those in the frequency-wavenumber domain. The quantity $\omega$ can be interpreted as the circular frequency, $k_x$ and $k_y$ are interpreted as the wavenumbers in the $x$ and $y$ directions, respectively. $\mathrm{i}=\sqrt{-1}$ is the unit imaginary. The $z$-dependent variable in the frequency-wavenumber domain is indicated by an over caret. Adopting the state space formalism (Tarn, 2002a, 2002b), we can reduce the transformed dynamic governing equations of a material layer corresponding to Eqs. (27) and (28) in right-handed coordinate systems, by eliminating the second group of generalized stresses, to the state equation as follows

$$\frac{\mathrm{d}\hat{\mathbf{v}}(z)}{\mathrm{d}z}=\mathbf{A}\hat{\mathbf{v}}(z)\tag{30}$$

It is noted that the transformed state vector $\hat{\mathbf{v}}(z)$ contains $n_v/2$ generalized displacement components and $n_v/2$ generalized stress components, with $n_v=6$ and $n_v=8$ for a non-piezoelectric (elastic) layer and a piezoelectric layer, respectively. Thus, the state equation is a system of $n_v$ first-order ordinary differential equations. The $n_v\times n_v$ coefficient matrix $\mathbf{A}$ of the state equation can be written in a blocked form

$$\mathbf{A}=\begin{bmatrix}-\mathrm{i}\mathbf{G}_{33}^{-1}\mathbf{W} & \mathbf{G}_{33}^{-1}\\ -\rho\omega^2\mathbf{M}+k_x^2\mathbf{G}_{11}+k_y^2\mathbf{G}_{22}+k_xk_y\left(\mathbf{G}_{12}+\mathbf{G}_{21}\right)-\mathbf{W}^{\mathrm{T}}\mathbf{G}_{33}^{-1}\mathbf{W} & -\mathrm{i}\mathbf{W}^{\mathrm{T}}\mathbf{G}_{33}^{-1}\end{bmatrix}\tag{31}$$

where $\mathbf{W} = -(k_x \mathbf{G}_{31} + k_y \mathbf{G}_{32})$. Assuming that the correspondence between the digital and coordinate indices follows $1 \to x$, $2 \to y$ and $3 \to z$, we have

$$\mathbf{G}_{kl} = \begin{bmatrix} c_{1k1l} & c_{1k2l} & c_{1k3l} \\ c_{2k1l} & c_{2k2l} & c_{2k3l} \\ c_{3k1l} & c_{3k2l} & c_{3k3l} \end{bmatrix}, \ \mathbf{M} = \mathbf{I}_3 \tag{32}$$

for a layer of arbitrarily anisotropic elastic material, and

$$\mathbf{G}_{kl} = \begin{bmatrix} c_{1k1l} & c_{1k2l} & c_{1k3l} & e_{l1k} \\ c_{2k1l} & c_{2k2l} & c_{2k3l} & e_{l2k} \\ c_{3k1l} & c_{3k2l} & c_{3k3l} & e_{l3k} \\ e_{k1l} & e_{k2l} & e_{k3l} & -\beta_{kl} \end{bmatrix}, \mathbf{M} = \begin{bmatrix} \mathbf{I}_3 & 0 \\ 0 & 0 \end{bmatrix} \tag{33}$$

for a layer of arbitrarily anisotropic piezoelectric material, with $\mathbf{I}_3$ the identity matrix of order 3.

According to the theory of ordinary differential equation (Coddington & Levinson, 1955), the solution to the state equation (30) can be expressed, in a form of traveling waves, as

$$\begin{aligned} \hat{\mathbf{v}}(z) &= \mathbf{\Phi} \exp(\mathbf{\Lambda} z) \mathbf{w} = \begin{bmatrix} \mathbf{\Phi}_- & \mathbf{\Phi}_+ \end{bmatrix} \begin{bmatrix} \exp(\mathbf{\Lambda}_- z) & \mathbf{0} \\ \mathbf{0} & \exp(\mathbf{\Lambda}_+ z) \end{bmatrix} \begin{Bmatrix} \mathbf{a} \\ \mathbf{d} \end{Bmatrix} \\ &= \begin{Bmatrix} \hat{\mathbf{v}}_u(z) \\ \hat{\mathbf{v}}_\sigma(z) \end{Bmatrix} = \begin{Bmatrix} \mathbf{\Phi}_u \\ \mathbf{\Phi}_\sigma \end{Bmatrix} \exp(\mathbf{\Lambda} z) \mathbf{w} = \begin{bmatrix} \mathbf{\Phi}_{u-} & \mathbf{\Phi}_{u+} \\ \mathbf{\Phi}_{\sigma-} & \mathbf{\Phi}_{\sigma+} \end{bmatrix} \begin{bmatrix} \exp(\mathbf{\Lambda}_- z) & \mathbf{0} \\ \mathbf{0} & \exp(\mathbf{\Lambda}_+ z) \end{bmatrix} \begin{Bmatrix} \mathbf{a} \\ \mathbf{d} \end{Bmatrix}, \end{aligned} \tag{34}$$

where $\exp(\cdot)$ denotes the matrix exponential function, $\mathbf{\Lambda}$ and $\mathbf{\Phi}$ are respectively the $n_v \times n_v$ diagonal eigenvalue matrix and square eigenvector matrix of the coefficient matrix $\mathbf{A}$, $\mathbf{w}$ is the vector of undetermined wave amplitudes with $n_v$ components. $\mathbf{\Lambda}_-$ ($n_a \times n_a$) and $\mathbf{\Lambda}_+$ ($n_d \times n_d$) are the diagonal sub-matrices of $\mathbf{\Lambda}$ corresponding respectively to the arriving wave vector $\mathbf{a}$ with $n_a$ wave amplitudes and the departing wave vector $\mathbf{d}$ with $n_d$ wave amplitudes. $\mathbf{\Phi}_-$ ($n_v \times n_a$) and $\mathbf{\Phi}_+$ ($n_v \times n_d$) are the corresponding sub matrices of $\mathbf{\Phi}$. $\mathbf{\Phi}_u$ and $\mathbf{\Phi}_\sigma$ are the $n_v / 2 \times n_v$ sub-matrices of $\mathbf{\Phi}$ corresponding to the generalized displacement and stress vectors, respectively. The sub-matrices $\mathbf{\Phi}_{u-}$, $\mathbf{\Phi}_{\sigma-}$, $\mathbf{\Phi}_{u+}$ and $\mathbf{\Phi}_{\sigma+}$ are defined accordingly. It should be noted that $\mathbf{a}$ consists of those wave amplitudes $w_i$ in $\mathbf{w}$, which correspond to the eigenvalues $\lambda_i$ satisfying $\mathrm{Re}(-\lambda_i) < 0$ or $\mathrm{Re}(-\lambda_i) = 0, \mathrm{Im}(-\lambda_i) < 0$, and the remaining wave amplitudes in $\mathbf{w}$ form $\mathbf{d}$. Obviously, we always have $\mathbf{w} = [\mathbf{a}^{\mathrm{T}}, \mathbf{d}^{\mathrm{T}}]^{\mathrm{T}}$ and $n_a + n_d = n_v$.

### 3.4. Reverberation-ray matrix analysis of integrated multi-layered structures

#### 3.4.1. Description of the structural system

Within the framework of MRRM, the physical variables associated with any surface/interface $J$ ($J=1,2,\cdots,N+1$) will be described in the global coordinate system $(X,Y,Z)$ as shown in Fig. 5 for the convenience of system analysis, and will be affixed with single superscript $J$ to indicate their affiliation. The physical variables associated with any layer $j$ (i.e. $JK$ or $KJ$, $j=1,2,\cdots,n$) will be described in the local dual coordinates $(x^{JK},y^{JK},z^{JK})$ or $(x^{KJ},y^{KJ},z^{KJ})$ as shown in Fig. 6 for the sake of member analysis, and will be affixed with double superscripts $JK$ or $KJ$ to denote the corresponding coordinate system and the pertaining layer. To make the sign convection clear, physical variables are deemed to be positive as it is along the positive direction of the pertinent coordinate axis.

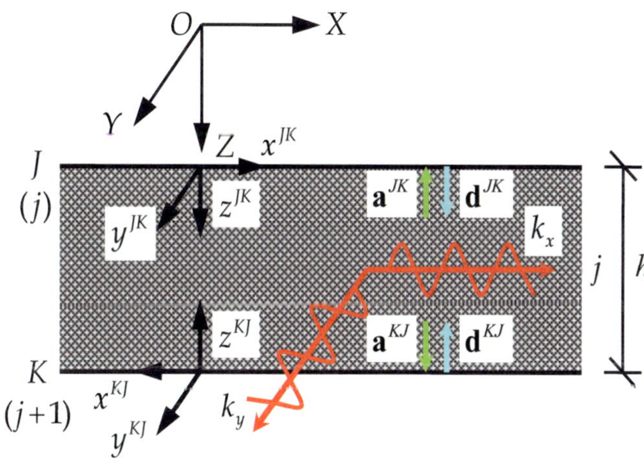


**Figure 6.** Description of a typical layer $j$ within the multi-layered model in local dual coordinates

#### 3.4.2. Traveling wave solutions to the state variables

It is seen from Fig. 6 that the local dual coordinates are both right-handed, thus the state equations for an arbitrary layer $j$ (i.e. $JK$ or $KJ$) in $(x^{JK},y^{JK},z^{JK})$ and $(x^{KJ},y^{KJ},z^{KJ})$ have the same form as Eq. (30). The traveling wave solutions to them can be written according to Eq. (34) as follows

$$\begin{Bmatrix} \hat{\mathbf{v}}_u^{JK}(z^{JK}) \\ \hat{\mathbf{v}}_\sigma^{JK}(z^{JK}) \end{Bmatrix} = \begin{Bmatrix} \mathbf{\Phi}_u^{JK} \\ \mathbf{\Phi}_\sigma^{JK} \end{Bmatrix} \exp(\mathbf{\Lambda}^{JK} z^{JK})\mathbf{w}^{JK} = \begin{bmatrix} \mathbf{\Phi}_{u-}^{JK} & \mathbf{\Phi}_{u+}^{JK} \\ \mathbf{\Phi}_{\sigma-}^{JK} & \mathbf{\Phi}_{\sigma+}^{JK} \end{bmatrix} \begin{bmatrix} \exp(\mathbf{\Lambda}_-^{JK} z^{JK}) & \mathbf{0} \\ \mathbf{0} & \exp(\mathbf{\Lambda}_+^{JK} z^{JK}) \end{bmatrix} \begin{Bmatrix} \mathbf{a}^{JK} \\ \mathbf{d}^{JK} \end{Bmatrix}, \quad (35)$$

$$\begin{Bmatrix} \hat{\mathbf{v}}_u^{KJ}(z^{KJ}) \\ \hat{\mathbf{v}}_\sigma^{KJ}(z^{KJ}) \end{Bmatrix} = \begin{Bmatrix} \mathbf{\Phi}_u^{KJ} \\ \mathbf{\Phi}_\sigma^{KJ} \end{Bmatrix} \exp(\mathbf{\Lambda}^{KJ} z^{KJ})\mathbf{w}^{KJ} = \begin{bmatrix} \mathbf{\Phi}_{u-}^{KJ} & \mathbf{\Phi}_{u+}^{KJ} \\ \mathbf{\Phi}_{\sigma-}^{KJ} & \mathbf{\Phi}_{\sigma+}^{KJ} \end{bmatrix} \begin{bmatrix} \exp(\mathbf{\Lambda}_-^{KJ} z^{KJ}) & \mathbf{0} \\ \mathbf{0} & \exp(\mathbf{\Lambda}_+^{KJ} z^{KJ}) \end{bmatrix} \begin{Bmatrix} \mathbf{a}^{KJ} \\ \mathbf{d}^{KJ} \end{Bmatrix}. \quad (36)$$

### 3.4.3. Scattering relations from coupling conditions on surfaces and at interfaces

Consider the compatibility of generalized displacements and the equilibrium of generalized stresses on surfaces and at interfaces. The spectral coupling equations on the top surface $1$, at any interface $J$ and on the bottom surface $N+1$ are expressed respectively as

$$\hat{\mathbf{v}}_u^{12}(0) = \hat{\mathbf{v}}_{uE}^1, \quad \hat{\mathbf{v}}_\sigma^{12}(0) + \hat{\mathbf{v}}_{\sigma E}^1 = \mathbf{0} \tag{37}$$

$$\mathbf{T}_u \hat{\mathbf{v}}_u^{JI}(0) = \hat{\mathbf{v}}_u^{JK}(0) = \hat{\mathbf{v}}_{uE}^J, \quad \mathbf{T}_\sigma \hat{\mathbf{v}}_\sigma^{JI}(0) + \hat{\mathbf{v}}_\sigma^{JK}(0) + \hat{\mathbf{v}}_{\sigma E}^J = \mathbf{0} \tag{38}$$

$$\mathbf{T}_u \hat{\mathbf{v}}_u^{(N+1)N}(0) = \hat{\mathbf{v}}_{uE}^{(N+1)}, \quad \mathbf{T}_\sigma \hat{\mathbf{v}}_\sigma^{(N+1)N}(0) + \hat{\mathbf{v}}_{\sigma E}^{(N+1)} = \mathbf{0} \tag{39}$$

where $\hat{\mathbf{v}}_{uE}^1$, $\hat{\mathbf{v}}_{uE}^J$ and $\hat{\mathbf{v}}_{uE}^{(N+1)}$ are the generalized displacement vectors of top surface $1$, interface $J$ and bottom surface $N+1$, respectively, $\hat{\mathbf{v}}_{\sigma E}^1$, $\hat{\mathbf{v}}_{\sigma E}^J$ and $\hat{\mathbf{v}}_{\sigma E}^{(N+1)}$ are the corresponding generalized stress vectors, $\mathbf{T}_u = \mathbf{T}_\sigma$ are the coordinate transformation matrix that equal to $<-1,1,-1>$ for non-piezoelectric (elastic) layers and $<-1,1,-1,-1>$ for piezoelectric layers. Here and after $<\cdot>$ denotes the (block) diagonal matrix with elements (or sub-matrices) only on the main diagonal.

It should be noticed that halves of all the components in vectors $\hat{\mathbf{v}}_{uE}^1$ and $\hat{\mathbf{v}}_{\sigma E}^1$, in vectors $\hat{\mathbf{v}}_{uE}^J$ and $\hat{\mathbf{v}}_{\sigma E}^J$, and in vectors $\hat{\mathbf{v}}_{uE}^{(N+1)}$ and $\hat{\mathbf{v}}_{\sigma E}^{(N+1)}$ are known, which are denoted by vectors $\hat{\mathbf{v}}_K^1$, $\hat{\mathbf{v}}_K^J$ and $\hat{\mathbf{v}}_K^{(N+1)}$, respectively. Substituting the solutions to the state variables of layers as given in Eqs. (35) and (36) into those coupling equations containing $\hat{\mathbf{v}}_K^1$, $\hat{\mathbf{v}}_K^J$ and $\hat{\mathbf{v}}_K^{(N+1)}$, we can obtain respectively the local scattering relations of top surface $1$, interface $J$ and bottom surface $N+1$ as follows

$$\mathbf{A}^1\mathbf{a}^1 + \mathbf{D}^1\mathbf{d}^1 = \mathbf{T}_K^1\hat{\mathbf{v}}_K^1 = \mathbf{s}_0^1 \tag{40}$$

$$\mathbf{A}^J\mathbf{a}^J + \mathbf{D}^J\mathbf{d}^J = \mathbf{T}_K^J\hat{\mathbf{v}}_K^J = \mathbf{s}_0^J \tag{41}$$

$$\mathbf{A}^{N+1}\mathbf{a}^{N+1} + \mathbf{D}^{N+1}\mathbf{d}^{N+1} = \mathbf{T}_K^{N+1}\hat{\mathbf{v}}_K^{N+1} = \mathbf{s}_0^{N+1} \tag{42}$$

where $\mathbf{a}^1 = \mathbf{a}^{12}$ ($\mathbf{d}^1 = \mathbf{d}^{12}$), $\mathbf{a}^J = [(\mathbf{a}^{JI})^{\mathrm{T}}, (\mathbf{a}^{JK})^{\mathrm{T}}]^{\mathrm{T}}$ ($\mathbf{d}^J = [(\mathbf{d}^{JI})^{\mathrm{T}}, (\mathbf{d}^{JK})^{\mathrm{T}}]^{\mathrm{T}}$) and $\mathbf{a}^{N+1} = \mathbf{a}^{(N+1)N}$ ($\mathbf{d}^{N+1} = \mathbf{d}^{(N+1)N}$) are the arriving (departing) wave vectors of top surface $1$, interface $J$ and bottom surface $N+1$, respectively, the corresponding coefficient matrices $\mathbf{A}^1$ ($\mathbf{D}^1$), $\mathbf{A}^J$ ($\mathbf{D}^J$) and $\mathbf{A}^{N+1}$ ($\mathbf{D}^{N+1}$) have components extracted, in accordance with $\hat{\mathbf{v}}_K^1$, $\hat{\mathbf{v}}_K^J$ and $\hat{\mathbf{v}}_K^{(N+1)}$, from respectively $\mathbf{\Phi}_-^{12}$ ($\mathbf{\Phi}_+^{12}$), $[\mathbf{T}_v\mathbf{\Phi}_-^{JI}, \mathbf{\Phi}_-^{JK}]$ ($[\mathbf{T}_v\mathbf{\Phi}_+^{JI}, \mathbf{\Phi}_+^{JK}]$) and $\mathbf{T}_v\mathbf{\Phi}_-^{(N+1)N}$ ($\mathbf{T}_v\mathbf{\Phi}_+^{(N+1)N}$), $\mathbf{s}_0^1$, $\mathbf{s}_0^J$ and $\mathbf{s}_0^{(N+1)}$ are the excitation source vectors of top surface $1$, interface $J$ and bottom surface $N+1$, respectively, the corresponding coefficient matrices $\mathbf{T}_K^1$, $\mathbf{T}_K^J$ and $\mathbf{T}_K^{N+1}$ consist of the components from $\mathbf{T}_v =< \mathbf{T}_u, \mathbf{T}_\sigma >$ in accordance with $\hat{\mathbf{v}}_K^1$, $\hat{\mathbf{v}}_K^J$ and $\hat{\mathbf{v}}_K^{(N+1)}$.

The local scattering relations of top surface, interfaces and bottom surface are grouped together from up to down to give the global scattering relation

$$\mathbf{A}\mathbf{a} + \mathbf{D}\mathbf{d} = \mathbf{s}_0, \tag{43}$$

where the global arriving and departing wave vectors $\mathbf{a}$ and $\mathbf{d}$ are

$$\begin{aligned} \mathbf{a} &= \left[ (\mathbf{a}^{12})^{\mathrm{T}}, (\mathbf{a}^{21})^{\mathrm{T}}, (\mathbf{a}^{23})^{\mathrm{T}}, \cdots, (\mathbf{a}^{JI})^{\mathrm{T}}, (\mathbf{a}^{JK})^{\mathrm{T}}, \cdots, (\mathbf{a}^{N(N+1)})^{\mathrm{T}}, (\mathbf{a}^{(N+1)N})^{\mathrm{T}} \right]^{\mathrm{T}}, \\ \mathbf{d} &= \left[ (\mathbf{d}^{12})^{\mathrm{T}}, (\mathbf{d}^{21})^{\mathrm{T}}, (\mathbf{d}^{23})^{\mathrm{T}}, \cdots, (\mathbf{d}^{JI})^{\mathrm{T}}, (\mathbf{d}^{JK})^{\mathrm{T}}, \cdots, (\mathbf{d}^{N(N+1)})^{\mathrm{T}}, (\mathbf{d}^{(N+1)N})^{\mathrm{T}} \right]^{\mathrm{T}} \end{aligned} \tag{44}$$

the corresponding coefficient matrices $\mathbf{A}$ and $\mathbf{D}$ are

$$\mathbf{A} =< \mathbf{A}^1, \mathbf{A}^2, \cdots, \mathbf{A}^J, \cdots, \mathbf{A}^{N+1} >, \quad \mathbf{D} =< \mathbf{D}^1, \mathbf{D}^2, \cdots, \mathbf{D}^J, \cdots, \mathbf{D}^{N+1} > \tag{45}$$

and $\mathbf{s}_0 = \left[ (\mathbf{s}_0^1)^{\mathrm{T}}, (\mathbf{s}_0^2)^{\mathrm{T}}, \cdots, (\mathbf{s}_0^J)^{\mathrm{T}}, \cdots, (\mathbf{s}_0^{(N+1)})^{\mathrm{T}} \right]^{\mathrm{T}}$ is the global excitation source vector.

It is noticed that the exponential functions in the solutions to the state variables of layers as shown in Eqs. (35) and (36) disappear in the scattering relations, since the thickness coordinates on the surfaces and at the interfaces are always zero in the corresponding local coordinates. This is the main advantage of introducing the local dual coordinates.

### 3.4.4. *Phase relations from compatibility conditions of layers*

Considering the compatibility between generalized displacements (generalized stresses) represented in local coordinates $(x^{JK}, y^{JK}, z^{JK})$ and the corresponding ones represented in $(x^{KJ}, y^{KJ}, z^{KJ})$ of any layer $j$ (i.e. $JK$ or $KJ$), we have

$$\hat{\mathbf{v}}_u^{JK}(z^{JK}) = \mathbf{T}_u \hat{\mathbf{v}}_u^{KJ}(h^{JK} - z^{KJ}), \; \hat{\mathbf{v}}_\sigma^{JK}(z^{JK}) = -\mathbf{T}_\sigma \hat{\mathbf{v}}_\sigma^{KJ}(h^{JK} - z^{KJ}) \tag{46}$$

where $h^{JK}$ $(= h^{KJ})$ denotes the thickness of layer $JK$ ($KJ$). Substitution of Eqs. (35) and (36) into Eq. (46), one obtains, by noticing $\mathbf{\Lambda}^{JK} = -\mathbf{\Lambda}^{KJ}$, $\mathbf{\Phi}^{JK} = \mathbf{T}_v \mathbf{\Phi}^{KJ}$, $\mathbf{T}_v = \mathbf{T}_v^{-1}$, $n_a^{JK} = n_d^{KJ}$ and $n_d^{JK} = n_a^{KJ}$, the local phase relation of a typical layer $j$ (i.e. $JK$ or $KJ$)

$$\begin{Bmatrix} \mathbf{a}^{JK} \\ \mathbf{a}^{KJ} \end{Bmatrix} = \begin{bmatrix} \exp(-\mathbf{\Lambda}_-^{JK} h^{JK}) & \mathbf{0} \\ \mathbf{0} & \exp(\mathbf{\Lambda}_+^{JK} h^{JK}) \end{bmatrix} \begin{Bmatrix} \mathbf{d}^{KJ} \\ \mathbf{d}^{JK} \end{Bmatrix} = \begin{bmatrix} \mathbf{P}^{JK} & \mathbf{0} \\ \mathbf{0} & \mathbf{P}^{KJ} \end{bmatrix} \begin{bmatrix} \mathbf{0} & \mathbf{I}_a^{JK} \\ \mathbf{I}_d^{JK} & \mathbf{0} \end{bmatrix} \begin{Bmatrix} \mathbf{d}^{JK} \\ \mathbf{d}^{KJ} \end{Bmatrix} \tag{47}$$

where $\mathbf{P}^{JK} = \exp(-\mathbf{\Lambda}_{-}^{JK} h^{JK})$ and $\mathbf{P}^{KJ} = \exp(\mathbf{\Lambda}_{+}^{JK} h^{JK})$ are respectively the $n_a^{JK} \times n_a^{JK}$ and $n_d^{JK} \times n_d^{JK}$ diagonal local phase matrices, and $\mathbf{I}_a^{JK}$ and $\mathbf{I}_d^{JK}$ are identity matrices of order $n_a^{JK}$ and $n_d^{JK}$, respectively.

Grouping together the local phase relations for all layers from up to down, one obtains the global phase relation

$$\mathbf{a} = \mathbf{PUd} \tag{48}$$

where $\mathbf{P}$ and $\mathbf{U}$ are respectively the blocked diagonal global phase matrix and global permutation matrix composed of

$$\mathbf{P} =< \mathbf{P}^{12}, \mathbf{P}^{21}, \mathbf{P}^{23}, \cdots, \mathbf{P}^{JJ}, \mathbf{P}^{JK}, \cdots, \mathbf{P}^{N(N+1)}, \mathbf{P}^{(N+1)N} > \tag{49}$$

$$\mathbf{U} =< \mathbf{U}^{12}, \mathbf{U}^{23}, \cdots, \mathbf{U}^{JK}, \cdots, \mathbf{U}^{N(N+1)} >, \quad \mathbf{U}_{n_v \times n_v}^{JK} = \begin{bmatrix} \mathbf{0} & \mathbf{I}_a^{JK} \\ \mathbf{I}_d^{JK} & \mathbf{0} \end{bmatrix} \tag{50}$$

### *3.4.5. System equation and dispersion equation*

The global scattering relation in Eq. (43) and the global phase relation in Eq. (48) both contain $n_v \times N$ equations for the $n_v \times N$ unknown arriving wave amplitudes (in $\mathbf{a}$) and $n_v \times N$ unknown departing wave amplitudes (in $\mathbf{d}$). Thus the wave vectors $\mathbf{a}$ and $\mathbf{d}$ can be determined accordingly. Substitution of Eq. (48) into Eq. (43) gives the system equation

$$(\mathbf{APU} + \mathbf{D})\mathbf{d} = \mathbf{Rd} = \mathbf{s}_0 \tag{51}$$

where $\mathbf{R} = \mathbf{APU} + \mathbf{D}$ is the system matrix.

If there is no excitation ($\mathbf{s}_0 = \mathbf{0}$), i.e. the free wave propagation problem is considered, the vanishing of the system matrix determinant yields the following dispersion equation

$$\left|\mathbf{R}(k_x, k_y; \omega)\right| = \mathbf{0} \tag{52}$$

which may be solved numerically by a proper root searching technique (Guo, 2008). Thus, the complete propagation characteristics of various waves can be obtained. In particular, the resonant frequency of the multi-layered structures can be obtained as $k_x = k_y = 0$.

It should be noted that the above proposed formulation of MRRM (Guo & Chen, 2008a, 2008b; Guo, 2008; Guo et al., 2009) excludes any exponentially growing function and matrix inversion, therefore possesses unconditionally numerical stability and enables inclusion of surface and interface wave modes.

### 3.5. Numerical examples

In this section, the above proposed formulation of MRRM for analyzing the propagation characteristics of various waves in the integrated acoustic wave devices are validated by a bulk acoustic resonator (BAR) consisting of 0.3μm Al film as the top electrode, 3.0μm AlN film as the propagation medium, 0.3μm Al film as the bottom electrode, alternate 0.81μm $SiO_2$ and 1.76μm AlN layers as the Bragg Cell, 0.81μm $SiO_2$ layer as the support medium and 42.6μm Si layer as the substrate. The material parameters of the exemplified BAR used in the calculation are given in Table 1.

| Type of material | Material | Material parameters: Elastic constants (GPa) | Mass density (kg/m³) | Dielectric constant ($\times 8.854\times 10^{-12}$ F/m) | Piezoelectric constants (C/m²) |
|---|---|---|---|---|---|
| Isotropic elastic material | Al | $E=69$, $G=26$ | $\rho=2700$ | $\varepsilon=1.6$ | — |
| | $SiO_2$ | $E=70$, $G=29.915$ | $\rho=2200$ | $\varepsilon=3.9$ | — |
| Transversely isotropic elastic material | Si | $c_{11}=c_{22}=c_{33}=164.8$ $c_{12}=c_{13}=c_{23}=63.5$, $c_{44}=c_{55}=79$, $c_{66}=50.65$ | $\rho=2330$ | $\varepsilon_{11}=\varepsilon_{22}=\varepsilon_{33}=11.8$ | — |
| Transversely isotropic piezoelectric material | AlN | $c_{11}=c_{22}=345$, $c_{33}=395$, $c_{12}=125$, $c_{13}=c_{23}=120$, $c_{44}=c_{55}=118$, $c_{66}=110$ | $\rho=3512$ | $\varepsilon_{11}=\varepsilon_{22}=9$, $\varepsilon_{33}=11$ | $e_{15}=e_{24}=-0.48$, $e_{31}=e_{32}=-0.58$, $e_{33}=1.55$ |

**Table 1.** Material properties of the exemplified bulk acoustic resonator

The resonant frequencies, represented by the engineering frequency $f=\omega/2\pi$ for the convenience of engineering application, of various waves in the multi-layered BAR are calculated by the formulation presented in Section 3.4 as the wavenumbers $k_x$ and $k_y$ are set to be zero. In order to show the influence of the number of unit cells in the Bragg Cell on the wave characteristics, Bragg Cells with 2 and 5 unit cells are respectively considered. Moreover, for sake of exploring the effects of electrodes, Bragg Cell and substrate on the wave characteristics in the propagation medium, the resonant frequencies of the 3.0μm AlN film, 3.0μm AlN film with top and bottom electrodes, and the bulk acoustic resonator without substrate and with 5 unit cells in the Bragg Cell are also calculated. The obtained first fifteen resonant frequencies of these multi-layered structures are listed and compared in Table 2.

| Order | 3.0μm AlN film | 3.0μm AlN film with electrodes | The whole BAR with 2 unit cells in the Bragg Cell | The whole BAR with 5 unit cells in the Bragg Cell | The whole BAR without substrate and with 5 unit cells in the Bragg Cell |
|---|---|---|---|---|---|
| 1 | 0.9750 | 0.8380 | 0.0523 | 0.0441 | 0.1275 |
| 2 | 1.8217 | 2.2329 | 0.1038 | 0.0655 | 0.2517 |
| 3 | 1.9328 | 3.3316 | 0.1564 | 0.0885 | 0.3810 |
| 4 | 2.9014 | 4.8778 | 0.2096 | 0.1332 | 0.5130 |
| 5 | 3.6438 | 5.2997 | 0.2608 | 0.1755 | 0.6320 |
| 6 | 3.8643 | 5.7110 | 0.3106 | 0.2031 | 0.8749 |
| 7 | 4.8326 | 6.5112 | 0.3645 | 0.2206 | 1.0236 |
| 8 | 5.4657 | 8.9604 | 0.4771 | 0.2652 | 1.2409 |
| 9 | 5.7971 | 9.7544 | 0.5747 | 0.2668 | 1.3322 |
| 10 | 6.7639 | 10.5897 | 0.6334 | 0.3066 | 1.4596 |
| 11 | 7.2868 | 12.1454 | 0.6964 | 0.3350 | 1.7058 |
| 12 | 7.7296 | 13.2501 | 0.7610 | 0.3542 | 1.7985 |
| 13 | 8.6956 | 14.6379 | 0.8262 | 0.3984 | 1.9343 |
| 14 | 9.1095 | 15.4651 | 0.8727 | 0.4394 | 2.1422 |
| 15 | 9.6604 | 16.2959 | 0.8943 | 0.4679 | 2.2865 |

**Table 2.** Effects of components on the first fifteen frequencies of the exemplified BAR (GHz)

From Table 2, it is seen that all component layers in the multi-layered bulk acoustic wave device have obvious influence on the wave propagation characteristics, which validates the necessity to model the multi-layered acoustic wave devices by an integrated model with all components considered. The electrodes generally raise the resonant frequencies in the propagation medium except for the first mode. Adding unit cells of the Bragg Cell and the appending of substrate in the multilayered BAR will reduce the resonant frequencies and increase the number of wave modes in a given frequency range. These findings about the effects of electrodes, Bragg Cell and substrate on wave characteristics in the multilayered acoustic wave devices can be used in the design of these devices.

## 4. Conclusion

The accurate analysis and design of layered Bragg Cell and of multi-layered acoustic wave devices with Bragg Cell are studied by the method of reverberation-ray matrix in this chapter. We obtain the analysis formulation, the features and the formation of SH-wave band structures in layered Bragg Cell and the design rules of layered Bragg Cell according to SH-wave band requirements. A unified formulation of MRRM is attained for the analysis of multi-layered acoustic wave devices modeled by integrated multi-layers consisting of working media, electrodes, Bragg Cell, support layer and substrate. The effects of other components on the resonant characteristics in the working media are gained. All findings are validated by numerical examples. The study in this chapter leads to the following conclusions:

(1) In the SH-wave band structures of layered Bragg Cell, the phase constant spectra in pass-bands and the attenuation constant spectra in stop-bands occur alternately. The phase constant spectra of characteristic SH waves are formed from the dispersion curves of equivalent SH waves due to the zone folding effect and wave interference phenomenon. All the attenuation constant loops as $k = 0$ and the second and upper attenuation constant loops as $k \neq 0$ of characteristic SH waves are formed due to the separation of the dispersion curves of equivalent SH waves with respect to frequency during the forming of the phase spectra. The first attenuation constant loop as $k \neq 0$ of characteristic SH wave is formed due to the cutoff property of SH waves in constituent layers. The contrasts of SH-wave characteristic impedances of the constituent layers, the characteristic time of the unit cell and the characteristic times of the constituent layers are three kinds of essential parameters determining the formation of the band structures. The contrasts of SH-wave characteristic impedances decide whether the stop-bands due to periodicity of the periodic layered media exist or not. If yes, it further decides the widths of the frequency bands. The characteristic time of the unit cell decides how many pass-bands/stop-bands exist in a specified frequency range. The characteristic times of the constituent layers mainly decides the mid-frequencies of the frequency bands. These rules can be used for the design of the layered Bragg Cell according to SH-wave bands requirements.

(2) The proposed MRRM for integrated multi-layered acoustic wave devices is analytical based on distributed-parameter model, yields unified formulation, includes all wave modes and possesses unconditionally numerical stability. It therefore leads to high accurate results at small computational cost and is applicable to complex multilayered acoustic wave devices while combined with a uniform computer program.

(3) The integrated model considers nearly all the components in practical multi-layered acoustic wave devices, which definitely renders accurate wave propagation characteristics for guiding the proper design of and suppresses unfavorable spurious modes in the devices. Generally, the electrodes raise the resonant frequencies, while the Bragg Cell and the substrate reduce the resonant frequencies.

In summary, the MRRM, the understanding of SH wave bands in the Bragg Cell and the integrated modeling of multi-layered acoustic wave devices with Bragg Cell in this chapter will push forward the design of high-performed acoustic wave devices.

## Author details

Yongqiang Guo
*Key Laboratory of Mechanics on Disaster and Environment in Western China, Ministry of Education, and School of Civil Engineering and Mechanics, Lanzhou University, P.R.China*

Weiqiu Chen
*Department of Engineering Mechanics, Zhejiang University,P.R.China*

## Acknowledgement

This study was financially supported by the National Natural Science Foundation of China (Nos. 10902045 and 11090333), the Postdoctoral Science Foundation of China (Nos. 20090460155 and 201104019) and the Fundamental Research Funds for the Central Universities of China (Grant No. lzujbky-2011-9).

## 5. References

Adler, E.L. (2000). Bulk and surface acoustic waves in anisotropic solids, *International Journal of High Speed Electronics and Systems*, 10(3): 653-684.

Ballandras, S., Reinhardt, A., Laude, V., Soufyane, A., Camou, S., Daniau, W., Pastureaud, T., Steichen, W., Lardat, R., Solal, M. & Ventura, P. (2004) Simulations of surface acoustic wave devices built on stratified media using a mixed finite element/boundary integral formulation. *Journal of Applied Physics*, 96(12): 7731-7741.

Benetti, M., Cannata, D., Di Pietrantonio, F., Verona, E., Almaviva, S., Prestopino, G., Verona, C. & Verona-Rinati, G. (2008). Surface acoustic wave devices on AlN/single-crystal diamond for high frequency and high performances operation, *Proceedings of IEEE Ultrasonics Symposium, 1924-1927*, Beijing, China, Nov. 2008.

Benetti, M., Cannatta, D., Di Pietrantonio, F. & Verona, E. (2005). Growth of AlN piezoelectric film on diamond for high-frequency surface acoustic wave devices, *IEEE Transactions on Ultrasonics, Ferroelectrics, and Frequency Control*, 52: 1806–1811.

Brillouin, L. (1953). *Wave Propagation in Periodic Structures: Electric Filters and Crystal Lattices*, Dover Publications, New York.

Brizoual, L.L., Sarry, F., Elmazria, O., Alnot, P., Ballandras, S. & Pastureaud, T. (2008). GHz frequency ZnO/Si SAW device, *IEEE Transactions on Ultrasonics, Ferroelectrics, and Frequency Control*, 55(2): 442–450.

Chung, C.-J. Chen, Y.-C., Cheng, C.-C., Wang, C.-M. & Kao, K.-S. (2008). Superior dual mode resonances for $1/4\lambda$ solidly mounted resonators, *Proceedings of IEEE Ultrasonics Symposium, 250-253*, Beijing, China, Nov. 2008.

Coddington, E. A., Levinson, N. (1955). *Theory of Ordinary Differential Equations*, McGraw-Hill, New York.

Ding, H.J. & Chen, W.Q. (2001). *Three Dimensional Problems of Piezoelasticity*. Nova Science Publishers, New York.

Eringen, A.C. & Suhubi, E.S. (1975). *Elastodynamics II: Linear Theory*, Academic Press, New York.

Guo, Y.Q. & Chen, W.Q. (2010). Reverberation-ray matrix analysis of acoustic waves in multilayered anisotropic structures. *Acoustic Waves* (Edited by Don W. Dissanayake, ISBN: 978-953-307-111-4): 25-46.

Guo, Y.Q. & Chen, W.Q. (2008a). On free wave propagation in anisotropic layered media, *Acta Mechanica Solida Sinica*, 21: 500-506.

Guo, Y.Q. & Chen, W.Q. (2008b). Modeling of multilayered acoustic wave devices with the method of reverberation-ray matrix, *Symposium on Piezoelectricity, Acoustic Waves, and Device Applications* (SPAWDA 2008), 105-110.

Guo, Y.Q. (2008). *The Method of Reverberation-Ray Matrix and its Applications*, Doctorial dissertation. Zhejiang University, Hangzhou, China. (in Chinese)

Guo, Y.Q., Chen, W.Q. & Zhang, Y.L. (2009). Guided wave propagation in multilayered piezoelectric structures, *Science in China, Series G: Physics, Mechanics and Astronomy*, 52(7): 1094-1104.

Hashimoto, K., Omori, T. & Yamaguchi, M. (2009). Characterization of surface acoustic wave propagation in multi-layered structures using extended FEM/SDA software, *IEEE Transactions on Ultrasonics, Ferroelectrics, and Frequency Control*, 56(11): 2559–2564.

Hashimoto, K.Y. (2000). *Surface Acoustic Wave Devices in Telecommunications: Modeling and Simulation*, Springer, Berlin.

Kirsch, P., Assouar, M.B., Elmazria, O., Mortet, V., & Alnot, P. (2006). 5GHz surface acoustic wave devices based on aluminum nitride/diamond layered structure realized using electron beam lithography, *Applied Physics Letters*, 88: 223504.

Lakin, K.M. (2005). Thin film resonator technology, *IEEE Transactions on Ultrasonics, Ferroelectrics, and Frequency Control*, 52(5): 707-716.

Lowe, M.J.S. (1995). Matrix techniques for modeling ultrasonic waves in multilayered media, *IEEE Transactions on Ultrasonics, Ferroelectrics, and Frequency Control*, 42: 525-542.

Makkonen, T. (2005). *Numerical Simulations of Microacoustic Resonators and Filters*, Doctoral Dissertation. Helsinki University of Technology, Espoo, Finland.

Marechal, P., Haumesser, L., Tran-Huu-Hue, L.P., Holc, J., Kuscer, D., Lethiecq, M. & Feuillard, G. (2008). Modeling of a high frequency ultrasonic transducer using periodic structures, *Ultrasonics*, 48: 141–149.

Mead, D.J. (1996). Wave propagation in continuous periodic structures: Research contributions from Southampton, 1964–1995. *Journal of Sound and Vibration*, 190(3): 495–524.

Nakanishi, H., Nakamura, H., Hamaoka, Y., Kamiguchi, H. & Iwasaki, Y. (2008). Small-sized SAW duplexers with wide duplex gap on a $SiO_2/Al/LiNbO_3$ structure by using novel Rayleigh-mode spurious suppression technique, *Proceedings of IEEE Ultrasonics Symposium*, 1588-1591, Beijing, China, Nov. 2008.

Naumenko, N.F. (2010). A universal technique for analysis of acoustic waves in periodic grating sandwiched between multi-layered structures and its application to different types of waves, *Proceedings of IEEE Ultrasonics Symposium, 1673-1676*, San Diego, USA, Oct. 2010.

Pao, Y.H., Chen, W.Q. & Su, X.Y. (2007). The reverberation-ray matrix and transfer matrix analyses of unidirectional wave motion, *Wave Motion*, 44: 419-438.

Pastureaud, T., Laude, V. & Ballandras, S. (2002). Stable scattering-matrix method for surface acoustic waves in piezoelectric multilayers, *Applied Physics Letters*, 80: 2544-2546.

Royer, D. & Dieulesaint, E. (2000). *Elastic Waves in Solids I: Free and Guided Propagation*, Springer, Berlin.

Shen, M.R. & Cao, W.W. (2000). Acoustic bandgap formation in a periodic structure with multilayer unit cells. *Journal of Physics D: Applied Physics*, 33: 1150–1154.

Stroh, A.N. (1962). Steady state problems in anisotropic elasticity, *Journal of Mathematics and Physics*, 41: 77-103.

Su, X.Y., Tian, J.Y. & Pao, Y.H. (2002). Application of the reverberation-ray matrix to the propagation of elastic waves in a layered solid, *International Journal of Solids and Structures*, 39: 5447-5463.

Tajic, A., Volatier, A., Aigner, R., Solal, M. (2010). Simulation of solidly mounted BAW resonators using FEM combined with BEM and/or PML, *Proceedings of IEEE Ultrasonics Symposium*, 181-184, San Diego, USA, Oct. 2010.

Tan, E.L. (2007). Matrix Algorithms for modeling acoustic waves in piezoelectric multilayers, *IEEE Transactions on Ultrasonics, Ferroelectrics, and Frequency Control*, 54: 2016-2023.

Tarn, J. Q. (2002a). A state space formalism for anisotropic elasticity. Part I: Rectilinear anisotropy, *International Journal of Solids and Structures*, 39: 5143-5155

Tarn, J. Q. (2002b). A state space formalism for piezothermoelasticity, *International Journal of Solids and Structures*, 39: 5173-5184

Wang, G., Yu, D.L., Wen, J.H., Liu, Y.Z., Wen, X.S. (2004). One-dimensional phononic crystals with locally resonant structures. *Physics Letters A*, 327: 512–521.

Wang, L. & Rokhlin, S.I. (2002). Recursive asymptotic stiffness matrix method for analysis of surface acoustic wave devices on layered piezoelectric media, *Applied Physics Letters*, 81: 4049-4051.

Wu, T.T. & Chen, Y.Y. (2002) Exact analysis of dispersive SAW devices on ZnO/diamond/Si-layered structures, *IEEE Transactions on Ultrasonics, Ferroelectrics, and Frequency Control*, 49: 142-149.

Yoon, G., & Park, J.-D. (2000). Fabrication of ZnO-based film bulk acoustic resonator devices using W/$SiO_2$ multilayer reflector, *IEEE Electronics Letters*, 36(16): 1435-1437.

Zhang, V.Y., Dubus, B., Lefebvre, J.E. & Gryba, T. (2008). Modeling of bulk acoustic wave devices built on piezoelectric stack structures: Impedance matrix analysis and network representation, *IEEE Transactions on Ultrasonics, Ferroelectrics, and Frequency Control*, 55(3): 704–716.

Zhang, V.Y., Lefebvre, J.E. & Gryba, T. (2006). Resonant transmission in stop bands of acoustic waves in periodic structures, *Ultrasonics*, 44(1): 899–904.

Section 4

# Design and Fabrication of Microdevices

Chapter 13

# High-Overtone Bulk Acoustic Resonator

T. Baron, E. Lebrasseur, F. Bassignot, G. Martin, V. Pétrini and S. Ballandras

Additional information is available at the end of the chapter

http://dx.doi.org/10.5772/56175

## 1. Introduction

Piezoelectricity has been used for the development of numerous time&frequency passive devices [1]. Among all these, radio-frequency devices based on surface acoustic waves (SAW) or bulk acoustic waves (BAW) have received a very large interest for bandpass filter and frequency source applications. Billions of these components are spread each year around the world due to their specific functionalities and the maturity of their related technologies [2]. The demand for highly coupled high quality acoustic wave devices has generated a strong innovative activity, yielding the investigation of new device structures. A lot of work has been achieved exploiting thin piezoelectric films for the excitation and detection of BAW to develop low loss RF filters [3]. However, problems still exist for selecting the layer orientation to favor specific mode polarization and select propagation characteristics (velocity, coupling, temperature sensitivity, *etc.*). Moreover, for given applications, deposited films reveal incapable to reach the characteristics of monolithic substrates [4].

For practical implementation, BAW is applied for standard low frequency (5 to 10MHz) shear wave resonators on Quartz for instance. SAW, Film Bulk Acoustic Resonator (FBAR) and High overtone Bulk Acoustic Resonator (HBAR) devices are applied for standard radio-frequency ranges and more particularly in S band (2 to 4GHz). HBAR have been particularly developed along different approaches to take advantage of their extremely high quality factor and very compact structure. Until now, many investigations have been carried out using piezoelectric thin films (Aluminum Nitride – AlN, Zinc Oxide – ZnO) atop thick wafers of silicon or sapphire [5] but recent developments showed the interest of thinned single-crystal-based structure in that purpose [6]. Although marginal, their application has been mainly focused on filters and frequency stabilization (oscillator) purposes [7], but the demonstration of their effective implementation for sensor applications has been achieved recently [8]. These devices maximize the $Q$ factor that can be obtained at room temperature using elastic waves, yielding quality factor times Frequency products ($Q.f$) close or slightly above $10^{14}$, *i.e.* effective $Q$ factors of about 10,000 at 1GHz in theory (practically, $Q$ factors in excess of 50,000 between 1.5 and 2GHz were experimentally achieved [9])

HBAR-based sensors exploit two principal features yielding notable differences with other sensing solutions. The first one is related to the anisotropy of piezoelectric crystals on which these devices are built, which allows one for selecting crystal cut angles to optimize their physical characteristics. It is subsequently possible to choose cut angles to favor or minimize the parametric sensitivities of the considered wave propagation. The second remarkable feature of these devices concerns the use of piezoelectric excitation/detection of acousto-electric waves which allows for wireless interrogation in radio-frequency ranges such as ISM bands centered at 434MHz, 868MHz, 915MHz or even 2.45GHz.

This chapter presents HBAR principles and related applications. Specific acoustic and electrical behaviors of HBAR are discussed and the different ways devoted to the manufacture of these devices also are presented. Applications of HBAR such as oscillator stabilization, intrinsically temperature-compensated devices and sensors are finally reported. Further developments required to promote the industrial exploitation of HBAR are discussed to conclude this article.

## 2. HBAR principles

HBARs are constituted by a thin piezoelectric transducer above a high-quality acoustic substrate, as shown in figure 1. The piezoelectric transducer generates acoustic waves in the whole material stack along its effective electromechanical strength. Stationary waves are established between top and bottom free surfaces according to normal stress-free boundary conditions. As no electrical boundary condition arises at this surface, all the possible harmonics of the fundamental mode can exist. However, only the even modes of the transducer are excited as the only ones meeting the electrical boundary conditions applied to the transducer.

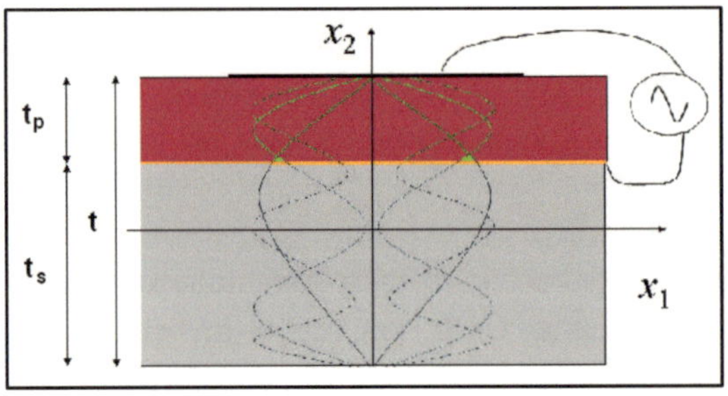


**Figure 1.** Schematic of HBAR

The electrical response of a HBAR can thus be interpreted as the modulation of the transducer resonance by the whole-stack bulk modes, presenting a dense spectrum of discrete modes localized around the resonance frequencies of the only piezoelectric transducer, as shown in figure 2. Since the substrate thickness is much larger than that of the piezoelectric film, most energy is stored in the substrate, and thus, the quality ($Q$) factor is dominated by the acoustic property of the substrate. When the thickness of the substrate decreases, the device tends to behave as a FBAR (corresponding to $t_s$=0 in fig.1). Depending on both the material and the cut orientations of piezoelectric transducer, pure longitudinal or pure shear waves or combinations of these basic polarizations can be excited.

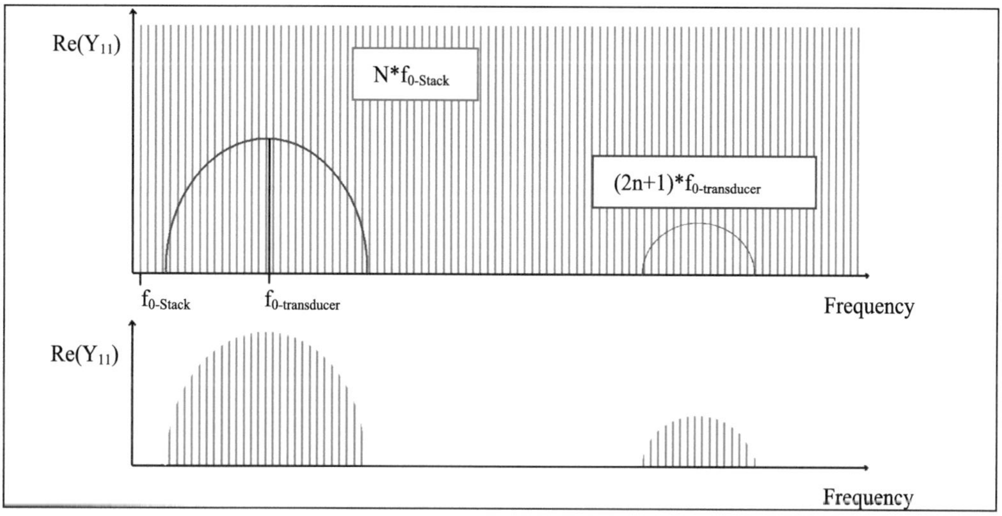


**Figure 2.** Schematic representation of the typical electrical response of HBARs.

Single port resonator structures can be easily achieved using HBARs, the use of two series devices being generally adopted to avoid etching the piezoelectric layer to reach the back electrode. Despite this favorable aspect, the exclusive use of single-port resonators limits HBAR applicability fields. Therefore, the possibility to fabricate four-port devices has been considered and experimentally tested (as shown in section 3.2). The leading idea consisted in transversely (or laterally) coupling acoustic waves between two adjacent resonators. The principle of such devices was inspired from the so-called monolithic filters based on coupled bulk waves in single crystals [10]. This is achieved by setting two resonators very close to one another. The gap between these resonators must be narrow enough to allow for the evanescent waves between the resonator electrodes to overlap and hence to yield mode coupling conditions. This system exhibits two eigenmodes with slightly different eigen-frequencies: a symmetric mode in which the coupled resonators vibrate in phase and an anti-symmetric mode in which they vibrate in phase opposition, as shown in figure 3. The gap between the two electrodes controls the spectral distance between the two coupled modes.

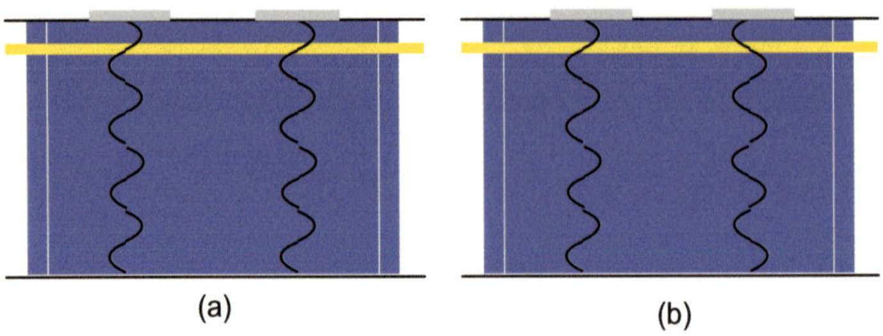


**Figure 3.** Principle scheme of the laterally-coupled-mode HBAR filter (a) symmetrical mode (b) anti-symmetrical mode

# 3. HBAR devices

## 3.1. Electrical and acoustic behavior

### *3.1.1. Single-port resonator*

As explained above, the electrical response spectrum of such HBAR presents a large number of overtones. A large band representation allows for the observation of several envelopes themselves composed of several overtones. The central frequencies of these envelopes correspond to fundamental and even overtone resonances of the only transducer and therefore are mainly controlled by the transducer thickness.

Figure 4 shows the $S_{11}$ response for the considered structure for different substrate thicknesses, illustrating the impact of this parameter on the overtone characteristics. The highest electromechanically-coupled overtone corresponds to the mode matching at best maximum energy location within the transducer, whereas the other overtones do exhibit smaller coupling factor proportionally to their spectral distance with the central overtone. The case of FBAR ($t_s$=0 μm in figure 1) is reported on this graph to effectively localize the central frequency of resonance and anti-resonance of the only transducer. In presence of a substrate, mode coupling between the two layers is made possible and several overtones appear for substrate thickness larger than the transducer one, as illustrated in figure 4. As suggested previously, the spectral distance between two overtones is mainly due to the substrate properties (velocity and thickness) when the later exhibits a thickness much larger than the other layers of the whole stack. Figure 4 also shows the evolution of the electromechanical coupling coefficient (generally noted $k_s^2$ for radio-frequency acousto-electric devices) when increasing the substrate thickness. The reduction of $k_s^2$ when increasing the substrate thickness is directly related to the energy ratio within the transducer and in the whole HBAR stack. Increasing the substrate thickness yields more energy in the whole HBAR structure and less energy within the transducer. Another interpretation consists in considering that the coupling of the transducer alone is spread on all the modes of the structure near the transducer resonance. Increasing the number of modes yields a reduction of the electromechanical of each mode coupled to the transducer mode. A trade-off between the mode density and the stack thickness therefore is mandatory to optimize the HBAR response. Increasing the number of modes experimentally tends to provide higher quality coefficients for modes close to the transducer one but also reduces the corresponding coupling and significantly impact the device spectral density, yielding more difficulty to exploit well defined resonance.

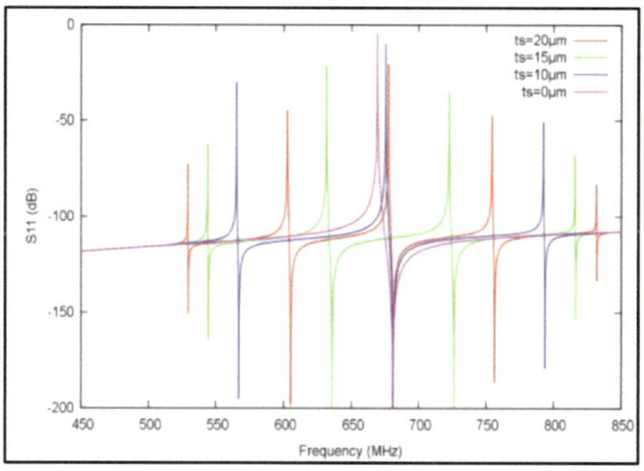


**Figure 4.** Impact of acoustic substrate. The reflection parameter-S11 with respect to a 50Ω load is measured for different substrate thicknesses. A material stack consisting of an acoustic substrate of $LiNbO_3$ (YX*l*)/163°, an aluminum electrode of 10nm thick, a 10µm thin piezoelectric layer of $LiNbO_3$ (YX*l*)/163° and a 10nm thick bottom aluminum electrode is considered here for a theoretical description of the HBAR characteristics. Electrode thickness are chosen extremely thin to neglect their acoustic influence. Acoustic and dielectric losses are only consider in $LiNbO_3$ layers for scaling the maximum achievable quality factors. For all computations, an active electrode surface of 100x100µm² has been considered for normative purposes.

For a given stack, the coupling coefficient of each group of overtones (these groups being defined by fundamental and even overtones of the transducer alone) depends on the material coupling coefficient of the transducer and on the order of the considered group. Indeed, the third order group presents a coupling coefficient divided by 9 compared to the fundamental group (one third of the fundamental mode coupling at excitation times one third at detection), the fifth is divided by 25, and so on. $LiNbO_3$ presents material coupling coefficient noticeably higher (3 to 7 times larger) than other material generally used for HBAR fabrication, such as AlN, ZnO. As a consequence, even transducer overtone groups can be effectively used with such a material and more especially when exciting pure shear waves as exposed further. Figure 5 shows the electrical response of a single-port HBAR built with (YX*l*)/163° $LiNbO_3$ piezoelectric layer and substrate. Only shear waves are excited and all even group can be visible from the fundamental to the 11$^{th}$ harmonic of the layer alone near 2GHz.

Each overtone in a given group presents a specific coupling coefficient $k_s^2$ and a specific quality factor *Q*. Central overtones present the highest coupling coefficients within a group, but not always the highest quality factors. Indeed, the substrate (Sapphire for instance) is usually chosen with acoustic quality better than the transducer material (Aln, ZnO) as it is expected to act as the effective resonant cavity, whereas the transducer material is selected for its piezoelectric strength. As explained above, the energy ratio within the transducer and the substrate is higher for the central overtones than for the overtones located at the edge of the group.

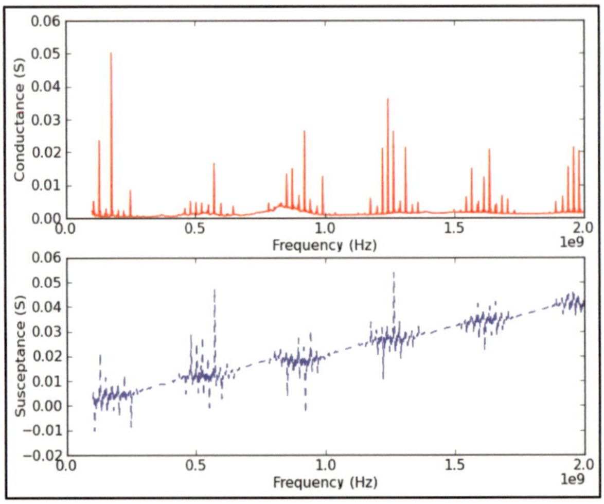


**Figure 5.** Single-port HBAR device built using $LiNbO_3$/$LiNbO_3$ (YX*l*)/163° cut.

In figure 6, a HBAR is constructed with $LiNbO_3$ material for the transducer. As shown further, $LiNbO_3$ presents better acoustic quality than Quartz, which is used for the HBAR substrate here to improve the device temperature stability (see section 4.2). In that example the overtone at 433.3MHz exhibits the best coupling coefficient $k_s^2$ as well as the best quality factor *Q*, due to the acoustic quality of $LiNbO_3$ compared to Quartz (see section 4.1).

According the above assumptions concerning material quality selection, the quality factor of overtones located at the edge of group is generally higher than the ones in the center of the group. Practically, it turns out that small-coupling overtones always exhibit better Q than the central overtones in a given group. One explanation of this objective result can be related to electrically generated losses (losses related to electrode resistivity and series resistance tends to increase with current).

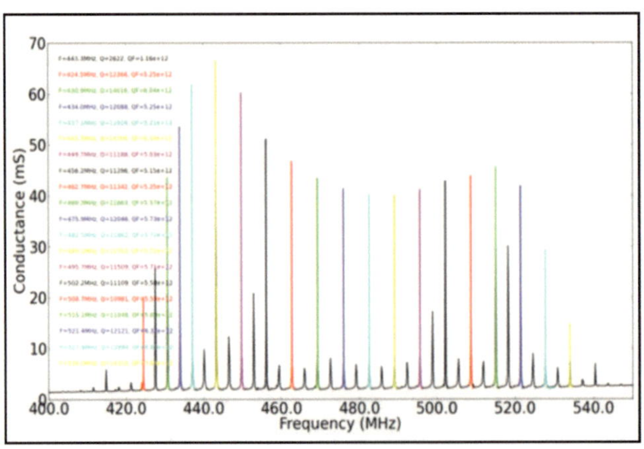


**Figure 6.** 5[th] Envelope of single-port HBAR device constituted by $LiNbO_3$ (YX*l*)/163°/Quartz.

### *3.1.2. Transversely-coupled HBAR*

As explained above, the possibility to fabricate four-port devices has been considered and experimentally tested. Two HBAR resonators were fabricated on a $LiNbO_3$ (34µm) / Au (300nm) / $LiNbO_3$ (350µm) stack. Two 145x200µm$^2$ surface aluminum electrodes, were patterned upon the stack and separated by a gap of 10µm. Figure 7 shows a typical coupled-mode filter response for a device manufactured atop a $LiNbO_3$/$LiNbO_3$ structure. Rejection in excess of 20dB is demonstrated at 720MHz with a single filter cell. Insertion losses of about 15dB are emphasized and could be easily improved by impedance matching. The measured transfer function actually exhibits a double mode response, providing a first evidence of the device operation according to the above assumptions.

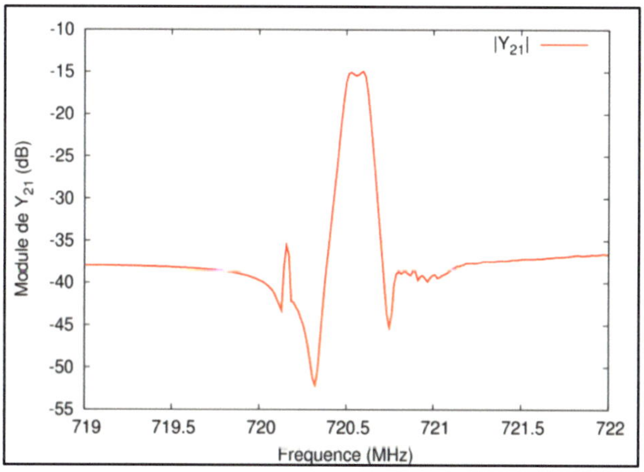


**Figure 7.** Four-port laterally coupled HBAR devices 0.1% band 720MHz $LiNbO_3$/$LiNbO_3$ filter.

Furthermore, the following experiment was applied for definitely validating the lateral mode coupling. The admittance of the first resonator was measured for two different loading conditions applied to the second resonator (Figure 8). For open circuit conditions, a main peak corresponding to the first resonator contribution is observed (Figure 8, left) together with a weaker contribution near the main resonance. For short-circuit conditions, the admittance measured on the first resonator shows two almost-balanced resonance peaks, (Figure 8, right). This behavior is explained by the fact that no current crosses the second resonator when in open condition, yielding a small contribution of the anti-symmetrical mode (due to poor boundary condition matching) whereas loaded electrical condition allows for an effective excitation of the later mode, yielding almost balanced contributions of symmetrical and anti-symmetrical modes as experimentally observed.

## 3.2. HBAR micro-fabrication

Two main approaches can be implemented to manufacture HBAR devices. The first approach consists in physical or chemical deposition of thin piezoelectric layers (such as ZnO, PZT, AlN and so on) onto the chosen substrate. The first HBAR was manufactured along this approach [11]. The main advantage of this kind of HBAR is the capability of the related techniques (sputtering, epitaxy, sol-gel spinning/firing, pulsed laser ablation, *etc.*) to deposit thin layers which allow for achieving device naturally operating at high frequency (for instance in the vicinity of the 2.45GHz ISM Band, or even more). This approach also did provide among the highest $Q$ factor ever measured for an acoustic-based resonator at room temperature [3], with $Q.f$ product values in excess of $10^{14}$ at parallel resonance ($7.10^{13}$ at series resonance).

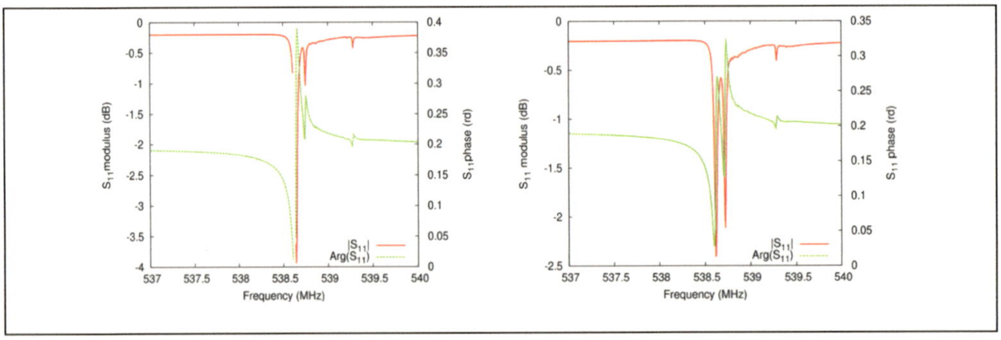


**Figure 8.** Admittances of one of the two resonators of the laterally-coupled structure as a function of the electrical conditions applied to the associated resonator (left) open circuit (right) 50Ω loading

Although this approach revealed efficient for operational device manufacturing, some drawbacks can be identified which limit the interest of the related resonators. Among these, one of the most problematic concerns the electromechanical coupling coefficient one can obtain particularly with AlN and ZnO, the most used thin piezoelectric layer for RF acoustic devices. As deposition techniques (principally reactive sputtering but also pulsed laser deposition (PLD)) generally allows for depositing well-controlled homogeneous C-oriented layers (i.e. with the C crystal axis oriented along the normal of the coated surface), the maximum accessible coupling remains much below 10%. Also the corresponding modes are purely longitudinal, with reduced degree-of-freedom for effectively controlled layer orientation suited for shear wave excitation/detection. Thin layers such as PZT can overcome this limitation but they generally exhibit notably high visco-elastic coefficients and significant dielectric loss which again limit their interest for high frequency (above 1GHz) applications. More generally, acoustic losses of most piezoelectric layers obtained by sputtering, sol-gel and techniques providing poly-crystalline materials always reveal larger than single-crystal ones. As explained before, the coupling coefficient of each high-overtone resonance depends on the number of overtones and on the intrinsic material electromechanical coupling coefficient. Poor material coupling coefficients prevent the use of overtones modulating the third (and therefore higher order) overtone of the piezoelectric transducer. Finally, compensating longitudinal modes thermal drift is particularly difficult as most of the high acoustic quality materials exhibit negative temperature coefficients of the corresponding phase velocity (as well as the transducer materials, ranging from -20 to -60ppm.$K^{-1}$). These negative aspects pushed to seek for other manufacturing approaches.

The opportunity to use single crystal layers for acoustic transduction therefore appears as an alternative solution. Assuming the possibility for manufacturing thin single-crystal films atop any material stack makes possible the use of specific crystal cut to select the polarization of the excited acoustic waves as well as its electromechanical coupling coefficient.

The development of the so-called Silicon On Insulator (SOI(TM)) wafers has demonstrated the huge opportunities offered by the Smart Cut(TM) approach [12]. Moreover, its application for transferring single crystal Lithium Niobate thin layer into silicon proved to be effective for SAW device development [13]. As this technology requires a severe know-how and complex technological facilities and environment, an alternative fabrication technique based on metal diffusion at the interface between the materials to be bonded together [14] has been developed together with a lapping/polishing technique for HBAR manufacturing [15].

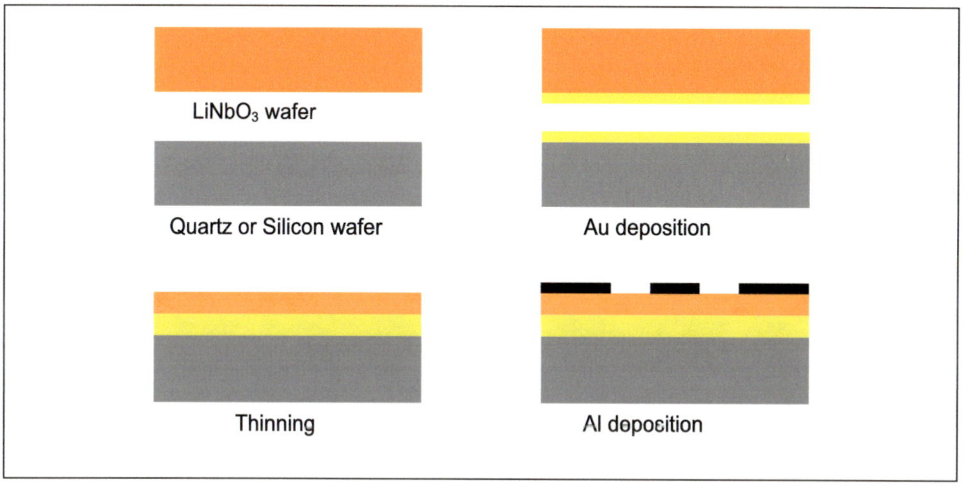


**Figure 9.** Process flow chart for the fabrication of the HBAR based on bonding and lapping technology.

In this particular approach, contrary to the sputtering method, thermal process forbids to stack materials presenting notably differential thermal expansion coefficients. Smart Cut(TM) approach allows one to produce thin single crystal layers (such as $LiNbO_3$ for instance, or even $LiTaO_3$ [16]) with typical thickness below 1µm. Along this approach, embedded metal electrodes are fabricated using the Smart Cuttrade technology [16].

The above-mentioned bonding and lapping technology has specifically been developed to allow for material stacking at room temperature, for exploiting any material of any crystal orientation. The process flow-chart reported in figure 9 allows one for a collective manufacturing of HBAR devices. This process is based on the mechanical diffusion of sub-micron gold layers, providing an effective acoustic liaison of the chosen material as well as the HBAR back electrode at once. As the bonding operation is achieved at room temperature, no significant thermal differential effects are observed and the resulting wafer can be handled and further processed provided thermal budget remains smaller than 100°C (as experimentally observed).

Along the proposed approach, optical quality polished surfaces are preferred to favor the bonding of the wafers. A Chromium and Gold thin layer is first deposited by sputtering on both wafers to bond ($LiNbO_3$ and Quartz in the example of figure 6 and 9). The $LiNbO_3$ wafer is then bonded onto the substrate by mechanical compression of the 200nm thick gold layers into an EVG wafer bonding machine as shown in figure 10. During the bonding process, the material stack is kept at a temperature of 30°C and a pressure of 65N.cm$^{-2}$ is applied on the whole contact surface. The bonding can be particularly controlled by adjusting the process duration and various parameters such as the applied pressure, the process temperature, the quality of the vacuum during the process, *etc.* In the reported development, the process temperature is kept near a value close to the final thermal conditions seen by the device in operation. Since substrate and piezoelectric materials have different thermal expansion coefficients, one must account for differential thermo-elastic stresses when bonding both wafers and minimize them as much as possible.

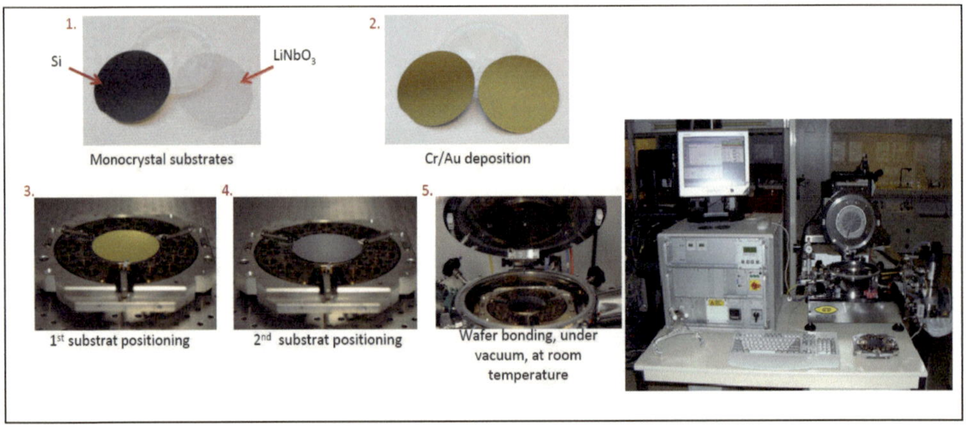


**Figure 10.** EVG wafer Bounder and illustration of Gold bonding process.

Once the bonding achieved, it is necessary to characterize the quality of the bonding. Due to the thickness of the wafers and the opacity of the stack (metal layers), optical measurements are poorly practicable. To avoid destructive controls of the material stack, ultrasonic techniques have been particularly considered here. The reliability of the bonding then is analyzed by ultrasonic transmission in a liquid environment. The bonded wafers are immersed in a water tank and the whole wafer stack surface is scanned. Figure 11 presents photography of the bench. Two focused transducers are used as acoustic emitter and receiver. They are manufactured by SONAXIS with a central frequency close to 50MHz, a 19mm active diameter and a 30mm focal length. The beam diameter at focal distance at -6dB is about 200μm.

Such a lateral resolution enables one to detect very small defects. The principle of the characterization method is based on the measurement of the received acoustic amplitude which depends on the variation of the acoustic impedance of the bonding area. If the bonding presents a defect at the interface between the two wafers, a dust or an air gap in most cases, the reflection coefficient of the incident wave is then nearly 1. The amplitude of the received wave is strongly reduced or even vanishes. Figure 12 shows a C-Scan of a Silicon/$LiNbO_3$ wafer bonding characterization. The blue color corresponds to bonded surfaces, whereas yellow and green regions indicate bonding defects.

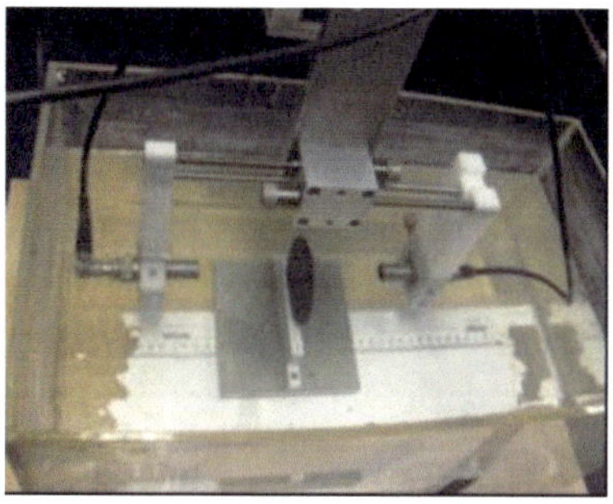

**Figure 11.** Ultrasonic characterization bench dedicated to non destructive control of the bonding interface.

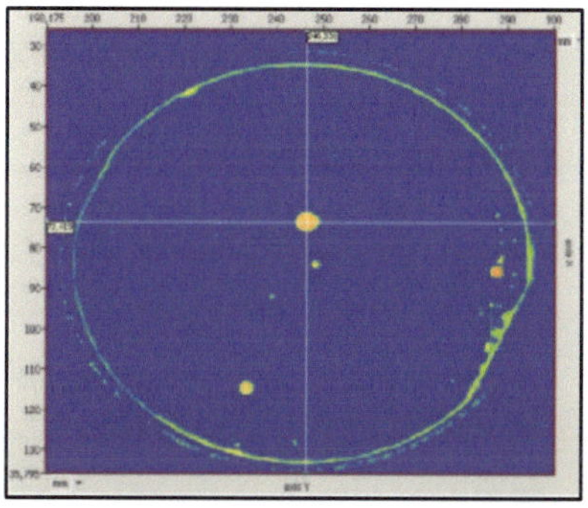

**Figure 12.** Characterization of a Silicon/LiNbO3 bonding – surfaces are bonded at 95%.

This method presents three major advantages:

- The control of the bonding can be made during the polishing steps without destruction; or the control can be done at the end of the process, indeed, the different layers obtained by sputtering do not disturb the measure.
- There is no constraint related to time resolution as in pulse-echo method, as the wafer thickness is not dramatically larger than the wavelength.
- The analysis of the ultrasonic transmitted signals is very simple because only the amplitude of the first detected signal contains the useful information.

**Figure 13.** SOMOS lapping/polishing machine.

The piezoelectric wafer is subsequently thinned by lapping step to an overall thickness of 20μm. The lapping machine used in that purpose and shown in figure 13 is a SOMOS double side lapping/polishing machine based on a planetary motion of the wafers (up to 4" diameter) to promote abrasion homogeneity. An abrasive solution of silicon carbide is used here. The speed of the lapping is controlled by choosing the speed of rotation of the lapping machine stages, the load on the wafer, the rate of flow or the concentration of the abrasive. Once close to the expected thickness, the lapping process is followed by a micro-polishing step. This step uses similar equipments dedicated to polishing operation and hence using abrasive solution (colloidal silica) with smaller grain. This polishing step is applied until the average surface roughness $r_a$ remains larger than 3nm. Afterward, the wafer is considered ready for surface processing.

The final step of the HBAR fabrication is the deposition and patterning of the top-side electrode. Aluminum electrodes are then deposited on the thinned $LiNbO_3$ plate surface with a lift-off process. This top electrode allows for connecting the HBAR-based sensor and for characterization operations.

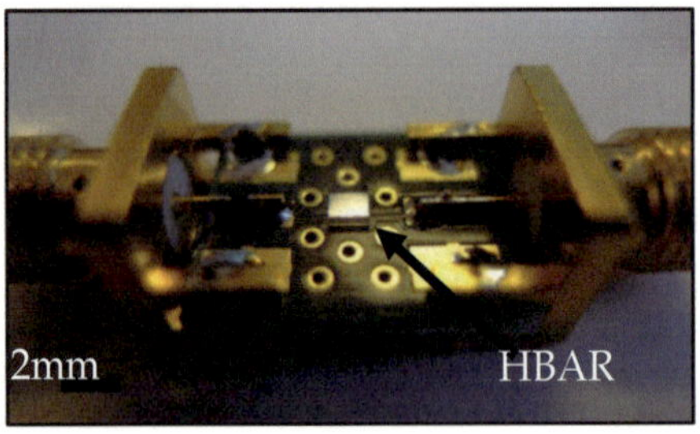


**Figure 14.** Flip chip of HBAR resonator on PCB substrate.

For all HBAR device, one technological problematic concerns packaging. Due to the HBAR operation, both sides must be kept free of any stress or absorbing condition. HBAR packaging therefore requires specific developments to meet such conditions. Experimental developments reveal that flip-chip techniques are the most appropriate approach in that purpose (as shown in figure 14).

## 4. HBAR optimization

### 4.1. Minimizing losses in HBARs

Since the 80's, HBAR devices have demonstrated high quality factor at high frequencies compared to other acoustic devices such as BAW, SAW. $Q.f$ products around $1.1\times10^{14}$ have already been obtained for high overtones using aluminum nitride (AlN) thin films deposited onto sapphire [3]. Hongyu Yu *and al.* showed HBAR with a structure of 0.10μm Al /0.88μm ZnO /0.10μm Al /400μm Sapphire which was measured to have a loaded $Q$ of respectively 15,000 and 19,000 for series and parallel resonant frequencies around 3.7GHz. The temperature coefficient of the resonant frequency is -28.5ppm/°C [17]. Resonators obtained by $LiNbO_3$ wafer as a transducer bonded on another $LiNbO_3$ wafer used as the HBAR substrate exhibit $Q$ factors of 53,000 at 1.5GHz using the Gold bonding technique [5] and $Q.f$ product above $8.10^{13}$ with an 800nm thickness for the piezoelectric layer by Smart Cut approach [18]. Understanding losses phenomena helps to design high quality factor devices. Loss origins can be classified into two categories: material (intrinsic) and geometry (technology-related). Due to the architecture of HBAR, the quality factor of such devices depends on the crystalline losses and on the material isotropy, on the surfaces parallelism and any loading due to the electrodes.

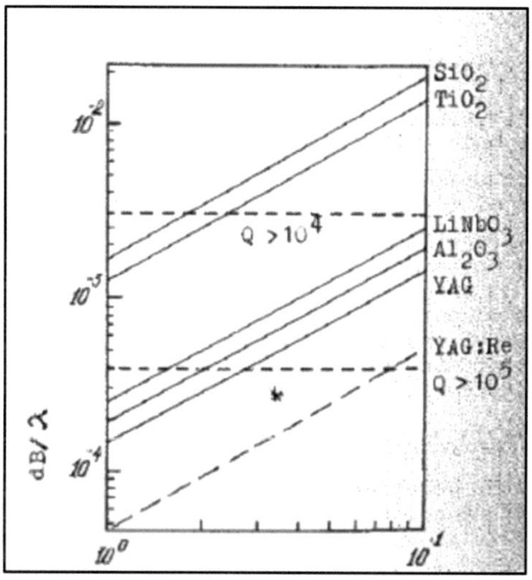


**Figure 15.** Losses per wave length and the resonator's quality $Q$ as a function of frequency (GHz) for various materials [19].

As explained before, the quality factor is directly link to the acoustic quality of the substrate. Some works have already been done to compare and improve materials to favor high acoustic resonance quality [19], [11]. Figure 15 shows an example of these works [19].

The polishing process providing damaged-free ultra-smooth surfaces is essential, as well as checking the substrate quality by X-ray topography. To take into account current industrial needs, using the technology of material crystal growth is crucial to obtain large wafers. In this context, $LiNbO_3$, $LiTaO_3$ Sapphire, and YAG are the preferred candidates as they do present effective acoustic quality (*i.e.* reduced visco-elastic and dielectric damping properties) and available as 4 inch wafers, excepted for YAG substrates.

The defect of parallelism between two surfaces of HBAR devices dramatically limits the quality factor [11]. Figure 16 shows the quality factor of HBAR modes on Sapphire-base structures versus the plate tilt. As clearly highlighted by this graph, the parallelism must be perfect to prevent power flow yielding $Q$ factor limitations. For example, a HBAR built on a 4 inch wafer with a total thickness variation (TTV) of 3μm (commercially accessible for Silicon) does not suffer from any parallelism defect and therefore the quality of its resonances is almost not limited by this effect ($Q>10^5$). However, one can see that a thickness variation of 3μm on 1cm yields effective limitation of the quality factor ($Q<10^4$).

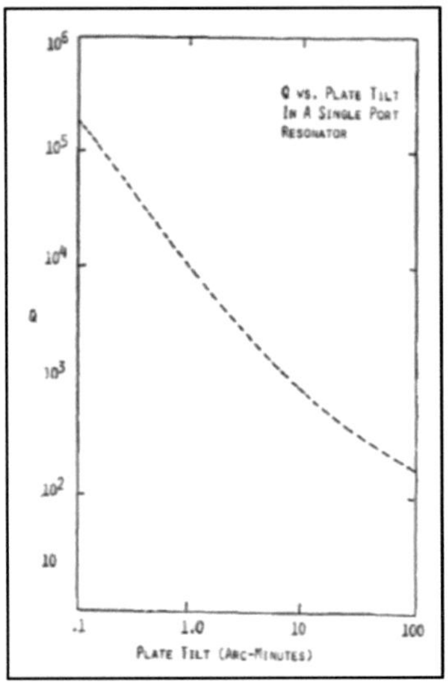


**Figure 16.** Parallelism-limited $Q$ in a single-port resonator built on Z cut Sapphire substrates [11].

The shape, size and nature of the electrodes can be also important to manufacture high $Q$ HBAR devices. Some works have been done on electrodes of HBAR devices [20], [21], [22], [23].

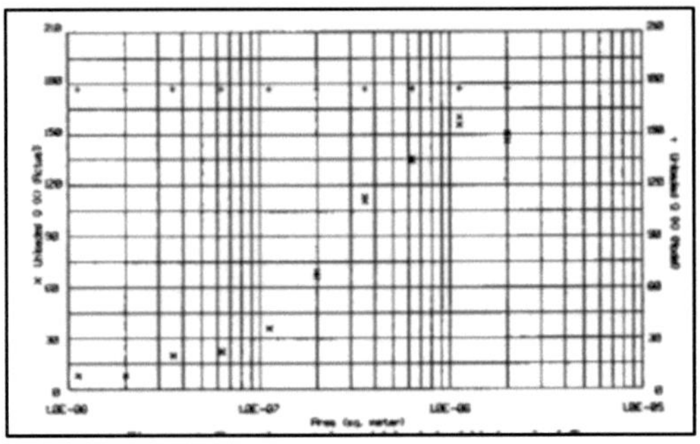

**Figure 17.** Experimental and Modeled Unloaded Q versus aperture [20].

D. S. Bailey *and al.* showed that the HBAR does not follow the one dimensional computer model [20]. Indeed, figure 17 shows the difference between the experimental $Q$ and the theoretical $Q$ versus the aperture of the electrode. The difference is due to the diffraction effect. The optimum electrode area can depend on two main parameters: the clamp capacitance $C_0$ and the geometry. This capacitance $C_0$ is proportional to the surface area and influences other parameters of resonator such as difference in impedance at series and resonance frequency. With a large active area, defects in transducer crystal or of geometry can happen more easily. The optimization of the area shape and surface to limit the diffraction effect and improve the quality factor is an area of ongoing works.

Furthermore, for ultra-high frequency HBAR devices, the electrodes are not thin compared to transducer layer. The thickness and the nature of electrodes have an influence on the quality factor and the other parameters such as the electromechanical coupling coefficient and the resonance frequency. Many works have been done on this subject [23], [24], [21].

The conditions of metal sputtering can influence the nature of the metallic electrode. Indeed, the conditions of metal sputtering for thin layers modify the density and the rate of impurity of the layer. The optimum must be found to have the highest metal density with the lowest impurities. Furthermore, some works compare the influence of different metallic layers (Al, Au, W, Ag) on the quality factor. If we consider the modified Butterworth-Van Dyke (MBVD) model, the best electrode is constituted with the lowest resistivity (Au), but experimentations also show the influence of other parameters. Thus, a Molybdenum layer used as an electrode shows better results due to better acoustic impedance [23].

Generally speaking, low losses applications also require a temperature compensation for the resonator. One solution is to have intrinsic compensation of temperature and it is the purpose of the next paragraph. Another solution is to control frequency by measuring temperature.

## 4.2. Temperature compensation

One challenge of the radio-frequency bulk acoustic devices is the temperature stability of their resonance frequency. A lot of work has been achieved exploiting thin piezoelectric films for developing temperature-compensated HBARs, with various successes. The possibility to use single crystal thinned films appears as an alternative to control the piezoelectric film properties (velocity, coupling, temperature sensitivity, and so on.) and to globally reconsider material association according to the technological assembly process previously presented.

The celebrated Campbell&Jones method [25] is used here for predicting the Temperature Coefficient of Frequency (TCF) of any mode of a given HBAR. As it has been reported hundred times in previous papers, only the main basic equation is reported below:

$$f=\frac{v}{2e}\rightarrow\frac{df}{f}(T)=\frac{dv}{v}(T)-\frac{de}{e}(T) \tag{1}$$

$f$, $e$, $v$ and $T$ are respectively frequency, thickness of resonator, wave velocity and temperature.

Which means that the frequency changes due to temperature variations is computed as the difference between the development of the velocity and of the stack thickness versus temperature. Theoretically using a standard anisotropic 1D model reveals that zero temperature coefficients of frequency (TCF) can be obtained and optimized along the mode order. It is well-known that Quartz and fused Silica (glass) do exhibit positive TCFs. So the use of the other temperature-compensated Quartz orientations, and hence of any other material sharing such property, has been checked theoretically and reveals applicable as well.

As example, Lithium of Niobate and Quartz have been associated for the fabrication of shear-wave based HBARs. $LiNbO_3$ provides crystal orientations for which very strongly coupled shear waves exist ($k_s^2$ in excess of 45%) whereas *AT* cut of Quartz allows for compensating second order frequency-temperature effects [WO2009156658 (A1)]. Although this idea was already proposed using other material combinations [US Patent #3401275A], no real design process was presented until now and therefore the possibility to actually determine structures allowing for high frequency operation with first order TCF smaller than 1ppm.$K^{-1}$ was quite hypothetical, but improvement of numerical tools allows this design.

Nevertheless, some works show the possibility to have an intrinsic compensation of the temperature for HBAR devices [8], [26]. Figure 18 shows the temperature dependence for different configuration of HBAR devices constituted by $LiNbO_3$ and Quartz layers with different cut orientations. This work shows clearly that the choice of materials and the cut orientation of these materials have a direct impact on the frequency shift with temperature variations [26].

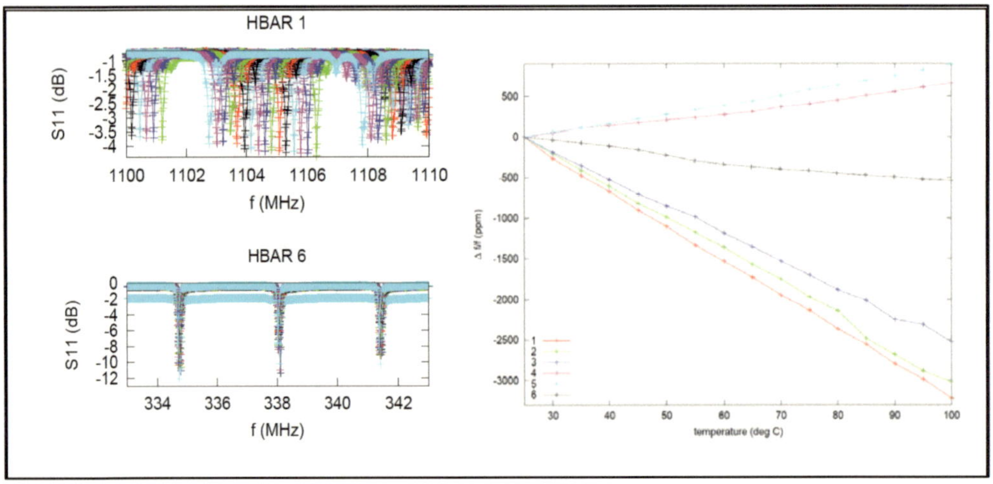


**Figure 18.** Electrical responses for two different configurations of HBAR (left), frequency variation versus temperature for six configurations of HBAR (right) [26].

Moreover, the frequency dependence on temperature is different for each overtone of HBAR devices. The thickness ratio between the transducer layer and the substrate also influences the frequency variations with different temperatures. Figure 19 shows the computation of the temperature coefficient of frequency (TCF) of a HBAR for various Lithium of Niobate / Quartz thickness ratios. This HBAR device is built on a (YX*l*)/163° $LiNbO_3$ thinned plate bonded on (YX*lt*)/36°/90° Quartz substrate of 50µm [27]. One can see that depending on the harmonic number, the $TCF_1$ changes from +1 to -14ppm.$K^{-1}$. Furthermore, depending on the harmonic of the transducer alone, the $TCF_1$ may notably change and thus it cannot be considered as a simple periodic function versus harmonic number. Therefore, it is mandatory to accurately consider all the actual features of the structure for an accurate design of a resonator, *i.e.* the operation frequency, the harmonic number and the thickness ratio for a given structure. To complete this, one should also account for the actual thickness of the device as this parameter will control the possibility to select one (frequency/harmonic number) couple. Finally, it clearly appears that the analysis of such HBAR TCF requires a numerical analysis and that if an intuitive approach allows for a first order definition of crystal orientations, the complicated distribution of energy within the stack versus all the structure parameters induces more intrication in the design process.

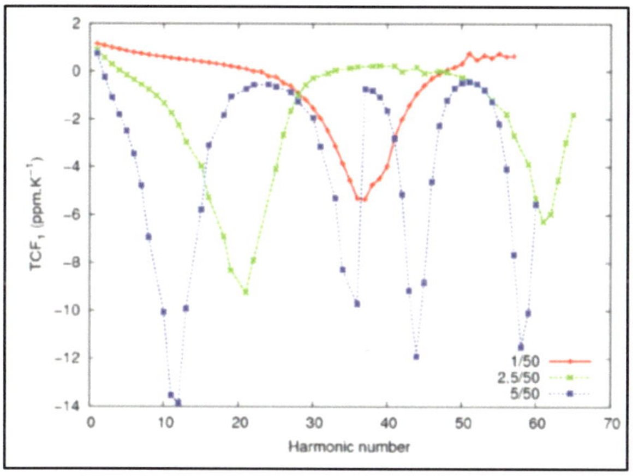


**Figure 19.** Plot of TCF of a HBAR built on a (YX*l*)/163° LiNbO3 thinned plate bonded on (YX*lt*)/36°/90° Quartz substrate for various Lithium of Niobate/Quartz thickness ratio (Quartz thickness arbitrary fixed to 50 μm) [27].

# 5. HBAR applications

## 5.1. HBAR sensors

Industrial acoustic-resonator-based sensors require adapted electronics to be efficiently operated. Two main approaches have been developed in that purpose:

- The first way is to use the acoustic resonator in an oscillator loop. Compared to normal oscillator use in frequency/time applications, some specific operation regimes must be considered for sensors [28]. Particularly, the resonance frequency is assumed to drift along the measured parameter on a large frequency domain. The *Q*-value of the resonator may notably vary as well as other parameters like electromechanical coupling and the overall electrical conductance (connected one another) yielding the need for improved electrical robustness of the circuit. The electronics therefore must be able to adapt its operation parameters. Distance between electronic and sensor is often important compare to classic oscillator. Due to these reasons, resolution of the system is limiting to $1.10^{-8}$. As example, sensors at 434MHz have resolution limitation of 5Hz due to oscillator loop measurement method. Finally, to use this method, we need resonator with low harmonic generation. BAW, SAW and FBAR can use this electronic.
- The second way is to have electronic interrogation which finds frequency resonance in determined range of frequency. With classic method it is possible to obtain 100Hz of resolution for 434MHz sensors [28]. If electronics is improving, we can achieve 5Hz of resolution for 434MHz sensors [29]. In this case, clock of electronic is really important for performance. All kind of resonators sensors can be interrogated by this technique, especially HBAR device which present high overtone generation.

Acoustic sensor is passive sensor. Device combine with antenna could be having great interest. Indeed, electromagnetic waves can be changed on electrical waves on electrodes, which can excite acoustic waves by piezoelectric effect. Furthermore this phenomenon is linear and invertible. So, wireless interrogation is possible with acoustic sensors. Wireless communication presents great interest for all hard environments. In that way, acoustic sensors can be use in engine, close environment and more generally in all environments where wire can not be employed.

With wireless interrogation, antenna size, quality factor of resonator, frequency have a strong impact. With increasing of frequency, antenna size decrease. Indeed, the size of antenna is equal to the quarter of wavelength. When higher ISM band is used, quality factor need to be increase to give the same obstruction of the ISM bandwidth. At -3dB, the bandwidth of resonator is proportional to frequency divided by quality factor. And finally, the flight time is proportional to quality factor divided by frequency. They are two consequences of this flight time. Firstly, to have enough energy when frequency increase, the quality factor need to increase. As example, SAW resonators at 434MHz ISM band have quality factor of 10,000 and can be interrogated by wireless approach. To pass at 2.45GHz ISM band, a quality factor equal to 20,000 is required. HBAR devices achieve these characteristics. Secondly, refresh rate increases with frequency. With bandwidth of few kHz the refresh rate is around one millisecond. In this case, quality factor of sensor could not be too higher. So, quality factor of HBAR device need to be optimize for wireless sensor application.

HBAR devices present a great advantage for achieving sensors device. As previously discussed, frequency shift due to temperature effects can be minimized and even compensated, but also magnified as well. As a consequence, HBAR temperature sensors are considered first. Moreover, due to high number of overtones of such devices, it is also possible to develop sensors exhibiting different sensitivity to a given parametric effect at different frequencies. Acoustic devices can also be effectively exploited as stress sensor or pressure sensors. The fabrication of SAW pressure sensor based on thinned Quartz membrane (for instance) was strongly investigated due to the dependence of the wave velocity versus tensile stress at the surface of the membrane when bent by pressure. In the case of bulk wave propagating in such a membrane, the strain variations across the membrane thickness forbid the use of such an approach to develop pressure sensor applications. This can be easily demonstrated using for instance static finite element analysis with a very simple mesh. Indeed, the strain and hence the stress change their signs along the membrane thickness when submitted to pressure. As a consequence, the strain variation across the HBAR generates equilibration of the velocity variations. On the one hand, the strain below the membrane neutral line yields an increase of resonant frequency of the HBAR; on the other hand, the strain above the neutral line yields a decrease of this frequency. Consequently, the resulting frequency shift is negligible. One solution consists in the fabrication of a micro-cavity within the HBAR stack near the neutral line. If the transducer of the HBAR structure is straight above this micro-cavity, the emitted bulk waves are reflected by this micro-cavity and hence confined in this membrane location. The micro-cavity then plays the role of a mirror for the waves. The structure of such device is shown in figure 20. The surface of the cavity should at minimum coincide strictly to the surface of the transducer, but to ease the fabrication (particularly to manage alignment issues) the cavity largely overlaps the transducer aperture. The micro-cavity/micro-mirror could be placed at different deph into the HBAR stack. Its location will define the HBAR sensibility to stress [8].

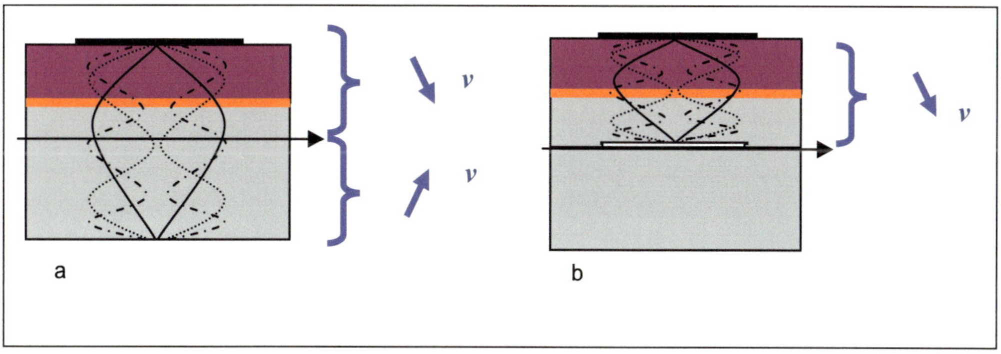


**Figure 20.** HBAR structures presented low frequency sensitive to stress (a), and highly frequency sensitive to stress with the realization of micro-mirror under the transducer aperture [8].

A lot of works has been done on liquid or gaz HBAR / FBAR sensor and as most representative example, gravimetric sensor. The basic principle of gravimetric acoustic wave sensors is the measurement of the phase velocity variations due to an adsorbed mass or a layer thickness change atop the device during a chemical reaction: this phase velocity is dependent on the boundary conditions of the propagating acoustic wave and is affected either by the layer properties or its thickness. The usual principle exploits bulk acoustic waves, yielding the well-known concept of Quartz Crystal Micro-balance (QCM). The gravimetric sensitivity of the QCM is directly related to its thickness and as a consequence to its fundamental frequency $f_0$. Adsorption on one side of the resonator modifies its resonance conditions and thus allows for a gravimetric detection. Furthermore, it is possible to functionalize the surface with specific reactants to provide information on the concentration of the adsorbed (target) species in the medium surrounding the sensor [31]. The case of HBAR is particularly attractive as one can expect probing the adsorbed material at various frequencies, providing frequency-dependent information such as viscosity for instance.

Copper electro-deposition on the back side of a HBAR has been used for calibrating the gravimetric sensitivity of its overtones. This approach was particularly implemented as it provides an independent estimate of the deposited metal mass through the measurement of the current. A negative current indicates copper reduction (deposition on the working electrode) whereas a positive current indicates oxidation (copper removal from the working electrode). Simultaneous to the current monitoring, the acoustic phase and magnitude at fixed frequency are recorded [30]. The figure 21.a shows four different overtone frequencies (red dot) recorded. Figure 21.b. shows relative frequency variations and clearly shows the sensitivity difference of the four HBAR overtones. Sensitivity of gravimetric HBAR directly depends on the stack thickness and more precisely on copper thickness versus transducer thickness. The best gravimetric HBAR sensor is constituted by the thinnest stack with metallic thickness equal to quarter of wavelength [31].

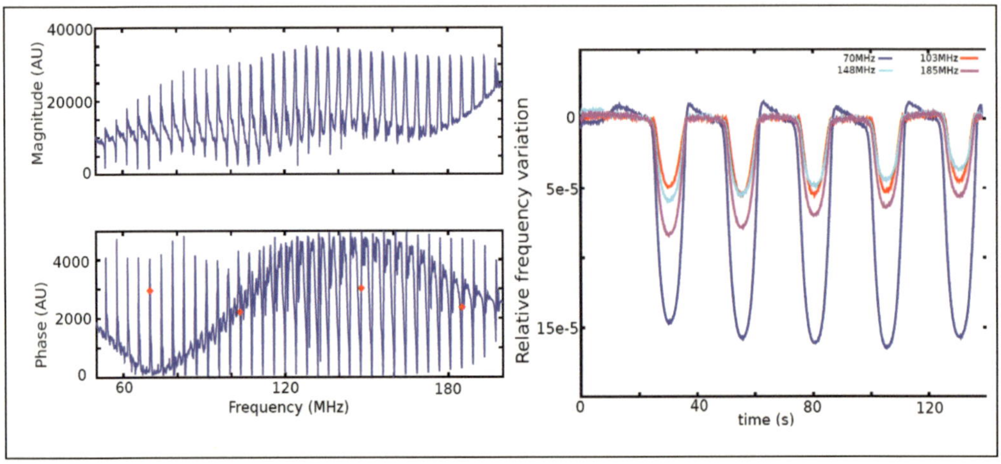


**Figure 21.** (a) Overtones of the fundamental transducer layer mode of a gravimetric HBAR sensor with four probed frequencies (red dot), (b) relative frequency variations of these four frequencies with their relative sensitivities [30].

## 5.2. HBAR-stabilized oscillators

Radio-Frequency oscillators can be stabilized by various resonating devices. Their stability is mainly conditioned by the spectral quality of the resonator even if the oscillator loop electronics must be optimized to lower the generated noise as much as possible. For mid-term stability, temperature compensation is a key point and allows for notably improving the corresponding figures of merit. The possibility to build temperature compensated HBARs has been shown in paragraph 4.2 and is a key-point for the fabrication of oscillator exhibiting short-term stability compatible with practical applications.

Moreover, the frequency stability of an oscillator can be characterized by its single-sideband phase noise, $L\{f_m\}$. Leeson's equation [33] shows that low phase noise operation can be achieved by increasing the loaded quality factor $Q_{load}$ of the resonator. According to Leeson's model, a high resonator quality factor ($Q$) or circulating power level improves the phase noise and, therefore, the short-term stability of the oscillator. Considering these aspects, HBAR device features for the frequency stability reveal more favourable than FBAR or SAW device ones. Therefore HBAR should allow for notably improving oscillator performances. However, multi-overtone features of HBAR do not facilitate resonance lock for oscillator applications. Therefore SAW or FBAR device are generally used to filter the frequency of HBAR. Consequently, the compactness of HBAR is deteriorated due to the need for this filter. Optimizing HBAR spectral response then is still an open question and should receive more attention in future developments. One can note that using single port HBARs with optimized frequency separation between the overtones [32] may allow to get rid of this filtering operation.

HBAR then are capable to address high frequency source applications without requiring multiplication stages as usually achieved. The idea then is to evaluate the effective interest of HBAR for direct frequency synthesis, reducing the oscillator architecture complexity and potentially improving the corresponding operational features.

Hongyu Yu and *al.* have presented a local oscillator based on a HBAR resonator associated to an atomic clock [34]. Atomic clocks are used for embedded applications which need high stability performance such as GPS station. The atomic transition allows having long-term stability in this application, but it presents poor short-term stability. To success oscillator based on this atomic transition, local oscillator is needed. This local oscillator stabilizes the short-term variation of the global oscillator with its short-term performance. The local oscillator need to have good phase noise (better than -70dBc/Hz at 1kHz for instance) to prevent global degradation of clock stability. Figure 22 shows the phase noise measurement data of the 3.67GHz Pierce oscillator and the 1.2GHz Colpitts oscillator, and the Allan deviation of the free-running 3.67GHz oscillator that consumes only about 3mW [34]. Local oscillator based on HBAR resonator need frequency control to achieve the atomic transition frequency. With the modulation of the HBAR frequency with an external synthesizer and FBAR filters, the local oscillator locked to the coherent population trapping resonance.

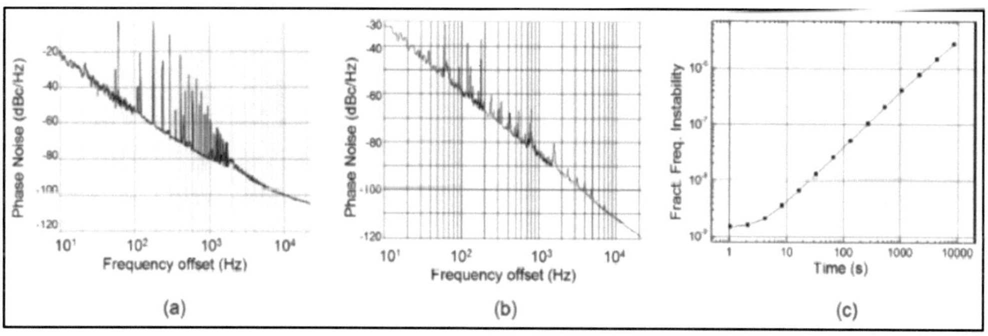


**Figure 22.** Phase noise measurement data of (a) the 3.67GHz Pierce oscillator and (b) the 1.2GHz Colpitts oscillator. (c) Allan deviation of the free-running 3.67GHz oscillator that consumes only about 3mW [34].

Other applications require low phase noise oscillator such as embedded RADAR. A radio-frequency oscillator operating near the 434MHz-centered ISM band validates the capability of the above-mentioned HBAR for such purposes. The composite substrates have been built using 3-inches (YX*l*)/163° $LiNbO_3$ cut wafer bonded and thinned down to 15µm onto a 350µm thick (YX*lt*)/34°/90° Quartz base. Single-port resonators operating near 434MHz (exploiting the third harmonic of the thinned Lithium of Niobate plate as the HBAR "motor") have been then manufactured. Electrical and thermoelectric characterizations have shown quality factor of the resonance in excess of 20,000, yielding a *Q.f* product of about $10^{13}$ and a third order frequency-temperature behavior. A SAW filter was used to select the ISM band and to filter the high spectral density HBAR response (figure 23). The oscillator then has been measured using a phase noise automatic bench. A phase noise better than --160dBc/Hz at 100kHz has been measured as well as a -165dBc/Hz level at 1MHz from the career (figure 23). Short-term stability characterizations show that the resonator stability is better than $10^{-9}$ at room conditions (no temperature stabilization).

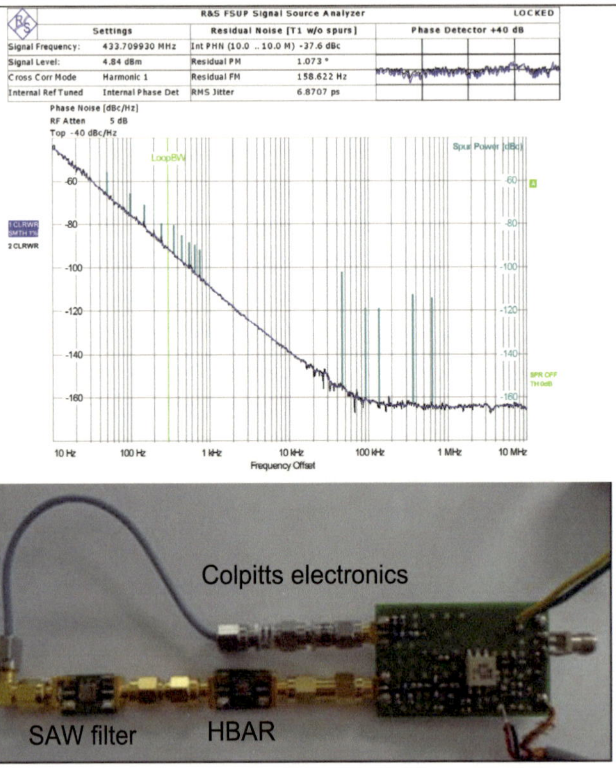


**Figure 23.** Phase noise curves for oscillators at 434MHz constituted by High Overtone Bulk Resonator, SAW filter and Colpitts electronics.

To achieve higher frequency (above 1GHz), four-port resonators are mandatory due to the difficulty to adjust the oscillator tuning elements. For this application, the temperature stability is required and therefore the resonator exploits a (YX*l*)/163° $LiNbO_3$ thinned layer atop a (YX*lt*)/36°/90° Quartz substrate. The electrodes defining the coupled transducers (four-port resonator) are two half-circles (300µm diameter) separated by a gap of 20µm, yielding favorable conditions for using the resonator to stabilize an oscillator loop at 935MHz and 1.636GHz. The device were cut and packaged, mounted in an oscillator loop, and measurements of phase noise were performed. Figure 24 shows the phase noise of the two corresponding oscillators compared with the phase noise of an oscillator stabilized by a classical BAW resonator at 100MHz, "octar 507X100" from AR-Electronics. The oscillator near 1.6GHz clearly shows better performances than the one at 935MHz, with a noise level lower than -130dBc/Hz at 10kHz from the carrier. In order to compare the 100MHz oscillator with the 1.6GHz one, the low frequency source has to be multiplied by 16, *i.e.* +12dB must be added to the noise level. It gives a level of -140dBc/Hz at 10kHz which is not far from HBAR solution.

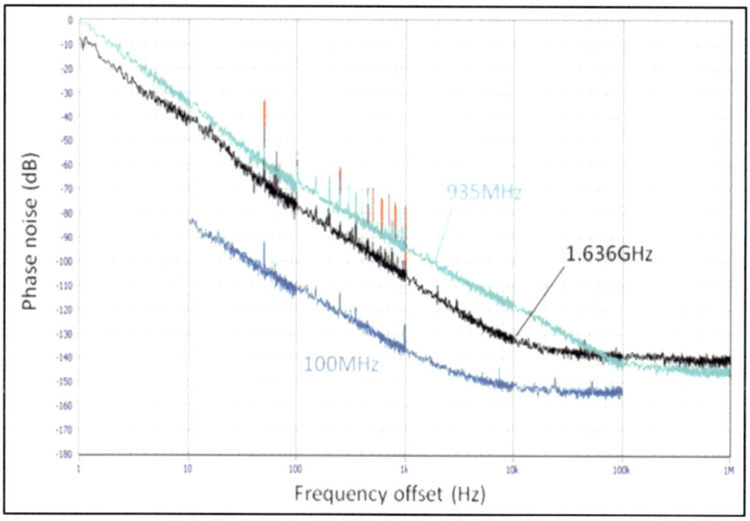


**Figure 24.** Phase noise curves for oscillators at 935MHz, 1,636MHz compared with phase noise of an oscillator stabilized by a classical BAW resonator at 100MHz [33].

## 6. Conclusions and perspectives

HBAR have been developed for the fabrication of passive radio-frequency devices capable to overcome standard SAW and BAW limitations considering the quality of the resonance, complexity of technological fabrication and operation frequencies.

These devices actually maximize the *Q* factor that can be obtained at room temperature using elastic waves, yielding quality factor times frequency products (*Q.f*) close or slightly above $10^{14}$, *i.e.* effective *Q* factors of about 10,000 at 10GHz in theory (practically, *Q* factors in excess of 50,000 between 1.5 and 2GHz were experimentally achieved). Single-port or four-port resonator has been described and the different approach of manufacturing as been explained. The choice of acoustic wave and acoustic substrate permits to address large range of application. First one is chose value of the quality factor and the electromechanical coupling coefficient of overtone frequencies. Second one is to minimize frequency shift due to temperature variation for chosen frequency. All these possibilities allow us to address different applications such as sensor or low phase noise applications.

Although HBAR device knows since several decades, HBAR device has not yet achieve is development maturity. Futures works will concern improvement of fabrication, frequency control, and wireless sensors. The large number of the parameters for optimizing HBAR in function of the applications requires well-control generic process of fabrication. The selection of the frequency resonance is also a key point for the emergence of HBAR devices, as the frequency tuning. And finally, wireless HBAR sensor need strong effort of development.

Some simple applications and large potentiality of HBAR conception has been presented. Two main approaches exist for the realization of HBAR devices. The first one based on piezoelectric deposition gives easily high quality factor and frequency device. The largest potentiality of conception to address different applications is obtained by the second approach based on mono-crystal wafer assembled. Further developments required to promote the industrial exploitation.

All previous technological industrial development for solidly mounted resonator for instance could be easily use for fabrication of HBAR based on piezoelectric deposition. More technological development is required to control thickness, repeatability and so on (see section 4.1) for the second approach. In both case, packaging aspect is a key-point. Both side of HBAR need to be free for acoustic reason. Flip-chip approach seems to give the best result for industrial needs.

In both cases, design tool needs to be developed to realize conception of all HBAR devices. SAW design tool could be a good base to develop such software.

More works also need to improve performances or to fixe limits of all different HBAR devices which are specialized for each application. These developments are the precondition for industrial actions.

## Author details

T. Baron, E. Lebrasseur, F. Bassignot, G. Martin, V. Pétrini and S. Ballandras
*FEMTO-ST, université de Franche-Comté, CNRS, ENSMM, UTBM, Département Temps-Fréquence, France*

## Acknowledgement

This work was partly supported by the Centre National d'Etudes Spatiales (CNES) under grant #04/CNES/1941/00-DCT094 and still supported by the CNES under grant #R-S08/TC-0001-026, and by the Direction Generale pour l'Armement (DGA) under grant #05.34.016.

## 7. References

[1] K. M. Lakin and J. S. Wang, UHF composite bulk wave resonators. *IEEE Ultrasonics Symposium Proceedings*, pp. 834-837, 1980

[2] K.M. Lakin, Thin film resonator technology. *IEEE Trans. on UFFC*, Vol.52, pp.707-716, 2005

[3] K.M. Lakin, G.R. Kline, K.T. McCarron, High Q microwave acoustic resonators and filters. *IEEE Trans. on Microwave Theory and Techniques, Vol. 41, n°12,* pp,2139-2146, 1993

[4] S.P. Caldwell, M.M. Driscoll, D. Stansberry, D.S. Bailey, H.L. Salvo, High overtone bulk acoustic resonator frequency stability improvements. *IEEE Trans. on FCS*, pp. 744-748, 1993

[5] D. Gachon, E. Courjon, G. Martin, L. Gauthier-Manuel, J.-C. Jeannot, W. Daniau and S. Ballandras, Fabrication of high frequency bulk acoustic wave resonator using thinned single-crystal Lithium Niobate layers. *Ferroelectrics, Vol. 362*, pp. 30-40, 2008

[6] Curran Daniel R & Al, US Patent #3401275A, 1968-09-10

[7] M. Pijolat, D. Mercier, A. Reinhardt, E. Defaÿ, C. Deguet, M. Aïd, J.S. Moulet, B. Ghyselen, S. Ballandras, Mode conversion in high overtone bulk acoustic wave resonators. *Proc. of the IEEE International Ultrasonics Symp.*, pp.201-204, 2008

[8] T. Baron, E. Lebrasseur, J.P. Romand, S. Alzuaga, S. Queste, G. Martin, D. Gachon, T. Laroche, S. Ballandras, J. Masson, Temperature compensated radio-frequency harmonic bulk acoustic resonators pressure sensors. *Proc. of the IEEE International Ultrasonics Symposium*, pp. 2040-2043, 2010.

[9] D. Gachon, T. Baron, G. Martin, E. Lebrasseur, E. Courjon, F. Bassignot, S. Ballandras, Laterally coupled narrow-band high overtone bulk wave filters using thinned single crystal lithium niobate layers. *Frequency Control and the European Frequency and Time Forum (FCS)*, 2011

[10] D. Royer & E. Dieulesaint. Elastic Waves in Solids II. *Springer*, 2000.

[11] R.A. Moore, J.T. Haynes, B.R. McAvoy. High Overtone Bulk Resonator Stabilized Microwave Sources. *1981 IEEE Ultrasonics Symposium*, pp.414-424, 1981

[12] B. Aspar, H. Moriceau, *et al.*. The generic nature of the Smart-Cut process for thin film transfer. *Journal of Electronic Materials*, Vol. 30, n°7, 2001, pp. 834-840

[13] T. Pastureaud, M. Solal, B. Biasse, B. Aspar, J.B. Briot, W. Daniau, W. Steichen, R. Lardat, V. Laude, A. Laëns, J.M. Frietd, S. Ballandras, High Frequency Surface Acoustic Waves Excited on Thin Oriented LiNbO3 Single Crystal Layers Transferred Onto Silicon. *IEEE Trans. on UFFC, Vol.54, n°4*, pp 870-876, 2007

[14] J.C. Ponçot, P. Nyeki, Ph. Defranould, J.P. Huignard. 3 GHz Bandwidth Bragg Cells. *IEEE 1987 Ultrasonics Symposium*. pp. 501-504, 1987

[15] T. Baron, E. Lebrasseur, *et al.*. Development of Composite Single Crystal Wafers for Sources and Sensor applications exploiting High-overtone Bulk Acoustic Resonators. *JNTE* 2010

[16] J.-S. Moulet, M. Pijolat, J. Dechamp, F. Mazen, A. Tauzin, F. Rieutord, A. Reinhardt, E. Defay, C. Deguet, B. Ghyselen, L. Clavelier, M. Aid, S. Ballandras, C. Mazure. High piezoelectric properties in LiNbO3 transferred layer by the Smart Cut™ technology for ultra wide band BAW filter applications. *Electron Devices Meeting, 2008. IEDM 2008. IEEE International.* pp. 1-4. 2008

[17] Hongyu Yu, Chuang-Yuan Lee, Wei Pang, Hao Zhang and Eun Sok Kim. Low Phase Noise, Low Power Consuming 3.7 GHz Oscillator Based on High-overtone Bulk Acoustic Resonator. *2007 IEEE Ultrasonics Symposium*, pp.1160-1163, 2007

[18] M. Pijolat, A. Reinhardt, *et al.*. Large Qxf Product for HBAR using Smart Cut (TM) transfer of LiNbO3 thin layers onto LiNbO3 substrate. *IEEE Ultrasonics Symposium 1-4 (2008)* pp. 201 – 204, 2008

[19] S. Ivanov, I. Koelyansky, G. Mansfeld and V. Veretin. Bulk Acoustic Wave High overtone resonator. *1990 1er congrès Français d'acoustique*, pp 599-601, 1990

[20] D. S. Bailey, M. M. Driscoll, and R. A. Jelen. Frequency Stability Of High Overtone Bulk Acous Tic Resonators. *1990 ultrasonics symposium.* pp. 509-512, 1990

[21] Lukas Baumgartel and Eun Sok Kim. Experimental Optimization of Electrodes for High Q, High Frequency HBAR. *2009 IEEE International Ultrasonics Symposium Proceedings.* pp. 2107-2110, 2009

[22] S.G.Alekseev, I.M.Kotelyanskii, G.D.Mansfeld, N.I.Polzikova. Energy Trapping in HBARs Based on Cubic Crystals. *2006 IEEE Ultrasonics Symposium.* pp. 1478-1480, 2006

[23] Hui Zhang, Zuoqing Wang, and Shu-Yi Zhang. Electrode Effects on Frequency Spectra and Electromechanical Coupling Factors of HBAR. *IEEE transactions on ultrasonics, ferroelectrics, and frequency control, vol. 52, no. 6,* june 2005

[24] Hui Zhang, Shu-Yi Zhang, Kai Zheng. Parameter characterization of high-overtone bulk acoustic resonators by resonant spectrum method. *Ultrasonics 43 (2005)* pp. 635–642. 2005

[25] J.J. Campbell, W.R. Jones, A method for estimating crystals cuts and propagation direction for excitation of piezoelectric surface waves. *IEEE Trans. On Sonics and Ultrasonics,* Vol. 15, pp. 209-217, 1968

[26] T. Baron, D. Gachon, *et al.*. Temperature Compensated Radio-Frequency Harmonic Bulk Acoustic Resonators. Proc of the IEEE IFCS, pp 625-655, 2010

[27] T. Baron, G. Martin, E. Lebrasseur, B. Francois, S. Ballandras, P.-P. Lasagne, A. Reinhardt, L. Chomeloux, D. Gachon, J.-M. Lesage. RF oscillators stabilized by temperature compensated HBARs based on LiNbO3/Quartz combination. *Frequency Control and the European Frequency and Time Forum (FCS), 2011 Joint Conference of the IEEE International.* pp. 1-4, 2011

[28] J.-M Friedt, C. Droit, G. Martin, and S. Ballandras, "A wireless interrogation system exploiting narrowband acoustic resonator for remote physical quantity measurement", *Rev. Sci. Instrum. vol. 81, 014701* (2010)

[29] C. Droit, G. Martin, S. Ballandras, J.-M Friedt, "A frequency modulated wireless interrogation system exploiting narrowband acoustic resonator for remote physical quantity measurement", *Rev. Sci Instrum. Vol 81, Issue 5, 056103* (2010)

[30] D. Rabus, G. Martin, E. Carry and S. Ballandras. Eight channel embedded electronic open loop interrogation for multi sensor measurements. *European Frequency and Time Forum (EFTF),* 2012

[31] G.D. Mansfeld. THEORY OF HIGH OVERTONE BULK ACOUSTIC WAVE RESONATOR AS A GAS SENSOR. *13th International Conference on Microwaves, Radar and Wireless Communications.* vol.2, pp. 469 – 472, 2000

[32] Jérémy Masson. Étude de capteurs résonants acoustiques interrogeables à distance à base de films minces micro-usinés sur silicium. *Mémoire de thèse,* 2007

[33] E. Lebrasseur *et al.*. A Feedback-Loop Oscillator Stabilized Using Laterally-coupled-mode Narrow-band HBAR Filters. *2011 IEEE International Ultrasonics Symposium Proceedings,* 2011

[34] Hongyu Yu, Chuang-yuan Lee, Wei Pang, Hao Zhang, Alan Brannon, John Kitching, and Eun Sok Kim. HBAR-Based 3.6 GHz Oscillator with Low Power Consumption and Low Phase Noise. *IEEE Transactions on Ultrasonics, Ferroelectrics, and Frequency Control, vol. 56, no. 2, February 2009.* pp.400-403, 2009

Chapter 14

# Modeling and Design of BAW Resonators and Filters for Integration in a UMTS Transmitter

Matthieu Chatras , Stéphane Bila, Sylvain Giraud, Lise Catherinot, Ji Fan, Dominique Cros, Michel Aubourg, Axel Flament, Antoine Frappé, Bruno Stefanelli, Andreas Kaiser, Andreia Cathelin, Jean Baptiste David, Alexandre Reinhardt, Laurent Leyssenne and Eric Kerhervé

Additional information is available at the end of the chapter

http://dx.doi.org/10.5772/56026

## 1. Introduction

Bulk-Acoustic Wave (BAW) resonators and filters are highly integrated devices, which represent an effective alternative for narrow-band components (up to 5% fractional bandwidth) up to few GHz [1].

This chapter presents the integration of a BAW filter and of a BAW duplexer in a UMTS transmitter. The first section details one dimensional and three-dimensional techniques for the modeling and the design of BAW resonators. The second section proposes a synthesis approach for dimensioning BAW filters and the third section illustrates the approach with the characterization of several fabricated prototypes. Finally, The UMTS transmitter incorporating a BAW filter and a BAW duplexer is described with a particular emphasis on the performances of these devices.

## 2. Model and design of bulk acoustic wave resonators

### 2.1. Modeling of a BAW resonator in 1 dimension

The proposed method compares the impedance of a piezoelectric resonator obtained by both an electrical equivalent model and a piezoelectric model. By this way, it is possible to obtain the values of the electrical model as functions of all geometrical and material characteristics. The two models and their relation are described in the following sections.

### 2.1.1. Resonator impedance based on electrical (BVD) model

A piezoelectric resonator can be modeled by the lossless BVD (Butterworth Van Dick) model as shown in Figure 1(a)) [2], [3]. $C_o$ is the geometric capacitance of the structure and the $L_m$ $C_m$ series circuit (called the motional arm) represents the mechanical resonance (motional behavior). According to the circuit in Figure 1 (a), two resonances are obtained, and the equivalent impedance $Z_{BVD}$ of this circuit can be easily derived:

$$f_s = \frac{1}{2\pi\sqrt{L_m C_m}} \tag{1}$$

$$f_p = f_s\sqrt{\frac{C_m + C_o}{C_o}} \tag{2}$$

$$Z_{BVD} = \frac{j(\omega L_m + 1/\omega C_m)}{1-\omega^2 C_o L_m + C_o / C_m} \tag{3}$$

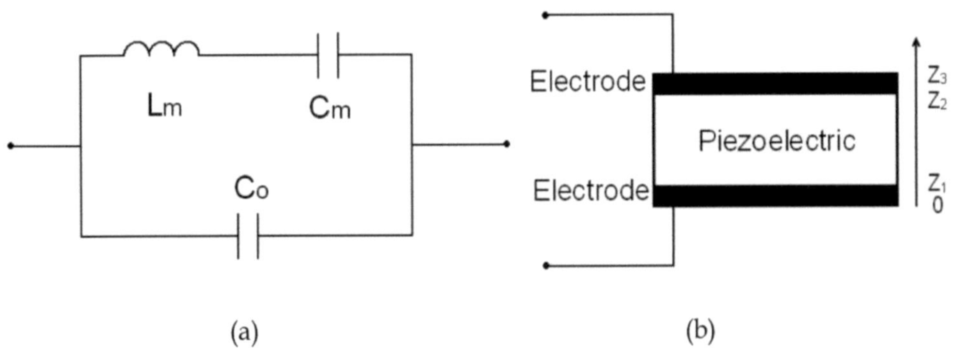


**Figure 1.** (a) Equivalent circuit of the lossless BVD model. (b) One-dimensional structure piezoelectric resonator.

### 2.1.2. Resonator impedance based on piezoelectric equations

The description of a piezoelectric resonator, made of a single piezoelectric layer and two thin electrodes as depicted in Figure 1 (b), is considered. Using piezoelectric fundamental equations [4], one can write:

$$T(z) = cS(z) - eE(z) \tag{4}$$

$$D = eS(z) + \varepsilon E(z) \tag{5}$$

$$S(z) = \frac{\partial u(z)}{\partial z} \tag{6}$$

$$-\rho^2 u(z) = \frac{\partial T(z)}{\partial z} \tag{7}$$

In the above equations, T is the mechanical stress tensor, S is the strain tensor, E is the electric field, D is the electrical displacement vector (C/m2), ϱ is the density, u is the mechanical displacement vector, c is the elastic stiffness tensor calculated at constant electric field, e is the piezoelectric tensor (C/m2), and ε is the relative permittivity.

Using the boundaries conditions for T (z) and u(z), we have:

$$T_1(0) = 0 \; ; \quad T_3(z_3) = 0 \tag{8}$$

$$T_1(z_1) = T_2(z_1) \; ; \quad u_1(z_1) = u_2(z_1) \tag{9}$$

$$T_2(z_2) = T_3(z_2) \; ; \quad u_2(z_2) = u_3(z_2) \tag{10}$$

The potential difference u on the piezoelectric layer can be obtained by integrating Ez (the electric field) on the thickness of the considered layer:

$$U = -\int_{z2}^{z1} Ez(z)dz \tag{11}$$

By definition, the current I is the temporal derivation of the charge Q at the surface of the electrodes, which in sinusoidal mode is equivalent to:

$$I = jwQ \tag{12}$$

The continuity of the normal component of the electrical displacement vector D at the interface piezoelectric-metal makes it possible to express the charge Q as a function of D and of the surface of the metal electrodes:

$$Q = Dz\ S \tag{13}$$

The expression of the current becomes then:

$$I = jwS\,Dz \tag{14}$$

Consequently, the impedance of the piezoelectric layer can be obtained in function of the thickness of the used materials and the dimensions of the resonator, as written in equation (15):

$$Z_{eq} = \frac{U}{I} = \frac{[(z_2 - z_1) - \alpha e_2(r_{21} - r_{11}) - \beta e_2(r_{22} - r_{12})] / \varepsilon_2}{j\omega A} \tag{15}$$

Where $e_2$ is the piezoelectric tensor of the piezoelectric layer, $A$ is the surface (area) of the electrodes, $\varepsilon_2$ is the permittivity of the piezoelectric layer, and $r_{11}$, $r_{12}$, $r_{21}$, $r_{22}$, $\alpha$, and β are expressions in function of known constants.

The impedance can be evaluated knowing the material properties and dimensions. This method can also be used for more complex structures, such as SCF (Stacked Crystal Filters), or CRF (Coupled Resonator Filters) resonators or filters, accounting for all the layers.

### *2.1.3. Equivalence of one dimensional models*

Using a least squares method, the two expressions in equations (3) and (15) can be equated. By this way, the values of $C_o$, $C_m$ and $L_m$ for different resonator areas $A$ and for the thicknesses of each layer can be obtained.

For example, the expression of $L_m$ as a function of the surface $A$ and the thickness $t$ of the top electrode is presented. For fixed values, $[A_1\ A_2\ A_3 \cdots\cdots A_i]$ and $[t_1\ t_2\ t_3 \cdots\cdots\ t_i]$, we obtain corresponding values of $L_m$. For each same thickness $t_i$, we can consider that $L_m$ is only varying according to the surface $A$. As shown in equation (16), we can find a polynomial function in the variable $A$ fitting the values of $L_m$. The polynomial coefficients ($W_i$, $X_i$, $Y_i$ and $Z_i$) are now independent of the surface $A$ and depend only on the thickness $t$. In a second time, using the same method, the coefficients can be expressed as a function of variable $t$.

$$\begin{aligned} L_m &= W_i A^n + X_i A^{n-1} + \cdots\cdots + Y_i A + Z_i \\ &= \left( \sum_{j=1}^{N} a_j t^{j-1} \right) A^n + \left( \sum_{j=1}^{N} b_j t^{j-1} \right) A^{n-1} + \cdots\cdots + \left( \sum_{j=1}^{N} c_j t^{j-1} \right) A + \left( \sum_{j=1}^{N} d_j t^{j-1} \right) \end{aligned} \tag{16}$$

This identification method can also be applied to the other variables, such as the thickness of the loading layer. The three elements of the BVD model ($C_o$, $C_m$ and $L_m$) can be determined as functions of the piezoelectric resonator dimensions (thickness of each layer and surface of resonator). According to the expressions of $C_o$, $L_m$, and $C_m$, surfaces and thicknesses can be optimized with an electrical software for designing a BAW resonator or a BAW filter.

## 2.2. Model in 3 dimensions

Even though Film Bulk Acoustic Resonator simulation with 1D model enables to quickly compute complex frequency response as filters, it becomes too restrictive when spurious modes in lateral dimensions have to be predicted. The 3D Finite Elements Method (FEM) enables to investigate the effect of the electrode shape on the spurious modes that are present in the electrical impedance. In order to reduce or to suppress these modes, solutions have to be investigated.

### *2.2.1. Finite element model*

Piezoelectricity is a phenomenon which couples electrical and mechanical domains. It can be modeled into coupled equations:

$$T_{ij} = c_{ijkl} * S_{kl} - e_{ijk} * E_k \tag{17}$$

$$D_k = e_{kl} * E_l + e_{ijk} * S_{ij} \tag{18}$$

With: T: mechanical stress (Pa), E: electric field (V/m), S: mechanical strain, D: electric displacement($C/m^2$), c: stiffness tensor (Pa), ε : permittivity tensor (F/m) and e: piezoelectric tensor ($C/m^2$)

Unfortunately, one-dimensional approximation becomes too restrictive when we need to predict spurious modes that may appear with lateral direction mode coupling or with resonators electrical or mechanical cross coupling. Thus a 3D simulation tool is needed to compute the resonator in three dimensions.

In the following sections, examples of 3D FEM computation will be proposed and described highlighting the advantages of 3D simulations.

### *2.2.2. Mechanical displacement modes*

We have analyzed the suspended resonator structure presented in Figure 2. The structure is clamped on lateral sides (no mechanical displacement in space directions), the bottom electrode is grounded and a potential constraint is applied to the top electrode.

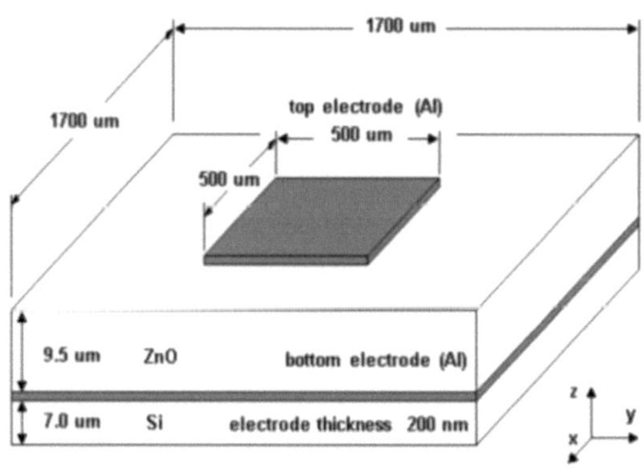


**Figure 2.** 3D suspended resonator structure.

In order to identify spurious modes, the mechanical displacement for four modes has been calculated, as displayed in Figure 3.

The first mode is the so-called thickness extensional mode. It corresponds to the maximum energy coupling and to minimum mechanical losses. This mode is the one taken into account in 1D analytical models.

The second mode is a cavity mode. In FEM, in order to keep the problem size finite, the physical domain needs to be truncated. This truncation introduces artificial boundaries where artificial boundary conditions are considered. Domain truncation causes reflection of the waves on clamped lateral sides. Then a standing wave may appear at certain frequencies. Those spurious modes are modeling errors and will be suppressed by mesh apodization.

Standing wave modes are harmonic thickness extensional modes due to acoustic wave reflection on top electrode edges. They appear as spurious modes in the electrical response. In order to use this electrical response for oscillator or filter applications, we need to suppress these modes. Different solutions can be considered.

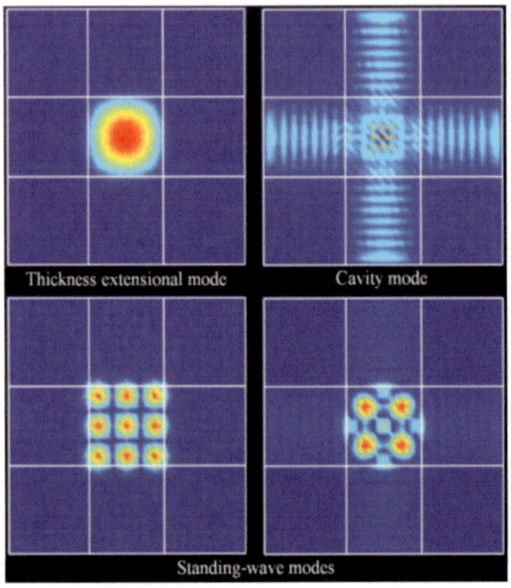


**Figure 3.** Top view of the suspended resonator mode shapes at resonant frequencies (The only top electroded region is the square in the middle).

### 2.2.3. *Suppression of standing wave modes*

#### 2.2.3.1. *Apodization*

The first solution consists of cut a triangular part of the square top electrode and to paste it on another edge as shown in Figure 4. Then a quadrilateral electrode is obtained with no parallel sides [5].

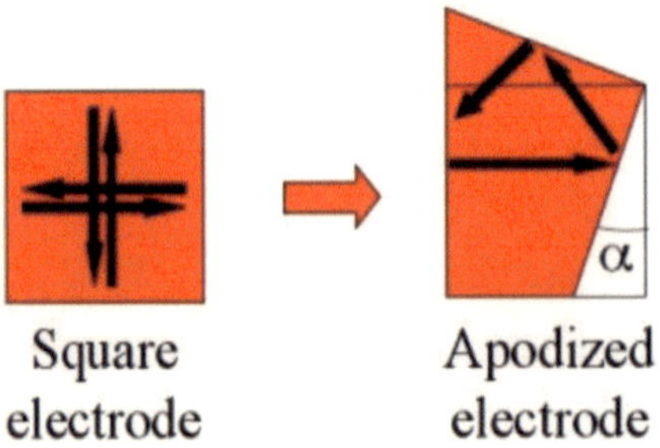


**Figure 4.** Apodization applied to a square electrode

Standing waves cannot appear since multiple reflections are not constructive. One can notice from Figure 5 that standing wave modes are not or weakly coupled as the apodization angle increases. A 3D simulation software enables the analysis of an entire apodized resonator and is very useful in this case for determining an optimum apodization angle.

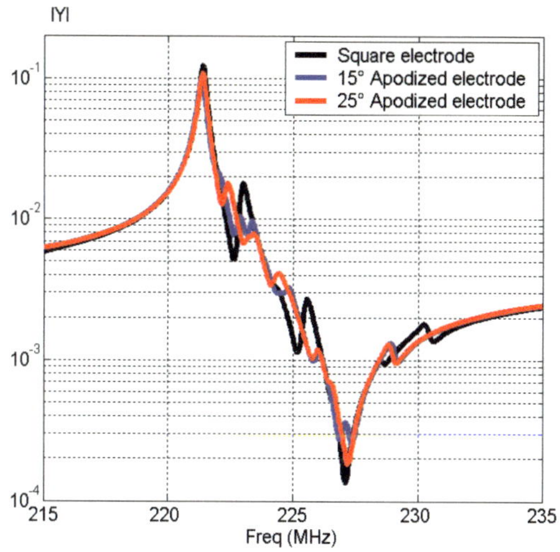


**Figure 5.** Apodized top electrode electrical admittance for two apodization angle values

### 2.2.3.2. *Edge loading*

Another solution proposed in [6] is considered. A narrow region is deposited at the edge of the suspended resonator top electrode as described in Figure 6. This thickened region constitutes a frame that matches the acoustic impedance and suppresses reflection on top electrode edges. Within certain optimum range for the edge region width, the resonator operates in a mode where the mechanical displacement is constant on the top electrode surface (Figure 7).

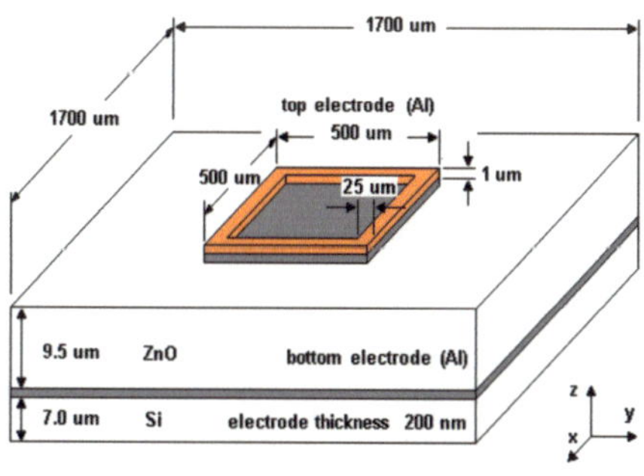


**Figure 6.** Framed top electrode suspended resonator 3D structure

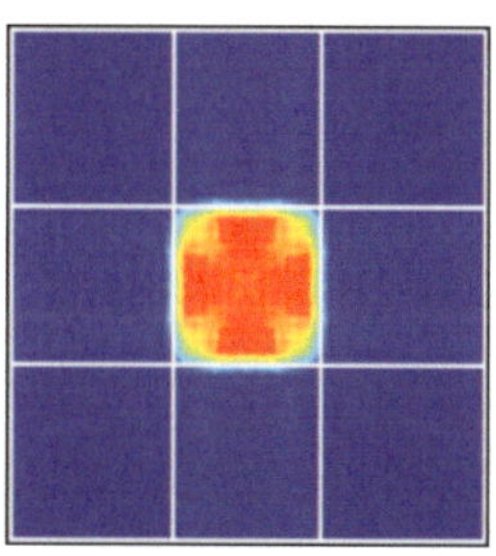

**Figure 7.** Framed top electrode suspended resonator thickness extensional mode

It can be observed, in Figure 8, that the electrical response is standing wave modes free. Standing waves on the top electrode are not coupled. Moreover, 3D resonant frequency tends toward 1D result.

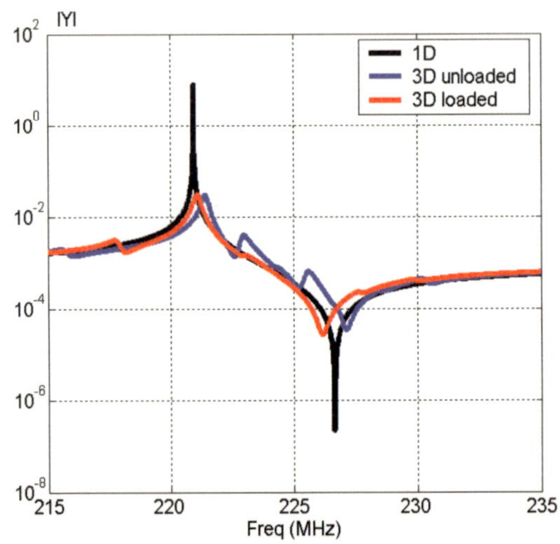


**Figure 8.** Framed top electrode suspended resonator admittance

*2.2.3.3. Cavity modes suppression by resonator apodization*

In order to identify spurious modes due to domain truncation, we have changed the distance from the top electrode edge to the structure edge (A=500 μm or 400 μm in Figure 9). We have found that resonant conditions and resonant frequencies for cavity modes change with cavity dimensions.

The mechanical displacement for those modes is displayed in Figure 9.

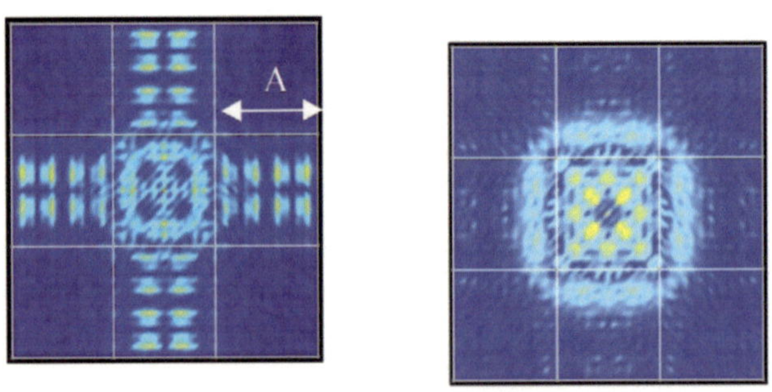


***f* = 224,5 MHz** ***f* = 228,4 MHz**

**Figure 9.** Mechanical displacement for two cavity modes (Top view)

While the mode which appears at *f* = *224,5 MHz* is obviously a cavity mode, we can have doubts about the one at *f* = *228,4 MHz*. In order to remove all doubts, we have applied the apodization technique to the resonator edges (Figure 10).

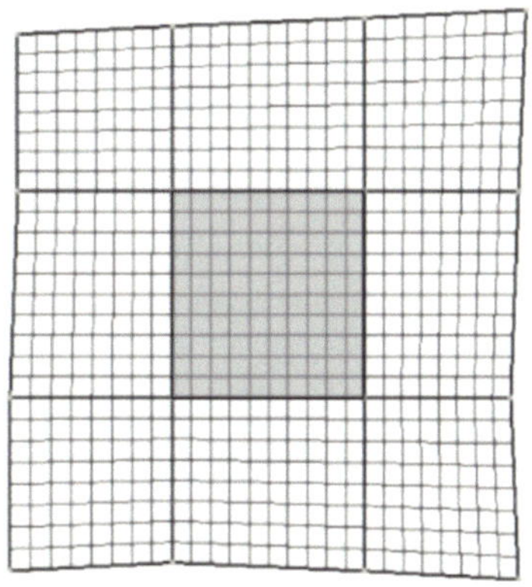

**Figure 10.** Apodized cavity (Top view) and mesh

We obtained an asymmetrical structure with no parallel sides. The irregular shape avoids phase reflections on the structure edges with constructive interferences, reducing the occurrence of standing waves.

One can observe in Figures 11, 12 and 13 that this technique enables to obtain an electric response free of spurious cavity modes.

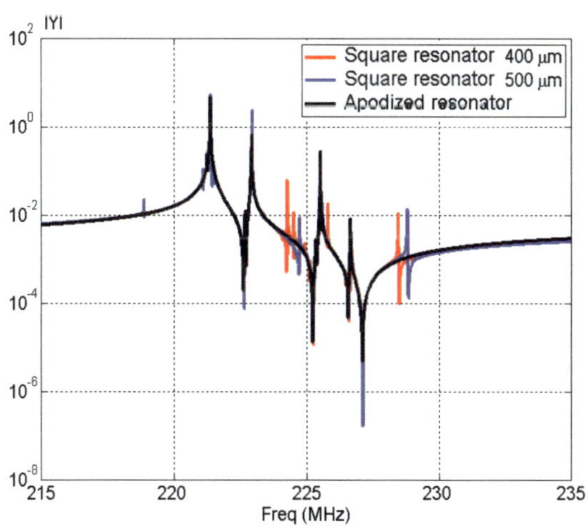


**Figure 11.** Suspended resonator electrical admittance

A full 3D FEM tool is very useful to analyze and to predict the behavior of complex structures. One can take into account non-homogenous structure and non-linear materials. Obviously it is the only method to compute 3D geometries. The shape of the mechanical displacement can be obtained to identify real or non-physical modes. Solutions have been proposed with apodized shape or with edge loading to remove parasitic modes.

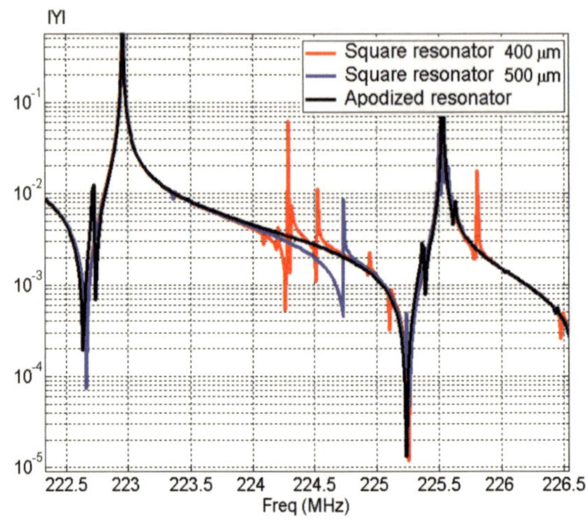


**Figure 12.** Suspended resonator electrical admittance around 224,5 MHz

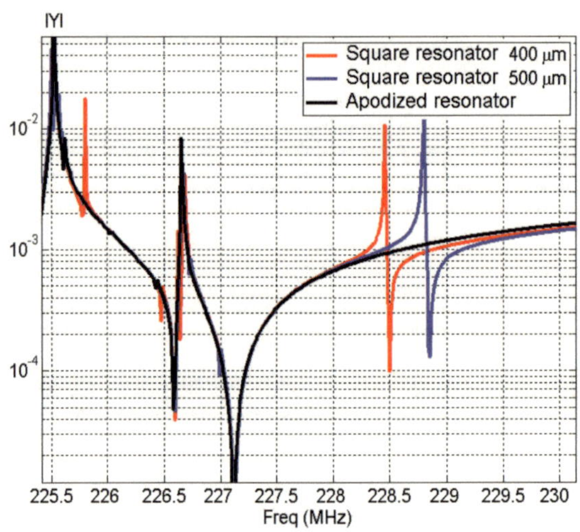


**Figure 13.** Suspended resonator electrical admittance around 228,4 MHz

# 3. Design of bulk acoustic wave filters

## 3.1. Implementation

The MBVD (Modified Butterworth Van Dyke) model, presented in Figure 14, can been used to compute the behavior of a BAW resonator [7]. Compared to the BVD model, the MBVD model incorporates resistances, which take into account losses in the piezoelectric material and in the electrodes.

This equivalent circuit resonates for two particular frequencies:

$$f_s = \frac{1}{2\pi\sqrt{Lm.Cm}} \tag{19}$$

$$f_p = f_s \sqrt{1 + \frac{Cm}{Co}} \tag{20}$$

*fs* and *fp* are known as the series and parallel resonant frequencies and correspond respectively to a minimum and to a maximum of the electrical impedance. *fs* and *fp* are related to the electromechanical coupling coefficient $k_t^2$ by:

$$k_t^2 = \frac{\pi^2}{4} \frac{f_p - f_s}{f_p} \tag{21}$$

Moreover, one can define $Q_s$ and $Q_p$ the quality factors of series and parallel resonances:

$$Q_s = \frac{2\pi f_s Lm}{Rm} \tag{22}$$

$$Q_p = \frac{1}{2\pi f_p Co Ro} \tag{23}$$

For dimensioning a BAW filter, each resonator is represented by its MBVD model and the synthesis of the BAW filter is carried out by optimizing each resonator with respect to a specified target.

BAW filters are usually implemented arranging series and shunt resonators. Basic resonator arrangements, namely ladder and lattice configurations, are defined as shown in Figures 15 and 16.

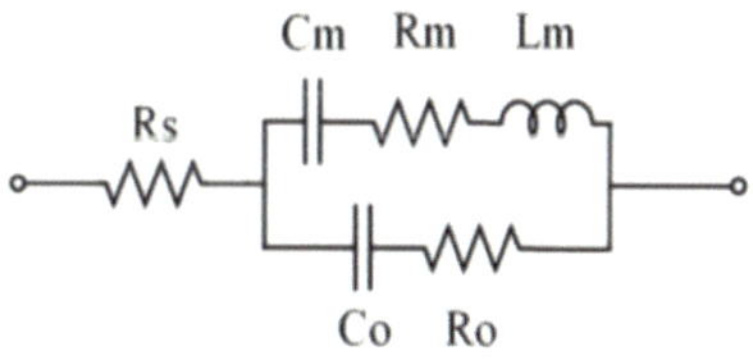


**Figure 14.** MBVD model of a BAW resonator

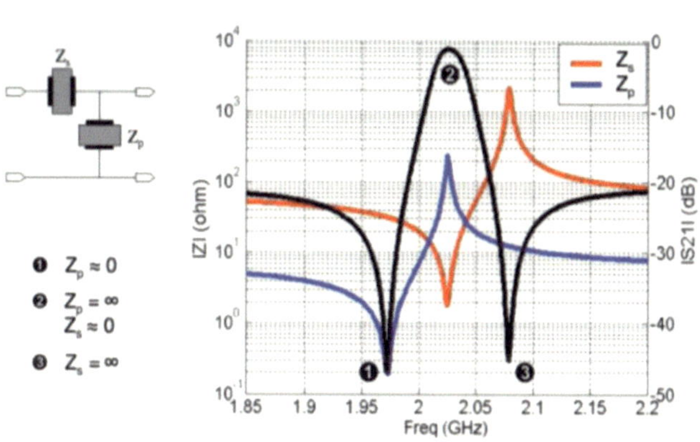


**Figure 15.** Electrical impedance of shunt and series resonators for providing a band-pass filter with a ladder network

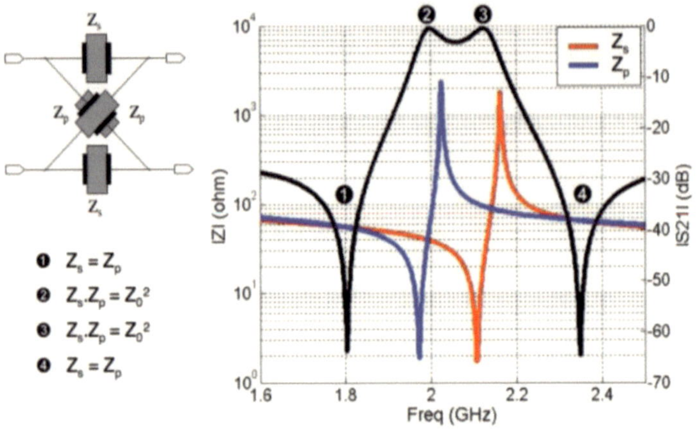


**Figure 16.** Electrical impedance of shunt and series resonators for providing a band-pass filter with a lattice network

The ladder configuration presents a high rejection close to the filter passband but a poor out of band rejection. On the other hand, the lattice filter exhibits higher out of band rejection but a poor rejection close to the filter passband. Combining these two configurations, one can obtain a mixed ladder-lattice filter with very good properties, as shown in Figure 17. $S_{21}$ parameter is the forward transmission coefficient of the filter.

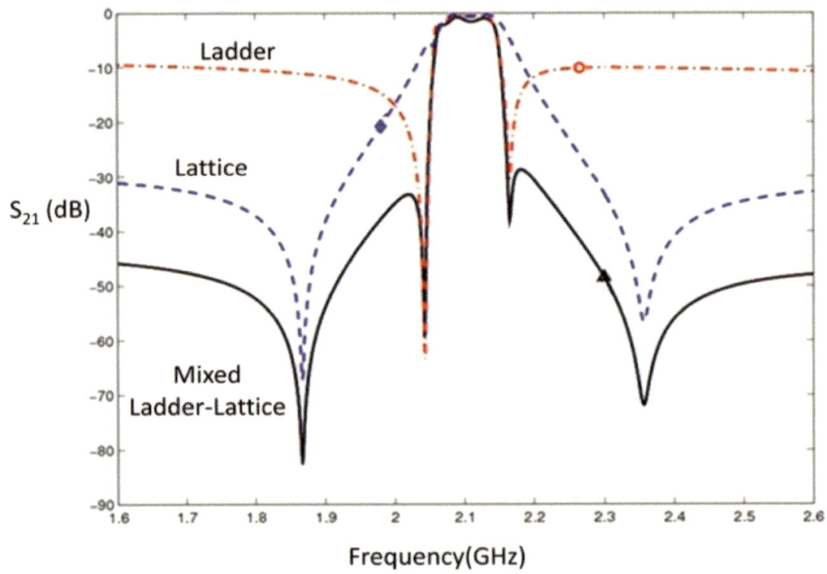


**Figure 17.** Behavior of the mixed ladder lattice filter.

## 3.2. Synthesis

Regarding the synthesis, each resonator is characterized by several fixed technological parameters obtained by electromechanical modeling or process characterization:

- $k_t^2$, the electromechanical coupling coefficient
- $Q_s$, the quality factor at series resonant frequency
- $Q_p$, the quality factor at parallel resonant frequency
- $\varepsilon_r$, the piezoelectric material permittivity
- $R_{[]}$, the square resistance of electrodes

The remaining variable parameters are:

- *A*, the surface of the top electrode (*A* = *L*.*W* where *L* and *W* are the length and the width of the top electrode respectively, the aspect ratio *L*/*W* remaining generally the same for all electrodes)
- *d* and *l*: the thickness of the piezoelectric layer and the thickness of the loading layer respectively, which control the series resonant frequencies ($f_s$) of series and parallel resonators.

The lumped elements of the MBVD model are related to these parameters through the following expressions:

$$C_0 = e_0 e_r \frac{A}{d} \tag{24}$$

$$C_m = C_o \left[ \left( \frac{f_s}{f_p} \right)^2 - 1 \right] \tag{25}$$

With

$$f_p = \frac{f_s}{1 - \frac{4}{\pi^2} k_t^2} \tag{26}$$

$$L_m = \frac{1}{C_m (2\pi fs)^2} \tag{27}$$

$$R_m = \frac{2\pi L_m fs}{Q_s} \tag{28}$$

$$R_o = \frac{1}{Q_p C_0 2\pi fp} \tag{29}$$

$$R_s = R_{[]} \cdot \frac{L}{\omega} \tag{30}$$

One can note that Lm, Cm and Co can also be approximated directly by polynomial expressions of the layer thicknesses and resonator area as explained in section 2.1.3.

The thicknesses of the piezoelectric and loading layers and the area of each resonator are optimized with respect to the specifications. This optimization can be driven by the minimization of a cost function defined by the filtering pattern [8].

### 3.3. Electromagnetic co-simulation

The previous synthesis relies on MBVD models of BAW resonators, which do not take into account metallic losses or couplings due to interconnections and access ports. Since metallic lines used for connecting resonators have irregular geometries depending on the arrangement of resonators, models for such elements cannot be implemented in a synthesis tool. Nevertheless a simulation is possible a posteriori with the layout of the filter in order to estimate additional losses or to check eventual couplings due to metallic lines.

The layout of the filter to be realized can be drawn with an electromagnetic (EM) software, e.g. Momentum included in Agilent ADS [9]. Using such a EM software, all electrostatic and electromagnetic phenomena are characterized considering the geometry and the physical characteristics of stacked layers.

The electrostatic part of MBVD resonators (*Rs*, *Ro* and *Co*) is directly taken into account in the distributed model (related particularly to the area of resonators). However, the motional part (*Rm*, *Cm* and *Lm*) is modeled by lumped elements connected through internal ports as shown in the example given in Figure 18.

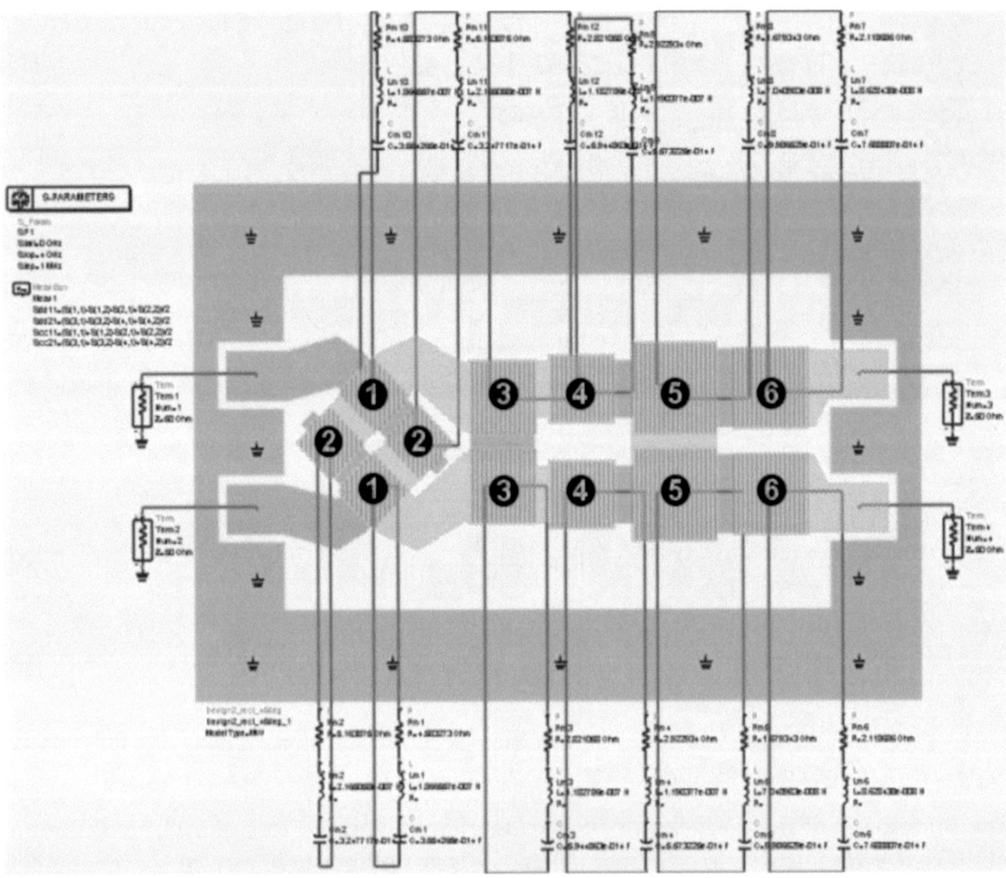


**Figure 18.** UMTS filter co-simulation

Figure 18 shows the co-simulation of a mixed ladder lattice filter in Agilent ADS/Momentum environment, including the layout and the motional parts directly derived from the previous synthesis.

# 4. Fabrication and characterization of BAW resonators and filters

## 4.1. Single resonator

Several BAW devices have been designed using the method proposed previously. Molybden is chosen as electrode material and the sputtering method is used. Therefore, AlN films have been deposited in (002) direction with the c-axis perpendicular to the substrate surface [10-13]. The optimized resonators and filters with the proposed method have been fabricated by CEA-Leti [14]. Figures 19 and 20 show a Solidly Mounted Resonator (SMR) resonator and a comparison between the simulated result and the measurement data respectively. The apodization of the top electrode was used to avoid parasite modes. One can observe a good agreement between simulations and measurements. $S_{21}$ and $S_{11}$ parameters are respectively the forward transmission and the reflection coefficients of the filter.

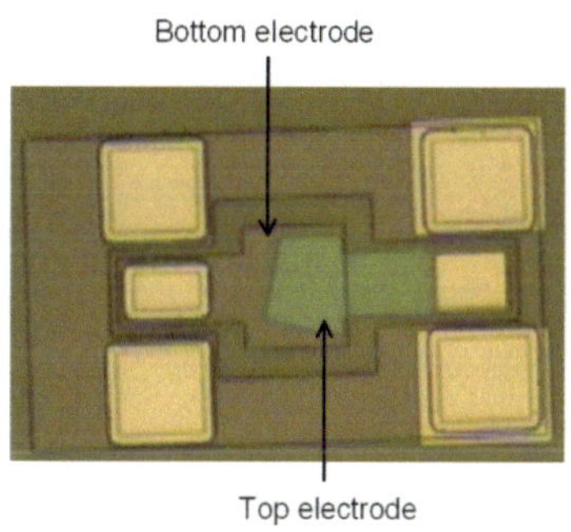


**Figure 19.** Structure of SMR resonator

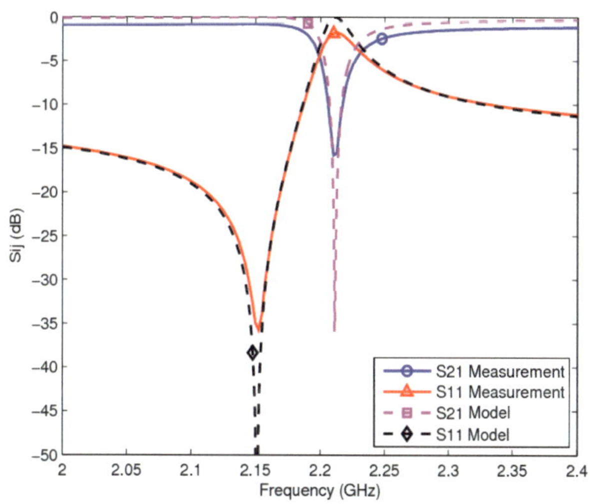


**Figure 20.** Comparison between the simulated result and the measurement data for SMR resonator

The resonant frequency is 2.212 GHz and anti-resonant frequency is 2.152 GHz, which leads to a coupling coefficient $k_t^2$ equal to 6.69%. The quality factor of this resonator is around 200.

### 4.2. 2-pole and 3-pole ladder filters

2-pole and 3-pole filters shown in Figure 21 were fabricated by CEA-Leti. Measurements and simulations for each filter are presented respectively in Figures 22 and 23. One can observe a good agreement between the measured responses and the simulated ones. The measured response of the 2-pole filter presents a passband of 55 MHz centered at 2.13 GHz (fractional bandwidth of 2.58 %) and the insertion losses are about 1.5 dB. Similarly, the 3-pole filter shows a 54 MHz passband at 2.13 GHz (FBW: 2.53 %) and insertion losses of approximately 2 dB.

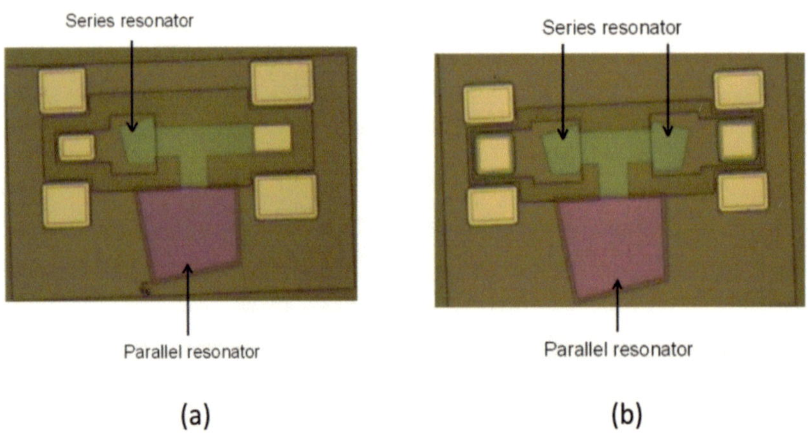


**Figure 21.** (a) Fabricated 2-pole filter. (b) Fabricated 3-pole filter (top view).

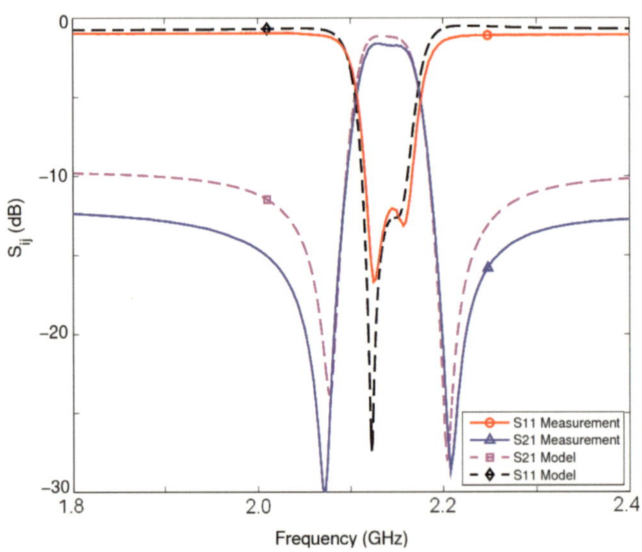


**Figure 22.** Measured and computed responses (model with losses) of the 2-pole filter

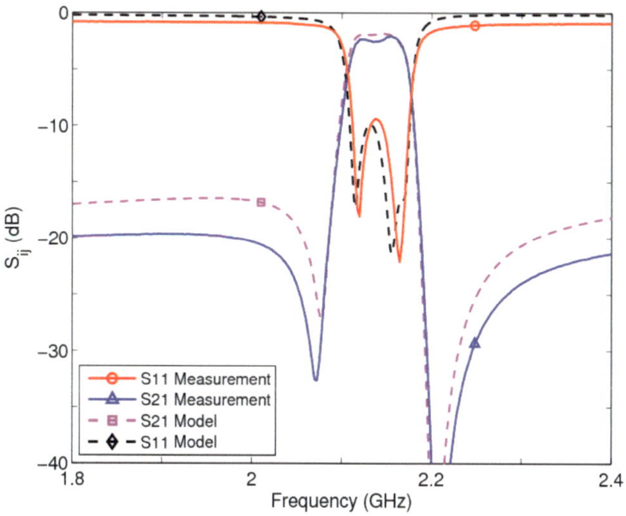


**Figure 23.** Measured and computed responses (model with losses) of the 3-pole filter

## 4.3. Differential ladder and lattice filters

Bandpass filters have been synthesised with differential structures (100Ω input/output impedance). These filters have been fabricated by CEA-LETI and UPM (Universidad Politécnica de Madrid) [15]. Each resonator is deposited on a Bragg mirror (SiN/SiOC) and is built with an Aluminium Nitride (AlN) piezoelectric layer and two Iridium (Ir) electrodes. With this technology, top electrode thickness is defined in order to act like a loading layer [16, 17] and to reach desired resonant frequency. Ir electrodes are utilized in order to enhance electromechanical coupling [18]. Iridium is a metal that presents a high-density [19] which leads to a high acoustic impedance, a low electric resistivity, and a specific crystal structure that promotes the growth of AlN films of excellent piezoelectric activity [20].

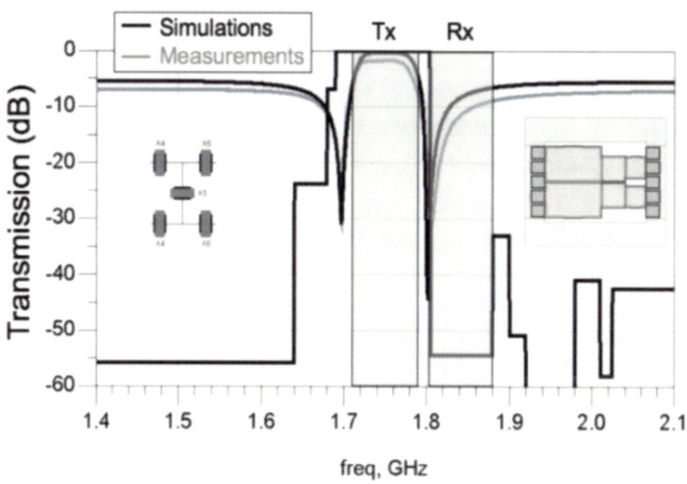


**Figure 24.** 1.5-stage ladder filter compared to UMTS standard

One can notice that there is a good agreement between simulated and measured responses in Figures 24 and 25. As electrical losses of Iridium were not characterized before simulation, they were not taken into account. Measured insertion losses are about 3dB for ladder structure and 6dB for lattice structure. A good out of band rejection is obtained, only limited by measurement noise.

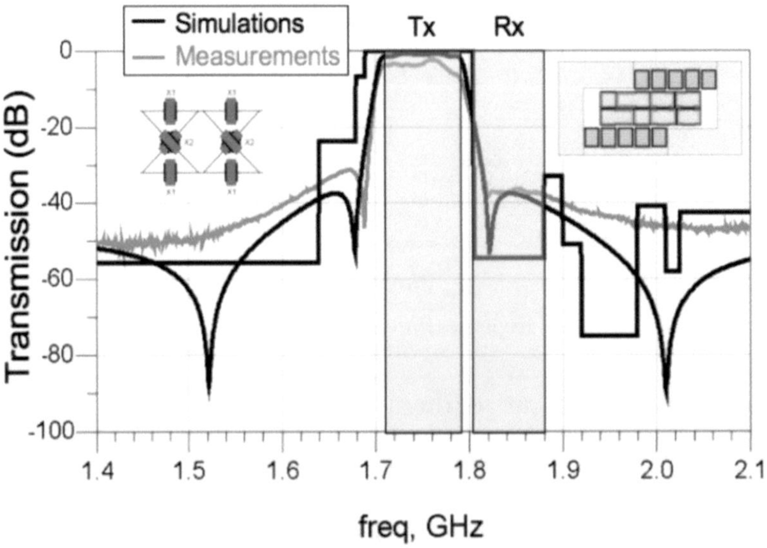


**Figure 25.** 2-stage lattice filter compared to UMTS standard

## 5. Integration of BAW components in a transmitter module

This section deals with the development of a BAW-SMR technology for application to mobile multi-standard communication terminals. UMTS and DCS1800 were chosen as targeted standards to validate a bi-standard BAW-based RF architecture, presenting both reconfigurability of active parts and integration of BAW technology.

This section presents the first demonstration of a complete UMTS transmitter including a CMOS 90nm digital RF signal generator connected to a BAW filter, a Power Amplifier (PA) stage and a BAW duplexer. Although the active part was originally designed for multi-mode purposes, UMTS standard was chosen to illustrate the proof of concept. The purpose is to emphasize the use of BAW-based structure for wireless mobile communications.

In the first section the proposed multi-standard transmission chain will be detailed, and a focus on the transmit BAW filter will be discussed. The design and measurement results of the stand-alone filter will then be presented before talking about the advantages of this filter with the RF signal generator. The BAW duplexer will be described and the measurement results of this filter will be detailed. Finally, the full transmitter test bench will be highlighted and the measurement results commented on.

## 5.1. RF signal generator

### *5.1.1. Third-order delta-sigma modulator*

The architecture presented in Figure 26 is clearly aimed at future software defined terminals by pushing the band or standard specific components as close as possible to the antenna. The targeted standard is UMTS, one of the main mobile communication standards in Europe using 1920 – 1980 MHz frequency band for TX and 2110-2170 MHz frequency band for RX. The architecture could easily be extended to additional standards such as DCS or PCS at the cost of extra BAW filters for the appropriate frequency bands.

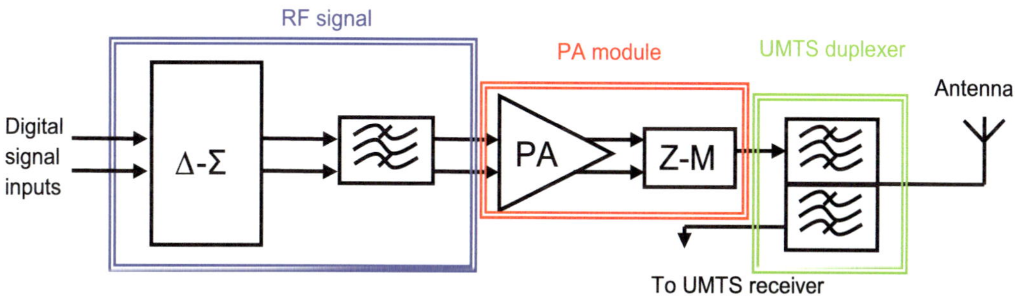


**Figure 26.** Modules definitions for first single standard UMTS transmitter

Two oversampled low-pass ΔΣ stages representing I and Q channels, work synchronously to generate a high sample rate 1-bit output stream [21] that feeds a digital image-reject mixer [22]. This latter stage produces a high frequency sampled two-level RF signal which exhibits very good in-band performances and a quantization noise shaping due to ΔΣ modulation [21]. A first stage of filtering precedes the power amplifier whose matching network converts a nominal single ended load impedance of 50 Ω to the optimal impedance for the active stage. A duplexer is inserted in order to isolate the receiver from the transmitter while enabling them to share a common antenna.

The UMTS transmit filter is designed on a ladder-lattice topology with 50 Ω differential input - 100 Ω differential output ports to benefit from advantages of both structures. This filter aims at meeting the standard emission mask by lowering the out-of-band quantization noise. Figure 27 shows the output spectrum of the digital RF signal generator and indeed justifies that the transmitter requires a large amount of out-of-band filtering at the output of the ΔΣ modulators. The most stringent specification for UMTS is the required rejection (50 dB) in the RX band (2110 MHz - 2170 MHz). It is very difficult to reach such a high rejection at a very close bandwidth from the center frequency band. A typical lattice BAW based filter exhibits a typical attenuation of roughly 40-45dB. Consequently, the insertion of a duplexer in the UMTS transmission chain is required to achieve this high attenuation in the RX band as well as its isolation between the TX and RX paths. The design and measurement results of these BAW-based structures are presented in the following sections.

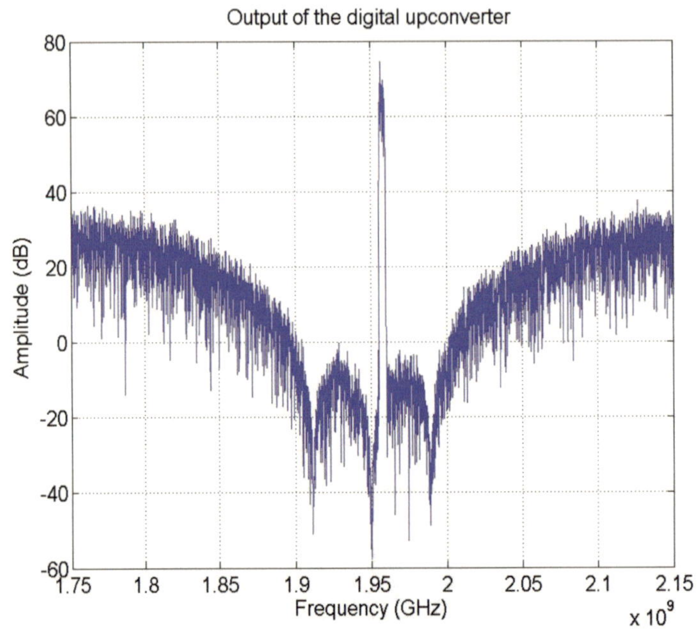


**Figure 27.** Spectrum of a ΔΣ modulated signal, up-converted around the RF carrier.

### 5.1.2. *BAW filter*

#### *5.1.2.1. Filter synthesis*

The synthesis of a UMTS filter has been performed [23], where a passband is needed for the 1920 – 1980 MHz range. Important constraints of the UMTS standard [24] are a high rejection and a high selectivity. To achieve these performances, a differential mixed ladder-lattice filter topology [25] with 100Ω differential impedance has been proposed. As explained in section 3, a mixed ladder-lattice filter as shown in Figure 28, provides sharp band edges and a good rejection in the stopband.

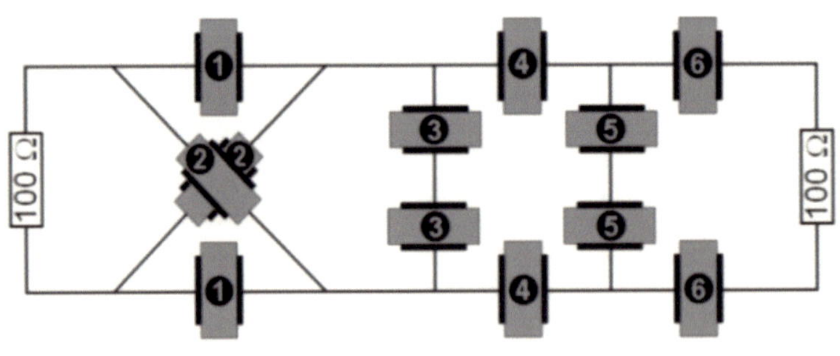


**Figure 28.** UMTS filter topology

After synthesis, optimal technological data for unloaded and loaded resonators are found as listed in Table 1. Related MBVD elements of each resonator are listed in Table 2 and the scattering parameters are given in Figure 29.

| | *fs* (MHz) | *fp* (MHz) | *K²eff* (%) | *Qs* | *Qp* |
|---|---|---|---|---|---|
| **Unloaded** | 1954 | 2004 | 6.2 | 500 | 300 |
| **Loaded** | 1898 | 1946 | 6.1 | 500 | 300 |

**Table 1.** Unloaded and loaded resonators given technological data

| | | X1 | X2 | X3 | X4 | X5 | X6 |
|---|---|---|---|---|---|---|---|
| **Elements values** | *Rm* (Ω) | 4.58 | 5.16 | 2.82 | 2.92 | 1.68 | 2.12 |
| | *Lm* (nH) | 186.6 | 216.5 | 118.3 | 119 | 70.4 | 86.3 |
| | *Cm* (fF) | 35 | 32 | 59 | 56 | 100 | 77 |
| | *R0* (Ω) | 0.39 | 0.43 | 0.23 | 0.25 | 0.14 | 0.18 |
| | *Co* (pF) | 0.68 | 0.63 | 1.16 | 1.07 | 1.94 | 1.47 |
| | *Rs* (Ω) | 0.19 | 0.17 | 0.19 | 0.20 | 0.23 | 0.22 |

**Table 2.** MBVD elements values of optimised UMTS filter

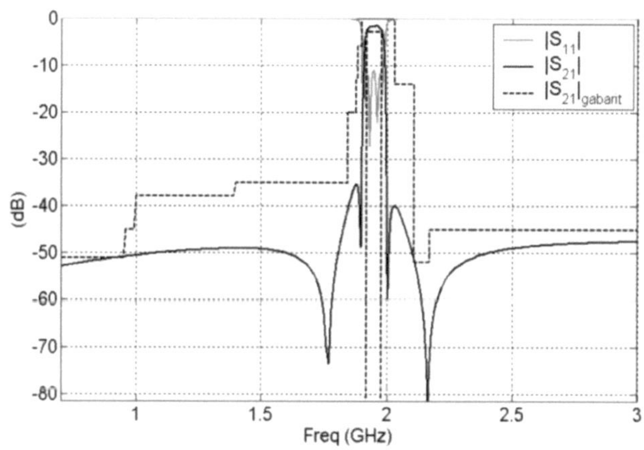


(a)

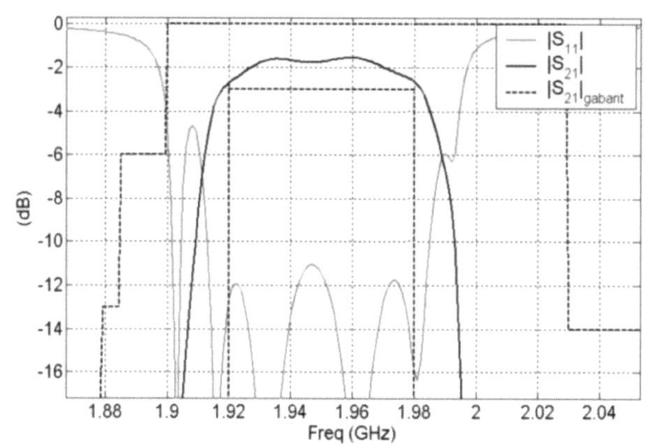


(b)

**Figure 29.** Scattering parameters of the UMTS filter: (a) wide band performance, (b) in band performance, compared to UMTS standard

One can observe, in Figure 30, that scattering parameters completely fulfill the specifications. Moreover, the layout of the filter has been co-simulated with a 2.5D electromagnetic software in order to estimate additional losses and to characterize eventual couplings due to metallic lines. As shown in Figure 30, the co-simulation is still in line with the specifications.

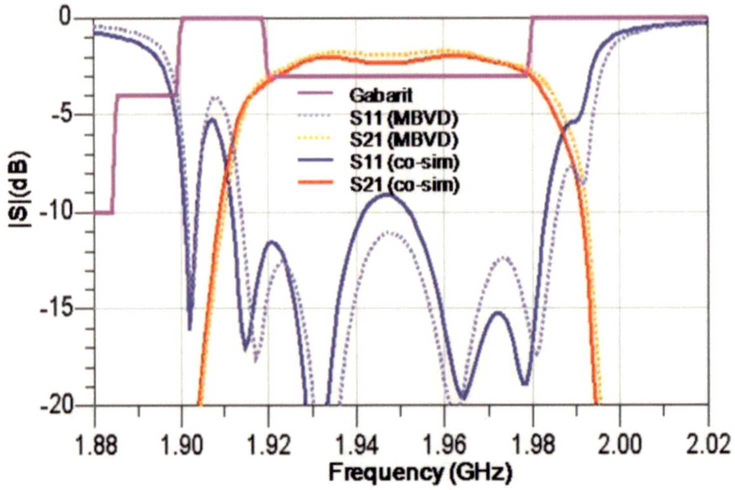


**Figure 30.** EM co-simulation of the UMTS filter, compared to UMTS standard

*5.1.2.2. Fabrication and measurements*

The filter has been fabricated by CEA-Leti. Each resonator is deposited on a Bragg mirror (SiN/SiOC), using an Aluminium Nitride (AlN) piezoelectric layer, two Molybdenum (Mo) electrodes and a Silicon Oxide (SiO2) loading layer. A photo of the filter is presented in Figure 31. The measured unpackaged quality factor at resonance frequency is close to 800 for the series resonators and close to 300 for the parallel ones. The measured response of the filter is presented in Figure 32. Insertion losses are around 3 dB and the required rejection and selectivity are fulfilled. The bandwidth is slightly reduced due to a lower resonant frequency for series resonators during fabrication

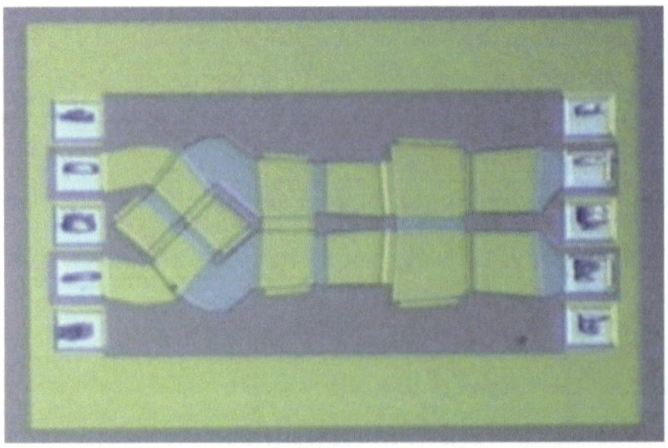

**Figure 31.** Fabricated filter with differential accesses.

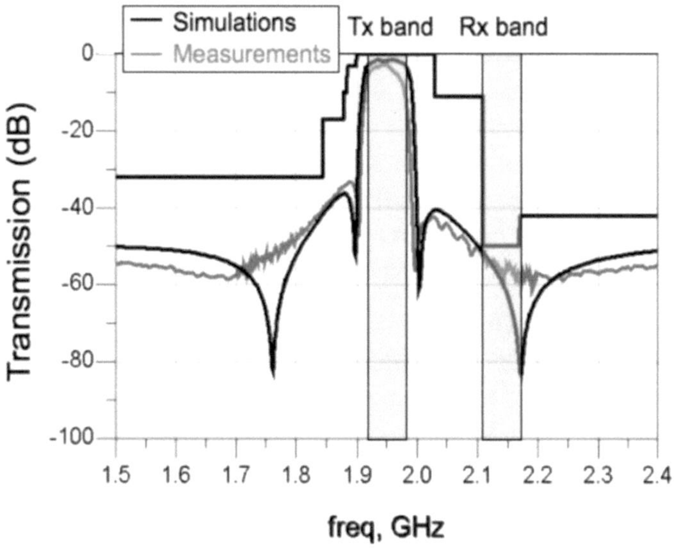


**Figure 32.** Measurement of the UMTS filter, compared to UMTS standard

## 5.2. Complete UMTS transmitter

### *5.2.1. RF Signal generator*

The all digital RF signal generator architecture is presented in Figure 33. A first stage oversamples the baseband I and Q signals to reach the modulator frequency. The principle of ΔΣ modulation and digital mixing has been explained in section 5.1 and one can find further explanations in [21]. The generator output signal is a very high speed 1-bit signal with bandpass shaped quantization noise centered on the standard transmit band. The differential output buffer is able to drive a differential 50Ω load with a 1V power supply, at very low output impedance of 0.6Ω. As the BAW filter needs a 50Ω differential input impedance match, 25Ω resistors need to be inserted between the signal generator and the BAW filter. This very simple passive and unfortunately unmatched matching network degrades the filter transfer function, introducing ripple in the pass band and lowering out-of-band attenuation. Both the RF signal generator and the BAW filter are molded with a resist on a module which is then soldered on a larger PCB. The output of this subset circuit is 100 Ω differential [26].

The baseband input signals are generated by an external Arbitrary Waveform Generator (Tektronix AWG 420), in which a Matlab sequence has been programmed to generate a WCDMA modulated signal on I and Q channels. The external clock reference comes from a synthesizer delivering a clock signal at 2.6GHz with an output level of –5dBm. This clock is the nominal frequency clock of the modulators and of the digital image-reject mixer and fully determines the center frequency of the RF signal. Using a 2.6GHz, the ΔΣ generator outputs an RF modulated signal at 1.95GHz using the first image-band.

The measurement results (Figure 34) consist of a spectrum analysis of output signals to evaluate the benefits of using BAW filtering in a ΔΣ modulation approach. It clearly shows that the out-of band quantization noise nearby the central bandwidth has been reduced below the noise level of the spectrum analyzer in this setup, due to the near-band high rejection of the ladder BAW filter configuration. Emission specifications in DCS are satisfied whereas more than 20dB of attenuation is still needed in UMTS RX band. Far-band filtering (thanks to lattice configuration) is also efficient, resulting in more than 35dB attenuation at low frequencies. In the first image band, the measured ACLR are 43 and 42dB, respectively at 5MHz and 10MHz offsets. The EVM is 3.7% and the measured channel power is -27dBm. This low value is due to the use of the first image and the loss in the series resistors.

The strongest constraint in UMTS transmission architecture is the very high level of rejection needed in the UMTS reception band which is very close to the transmission band. The measurement results show that a ladder-lattice BAW filter is not sufficient to fulfill this specification. A BAW-based duplexer, providing further isolation between both signal paths is consequently mandatory and will be described in the following section.

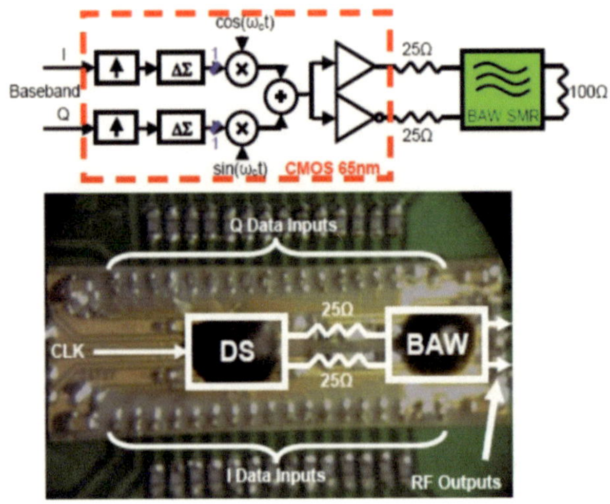


**Figure 33.** Signal generator and BAW filter architecture (left), photograph of the DS modulator and BAW assembly

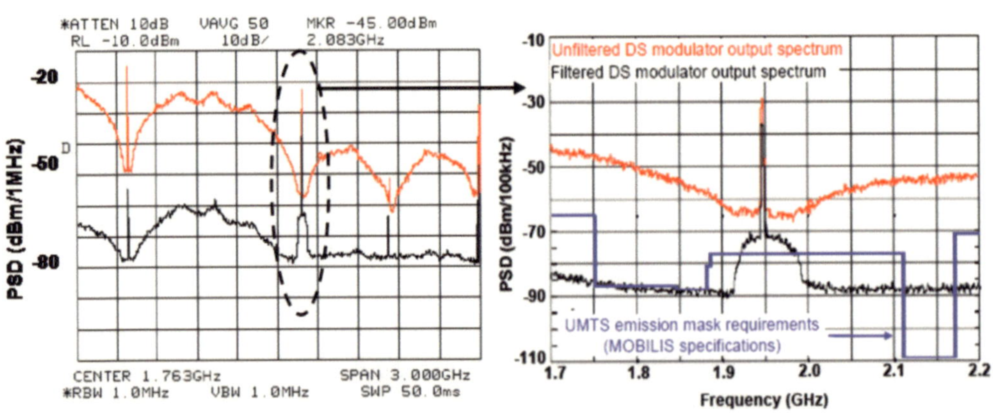


**Figure 34.** Delta-sigma digital generator output spectrum before (red) and after (black) BAW filtering

### 5.2.2. BAW duplexer

The BAW duplexer has been designed using a mix of the exposed BAW filter methodology and co-simulation with electromagnetic tools [27], [28], [29]. The topology of this duplexer is shown in Figure 35.

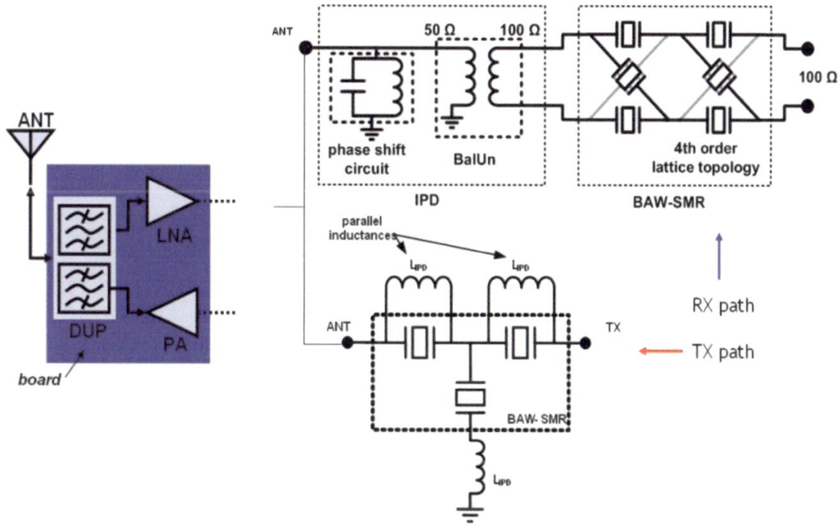


**Figure 35.** BAW duplexer topology

The BAW duplexer was made with two BAW filters and a glass substrate containing high quality passive elements (IPD from ST Microelectronics [30]). Each BAW filter occupies 1mm². The TX and RX filters were flip-chipped on the 3.9x3.9mm² substrate, as shown in Figure 36.

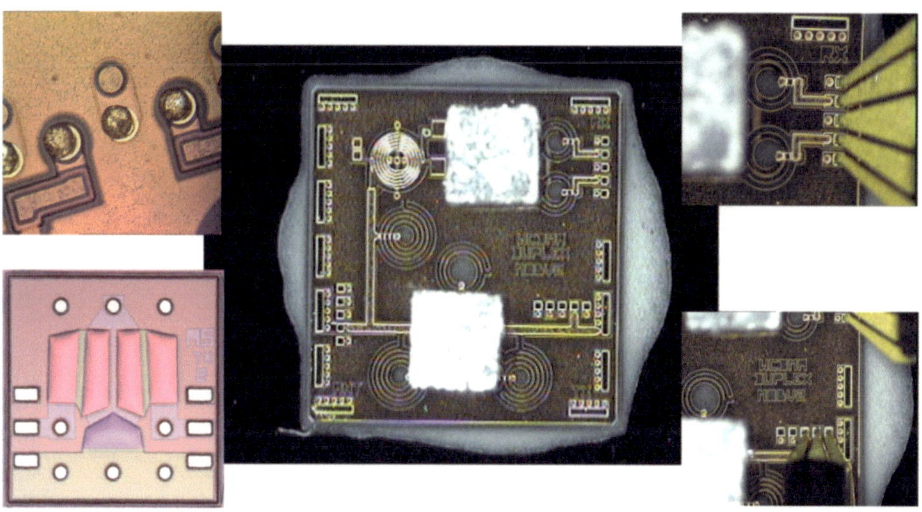

**Figure 36.** Left: Photograph of bumps and micropackaged TX filter. Center and right: On probe measurement of the BAW duplexer

The on-probe measurement of the duplexer is performed in two steps, for TX and RX paths respectively. This is due to calibration restrictions: two ports are 50 Ω single ended and the third one is a 100 Ω differential port, and on probe calibration of the vectorial analyzer is not guaranteed in these conditions. Figure 37 shows the comparison between probe measurement and backward simulation (taking into account of shifted components compared to expected elements in initial simulation). It shows globally a good fitting. The

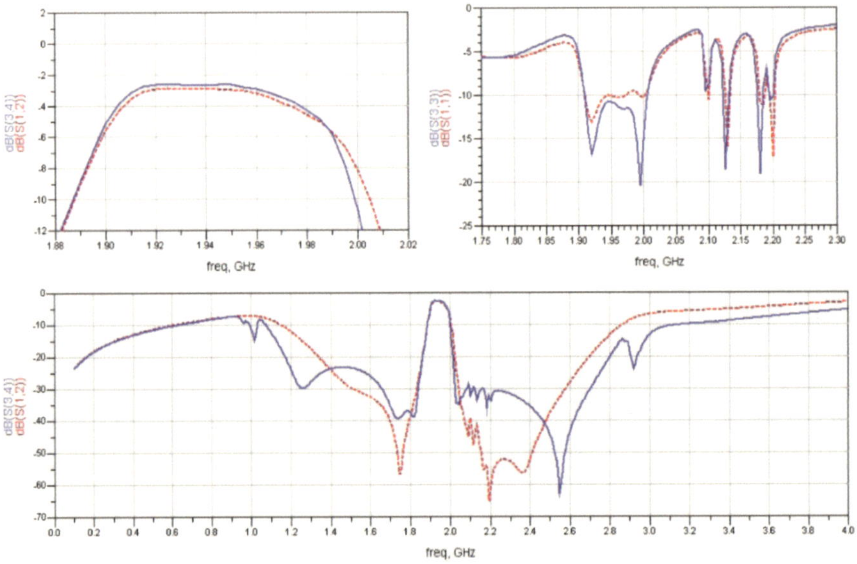


(a) BAW filter behavior (Tx path)

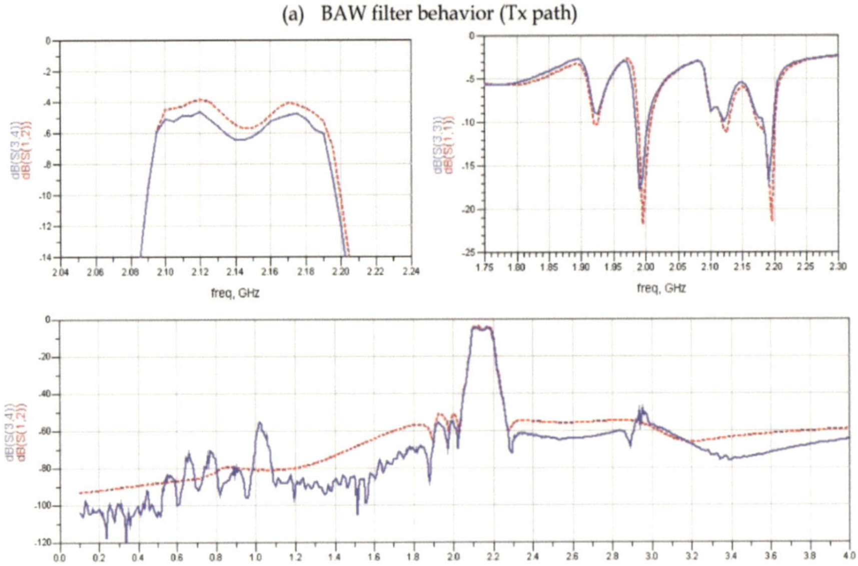


(b) BAW filter behavior (Rx path)

**Figure 37.** On probe measurement (blue) and simulation (BAW) – (a) TX path, (b): RX Path only strong difference is the TX isolation at the RX frequency, which is due to grounding effects; it disappears when the ground is bounded. It should be noticed that because of the one-path measurement conditions, the other path is open, leading to a strong mismatch in the adjacent band. Moreover, the unexpected low coupling factor of the resonators and the especially balun variations explain the relatively low performances of this duplexer.

When mounted on board as shown in Figure 38, the BAW duplexer exhibits approximately the same performances as on probe, with a better out-of-band rejection. The RX rejection in the TX path is better than 40dB, the TX rejection in the RX path is better than 45dB, the TX insertion loss evolves from 2.6dB to 4.6dB in the upper border of the band, and the RX insertion loss falls from 4.5 to 6dB mainly due to the balun mismatch.

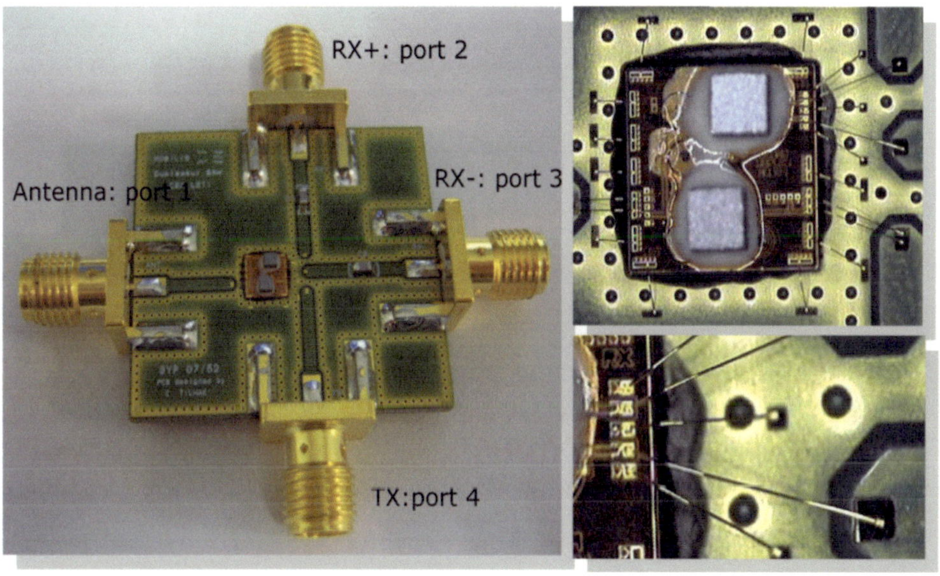


**Figure 38.** On board mounting of the BAW duplexer.

### *5.2.3. Complete transmitter test bench and measurement results.*

Figures 39 and 40 present the measurement setup. Two test benches have been implemented: one with the RF signal generator feeding the transmit BAW filter and the PA module, and the second one is made of the latter test bench which has been completed by the BAW duplexer to get the complete transmitter.

At the differential output of the RF signal generator board, the BiCMOS7RF differential input – single ended output PA is connected. Isolators are placed on each signal path between the two boards. The PA exhibits a peak power gain of 13dB at 1.7GHz and almost 500MHz of -3dB bandwidth with an output compression point of 27.5dBm showing that this stage will not contribute to non-linearity with output power below 20dBm. The power gain at 1.95GHz is 10.3dB. Figure 41 presents the measured output spectrum for the signal at the output of the PA module.

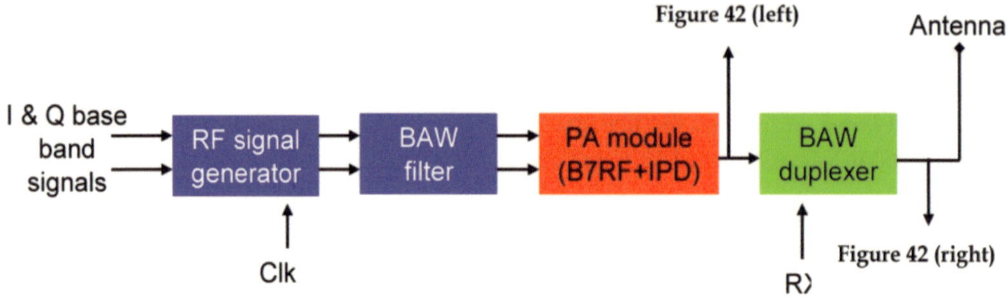


**Figure 39.** Test fixture for the full transmitter

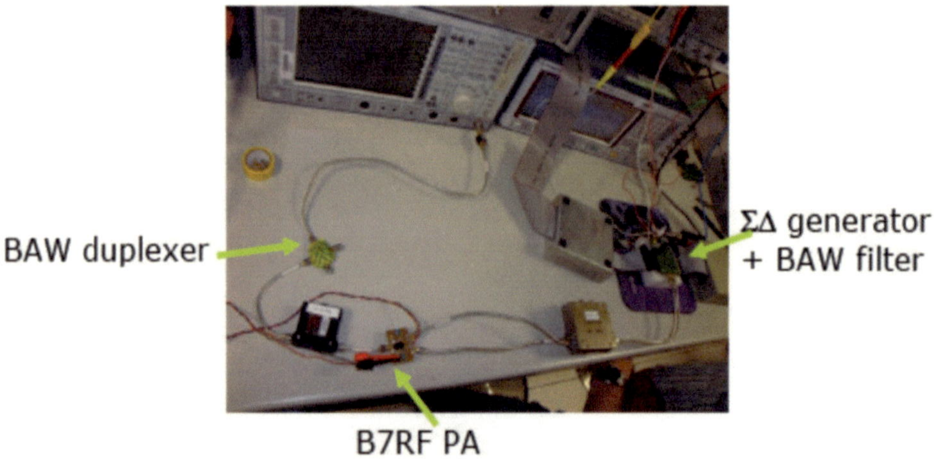


**Figure 40.** Measurement setup for filtered ΔΣ+ B7RF PA + BAW duplexer (full transmitter)

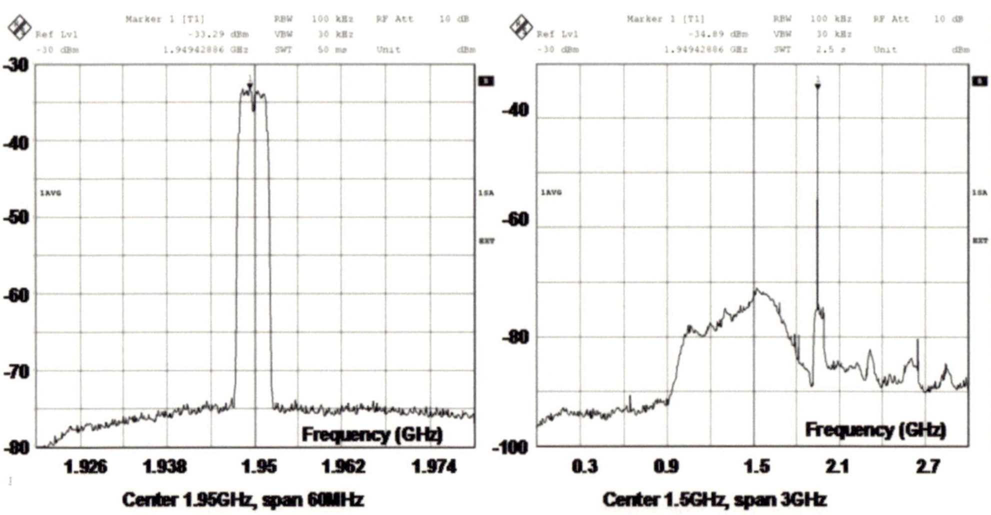


**Figure 41.** Output spectrum of the power gain stage, zoomed on the bandwidth (left) and full span (right)

One can observe that as the signal level at the input of the PA is low, there is no "visible" distortion on the signal (when looking in-band), and the ACLR level is preserved. When investigating a wider frequency window, we see the Sigma-Delta like lower side skirts re-growing in the band due to the wide-band amplification of the B7RF PA. Finally, a 42dB ACLR at 5 and 10MHz from the carrier for a -17.3dBm integrated power is obtained. EVM has been measured to be 4%.

At the single-ended output of the B7RF power amplifier, an isolator and the BAW duplexer are connected on the Tx input. The output chain measurements are done on the antenna pin of the duplexer demo-board, while the differential output toward the Rx path is shunted to 50 Ω loads. In figure 42, the filtering role of the duplexer Tx path on the overall transmitted signal is clearly shown. The ACLR constraints are generally preserved through the complete transmission chain. Moreover, the low-side spectrum re-growth observed in figure 42 is swept over by the BAW duplexer Tx filtering path. Finally, a 41dB and 41.7dB ACLR respectively at 5 and 10MHz from the carrier for a -20.8dBm integrated power is obtained. EVM has been measured to be 5%.

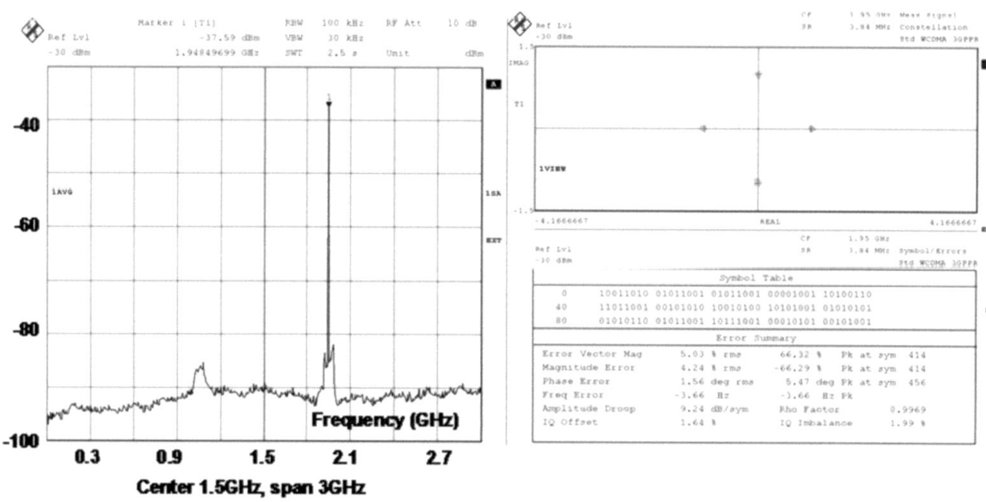


**Figure 42.** Measured wide-band output spectrum and EVM feature for the full transmitter (filtered ΔΣ + B7RF PA + BAW duplexer)

The complete transmitter is in line with the ACLR and EVM specifications. The transmitter was not able to fulfill the transmit power specifications due to some discrepancies on the distributed gain over the chain. Nevertheless, spurious emission requirements are almost fulfilled, thanks to the BAW high out-of-band rejection. The goal of the presented results was to demonstrate a novel type of Software Defined Radio transmitter architecture for W-CDMA (and DCS) standard(s) with BAW based filter and W-CDMA duplexer. Extra gain should be inserted into the whole transmission chain at the price of an increase in EVM and non-linearity degradation (lowering the ACLR).

## 6. Conclusion

In this chapter several methods for BAW devices simulation have been presented and investigated. Regarding the efficiency and the rapidity of the needed computation, designers could choose the 1D, the co-simulation piezoelectric equations and 2.5 D simulations or the full 3D Finite Element Method. The co-simulation equations and 2.5 D seems to be the best compromise between simulation time and quality of results. The detailed methods have been validated on several types of filters (Ladder, Lattice and mixed Ladder-Lattice). Optimization method and tools for filters have also been proposed to compute high quality filters (low insertion losses, high rejection, high pole number…).

BAW based filters and duplexers have then been presented and integrated in a complete UMTS transmitter highlighting the advantages of BAW devices.

Other piezoelectric materials are investigated and appear as promising solutions for realizing wide-band filters or sensors [31], [32].

## Author details

Matthieu Chatras, Stéphane Bila, Sylvain Giraud, Lise Catherinot,
Ji Fan, Dominique Cros and Michel Aubourg
*XLIM, UMR CNRS 7262, University of Limoges, Limoges, France*

Axel Flament , Antoine Frappé , Bruno Stefanelli and Andreas Kaiser
*IEMN, UMR CNRS 8520, Villeneuve d'Ascq, France*

Andreia Cathelin
*STMicroelectronics, TR&D, Crolles, France*

Jean Baptiste David and Alexandre Reinhardt
*CEA-LETI, Grenoble, France*

Laurent Leyssenne and Eric Kerhervé
*IMS, UMR CNRS 5818, Université de Bordeaux, Talence, France*

## 7. References

[1] K.M.Lakin, G.R. Kline, K.T. Mccarron, "High Acoustic Resonators and Filters", 1993 IEEE Microwave Symp. Digest, 3, pp 1517-1520

[2] K. Y. Hashimoto, "RF Bulk Acoustic Wave Filters for Communications", Artech House, 2009.

[3] A. Shirakawa, JM. Pham, P. JarrY, E. Kerherve, "FBAR Filters Synthesis Method and Reconfiguration Trends", Chapter 3 of Microwave Filters and Amplifiers book, pp.19-47, Research Signpost, 2005.

[4] Lakin, K. M., Kline, G. R., and Mccarron, K. T., "Development of Miniature Filters for Wireless Application," *IEEE Transaction on Microwave Theory and Techniques*, Vol.43, 2933-2939, 1995
[5] K. M. Lakin and K. G. Lakin, « Numerical Analysis of Thin Film BAW Resonators », *Proceedings of Ultrasonics Symposium*, Vol. 1, pp. 74-79, October 2003.
[6] R. Thalhammer, R. Aigner, « Energy loss mechanisms in SMR–type BAW devices », *Microwave Symposium Digest*, 2005 IEEE MTT-S International 12-17 June 2005
[7] J. F. Rosenbaum "Bulk Acoustic Wave Theory and Devices", Boston: Artech House, 1988
[8] L. Catherinot, S. Giraud, M. Chatras, S. Bila, D. Cros, T. Baron, S. Ballandras, P. Monfraix, L. Estagerie "A general procedure for the desing of BAW filters", International Journal of RF and Microwave Computer-Aided Engineering, September 2011, Vol. 21, Issue 5, pp 458-465.
[9] http://www.home.agilent.com
[10] Lanz, R.; Muralt, P. "Solidly mounted BAW filters for 8 GHz based on AlN thin films", IEEE Symp. on ultrasonics, 2003, Vol. 1, pp. 178-181.
[11] C. Cibert, C. Champeau, M. Chatras, D. Cros and A. Catherinot « Pulsed laser deposition of aluminum nitride thin films for FBAR application » Applied Surface Science 253 (2007) pp. 8151-8154
[12] H.P.Loebl, M.Klee, C.Metzmacher, W.Brand, R.Milson And P.Lok "Piezoelectric Thin AlN Film for Bulk Acoustic Wave (BAW) Resonators" Materials Chemistry and Physics, Vol. 79, pp. 143-146, 2003
[13] Ylilammi, J.Ella, M.Partanen and J.Kaitila "Thin Film Bulk Acoustic Wave Filter" IEEE Transactions on Ultrasonics, Ferroelectrics, and Frequency Control, vol. 49, No. 4, pp. 535-539, April 2002
[14] http://www.leti.fr/en
[15] http://www.upm.es/internacional
[16] E. Iborra, M. Clement, J. Olivares, S. Gonzalez-Castilla, J. Sangrador, N. Rimmer, A. Rastogi, B. Ivira, and A. Reinhardt, "BAW resonators based on AlN with Ir electrodes for digital wireless transmissions." 2008 IEEE Ultrason. Symp. Proc., (2008) pp. 2189-2192.
[17] B. Ivira, P. Benech, R. Fillit, F. Ndagijimana, P. Ancey, and G. Parat, "Self-Heating Study of Bulk Acoustic Wave Resonators Under High RF Power" IEEE Trans. Ultrason. Ferr. Freq. Control, 55, (2008). pp. 139-147.
[18] A. Reinhardt, F. de Crécy, M. Aïd, S. Giraud, S. Bila, and E. Iborra, "Design of Computer Experiments: A powerful tool for the numerical design of BAW filters" 2008 IEEE Ultrason. Symp. Proc., (2008) pp. 2185-2188.
[19] A. Devos, E. Iborra, J. Olivares, M. Clement, A. Rastogi, and N. Rimmer, "Picosecond Ultrasonics as a Helpful Technique for Introducing a New Electrode Material in BAW Technology: The Iridium Case", 2007 IEEE Ultrason. Symp. Proc., (2007) pp. 1433-1436.
[20] J. Olivares, M. Clement, E. Iborra, S. González-Castilla, N. Rimmer, and A. Rastogi, "Assessment of Aluminum Nitride Films Sputtered on Iridium Electrodes ", Ultrasonics Symposium 2007, pp 1401-1404.

[21] A. Frappé, A. Flament, B. Stefanelli, A. Kaiser, A. Cathelin, "An all-digital RF signal generator using high-speed ΔΣ modulators", IEEE Journal of Solid-State Circuits, art. No. 15, Vol.44, Oct. 2009

[22] Vankka, J. Sommarek, J. Ketola, I. Teikari, M. Kosunen and K. Halonen, "A Digital Quadrature Modulator with on-chip D/A Converter," IEEE Journal of Solid-State Circuits, Vol. 38, No. 10, pp. 1635-1642, Oct. 2003.

[23] S. Giraud, S. Bila, M. Chatras, D. Cros, M. Aubourg, "Bulk acoustic wave filter synthesis and optimisation for UMTS application", EuMW 2009, Rome

[24] 3GPP UE Radio Transmission and Reception (FDD) TS 25.101. Available at http:\\www.3gpp.org\

[25] A. Shirakawa, P. Jarry, J.-M. Pham, E. Kerhervé, F. Dumont, J.-B. David, A. Cathelin, "Ladder-Lattice Bulk Acoustic Wave Filters: Concepts, Design, and Implementation", International Journal of RF and Microwave Computer-Aided Engineering, 5 June 2008, pp.476

[26] A. Flament, S. Giraud, S. Bila, M. Chatras, A. Frappe, B. Stefanelli, A. Kaiser, A. Cathelin, *"Complete BAW filtered CMOS 90nm digital RF signal generator",* Joint IEEE North-East Workshop on Circuits and Systems and TAISA Conference, 2009

[27] E. Kerhervé, J.B. David, A. Shirakawa, M. El Hassan, K. Baraka, P. Vincent, A. Cathelin, "SMR-BAW duplexer for W-CDMA application", Journal of Analog Integrated Circuits and Signal Processing,2010.

[28] A. Shirakawa, P. Jarry, J.M. Pham, E. Kerherve, F. Dumont, J.B. David, A. Cathelin, "Ladder-Lattice BAW Filters: Concepts, Design and Implementation" , International Journal of RF & Microwave Computer Aided Engineering (RFMiCAE), 2008

[29] P.Bradly, R.Ruby, J.Larsoniii, Y.Oshmyansky and D. Figueredo "A Film Bulk Acoustic Resonator (FBAR) Duplexer for USPCS Handset Applications" IEEE International Microwave Symposium Digest 2001, Vol.1, P367-370

[30] Calvez, C.; Person, C.; Coupez, J.; Gallée, F.; Gianesello, F.; Gloria, D.; Belot, D.; Ezzeddine, H. "Packaged hybrid Si-IPD anrenna for 60 GHz applications", EuMW 2010, pp683-686.

[31] M. Chatras, L. Catherinot, S. Bila, D. Cros, S. Ballandras, T. Baron, P. Monfraix, L. Estagerie "Large Band Pass BAW Filter for Space Applications"IEEE, IFCS, San Francisco, May 2011

[32] T. Baron, J. Masson, D. Gachon, J.P. Romand, S. ALzuaga, L. Catherinot, M. Chatras, S. Ballandras. "A Pressure Sensor based on HBAR micromachined structure" IEEE IFCS, New Port(USA), june 2010.

Chapter 15

# Surface Acoustic Wave Based Magnetic Sensors

Bodong Li, Hommood Al Rowais and Jürgen Kosel

Additional information is available at the end of the chapter

http://dx.doi.org/10.5772/55220

## 1. Introduction

Since the radar system was invented in 1922, the development of devices communicating by means of reflected power has experienced a continuously growing interest. In 1948, Harry Stockman published a paper [1] in which he laid the basis for the idea of radio frequency identification (RFID), and the first patent had been filed in 1973 by Charles Walton. After decades of research and commercialization, RFID products became a part of everyday life (e.g. logistics, access control, security). With the growing interest in remote and battery-free devices, researchers are pushing the boundaries of RFID technology to find solutions in new fields like sensing applications.

For many sensors such as those operated in remote or harsh environments, the sensitivity is not the only evaluation criteria. The lifetime, especially of the power source, and the complexity added by wiring often demand wireless and passive operation. Batteries have limited lifetime and also add to the size and mass of the sensors. Alternatively, energy harvesting or an RF-based wireless power supply can be employed [2, 3]. The former method depends on environmental conditions such as solar radiation, temperature change, chemical reagents, vibration etc., which are often not constantly or not sufficiently available. RF power sources, on the other hand, transmit power wirelessly and with full control over amount and timing.

Passive and remote sensors utilizing SAW transponders are devices, which are powered by an RF source. These systems require an interrogation device that requests the sensor signal, a SAW transponder plus a sensing element and two antennas. The basic idea is that an RF signal of certain frequencies generated by the interrogator is received by the SAW transponder, which reflects back a signal modified by the sensing element. This signal contains the environmental information in an amplitude and phase change, which is converted into the physical parameters by the interrogator. In most cases, SAW sensors are coded by having different reflector designs in order to have multiple measurement capabilities from sensors located in the same interrogation area. A great amount of research has been carried out in the past decades in this field, and, as a result, different wireless SAW sensors have been developed to measure a variety of physical and chemical parameters including temperature, stress, torque, pressure, humidity, magnetic field, chemical vapor etc. [4-9]. Several devices are already commercialized [10-12].

SAW-based magnetic sensors have, so far, not been studied in detail. Magnetic sensors are one of the most pervasive kinds of sensors for a large number of applications and are employed in different fields like automotive, biomedical or consumer electronics. Integrating magnetic sensors with SAW transponders enables remote and passive operation, thereby, opens a door for further applications. Early in 1975, a magnetically tuned SAW phase shifter was proposed by Ganguly et al [13]. A thin film of magnetostrictive material was fabricated on the delay line of a SAW device. A phase shift was observed due to the dependence of the wave propagation velocity on the external magnetic field. Recently, a new concept of a magnetic sensor based on a SAW resonator has been published [8]. A magnetostrictive material was used to fabricate the interdigital transducers (IDT) of the SAW device. The resonant frequency of the device changes with an external magnetic field. A different idea was put forth by Hauser and Steindl [14-16] combining a SAW transducer with a giant magnetoimpedance (GMI) microwire sensor. The GMI sensor has a magnetic sensitivity, at least, one order higher than contemporary giant magnetoresistance (GMR) sensors and can be used to measure very low magnetic fields such as those generated by the human heart or muscles. A GMI sensor is operated by an ac current and the impedance changes upon changes of a magnetic field. This makes it a suitable load for a SAW transponder, which converts this impedance change into a magnitude and phase change of the reflected acoustic waves. In order to reduce the size and improve the level of integration of the senor, a new design of an integrated SAW transponder and thin film GMI sensor has been proposed and developed recently by the authors [17, 18]. The SAW transponder and GMI thin film were integrated on the same chip using standard micro-fabrication technology suitable for mass fabrication.

The ideal SAW-based magnetic sensor is small and highly integrated, inexpensive, passive, remotely controlled and have a high magnetic sensitivity together with a large linear range. With regard to these criterions, an SAW-GMI sensor is a very promising candidate.

SAW-based magnetic sensors have been studied for several years. However, this topic has yet not been comprehensively summarized, and the aim of this chapter is to provide a systematic review of the past research as well as the latest results. The performance of the devices crucially depends on different design parameters in a complex fashion. This will be shown by a detailed description and analysis for a device consisting of a SAW transponder and GMI thin film sensor.

## 2. SAW based passive sensors

A basic SAW device consists of an input interdigital transducer (IDT) and an output (or reflector) IDT, which are fabricated on a piezoelectric substrate. The area between the input IDT and output IDT is called the delay line. The IDT is made of two metallic, comb-like structures arranged in an interdigital fashion, whereby the distance between two fingers of a comb defines the periodicity ($p$) (Fig. 1). Upon application of a voltage, charges accumulate at the fingers of the IDT depending on the capacitance of the structure. The resulting electric field generates stress in the substrate due to the piezoelectric effect. If an ac input voltage is applied, the continuously changing polarity of the charges will excite an SAW (Rayleigh wave) traveling through the substrate. At the operating (resonant) frequency of the SAW device, the value of $p$ equals the wavelength of the SAW, and the SAW amplitude shows a maximum value due to constructive superposition.

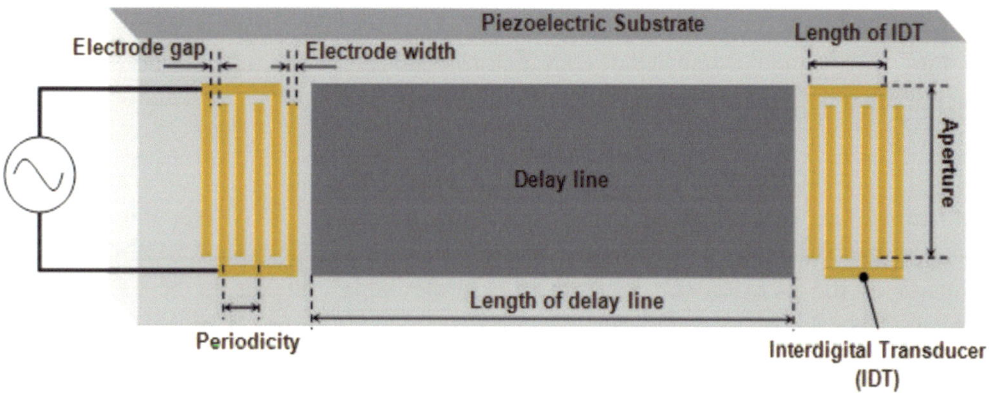


**Figure 1.** Schematic of a SAW device.

## 2.1. Basics of SAW devices

**SAW Resonant Frequency:** The resonant motion of an acoustic resonator is caused by the coupling between the transducer (IDT) and the acoustic medium. The resonant frequency, or operating frequency, is determined by the periodicity and the acoustic wave velocity ($v$)

$$f = \frac{v}{p}. \tag{1}$$

The value of $v$ mainly depends on the substrate's material. A typical SAW velocity for piezoelectric materials is several thousand meters per second. Due to the intrinsic anisotropy of piezoelectric materials, $v$ is dependent on the direction of propagation. Since different acoustic modes have different wave velocities, a device can resonate at different frequencies. The SAWs are Rayleigh waves, which have a longitudinal and a vertical shear component that can couple with any media in contact with the surface. This coupling strongly affects the amplitude and velocity of the wave allowing SAW sensors to directly sense, e.g., mass loads.

**Electro-Mechanical Coupling Coefficient:** The electro-mechanical coupling coefficient ($\kappa$) defines the conversion efficiency of the piezoelectric material between the electrical and mechanical energies, determined by

$$\kappa^{-2} = \frac{input\ energy}{converted\ energy}. \tag{2}$$

A high coupling coefficient reduces the insertion loss caused by the energy conversion, which results in smaller energy consumption as well as larger effective readout distance of a SAW-based wireless sensor.

**SAW Delay Line:** The SAW delay line refers to the area between the input IDT and output IDT on the substrate (Fig. 1). It creates a time delay between the input signal and the output signal depending on the SAW velocity and the length of the delay line. Due to this feature, SAW devices are widely used in RF electronics. It is also used in sensing applications, where the measurand causes, e.g., a change in the SAW.

**Temperature Coefficient of Delay (TCD):** The TCD reflects the temperature dependence of the time delay and is connected with the thermal expansion coefficient ($\alpha$) and the temperature coefficient of the phase velocity (TCV) by

$$TCD = \alpha - TCV \ . \quad (3)$$

The temperature dependence of the time delay is the basis of SAW temperature sensors, where higher TCD values yield higher sensitivity. However, for other SAW devices, the influence of the TCD on the time delay is undesirable and has to be minimized or eliminated. For this purpose, temperature compensated cuts of the crystalline substrates are employed, where the TCD is minimized over certain temperature ranges [19-20]. Piezoelectric bi-layers are another concept that has been utilized in order to compensate the TCD in sensing applications [21, 22].

## 2.2. Basic design concepts of passive SAW sensors

Passive SAW sensors typically operate as resonators, delay lines or loaded transponders. In case of resonators, the reflection of the interrogation signal from the SAW device is a function of the SAW device's resonant frequency, which depends on the measurand. In case of delay lines, the request signal is separated from the response signal by a time difference, whereby this time difference depends on the measurand. Similarly, the request signal and response signal are separated by a time difference in case of a loaded transponder. However, the time difference is constant and the measurand affects the signal amplitude. Intrinsic SAW sensors utilize a change of the substrate's properties. For example, intrinsic temperature sensors were realized by detecting the resonant frequency or phase change of the SAW in materials with large TCD [23]. Intrinsic stress sensors utilize the length change of the delay line caused by mechanical strain applied to the substrate. The stress can be evaluated by measuring the SAW phase shift [24]. Extrinsic SAW sensors can be realized by integrating a SAW device and an additional sensing element. A common extrinsic sensor concept utilizes selective thin films on top of the delay line leading to a change in mass by the measurand [13, 25]. This can be, for example, a thin film with high $CO_2$ solubility and selectivity [26]. As $CO_2$ dissolves into the film, the additional mass load causes a detectable phase shift in the SAW. Another extrinsic concept utilizes a sensitive IDT. For example, in case of a magnetostrictive IDT, a magnetic field applied to the sensor causes a change of the resonant frequency [8]. A loaded transponder is another extrinsic design, where the output

IDT is connected to a sensor, which changes the IDT's electrical characteristics as a function of the measurand. An example for a load sensor is a pair of conducting rods placed in the earth with a certain distance from each other. As the water level changes, the resistance between the rods changes, which can be detected as a magnitude and phase change of the signal reflected from the load IDT [27]. Another example for a load is a giant magnetoimpedance sensor [16, 17]. A change in the magnetic field yields a change in the sensor's impedance. Consequently, changes the reflectivity of the output IDT.

Some SAW sensors, their classification and method of detection are presented in Table 1.

| Sensor Type | Commercialization | Year | Intrinsic/ Extrinsic | Design | Detection Method | Access | Paper |
|---|---|---|---|---|---|---|---|
| Temperature | Yes | 1990 | Intrinsic | Resonator | Frequency | None | [23] |
| | | 2003 | Intrinsic | Delay line | Phase velocity | None | [28] |
| Pressure | Yes | 2001 | Extrinsic | Loaded Transponder | Phase | Capacitive pressure sensor | [25] |
| | | 2007 | Intrinsic | Delay line | Phase | None | [24] |
| Bio/Chem | No/Yes | 2006 | Extrinsic | Resonator | Frequency | Thin film | [29] |
| | | 2011 | Extrinsic | Delay line | Phase | Thin film | [26] |
| | | 2001 | Extrinsic | Loaded Transponder | Amplitude/ Phase | Conducting rods | [30] |
| Magnetic | No | 1975 | Extrinsic | Delay line | Phase | Thin film | [13] |
| | | 2011 | Extrinsic | Resonator | Frequency | Magnetostrictive IDT's | [8] |
| | | 2006/11 | Extrinsic | Loaded Transponder | Amplitude | GMI wire/ thin film | [16,17] |
| Sound | No | 2005 | Extrinsic | Loaded Transponder | Phase | Capacitive pressure sensor | [31] |
| Torque | Yes | 1996 | Intrinsic | Delay line | Phase | None | [32] |

**Table 1.** SAW-based passive sensors.

## 3. Passive and remote SAW-based magnetic sensors

Magnetic sensors are one of the most versatile sensors employed not only for the task of measuring magnetic fields but for a large number of different applications, thereby detecting the measurand indirectly, e.g., via a change of material parameters in construction monitoring or a change of distance in position monitoring. A passive and remote operation of magnetic sensors can be advantageous in many cases and considerably increase their applicability.

A SAW-based passive magnetic sensor can be realized either by adding an additional material layer, which is sensitive to magnetic fields, or by loading the output IDT with a magnetic sensor. In the first case, the magnetic layer changes the delay line or the resonant frequency of the SAW device. While in the second case, the sensor changes the reflection signal of the output IDT. Since SAW devices are operated by RF power, the sensor element has to work at the operation frequency of the SAW device. Among the available magnetic sensors, GMI sensors are the most suitable candidates as they have a high magnetic sensitivity as well as a high operating frequency.

### 3.1. Magnetostrictive SAW devices

Magnetostriction defines the relationship between the strain and the magnetization states of a material. It is an important property of ferromagnetic materials and was first observed by James Joule in 1842 in nickel samples. For a positive/negative magnetostrictive material, an applied magnetic field causes the material to expand/shrink in the field direction. Inversely, when a stress is applied to the magnetostrictive material, its magnetic anisotropy will change accordingly.

A magnetostrictive-piezoelectric resonator consists of amorphous magnetostrictive material layers as the electrodes sandwiching a piezoelectric core (Fig. 2). An ac signal applied to the electrodes causes the quartz layer to oscillate. The resonant frequency of this oscillation depends on the thickness of the piezoelectric material, the crystal orientation, temperature and mechanical stress, etc.

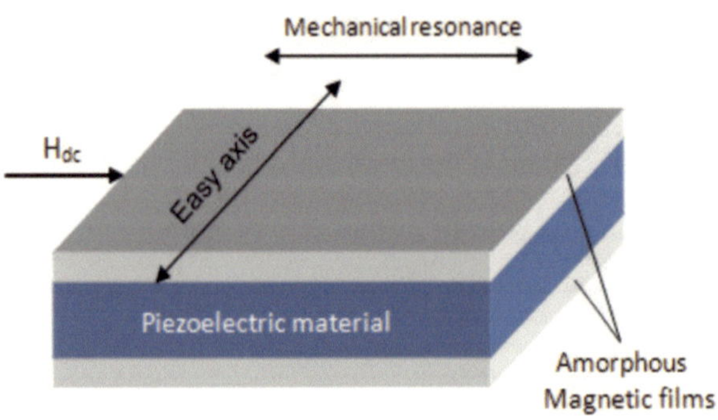


**Figure 2.** Structure of a composite magnetostrictive/piezoelectric resonator. The magnetic anisotropy is perpendicular to the external magnetic field $H_{dc}$.

When a magnetic field $H_{dc}$ is applied, the length change induced in the magnetostrictive film exerts stress to the piezoelectric material and, consequently, shifts the resonant frequency of the device. Utilizing this concept, a magnetic sensitivity high enough to detect the terrestrial field has been achieved [33]. In a similar work, a magnetostrictive-piezoelectric tri-layer structure has been embedded in a coil. The dc magnetic field sensitivity was as high as $10^{-8}$ T [34].

#### *3.1.1. Passive resonator*

A SAW-based, passive resonator for magnetic field detection was developed recently by Kadota et al [8]. Nickel, which is a negative magnetostrictive material, was used to fabricate the sensing IDT on a quartz substrate (Fig. 3). Upon the application of a magnetic field, stress will be induced to the substrate by the IDT change causing a change in the resonant frequency. This sensor showed a frequency change of 200 ppm for a magnetic field of 100 mT applied perpendicularly to the direction of SAW propagation.

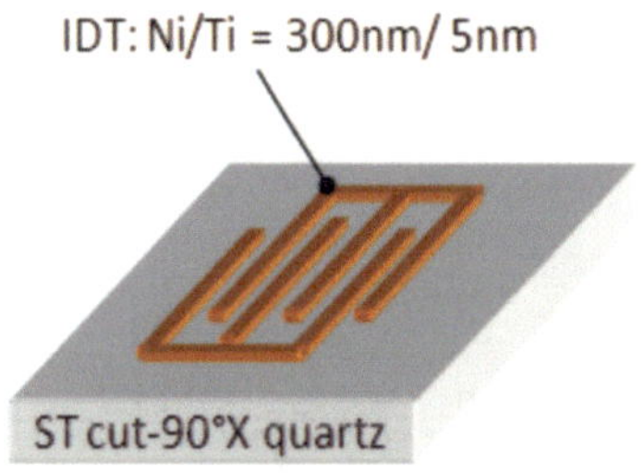


**Figure 3.** Schematic of a magnetic sensor device using a magnetostrictive IDT on a SAW substrate.

### *3.1.2. Passive phase shifter*

A magnetically tuned SAW phase shifter is a one-port SAW structure with a magnetic sensing functionality achieved through a delay line sensitive to magnetic fields. This idea was first introduced by Ganguly et al in 1975 [13]. In their device, the acoustic velocity is varied by an external magnetic field. This functionality is facilitated by a magnetostrictive thin film deposited on top of the delay line (Fig. 4). The propagation velocity of the SAW in the film region depends on the magnetic field. Hence, there is a correlation between the time shift of the reflected signal and the magnetic field.

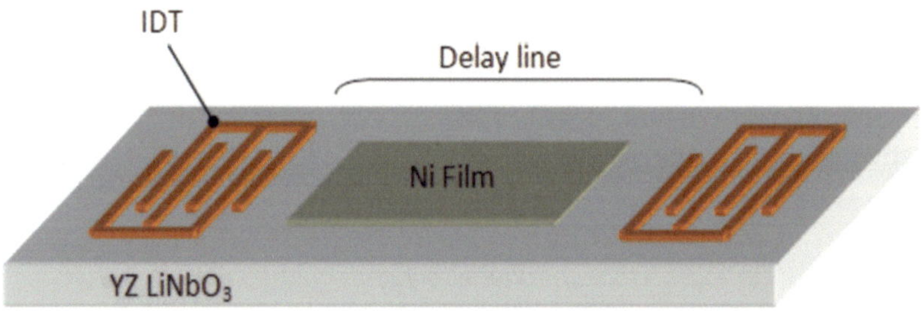


**Figure 4.** Schematic of a magnetically tuned SAW wave phase shifter.

Later, research efforts focused on different magnetostrictive materials and measurement methods [9, 35, 36], and a magnetic sensitivity of $10^{-4}$ %/kOe was achieved.

## 3.2. SAW-GMI sensors

A SAW-based, magnetic and passive sensors comprises a two-port SAW transponder and a magnetic sensor acting as a load at the output IDT. Among the available magnetic field sensors, giant magnetoimpedance (GMI) sensors offer favorable characteristics like high sensitivity to magnetic fields and high operation frequency (compatible with SAW transponders) making them a very suitable load. SAW-GMI sensors have been fabricated by combining SAW transponders with GMI wire sensors as well as thin film GMI sensors. Both of these methods have shown a higher magnetic sensitivity than direct designs.

### 3.2.1. *Basics of GMI sensors*

The GMI effect was first observed in Co-based amorphous wires by Panina and Mohri in 1994 [37] and has since attracted strong interest due to its sensitivity enabling magnetic field measurement with a nT resolution. The GMI effect is the impedance change of an ac-powered ferromagnetic conductor upon the change of a magnetic field. The relative impedance change, also called GMI ratio, is expressed as

$$\text{GMI Ratio (\%)} = 100\% \times \frac{Z(H)-Z(H_0)}{Z(H_0)} \quad \text{or} \quad \text{GMI Ratio (\%)} = 100\% \times \frac{Z(H)-Z(H_{max})}{Z(H_{max})}, \tag{4}$$

where $Z(H_0)$ is the impedance at zero magnetic field and $Z(H_{max})$ is the impedance at saturation field. Both definitions have particular aspects that should be considered. In case of the first expression, $Z(H_0)$ depends on the remanent state of the magnetic material while, in the second case, $Z(H_{max})$ is not always achievable and equipment dependent.

The GMI effect is explained by classical electromagnetism. The change of the complex impedance mainly originates from the skin effect in conjunction with a change of the complex permeability. Analytically, the complex impedance ($Z$) of a conductor is defined by

$$Z = \frac{U_{ac}}{I_{ac}} = \frac{\int_L \frac{1}{\sigma} J_z(S)dz}{\iint_q J_z dq}, \tag{5}$$

where $U_{ac}$ is the applied ac voltage, $I_{ac}$ is the current, $L$ is the length and $\sigma$ the conductivity. $S$ and $q$ refer to the surface and the cross section of the conductor, respectively. $J_z$ is the current density in the longitudinal direction obtained by solving Maxwell's equations. In ferromagnetic materials, by neglecting displacement currents ($\dot{\boldsymbol{D}} = \mathbf{0}$), Maxwell's equations can be written as follows:

$$\nabla \times \boldsymbol{H} = \boldsymbol{J}, \tag{6}$$

$$\nabla \times \boldsymbol{J} = -\frac{\mu_0}{\rho_f}(\dot{\boldsymbol{H}} + \dot{\boldsymbol{M}}), \tag{7}$$

$$\nabla \cdot (\boldsymbol{H} + \boldsymbol{M}) = 0, \tag{8}$$

J is the current density, H is the applied magnetic field, M is the magnetization of the ferromagnetic material, $\rho_f$ is the free charge density and $\mu_0$ is the permeability of vacuum. From Equ. (6) to (8), the expression

$$\nabla^2 \boldsymbol{H} - \frac{\mu_0}{\rho_f}\dot{\boldsymbol{H}} = \frac{\mu_0}{\rho_f}\dot{\boldsymbol{M}} - \nabla(\nabla \cdot \boldsymbol{M}), \tag{9}$$

can be derived. Equ. (9) can be solved using the Landau-Lifshitz equation, which relates $\boldsymbol{M}$ and $\boldsymbol{H}$

$$\dot{\boldsymbol{M}} = \gamma \boldsymbol{M} \times \boldsymbol{H}_{eff} - \frac{\alpha}{\boldsymbol{M}_s} \boldsymbol{M} \times \dot{\boldsymbol{M}}, \tag{10}$$

where $\gamma$ is the gyromagnetic ratio, $M_s$ is the saturation magnetization, $\alpha$ is the damping parameter and $H_{eff}$ is the effective magnetic field expressed as [30]

$$\boldsymbol{H}_{eff} = \boldsymbol{H} + \boldsymbol{H}_{\mathrm{a}} + \frac{2A}{\mu_0 M_s} \nabla^2 \boldsymbol{M} \tag{11}$$

where $\boldsymbol{H}$ is the internal magnetic field that includes the applied field and demagnetizing field, $\boldsymbol{H}_a$ is the anisotropy field and $A$ is the exchange stiffness constant..

By combining Equ. (5) to (11), a theoretical impedance model can be evaluated for GMI sensors with different geometries [37-40].

Although the experimentally obtained GMI effect shows a large sensitivity compared to other effects exploited for magnetic sensors, the theoretically estimated values have not been achieved yet. Therefore, a lot of effort has been put into improving the magnetic properties of GMI materials [41-44]. At the same time, GMI sensors of different structures have been developed such as glass-coated wires, thin films, multi layer thin films, meander structures, ribbons, etc. [45-47]

### 3.2.2. *Wire GMI sensors*

As the first discovered GMI sensor structures, GMI wire sensors have been extensively studied. Based on the classical electromagnetism, the theoretical model of the GMI wire is (Panina et al, 1994) [37]

$$Z = \frac{R_{dc} kr \zeta_0(kr)}{2\zeta_1(kr)}, \tag{12}$$

where

$$k = \frac{1+j}{\delta_m} \tag{13}$$

and

$$\delta_m = \frac{c}{\sqrt{4\pi^2 f \sigma \mu_\emptyset}}. \tag{14}$$

$R_{dc}$ is the dc resistance of the wire, $\zeta_0$, $\zeta_1$ are the Bessel functions, $r$ is the radius of the wire, $j$ is the imaginary unit, $\delta_m$ is the penetration depth, $c$ is the speed of light, $f$ is the frequency of the ac current, $\mu_\emptyset$ is the circumferential magnetic anisotropy. The origin of the GMI effect lies in the dependence of $\mu_\emptyset$ on an axial magnetic field resulting in a change of $\delta_m$. In order to obtain a high GMI ratio, the value of $\delta_m$ has to be close to the thickness of the conductor. Hence, the thinner a ferromagnetic conductor and the lower its permeability, the higher the operation frequency required. A well-defined circumferential magnetic anisotropy in combination with a soft magnetic behavior is desirable, since it will provide a large permeability change for small magnetic fields.

Different amorphous and ferromagnetic materials were used to fabricate GMI wires [48], and various fabrication methods were developed such as melt spinning, in-rotating water spinning, glass-coated melt spinning etc. [45, 49, 50]. Glass-coated micro-wires (Fig. 5) present outstanding properties in terms of the magnetic anisotropy distribution, which is reinforced by the strong mechanical stress induced by the coating. $(Co_xFe_{1-x})_{72.5}Si_{12.5}B_{15}$ is one of the most typical materials. By adjusting x from 0 to 1, the magnetostriction of the material changes from positive at high Fe content to negative at high Co content. Negative magnetostrictive compositions in combination with the compressive, radial stress induced by quenching and the glass coating provide the best results, since it supports a strong circumferential anisotropy.

**Figure 5.** SEM image of a glass-coated amorphous micro-wire (Courtesy of M. Vazquez, Inst. Materials Science of Madrid, CSIC).

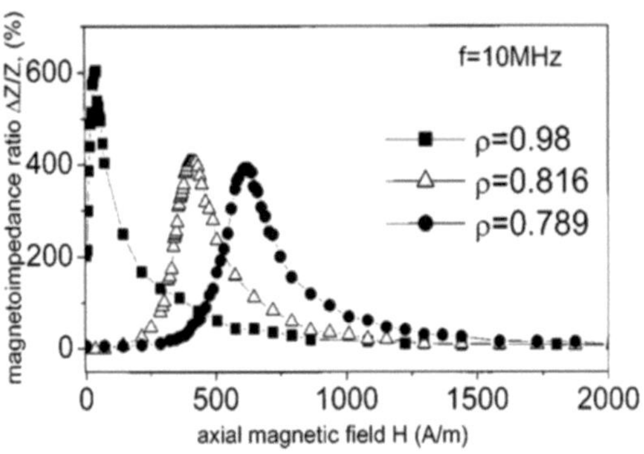


**Figure 6.** GMI ratio of $Co_{67}Fe_{3.85}Ni_{1.45}B_{11.5}Si_{14.5}Mo_{1.7}$ glass-coated wires with different geometric ratio $\rho$ (the metallic nucleus diameter to the total microwire diameter) at 10 MHz.

Wire-type GMI sensors provide the best performance in terms of the GMI ratio with values as high as 615% (Fig. 6) achieved with optimized glass coated microwires (Zhukova et al, 2002) [43]. The value of the magnetic field at which the maximum GMI ratio is obtained increases as the diameter of the magnetic nucleus decreases compared to the diameter of the glass coating. This is attributed to the different anisotropies induced by the stress from the coating. Due to the high sensitivity provided by GMI wires they have been commercialized despite the facts that fabrication is not silicon based, does not use standard microfabrication methods and, as a consequence, integration with electronics is complex.

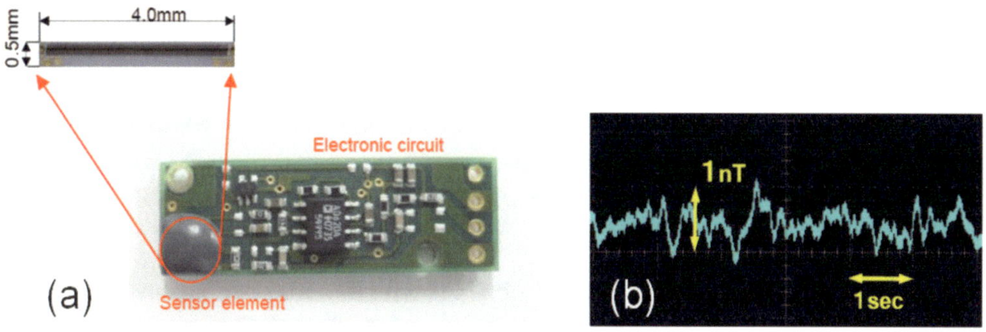


**Figure 7.** (a) Layout of the commercialized GMI sensor from Aichi Steel Co. (b) Noise output of the GMI sensor.

Fig. 7 shows a GMI sensor developed by Aichi Steel Co., which has a very high sensitivity of 1 V/μT and a noise level of 1 nT [51].

### *3.2.3. Ribbon GMI sensors*

Magnetic ribbons discussed in this section are planar structures of rectangular shape with a thickness of a few tens of micrometers and a length and width from several millimeters to centimeters. Similar to the micro-wires, magnetostriction is utilized in order to create certain anisotropies during the ribbon's fabrication. Magnetic ribbons that exhibit a strong GMI effect have a high permeability as well as a transversal magnetic anisotropy.

For a planar film of infinite width, the impedance is given by

$$Z = R_{dc} \cdot \frac{jka}{2} coth(\frac{jka}{2}), \tag{15}$$

where $R_{dc}$ is the dc resistance, $a$ is the thickness of the ribbon, $k$ and $\delta_m$ can be obtained from Equ. (14) with the only difference that $\mu_\phi$ represents the transversal permeability instead of the circumferential one [52].

Again, Fe- and Co-based amorphous alloys are preferably used as the magnetic material. The standard fabrication method for the ribbons is melt spinning, where a rotating copper wheel is used to rapidly solidify the liquid alloy. This method produces magnetic ribbons with a thickness of about 25 μm and a width of several mm. With this thickness, ribbon GMI sensors operate at comparably low frequencies of hundred kHz up to a few MHz. A GMI ratio of, e.g., 640% has been obtained with a GMI ribbon made of $Fe_{71}Al_2Si_{14}B_{8.5}Cu_1Nb_{3.5}$ at 5 MHz [53].

### *3.2.4. Thin film GMI sensors*

In theory, a single layer magnetic thin film is similar to a magnetic ribbon, and the same analytical expressions are applied for modeling the GMI effect. Practically, the main difference is the fabrication method. Thin film fabrication is a standard micro-fabrication technology producing a film thickness of some nanometers up to a few micrometers. Thin film GMI sensors are of great interest due to the advantages arising from the fabrication in terms of the flexibility in design and integration. They can easily be fabricated on the same substrate as the electronic circuit and other devices. In the context of passive and remote sensors, this is particularly relevant, since the GMI element can be easily integrated with an SAW device. For this reason, GMI thin film sensors will be discussed in more detail and our recent results will be presented.

Compared to wires and ribbons, the results obtained with thin film sensors have not been as good, and the highest GMI ratios reported are around 250% [42]. This may be due to the differences in the magnetic softness as well as the magnetic anisotropy, which is very well established in circumferential and transversal direction in wires and ribbons, respectively, and is difficult to control in thin films. In thin films transverse anisotropy is mainly realized through magnetic field deposition or field annealing, Fig. 8 (a) and (b) show the magnetization curve and domain structure of a $Ni_{80}Fe_{20}$ thin film (100nm thick) fabricated under a magnetic field of 200 Oe during deposition. A magnetic easy axis and domain structures in transverse direction are observed. Due to the small thickness, thin film GMI sensors normally operate at a higher frequency from hundred MHz to several GHz where the penetration depth is in the range of the film thickness.

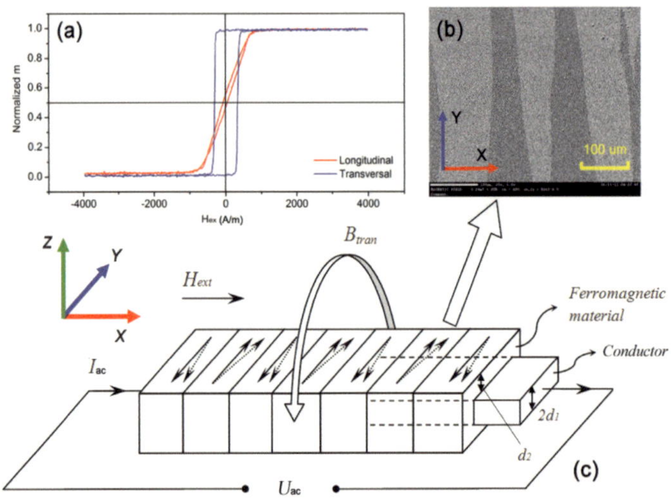


**Figure 8.** (a) Magnetization curves obtained by vibrating sample magnetometry of a magnetic thin film (100nm of $Ni_{80}Fe_{20}$) in transversal and longitudinal directions. (b) Domain pattern of the magnetic layer. (c) Schematic of a typical multi layer GMI structure. The arrows in the ferromagnetic material indicate the magnetization of individual domains (simplified). Upon application of an external magnetic field $H_{ext}$, the magnetization rotates into the direction of $H_{ext}$ (dotted arrows).

In general, a GMI sensor with high sensitivity consists of a stack of several material layers. In case of a tri-layer element, one conducting layer is sandwiched between two magnetic layers as shown in Fig. 8 (c). The conducting layer ensures a high conductivity and, in combination with the highly permeable magnetic layers, a large skin effect is obtained [51,

54]. An alternating current $I_{ac}$ mainly flowing through the conductor generates a transversal flux $B_{tran}$, which magnetizes the magnetic layers. Upon the application of an external field $H_{ext}$ in longitudinal direction, the magnetization caused by $I_{ac}$ will be changed. This is equivalent to a change of the transversal permeability of the magnetic layers and is reflected by an impedance change.

The analytical model of the impedance for a magnetic/conducting/magnetic tri-layer structure is given by

$$Z = R_{dc}(1 - 2j\mu \frac{d_1 d_2}{\delta_c^2}) , \quad (16)$$

where $R_{dc}$ is the dc resistance of the inner conductor, $2d_1$ is the thickness of the conductor, $d_2$ is the thickness of the magnetic layers as shown in Fig. 8 (c) and $\delta_c$ is the penetration depth of the conducting layer [39].

Analytical solutions for the impedance of thin film GMI sensors can only be found for rather simple structures. In order to calculate the impedance of more complicated geometries, for example, a sandwich structure with isolation layers between the conductor and the magnetic layers [41], a meander structure multilayer [46] or to take into account edge effects, the finite element method (FEM) provides a viable solution [55].

Fig. 9 shows the comparison of the GMI ratios simulated for a single magnetic layer, a tri-layer structure made of a magnetic/conducting/magnetic stack and a five-layer structure with isolation layers between the conducting and magnetic layers using the FEM. The simulated GMI sensors have a width of $w$ = 50 μm and length of $l$ = 200μm. The magnetic layers have a thickness of $t_{mag}$ = 1 μm and the conducting layer has a thickness of $t_{met}$ = 4 μm. The material of the isolation layer is $SiO_2$ with a thickness of 1μm. The conductivity of the ferromagnetic and conducting layers are $7.69\times10^5$ S/m ($(CoFe)_{80}B_{20}$) and $4.56\times10^7$ S/m (Gold), respectively. All parameters including $M_s$ = $5.6\times10^5$ A/m, $\gamma$ = $2.2\times10^5$ m/(A·s), $\alpha$ = 0.3, $H_a$= 1890 A/m and are taken from literature [56].

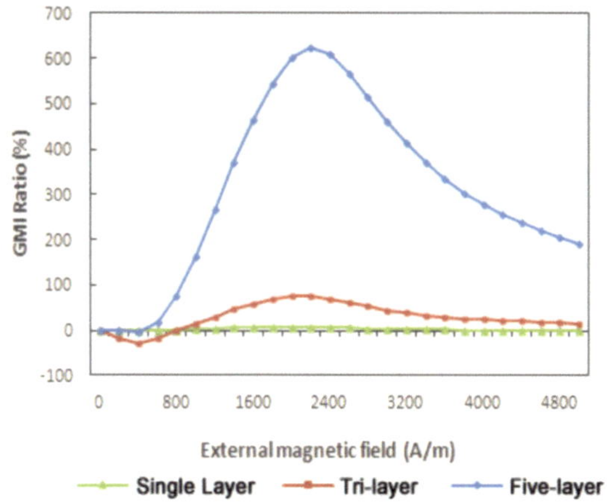


**Figure 9.** Simulated GMI ratios of single layer, sandwiched multilayer and isolated sandwiched multilayer structures.

The results clearly show the performance increase achieved with the multilayer structures. Specifically, the isolated sandwich structure has a superior performance, which is due to preventing the current from flowing in the magnetic layer.

For the fabrication of thin film GMI sensors, Co-based and Fe-based amorphous magnetic alloys were used in earlier studies. Recently, permalloy, which is a NiFe compound, became popular as it provides very high permeability, zero magnetostriction and simple fabrication. Meander shaped multilayers and different stacks of magnetic and conductive layers using permalloy were developed. Some results are summarized in Table 2.

| Year | Material | Frequency | GMI Ratio (%) | Sensitivity (%/Oe) | Reference |
|---|---|---|---|---|---|
| 1999 | FeNiCrSiB/Cu/FeNiCrSiB | 13MHz | 77 | 2.8 | [57] |
| 2000 | FeSiBCuNb/Cu/FeSiBCuNb | 13MHz | 80 | 2.8 | [58] |
| 2004 | $Ni_{81}Fe_{19}$/Au/$Ni_{81}Fe_{19}$ | 300MHz | 150 | 30 | [59] |
| 2004 | ($Ni_{81}Fe_{19}$/Ag)n | 1.8GHz | 250 | 9.3 | [42] |
| 2005 | FeCuNbSiB/$SiO_2$/Cu/$SiO_2$/FeCuNbSiB | 5.45MHz | 33 | 1.5 | [60] |
| 2011 | NiFe/Ag/NiFe | 1.8GHz | 55 | 1.2 | [61] |
| 2011 | NiFe/Cu/NiFe | 20MHz | 166 | 8.3 | [62] |

**Table 2.** Recent results on thin film GMI sensors.

GMI thin film sensors not only offer the advantages of standard microfabrication and straight-forward integration with SAW devices, but, as can be seen from Table 2, the operation frequency of GMI thin film sensors is also compatible with the one of SAW devices (usually from hundred MHz to several GHz and can be adjusted within a wide range.

### *3.2.5. Integrated SAW-GMI sensor*

In the first studies, SAW transponders and GMI wire sensors were combined to form remote devices [14-16]. GMI wires were selected for their high sensitivity, and they were bonded to the output IDT of the SAW device, which operated as a reflector, in order to act as load impedance. The strong dependence of the impedance on magnetic fields causes a considerable amplitude dependence of the reflected signal on magnetic fields. Even though these studies provided good results for passive and remote magnetic field sensors, the fabrication method for the GMI wires, which is not compatible with standard microfabrication, is a considerable problem with respect to reproducibility and costs, hence, hindering commercial success of such devices. In order to conquer this problem, a fully integrated SAW-GMI design utilizing standard microfabrication processes is required. The most viable option is a thin film GMI sensors for the following reasons:

1. Thin film GMI sensors can be produced by the same metallization processes as the SAW transponders and on the same substrate.

2. Standard photolithography technique guarantees an accurate and reproducible alignment of the two devices.
3. Thin film GMI sensors provide a wide range of working frequencies up to GHz, which matches the high frequency requirement of the SAW transponders.
4. Thin film GMI sensor can have a minimized and flexible design as well as large magnetic field sensitivity.

In this section, a detailed description of our recent work on the design, fabrication and testing of an integrated SAW-GMI sensor is presented.

**Design**

Fig. 10 shows a schematic of a GMI thin film sensor integrated with a SAW transponder. A wireless signal applied to the source IDT (IDT1) is converted to an SAW and propagates towards the other end of the substrate, where it is reflected from the reference IDT (IDT2) and the load IDT (IDT3). The reflected waves containing the reference and load information are received by IDT1 at different time instants and reconverted to a wireless electrical signal sent out via the antenna.

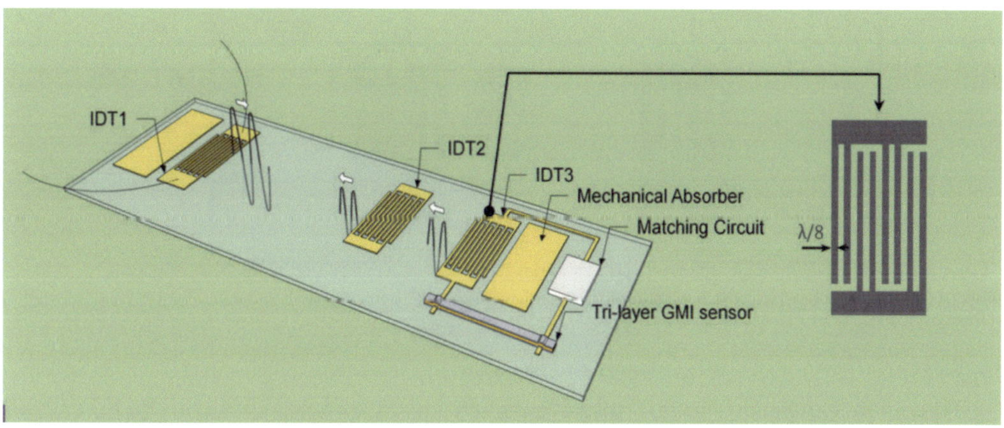


**Figure 10.** Schematic of an integrated passive and remote magnetic field sensor consisting of a SAW transponder and thin film GMI sensor.

In order to obtain high magnetic field sensitivity, the GMI sensor is matched to the output port (IDT3) at the working frequency of the SAW device. As the impedance of the GMI sensor changes with an applied magnetic field, the matching deteriorates, which causes the amplitude of the signal reflected from IDT3 to change. Since the piezoelectric material is sensitive to environmental changes, e.g. temperature, a reference IDT is used to provide a signal that enables the suppression of such noise by means of signal processing. Two metallic pads next to the input and output IDTs act as mechanical absorbers and suppress reflections from other structures on the substrate or the edge of the substrate.

Matching the sensor load to the optimal working point of IDT3 is a crucial aspect in the device design. Therefore, the influence of the load on the signal reflected from IDT3 is simulated. The interaction of a SAW with an IDT can be described by the P-matrix model introduced by Tobolka [63]. As shown in Fig. 11, $P_{11}$ is the acoustic wave reflection at the output IDT [24]. Specifically, the dependence of $P_{11}$ on the load impedance $Z = Z(H_{ext}) + Z_m$, where $Z(H_{ext})$ is the impedance of the GMI element and $Z_m$ is the matching impedance, is expressed as

$$P_{11}(Z) = P_{11,sc} + \frac{2 \cdot P_{13}^2}{P_{33} + \frac{1}{Z}}, \tag{17}$$

where $P_{11,sc}$ is the short circuit reflection coefficient, $P_{13}$ is the electro-acoustic transfer coefficient and $P_{33}$ is the input admittance of the transducer. In order to have a large change of $P_{11}$, which is equivalent to the sensitivity of the SAW device loaded by an impedance sensor, the influence of Z in equation (17) needs to be large. Therefore, a SAW transducer with a small $P_{11,sc}$ and large $P_{13}$ will provide a large sensitivity. $P_{11,sc}$ can be minimized by using a double electrode IDT design as shown in Fig. 10, which provides cancelation of the internal mechanical reflections of the IDT.

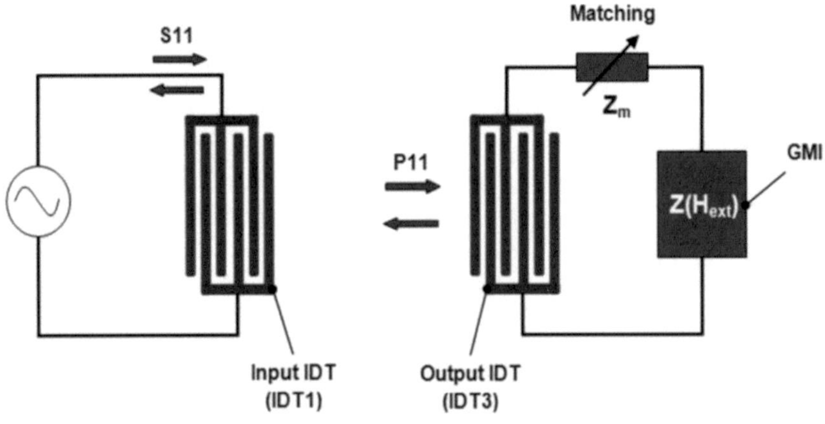


**Figure 11.** Electric and acoustic ports of the SAW sensor

The electro-acoustic transfer coefficient $P_{13}$ and input admittance $P_{33}$ can be obtained by,

$$P_{13} = \frac{1}{2\sqrt{2Z_a}} r_m (1 - e^{-j\varphi_m}) \tag{18}$$

$$P_{33} = j\omega C_{IDT} + \frac{r_m^2}{Z_a}(1 - e^{-j\varphi_m}) \tag{19}$$

Where $r_m$ is the ratio of the electrical to acoustical transformer, $C_{IDT}$ is the capacitance of the IDT, Za is the acoustic impedance and $\varphi_m$ is the transit angle [63].

Since the GMI sensor is an inductive element, matching is accomplished by a series capacitance resulting a load impedance

$$Z = 1/j\omega C_m + R + j\omega L(H_{ext}), \tag{20}$$

where $C_m$ is the matching capacitance, $R$ is the average resistance (over the considered magnetic field range) of the GMI sensor and $L(H_{ext})$ the inductance of the GMI sensor.

Fig. 12 (a) shows the simulation result of the IDT's reflectivity as a function of the load. The slope of this plot corresponds to the magnetic field sensitivity. Therefore, the optimum matching capacitance can be determined. Fig. 12 (b) presents the rate of change of $P_{11}$ for 1nH inductance changes (corresponding to a field change of approximately 50A/m). The result shows that with the optimum matching capacitance a maximum reflectivity change rate of 0.3dB/nH can be achieved. As the fabricated GMI sensor has an inductance change from 5nH to 15nH, a reflectivity change of 3dB can be expected.

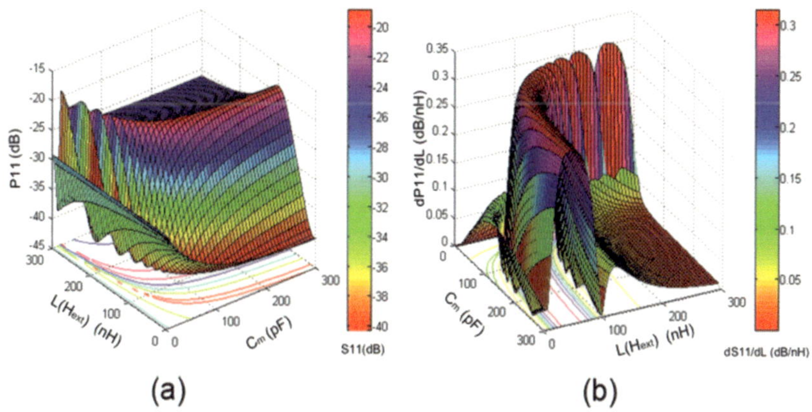


**Figure 12.** (a) Magnitude $P_{11}$ as a function of the matching capacitance and sensor inductance. (b) Rate of change (absolute value) of $P_{11}$ for 1nH load inductance change.

The piezoelectric substrate chosen for this application is $LiNbO_3$ as it provides a strong electromechanical coupling corresponding to a high value of $P_{13}$. The detailed design parameters of the SAW transponder are shown in Tab. 3. The working frequency of the device is 80MHz, resulting in a periodicity $p$ of 50μm (Equ. (1). The value of $p$ determines the electrode width and gap. The distances between the IDTs yield a 1.25μs delay between IDT1 and IDT2 and a 0.625μs delay between IDT2 and IDT3.

| Design parameter | | Design parameter | |
|---|---|---|---|
| Substrate material | $LiNbO_3$ (128 deg. Y-X cut) | Electrode material | Gold |
| Center frequency | 80MHz | Aperture | 30λ |
| Periodicity | 50μm | Electrode/gap width | 6.25μm |
| Electrodes per segment | 4 | IDT segment number | 30 |

**Table 3.** Design parameters for the SAW device.

The GMI sensor consists of a tri-layer structure with two ferromagnetic layers of 100nm in thickness made of $Ni_{80}Fe_{20}$ and a conducting copper layer with a thickness of 200nm. The sensor has a rectangular geometry of 100 μm× 4000 μm. The conducting layer is connected to the IDT3 [18].

**Fabrication**

The fabrication of the combined device is accomplished in several steps as shown in Fig. 13. On a $LiNbO_3$ wafer, a 40 nm Ti adhesion layer and 200 nm gold layer are sputter deposited and patterned by ion milling into individual SAW devices. The leads and SMD footprints are designed together with the SAW device to facilitate an on-chip impedance matching circuit, which was accomplished by a 150pF capacitor connected in series with the GMI element. The GMI element comprises a tri-layer structure ($Ni_{80}Fe_{20}$(100nm) / Cu(200nm) / $Ni_{80}Fe_{20}$(100nm)) deposited at room temperature with a uniaxial magnetic field of 200 Oe applied in the transversal direction.

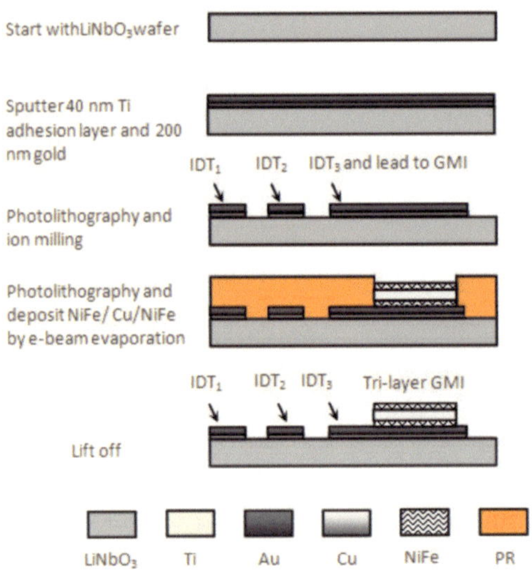


**Figure 13.** Fabrication flow chart of the integrated SAW-GMI device.

**Results**

A network analyzer (Agilent E8363C) is used to apply an RF signal to IDT1 and measure the electric reflection coefficient ($S_{11}$) of the input IDT, which is related to the admittance matrix of the whole device and $P_{11}$. The time domain signal of $S_{11}$ is converted from the frequency domain using fast Fourier transform. As shown in Fig. 14 (a), two reflection peaks at 2.45μs and 3.55μs are observed indicating the reflections from the reference IDT and the load IDT accordingly. The magnetic response of the integrated device is determined by applying a variable magnetic field in longitudinal direction to the device. A 2.4dB amplitude change of the reflection signal can be observed. A comparison of the simulated and experimental results together with the measured GMI ratio curve is shown in Fig. 14 (b).

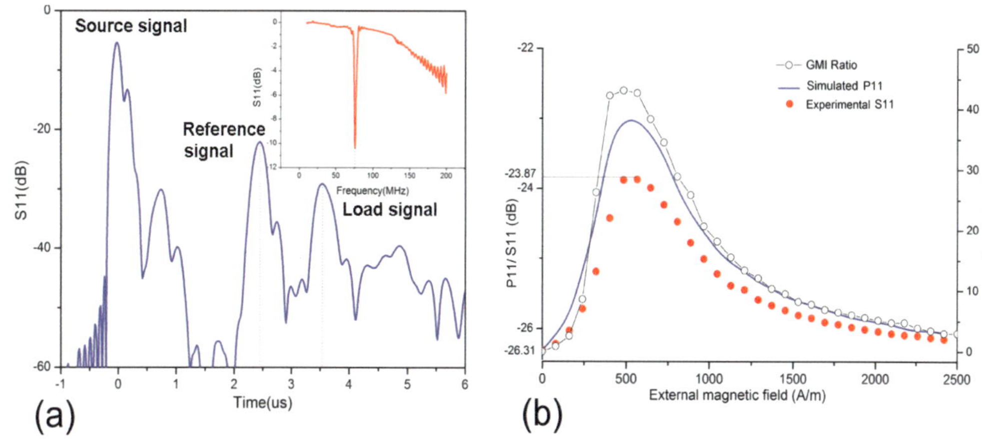


**Figure 14.** (a) Time domain measurement of the SAW-GMI device. Inset: Frequency domain measurement. (b) Comparison of the simulated and experimental device response together with the measured GMI ratio curve.

# 4. Potential applications

Magnetic field sensors are one of the most widely used sensors and employed for many different applications. Current commercial magnetic sensors are wire connected to a circuit providing power and readout. These wire connections prevent the sensors from being used for certain applications. In addition, as the complexity and the number of devices, increases in modern systems such as automobiles, wire connections are becoming an increasing problem due to limited space. For those and other reasons, wireless solutions are being much sought after.

As pointed out in the previous sections, SAW-based sensors have been developed for different applications, and this technology also provides a platform for realizing wireless and passive magnetic sensors. They can provide solutions for various applications, for example, where the sensors have to withstand harsh environmental conditions or are required to have a long lifetime without maintenance.

Out of the countless applications for SAW-based passive and remote magnetic field sensors, a few will be highlighted in the following.

Nanotechnology and miniaturized systems are becoming increasingly popular in the biomedical field. Technologies based on magnetic effects are of particular interest since they can be controlled remotely via magnetic fields. For example, NVE Corporation recently developed a battery operated magnetic sensor to be used as a magnetic switch for implantable devices. When a magnetic field is applied, the sensor turns on triggering a specified action. It turns off when the field is removed. The sensor works at a stable operating point of 15 Oe [64]. Magnetic beads have been extensively used in many biomedical applications. These magnetic beads are made of ferromagnetic material ranging in size between 5 nm to 500 um. A new application of such particles promises benefits in cancer therapy by employing the particles either as drug carriers or heat sources (hyperthermia) [65]. In order to have better control of the treatment, magnetic sensors are considered to measure and detect the concentration of these magnetic particles.

The automotive industry extensively uses magnetic sensors for different purposes, for example, to measure current in electric vehicles [66] or the rotation speed of gears [67]. Another application employs magnetic sensors to detect passing vehicles using lane markers [68]. Such a system could also be used to detect vehicle speed by measuring the time between two markers of a fixed distance. In yet another application, developed by Stendl et al [69], the wear of a vehicle's tire is detected by measuring the field of magnetic beads embedded in the rubber of tire treads. As the tread size decreases, the magnetic field also decreases. A wireless magnetic sensor is placed just below the threads.

Construction monitoring is an upcoming application for wireless sensor. Long-term monitoring of metallic reinforcements in, e.g., bridges or buildings requires passive and remote sensors, which are capable of detecting changes of the metal. Similarly, the detection of internal defects or corrosion of pipelines is of great interest. Gloria et al [70] developed an Internal Corrosion Sensor (ICS) consisting of a magnet and a Hall sensor. A disturbance in the magnetic field caused by changes of the metal changes the sensor readout. This information is used for both to detect and size the defects.

## 5. Conclusion

In this chapter we discussed different types of SAW-based, magnetic sensors including resonators, phase shifters and loaded transponders. Sensitivity to magnetic fields can be achieved by either changing the properties of the IDT or delay line utilizing magnetostrictive materials or loading the output IDT with a magnetic field sensor. GMI sensors feature a very high sensitivity and wide range of operating frequencies and, therefore, constitute an especially suitable load. The principle of GMI sensors is described in detail and different GMI structures are discussed. While the highest sensitivity has been obtained with GMI microwires, thin film GMI sensor are advantageous because they can be produced using standard microfabrication methods, and they can be easily integrated with a SAW transponder on the same substrate. These features are crucial with respect to production complexity and costs.

A SAW transponder combined with a GMI element connected to the output IDT is a passive and remote magnetic field sensor, which responds to an interrogation signal with a delayed response signal. The design of such a device needs to take into account different aspects like operation frequency, dimensions of IDTs and delay line or matching the load with the output IDT. In order to obtain a high sensitivity, an impedance change of the GMI element caused by a magnetic field, has to yield a large change in the SAW reflected from the output IDT. A model is presented to simulate the electro-acoustic interaction of the output IDT with the GMI sensor's impedance and the impedance matching capacitance. The simulation results provide information regarding the matching parameters and are invaluable for obtaining an optimized performance. A detailed description of the fabrication of an integrated SAW-GMI sensor is provided using standard microfabrication technologies. The GMI ratio of the fabricated sensor is 45 % The SAW-GMI sensor provides a sensitivity of 3 dB/mT, and its output corresponds well with the simulation results.

Magnetic field sensors have countless applications and are widely used in many different fields. The trend towards wireless operation, which is generally observed nowadays, drives the development of passive and remote magnetic field sensors. Several concepts of such sensors employing SAW devices have been presented in this chapter. The most promising one is a SAW-GMI sensor, which has been discussed in detail and which features wireless and battery less operation as well as durability and the ability to withstand harsh environments. This kind of sensor is considerable not only for providing existing applications with a wireless mode; it also largely expands the potential applications of magnetic field sensors.

## Author details

Bodong Li and Jürgen Kosel
*Electrical Engineering Department, King Abdullah University of Science and Technology, Thuwal, Saudi Arabia*

Hommood Al Rowais
*Electrical Engineering Department, Georgia Institute of Technology, Atlanta, Georgia, USA*

## 6. References

[1] Stockman H (1948) Communication by means of reflected power, Proceedings of IRE. 36: 1196-1204

[2] Plath F, Schmeckebier O, Rusko M, Vandahl T, Luck H, Moller F, Malocha D C (1994) Remote sensor system using passive SAW sensors. ULTRASON. 1: 585-588

[3] Reindl L, Scholl G, Ostertag T, Scherr H, Wolff U, Schmidt F (1998) Theory and application of passive SAW radio transponders as sensors. IEEE Trans. Ultrason., Ferroelectr., Freq. Control. 45: 1281-1292

[4] Reeder T M and Cullen D E (1976) Surface-acoustic-wave pressure and temperature sensors. Proceedings of the IEEE. 64: 754-756

[5] Lee K, Wang W, Kim G and Yang S (2006) Surface Acoustic Wave Based Pressure Sensor with Ground Shielding over Cavity on 41° YX $LiNbO_3$. Jpn. J. Appl. Phys. 45: 5974–5980

[6] Pohl A (1997) Wirelessly interrogable surface acoustic wave sensors for vehicular applications. IEEE T Instrum Meas. 46: 1031-1038

[7] Calabrese G S, Wohltjen H, Roy M K (1987) Surface acoustic wave devices as chemical sensors in liquids: Evidence disputing the importance of Rayleigh wave propagation, Anal. Chem. 59: 833–837

[8] Kadota M, Ito S, Ito Y, Hada T, And Okaguchi K (2011) Magnetic Sensor Based on Surface Acoustic Wave Resonators. Jpn. J. Appl. Phys. 50: 07HD07

[9] Hanna S M (1987) Magnetic Field Sensors Based on SAW Propagation in Magnetic Films. IEEE Trans. Ultrason., Ferroelectr., Freq. Control. UFFC-34: 191-194

[10] http://www.asrdcorp.com

[11] http://www.senseor.com

[12] http://www.transense.co.uk

[13] Ganguly A K, Davis K L, Webb D C, Vittoria C, Forester D W (1975) Magnetically tuned surface-acoustic-wave phase shifter. *Electronics Letters*. 11: 610-611

[14] Hauser H, Steindl R, Hausleitner Ch, Pohl A, Nicolics J (2000) Wirelessly Interrogable Magnetic Field Sensor Utilizing Giant Magneto-Impedance Effect and Surface Acoustic Wave Devices. IEEE T INSTRUM MEAS. 49: 648-652

[15] Steindl R, Hausleitner Ch, Pohl A, Hauser H, Nicolics J (2000) Passive wirelessly requestable sensors for magnetic field measurements. Sens Actuators A. 85: 169-174

[16] Hauser H, Steurer J, Nicolics J, Musiejovsky L, Giouroudi I (2006) Wireless Magnetic Field Sensor. Journal of Electrical Engineering 57: 9-14

[17] Al Rowais H, Li B, Liang C, Green S, Gianchandani Y, Kosel J (2011) Development of a Passive and Remote Magnetic Microsensor with Thin-Film Giant Magnetoimpedance Element and Surface Acoustic Wave Transponder. J. Appl. Phys. 109: 07E524

[18] Li B, M H. Salem N P, Giouroudi I, Kosel J (2011) Integration of Thin Film Giant Magneto Impedance Sensor and Surface Acoustic Wave Transponder. J. Appl. Phys. 111: 07E514

[19] Sinha B K, Tiersten H F (1979) Zero temperature coefficient of delay for surface waves in quartz. Appl. Phys. Lett. 34: 817

[20] Ebata Y, Suzuki H, Matsumura S and Toshiba K F (1982) SAW Propagation Characteristics on $Li_2B_4O_7$. Jpn. J. Appl. Phys. 22: 160-162

[21] Tsubouchi K.Sugai K, Mikoshiba N (1982) Zero Temperature Coefficient Surface-Acoustic-Wave Devices Using Epitaxial AlN Films. Ultrasonics Symposium, 340 – 345

[22] Dewan N, Tomar M, Gupta V, and Sreenivas K (2005) Temperature stable LiNbO3 surface acoustic wave device with diode sputtered amorphous TeO2 over-layer. Appl. Phys. Lett. 86: 223508

[23] Viens M, Cheeke J D N (1990) Highly sensitive temperaturesensor using SAW resonator oscillator. Sensors and Actuators A: Physical 24: 209-211

[24] Wang W, Lee K, Woo I, Park I, Yang S (2007) Optimal design on SAW sensor for wireless pressure measurement based on reflective delay line. Sensors and Actuators A: Physical. 139: 2-6

[25] Schimetta G, Dollinger F, Scholl G, Weigel R (2001) Optimized design and fabrication of a wireless pressure and temperature sensor unit based on SAW transponder technology. Microwave Symposium Digest, 2001 IEEE MTT-S International. 1: 355-358

[26] Lim C, Wang W, Yang S, Lee K (2011) Development of SAW-based multi-gas sensor for simultaneous detection of CO2 and NO2, Sensors and Actuators B: Chemical. 154: 9-16

[27] Reindl L, Ruppel C C W, Kirmayr A, Stockhausen N, Hilhorst M A, and Balendonck J (2001) Radio-Requestable Passive SAW Water-Content Sensor, IEEE Transactions on Microwave Theory And Techniques. 49: 803

[28] Wang S Q, Harada J and Uda S (2003) A Wireless Surface Acoustic Wave Temperature Sensor Using Langasite as Substrate Material for High-Temperature Applications. Jpn. J. Appl. Phys. 42: 6124

[29] Wu C (2006) Fabrication of Surface Acoustic Wave Sensors for Early Cancer Detection, Electrical Engineering, University of California, Los Angeles

[30] Knobel M, Vazquez M, and Kraus L, Buschow ed. K.H.J. (2003) Giant Magnetoimpedance, Handbook of magnetic materials. 15: 497-563

[31] Sezen A S, Sivaramakrishnan S, Hur S, Rajamani R, Robbins W, and Nelson B J (2005) Passive Wireless MEMS Microphones for Biomedical Applications, J. Biomech. Eng. 127: 1030

[32] Wolff U, Schmidt F, Scholl G, Magori V (1996) Radio accessible SAW sensors for non-contact measurement of torque and temperature. Ultrasonics Symposium Proceedings. 1: 359-362

[33] Yoshizawa N, Yamamoto I, and Shimada Y (2005) Magnetic Field Sensing by an Electrostrictive/Magnetostrictive Composite Resonator. IEEE TRANSACTIONS ON MAGNETICS 41: 11

[34] Dong S, Zhai J, Li J, and Viehland D (2006) Small Dc Magnetic Field Response Of Magnetoelectric Laminate Composites, Applied Physics Letters 88: 082907

[35] Koeninger V, Matsumura Y, Uchida H H, Uchida H (1994) Surface acoustic waves on thin films of giant magnetostrictive alloys. J ALLOY COMPD 211/212: 581-584

[36] Uchida H, Wada M, Koike K, Uchida H H, Koeninger V, Matsumura Y, Kaneko H, Kurino T (1994) Giant magnetostrictive materials: thin film formation and application to magnetic surface acoustic wave devices. J ALLOY COMPD 211/212: 576-580

[37] Panina L V and Mohri K (1994) Magneto-impedance effect in amorphous wires. Appl.Phys.Lett. 65: 1189-1191

[38] Machado F L A, Rezende S M (1996) A theoretical model for the giant magnetoimpedance in ribbons of amorphous soft-ferromagnetic alloys. Journal of Applied Physics. 79: 6558 - 6560

[39] Hika K, Panina L V, Mohri K (1996) Magneto-Impedance in Sandwich Film for Magnetic Sensor Heads. IEEE Trans Magn. 32: 4594-4596.

[40] Panina L V, Makhnovskiy D P, Mapps D J, and Zarechnyuk D S (2001) Two-dimensional analysis of magnetoimpedance in magnetic/metallic multilayers. J. Appl. Phys. 89: 7221

[41] Morikawa T, Nishibe Y, Yamadera H, Nonomura Y, Takeuchi M, Taga Y (1997). Giantmagneto-Impedance Effect in Layered Thin Films. IEEE Trans Magn. 33: 4367-4372

[42] de Andrade A M H, da Silva R B, Correa M A, Viegas A D C, Severino A M, Sommer R L (2004) Magnetoimpedance of NiFe/Ag multilayers in the 100 kHz–1.8 GHz range. Journal of Magnetism and Magnetic Materials 272–276: 1846–1847

[43] Zhukova V, Chizhik A, Zhukov A, Torcunov A, Larin V and Gonzalez J (2002) Optimization of giant magnetoimpedance in Co-rich amorphous microwires. IEEE Trans. Magn. 38: 3090-3092

[44] Le A T, Phan M H, Kim C O, Vázquez M, Lee H, Hoa N Q and Yu S C (2007) Influences of annealing and wire geometry on the giant magnetoimpedance effect in a glass-coated microwire LC-resonator, J. Phys. D: Appl. Phys. 40: 4582–4585

[45] Vázquez M, Adenot-Engelvin A L (2009) Glass-coated amorphous ferromagnetic microwires at microwave frequencies. Journal of Magnetism and Magnetic Materials. 321: 2066–2073

[46] Zhou Z, Zhou Y, Chen L (2008) Perpendicular GMI Effect in Meander NiFe and NiFe/Cu/NiFe Film. IEEE Transactions on Magnetics. 44: 2252 - 2254

[47] Pompéia F, Gusmão L A P, Hall Barbosa C R, Costa Monteiro E, Gonçalves L A P and Machado F L A (2008) Ring shaped magnetic field transducer based on the GMI effect. Meas. Sci. Technol. 19: 025801

[48] Phan M H , Peng H X (2008) Giant magnetoimpedance materials: Fundamentals and applications. Progress in Materials Science. 53: 323–420

[49] Squire PT, Atkinson D, Gibbs M R J, Atalay S J (1994) Amorphous wires and their applications. J Magn MagnMater. 132: 10–21

[50] Ohnaka I, Fukusako T, Matui T (1981) Preparation of amorphous wires. J. Jpn. Inst. Met. 45: 751–62.

[51] http://www.aichi-steel.co.jp/ENGLISH/pro_info/pro_intro/elect_3.html

[52] Panina L V, Mohri K, Uchiyama T, Noda M (1995) Giant magneto-impedance in Co-rich amorphous wires and films. IEEE Trans Magn. 31:1249–60

[53] Phan MH, Peng HX, Yu SC, Vazquez M (2006) Optimized giant magnetoimpedance effect in amorphous and nanocrystalline materials. J Appl Phys. 99: 08C505

[54] Sukstanskii A L and Korenivski V (2001) Impedance and surface impedance of ferromagnetic multilayers: the role of exchange interaction J. Phys. D 34: 3337

[55] Li B, Kosel J (2001) 3d Simulation of GMI Effect In Thin Film Based Sensors. J. Appl. Phys. 109: 07E519

[56] Dong C, Chen S, Hsu T Y (Xu Zuyao) (2003) A modified model of GMI effect in amorphous films with transverse magnetic anisotropy. J. Magn. Magn. Mater. 263: 78-82

[57] Xiao S, Liu Y, Dai Y, Zhang L, Zhou S, and Liu G (1999) Giant Magnetoimpedance Effect in Sandwiched Films. J. Appl. Phys. 85: 4127

[58] Xiao S, Liu Y, Yan S, Dai Y, Zhang L, and Mei L (2000) Giant Magnetoimpedance and Domain Structure in FeCuNbSiB Films and Sandwiched Films. Phys. Rev. B 61: 5734–5739

[59] de Cos D, Panina L V, Fry N, Orue I, Garcia-Arribas A, Barandiaran J M (2005) Magnetoimpedance in narrow NiFe/Au/NiFe multilayer film systems. IEEE Transactions on Magnetic. 41: 3697 - 3699

[60] Li X, Yuan W, Zhao Z, Ruan J and Yang X (2005) The GMI effect in Nanocrystalline FeCuNbSiB Multilayered Films with a SiO2 Outer Layer. J. Phys. D: Appl. Phys. 38: 1351–1354

[61] Corrêa M A, Bohn F, Escobar V M, Marques M S, Viegas A D C, Schelp L F, and Sommer R L (2011) Wide Frequency Range Magnetoimpedance in Tri-layered Thin NiFe/Ag/NiFe Films: Experiment and Numerical Calculation. J. Appl. Phys. 110: 093914

[62] Zhou Z, Zhou Y, Chen L and Lei C (2011) Transverse, Longitudinal and Perpendicular Giant Magnetoimpedance Effects in a Compact Multiturn Meander NiFe/Cu/NiFe Trilayer Film Sensor. Meas. Sci. Technol. 22: 035202

[63] Tobolka G (1979) Mixed matrix representation of SAW transducers. IEEE Trans. Sonics Ultrason. SU-26

[64] http://www.medicalelectronicsdesign.com/products/nanopower-magnetic-sensors-fit-implantable-devices

[65] Corchero J L, Villaverde A (2009) Biomedical Applications of Distally Controlled Magnetic Nanoparticles. Trends in Biotechnology 27: 468-476

[66] Ripka P (2008) Sensors based on bulk soft magnetic materials: Advances and challenges. Journal of Magnetism and Magnetic Materials. 320: 2466-2473

[67] Lenz J, Edelstein S (2006) Magnetic sensors and their applications. IEEE Sensors Journal. 6: 631-649

[68] Nishibe Y, Ohta N, Tsukada K, Yamadera H, Nonomura Y, Mohri K, Uchiyama T (2004) Sensing of passing vehicles using a lane marker on a road with built-in thin-film MI sensor and power source. IEEE Transactions on Vehicular Technology. 53: 1827- 1834

[69] Steindi R, Hausleitner C, Hauser H, Bulst W (2000) "Wireless magnetic field sensor employing SAW-transponder," Applications of Ferroelectrics. Proceedings of the 2000 12th IEEE International Symposium on. 2: 855-858

[70] Gloria N B S, Areiza M C L, Miranda I V J, Rebello J M A (2009) Development Of A Magnetic Sensor for Detection And Sizing of Internal Pipeline Corrosion Defects, NDT & E International. 42: 669-677

Chapter 16

# Electromagnetic and Acoustic Transformation of Surface Acoustic Waves and Its Application in Various Tasks

Sergey E. Babkin

Additional information is available at the end of the chapter

http://dx.doi.org/10.5772/56307

## 1. Introduction

### 1.1. Electromagnetic and acoustic transformation

Electromagnetic and acoustic transformation (EMAT) is transformation of high-pitched electromagnetic oscillations in the inductive sensor over a specimen into acoustic oscillations in the specimen. For the transformation to be performed the padding constant magnetic field is required. This process is referred to as a direct EMAT. Further oscillations are extended in the specimen in the shape of acoustic waves. Acoustic waves can be deduced outside by means of revertive EMAT when acoustic oscillations in the surface layer of a specimen will be transformed to electromagnetic oscillations in the receiving sensor. The overall process is as follows: a direct EMAT, a distribution of acoustic waves and a revertive EMAT, which is in practice referred to as a double EMAT or just EMAT [1, 2].

Materials in which EMAT is possible to occur:

- ferromagnetics, i.e. substances which possess electrical conductivity, magnetic properties, and magnetostriction (e.g. iron, nickel, steels);
- the conductors which have no essential magnetic properties (e.g. copper, aluminum);
- materials possessed at least one of following properties: magnetism, conduction, magnetostriction (ferrite, amorphous, rare-earth materials).

*EMAT mechanisms*

EMAT may proceed by three basic mechanisms.

1. A vortical current mechanism (Lorentz force mechanism). Electrons in a surface layer will fluctuate under the influence of Lorentz force:

$$F_L = f\left(J, H\right),$$

where J is a current density, H is a magnetic field.

In conductors this thin layer is called as a skin layer.

2. A magnetic mechanism. This mechanism defines power interaction of an alternative electromagnetic field of the sensor, h, and a constant magnetic field of a specimen, H.

$$F_m = f\left(h,\ H,\ \mu_{ij}\right),$$

where $\mu_{ij}$ is a tensor of magnetic conductivity of a material.

3. A magnetostriction mechanism. For the materials possessing magnetostriction, this mechanism is responsible for changing linear dimensions of microvolumes in the surface layer of a specimen under the sensor depending on the alternative field of the sensor.

$$F_\lambda = f\left(Q_{ij},\ M\right),$$

where Qij is a tensor of magnetoelastic communication, M is magnetization of a specimen.

The first and the second mechanisms are often considered as one which is referred to as an electromagnetic or electrodynamic mechanism of transformation.

*Technical realization of EMAT*

In technical realization terms EMAT is defined a non-contact method of generation and reception of acoustic waves (ultrasonic waves). The method of generation and reception of ultrasonic waves by means of piezoelectric transducers (PET) appears to be the nearest and best analog in this field. Therefore, the technical methods developed for PET methods generally are appropriate for EMAT methods as well.

There are two main techniques, i.e. a resonance technique and a pulse technique.

1. A resonance technique. A small specimen (a cylinder or a rectangular parallelepiped) is placed into solenoid in order to create a magnetic field. By means of a round wire EMA coil which is put on the specimen the loose oscillations are generated in it. A standing waves resonance is reached by changing the frequency of generation. Thus receiving EMA coil shows a signal maximum. A resonance technique is hard to realizing for a PET method, because the contact is required. These problems are easily solved by the non-contact EMAT method.

An amplitude of the received signal, a resonance frequency, a size of a magnetic field are the key information parameters. The attenuation and Q-factor of the system is possible to define.

2. A pulse technique. Short electric pulses of the generator are transformed to acoustic waves by means of EMAT. These pulses are of high-pitched filling. The generation

EMAT sensor can be to used as a revertive EMAT sensor, if the sensor receives a reflected signal.

Each technique developed for a PET method are suitable for pulse EMAT (an echo method, a shadow method etc.).

Key informational parameters of the pulse technique: amplitude of the received signal, speed of a wave, size of a magnetic field, wave attenuation.

## 2. The surface acoustic waves

Classically the term «the surface acoustic waves» (SAW) is considered to involve the extending of the waves along the surface of the solid and vacuum; waves poorly fade extending along the surface boundary, the waves quickly fade when moving away from surface boundary into a solid. It concerns only Rayleigh waves and the Guljaev-Blyushteyna waves but in piezocrystals.

There is an expanded treatment of the term «the surface acoustic waves» [3]. Firstly, surface waves at the boundary of the solid body and the gases are considered. It allows Stonly waves and the waves of leak being referred to SAW. They have vertical polarization and a quasyrayleigh structure. Secondly, the expanded treatment considers waves in a layer on the solid body surfaces, or waves with the inhomogeneous elastic properties which are observed in a closely to a surface layer. Loves waves and generalized Lamb waves are considered as well. Thirdly, if the wave extending along the surface is thought to be the main factor, the waves in plates, i.e. Lamb waves and horizontal polarization waves (SH-wave) can be referred to surface waves.

## 3. Electromagnetic and acoustic transformation of surface acoustic waves

The surface waves in a solid body can be generated in the different ways. The most widespread way is the way using PET. The main advantages of the EMA method of SAW generation in comparison with PET consist of a) not – contact generation and SAW receiving and b)availability to use different SAW which are hardly generated by PET. To reach this purpose it is enough to change a configuration of the wire EMA coil as well as its orientation regarding the magnetic field.

From the practical point of view Rayleigh and Lamb waves being waves vertical polarization as well as SH–waves and Love waves being waves of horizontal polarization are of great interest.

1. Waves of vertical polarization.

The Rayleigh wave can be generated and received by means of meandr EMAT at the normal and tangential magnetic field, $H_z$ and $H_x$ (fig. 1). The double EMAT effectiveness is characterized by the amplitude of the received signal in the receiving coil (E). He is measured in volt usually.

On tangential magnetizing ($H_x$) the amplitude of the received signal according to the mechanisms of transformation is described by the following approximate formulas[1, 4].

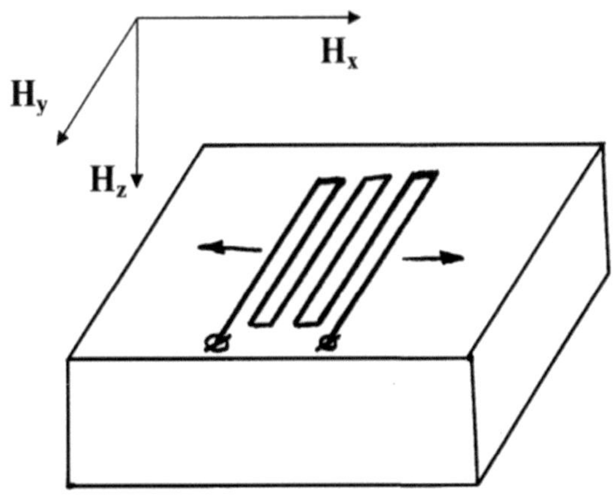


**Figure 1.** Position of the meandrovy coil on an specimen.

For the magnetostriction mechanism:

$$E^{ms} \approx \frac{1}{\mu_d}\left(\frac{d\lambda}{dH}\right)^2 P_1, \tag{1}$$

where $\mu_d$ is differential magnetic permeability, $\lambda$ is the linear magnetostriction ( $\lambda$ =Δl/l is specific elongation of the specimen), $P_1$ is constant factor which depends from parametres of the generating sensor, the reception sensor; physical constants of a material (for example, density, electroconductivity); strengthenings of a reception path, H is a magnetic intensity.

For the sum of vortical current and magnetic mechanisms (the electrodynamic mechanism):

$$E^i + E^m \approx (\mu_0 H\ )^2 P_2, \tag{2}$$

where $\mu_0$ is a permeability of vacuum, $P_2$ is constant factor the similar $P_1$.

At normal magnetization ($H_z$).

For the magnetic mechanism:

$$E^m \approx (\mu M)^2 P_3, \tag{3}$$

where, $\mu$ $M$ are magnetic conductivity and magnetization of the material in this orientation, $P_3$ is constant factor the similar $P_1$.

For the vortical current mechanism:

$$E^i \approx (B)^2 P_4, \quad (4)$$

where $B$ is a magnetic induction, $P_4$ is constant factor the similar $P_1$.

The dependence of the amplitude of the received signal (E) on the magnetic field size, the so-called «field curve EMAT» is the total characteristic EMAT SAW [5]. Two sensor (such as in fig. 1) place on a surface of the sample. One sensor generates the surface wave, the second sensor accepts a wave. The accepted signal strengthen and measure in volt. Often E normalized in relation to any value E (for example, to the maximum value) and receive abs. units. This size postpone on axis Y. The magnetic field size in the sample is shown on an axis X (B (T) or H (A/cm), as $B=\mu\mu_0 H$).

Fig. 2 shows a typical field curve for Rayleigh waves EMAT taking ARMCO iron as an example (curve 1). The curves for the transformation mechanisms calculated on the basis of equation (1) and (2) are presented in fig.2 as well (curve 2 and 3, accordingly). At calculation of the equations use known functions for a material: $B=\mu\mu_0 H$, $\lambda = f(H)$ [12].

The amplitude of the signal (E) is normalized concerning a maximum.

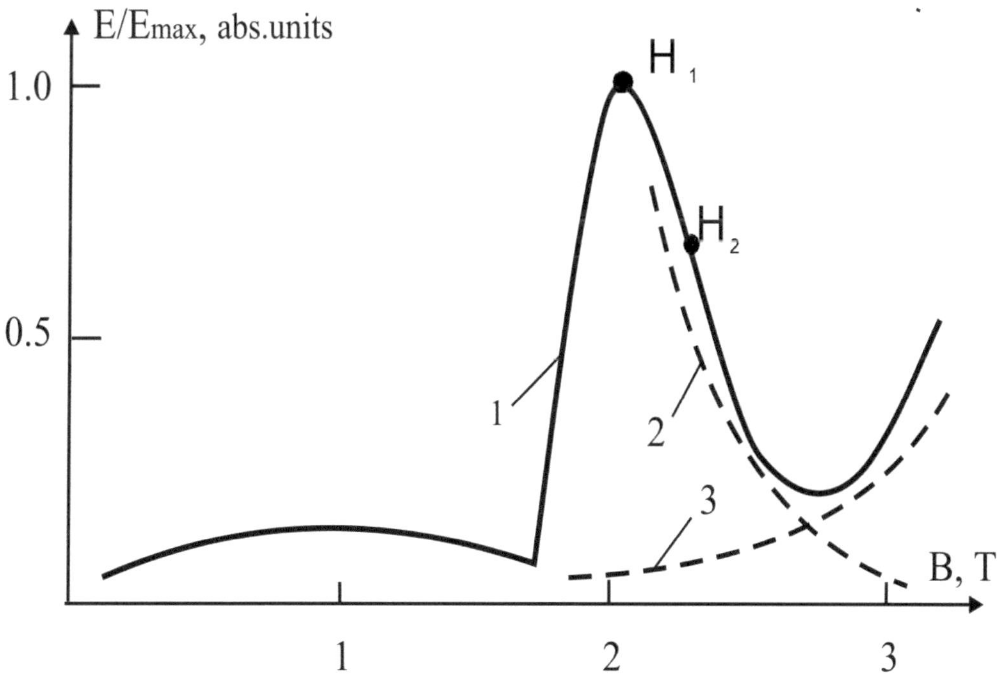


**Figure 2.** The experimental EMAT field curve of Rayleigh waves in a tangential field for iron (curve 1). The theoretical curve 2, 3 are presented equations (1) and (2).

An EMAT field curve of Rayleigh waves registered in a normal magnetic field(curve 1) and the curves 2, 3 calculated using (3) and (4) are shown in fig.3. As seen from fig.2 and fig.3 the experimental process is well described by the formulas.

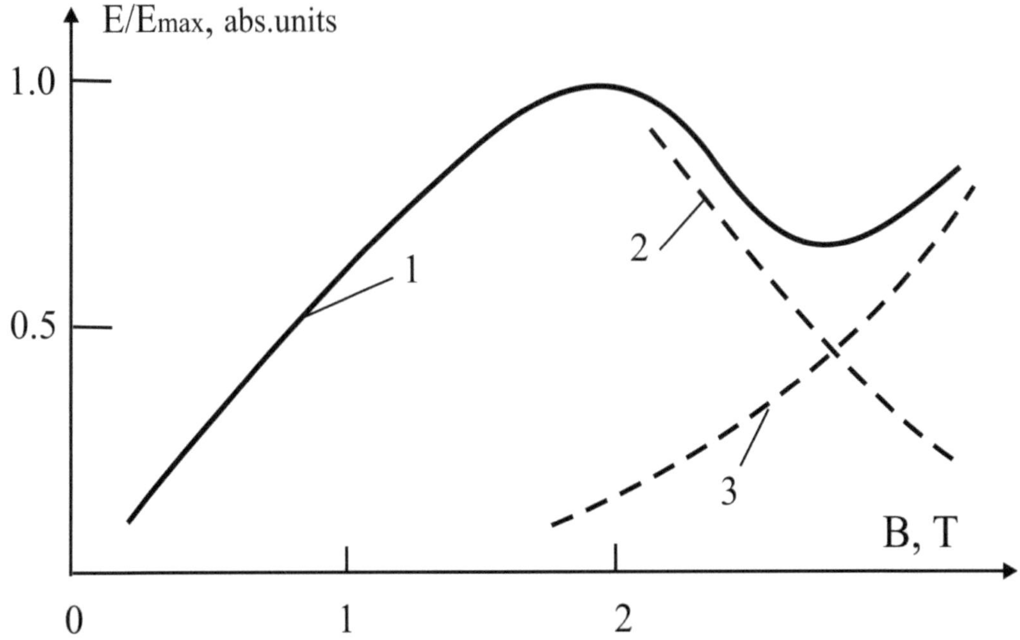


**Figure 3.** 1 – is the experimental EMAT field curve of Rayleigh waves in a normal field for iron. The theoretical curve 2, 3 are presented equations (3) and (4).

2. Waves of horizontal polarization.

Waves of horizontal polarization are generated in a case when current elements of the meander coil are parallel to a magnetic field. In this case the elementary volumes of the specimen under the EMA coil are in the cross fields, i.e. under the big constant magnetize field and the small variable magnetic field of the coil. Such an arrangement results in shear modes of horizontal polarization which are synchronized under EMA current elements of the coil (fig. 1 field $H_y$).

The following simplified formula [6] is derived based on the theoretical calculation for SH waves in a **d-** thick specimen [6]:

$$E_{SH} \approx \frac{1}{d}\left(\frac{\lambda}{H}\right)^2 P_5, \tag{5}$$

where $P_5$ is constant factor the similar $P_1$.

A structurally similar formula is deduced for the waves of horizontal polarization in the layer (Love waves), but a layer thickness and some known restrictions concerning speeds ratio should be taken into consideration.

The experimental field curve of EMAT of SH waves in ARMCO iron is presented in fig. 4.

If I construct the chart of the equation (5) on this coordinates, I also will see good coincidence to experiment.

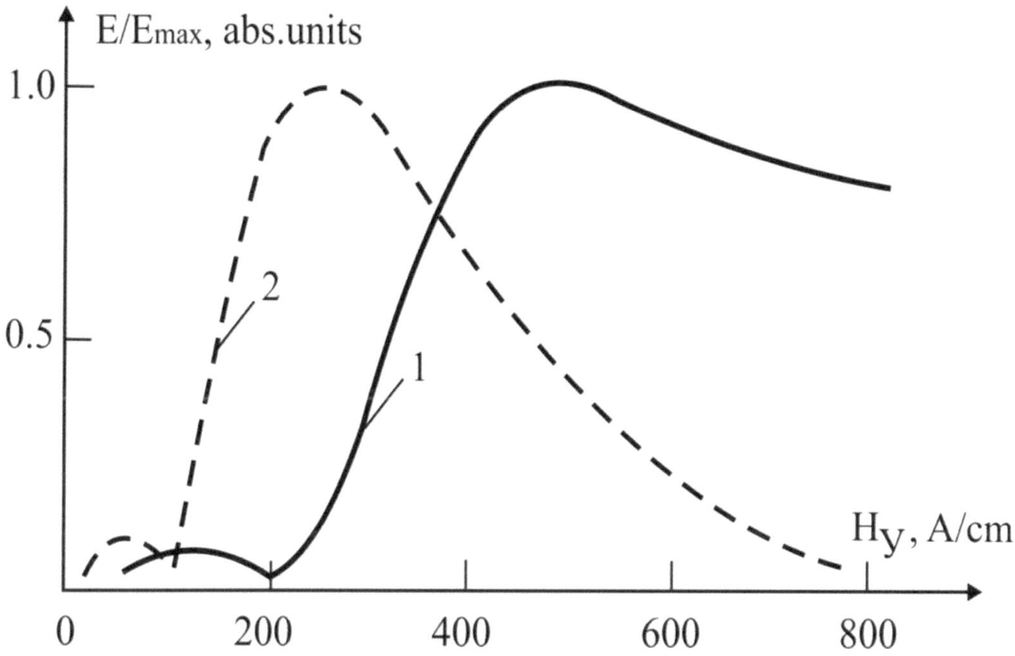


**Figure 4.** The experimental field curve of EMAT of SH waves in iron (1), in comparison with Rayleigh waves (2).

Love waves have been studied using 20-80 microns thick nickel films applied on the surface of an aluminum substrate. The field curve is presented on fig. 5.

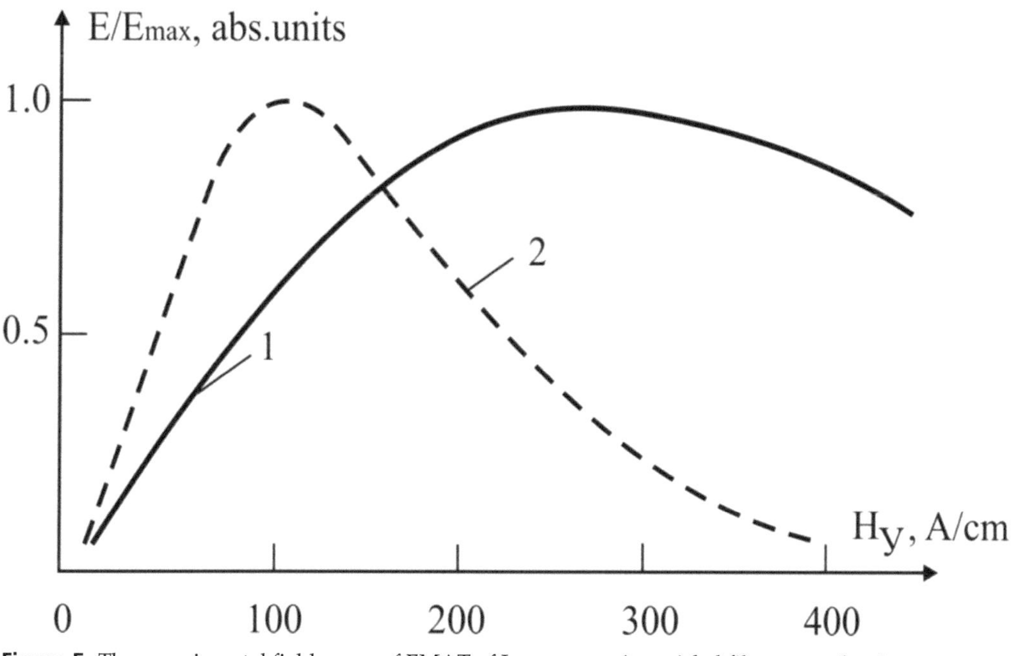


**Figure 5.** The experimental field curve of EMAT of Love waves in a nickel film on an aluminum substrate (1), in comparison with Rayleigh waves (2).

*SAW identification*

SAW identification is essential in studying EMAT of the surface wave experimentally.

SAW signals should be distinguished from the signals of other waves. E.g. from the volume waves signals starting from the surface of the specimen to its depth at different angles with interfering reflections in the receiving EMA sensor. It is also necessary for SAW to be recognized in between.

*Basic identification methods*

1. Pulp method. It involves simple damping by a finger of the surface of the specimen which strongly affects the SAW wave amplitudes but doesn't affect other waves. As long as it is a subjective method to make it being more objective a method of buttered drops is used. Using a pipette to drop the oil from the same height an identical damping is obtained. It allows distinguishing even the Rayleigh surface waves from SH waves. In the same conditions using a 6-drop method a 28 % decrease of Rayleigh wave amplitude and a 7 % decrease of SH waves are observed.
2. A Rayleigh wave $C_R$ strongly differs from longitudinal waves $C_l$ in speed, its speed being slightly differs from the speed of transverse waves $C_t$ ($C_R$= 0, 87 -0, 96 $C_t$). A zero mode of a SH wave speed is equal to $C_t$. To be more sure in distinguishing from a Rayleigh wave the accuracy of speed determination is be within 0, 5 %.
3. A sensor driving method is based on the following: if a generating sensor moves regarding to the receiving one, the signals on the a receiving oscilloscope screen move synchronously.

Thus, we can tell the following.

- We understand work electromagnetic and acoustic transformation of surface acoustic waves in the theory and in practice.
- It is understood how to choose operating mode electromagnetic and acoustic transformation of surface acoustic waves for the decision various applied problems.

# 4. Using EMAT surface acoustic waves

## 4.1. Non-destructive testing

### *4.1.1. Defectoscopy of the surface defects*

Using an ultrasonic method for testing materials and products EMAT incorporates all the advantages of ultrasonic testing methods. A piezoelectric transducers (PET) method is similar to EMAT. The main ultrasonic techniques developed for PET, are applied to EMAT.

In fig. 6 the design of EMA of the Rayleigh wave converter on the basis of П-shaped electromagnets [7] is shown.

The sensor consists of two identical half-cells: П-shaped electromagnet 3 and meandrovy coil 5, 6. Both converters are located in the case 2 and are filled in with a filler 4. One half-cell generates the surface wave in a specimen 1, another accepts the surface waves. It is a separate testing regime. It is possible to use the combined testing regime when one half-cell generates a wave and accepts the signal reflected from defect.

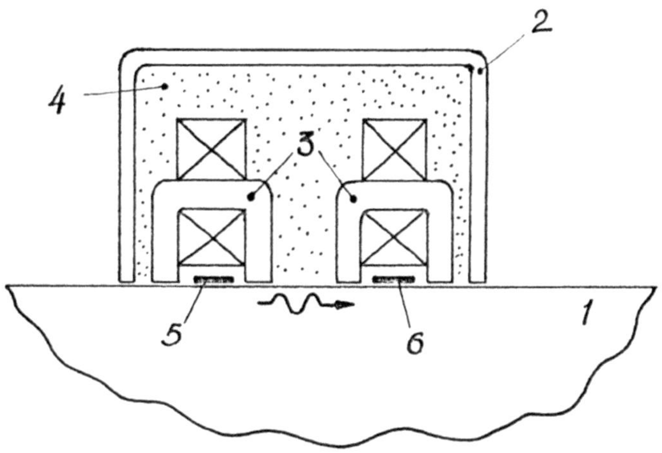


**Figure 6.** Design of the EMAT sensor of waves of Reley. 1-an specimen, 2 - a housing, 3 - electromagnets, 4 - filler, 5, 6 - generating and receiving EMA coils.

A magnetic field can be created the strong permanent magnets, but in this case it will be uncontrollable. Electromagnets permits operating a magnetic field, choosing the working point on field curve EMAT.

EMAT loses to a way PET in sensitivity, but in this case a contact is not required. It is meant it that using the EMAT method an immediate contact between EMA of the sensor and a specimen is not needed, and the sensor can operates through air or other gaps. For example, the layer of scale or paint isn't an absolute obstacle for EMAT. The amplitude of the received signal has been recognized to be strongly dependent on the gap size. It is due to two factors: 1) an electromagnetic field of EMA of the sensor in a gap called "the sensor – an specimen surface" decreases according to the exponential curve law, 2) a magnetic field of the specimen decreases as a gap occurs. For ferromagnetic materials the field curve of Rayleigh waves EMAT effectiveness has a characteristic maximum defined by the magnetostriction mechanism (fig. 2). The amplitude of the received signal depends on a working point choice in a field curve.

Usually the working point at non-destructive monitoring is chosen at top of the main maximum of the field curve (point $H_1$ in fiq.2). The maximal receiving signal is obtained in this way. For most ferromagnetic materials this maximum lies within 100 – 300A/cm magnetic fields. The same effectiveness using the vortical current mechanism is reached in magnetic fields about 5000A/cm in size.

If the working point of monitoring is chosen at the maximum top (H1 in fig. 2), the emergence of the gap in a magnetic circuit will result in decreasing the magneticfield in size and falling transformation effectiveness, i.e., the amplitude of the received signal. However, if the working point is chosen on the slope of the field curve (H2), a magnetic field decreases with the gap increasing which leads to the transformation effectiveness. Thus, both factors, i.e., the removal of EMA of the coil from a surface of the specimen and a magnetic field decrease in an specimen compensate each other. [8].

If the gap between the coil and the surface is introduced before the measurements are carried out the measurement results are more reliable. In this case the uncontrollable gap less influences on the received signal. If the working point is chosen in the slope of the field curve the signal received does not depend on the gap (fig. 7).

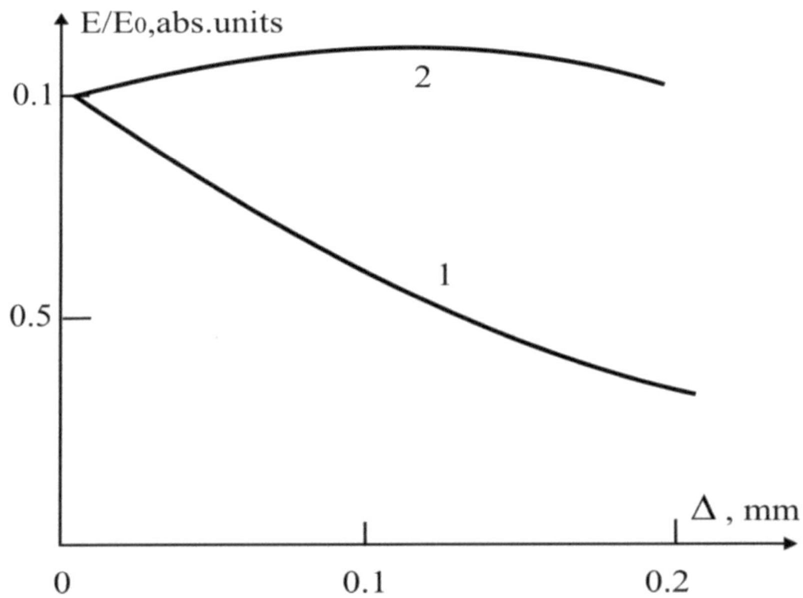


**Figure 7.** Dependence of the received signal from dimension of the gap between sensor and specimen at the working point $H_1$(1), at the working point $H_2$ and preliminary gap (2).

The experiments on detection of model defects showed that by means of EMA of the converter of Rayleigh waves presented in fig. 6, it is possible to reveal the surface defects such as cracks 5MM in length, 0, 5MM in depth. Such defects come to light both at reflection from the defect, and when weakening a signal passing the defect (weakening of 20 %).

### 4.1.2. *Structurescopy of metals*

Structural changes are well recognized by EMAT in ferromagnetic materials taking place in technological processes. This results from the fact that at EMAT four subsystems are involved in a material: magnetic, electric, magnetoelastic and elastic. All these factors influence informational parameters of the EMAT method.

Thus, if a dependence between the parameters of the technological process and the information parameters of the method of one subsystem is not a one- direction dependence the EMAT method can be used.

So the dependence between the amplitude of the received signal of Rayleigh waves sensor (fig. 6) and the hardness of the steel 38ХГС specimens being treated at different temperatures is practically linear. To increase the specimen hardness it was treated at high temperatures. The coefficient of correlation was equal to 0, 92. It allows EMAT controlling.

Another structural component is anisotropy of the material properties which can be also measured by the sensor described above. The sensor is rotated on the specimen surface showing the information parameters change. However, the data can be distorted due to different factors.

To overcome this trouble a meander -curve sensor was suggested to be curved into a ring (fig. 8) [9].

In this case each element of the dl coil generates SAW in K1 and K2 directions. In general two circular waves are observed: a coverging to the center wave and a diverging from the center one. A converging to the O point wave then turns into a diverging wave. It is received by the same circular EMA coil when the wave moves under it.

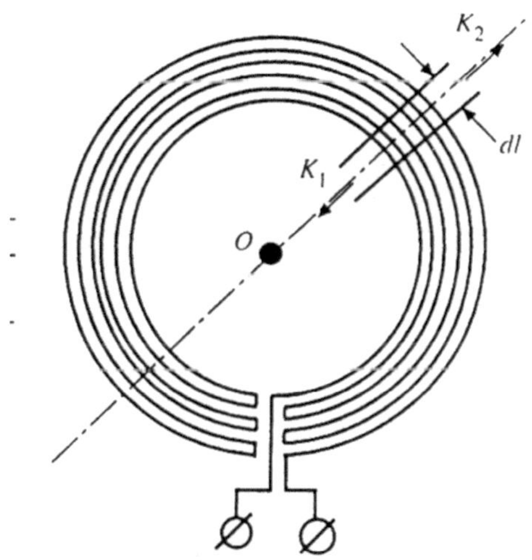


**Figure 8.** Circular EMAT sensor.

Two case of magnetizing are likely to observed for the circular EMA sensor. In case of normal magnetizing a rod electromagnet is used (fig. 9a), and in case of a tangential magnetizing it is realized by means of a cylindrical m-shaped electromagnet (fig. 9b). Cracks are revealed by such a sensor.

But the most interesting area for the sensor to be applied is in controlling anisotropy of sheet ferromagnetic materials. It allows revealing three types of anisotropy: elastic, magnetoelastic and magnetic.

At generation and reception of a circular surface wave in a completely isotropic metal we have the individual fine-bored received pulse which is a superposition of the signals from all directions (fig. 10a). In case of elastic anisotropy (anisotropy of sound speed) signals in the different directions take different time. And the received pulse extends, or splits into a number of pulses. An elastic anisotropy of the material is characterized by the pulse duration or the distance between extreme pulses (fig. 10b).

The amplitude of the obtained pulses (A1, A2) demonstrates the effectiveness of EMAT in different directions which is defined by magnetoelastic properties of a material. Magnetoelastic anisotropy of materials is defined by the amplitude distinction of pulses.

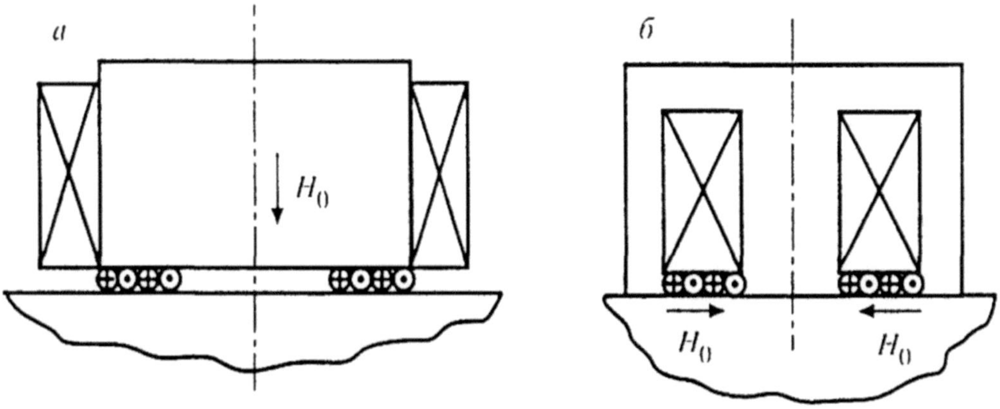


**Figure 9.** Tangential (a) and normal (в) magnetization of EMAT sensor.

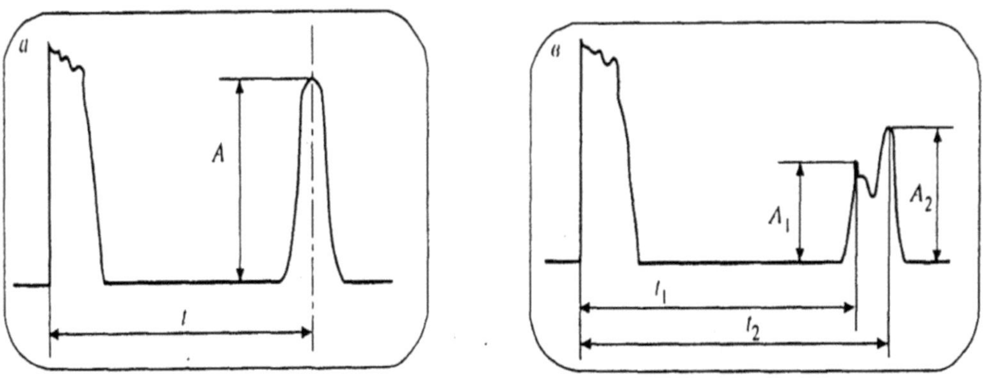


**Figure 10.** Oscillograms of signals, isotropic material (a), non-isotropic material (b), which are received by means circular EMAT sensor.

The magnetic anisotropy of a sheet is shown as follows. Effectiveness of EMAT from a magnetize field has a maximum. In case of a magnetic anisotropy the maximum of EMAT signal is reached in different directions at different currents of magnetization. Thus, on changing a magnetization current for an isotropic material the range of optimal currents is narrow, and for a magnetic anisotropic material the range is wider the one mentioned above.

The developed technique was tested on various materials and showed high effectiveness.

### 4.1.3. *Tension controlling*

EMAT SAW enables tension and operational loadings controlling. For this purpose the sensor (fig.6) was used. The sensor platform area is to be 30 x 60MM. Adjusting to obtain a magnetostriction maximum the amplitude of the received signal is an information parameter.

At stretching the specimen made of steel 25 in the elastic area it has been shown that the amplitude behaves depending on the positions of the sensor i.e. lengthwise and crosswise of the loading. Thus dependence is almost linear.

On measuring the sound speed, it has been found out that in the elastic area it changes less than 0, 5 %. Hence, at controlling tension concerning the sound speed more precise equipment and computer processing of signals should be applied.

The studying of the field curve of Rayleigh waves EMAT under «the sensor is along loading» shows that at specimen stretching in the elastic area and at the beginning of the area plastic deformation both the amplitude of the received signal and an optimal field of magnetizing (a field of the maximal signal in the EMAT curve) for a magnetostriction maximum have changed. It depends on whether magnetostriction is positive or negative [10].

For materials with the negative magnetostriction of saturation, $\lambda_s$, (nickel, constructional steel) the dependence is as follows.

Fig. 11 shows the dependence of the maximum point of the field curve of EMAT on loading, the maximum point being normalized in relation with its initial value. At the initial stage of a loading the amplitude of the maximum grows, and this maximum is displaced to the area of the bigger fields. Then, in the area of the inflection point a decrease of a maximum value is observed. The inflection point lies in the area of the elasticity limit (the conditional point where the size of permanent deformations makes 0, 05 %).

It can be explained from the physical point of view. In [4] it is noted that at $\lambda_s \sigma<0$ pulling stresses of $\sigma$ are considered to be positive, two factors being competitive: on the one hand the energy of a magnetic field, , on the other hand the crystallographic anisotropy and magnetoelastic energies. Tension in a metal makes a magnetization vector turn perpendicular to the magnetic field. As a result, it is more difficult to obtain an amplitude maximum of EMAT.

The amplitude of the EMA signal is described by formula (1). As the loading in the elastic area increases magnetic conductivity, $\mu$, decreases, and the parameter $\frac{d\lambda}{dH_0}$ grows. It results in the received signal growth. At larger loadings there is tension which leads to the signal decrease. So a maximum occurs in the elasticity limit.

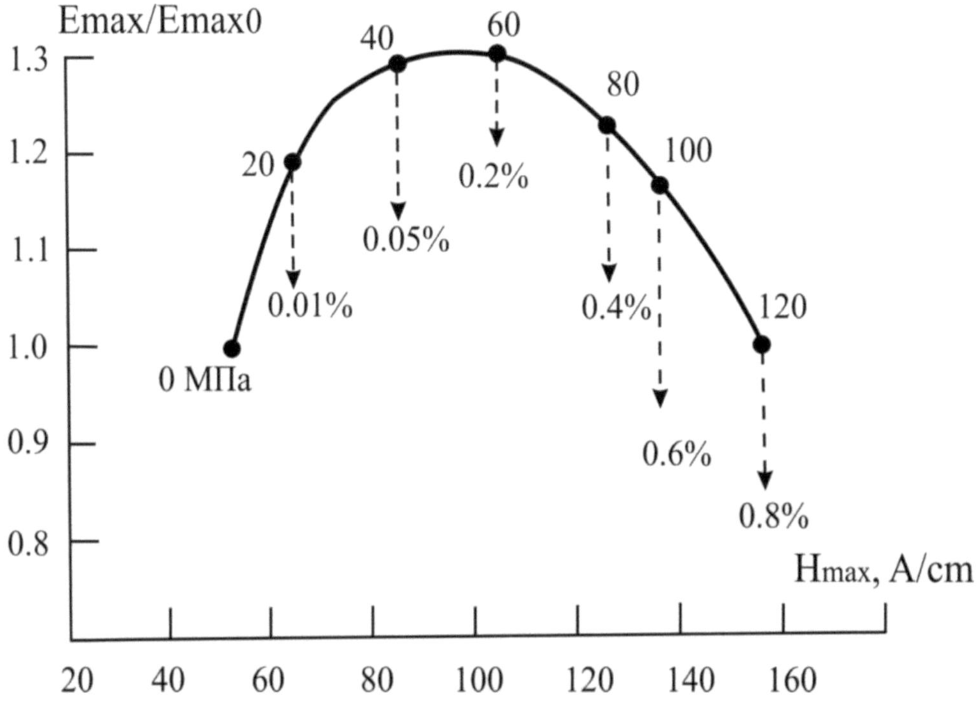


**Figure 11.** The shift of the maximum point of a field curve of Rayleigh waves EMAT in nickel from the loading (the loading is given in MPa, and the length change after loading is given in percent).

If materials have a positive magnetostriction of saturation, the behaviour of the maximum of the field curve for Rayleigh waves EMAT is essentially different. Using alloy H18 (18% Ni, 82 % Fe) $\lambda_s$› 0 as an example it is shown that a received EMA signal unequivocally falls (fig. 12). As to a magnetic field first an optimal magnetic field decreases and then it becomes larger, the change takes place at the point close to a limit of material elasticity. It can be also explained from the physical point of view.

Thus, it has been shown that using Rayleigh waves EMAT at least two information parameters are obtained which are used for the applied loadings to be characterized.

It can be carried out either by calibration curves at monitoring or without calibration curves if the tendency of change of parameters EMA is used.

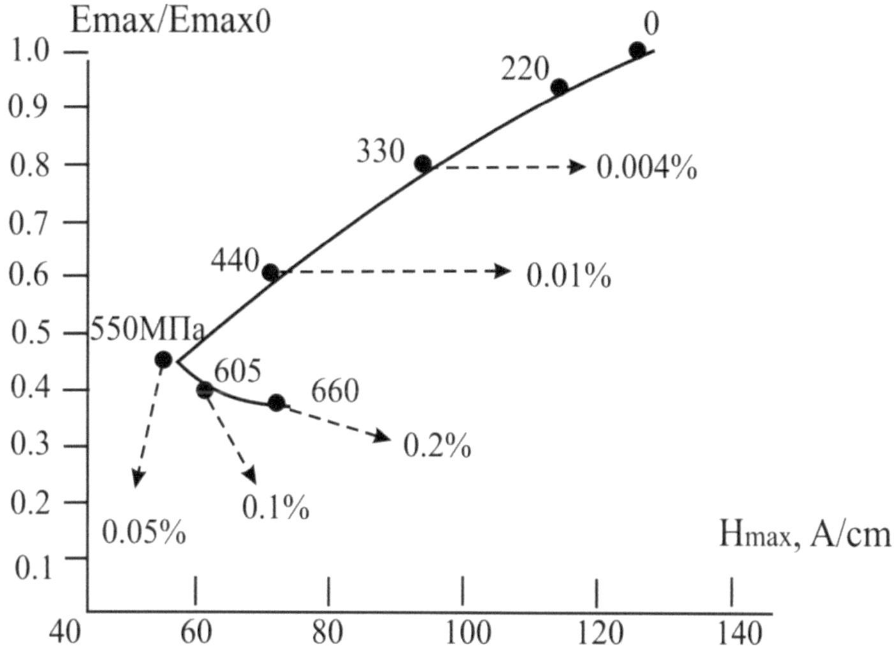


**Figure 12.** The shift of the maximum point of a field curve of Rayleigh waves EMAT in H18 alloy from the loading (the loading is given in MPa, and the length change after loading is given in percent).

## 4.2. Physical study of materials

It has been shown that the EMAT SAW method are defined by different subsystems of the material. Information on the parameters of the material related to these subsystems is revealed by the information parameters of the EMA method.

### *4.2.1. Magnetostriction assessment of a material*

Magnetostriction is one of the important characteristics of a material. Due to the wide use of these materials in science and industry the assessment of magnetostriction properties of a material is of great importance.

The main characteristic of magnetostriction properties of a material is the field curve of magnetostriction, i.e., dependence of the linear magnetostriction, $\lambda$, on a magnetic field. Though from the practical point of view it is enough to know magnetostriction of saturation, $\lambda_s$.

At magnetostriction measurement the method based on tension resistivity of sensors stuck by glue on an specimen is the most extensively used. There are some shortcomings in this method. Alternatively a quick test based on non-contact EMAT SAW [11] may be suggested.

So for Rayleigh waves EMAT formula (1) is used. The task is as following: based on the known curve of EMAT we are to obtain a curve of magnetostriction, $\lambda$ =f(H). It is clear that from curve (1) a field curve of magnetostriction can be obtained by the integration method only. This process is difficult and inaccurate.

Horizontal polarization waves EMAT (SH wave) is more preferable to be used to obtain λ =f(H). A field curve of SH waves EMAT is described by the simplified formula:

$$E_{SH} \approx \frac{1}{\mu d}\left(\frac{\lambda}{H}\right)^2 P \tag{7}$$

where - $P = \rho C_t / \sigma$ is a material parameter, $\mu$ is a static magnetic conductivity, $d$ is a sheet thickness, $\rho$ is material density, $C_t$ is a speed of shift waves, $\sigma$ is an electrical conductivity.

Thus we obtain :

$$\lambda = \sqrt{\frac{E_{SH} d}{P}} H\sqrt{\mu} \tag{8}$$

Knowing B=f (H) it is easy to calculate μ=f (H), and then knowing the material parameters it quite easy to calculate a curve λ=f (H).

The calculations above have been used to obtain λ=f (H) for four materials: iron, cobalt, nickel, permendur (brand 49KФ2). Fig. 1 presents the orientation scheme of the magnetic Hy field for EMAT of SH waves. Fig. 13 shows field characteristics of EMAT of the materials mentioned above. In fig. 14á field characteristics of the magnetostriction module calculated on the basis of formula (8) are presented. Fig. 14б shows the same curves based on data [12]. High curves coincidence is observed.

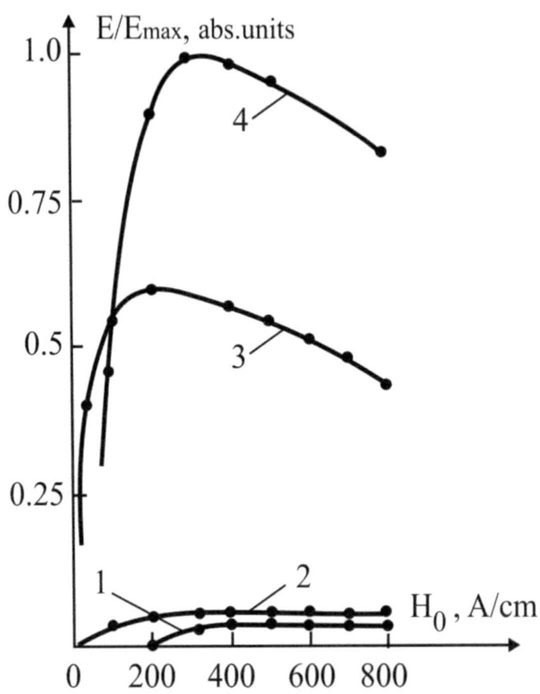


**Figure 13.** The experimental EMAT field curve of SH- waves (1-iron, 2 - cobalt, 3 – alloy 49KФ2, 4 – nickel).

In 800A/cm fields magnetostriction of materials approaches saturation. To obtain the quantitative value of saturation magnetostriction the required data were obtained by a method of tension resistors. The comparative table on the magnetostriction of saturation received by an EMAT method and a method of resistance strain gages is given in table 1.

As shown in the table, the relative ratios for the EMA method are obtained with an accuracy of 5-7 %.

It should be noted that EMA method allows defining the magnetostriction module $\lambda$ only. A sign of magnetostriction is complicated to be defined, but it is quite possible. To carry it out it is necessary to observe a phase of the received signal and then to compare it with a signal phase of the known material.

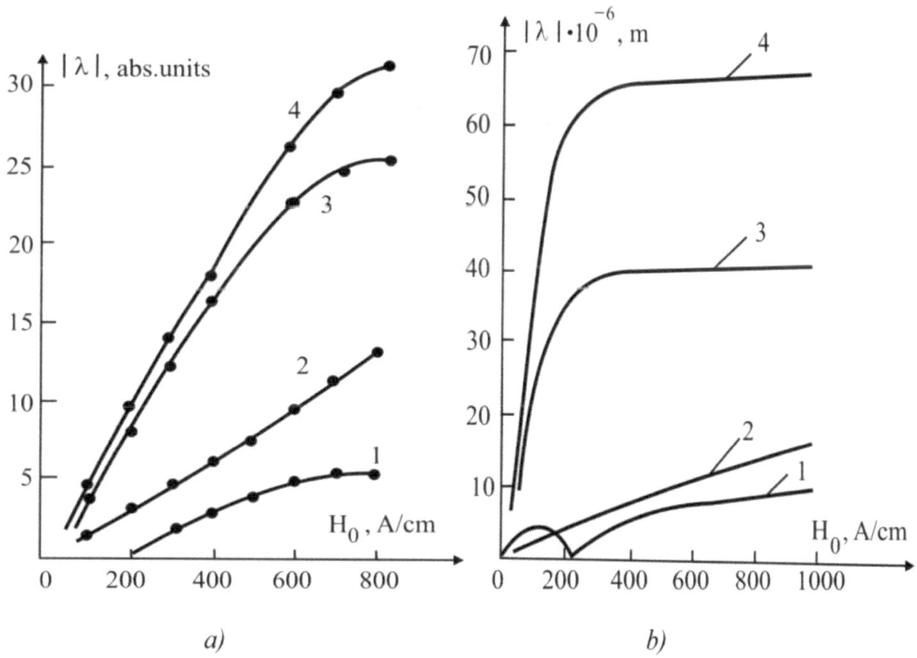


**Figure 14.** Field curves of the magnetostriction module calculated based on the experimental data (a) the data are given according to literature data [12] (b) (1-iron, 2 - cobalt, 3 - alloy 49KФ2 (Permendur), 4 – nickel).

| Material | $\lambda_{max}/\lambda_{max\ Fe}$ EMA method | $\lambda_{max}/\lambda_{max\ Fe}$ Tenzometod |
|---|---|---|
| Fe | 1 | 1 |
| Co | 2, 6 | 2, 8 |
| Ni | 5, 2 | 5, 0 |
| 49KФ2 | 6, 2 | 6, 5 |

**Table 1.** The comparative table on the magnetostriction of saturation received by an EMAT method and a method of resistance strain gages.

### *4.2.2. Assessment of other characteristics of a material*

Electric properties of ferromagnetic materials can be estimated using the amplitude of the received signal in large magnetic fields (more than 800A/cm). EMAT effectiveness depends on the electrical conductivity of a material under the same conditions.

Magnetic properties of a material can be estimated by the location of the magnetostriction maximum.

Two parameters are used to estimate elastic properties of a material:

Firstly, a sound speed in a material. To measure the speed of the Rayleigh surface wave, $C_R$, and the speed of horizontal polarization waves, $C_{SH}$, the techniques described are employed. Here, it is should be considered that $C_{SH}$ for a zero mode is equal to the speed of shift waves of the material, $C_{t,}$. These measurements enables estimating the elastic constants of a material such as shift modulus and a Poisson's ratio. The constants defined allows us to determine the remained elastic modules of a material.

Secondly, sound attenuation in a material is estimated by EMAT SAW methods. In this case two receiving EMA coils are used.

Having defined the attenuation parameter it is easy to find such important production characteristics as good quality of the system and an internal friction parameter of a material.

## 4.3. Prospects of use of EMAT SAW

The main examples of using the surface waves EMAT are presented in the paper only. The use of EMAT SAW in non-destructive monitoring as well as in studying new materials and the phenomena is possible in the nearest future.

Love waves EMAT may be used for non-destructive monitoring of coatings and thin films and for their thickness measurements as well. SH waves EMAT of can be applied in cylindrical specimens where such waves are known to be generated and received very efficiently.

One can find papers on the use of surface waves EMAT to study magnetic phenomena in rare-earth metals. At the increased or decreased temperatures of the specimens it is easier to apply the EMAT method than the method of piezoelements.

When studying various acoustic waves the EMAT method is more perspective due to easy generation of acoustic waves.

There are combined methods of generating and receiving acoustic waves where EMAT is used together with other methods, for example, with laser methods and with methods of an optical interferometry.

All we said above allows us to draw a conclusion that the EMAT methods are very promising in the future.

## Author details

Sergey E. Babkin
*Physical-Technical Institute, Ural Branch of Russian Academy of Sciences, Izhevsk, Russia*

## 5. References

[1] Комаров В.А. Квазистационарное электромагнитно-акустическое преобразование в металлах. – Свердловск: УНЦ АН СССР, 1986, 235с.

[2] Васильев А.В., Бучельников В.Д., Гуревич С.Ю., Каганов М.И., Гайдуков Ю.П. Электромагнитное возбуждение звука в металлах. – Челябинск. – М.: Издательство ЮурГУ. – 2001. –339 с.

[3] Викторов И.А. Звуковые поверхностные волны в твердых телах. М.: Наука, 1981.

[4] Ильясов Р.С., Бабкин С.Э., Комаров В.А. О механизмах ЭМА преобразования волн Рэлея при различных частотах. - Дефектоскопия, 1988, №10, с.77 – 82.

[5] Thompson R.B. A Model for the Electromagnetic Generation and Detection of Rayleigh and Lamb Waves. – IEEE Trans. Sonics and Ultras., 1973, v.20, N 4, p.340-346.

[6] Комаров В.А., Бабкин С.Э., Ильясов Р.С. ЭМА преобразование волн горизонта льной поляризации в магнитоупругих материалах. - Дефектоскопия, 1993, №2, с.11 – 17.

[7] Бабкин С.Э., Ильясов Р.С., Комаров В.А., Рубцов В.И. Устройство для бесконтактного возбуждения и приема волн Рэлея в ферромагнетиках. – Дефектоскопия, 1989, №6, с.93-94.

[8] Бабкин С.Э., Ильясов Р.С., Зверев Н.Н. Отстройка от зазора при электромагнитно-акустическом способе контроля. - Дефектоскопия, 1998, №4, с.19-22.

[9] Бабкин С.Э., Ильясов Р.С. Кольцевой электромагнитно-акустический преобразователь поверхностных волн. - Дефектоскопия, 2002, №1, с.78-82.

[10] Бабкин С.Э., Ильясов Р.С. О возможности использования параметров ЭМАП для оценки предела упругости и остаточных деформаций ферромагнитных материалов. - Дефектоскопия, 2010, №1, с.83-89.

[11] Бабкин С.Э., Ильясов Р.С., Комаров В.А. Бесконтактный метод оценки магнитострикции материалов. - Дефектоскопия, 1996, №1, с.41-45.

[12] Bozort R.M., Ferromagnetism, D.Van Nostrand Co., Princeton, 1951.

Chapter 17

# Acoustics in Optical Fiber

Abhilash Mandloi and Vivekanand Mishra

Additional information is available at the end of the chapter

http://dx.doi.org/10.5772/54477

## 1. Introduction

Optical filters are the heart of optical networks; without the wavelength selective device wavelength division multiplexing and dense wavelength division multiplexing network will not exist. As the networks are progressing towards closer wavelength spacing, performance requirement for filters are becoming more demanding. Currently, the popular filters include gratings, thin-film filters, and Fabry-Perot filters and acoustoi optic tunable filters (AOTFs).

Acousto-optic (AO) effect in fibers has been studied to produce tunable filters, gain flatteners, modulators, frequency shifters, and optical switches reported. Most AO devices work on coupling from the fundamental mode ($LP_{11}$) of light to a higher order asymmetrical ($LP_{1l}$, $LP_{12}$ .... $LP_{1n}$) modes. Acousto-optics is defined as the discipline devoted to the interactions between the acoustic waves and the light waves in a material medium. Acoustic (vibrational) waves can be made to modulate, deflect and focus light waves by causing a variation in the refractive index. Acousto optic tunable filters are a promising technology for dynamic gain equalization of optical fiber amplifiers [1]. By launching an acoustic wave directly on the fiber, the device combines the merits of fiber and AOTF devices namely the low insertion loss, low polarization dependence loss, wide tunability, fast tuning speed and ease of packaging. When a flexural acoustic wave is applied to a tapered single mode fiber, coupling takes place between the core mode and the cladding mode. The coupled energy in the cladding mode is essentially absorbed by the fiber jacket as reported so that the device is a notch filter. It means the centre frequency and the rejection efficiency can be tuned by adjustment of the frequency and voltage being applied. Varying the amplitudes and frequency of a RF generator can change the spectral profile of these filters.

To improve the rejection efficiency of the filters, the thickness of the fiber can be reduced. This is achieved through the heating and the acid-etching method. In the heating method, the ratio of cladding to core size is maintained while in the acid etching-method, the ratio between the cladding and core can be changed.

## 2. Acousto-optic tunable filter

### 2.1. Device design

The fiber used in our experiment is a Corning SMF-28, standard telecommunication single mode fiber. A region of SMF is etched by dipping the fiber in a hydrofluoric acid solution, which has a concentration of 40%. Etching rate controls the thickness of the SMF and the diameter reduction is observed using a CCD camera.

When the optical signal enters the fiber and interacts with the acoustic energy in a jacket stripped segment of the fiber, the core mode of the light is converted to a higher order cladding mode producing a notch filter like characteristics in the transmission spectrum. Core mode converting to various cladding modes will produce a few notch filters, with each having its peak notch at a separate wavelength [2-4]. A vibrating PZT transducer driven by a RF generator produces the acoustic energy as stated by Yun, Hwang and Kim, (1996). The acoustic energy is further amplified and concentrated to the fiber by a machined aluminium horn.

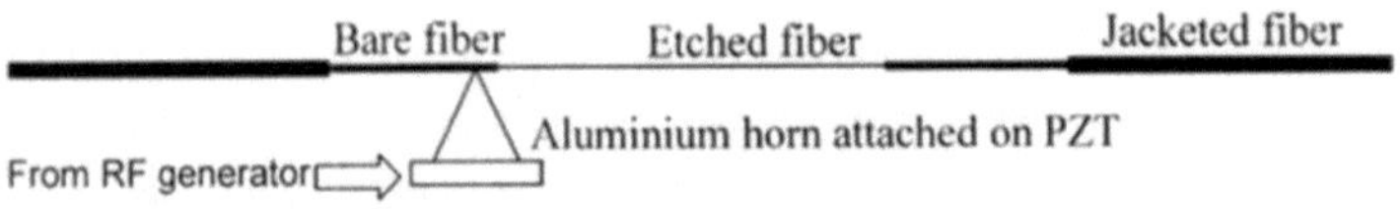


**Figure 1.** The setup to study AO interaction inside a fiber.

### 2.2. Horn design

An acoustic horn functions to transfer and amplify the surface acoustic wave to the fiber. All horns made were conical in shape, where the tip is narrow and the base is broad as described by Lee, Kim Hwang and Yun, (2003). All the horns fabricated for the AOTF experiment have a ratio of length to outer diameter ratio of 2. Length is defined as the length from the tip of the horn to the base of the horn. Horns taken are 1cm in length. Outer diameter of the horn is defined at the diameter at the base. The inside of the horns are made hollow. When the horn is made considerably small, the frequency dependence on the acoustic generator is low. In the experiments done, no transduction is observed when the fiber is not etched. Potential problems can be attributed to the size of the transducer and the adhesive used to bond the tip of the horn to the fiber. Solder acts as a strong, metallic, thermally stable, and acoustically transmitting joint. In these experiments however glue was chosen as the bonding agent. Of particular interest will be the horn tip size. Acoustic impedance at the horn tip is given by:

$$Z_r = c_a v_a A_r \tag{1}$$

where $c_a$ is the longitudinal velocity inside aluminium, $v_a$ is the density aluminium and $A_r$ is the cross section of the horn tip. Acoustic impedance at the bond junction along the fiber is given by:

$$Z_s = 2c_s v_s A_s \tag{2}$$

where $c_s$ is the longitudinal velocity inside silica, $v_s$ is the density silica and $A_s$ is the cross section of fiber. Acoustic impedance inside the fiber is counted twice because of bidirectional acoustic movement along the fiber. Optimum transduction occurs when $Z_r = Z_s$ and since acoustic impedance of silica is almost matching that of aluminium, according to engan et. al maximum acoustic wave transfer occurs when horn tip diameter is almost matching that of the fiber.

## 2.3. Tuning of peak wavelength

By driving a piezoelectric (PZT) device at an ultrasonic frequency the periodic perturbations can be created inside the fiber. In a phase-matched condition, where the momentum and energy conservation requirement ($L_B = \Lambda$) are met, the resonant frequency of an acoustic wave according to Birks, Russel and Culverhouse (1992) is given by

$$f = \frac{\pi b C_{ext}}{L_B^2} = \frac{\pi b C_{ext}}{\Lambda^2} \tag{3}$$

where $b$ is the radius of the fiber, $C_{ext}$ is the speed of fundamental acoustic mode, which for silica is 5760 $ms^{-1}$, $\Lambda$ is the period of the microbend[1].

Assuming a phase-matched condition, the frequency needed to transfer the modes from core to cladding mode for various thickness of the fiber is given in Fig 2 and Fig 3. As the fiber diameter is reduced, the values of $df/d\lambda$ *get* smaller. For unetched fibres, the frequencies used to create the micro bends and thus, convert the modes are from 1.75 MHz to 2.25 MHz. For thin diameter fibers (20 $\mu$ m, 30 $\mu$ m, 40 $\mu$ m), the frequencies are from 800 kHz to 1.1 MHz.

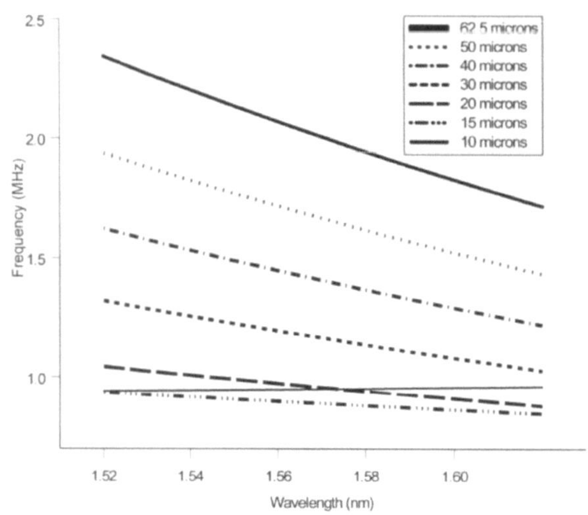


**Figure 2.** Calculated RF frequency to convert the LP01 mode to LP11 mode plotted against wavelength (for various thickness of fibre diameter).

[1] A microbend is the physical deformation of fiber achieved mechanically or chemically done to perturbing the optical modes to study mode coupling between core and cladding mode.

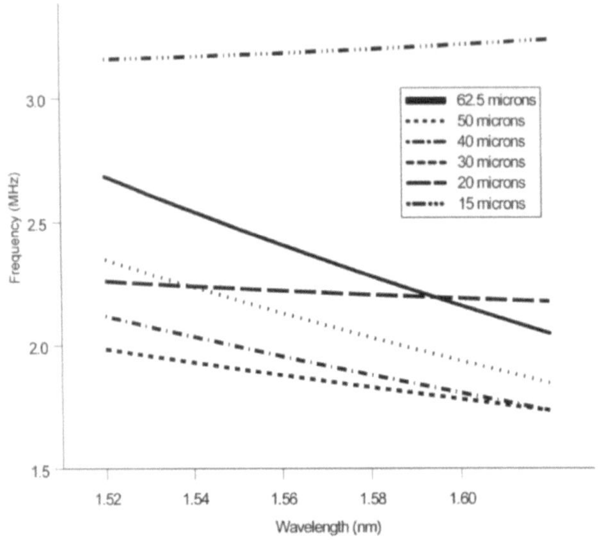


**Figure 3.** Calculated RF frequency to convert the LP01 mode to LP12 mode plotted against wavelength (for various thickness of fibre diameter).

Frequency from the RF generator can be used to control the peak wavelength tuning of the notch filters (Fig.4). The fiber used in the experiment has a diameter of 30 μm, and length of 17 cm. Higher frequencies of the RF generator will blue shift the peak wavelength of the filter [3-7]. The tuning range of the filter is slightly less than 300 nrn. From Eq.1.3, we deduce that, micro bend's period is inversely proportional to the frequency of the RF generator. For a larger value of period, the filter's peak is red shifted. Thin fibers have lower period values, thus etching the fibres will blue shift the peak wavelength of the notch filters. Frequency used to tune the peak wavelength as in for thin fibres is from 800 kHz to 1.1 MHz, which is in excellent agreement with the theory as in Fig.3 and Fig.4.

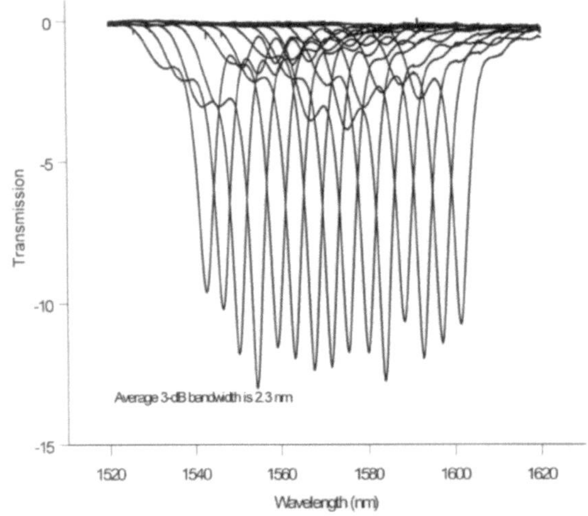


**Figure 4.** Measured peak wavelength tuning of the filter by changing the RF frequency. Frequency used is from 970 kHz to 1045 kHz. The fiber used has a thickness of 30 μm and length of 17 cm.

## 3. Tuning of attenuation depth

The RF generator's $V_{p\text{-}p}$ level will be used to control the attenuation depth of the filter. $V_{p\text{-}p}$ level is actually referring to the acoustic power transferred to the fiber. Increasing the $V_{p\text{-}p}$ level will generally increase the bottom level of the filter as seen from Fig. 5. However in some cases increasing the $V_{p\text{-}p}$ level will only distort the shape of the filter without increasing the notch's depth. For this strong over-coupled phenomenon, side lobes of the filter is actually increasing. One way to eliminate the problem is by limiting the interaction length of light inside the etched region. Here the power means RF generator's Vp-p level which will be used to control the attenuation depth of the filter [8-10]. Vp-p level is actually referring to the acoustic power transferred to the fiber. Increasing the Vp-p level will generally increase the bottom level of the filter as seen from Fig. 5. Here acoustic power supplied to PZT is 1.6 W to allow mode conversion.

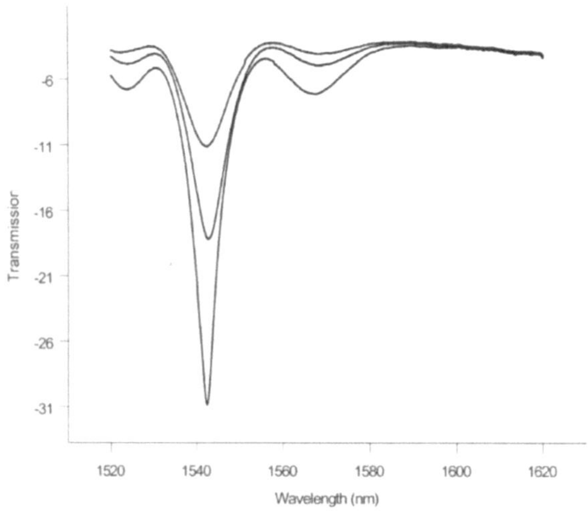


**Figure 5.** Measured attenuation variation of filter when the power of RF generator ($V_{p\text{-}p}$) is increased.

An effort to reduce the acoustic power fed into the fiber is by reducing the thickness of the fiber. The minimum acoustic power required by the device to operate or to allow mode conversion, is given by

$$P = 2\pi^3 \rho v_g (fR)^2 (u_t)^2 \tag{4}$$

where $\rho$ is the mass density of the fiber ($\rho$ =2200kg/m3 for fused silica), $V_g$ is the group velocity of the wave and R is is curvature of fiber, and $u_t$ is the transverse acoustic amplitude which is given by:

$$u_t = \frac{\pi}{2} L_B \frac{a}{L} \frac{1}{0.908} \tag{5}$$

where $L$ is referring to the interaction length of acoustic and light inside the fiber and $L_B$ is the optical beat length. Fig. 6 shows the calculated power required for mode conversion is lower for etched fibers. When the fiber is unetched the power required will be 287 mW. For a 20 μ m fiber, the power required for conversion is only 1.17 mW.

Experimentally, as seen from Fig. 7 for a 37 μm thick fiber, the acoustic power supplied to PZT is 1.6 W to allow mode conversion. Mode conversion was confirmed using far-field radiation pattern as reported by Doma, and Blake (1992). However, in a 26 μ m fiber, power requirement for mode conversion is reduced to a mere value of 42 mW. The difference in the power reduction with the calculated value, suggests that the loss at the point of contact is high[10]. It is believed that the horn design is still not optimized; nevertheless, this transduction is sufficient to demonstrate conversion between two modes. Typically, only the lowest order flexural acoustic mode should be made to travel inside the fiber, and this can be achieved by ensuring the horn tip's thickness is matching that of the fiber.

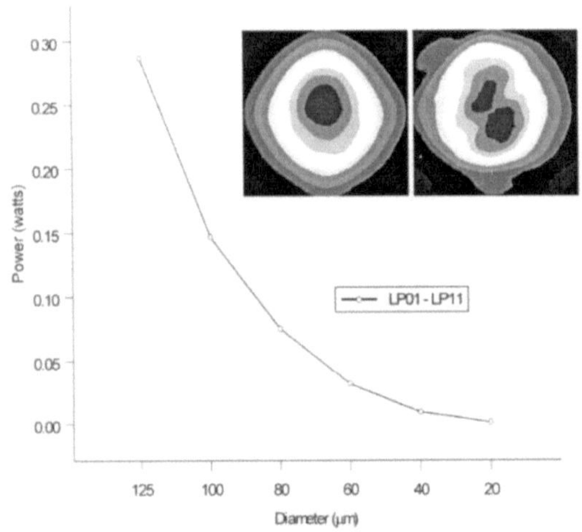


**Figure 6.** Calculated acoustic power required to allow mode conversion. Interaction length was set to 13 cm. Inset: Far field radiation pattern of modes involved in conversion. Left- LP01, Right- LP11.

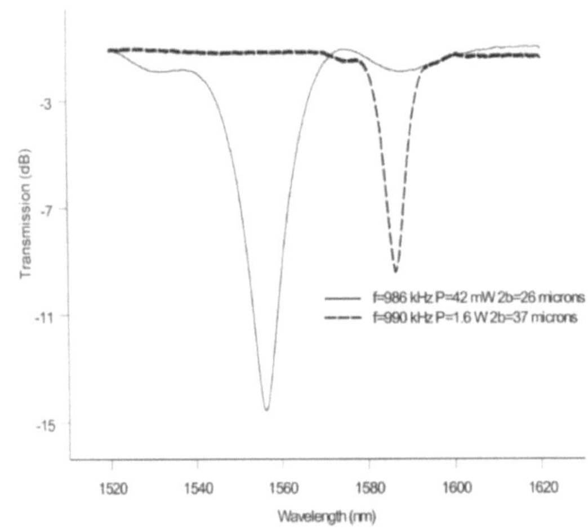


**Figure 7.** Measured transmission spectrum when fiber is etched. Significant reduction in acoustic power is observed.

## 4. Tuning of 3-dB bandwidth

The 3-dB bandwidth of the notch filter is given by the equation below as reported by D. Ostling, H.E. Engan (1995):

$$\Delta\lambda = \frac{0.8}{L}\left[\frac{\partial L_B(\lambda)}{\partial \lambda}\right]^{-1}\left[L_B(\lambda)\right]^2 \tag{6}$$

Where $\lambda$ is the wavelength of the light, L is the length of the coupling interaction, and $L_B$ is the optical beat length [11-14]. For a broadband filter, a short coupling length, a long beat length and small beat length dispersion is required. Without making the device short, only by etching the fiber to that thickness a broad filter can be obtained as reported. However this bandwidth is not tuneable and so is not suitable for spectral shaping. In this section, a similar achievement by only using a SMF to tune the 3-dB bandwidth of the filter is demonstrated. In this device, the notch filter's attenuation, peak wavelength tuning and 3-dB bandwidth can be simultaneously controlled in a single device.

To achieve this, a tunable acoustic absorber is added to the original AO setup as shown in Fig. 8. By moving the acoustic absorber along the etched region of the fiber, the interaction of light inside the acoustic region can be controlled. From Eq.6, we know that by controlling the coupling interaction length the 3-dB bandwidth of the filter can be controlled [15]. A strong acoustic absorbing material such as cotton or polystyrene can be used as the acoustic absorber [16]. The absorbing material functions to ensure no surface acoustic wave beyond the absorber's position are present. Since the interaction length of light inside the acoustic region can be controlled, over-coupling phenomenon can be monitored, to reduce the effects of undesirable side lobes. Broad filters require higher power to operate when the attenuation level is maintained the same as a narrow filter.

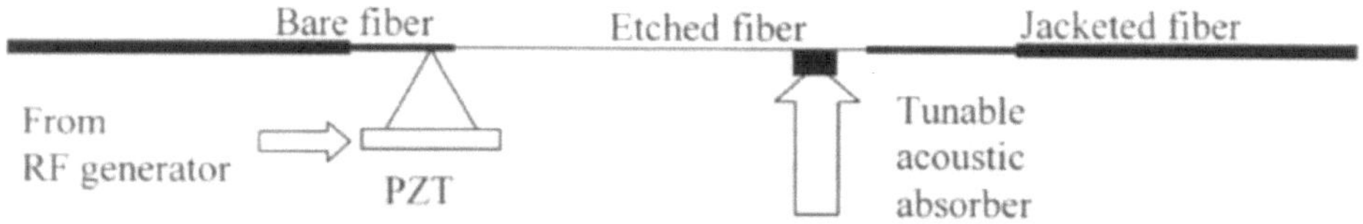


**Figure 8.** Setup to study the bandwidth variation using the AO interaction inside fiber.

From Fig. 9, the narrowest filter with a 3-dB bandwidth of 13 nm is obtained when maintaining the interaction length at 14 cm while the broadest filter with a 3-dB bandwidth of 28 nm is obtained when the interaction length is reduced to 7 cm. For an interaction length of 11.5 cm the spectral width is 16 nm and for 9 cm the spectral width will be 21 nm. The frequency used for wavelength tuning was from 960 kHz to 995 kHz and was sufficient to cover a wavelength span of 100 nm (1520 om-1620 nm).

For the broadest filter, the RF generator's $V_{P-P}$ needed to generate coupling between the modes, seem to be the highest at 14 V. Meanwhile, for the narrowest filter, the $V_{P-P}$ needed is only 6 V. Thus, we need a higher $V_{P-P}$ to generate filters for shorter interaction length of light inside the grating region [17]. The introduction of tuneable acoustic absorber will change the strain dependency on the device. To limit the strain change introduced in the device only the tip of the absorber is allowed to touch the fiber, in our case, the resonant frequency change corresponding to the strain change was maintained at +/- 0.7kHz. Throughout the experiment the total IL was maintained less than 0.1 dB and the PDL was less than 0.4 dB.

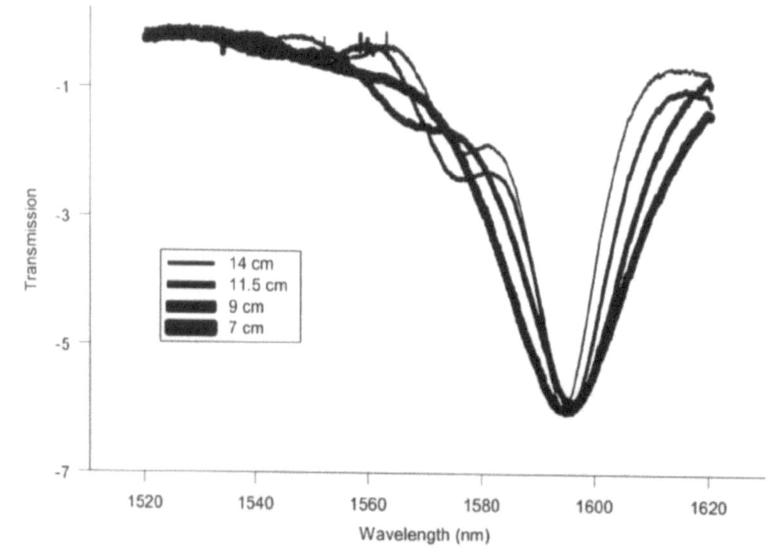


a)

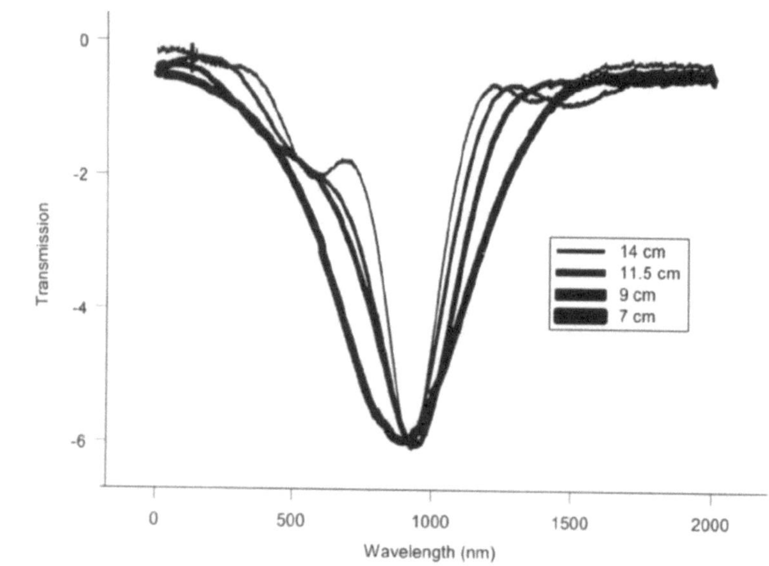


b)

a) 1596 nm b) 1566 nm

**Figure 9.** Measured bandwidth variation of filters at different peak wavelengths

## 5. Double-pass configuration

One of the key problems in fiber-based AOTF is the low attenuation level of the notch filter. Superposing two or more filters according to Yun, Lee, Kim and Kim (1999) produced by multiple transducers can increase the attenuation level. But this method introduces a very high crosstalk in the device especially when the filter's peak: wavelengths are very near to each another and prove [18] highly impractical. Alternatively to improve the attenuation level of the filter, a double pass AO setup reported by Satorious, Dimmick, Burdge (2002) and Culverhouse, Yun, Richardson, Birks, Farwell, Russell (1997) can be used. In the new setup as in Fig 1.10, a 3-port circulator is added before and after the AO device. Light comes in from port 1 of circulator 1 and goes through the acoustic region and experiences mode conversion. The LP 11 coupled mode is converted back to the fundamental mode at the jacket of the fiber. The light rounds circulator 2 and goes through the AO device and experiences mode conversion again. The produced notch filter is observed using the OSA connected to the port 3 of circulator 1. Since the period of the acoustic inside the fiber is not changed, the light going through this region experiences mode conversion at the same wavelength of the incident and returning light.

The insertion loss (IL) of a double-pass is increased to less than 3 dB and the Polarization Dependent Loss (PDL) was less than 0.6 dB. IL was not intentionally increased to a high value here, because 2 FC/FC connectors were introduced in the setup to connect port 2 of both circulators to the AO device. Splicing the ports to the device will reduce the IL loss to values less than 1 dB. Using higher quality circulators can further reduce PDL of the notch filters. The filters however will be more expensive to fabricate.

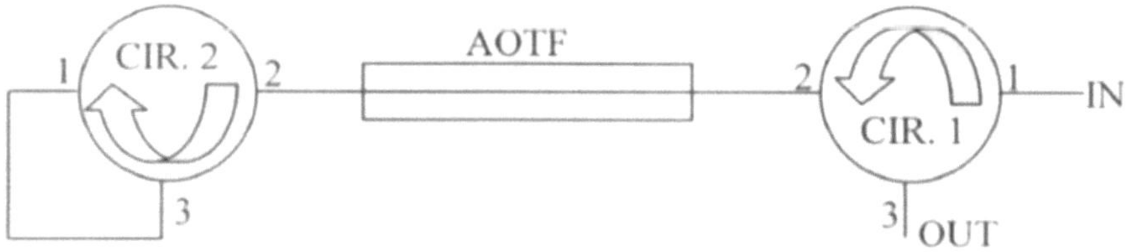


**Figure 10.** Double pass configuration to increase the maximum attenuation of the notch filter.

The AO band pass filter by Satorius *et. al.*, mentions side lobe suppression and maximum attenuation suppression using the double-pass configuration. Unlike the band pass filter, the notch filter will increase the side lobe level and maximum attenuation level using the double-pass configuration. The side lobe increment is not significant to exceed the bottom level of the main lobe of the notch filter.

From Fig. 11 the maximum attenuation of the notch filter was -28 dB for the double pass configuration and -12 dB for the single pass configuration. The maximum attenuation of the filter was increased to more than two times. The 3-dB bandwidth of the single pass AO device is 6.14 nm and the 3-dB bandwidth of the double pass AO device is 2.383 nm. This technique will be useful in producing narrow filters with high attenuation suitable in switching applications. However, there will be a frequency shift of 7 nm introduced using this setup.

The optical signal coupled from the slow mode (LP01) to the fast mode (LP11) will be downshifted in frequency when the acoustic wave is in the same direction as the optical signal. Frequency is shifted up when the fast mode is coupled to the slow mode for the same acoustic wave [19]. The frequency shift direction is reversed when the acoustic wave is in opposite direction with the optical signal as reported by Kim, Blake, Engan, and Shaw, (1986). In a double-pass setup, the optical signal is both the same and opposite direction to the acoustic wave, while in a single-pass setup, optical signal is maintained in the same direction as the acoustic wave. Thus, a frequency shift is observed in a double-pass setup.

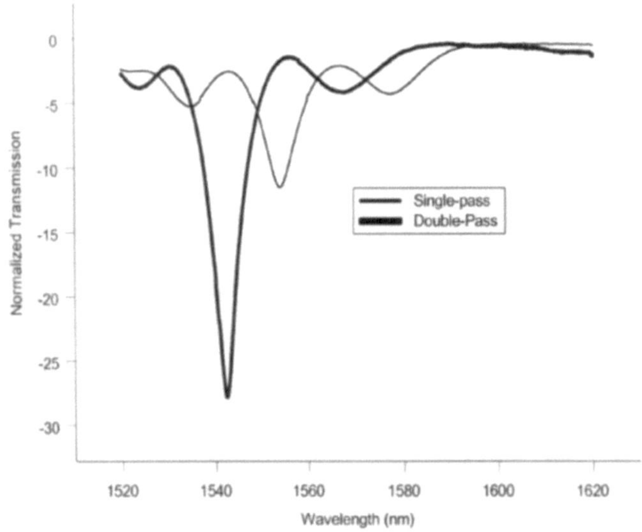


**Figure 11.** Measured normalized transmission spectrum using the double pass configuration. The result is compared with single pass configuration (refer to Fig4.1). There is a frequency shift of7 nm to the left using the double-pass setup.

## 6. Gain flattening filter

The technique to vary the 3-dB bandwidth of filter inside SMF is then extended as a dynamic gain equalizer for the gain profile of an Erbidium Doped Fiber Amplifier (EDFA). This is just one of the possible applications of AO interaction as efficient spectral shaping devices. Various efforts to dynamically control the gain flatness of the ASE spectrum using acousto-optic tuneable filters (AOTF) were well demonstrated. Passive gain equalization as reported by Vengsarkar, Pedrazzani, Judkins, Lemaire, Bergano, and Davidson (1996) is unable to encounter gain variations due to different input optical power of Wavelength Division Multiplexing (WDM) channels. Meanwhile, integrated AOTF as gain flattening filters have a serious limitation of high insertion loss and crosstalk problems. Fiber-based AOTF by H.S. Kim, Park, and B.Y. Kim (1998) and the setup by Feced, Alegria, and Zervas

(1999) uses two transducers with six synthesizers to obtain the desired spectral filters. In this technique, to shape the gain, the AOTF setup is using only one transducer and a single-taper. This is possible because the 3-dB [21-24] bandwidth of the filter we demonstrated can be varied on the same device. In our setup to flatten the gain profile of the Amplified Stimulated Emission (ASE) spectrum, an AOTF device with two frequency generators and a double-branched power combiner is used as in Fig 12. The power combiner typically introduces a 3-dB loss to the system, thus higher Vp-p from the RF generator is needed to produce the filters for spectral shaping. Total insertion loss of the setup is less than 0.2 dB. For the measurement, the EDF A was used as the ASE source and the output spectrum measured [20] using an Optical Spectrum Analyser (OSA).

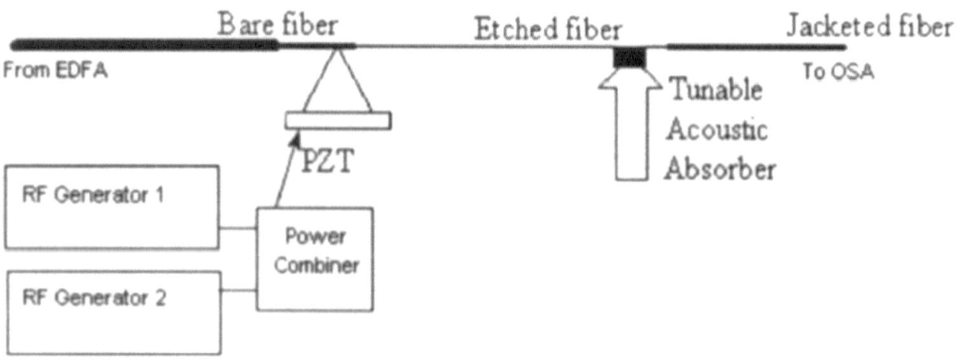


**Figure 12.** AOTF setup to flatten the gain of ASE spectrum.

The gain was flattened by changing the $V_{P-P}$ level of the RF generator, and moving the tuneable acoustic absorber along the etched region of the SMF. The degree of freedom to shape the filter is very high, thus the necessity of cascading another AOTF to the setup is not needed. Fig. 13 shows the effect of shaping the filter on the Amplified Stimulated Emission spectrum of EDFA. Typically has it Amplified Stimulated Emission s peaks at 1532 run and 1550 run. For low gain, however there is a single broad peak at 1560 run. By using this method we show that, the [26-28] ASE spectrum can be flattened regardless of the peak's position and bandwidth using the same device. Since the tuning range is about 300 run, any Amplified Stimulated Emission spectrum that is lying from 1350 run to 1630 run can be successfully flattened using the same device.

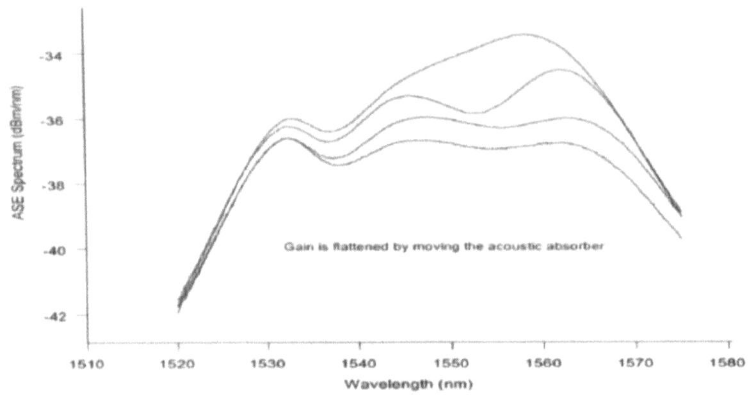


a)

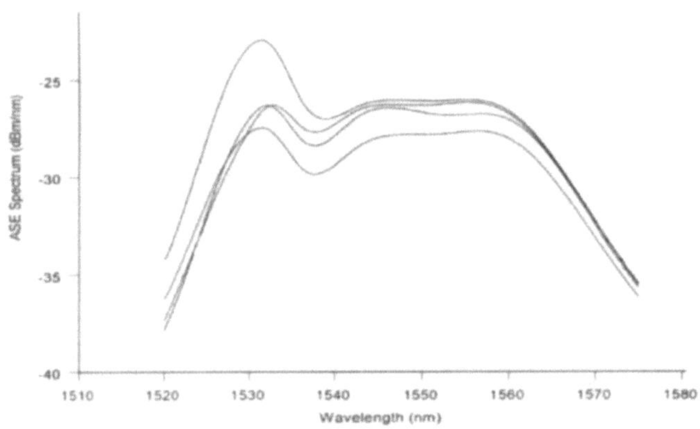


b)

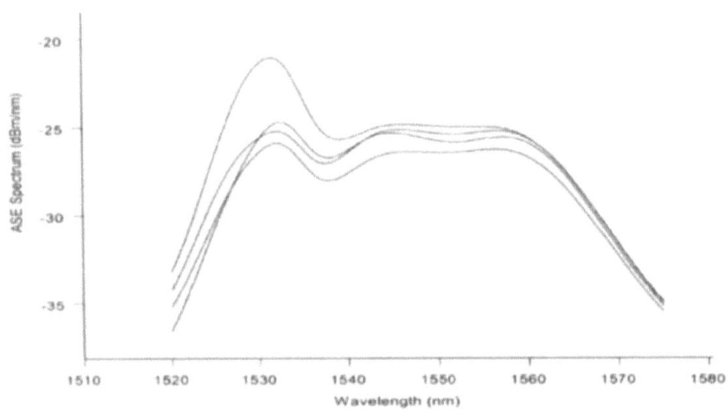


c)

**Figure 13.** The effect of moving the tuneable acoustic absorber on the Amplified Stimulated Emission spectrum at various gain levels: a) low gain single peak at 1560 nm b) and c) high gain two peaks at 1532 nm and 1550 nm.

Fig. 14 shows the flattened gain of ASE spectrum at various gain levels using this technique. For the lowest gain, at -30 dBm, which is achieved with a pump power of 96 mA, a broad filter is needed at 1545 nm; to obtain this; the tuneable acoustic absorber is positioned 14 cm after the AOTF device. The required resonant frequency to produce the coupling will be 990 kHz. A deeper notch is needed at 1532 nm; which is produced through the second frequency generator that is set at 993 kHz. The flattened gain is less than 0.8 dB. For gain at -25 dBm, which is achieved with a pump power of 150 mA, a filter is needed at 1556 nm; and a narrow deep notch is needed at 1532 nm; the required resonant frequency to produce the coupling respectively will be 986 kHz and 993 kHz. To obtain this, the tuneable acoustic absorber is positioned 16 cm after the AOTF device. The flattened gain is less than 0.9 dB.

Similarly for gain at -22 dBm, which is achieved with a pump power of 220 mA, a very deep filter is needed at 1532 nm and a small filter at 1545 nm. The resonant frequencies corresponding to these wavelengths are 990 kHz and 993 kHz respectively. To obtain the narrow filter, the tuneable acoustic absorber was set 17 cm after the AOTF device. And the measured flattened gain is less than 0.9 dB. Fig. 5 represents the notch filters obtained to flatten the gain of the Amplified Stimulated Emission (ASE) spectrum.

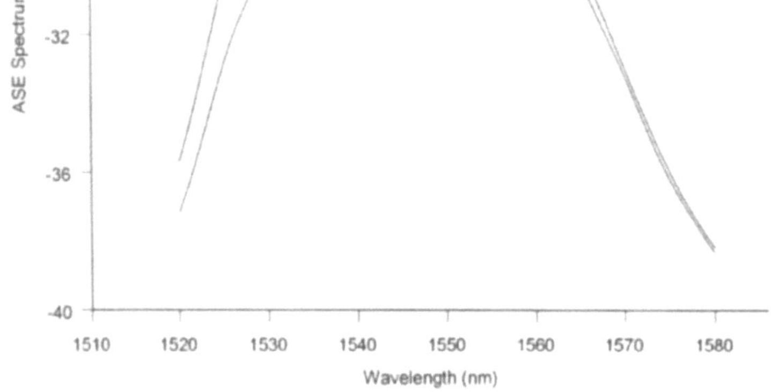


a)

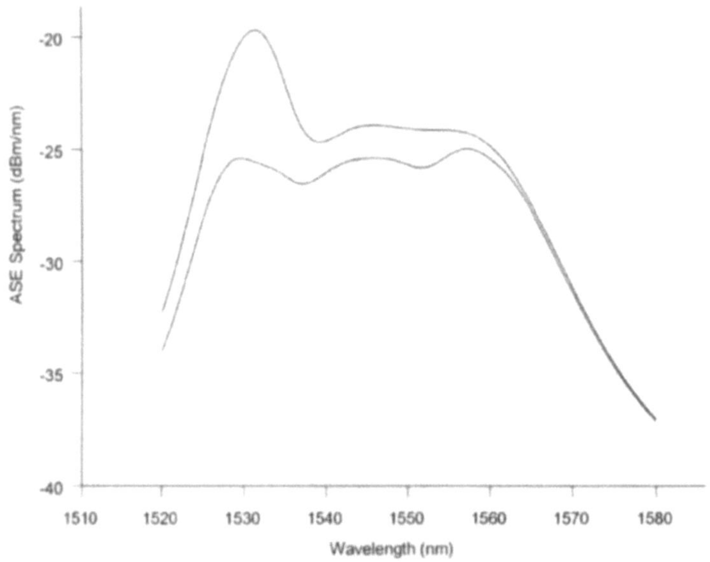


b)

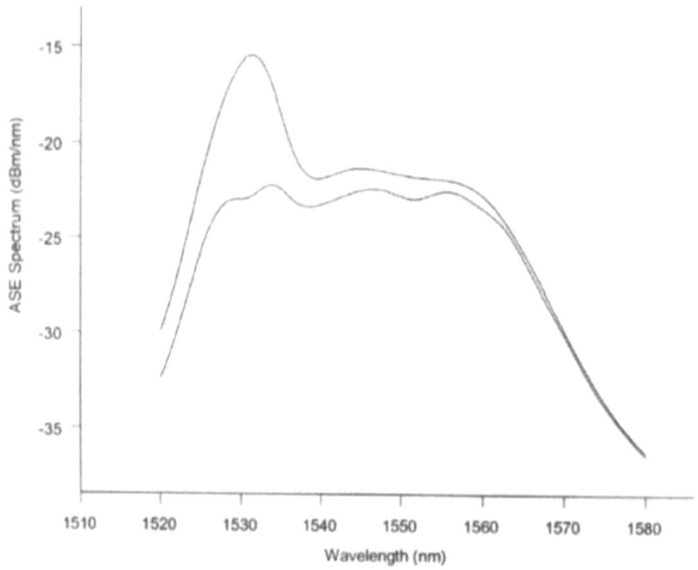


c)

**Figure 14.** Gain profiles of the ASE spectrum and the flattened gain at various pump powers: a) 96 mA which has a gain of -30 dBm b) 150 mA which has a gain of -25 dBm and c) 220 mA which has a gain of -22 dBm.

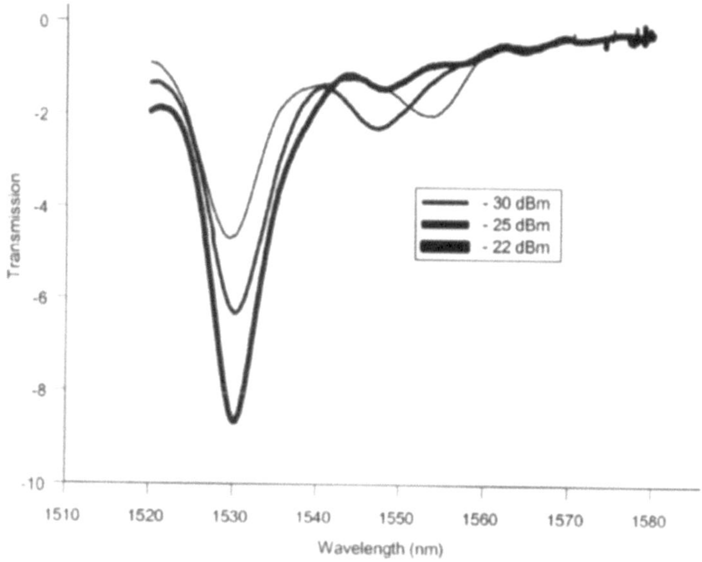


**Figure 15.** Corresponding filter spectrum to the flattened gain of various gain levels in Fig. 16.

## 7. Conclusion

The presence of acoustics inside the fiber will create a sequence of bends periodic in nature along the direction of its propagation. Core mode's energy is transferred to a cladding mode's, when it passes through the sequence of bends. The fiber jacket absorbs the coupled energy and this produces a notch filter observed using an optical spectrum analyzer.

Acoustic horn functions to transfer the acoustic wave of the transducer to the fiber. Aluminium horn is preferred over silica horn because it can be easily reproduced. Furthermore, its acoustic impedance almost matches that of silica's. To allow optimum transmission of acoustics to the fiber, the tip of the horn is made small, with its diameter matching that of silica's.

No resonance peaks were observed when the fiber is unetched First peaks are observed when the thickness of fiber is approximately 40 μ m. Overlap integral between the modes is not high in thicker fiber, meaning the transfer of acoustic wave to the fiber is not optimized. Thickness reductions in fibers are observed using a CCD camera. The characteristics of the resonance peaks can be controlled electrically using a RF generator. Voltage of the generator can be used to tune the attenuation depth of the resonance peaks. Frequency of the generator can be used to tune the peak wavelength. Frequency is inversely related to period of bends, thus higher frequencies will shift the peak to lower wavelengths. The 3-dB bandwidth of the resonance peaks can be adjusted by limiting the acoustic bend produced inside the fiber. Introducing a tunable acoustic absorber along the fiber can do this. Frequency used in all experiments was from 800 kHz to 1.1 MHz. All the coupled energy to produce the resonance peaks were to LP11 modes, mode conversion observed using a beam pro filer.

The power fed to cause resonance peaks can be reduced by reducing the thickness of the fiber to a value close to 20 $\mu$ m. Allowing light to pass through the acoustic bend region twice, as proposed in the double pass configuration, can increase the attenuation peaks. However, a frequency shift of 13 nm is observed because the light is passing through the bend in opposite directions.

As a spectral shaping tool, the attenuator is efficient as a gain flattening filter for an erbium doped amplifier. The peak of an amplified spontaneous emission at 1531 nm can be reduced to flat levels for various gains of the EDF A pump power. Insertion loss is less than 0.2 dB and polarization dependence loss is less than 0.4 dB.

## Author details

Abhilash Mandloi and Vivekanand Mishra
*Department of Electronics Engineering, S.V. National Institute of Technology, Surat 395007, Gujarat, India*

## 8. References

[1] Sutharsanan Veeriah, "Design and Characterisation of All Fiber Optical Filters", Master of Science Thesis, Faculty of Engineering, Multimedia University, Malaysia, Feb 2006.

[2] Abdulhalim 1., Pannell C.N., (1993). Acoustooptic in-fiber modulator using acoustic focusing. IEEE Photonics Technology Letters, 9 (5), 999-1002

[3] Au A. A., Liu Q., Lin C.H., Lee H.P., (2004). Effects of Acoustic Reflection on the Performance of a Cladding-Etched All-Fiber Acoustooptic Variable Optical Attenuator. IEEE Photonics Technology Letters, 16 (1), 150-152

[4] Birks T.A., Russell P.SJ., Culverhouse D.O., (1992). The Acousto-Optic Effect in Single-Mode Fiber Tapers and Couplers. Journal of Lightwave Technology, 14 (11), 2519-2529

[5] Culverhouse D.O., Yun S.H., Richardson D.J., Birks T.A, Farwell S.G., Russell P.StJ., (1997). Low-loss all- fiber acousto-optic tunable filter. Optics Letters, 22 (2), 96-98

[6] Engan H.E., (2000). Acousto-Optic Coupling In Optical Fibers. IEEE Ultrasonics Symposium, 625-629

[7] Engan H.E., Kim B.Y., Blake J.N., Shaw H.J., (1988). Propagation and Optical Interaction of Guided Acoustic Waves in Two-Mode Optical Fibers. Journal of Lightwave Technology, 6 (3), 428-436

[8] Engan H.E., Ostling D., Kval Per 0., Askautrud Jan 0 .. Wideband Operation of Horns for excitation of Acoustic Modes in Optical Fibers. 10th Optical Fibre Sensors Conference, 568-571

[9] Feced R., Alegria c., Zervas M.N., (1999). Acoustooptic Attenuation Filters Based on Tapered Optical Fibers. IEEE Journal of Selected Topics in Quantum Electronics, 5 (5), 1278-1288

[10] Jung Y., Lee S.B., Lee J.W., Oh K., (2005). Bandwidth control in a hybrid fiber acoustooptic filter. Optics Letters, 30 (1), 84-86

[11] Keiser G., (1991). Optical Fiber Communications. McGraw-Hill, Inc., (2nd Edition) Kim RY, Blake J.N., Engan H.E., Shaw H.J., (1986). All-fiber acousto-optic frequency shifter. Optics Letters, 11 (6),389-391

[12] Kim H.S., Yun S.H., Kim H.K., Park N., Kim B.Y., (1998). Actively Gain-Flattened Erbium-Doped Fiber Amplifier Over 35 nrn by Using All-Fiber Acoustooptic Tunable Filters. IEEE Photonics Technology Letters, 10 (6), 790-792

[13] Kim H.S., Yun S.H., Kwang LK., Kim B.Y., (1997). All-fiber acousto-optic tunable notch filter with electronically controllable spectral profile. Optics Letters, 22 (19), 1476-1478

[14] Lee S.S., Kim H.S., Hwang LK., Yun S.H., (2003). Highly-efficient broadband acoustic transducer for all-fibre acousto-optic devices. Electronics Letters, 39 (18)

[15] Li Q., Au A.A., Lin C.H., Lyons E.R, Lee H.P., (2002). An Efficient All-Fiber Variable Optical Attenuator via Acoustooptic Mode Coupling. IEEE Photonics Technology Letters, 14 (11),1563-1565

[16] Li Q., Liu X., Lee H.P., (2002), Demonstration of Narrow-Band Acoustooptic Tunable Filters on Dispersion-Enhanced Single Mode Fibers. IEEE Photonics Technology Letters, 14 (11), 1551-1553

[17] Li Q., Liu X., Peng J., Zhou B., Lyons E.R, Lee H.P., (2002). Highly Efficient Acoustooptic Tunable Filter Based on Cladding Etched Single-Mode Fiber. IEEE Photonics Technology Letters, 14 (3), 337-339

[18] Liu Q., Chiang K.S., Rastogi V., (2003). Analysis of Corrugated Long-Period Gratings in Slab Waveguides and Their Polarization Dependence. Journal of Lightwave Technology, 21 (12),3399-3405

[19] Love J.D., Henry W.M., Stewart W.J., Black RJ., Lacroix S., Gonthier F., (1991). Tapered Single-mode fibres and devices Part 1: Adiabaticity criteria. IEE Proceedings, 138, (5), 343-354

[20] Monerie M., (1982). Propagation in Doubly Clad Single-Mode Fibers. IEEE Transactions on Microwave Theory and Techniques, MTT-30 (4), 381-388

[21] Mononobe S., Ohtsu M., (1996). Fabrication of a Pencil-Shaped Fiber Probe for Near-Field Optics by Selective Chemical Etching. Journal of Lightwave Technology, 14 (10), 2231-2235

[22] Ostling D., Engan H.E., (1995). Narrow-band acousto-optic tunable filtering in a twomode fiber. Optics Letters, 20 (11), 1247-1249

[23] Ostling D., Engan H.E., (1995). Spectral Flattening by an All-Fib er Acousto-Optic Tunable Filter. IEEE Ultrasonics Symposium, 837-840

[24] Pannell C.N., Wacogne B.F., Abdulhalim 1., (1995). In-Fib er and Fiber Compatible Acoustooptic Components. Journal of Lightwave Technology, 13 (7),1429-1434

[25] Satorius D.A., Dimmick T.E., Burdge G.L, (2002). Double-Pass Acoustooptic Tunable Bandpass Filter With Zero Frequency Shift and Reduced Polarization Sensitivity. IEEE Photonics Technology Letters, 14 (9), 1324-1326

[26] Yun S.H, Lee B.W., Kim H.K., Kim B.Y., (1999). Dynamic Erbium-Doped Fiber Amplifier Based on Active Gain Flattening with Fiber Acoustooptic Tunable Filters. IEEE Photonics Technology Letters, 11 (10), 1229-1231

[27] Yun S.H., Kim B.Y., Jeong HJ., Kim B.Y., (1996). Suppression of polarization dependence in a two-mode-fiber acousto-optic device. Optics Letters, 21 (12), 908-910

[28] Yun S.H., Kim H.S., (2004). Resonance in Fiber-Based Acoustooptic Devices Via Acoustic Radiation to Air. IEEE Photonics Technology Letters, 16 (1), 147-149

Section 5

# Technological Systems

# Chapter 18

# Techniques for Tuning BAW-SMR Resonators for the 4th Generation of Mobile Communications

M. El Hassan, E. Kerherve, Y. Deval, K. Baraka, J.B. David and D. Belot

Additional information is available at the end of the chapter

http://dx.doi.org/10.5772/55131

## 1. Introduction

In telecommunication systems, all filters and resonators that constitute the RF part have the tendency to be integrated on the same chip that contains the information treatment.

In order to achieve miniaturization, bulk acoustic wave (BAW) technology is presented. BAW filters are very sensitive to surface contamination, and can exhibit very small sizes. In addition, BAW resonators could be fabricated using compatible material CMOS and BiCMOS [1]. In this context and in order to compensate the variation due to the fabrication process, the work presented in this paper focuses on the tuning of BAW-SMR resonators and filters.

This work is divided in two parts. The first part consists of designing BAW-SMR (Solidly Mounted Resonator) filters. In the second part, we propose the use of two methods to tune this type of filters. Thus, we present the design methodology, the study, and the experimental realization of the BAW-SMR tunable filters.

## 2. BAW impedance behavior

### 2.1. SMR impedance behavior

The bulk acoustic wave resonator is basically constituted by a piezoelectric layer sandwiched between two electrodes (Fig.1). The application of an electric field between the two electrodes generates a mechanical stress that is further propagated through the bulk of the structure (acoustic wave). The resonance condition is established when the acoustical path (in thickness direction) corresponds to odd integer multiples of the half acoustic wavelength. The bulk acoustic wave resonator is basically constituted by a piezoelectric layer sandwiched between two electrodes (Fig.1) [2, 3].

The bulk acoustic wave resonator fabrication over silicon substrates imposes its acoustical isolation, confining the acoustic waves into the main resonant structure. Two configurations are proposed: the membrane suspended structure (FBAR – Film Bulk Acoustic wave Resonator) [3], where the resonator is suspended by an air-bridge (Fig.1a); and the solidly-mounted structure (SMR – Solidly Mounted Resonator), where the resonator is mounted over a stack of alternating materials (Fig.1b). This stack is built on a Bragg reflector basis and it has an acoustic mirror behavior [2-3]. Both, air and acoustic mirror, present an optimum discontinuity for reflecting the acoustic waves at the interface with the bottom electrode, confining waves into the main resonant structure.

In the solidly mounted resonator (SMR), the piezoelectric is solidly mounted to the substrate (Fig.1.b). Some means must be used to acoustically isolate the piezoelectric from the substrate if a high quality factor (Q) resonance is to be obtained. In effect, the quarter wavelength layers act as a reflector to keep waves confined near the piezoelectric transducer film [4]. The effect of the reflector on mechanical displacement is to cause the wave amplitude to diminish with depth into the reflector. The number of layers required to obtain a satisfactory reflection coefficient is dependent on the mechanical impedances between layers and, to a lesser extent, the substrate [5]. The number of layers is best determined by an analysis of the resonance response as a function of the number of layers versus resonator 'Q' and coupling coefficient.

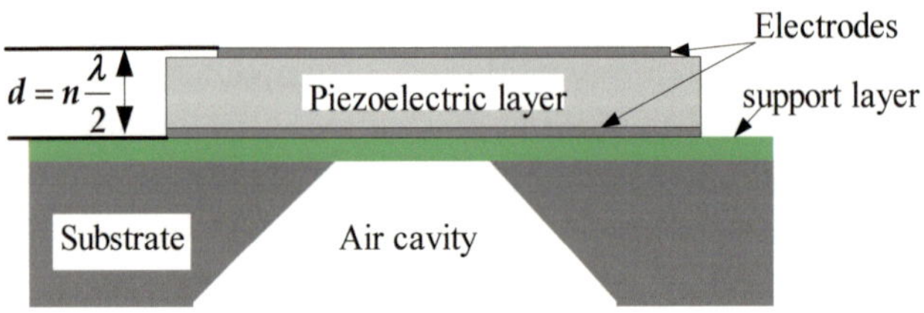


(a)

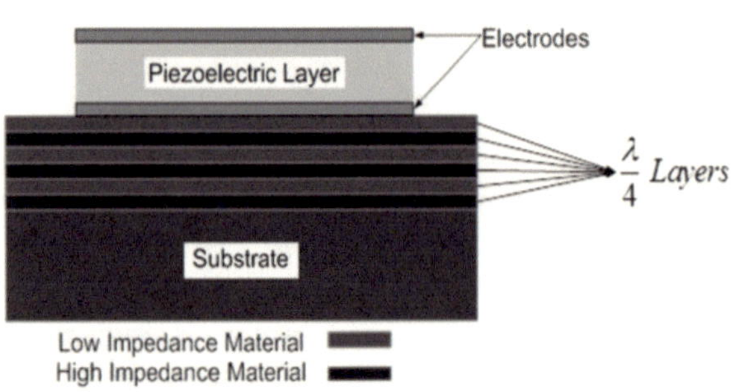


(b)

**Figure 1.** (a) Film Bulk Acoustic Resonator (FBAR). (b) Solidly Mounted Resonator (SMR).

An important effect of the reflector layers, as demonstrated by Newell [4], is a partial lateral stiffening of the piezoelectric plate that minimizes plate wave generation and spurious resonances normally observed in free plates. However, real resonator structures are inherently 3D and some form of radiation beyond the simple thickness dimension is to be expected. If energy leaves the resonator structure, through radiation, then it counts as a loss mechanism. Reflections of lateral waves at the edge of the resonator can lead to standing waves and spurious responses. The SMR approach requires that the substrate be smooth in order to proceed with the fabrication of reflectors, electrodes, and piezoelectric film [5].

## 2.2. The Butterworth-Van-Dyke (BVD) model

The Modified Butterworth-Van-Dyke (MBVD) model is an electrical schematic around resonance (Fig.2). The elements $R_a$, $L_a$, $C_a$ present the series resonance and the insertion losses. The capacity $C_0$ represents the piezoelectric material between the two electrodes.

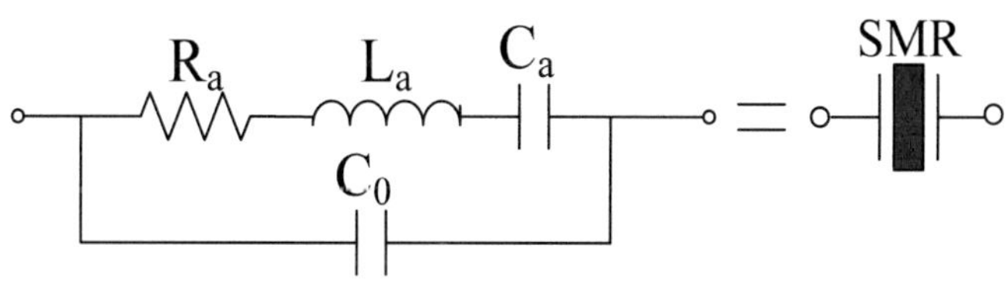


**Figure 2.** SMR-MBVD model.

The impedance characteristic of a measured BAW-SMR is shown in Fig.3. In this graph, it can be observed that the SMR presents mainly two resonance pulsations: the series resonance ($f_s$), when the electrical impedance approaches to zero, and the parallel resonance ($f_p$), when the electrical impedance approaches to infinity. For all other frequencies far from

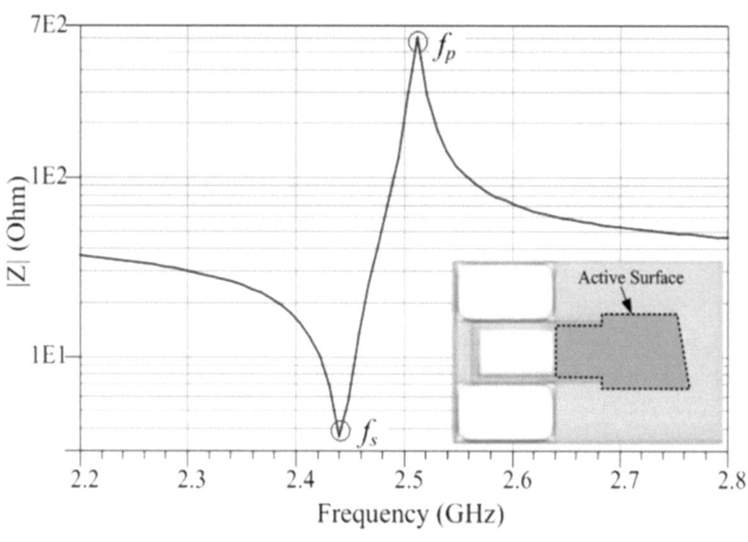


**Figure 3.** Impedance characteristic of a measured BAW-SMR.

the resonances, the SMR presents static capacitor behavior. $f_s$ is adjusted according to the thickness of the piezoelectric layer and it is spaced by the parallel resonance $f_p$. The instantaneous frequency deviation between the two resonances is determined by the electromechanical coupling coefficient of the piezoelectric layer. The quality factor of the measured BAW-SMR is 192.5 and its active area is 16800 μm². This relatively small quality factor is due to some technological problems.

The electrical impedance of an SMR is obtained by solving the acoustic boundary problem and applying the transmission line theory [6]. The electrical SMR impedance can be simplified and expressed by the following equation:

$$Z_{SMR} = \frac{1}{sC_0} \frac{(f^2 - f_s^2)}{(f^2 - f_p^2)} \quad (1)$$

where 's' is the complex variable: s = $jw$.

### 2.3. Addition of external passive elements

The association of external passive components (L, C) to the resonators was made in two parts. First, the resonators analyses have been carried out using on-wafer measurements. Then, these results have been associated with the capacitors and inductors. The combination of experimental results and modeling of passive elements constitutes the final response of the tuned resonators.

#### *2.3.1. Addition of capacitors*

The addition of the capacitors (having an intrinsic quality factor 'Q' > 140) to the SMR circuitry doesn't affect severely the quality factor of the overall design. The resonator's electromechanical coupling coefficient is described indirectly by the capacitor ratio $C_0/C_a$ as determined by the resonator physical configuration and piezoelectric material properties [7]. When changing the electromechanical coefficient of the piezoelectric material, the bandwidth changes. Our goal is to tune the capacitor ratio $C_0/C_a$ by adding series or parallel capacitors to the resonator. Thus, controlling this ratio will enable us to control the electromechanical coefficient, and as a sequence the bandwidth of the resonator.

##### *2.3.1.1. Series capacitor association*

The performance analysis of the association of series capacitor to the SMR is based on the BVD model (Fig.4). The analysis of the frequency response of the series capacitor associated to the SMR will lead us to (2) presented below:

$$Z_{in} = \frac{s^2(CC_aL_a + C_0C_aL_a) + C_0 + C_a + C}{sC(s^2C_0C_aL_a + C_0 + C_a)} \quad (2)$$

Where's' is the complex variable: s = $jw$.

From (2), it is possible to notice the insertion of a pole and zero to the frequency response of the simple resonator. The extraction of the poles and zeroes values of (2), will lead us to (3) & (4) shown below.

$$f_s = \frac{1}{2\pi}\sqrt{\frac{C_0 + C_a + C}{CC_aL_a + C_0C_aL_a}} \tag{3}$$

$$f_p = \frac{1}{2\pi}\sqrt{\frac{C_0 + C_a}{C_0C_aL_a}} \tag{4}$$

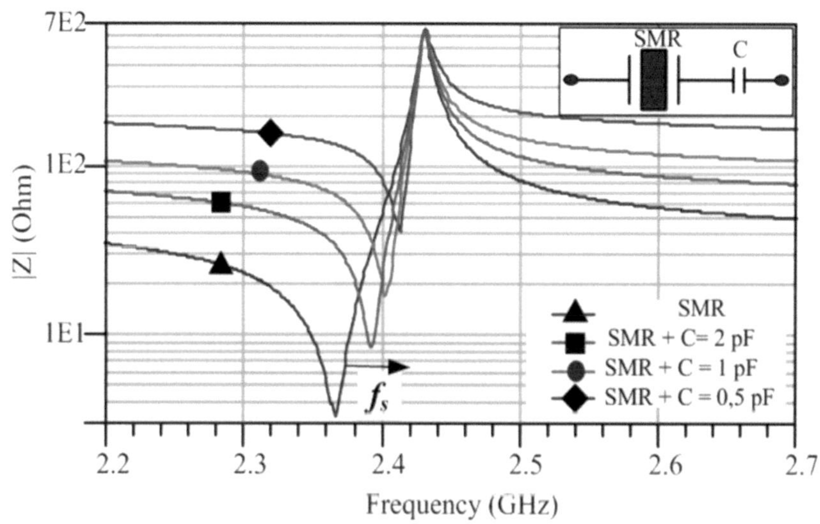


**Figure 4.** Impedance response of the measured BAW-SMR with series capacitor.

From (3) & (4), it may be noted that the capacitor added in series with the BAW resonator affects only the series resonance frequency. It is inversely proportional to $f_s$. To illustrate this theoretical study, responses with different values of capacitors added in series with the measured BAW-SMR are shown in Fig.4.

*2.3.1.2. Parallel capacitor association*

The analysis of the frequency response of the parallel capacitor associated to the SMR will lead us to (5), presented below.

$$Y_{in} = \frac{s(s^2L_aC_a(C_0 + C) + C_0 + C + C_a)}{s^2L_aC_a + 1} \tag{5}$$

$$f_s = \frac{1}{2\pi}\sqrt{\frac{1}{C_aL_a}} \tag{6}$$

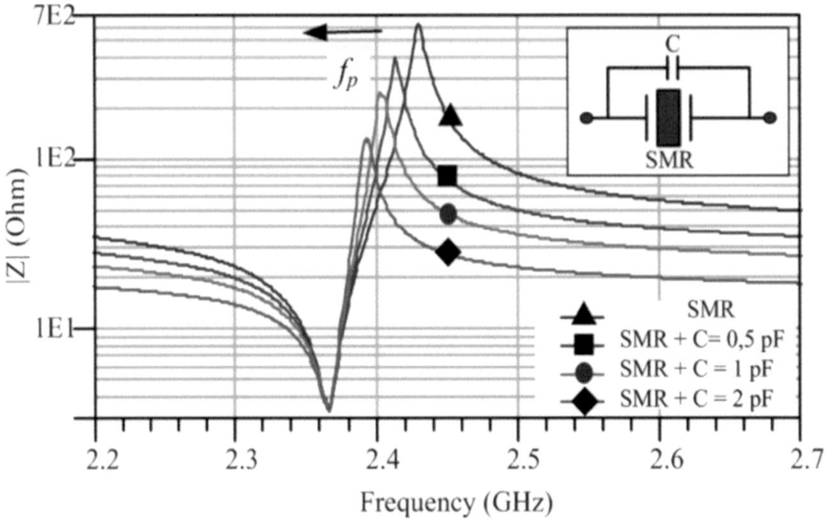


**Figure 5.** Impedance response of the measured BAW-SMR with parallel capacitors.

$$f_p = \frac{1}{2\pi}\sqrt{\frac{C_0 + C_a + C}{L_a C_a (C_0 + C)}} \tag{7}$$

From (5), it is possible to notice the displacement of the poles of the device's resonance frequency. The extraction of the poles and zeroes values of (5), will guide us to (6) & (7) as shown below.

It is noticed from (6) & (7) that the capacitor added in parallel with the BAW resonator affects only the parallel resonance frequency. It is inversely proportional to $f_p$. Fig.5 presents the final responses of the measured BAW-SMR with different values of capacitors added in parallel.

### 2.3.2. *Addition of inductors*

The association of inductors to the BAW resonators could be realized in series or parallel. We should note that in the case of VLSI-CMOS, these inductors are characterized by a small quality factor with respect to the BAW (400 to 1000) which degrades the quality factor of the overall circuitry, and they occupy a relatively large size.

#### 2.3.2.1. *Series inductor association*

The performance analysis of the assembly of the inductors in series with the SMRs is based on the BVD model. The analysis of the frequency response of the series inductors associated to the SMR will lead us to (8).

$$Z_{in} = \frac{s^4 C_0 C_a L_a L + s^2 (C_a L_a + C_0 L + C_a L) + 1}{s(s^2 C_a L_a C_0 + C_0 + C_a)} \tag{8}$$

From (8), it is possible to notice the insertion of a zero to the resonator frequency response. The extraction of the poles and zeroes values from (8), will bring (9) & (10). Based on these equations, we can see that the values of the poles are not deteriorated. However, from (9), we notice that the association of series inductors to the SMR will modify the zeroes, and only the series resonance frequency is affected. It is inversely proportional to fs.

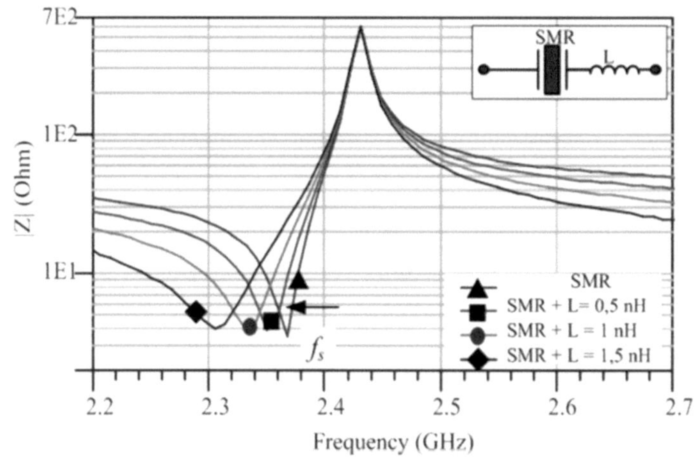


**Figure 6.** Impedance response of the measured BAW-SMR with series inductors.

To illustrate this theoretical study, impedance behavior of the measured BAW-SMR with different values of inductors associated in series is shown in Fig.6.

$$f_s = \frac{1}{2\pi}\sqrt{\frac{(C_aL_a + C_0L + C_aL) \pm \sqrt{\left(C_aL_a + C_0L + C_aL\right)^2 - 4C_0C_aL_aL}}{2C_0C_aL_aL}} \tag{9}$$

$$f_p = \frac{1}{2\pi}\sqrt{\frac{C_0 + C_a}{C_aL_aC_0}} \tag{10}$$

*2.3.2.2. Parallel inductor association*

Based on the same procedure used above, the analysis of the frequency response of the parallel inductors associated to the SMR will lead us to (11), presented below.

$$Y_{in} = \frac{s^4C_0C_aL_aL + s^2\left(C_0L + C_aL + C_aL_a\right) + 1}{sL(1 + s^2C_aL_a)} \tag{11}$$

From (11), it is possible to notice the insertion of a pole to the resonator frequency response and the displacement of another. Also, we can notice the composition of a double pole near the frequency response of the device. The extraction of the poles and zeroes values from (11), will lead us to (12) & (13).

From (12) & (13), we can see that that the inductor added in parallel with the BAW resonator affects only the parallel resonance frequency. It is inversely proportional to $f_P$. Fig.7 presents the final responses of the measured BAW-SMR with different values of inductors associated in parallel.

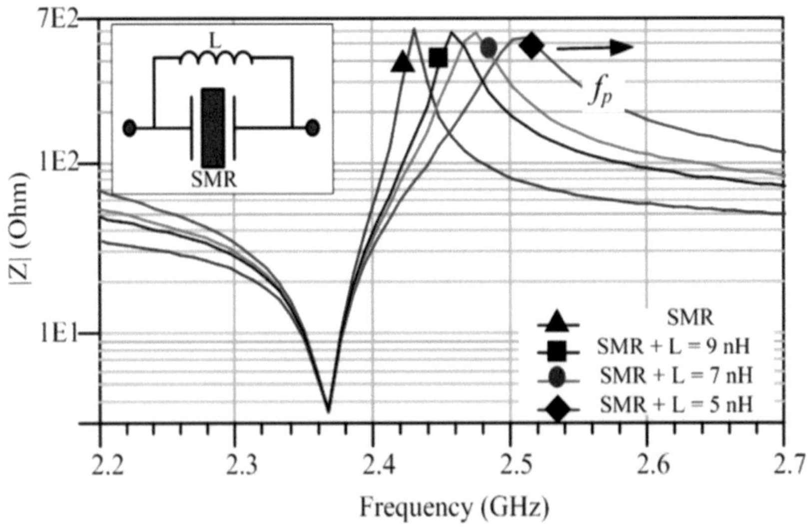


**Figure 7.** Impedance response of the measured BAW-SMR with parallel inductors.

$$f_p = \frac{1}{2\pi}\sqrt{\frac{\left(C_0 L + C_a L + C_a L_a\right) \pm \sqrt{\left(C_0 L + C_a L + C_a L_a\right)^2 - 4C_0 C_a L_a L}}{2C_0 C_a L_a L}} \tag{12}$$

$$f_s = \frac{1}{2\pi}\sqrt{\frac{1}{C_a L_a}} \tag{13}$$

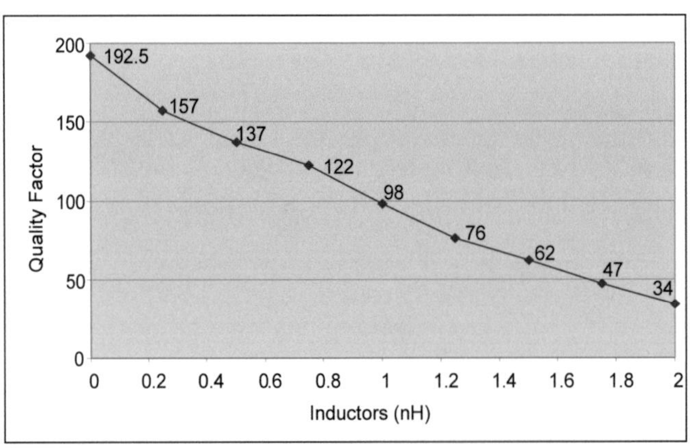


**Figure 8.** Impact of the quality factor of the inductors added in series with the BAW resonator on the series quality factor '$Q_s$' of the overall resonator circuitry (SMR+L).

The addition of inductors to the BAW-SMR increases the insertion losses and degrades the quality factor of the overall circuitry. Fig.8 illustrates the influence of inductors added in series with the SMR on the overall quality factor of the assembly.

# 3. BAW filter design

## 3.1. Bulk Acoustic Wave (BAW) filters topologies

Bulk acoustic wave filters are basically divided in three topologies: ladder, lattice and ladder-lattice (Fig.9) [3].

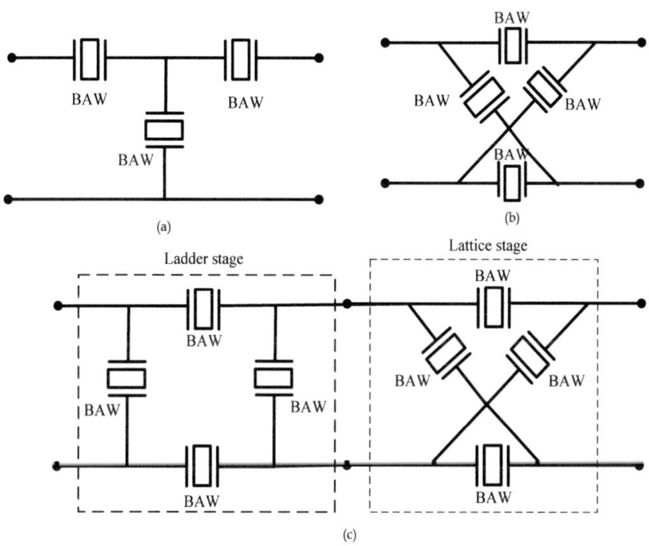


**Figure 9.** Bulk acoustic wave resonator filter topologies: (a) ladder stage, (b) lattice stage and (c) ladder-lattice stage.

Ladder filter are characterized by an unbalanced operation mode and very small size able to deliver very high selectivity filtering responses, however presenting low rejection or isolation out-of-band. Typically, ladder bulk acoustic wave filters are quite effective for blocking signals close to the passband, but poor at rejecting undesired bands [3-4].

On the other hand, lattice bulk acoustic wave filters are characterized by a balanced operation mode. In contrast to the first one, this topology presents typically a low selectivity close to the passband, followed by a high rejection out-of-band. They present slower roll off coefficient and higher rejection. Thus, lattice networks are not interesting to block signals close to the passband, but they are more effective for rejecting undesired bands [3].

Ladder-lattice bulk acoustic wave filters are also characterized by a balanced operation mode. Ladder-lattice filters ally advantages of both network types, enabling high isolation at undesired bands and steep responses close to the passband [3]. This topology is able to strongly reduce the linearity constraints of the receiver RF chain. Fig.10 shows a comparison between the typical theoretical transmission responses ($S_{21}$) of these three network types.

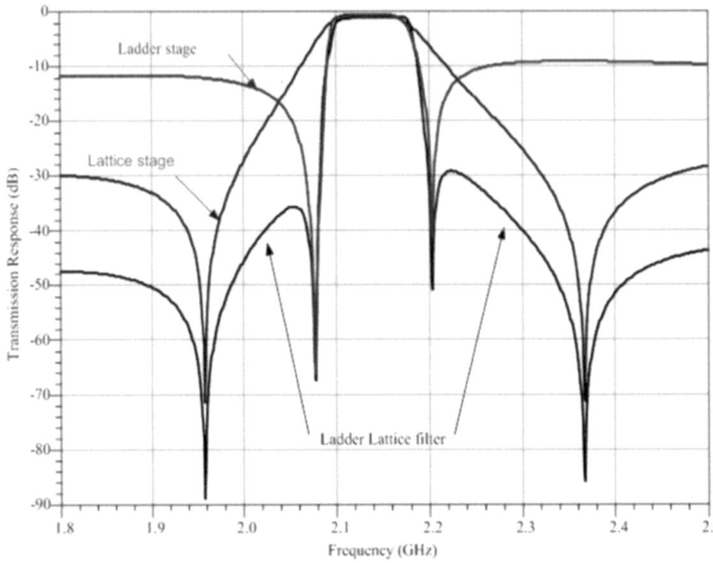


**Figure 10.** Theoretical transmission responses for main bulk acoustic wave (BAW) filter topologies.

## 3.2. Ladder BAW-SMR filter design

The ladder filter is an association of resonators in series and in parallel. The shunt resonators are loaded and their resonance frequencies are smaller than the series resonators. Ladder BAW topology presents a good selectivity which enables to block undesired signals near the pass band. In this context, and in order to support the theoretical study, a tunable BAW-SMR ladder filter was designed for the 802.11b/g standard (2.40 – 2.48 GHz). The resonators and filters were fabricated at the CEA-Leti in the framework of the project '*EPADIM*'. The filter is composed of five SMRs, associated in ladder topology. The filter stack can be divided into resonators' layers and Bragg reflector's layers. The resonator's layers are composed by the classical couple AlN-Mo [8]. However, in contrast to [8], the Bragg reflector was implemented using an exclusive dielectric stack composed by SiOC:H and $Si_xN_y$ [9].

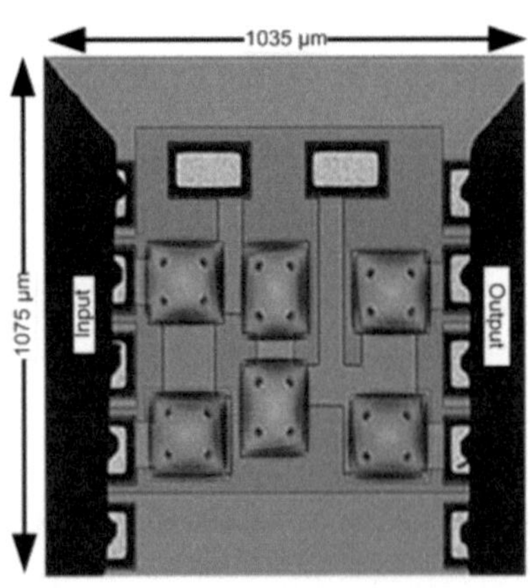


**Figure 11.** Microphotography of the Ladder BAW-SMR for the 802.11b/g standard.

The acoustical performance of the fully dielectric stack is comparable to the traditional SiO2-W reflectors; however, it strongly reduces the coupling between resonators through the Bragg reflector. Furthermore, the filter stack was realized on a high resistive silicon substrate in order to reduce losses due to the capacitive coupling [10].

In order to optimize the filter performance, a double resonator and apodized geometries have been used. Indeed, double resonators present large electrodes' areas, which results in lower resistive losses. Also, the filter resonators present apodized geometries in order to avoid spurious resonances caused by the parasitic lateral acoustic modes [11]. Fig.11 and 12 show the microphotography and a comparison between the measured and simulated results of the tunable ladder BAW-SMR filter, respectively. The filter occupies a small area and has reduced dimensions (1035*1075 μm²). Electromagnetic simulation of the overall filter structure has been performed using the ADS-Momentum software. Next, the acoustical effects have been considered using the Mason Model [12] and included in the simulations.

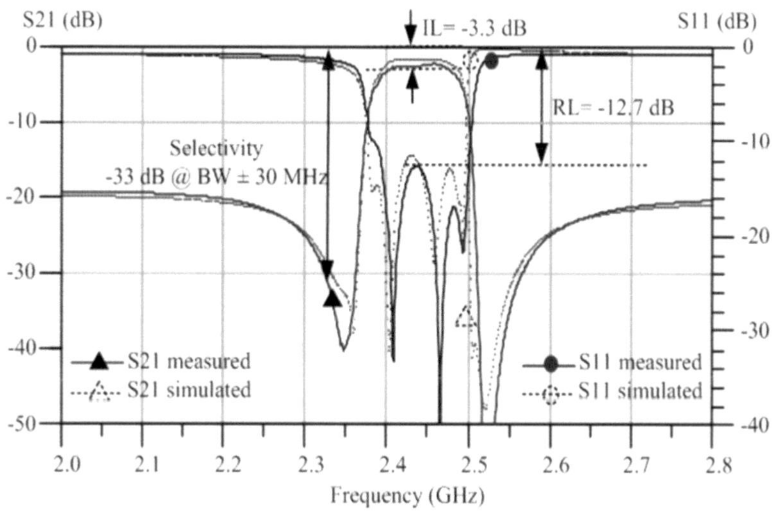


**Figure 12.** Comparison between measured and simulated results of the 802.11b/g ladder BAW-SMR filter.

The filter design was realized for implementation in SiP context. The performances of the tunable BAW-SMR filter are in concordance with the simulation results. Mainly, the filter fulfills the requirements for the WLAN 802.11 b/g standard, presenting -3.3 dB of insertion loss, -12.7 dB of return loss and a selectivity higher than 33 dB at ± 30 MHz of the bandwidth. The filter high insertion losses are mainly due to the low resonators quality factor obtained in the fabrication (Q = 200). Therefore, these losses can be strongly reduced using mechanical energy concentration techniques in the resonator acoustical cavity [13].

## 4. Tunability of the ladder filter

The shunt resonators of the ladder filter determine the position of the zeroes at the left of the center frequency and the series resonators determine the position of the zeroes at the right of the center frequency. Thus, changing the impedance of the parallel and series resonators leads to a change in the zeroes' positions.

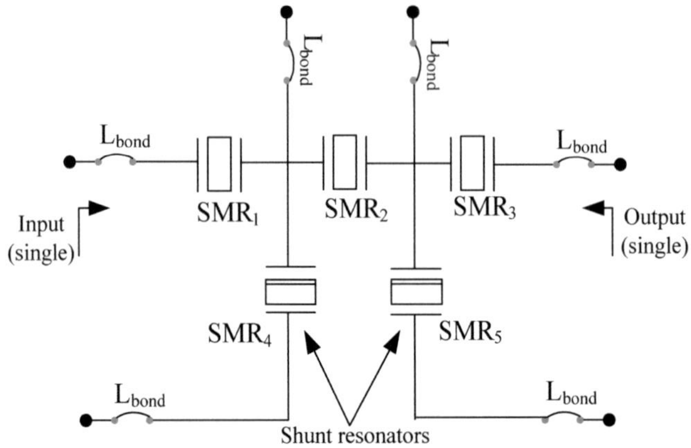


**Figure 13.** Ladder BAW-SMR filter.

Based on this theory and in order to tune the BAW-SMR filter (Fig.13), we propose to add passive elements to the shunt and series resonators that constitute the filter.

## 4.1. Shift towards higher frequencies

To shift the center frequency of the filter towards higher frequencies we have to move all the zeroes towards these frequencies. Thus, we have added inductors in parallel with the series resonators and capacitors in series with the shunt resonators that constitute the ladder BAW-SMR filter (Fig.14).

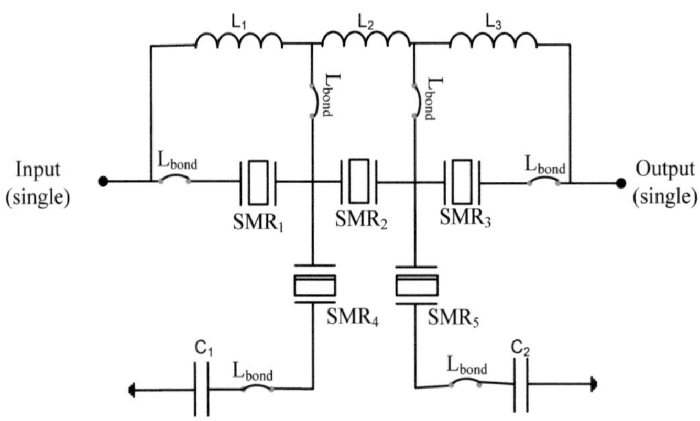


**Figure 14.** Tunable BAW-SMR filter with inductors added in parallel with the series resonators and capacitors added in series with the shunt resonators.

Table 1 presents the values and the quality factor of the external passive elements used in the circuitry designed to move the center frequency of the tunable BAW-SMR filter towards higher frequencies.

| | Value | Quality Factor 'Q' |
|---|---|---|
| Capacitors | $C_1$= $C_2$=2.2 pF | 140 |
| Inductors | $L_1$=3.9 nH | 47.7 |
| | $L_2$=5.6 nH | 40 |
| | $L_3$=4.7 nH | 45 |

**Table 1.** Values of the passive elements used to shift the center frequency of the tunable BAW-SMR filter towards higher frequencies and their quality factors.

The tunable BAW-SMR filter and the passive components are mounted on a FR4 substrate as shown in Fig.15.

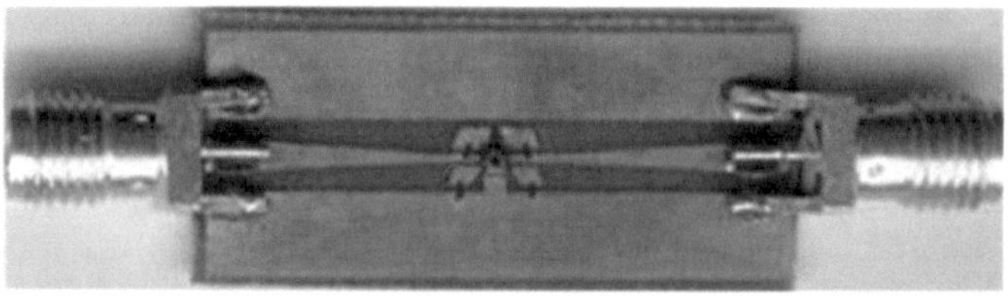

**Figure 15.** Tunable BAW-SMR filter and passive elements mounted on FR4 PCB.

Fig.16 shows the simulation results of the tunable filter. The insertion loss (IL = -2.4 dB) obtained in the simulation is due to quality factors of resonators (Q = 500). The return loss (RL) is -10 dB, and the out of band rejection is 26 dB at 2.0 GHz. The simulation results shows that a shift of +1% of the initial central frequency (2.44 GHz) is obtained.

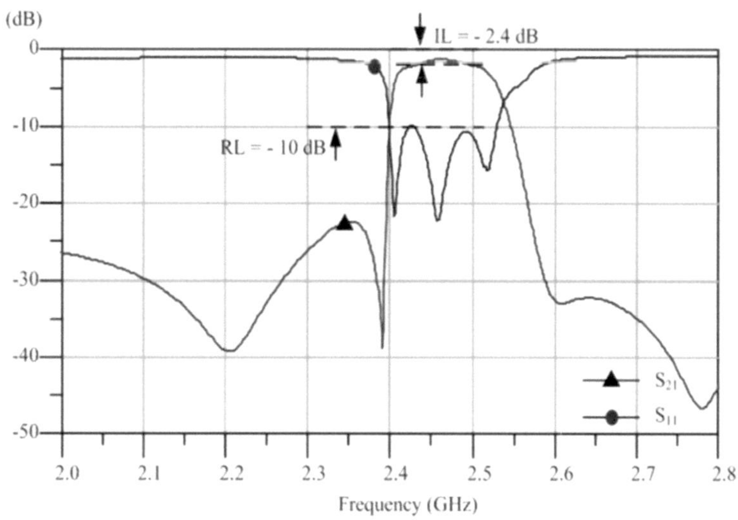


**Figure 16.** Simulation of tunable BAW filter with inductors added in parallel with the series resonators and capacitors added in series with the shunt resonators.

Comparisons between the measurements of the tunable filter with the original one are shown in Fig.17 and Fig.18.

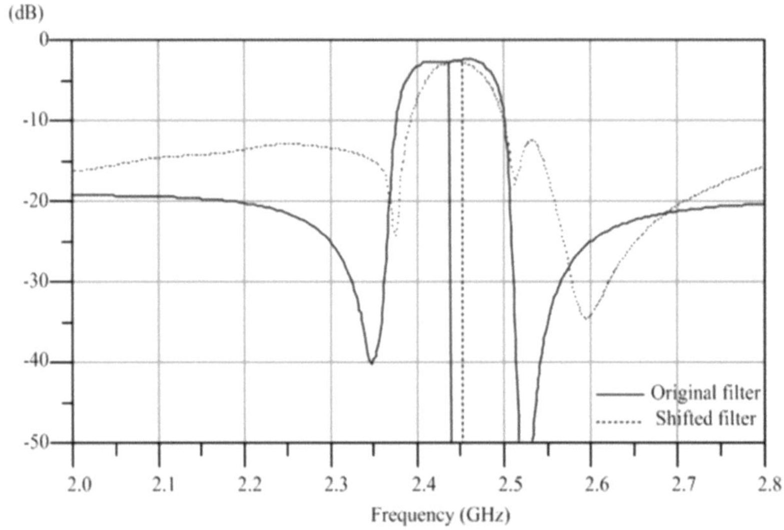


**Figure 17.** Shift towards higher frequencies: Comparison of transmission characteristic ($S_{21}$)between the tunable BAW-SMR filter (passive elements added) and the original one.

Based on the measurements of the tunable filter, we can note -4.5 dB of insertion losses and a shift of +0.6% of the center frequency (2.44 GHz) towards the higher frequencies (Fig.17). As well, a return loss of -7 dB is obtained (Fig.18). The filter high insertion losses are mainly due to the low resonators quality factor obtained in the fabrication (Q = 200) and to the low quality factor of the passive element used. Moreover, the parasitic capacitors generated by the FR4 PCB and the bonding wires used to connect the tunable filter with the passive elements caused a reduction of 15 MHz to the bandwidth of the tunable filter.

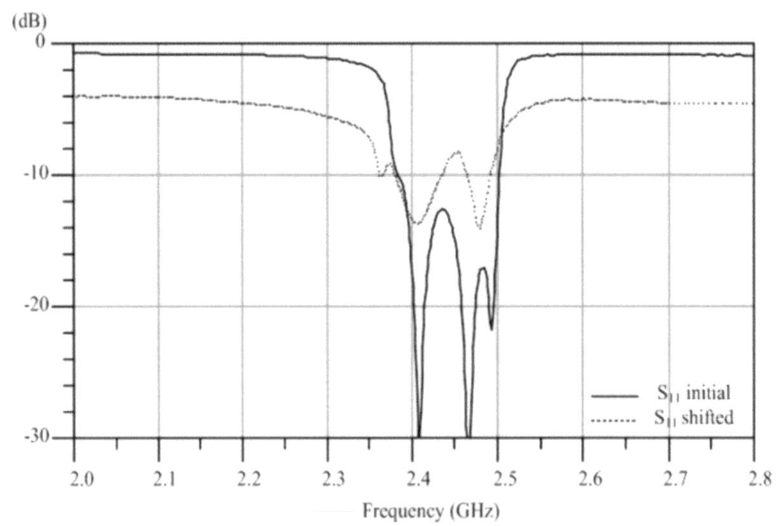


**Figure 18.** Shift towards higher frequencies: Comparison of reflexion characteristic ($S_{11}$) between the tunable BAW-SMR filter (passive elements added) and the original one.

## 4.2. Shift towards lower frequencies

This time and in order to shift the center frequency of the filter towards lower frequencies, we have to move all the zeroes towards these frequencies.

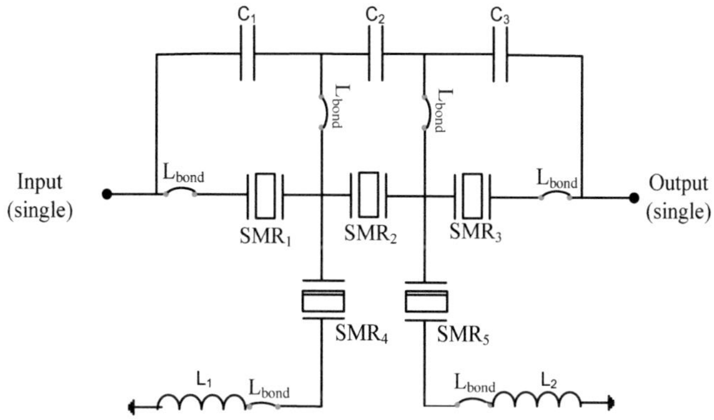


**Figure 19.** Tunable BAW-SMR filter with capacitors added in parallel with the series resonators and inductors added in series with the shunt resonators.

Thus, we have added capacitors in parallel with the series resonators and inductors in series with the shunt resonators that constitute the ladder BAW-SMR filter (Fig.19). Table 2 presents the values and the quality factor of the external passive elements used.

| | Value | Quality Factor 'Q' |
|---|---|---|
| Capacitors | $C_1$=2.2 pF | 140 |
| | $C_2$=1 pF | 200 |
| | $C_3$=2 pf | 150 |
| Inductors | $L_1$=$L_2$=2.2 nH | 57 |

**Table 2.** Values of the passive elements used to shift the center frequency of the tunable BAW-SMR filter towards lower frequencies and their quality factors.

As same as before, the tunable BAW-SMR filter and the passive components are mounted on a FR4 substrate. Fig.20 shows the simulation results of the tunable filter.

The insertion loss (IL = -2.4 dB) obtained by simulation is due to quality factors of resonators (Q = 500). In addition the return loss (RL) is -9.5 dB, and the out of band rejection is 16 dB at 2.0 GHz (Fig.20). The simulation results shows that a shift of -1% of the initial central frequency (2.44 GHz) is obtained. A comparison between the measurements of the tunable filter with the original one is shown in Fig.21.

Based on the measurements of the tunable filter, we can note -4.5 dB of insertion losses and a shift of -1.3% of the center frequency (2.44 GHz) towards the lower frequencies (Fig.21). In addition, the parasitic capacitors generated by the FR4 PCB and the bonding wires used to connect the tunable filter with the passive elements caused a reduction of 13 MHz to the bandwidth of the tunable filter.

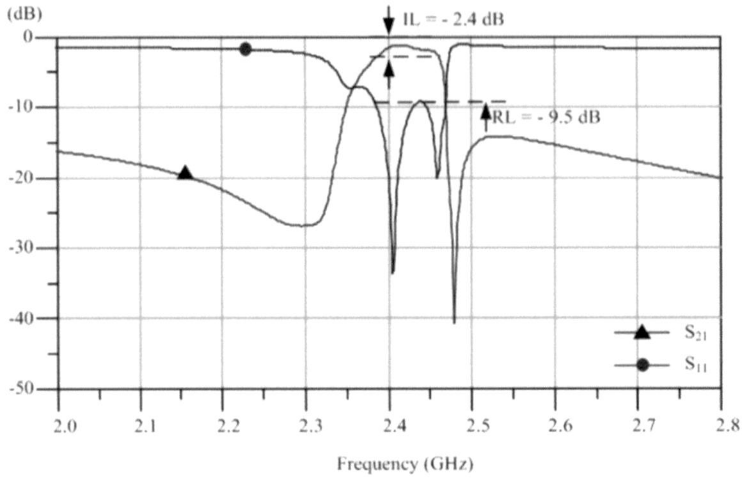


**Figure 20.** Simulation of tunable BAW filter with capacitors added in parallel with the series resonators and inductors added in series with the shunt resonators.

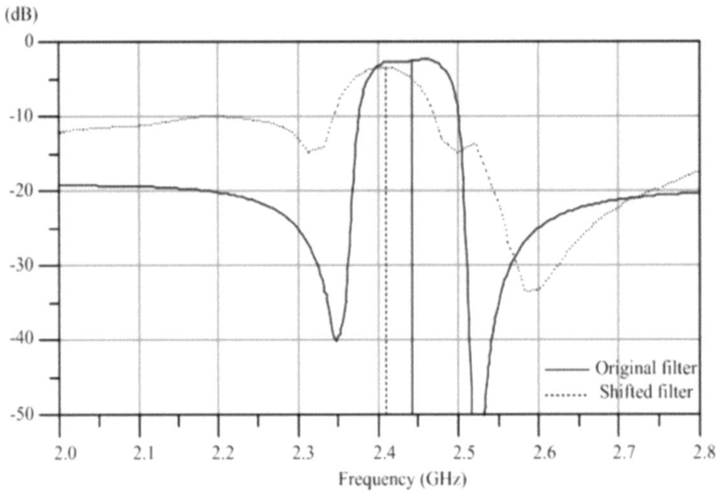


**Figure 21.** Shift towards lower frequencies: Comparison between the tunable BAW-SMR filter (passive elements added) and the original one.

As a conclusion, one should note that in contrast to [13], where lumped elements (inductors or capacitors) were proposed to be added at a time, in this paper the use of capacitors and inductors together have shown how to shift the center frequencies towards higher or lower frequencies.

## 5. Digitally tuning BAW filters

To validate the concept of digitally tuning BAW filters using passive elements controlled by CMOS transistors, we present in this part the use of CMOS switches at the terminals of capacitors (Fig.22) [6].

When a transistor is ON, the capacitor is short circuited, and when a transistor is OFF, the capacitor will be considered in series with the shunt resonator. Thus, the bandwidth and the characteristics of the filter will be modified. The circuitry of the filter, capacitors, transistors and the associated "bonding wires" are shown in Fig.22.

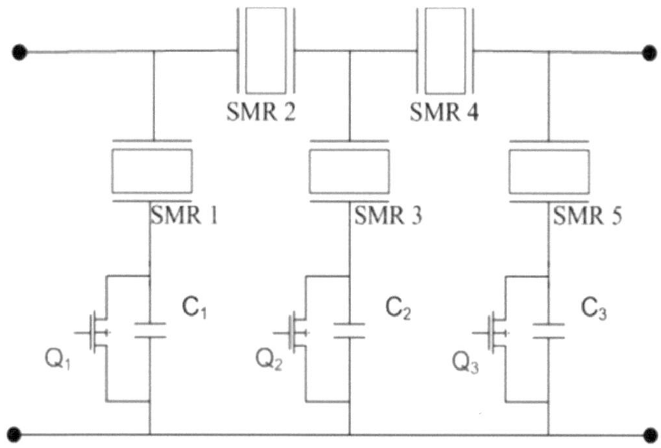


**Figure 22.** Tunable BAW-SMR using CMOS transistors.

The BAW-SMR filter used in this study is a fifth order filter designed for the W-CDMA standard in the ladder topology. This topology is composed by the resonator in series and parallel, the parallel resonators are loaded and their resonance frequencies are smaller than the series resonators. The die photography of the tunable BAW-SMR filter is shown in Fig.23. The filter has reduced size, and the die area is 1450*985 μm². Moreover, many passive pads connecting the filter with the active chip were taken in consideration.

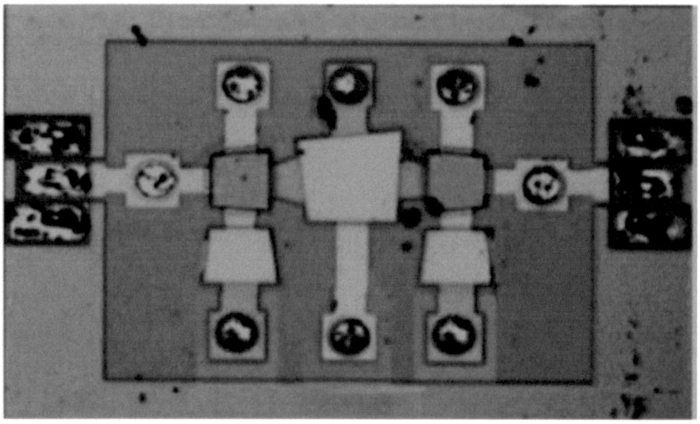

**Figure 23.** Microphotography of ladder BAW-SMR for the W-CDMA standard.

Fig.24 presents the comparison between the measurement and the simulation of the BAW-SMR filter. Electromagnetic simulation of the overall filter structure has been accomplished by using the ADS-Momentum software, where the acoustical effects have been included using the Mason Model. The filter is designed for implementation in SiP context. As shown in Fig.24, the measurements are in concordance with the simulations. However, the filter fulfills the requirements for the W-CDMA standard, exhibiting -2.77 dB of insertion loss, -8.75 dB of return loss and selectivity higher than 38 dB at 40MHz offset from the operating frequency.

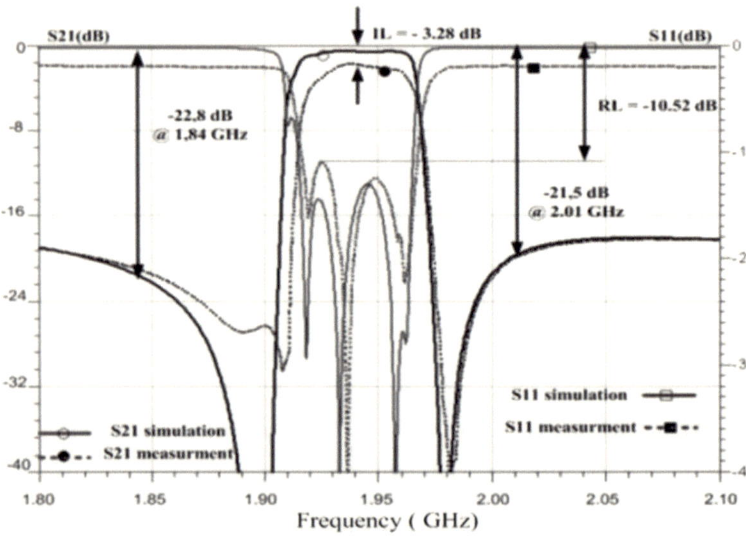


**Figure 24.** Simulation and measurement results of the W-CDMA ladder BAW-SMR filter.

## 5.1. Switches design

To adjust the bandwidth of the BAW filter, a chip is realized in 65nm CMOS technology. This chip is composed by the capacitors mounted in series with the MOS transistor, and these transistors are controlled by a 2 to 4 decoder (Fig.25).

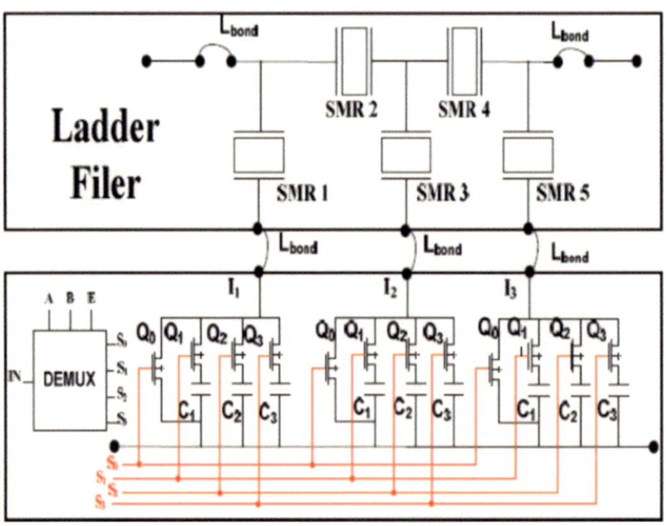


**Figure 25.** Circuitry of the tuning mechanism.

Fig.26 shows the layout of the tuning circuit. $I_n$ symbolize the pad for connection with ladder filter. $V_G$, $V_{dd1}$, $V_{dd2}$ and GND correspond to the gate voltage, the command of decoder and the ground respectively. The size of the Silicon area is 335*330μm².

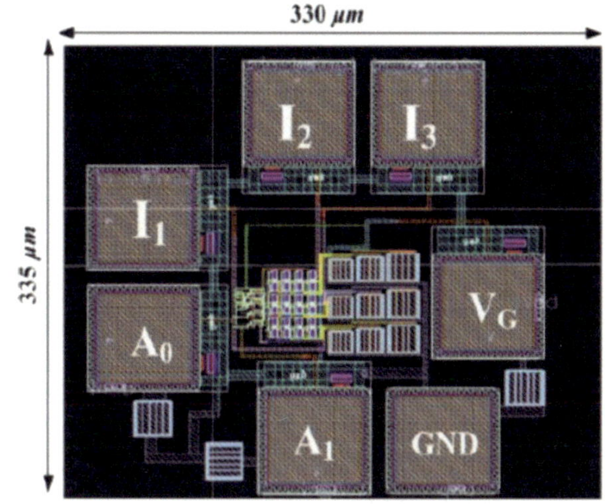


**Figure 26.** Layout of the tuning mechanism.

The tuning is attained by controlling the MOS transistors and capacitors in series with shunt resonators. Each transistor is open or short circuited by obtaining different outputs of the 2 to 4 decoder. Table 3 shows the truth table of the realized decoder. A, B, E symbolize the input of a decoder commanded by $V_{dd1}$, $V_{dd2}$ and $V_G$ respectively, $S_n$ (n = 0, 1, 2 or 3) represent the output of this decoder used to control $Q_n$. All transistors used in the tuning mechanism are provided by STMicroelectronics (CMOS 65 nm). The width and length of the gate are: W = 50 μm and L = 0.06 μm. The main parasitic elements are taken into account in the simulation ($C_{gs}$, $C_{gd}$, $C_{ds}$ and $R_{on}$). The length of bonding wire is 2 mm. It represents an inductive effect of approximately 2 nH at 2 GHz. $R_{on}$ of the MOS transistor is function of its dimensions and of the gate voltage ($V_G$). Thus, with an external adjustment of $V_G$, $R_{on}$ value is regulated.

| A | B | E | $S_0$ | $S_1$ | $S_2$ | $S_3$ |
|---|---|---|---|---|---|---|
| | 0 | 1 | 0 | 0 | 0 | 1 |
| 0 | 1 | 1 | 0 | 0 | 1 | 0 |
| 1 | 0 | 1 | 0 | 1 | 0 | 0 |
| 1 | 1 | 1 | 1 | 0 | 0 | 0 |
| X | X | 0 | 0 | 0 | 0 | 0 |

**Table 3.** Truth table of the 2 to 4 decoder.

## 5.2. Co-design: BAW filter- 65nm CMOS chip

Fig.27 shows the microphotography of the association of the ladder filter with the active chip. The devices are connected with bonding wires. The capacitors $C_1$, $C_2$ and $C_3$ values are fixed to achieve 12, 9 and 6 MHz tuning range, respectively. When the output $S_0$ of the decoder is ON, the transistor $Q_0$ is ON and all of the capacitors are short circuited. When the output $S_n$ (n = 1, 2 or 3) is ON, the transistor $Q_n$ is ON and the capacitor $C_n$ will be considered in series with the shunt resonators.

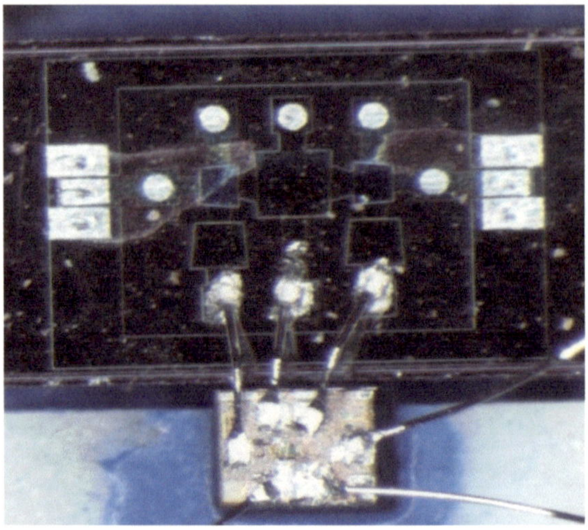

**Figure 27.** Microphotography of the digitally tunable ladder BAW-SMR for the W-CDMA standard.

The comparison between the simulation and the measurement results is shown in Fig.28. of the tunable BAW-SMR filter combined with the active chip presents a tuning range of 12 MHz, when the output $S_3$ of the decoder is ON. It show also -1.52 dB of insertion loss and 12 dB of the out-band rejection at 1.85 GHz. This out-band rejection is improved by 3 dB. This degradation is due to the length of bonding wire associating the active chip and PCB [10].

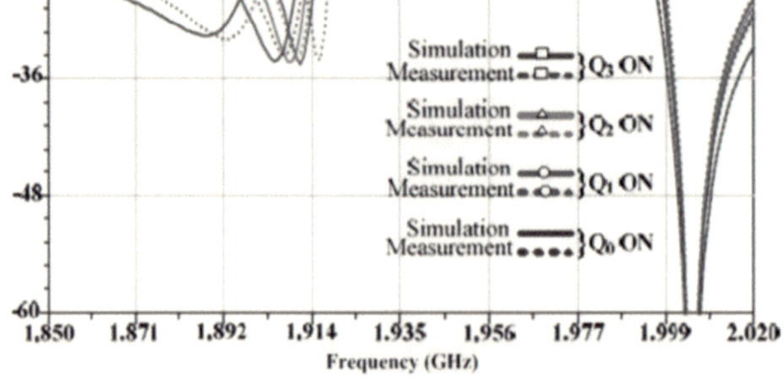


**Figure 28.** Comparison between the simulation and the measurement of the digitally tunable BAW filter.

## 6. Conclusion

In this paper, the impedance behavior of the BAW-SMR has been shown. Also, the effects of the addition of passive elements (L, C) to this type of resonators have been illustrated. In addition, a tunable BAW-SMR filter realized in a ladder topology used for the 802.11b/g standard (2.40 - 2.48 GHz) was shown. Mainly, the filter fulfilled the requirements for the WLAN 802.11 b/g standard, presenting a measured -3.3 dB of insertion loss, -12.7 dB of return loss and selectivity higher than 33 dB @ ± 30 MHz of the bandwidth. This tunable BAW-SMR filter has reduced dimensions (1035*1075 μm2). Moreover, the center frequency of this tunable filter was shifted towards higher and lower frequencies by adding passive elements. Measured shifts of -1.3% of the center frequency (2.44 GHz) towards lower frequency and +0.6% of the center frequency towards higher frequencies were obtained. Furthermore, digitally tunable BAW-SMR filter implementation was shown. The tunable filter was designed for the W-CDMA standard. The filter fulfilled the requirements for the WCDMA standard, presenting a measured -2.77 dB of insertion loss, -8.75 dB of return loss and selectivity higher than 38 dB @ ±40 MHz of the bandwidth. Moreover, the center frequency of this tunable filter is digitally shifted towards higher frequencies by adding capacitors in series with transistors that act as switches. These switches are controlled by a 2to4 decoder, and they are added to the shunt resonators.

## Author details

M. El Hassan
*University of Balamand –Al Kura, Lebanon*

E. Kerherve, Y. Deval and K. Baraka
*IMS Laboratory – UMR 5218 CNRS – University of Bordeaux, France*

J.B. David
*CEA-Leti – Minatec – Grenoble, France*

D. Belot
*ST Microectronics – Crolles, France*

## Acknowledgement

IMS laboratory is acknowledged for all facilities offered and the access to obtain the filter measurements. Also, CEA-LETI (Grenoble, France) and STMicroelectronics (Crolles, France) are acknowledged for the technology access and filter fabrication.

## 7. References

[1] Bradley, P. et al. "A Film acoustic bulk resonator (SMR) duplexer for USPCS Handset Applications", *IEEE MTT-S*, pp.367–370, 2001.

[2] Carpentier, J. F. et al. "A SiGe:C BICMOS WCDMA zero-IF RF front-end using an above-IC BAW filter", *IEEE ISSCC,* 2005, pp. 394-395.

[3] Shirakawa, A. A. Jarry, P. Pham, J-M. Kerherve, E. Dumont, F. David, J-B. Cathelin, A. "Ladder-Lattice Bulk Acoustic Wave Filters: Concepts, Design and Implementation", *RF and Microwave Computer Aided Engineering,* 2007.Newell, "Face-mounted piezoelectric resonators", Proc. IEEE, Vol 53, June 1965, pp.575-581.

[4] Newell, "Face-mounted piezoelectric resonators", Proc. IEEE, Vol 53, June 1965, pp.575-581.

[5] Lakin, K.M. Kline, G. McCarron, K.T. "High-Q microwave acoustic resonators and filter", *IEEE Trans. On Microwave Theory and Techniques,* vol. 41, pp. 2139-2146, Dec. 1993.

[6] Lakin, K.M. Belsick, J. McDonald, J.F. McCarron, K.T. "Improved bulk wave resonator coupling coefficient for wide bandwidth filters", *IEEE Ultrasonics Symposium,* Atlanta, GA, USA, pp.827-831, October 9, 2001.

[7] Fattinger G. G. et al. "Thin Film Bulk Wave Devices for Applications at 5.2 GHz", *IEEE UFFC Symposium,* Honolulu, Hawaii, pp. 174-177, 2003.

[8] Ancey, P. "Above IC RF MEMS and BAW filters: fact or fiction", *IEEE BCTM Proceedings,* Maastricht, Netherlands, pp. 186-190, 2006.

[9] Shirakawa, A. A. Pham, J-M. Jarry, P. Kerherve, E. Dumont, F. David, J-B. Cathelin, A."A High Isolation and High Selectivity Ladder-Lattice BAW-SMR Filter", *36th European Microwave Conference,* 2006, Manchester, UK, pp. 905 – 908, 10-15 September.

[10] Tsutsumi, J. Iwaki, M. Iwamoto, Y. Yokoyama, T. Sakashita, T. Nishihara, T. Ueda, M. Satoh, Y. "A Miniaturized FBAR Duplexer with Reduced Acoustic Loss for the W-CDMA Application", *IEEE Ultrasonics Symposium Proceedings,* Rotterdam, Netherlands, pp. 93-96, 2005.

[11] Lakin, K. M. Lakin, K. G. "Numerical Analysis of Thin Film BAW Resonators", *Proceedings of Ultrasonics Symposium,* Honolulu, Hawaii, Vol. 1, pp. 74-79, 2003.

[12] Mason, W. P. "Electromechanical Transducers and Wave Filters", Princeton, New Jersey, Van Nostrand, 1948.

[13] El Hassan, M. Kerherve, E. Deval, Y. David, J.B. Belot, D. "Reconfiguration of Bulk Acoustic Wave Filters Using CMOS Transistors: Concept, Design and Implementation", 2010 IEEE RFIC Symposium, Anaheim, CA, USA, 23 - 25 May, 2010. pp. 241 – 244.

# Chapter 19

# Seismic Vibration Sensor with Acoustic Surface Wave

Jerzy Filipiak and Grzegorz Steczko

Additional information is available at the end of the chapter

http://dx.doi.org/10.5772/55445

## 1. Introduction

Mechanical vibrations are a movement of particles around the state of equilibrium in a solid environment. Vibrations are a common phenomenon in our daily life. These vibrations are often parasite effects threatening our existence. Vibrations of the ground, machines, or a number of technical devices present a process, which require a continuous or a long-term monitoring. In many sectors vibrations are a working factor in a production process.

Mechanical vibrations serve as a source of information in medicine, diagnostics of the structure of many machines and in perimeter protection (monitoring). The knowledge of vibration parameters allows evaluating the technical condition of machines, the quality of their design and manufacture and their reliability. Early detection of ground vibrations serves to predict and warn of earthquakes. Ground vibrations serve to monitor explosions and are used in the reflexive seismology (prospecting for mineral deposits). Detection of ground vibrations in systems of perimeter protection allows detecting an intrusion into an area under surveillance. Mechanical vibrations are characteristic for their differing frequencies and amplitude. The frequency of mechanical vibrations ranges usually from a hundredth of Hz to a dozen or so kHz. Parameters of mechanical vibrations are measured with vibration sensors.

At present practically three types of seismic sensors are used:

- geophones,
- piezoelectric acceleration sensors,
- micro-mechanical silicone acceleration sensors.

Geophones belong to the simplest and most inexpensive vibration sensors. They feature a low mechanical resonance frequency, which ranges usually between 4Hz and 14Hz. They are used in mining, for safety perimeter protection [1] and in the reflexive seismology (Figure 1).

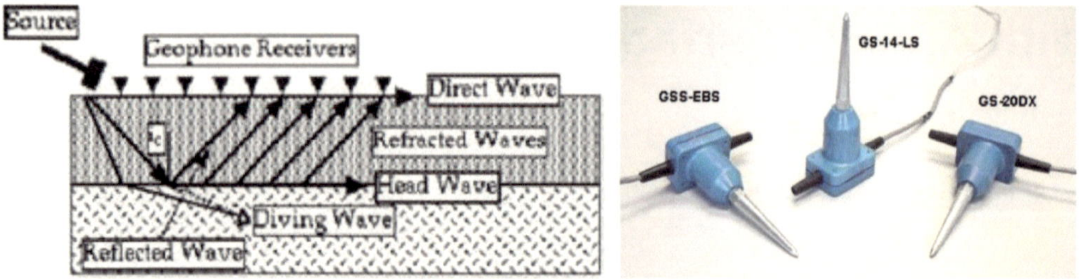


**Figure 1.** Reflexive seismology and geophones [1]

Micro mechanical silicone acceleration sensors (MEMS) [2-3] are mass-produced and used in many fields (e.g. deploying air bags, laptops). They are characterised by small dimensions. Due to their miniature dimensions their mechanical resonance frequency is very high, usually a dozen or so kHz or a few dozen kHz. With piezoelectric acceleration sensors [4] one can measure variable accelerations. Their mechanical resonance frequency is higher than that of MEMS.

**Figure 2.** Seismometers offered by MEROZET 1 - portable broadband seismometer, 2 - very broadband seismometer 3 - triaxal seismic accelerometer.

Figure 2 shows seismometers offered by MEROZET: a portable one, a wide-band and a triaxial one; all these 3 models are based on and include piezoelectric acceleration sensors. Vibration sensors can be built using sensors with the acoustic surface wave (SAW).

These SAW-based sensors are used to measure a number of physical quantities: gas concentration [5-8], temperature, pressure [9-13], and mechanical quantities: torque of a rotating shaft [14], stress [15-17], acceleration [18-19] and vibrations [20-21]. All these SAW–based sensors work on the basis of measuring changes in the delay of a surface wave due to the impact of a physical quantity being measured on its speed and the propagation path. However, depending on the kind of a measured physical quantity, a number of problems occur, which are characteristic for the group of sensors used for measuring that quantity.

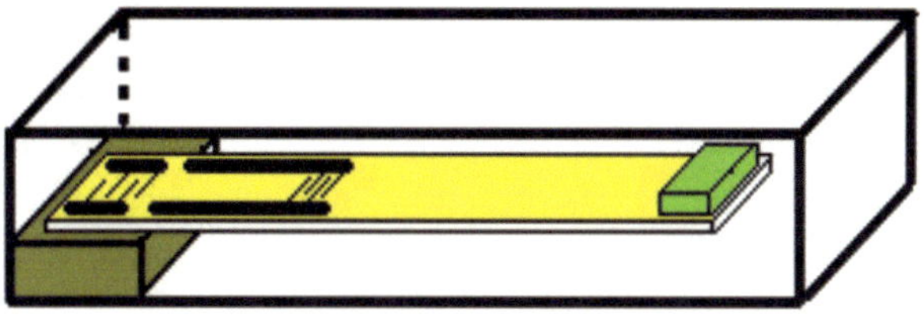

**Figure 3.** Basic structure of a SAW vibration sensor

Figure 3 presents the idea of the structure of an SAW-based sensor. The main element of this sensor is an anisotropic plate of a piesoelectric material. One end of the piesoelectric plate is made fast to the sensor housing, while on the other free end a seismic mass can be placed. An SAW-delaying line in the form of a four-terminal-network is made on the top surface of the sensor. The movement of the sensor housing causes its plate to vibrate and the SAW-based delaying line delay to change. This is why the phase of a high frequency signal passing thru such a line changes. The magnitude of a signal pahse change will be proportional to the change in the delay of an SAW-based delaying line.

The presented sensor design presents three different issues, which must be solved:

- modeling of the sensor mechanical system, which amounts to the description of plate deformations and stresses occurring in it.
- modeling of the sensor mechanical-electrical converter – delaying line with the SAW, which amounts to the description of the change in parameters of a delaying line with the SAW (first of all of the delay) due to distortions and stresses in the plate
- modeling of the electric circuit cooperating with the sensor, which amounts to the analysis and synthesis of an electric circuit measuring the changes in the delay of the surface wave in a delaying line including the SAW.

The work presents a solution of the mentioned problems, which were further analysed. The work presents executed models of SAW-comprising sensors and the results of a study of their parameters. The use of realised sensors in a system of perimeter protection is described. The structure of an SAW-comprising sensor (Fig. 3) is a combination of a continous system in the form of a piesoelectric, anisotropic support plate and a discrete system in the form of a concentrated mass. In theory such a system can feature an infinite number of free vibration frequencies. Writing a description of the mechanics of the plate of an SAW-based vibration sensor is a complicated proces; what makes it dificult is the tensor description of the plate mechanical properties. The knowledge of the value of an attenuation tensor (viscosity tensor) poses a problem. Therefore an analysis was conducted, which allows to simplify the sensor model presented in Figure 3 and next as a result of this analysis the movement of a piesoelectric, anisotropic plate with a concentrated mass was described with the aid of a discrete system of one degree of freedom. Elastic and viscous properties of the plate material were taken into account. This model was introduced by way of an isotropic descritpion of anisotropic material properties. The model accuracy was evaluated. Explicit realtionships between sensor plate movement parameters and its geometey and parameters describing its elastic an viscous properties were determined, thus a simple analysis and synthesis of the sensor plate movements were possible. The main feature of the sensor mechanical system (a continous system combined with a discrete one) is the occurence of pratically only one resonance frequency. A simple description of this magnitude by design parameters of the system and elastic and viscous plate parameters allow a simple modelling, how these sensors function, and as well to determine elastic and viscous parameters of a plate empirically. These parameters in the form : an equivalent Young's modulus and an equivalent material damping coefficient for a selected direction of a piezoelctric substrate (i.e. the direction of the surface wave propagation) were determined in works [22][23]. In available bibliography they are not knowen or utilised. The above considerations are presented in the work [21]. For a full description of the SAW-comprising vibaration sensor designing process in Section 2 we present a modelling process of its mechanical system.

## 2. Model of mechanical unit for SAW vibration sensor

The object of consideration has been presented in Figure 4. One end of the plate is stiffly attached, and the other is free and without any concentrated mass. The piezoelectric properties of sensor plate will be omitted in the analysis.

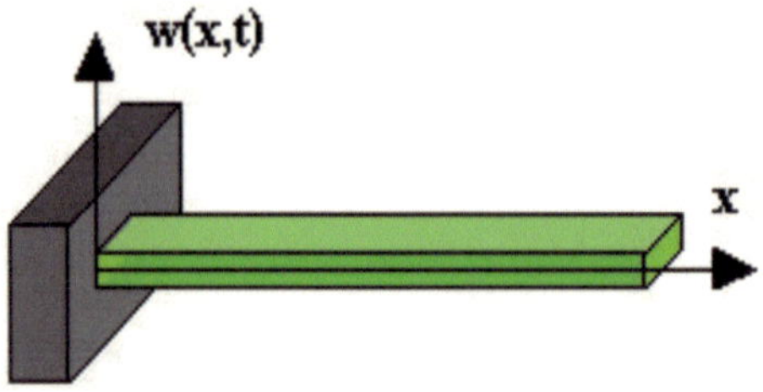


**Figure 4.** The plate of a vibration sensor.

The equation of a movement of an anisotropic body with the mass density ϱ is:

$$\frac{\partial \sigma_{ij}}{\partial x_j} + \rho \frac{\partial^2 u_i}{\partial t^2} = 0 \tag{1}$$

The stress tensor $\sigma_{ij}$ depends on the strain tensor $\varepsilon_{kl}$ through Hook-Voight equation:

$$\sigma_{ij} = C_{ijkl}\varepsilon_{kl} + \Sigma_{ijkl}\varepsilon_{kl,t} \tag{2}$$

where: $C_{ijkl}$—is a elasticity tensor, $\Sigma_{ijkl}$—is a material damping tensor, $u_i$—is a displacement vector,

$$\varepsilon_{kl} = \frac{1}{2}\left(\frac{\partial u_k}{\partial x_l} + \frac{\partial u_l}{\partial x_k}\right) \quad \text{– is a strain tensor} \tag{3}$$

The mathematical description of this issue will be closed if the initial and boundary conditions are added to the aforementioned equations. It is complicated to solve the problem. The causes of the complications are huge number of non-vanishing modules of an elasticity and material damping tensor. The material damping tensor $\Sigma_{ijkl}$ has an identical symmetry as the elasticity tensor $C_{ijkl}$. For materials used in SAW devices (quartz, lithium niobate), the value of damping constants are difficult to experimental verification. For higher class of symmetry of an anisotropic materials, equations (2, 3) are simple. For the isotropic material the elasticity tensor has only two independent components $C_{1111}=\lambda+2\mu$, $C_{1122}=\lambda$, $C_{2323}=\mu$. Therefore the elastic properties of an isotropic substance describe two quantities ($\lambda$, $\mu$). They are often shown in form of a Young's modulus $E$ and a Poisson ratio $\nu$. The following relations occur between quantities $E, \nu$, and elasticity tensor components [24]:

$$E = c_{2323}\left(2+\frac{c_{1122}}{c_{1122}+c_{2323}}\right), \quad \nu=\frac{c_{1122}}{2\left(c_{1122}+c_{2323}\right)} \tag{4}$$

The description of viscous properties of an isotropic body done by material damping tensor is analogical. It is usually described by two quantities [24]:

$$\tau = \frac{\Sigma_{2323}}{E}\left(2+\frac{\Sigma_{1122}}{\Sigma_{1122}+\Sigma_{2323}}\right), \quad \eta=\frac{\Sigma_{1122}}{2\left(\Sigma_{1122}+\Sigma_{2323}\right)} \tag{5}$$

The Young's modulus $E$ is described as a proportion of a longitudinal stress to longitudinal strain for the direction of the functioning of a stress. To describe the mechanical properties of anisotropic materials taking into account a particular direction of the stress an effective Young's modulus may be used $E$ [24][25](Its magnitude is described by overt dependence). An exemplary expression for a inverse effective Young's modulus for a trigonal unit (lithium niobate, quartz) is:

$$1/E=(1-l_3^2)^2 s_{11}+l_3^4 s_{33}+l_3^2(1-l_3^2)(2s_{13}+s_{44})+2l_2 l_3(3l_1^2-l_2^2)s_{14} \tag{6}$$

where: $s_{ij}$—is an element of an compliance matrix, $l_j$—is a cosine of an angle between the chosen direction and the axis – $j$, in Cartesian coordinates. The compliance matrix $s_{ij}$ is reverse to the stiffness elasticity matrix $c_{ij}$.

It is possible to calculate the values of material damping coefficients in a chosen crystallographic direction, too. The presented approach allows to model the anisotropic material by the isotropic model. In a such model the stresses are the sum of elastic and dissipative components:

$$\sigma = E\varepsilon + E\tau\frac{\partial\varepsilon}{\partial t} \tag{7}$$

We consider equivalent isotropic model of cylindrically bent plate [19]. Equation of free vibrations has the form:

$$\rho\frac{\partial^2 w(x,t)}{\partial t^2}+E_e\frac{h^2}{12}\left(1+\tau\frac{\partial}{\partial t}\right)\frac{\partial^4 w(x,t)}{\partial x^4}=0 \tag{8}$$

where: $\rho$—mass density, $h$—plate thickness, $L$—plate length, $\tau$—equivalent material damping coefficient, $E_e=\dfrac{E}{1-\nu^2}$

$E_e$ is an equivalent Young's modulus.

At the boundaries we have:

$$w(0,t)=0,\quad \frac{\partial w(0,t)}{\partial x}=0,\quad \frac{\partial^2 w(L,t)}{\partial x^2}=0,\quad \frac{\partial^3 w(L,t)}{\partial x^3}=0, \tag{9}$$

The solution to the boundary problem (8), (9) has the form:

$$w(x,t)=\sum_{n=1}^{\infty}W_n(x)\cdot A_n\, e^{-\frac{\varpi_n^2}{2}t}\cdot\sin(\bar{\omega}_n t+\varphi_n) \tag{10}$$

where: constants $A_n$ and $\phi_n$ are determined by initial conditions.

The angular frequency of non-damped vibrations is equal to:

$$\omega_n=k_n^2\frac{h}{l^2}\sqrt{\frac{E_e}{12\rho}}\,. \tag{11}$$

The angular frequency of damped vibrations is equal to:

$$\bar{\omega}_m=\omega_m\sqrt{1-\frac{\omega_m^2\tau^2}{4}} \tag{12}$$

$$\text{where}: \mathrm{k}_1=1{,}875,\ \mathrm{k}_2=4{,}694,\ \mathrm{k}_3=7{,}855 \tag{13}$$

The orthonormal set of function $W_n$ (eigenfunctions) is taken from [26]. Only some elements in the sum (10) represent vibrations. For $N < n$ where $N$ is the greatest natural number for chich $\omega_n < 2/\tau$ , the element of sum represents very strongly dumped movement and there is no resonance at this frequency. Each of the harmonics n=1,2,3… has part of energy. How great is the part depends on the unit (normal vibrations, $\omega_n$, $w(x)$ ) and depends on activation. In the paper [18][19][27][28] a simplified model with one degree of freedom was presented and it is shown in Figure 5. It has been used to describe the dynamics of sensor plate movement. It was derived according to the Rayleigh method. This method is based on a simplified modeling of a plate with the use of an equivalent circuit with one degree of freedom which is energetically equivalent. The free end of sensor plate has been taken as a point of reduction. The equivalent circuit present Figure 5

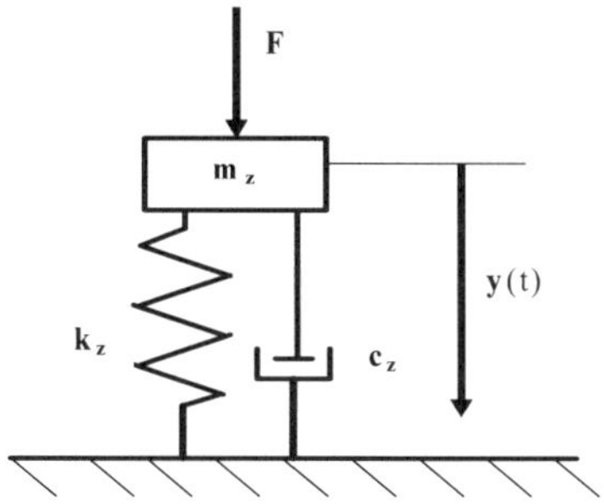


**Figure 5.** The equivalent circuit of sensor plate with one degree of freedom.

Parameters in the model are as follows [18]:

$$m_z = 0.25196\,\rho b h l, \quad k_z = 3.1169\frac{E_e b\, h^3}{l^3}, \quad c_z = \tau\, 3.1169\frac{E_e b h^3}{l^3} \tag{14}$$

where: b — plate width

The model with one degree of freedom has only one resonance frequency. The equation of mass movement is as follows:

$$\frac{d^2 y(t)}{dt^2} + \omega_0^2 \tau \frac{dy(t)}{dt} + \omega_0^2 y(t) = \frac{F(t)}{m_z}. \tag{15}$$

The solution of an equation for natural vibrations:

$$y(t) = Ae^{-\frac{\omega_0^2 \tau}{2} t} \sin(\omega_r t + \phi), \tag{16}$$

where:

$$\omega_0 = 3.5172\left(\frac{h}{l^2}\right)\sqrt{\frac{E_e}{12\rho}}, \tag{17}$$

$$\omega_r = \omega_0 \sqrt{1 - \frac{\omega_0^2 \tau^2}{4}} \tag{18}$$

It is analogical to the Equation (10) obtained with the use of an isotropic model of sensor plate. Comparison of the first frequency of damped vibrations of the plate obtained in an isotropic model (11) and the frequency of damped vibrations obtained with the use of a model with one degree of freedom (17) are in fulfils relation:

$$\omega_1 = 0.9996\omega_r, \tag{19}$$

First frequency of damped vibrations calculated in an isotropic model is 0.5 per cent lower than frequency calculated with the use of a discrete model. This difference could be smaller in case of a sensor construction with the concentrated mass attached to the movable end of the plate. That is why the model with one degree of freedom may be used to describe the movement of sensor plate. It allows relatively easy simulation of vibrations of the plate with the mass attached to its movable end. Free vibrations of sensor plate are a definite as a sum of damped harmonic frequency vibrations. But, in free vibration damped vibrations with first harmonic frequency will dominate. The amplitudes of the superior harmonic vibrations will be extremely small. As it is shown in [18] their quantity is 40 dB smaller than the first harmonic amplitude. This is the reason why a model with one degree of freedom [18][27][28] has been used to analyze the movement of the plate with concentrated mass. Vibrations of the plate have been activated by the movement of the sensor casing *Y(t)*. The equation of movement is as following:

$$\frac{d^2y(t)}{dt^2}+\omega_0^2\tau\frac{dy(t)}{dt}+\omega_0^2y(t)=\omega_0^2\tau\frac{dY(t)}{dt}+\omega_0^2Y(t)\,. \tag{20}$$

where:

$$\omega_0=3.5172\left(\frac{h}{l^2}\right)\sqrt{\frac{E_e}{12\rho}}\sqrt{\frac{1}{1+r\ 3.9689}} \tag{21}$$

r is a ratio between seismic mass and mass of sensor plate.

The solution of the Equation (20) is a function:

$$y(t)=A\exp\left[-\frac{\omega_0^2\tau t}{2}\right]\sin\left[\omega_r\left(t+\varphi\right)\right]- \\ -\frac{4\omega}{4-\omega_0^2\tau^2}\int_0^t\left(\tau\frac{dY(\xi)}{dt}+Y(\xi)\right)\cdot\exp\left[-\frac{\omega_0^2\tau}{2}\left(t-\xi\right)\right]\cdot\sin\left[\omega_r\left(t-\xi\right)\right]d\xi \tag{22}$$

where: constants $A$ and $\varphi$ are determined by initial conditions.

Relations between $\omega_0$ and $\omega_r$ are as in identity (18). In both components of the solution (22) appears the following function:

$$\varphi_\delta(t)=Ae^{-\frac{\omega_0^2\tau}{2}t}\sin(\omega_rt+\phi) \tag{23}$$

It is a product of a harmonic and damping (exponentially decay with time) function. The frequency of a harmonic function is the resonance frequency of the unit. This function describes sensor impulse response and its natural vibrations. It is a sum of:

- convolution of an impulse response of the plate and the component of describing movement of the sensor casing,
- damped vibrations with the resonance frequency of a sensor plate.

It will always have a factor in form of a harmonic function with the frequency equal to the resonance frequency of sensor plate and with variable amplitude. That is why the frequency response of sensor plate may be quantity identifying the sensor. Frequency response of the sensor plate is the ratio of the amplitude of the deflection plate sensor to the harmonic amplitude of its case The frequency response of the sensor plate calculated from the Equation (20) is as follow:

$$H(\omega)=\sqrt{\frac{1+(\omega\tau)^2}{[1-(\frac{\omega}{\omega_r})^2]^2+(\omega\tau)^2}} \qquad (24)$$

Parameters in relation (24) depends on mechanical properties of the sensor plate material. The quantities of elastic and viscous parameters for quartz are shown in Table 1 [22][23].

| | ST-cut quartz |
|---|---|
| Equivalent Young's modulus [GPa] | 76 |
| Dynamic critical compressive stress [MPa] | 80 |
| Equivalent material damping coefficient [$\mu s$] | 29.3 |
| Density [kg/m³] | 2650 |

**Table 1.** Material parameters of quartz.

Theoretical frequency response for plates made of ST-cut quartz with the resonance frequencies of 22 Hz and 100 Hz are shown in Figure 6. The most important is that for low frequency the frequency response has narrower band and higher magnitude so the selectivity of the sensor is high. It decreases with increased resonance frequency.

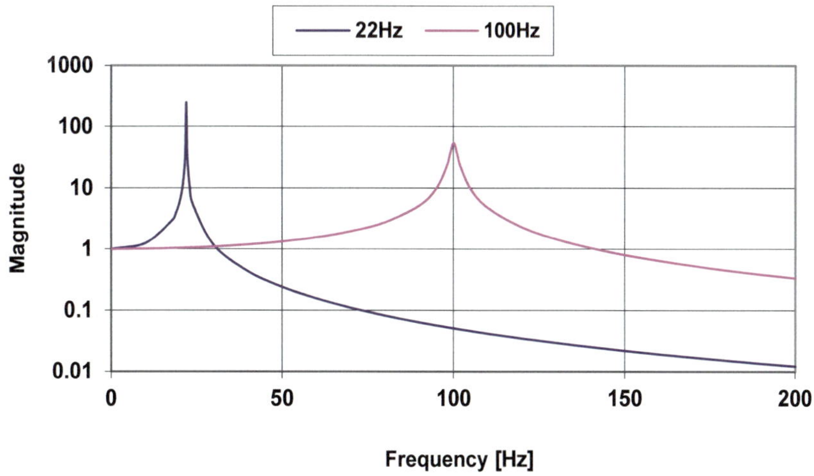


**Figure 6.** Theoretical resonance characteristics of plates with resonance frequencies of 22 Hz and 100 Hz.

The maximum value of the frequency response of the plate will occur for $\omega = \omega_r$. It has been described with the relationship:

$$H(\omega_r) = \sqrt{\frac{1+(\omega_r\tau)^2}{(0,5\omega_0\tau)^4 + (\omega_r\tau)^2}} \quad (25)$$

Its value exceeds repeatedly the value of static deflection. (e.g., for the resonance frequency of 22 Hz it is 246 times higher).

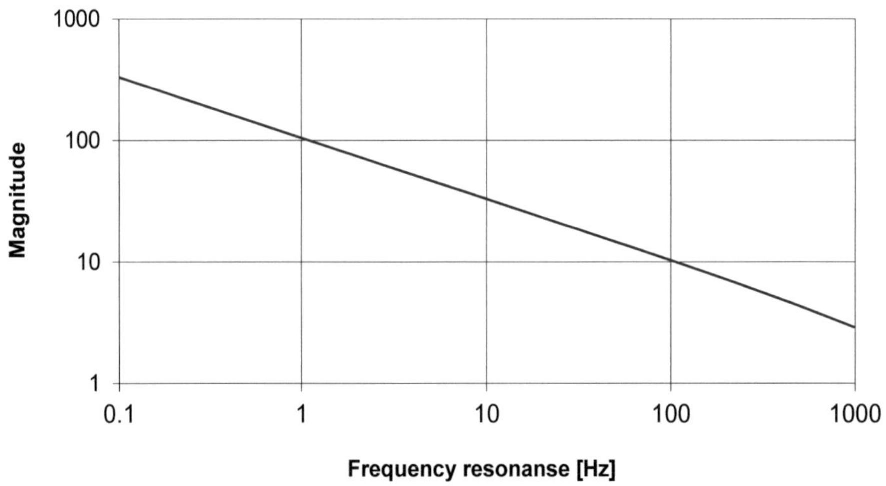


**Figure 7.** Maximum magnitude of frequency response versus the resonance frequency of the sensor plate.

The change in maximum magnitude of frequency response as a function of resonance frequency of the plate is shown in Figure 7 For the resonance frequency of a plate of 10Hz the vibrations amplitude multiplication is 1,600 higher than static deflection. This property may be used to construct sensors with high sensitivity level. But it is necessary to answer one question beforehand: what is the lowest possible resonance frequency of a plate that we can manufacture? The answer is accessible on the basis of the described model and the length of available plates. The resonance frequency of sensor plate is described by the relation (21). It depends on plate length (l) and on quantity of a concentrated mass (r) attached to the free end of sensor plate. The increase of the concentrated mass lowers resonance frequency of the plate, simultaneously increasing stresses of the plate.

The influence of a change of concentrated mass on resonance frequency of the sensor plate is shown in Figure 8.

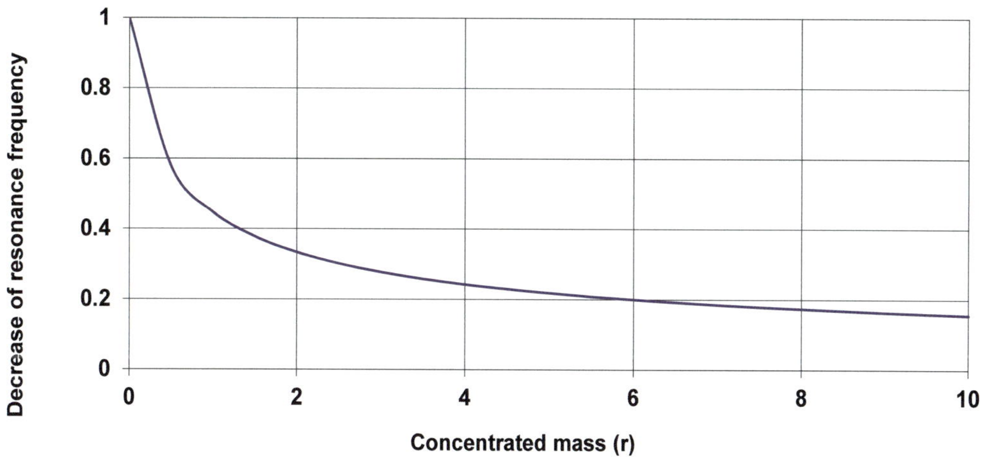


**Figure 8.** The influence of a change of a concentrated mass on sensor plate resonance frequency.

It is visible that the use of concentrated mass quantities exceeding two times the mass of the plate enables and triple decrease of resonance frequency of a plate. It is the most effective place to decrease the resonance frequency of a plate. Continuous increase of a concentrated mass does not substantially decrease the resonance frequency of sensor plate. The further analysis of sensor parameters will be limited to such range of concentrated mass quantities. The relationship between the value of resonance frequency of a plate made of ST-cut quartz and length of the plate determined by three different concentrated mass values is shown in Figure 9.

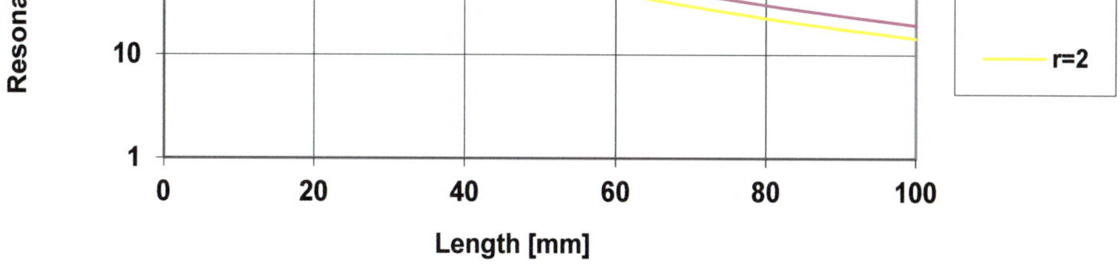


**Figure 9.** The relationship between sensor plate resonance frequency made of ST-cut quartz 0.5 mm thick, and plate length (l) and the quantity of concentrated mass (r).

From the figure presented above may conclude that it is relatively easy to create plates of resonance frequency form 20 Hz do 4 kHz. For the 0.5 mm thick plates it is necessary to use the concentrated mass up to 1.5 g. The relation between the concentrated mass and the plate length is shown in Figure 10. The sensor impulse response presented by the relation (23) has a damped character. Its fast fading can impose an upper limit of the resonance frequency. The damping value depends on the geometry of the plate and the equivalent damping coefficient. In order to simplify the illustration the impulse response damping measure has been introduced as a relative decrease of its quantity after one period.

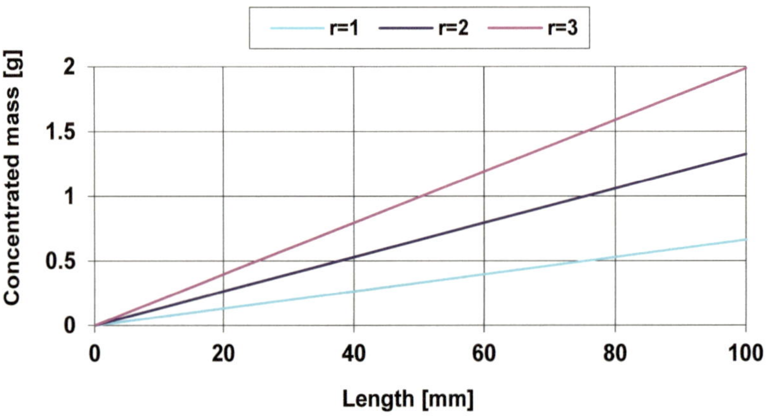


**Figure 10.** The concentrated mass quantities used in the considered sensor constructions.

The relation of impulse response damping in form of a function of length of ST-cut quartz plate for three different concentrated mass values is presented in Figure 11. For plates longer than 40 mm loaded by the concentrated mass equal to the mass of the plate (r = l) the

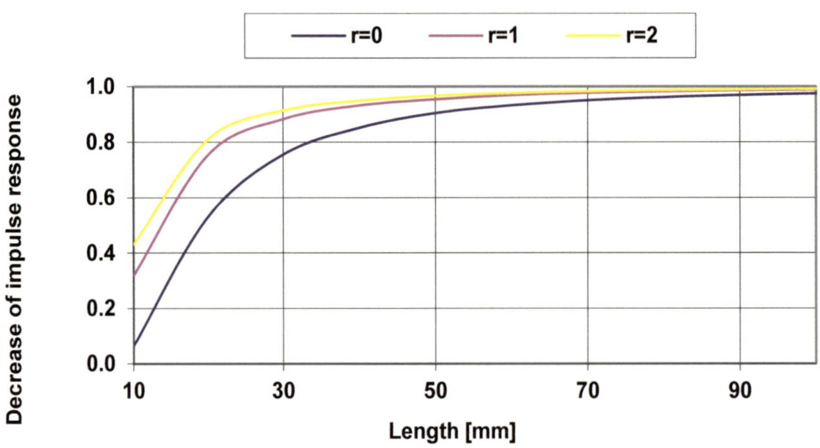


**Figure 11.** Relative decrease of impulse response amplitude after the time equal to its period in form of a function of plate length for different concentrated mass quantities (r).

damping of free vibrations of the plate is relatively slow. The impulse response of shorter plates is dampened relatively fast. This is why it seems to be beneficial to use possibly long plates loaded by concentrated mass equal to plate mass. The value of resonance frequencies of plates possible to manufacture has changed. It seems that the range of resonance frequency plates available to use is limited to the scope from 20 Hz to 250 Hz. The parameters of resonance frequencies of the plates in the aforementioned range are shown in Figure 12.

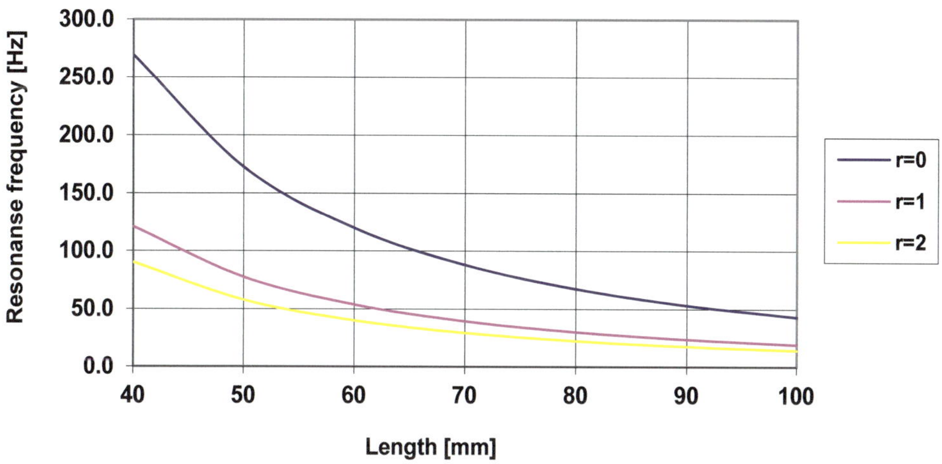


**Figure 12.** The relation between resonance frequency of a sensor plate and plate length and the value of concentrated mass.

From the above considerations one can draw a conclusion that working with SAW vibration sensors one can utilise their pulse responses or forced vibrations. The sensor resonance characteristics is a basic parameter of the first operation mode. On 0.5 mm thick quartz plates with a concentrated mass equalling the mass of a plate we can achieve plate own vibration frequencies from 20 Hz thru 250Hz. In the second operation mode the vibration sensors with SAW will operate like classical acceleration sensors. The plate resonance frequency should be above the measuring range of a sensor.

The knowledge of the resonance characteristic curve is required for both sensor operation modes. That characteristic can be easily calculated with the aid of the presented model. In the Section 4 we will present experimental examples of the operation of a seismic vibration sensor with SAW, which will enable to evaluate the precision of this model and its usefulness.

## 3. Vibration sensor electronic components

In order to ensure the transmission of test and supply signals through one coaxial cable there must be a system at the sensor input separating the test signal (74MHz) and supply signal 12VDC (separator). At the output there must be a system summing up the test signal with a constant supply voltage (adder). Test signal (high frequency) after going through SAW delay line must be amplified to input quantity. It will ensure loss compensation caused by SAW delay line. In the entire line of high frequency test signal (74MHz) a characteristic impedance of 50Ω should be retained. Input and output impedance must have the value of 50Ω. Figure 13 shows the basic functional elements of SAW vibration sensor. Depending on the function in the whole system the following components may be distinguished:

- system separating and summing up the test and supply signals;
- systems adjusting the impedance of SAW line to 50Ω;
- SAW delay line;
- amplifier compensating losses caused by SAW delay line.

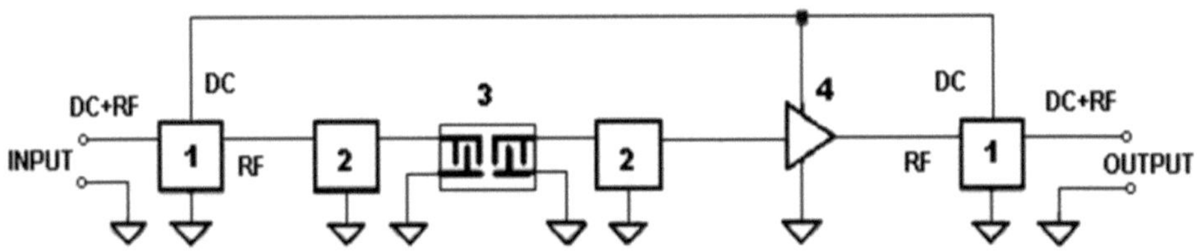


**Figure 13.** Block diagram of SAW vibration sensor.

The method of making the aforementioned components will be discussed in the next Section.

## 3.1. System separating or summing up electrical signals (separator/adder)

A system separating test and supply signals is placed at the sensor input. A system summing up these signals is placed at the sensor output. Figure 14 shows a system separating or summing up test and supply signals. The system is in a form of a circulator. It is connected to the line of high frequency signal with a characteristic impedance of 50Ω. A point of separation (or summation) of signals is the place where additional impedance is added to a line of characteristic impedance of 50Ω by inductance $L_1$. It may change the characteristic impedance of the line and be the reason of signal reflections. In order to avoid this the quantity of added impedance must be much larger than line characteristic impedance (50Ω). In order to fulfill this requirement inductance $L_1$=4,7uH of its own parallel resonance frequency of 74MHz has been chosen. Figure 15 shows inductance equivalent system diagram.

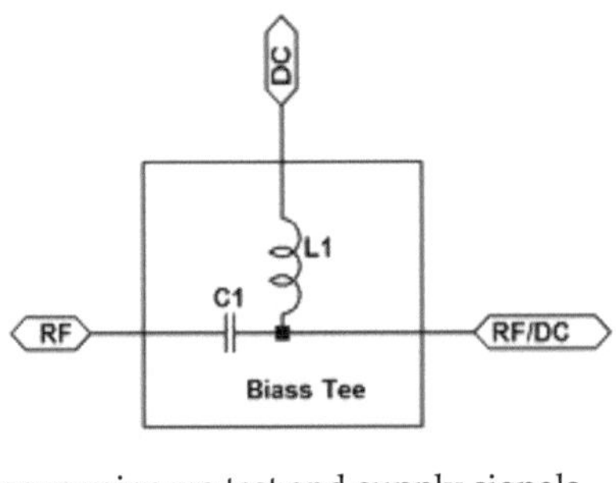


**Figure 14.** System separating or summing up test and supply signals.

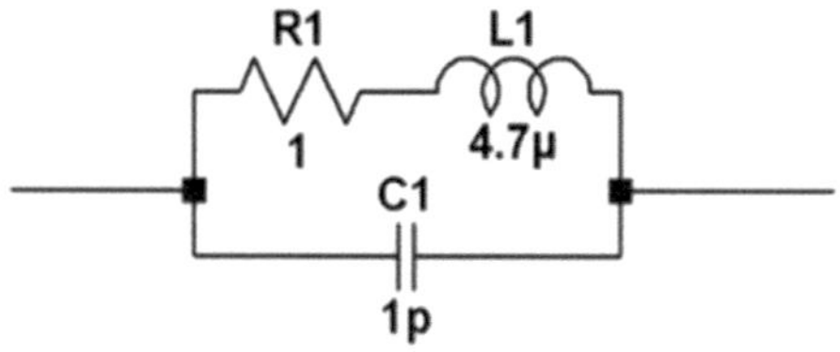


**Figure 15.** Inductance equivalent system diagram

In reality, the chosen inductance is a parallel resonant circuit. Figure 16 presents change in impedance of such a system in frequency function.

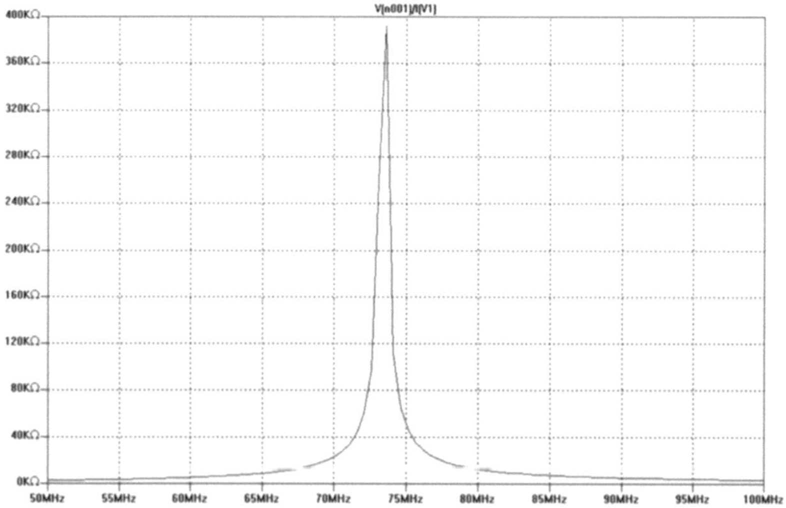


**Figure 16.** Relationship between impedance and frequency of a system presented in Figure 15

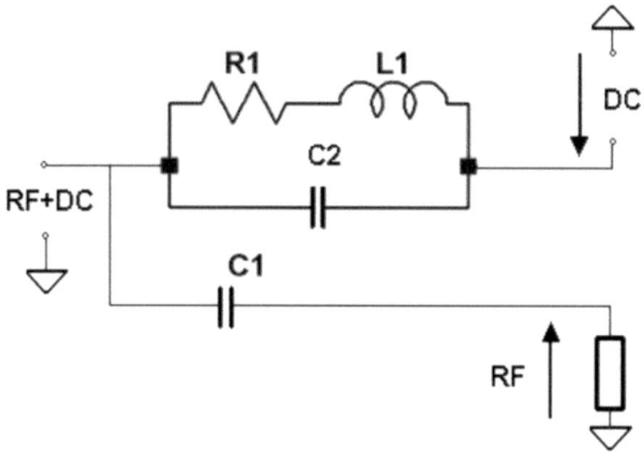


**Figure 17.** Actual system separating the test and supply signal.

For frequency equal to 74MHz the system impedance value amounts to 400kΩ. It is relatively high in comparison with characteristic impedance of the test signal transmission line (50Ω). It is then possible to obtain considerable attenuation of the test signal entering the supply circuit and it practically eliminates reflections at the point of signal separation or summation. Figure 17 presents an actual system separating the test and supply signals. Impedance of connected in series: C1 capacity and RF output impedance is equal to characteristic impedance of 50Ω. A diagram presented in Figure 17 enables to analyze test signal attenuation in supply circuit. It allows to calculate the change in line characteristic impedance made by the separating system.

Figure 18 shows attenuation of the test signal at DC output and change in line impedance in frequency function for R1=1Ω.

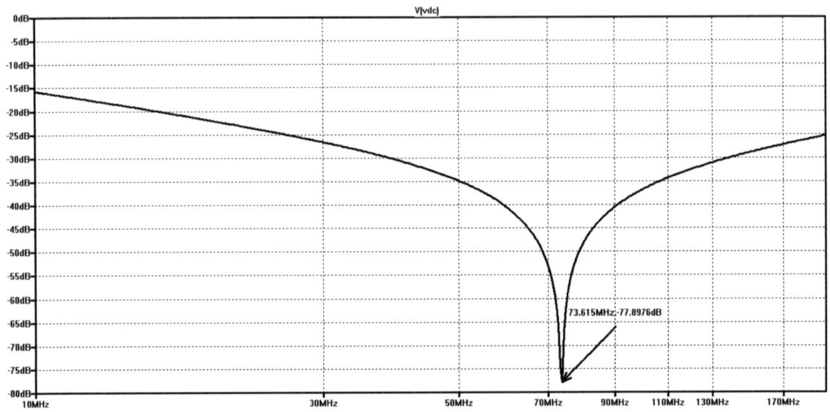


**Figure 18.** Attenuation of test signal at DC output vs. frequency

Calculations of transmission line impedance change have been done on the assumption that the test signal line has the impedance equaling 50Ω in the entire frequency range. This assumption is correct in the range of line frequency. It substantially simplifies modelling of the system. Figure 19 shows the change line characteristic impedance vs. frequency for R1 = 1Ω.

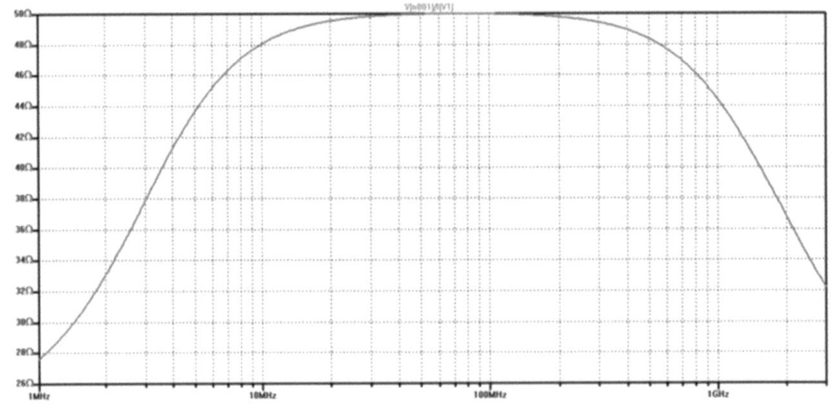


**Figure 19.** Change in test signal line characteristic impedance in frequency function.

This simplification does not influence the results in the system frequency range. For the sensor frequency equalling 74MHz the test signal attenuation at DC output equals –78dB, and line impedance equals 50Ω. A separating or summing up system prepared in such a manner does not influence the test signal transmission through SAW vibration sensor.

### 3.2. Delaying line and adapting system

The design of SAW-comprising delaying lines used for vibration sensors (Figure 20A) differ from those used for sensors of other physical quantities (Figure 20B). Since the sensor plate moves, electrodes applying electric signals to converters (bus bar) should be situated on the immobile part of the plate. (Thus a proper strength of electric contacts for these electrodes will be ensured.) Electrodes are long and their resistance is specific. As the plate moves its housing is greater than that of classic filters comprising the SAW. This causes the signal passing directly between the SAW-comprising delaying line inlet and outlet to increase.

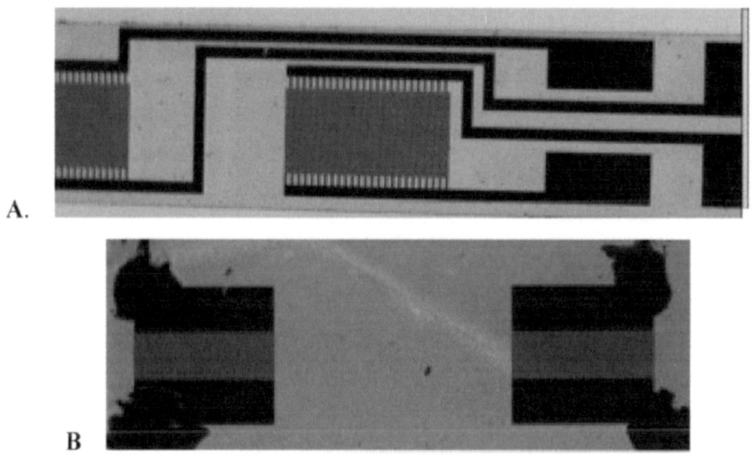


**Figure 20.** SAW delay lines: for vibration sensor (A), classic (B)

The line was designed in the form of two cooperating, identical, simple, periodical, double-electrode interdigital transducers. Figure 21 shows a system of converter electrodes. Such a structure of converters enables their operation on the third harmonic. The lines were designed to fabricate them with a ST-cut quartz.

**Figure 21.** Structure of delaying line converters

Due to a low value of the electro-mechanical coupling factor for a ST-cut quartz losses for a mismatch its inlet impedances to 50Ω are significant. In order to cut on these losses operation of interdigital transducers under conditions of matching to impedance of 50Ω at a frequency of 74MHz was selected [29-32]. Figure 22 shows a converter matching system. The element for matching a converter having the conductance of $G_P$ and the capacity $C_P$ to the impedance $R_g$ = 50Ω is the inductance $L_1$. The matching takes place on the condition that the available power derived from a voltage source $E_g$ of the internal impedance $R_g$ = 50Ω is distributed released/emitted on the converter conductance $G_P$.

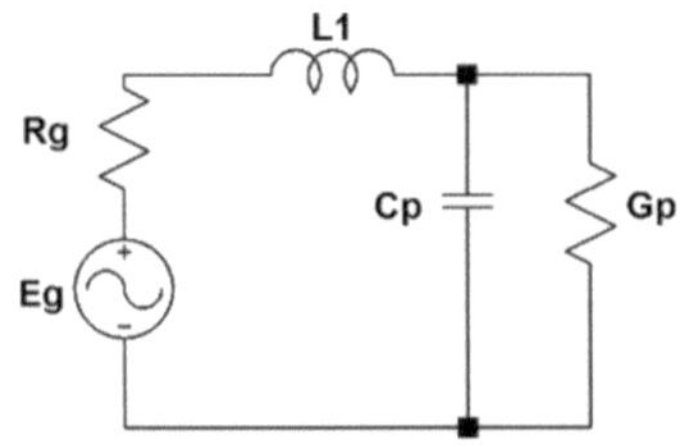


**Figure 22.** Interdigital transducer matching system to 50Ω impedance..

This condition is met for a conductance resulting from the relationship [29]:

$$L_1 = \frac{C_P}{G_P{}^2 + \omega^2 C_P{}^2} \tag{26}$$

With the help of the (26) relationship the converter geometry for a ST-cut quartz was determined. The aperture of the converter A= 2,5 mm was adopted. To fulfill the purpose converters operated on the third harmonic at 74MHz. For such parameters a converter consisting of N=25 pairs of electrodes was received. The electrode width and the gap between electrodes were 16 μm, and the surface length 37 μm. The results of theoretical calculations of conductance and converter capacity versus frequency are presented in Figure 23.

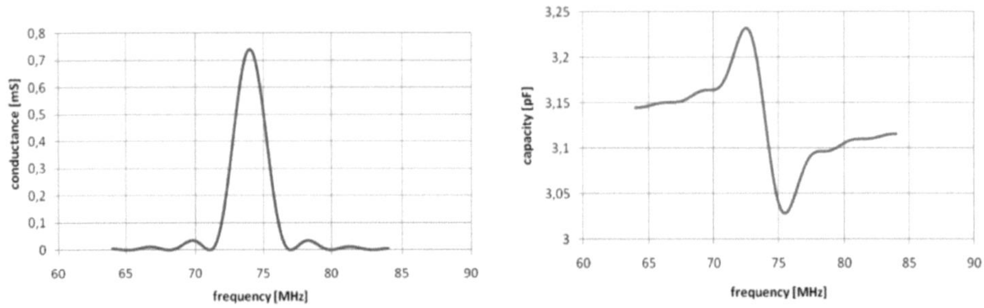


**Figure 23.** Theoretical dependence of conductance and capacity of simple transducer composed of 25 double electrode pairs on ST-cut quartz.

At 74MHz frequency the converter conductance is 0.74mS and the capacity 3.13pF. For these quantities the inductance $L_1$= 900nH was calculated, at which value the condition of matching the converter to the impedance 50 Ω is satisfied. Practically, the matching of interdigital transducers to the impedance of 50 Ω was carried out by measurement of the coefficient of reflection Figure 24 shows the change in the coefficient of reflection from a matched converter versus frequency in a system of an impedance of 50 Ω.

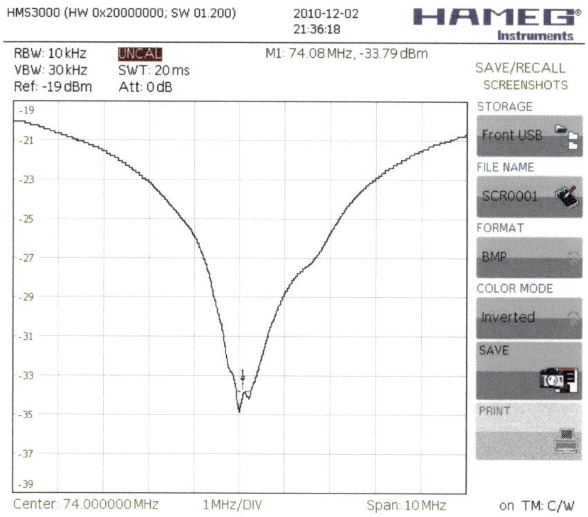


**Figure 24.** Coefficient of reflection from a matched converter versus frequency.

Figure 25 shows the mounting of the sensor plate.

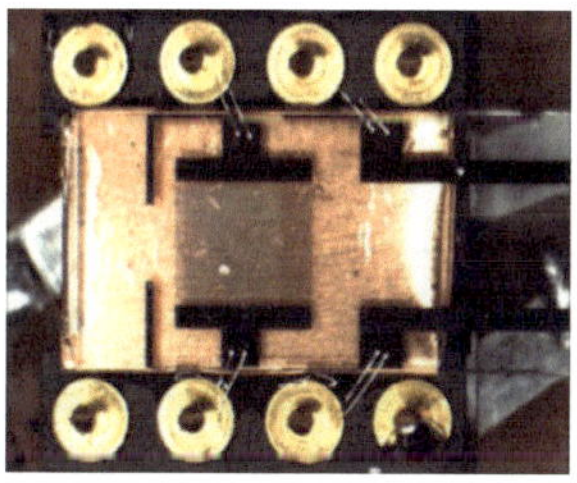

**Figure 25.** Mounting of the sensor plate

Figure 26 presents the frequency amplitude curve for the fabricated SAW delaying line. The measurement was conducted after the line was mounted in an SAW-based vibration converter.

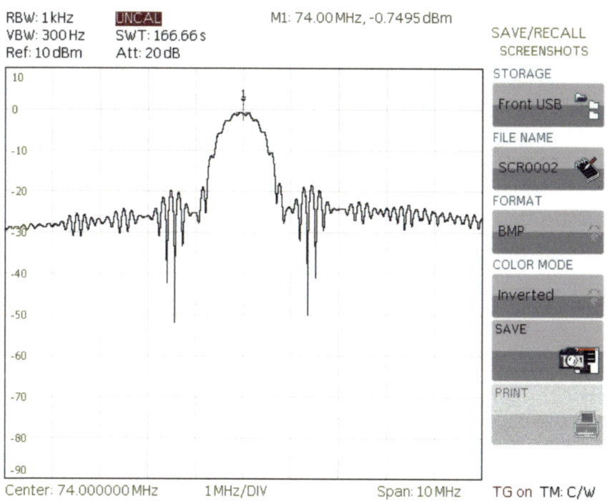


**Figure 26.** Attenuation frequency diagram of SAW vibration sensor.

## 3.3. Amplifier

The role of an amplifier is to compensate losses caused by SAW delay line. An amplifier has been built on a monolithic system MAR-6 manufactured by Mini-Circuits. Figure 27 shows a diagram of an amplifier being a part of an electronic system of SAW vibration sensor.

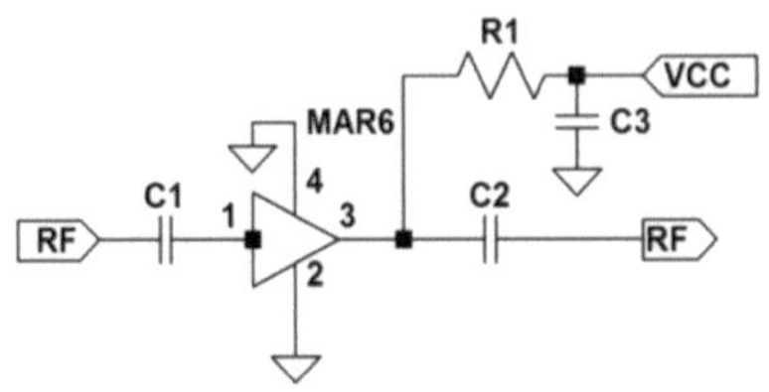


**Figure 27.** A diagram of an amplifier for SAW vibration sensor.

The amplifier is supplied at the output side by an R1 resistor. The R1 value is selected according to the DC supply voltage. For a supply voltage of 12V the resistance R1 equaled 560Ω. An amplification of 22dB was achieved for test signal frequency of 74MHz. It is the highest amplification value possible to achieve in this system. The value of current input equaled 16mA. Figure 26 shows an experimental frequency characteristics of SAW vibration sensor. Measurement of attenuation frequency diagram of SAW vibration sensor has been conducted on a spectrum analyzer HMS 1010. A supply voltage system has been put at line input. A supply voltage blocking system has been put before the analyzer at line output. Losses of 0.75dB consist of line losses and losses in the connection wiring and discussed separation systems. The value of these losses has been estimated at the level of 1dB. A conclusion may be drawn that an amplifier compensates the losses caused by SAW delay line. A theoretical shape of attenuation frequency diagram of a sensor should be described by the function (sin(x)/x). An experimental characteristic has high-frequency irregularities. Their reasons are the signals going from sensor output to input, omitting sensor electronic components. This signal amplitude is around –36dB lower than useful signal amplitude. The reason of occurrence of signal going from sensor input to output will be discussed in the next Section.

## 3.4. Parasitic signals

Parasitic signals are the signals going from electronic system input to output, omitting any component which is a part of the test signal transmission line. It is possible due to the occurrence of a parasitic coupling between any place of electronic system. There are two mechanisms leading to the occurrence of couplings [18][33]. The first one is electromagnetic coupling. The second one is ground current coupling. Figure 28 shows the mechanism of electromagnetic coupling. Red lines indicate paths of electromagnetic coupling which may occur in the electronic system of SAW vibration sensor. Electromagnetic couplings occur in all the electronic components constituting a sensor system. Paths of printed circuit are matched to the impedance of 50Ω. They are simultaneously transmitting and receiving aerials. Their efficiency depends on the path length. A similar role is played by inductances occurring in the system and capacities between paths. In order to reduce the electromagnetic coupling the inductances should be placed perpendicularly to each other and placed at a distance. These elements define the manner of making of the printed circuit plates. Any problems are solved individually, in accordance with a chosen construction.

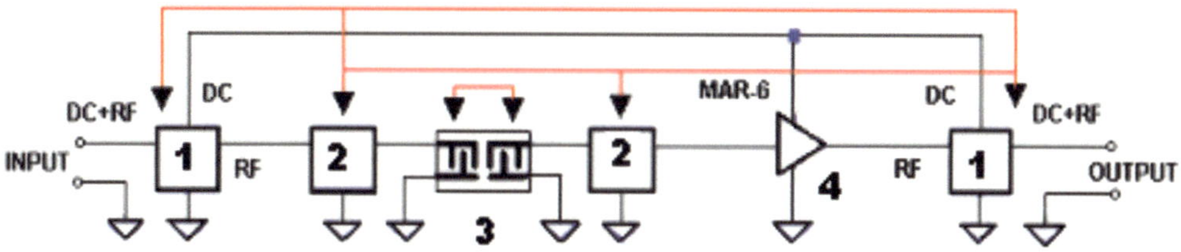


**Figure 28.** Electromagnetic coupling in electronic system of SAW vibration sensor.

The fundamental problem is an occurrence of electromagnetic coupling between the transducers and SAW delay line. Bus-bars delivering electric signal to transducers are placed on an immobile part of the plate. They are long and they are placed close to each other. It causes an increase in capacity between IDTs. The direct signal going through this way is also strengthened. Because of sensor plate motion its casing is larger than those used in traditional SAW filters. It also causes an increase in direct signal strength in the delay line. This problem and possible solutions are known in literature [29-31]. The most effective solution is symmetrical supply of one of transducers and their functioning in a bridge circuit [34]. Such a solution has been used in the presented SAW vibration sensor. The signal strength at the level of –35dB has been achieved. The second mechanism causing increase in direct signal strength is the ground current coupling [33]. Figure 29 shows the mechanism of this coupling.

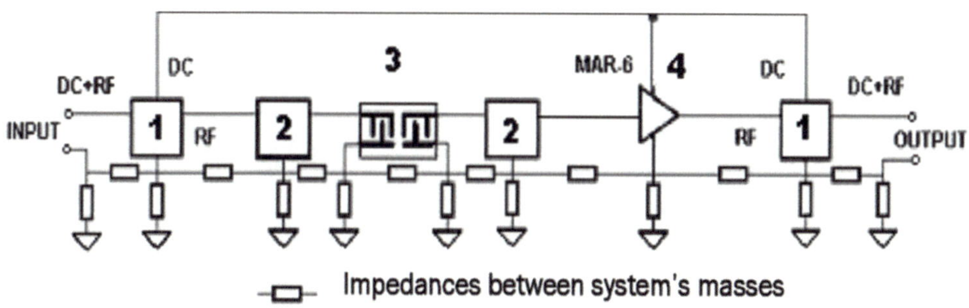


**Figure 29.** The mechanism of ground current coupling.

Ground current couplings occur only when the connection between component mass and joint mass is not perfect. A diagram in Figure 29 shows this effect by introduced impedances. An ideal connection is characterised by a null value of all impedances. Introduced impedances change current distribution in the entire system. Values of these impedances are small (fractions of ohm). That is why they are difficult to model. A physical making of component mass connection to the system joint mass must be considered during the design stage of the system. Reduction of this value by careful preparation of the system joint mass is a proper solution. Similarly to the first mechanism, SAW delay line plays an important part. Impedance of bus-bars and impedance of contacts leading the signal to transducer are crucial elements in SAW vibration sensor delay line. The joint mass of the discussed system has been made of 5 mm copper plate to which a printed-circuit board has been soldered. This side of the plate was completely bonded. The electronic system joint mass has been connected to component masses. Figure 30 shows SAW vibration sensor. Only supply voltage of the amplifier system goes through printed circuit paths. Longer segments of test signal line have been made by means of coaxial cables. It allowed to reduce an electromagnetic couplings value in the system. Attenuation frequency diagram of this sensor is presented in Figure 26.

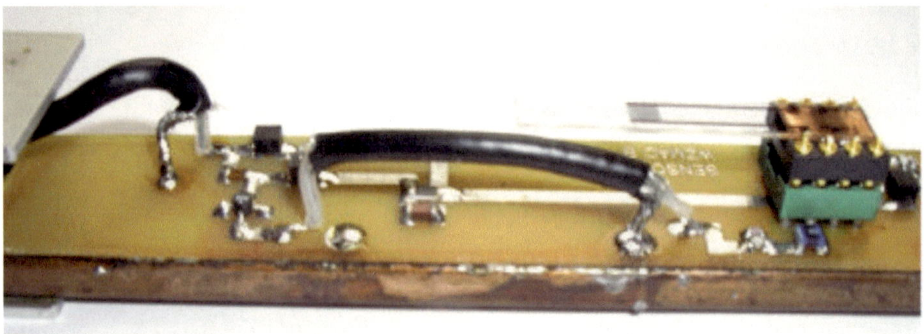

**Figure 30.** SAW vibration sensor

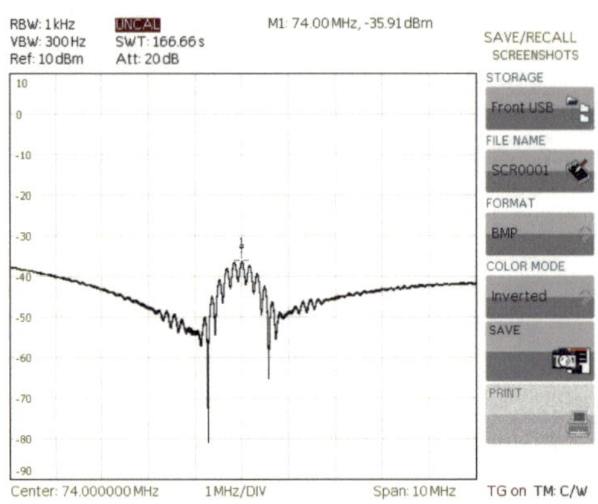


**Figure 31.** Attenuation frequency diagram of SAW vibration sensor after shut-down of amplifier supply.

Figure 31 presents this characteristic after a shut-down of amplifier supply. Shapes of characteristics in delay line operation band are similar in both figures. It suggests that signal source at the sensor output is situated behind SAW delay line. This signal strength is –35dB below the sensor frequency characteristic signal. Beyond the operation band the signal strength equals –40dB. Conducted measurements make use of the time trace of stationary signals. The lack of information about the delay of these signals does not allow to determine their source. IDTs are selectively matched to the impedance of 50Ω. It can be a signal going directly between interdigital transducers of SAW delay line. In order to explain the origin of the signal, it is important to learn about its delay time. It has been made in a system presented in Figure 32.

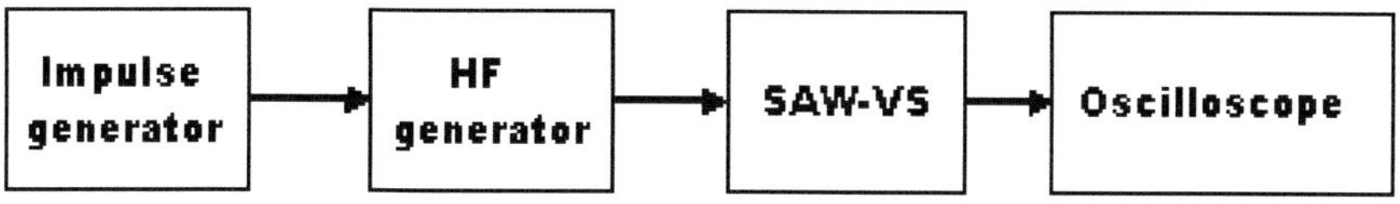


**Figure 32.** Meter circuit for parasitic signals in SAW delay line.

At sensor input a signal in form of wave packet at frequency of 74MHz has been delivered. The length of the packet is smaller than line delay. In this way a temporary separation of the direct and useful signals has been ensured.

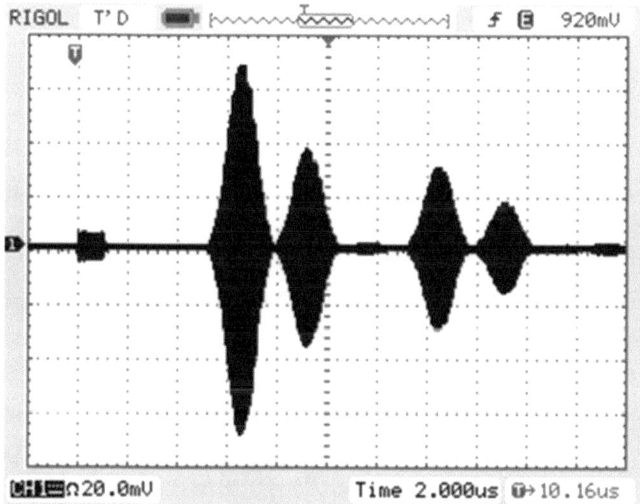


**Figure 33.** Time run of signals in SAW delay line.

Figure 33 presents timing of signals at the output of SAW vibration sensor. There are five signals at the sensor output. The first signal is going directly from sensor output to its input. The second one is a useful signal. Next three signals are reflected from the plate edge. Reflected signals will be attenuated by a damping paste. The amplitude of direct signal is –35dB below the useful signal. It leads to conclusion that the signal shown in Figure 31 is a signal going directly between SAW delay line IDTs. When the electronic system is properly made, the elimination of this signal is the most fundamental problem in SAW vibration sensor design.

## 4. Measurements of vibration sensor parameters

SAW-based vibration sensors with SAW were made as described in Section 3. Delaying lines of a 74MHz middle frequency and various delay quantities of 4.2μs, 6.2μs, 8.2μs were worked out. These delaying lines were made on ST-cut quartz plates of different lengths. These plates were 5.7 mm wide and 0.5 mm thick.

The signal passing thru a sensor is continuous one. In a sensor every parasite signal adds up to a useful signal and causes the amplitude of a useful signal to modulate. This causes the sensor sensitivity to decrease; therefore parasitic signals must be removed. An essential problem is the reduction of a signal passing directly from the delay line inlet to the delay line outlet. The magnitude of this signal for sensors with lines of different delays is shown in Figure 34. The larger the line delay the lower the direct signal level. This level depends on the outlay of electric inlets to the delaying line converters. This is illustrated in Figures 35 (oscilloscope signals A and C). In order to lower the direct signal level additional screening of interdigital transducers or a symmetric supply of one of the converters was used [34]. For executed sensors a direct signal at a level of –36dB was obtained.

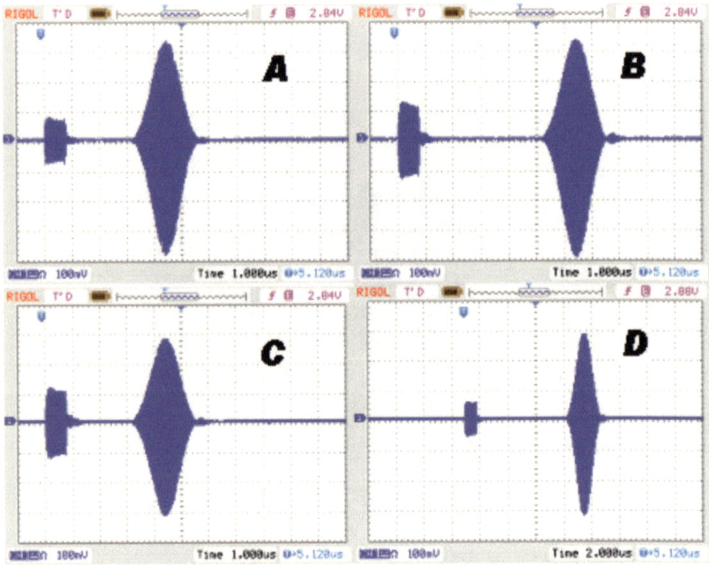


**Figure 34.** Direct and useful signal in delaying lines of different delays: A – 4,2μs, B – 6,2μs, C – 4,2μs, D – 8,2μs.

To basic parameters of vibration sensors belong their frequency characteristic curves and the static sensitivity. We will present the way they are measured. Determined experimental characteristics will be compared to the theoretical one. This will enable to estimate the model precision and to what extent it can help to model the parameters of SAW-based vibration sensors. The sensor frequency characteristics were determined in two stages. At the first stage the sensor pulse response was registered. At the second stage the spectrum of this pulse response was determined. Its shape corresponds with the sensor frequency characteristics. Pulse responses of sensors were measured and recorded as well as the spectrum of pulse responses from sensors were calculated with the aid of the system shown in Figure 35. Sensors were agitated for vibrations by an impact. Pulse responses were recorded with the help of the Agilent VEE Pro programme.

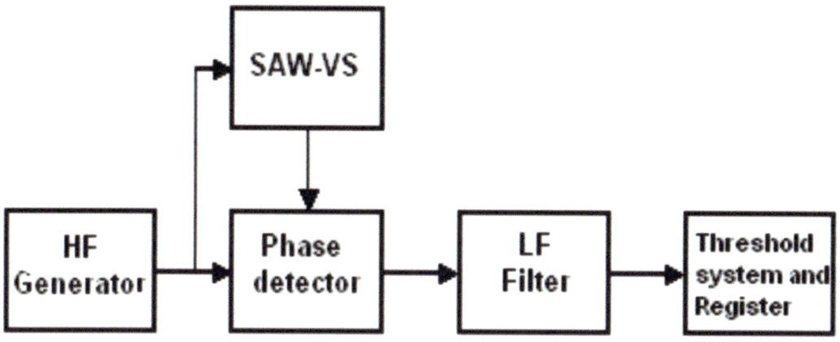


**Figure 35.** System to measure responses from SAW-based vibration sensor pulse plates.

Figure 36 presents a pulse response of a sensor where a delaying line with a 65 mm long plate was used. The sensor plate was not loaded with a seismic mass. The length of the pulse response was approximately 5s and its frequency was 92Hz.

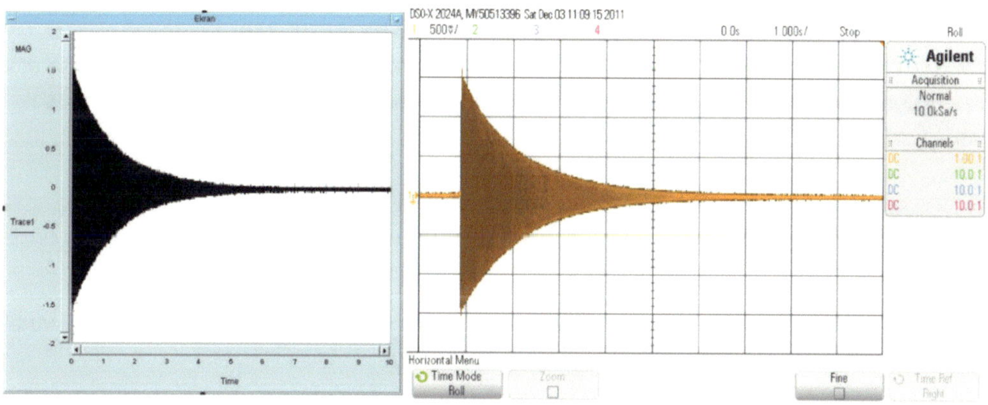


a. b.

**Figure 36.** Pulse response of sensor with 65 mm long plate without seismic mass recorded with VEE programme – (a) and oscilloscope – (b)

The spectrum of a pulse response (a Fourier transform) was calculated with the help the Agilent VEE Pro programme, but it can be achieved also directly on an oscilloscope. Figure 37 presents the amplitude of this spectrum. Its shape corresponds to the amplitude of the frequency characteristics of a tested sensor. The resonance frequency equals theoretical values calculated with the relationships presented in Section 2. The frequency characteristics shows a harmonic at a 400Hz frequency. Its level is –26dB below the level of the sensor characteristics for the resonance frequency. This level is higher than its theoretical estimate presented in Section 2. It is difficult to determine the reason for this difference. It can be the inaccuracy of the model. However, it can be also due to a differing effectiveness of incitation of the resonance frequency component and harmonic frequency component.

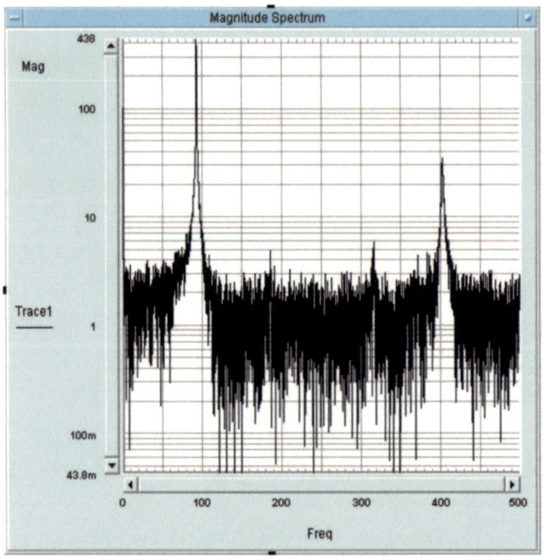


**Figure 37.** Spectrum of sensor with 65 mm long plate without seismic mass

Figure 38 presents the measurement of the statistical sensitivity of a sensor. It was conducted by recording the sensor output signal during its rotation by 180 degrees. This rotation causes the constant acceleration affecting a sensor to change by a value of two gravitational accelerations – i.e. 2 „g". An estimated static sensitivity of a sensor is 40mV/(„g").

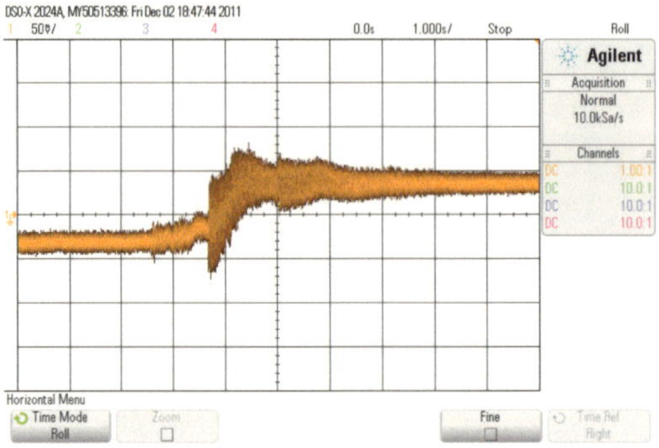


**Figure 38.** Method of determining static sensitivity of sensor

A 1.07g (r=2,27) seismic mass was placed on a sensor. This caused its resonance frequency to drop to 29Hz, and then for this sensor the above-presented tests were repeated. Fig. 39 shows the pulse response of a sensor with a 65mm long plate with a 1.07 g seismic mass. The pulse response is longer than 10s, with its frequency being 29Hz.

Figure 40 shows the frequency characteristics of the tested sensor. The value of the resonance characteristics equals the theoretical value. The characteristics shows a harmonic at 58Hz. Its level is –4dB below the sensor characteristic value for the resonance frequency, and being low, practically has no impact on the sensor function. Figure 41 shows the measurement result of the sensor static sensitivity as 100mV/(„g"). Against a sensor without a concentrated mass this value rose 2 ½ times. The length of the pulse response increased more than 2 times. These changes are obvious and their quantities determine explicit relationships describing the sensor model. The only difference between the results of experimental tests against the theoretical model is a higher level of the harmonic resonance frequency. To explain this difference classical measurements of the sensor resonance characteristics must be conducted. To conduct these measurements an exciter of stable mechanical vibrations of the sensor housing of an adjusted amplitude and frequency is required. It was not possible for the authors to carry out these tests.

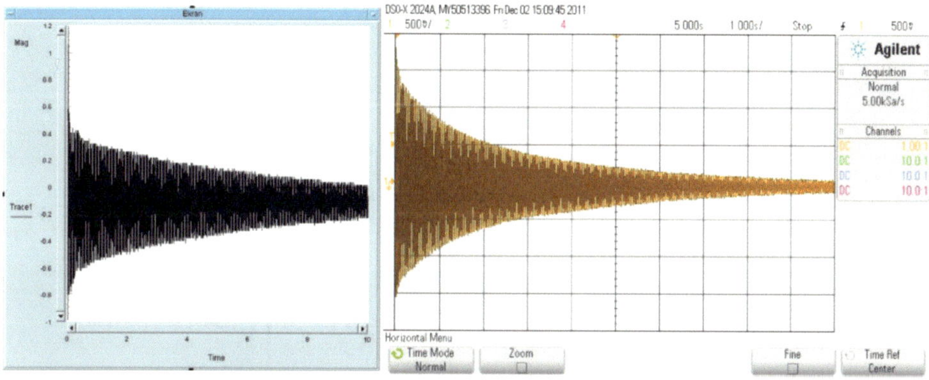


a. b.

**Figure 39.** Pulse response of sensor with 65 mm long plate with a 1.07g seismic mass recorded with VEE programme – (a) and oscilloscope – (b)

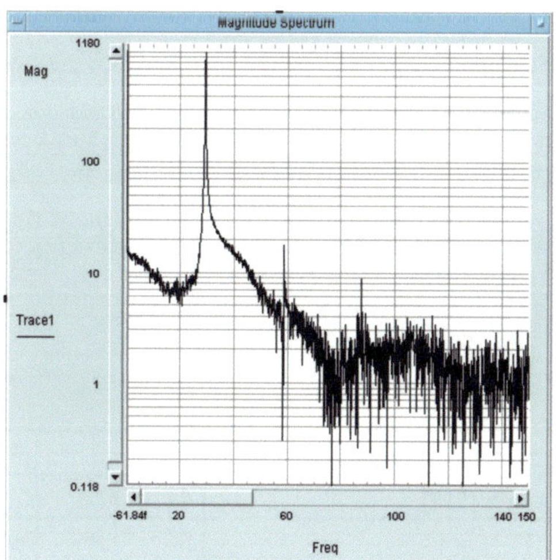


**Figure 40.** Spectrum of a 65mm long sensor with 1.07 g seismic mass

The test results demonstrated a good compatibility between the theoretical parameters of the sensor pulse response (resonance frequency and the decay time) and their experimental realization. The model presented in Section 2 was used to elaborate SAW-based sensors, which featured required parameters of the pulse response. Figure 42 shows a block diagram of an electronic warning system with SAW-based vibration sensors.

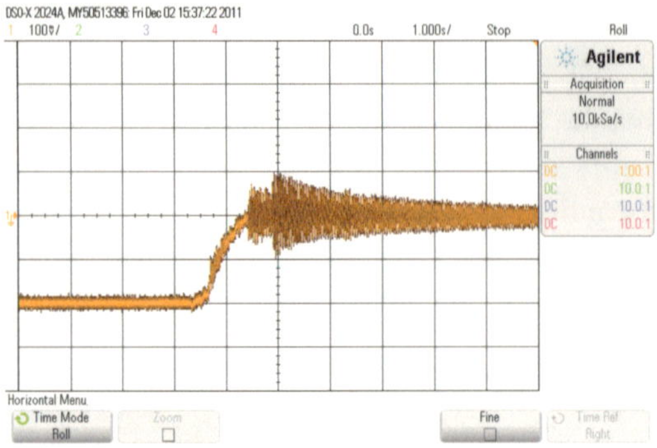


**Figure 41.** Method of determination the sensor static sensitivity

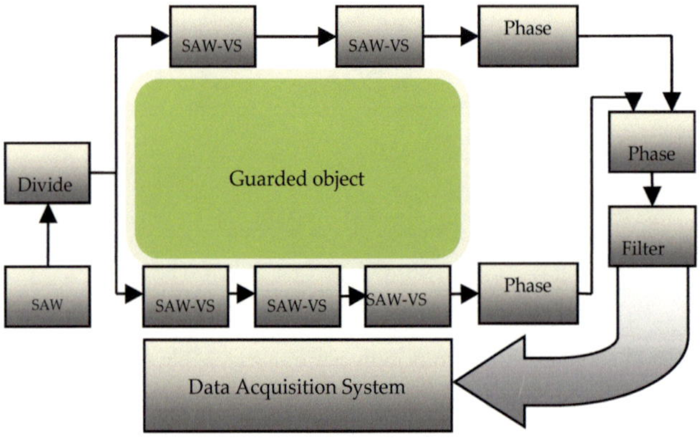


**Figure 42.** Block diagram of electronic warning system with SAW-comprising vibration sensors.

During operation of the system vibratons of the sensor plate change the phase of the measuring signal. The frequency of these changes equals the plate resonance frequency. Thus the signal from every sensor passing thru a set of filters at the phase detector inlet can be separated. The system operation is discussed further in the following works [35][36]. The said system required SAW-based vibration sensors having parameters given in the Table 2; the authors prepared these sensors.

| Sensor No. | 1 | 2 | 3 | 4 | 5 |
|---|---|---|---|---|---|
| Resonance frequency [Hz] | 41 | 56 | 73 | 90 | 159 |
| Plate length [mm] | 86 | 85 | 66 | 68 | 26 |
| Seismic mass - r | 0.22 | 0 | 0.14 | 0 | 1 |
| Line delay [μs] | 6.2 | 4.2 | 8.2 | 4.2 | 4.2 |

**Table 2.** Parameters of SAW-VS sensors made for electronic warning system

By selection of a seismic mass required resonance frequencies of sensors were achieved. In case of sensors No. 1 and 2 as well as No. 3 and 4 plates of similar lengths were used. Figure 43 presents characteristics of four assembled sensors.

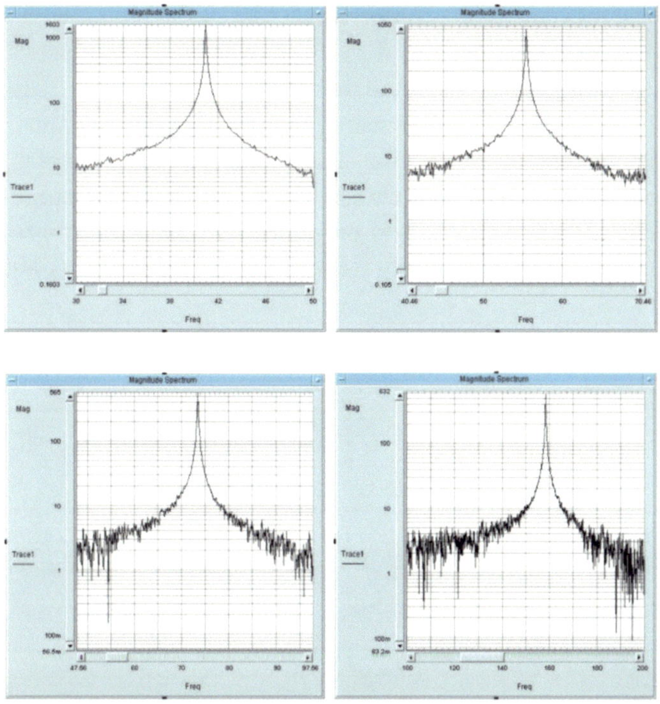


**Figure 43.** Frequency characteristics of sensors of resonance frequency of 41Hz, 56Hz, 73Hz, 159Hz.

These five SAW-based vibration sensors fabricated and used in an electronic warning system proved the efficiency of the presented modeling method. The system was tested on a stand shown in Figure 44. The sensors were attached to steel ropes tensioned as required. In order to describe the movement of sensors and ropes a model of a string loaded with a sensor mass taking into account its moment of inertia [37] was developed.

**Figure 44.** SAW-comprising vibration sensors attached to steel ropes

Vibrations of sensors were stimulated by deflecting them out of the state of equilibrium. The string vibration time was several times as long as that of the decay time of a senor plate pulse response. The movement of a sensor was a sum of the fading with time of the pulse response and vibrations enforced by a cyclic movement of the sensors housing. The frequency of the housing movements was that of the rope vibration frequency, which was selected so that it was lower than the resonance frequency of sensors. Thus it was possible to analyze every component of the sensor movement. For experimental testing a string vibration frequency of some 6 Hz was chosen. For every sensor the output signal from the phase detector (Fig. 42) was recorded and processed with the VEE program. Figure 45 and 46 show the course of signals of vibrating sensor of various resonance frequencies and their spectra.

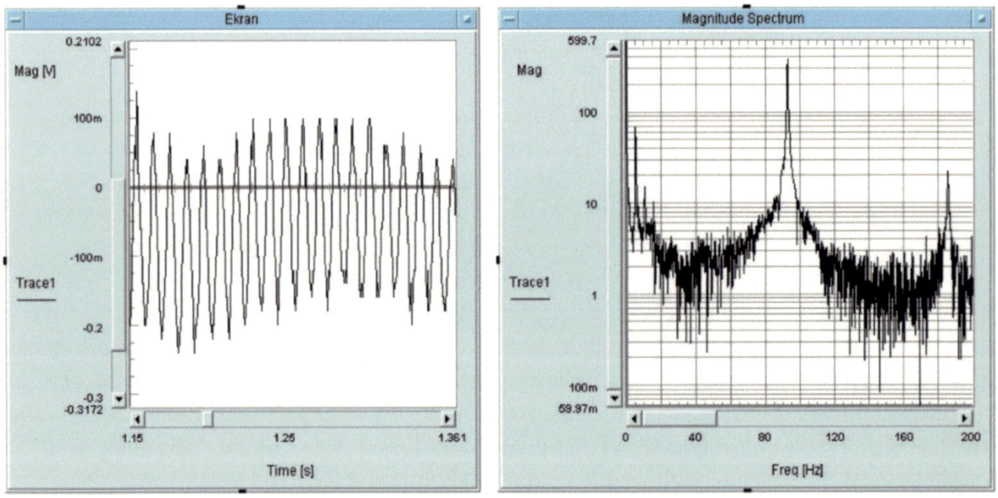


**Figure 45.** Part of output signal from sensor of plate resonance frequency of 91 Hz and string vibration frequency of 6.7 Hz and its spectrum

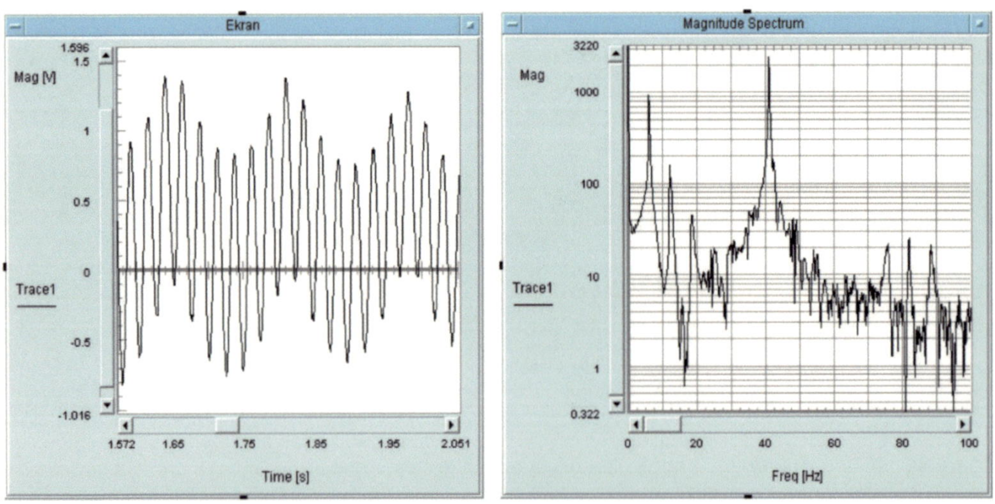


**Figure 46.** Part of output signal from sensor of plate resonance frequency of 41 Hz and string vibration frequency of 6.7 Hz and its spectrum

Pulse responses of sensors of an 91Hz (Fig. 45) or a 41Hz (Fig. 46) frequency can be easily discerned form signals derived from a string movement of a 6.7 Hz frequency, especially within the signal spectrum. Relations between their amplitudes are visible. After decay of pulse responses sensors measure the rope vibrations; they work then like classical vibration sensors.

In case of a sensor plate without the seismic mass (Fig. 45) the harmonic amplitude occurs at the – 34dB level below the resonance frequency amplitude. With the plate loaded with a small seismic mass (r=0,22, Fig. 46) the level is lower than -40dB. Harmonic vibrations had no impact on the operation of a system presented in Figure 43.

Depending on the use of a sensor by selection of its resonance frequency we can change its sensitivity and linearity. The practically prepared electronic warning system with SAW-based vibration sensors has fully proven the usefulness of the model presented in Section 2.

## 5. Conclusions

The work presents development, execution and parameters of SAW seismic vibration sensors. A sensor is a two-terminal pair network consisting of a delaying line with SAW and an amplifier compensating losses introduced by a ST-cut quartz. The delay line is fabricated on the CT-cut quartz surface.

A simple vibration model of an anisotropic plate was used to develop sensors. By way of successive simplifications of the description of vibrations of a viscous-elastic sensor plate a model of one degree of freedom was obtained. An explicit description of the movement parameters of the sensor plate was achieved. A material damping of the plate practically causes it to vibrate at only one resonance frequency, thus enabling to design SAW seismic vibration sensors. The results of experiments proved the effectiveness of using this model to design SAW-comprising seismic vibration sensors.

Basic parameters of realised vibration sensors (resonance characteristics, pulse responses, static sensitivity) are presented and analysed.

The range of resonance frequencies of plates made of ST-cut quartz, which were feasible, was determined. The plate length restricts the lower range of resonance frequencies. This range was determined on the basis of an available length of 100mm of a CT-cut quartz. The upper range of resonance frequencies is restricted by the speed of decay of a pulse response of a sensor plate. However, this restriction applies only to a sensor operating on its pulse responses. For the determination of this range the magnitude of stresses occurring in the plate was not taken into account. This magnitude must be smaller than the size of critical stresses presented in Table 1. In the subject work this element was not analysed. The magnitude of dynamic, critical stresses for a ST-cut quartz was determined in works [22] [23]. These values do not conform to standards and they apply to a series of plates cut out with a wire saw. In the course of that determination it turned out that the technology of plates production has a great impact on the value of dynamic critical stresses. The work [18] demonstrated that their value for a given design of a sensor does not restrict the determined range of the plate resonance frequencies.

As the plate resonance decreases the sensor sensitivity increases. Therefore high sensitivity sensors can be designed. Increasing its length can lower the resonance frequency of a plate. This is the most effective way to reduce the plate resonance frequency. It is possible to develop high-sensitivity vibration sensors of resonance frequencies in the order of a few Hz. To design such sensors one can use directly the presented model. Lets compare the SAW (SAW-VS) vibration sensor presented in the work to three kinds of sensors used at present. Resonance frequencies and basic applications of these sensors are presented in the introduction to this work.

The first kind of these sensors is geophones, where the sensor pulse response is utilized, which explains their low resonance frequency (several Hz) and a high sensitivity.

The two remaining kinds of vibration sensors are micro-mechanical silicone acceleration sensors (Micro Electro Mechanical Systems accelerometers - MEMS accelerometers) and piezoelectric acceleration sensors. Lets compare the basic parameters of these sensors: sensitivity, range of measured values and the frequency of acceleration changes, their structure (resonance frequency (natural frequency), weight and the manner of measuring the acceleration. We don't compare acceleration sensors reacting to impacts. They are very light and feature a very broad measuring range. Piezoelectric acceleration sensors work within a range between fractions of Hz thru a dozen or so kHz. The lower and upper range of the measurement dynamics is from 1 "mg" up to 100.000 „g". („g" is a unit of acceleration equal to the gravitational acceleration at sea level, i.e. $9.81 m/s^{-2}$ ). Depending on the design the sensor sensitivity is 0.2mV/g - 0.7V/g. The natural (resonance) frequency ranges from a few Hz to a few dozen of Hz. These sensors weigh from 3g to 500 g. The structure of these sensors is relatively simple, however, the measuring system is complicated (measuring of charge changes in the order of pC). MEMS accelerometers are characterized by small dimensions (an integrated circuit) and a low price. With these devices constant and variable acceleration up to a frequency of a few hundred Hz can be measured.

The lower and upper range of the measurement dynamics is from fractional „g" to 10 000 „g". Depending on the design the device sensitivity ranges from 0.2mV/g up to 10V/g (this applies to seismology sensors).

The device resonance frequency (natural frequency) is high: several Hz. These devices weigh from a few g up to 2 500 „g" (in case of seismology sensors). The sensor design is relatively simple; as well its measuring system is simple. (Measuring of changes in the charge in the order of aF's.)

The measuring system of a MEMS sensor is similar to a measuring system of an SAW-vibration sensor presented in the work (Figure 35). It consists of a measuring generator of a 1MHz frequency, two measuring paths and a phase detector.

The mechanical frequency of SAW-comprising vibration sensors can be changed within a range from several to a few hundred Hz. This is a significant difference between these both kinds of sensors. The sensitivity of an SAW vibration sensor depends on the resonance characteristics of the sensor plate, the length of the surface wave utilized in a sensor and on the sensor design. One can define here the sensitivity for a constant acceleration and the sensitivity for the resonance frequency of the sensor plate or some other frequency. For comparison's sake we can assume the sensitivity for a constant acceleration.

An example: for a SAW-VS with a plate 57.5 mm long and 0.5 mm thick, loaded with a seismic mass equal to 4 times the plate mass, a static sensitivity of 0.5V/„g" was achieved. The sensitivity of SAW-sensors for changing accelerations can be a couple of times greater than that for constant accelerations. The sensitivity degree is determined by the resonance curve of a sensor plate. The sensitivity of vibration sensors with SAW-VS can be increased by decreasing the length of the surface wave, or changing the sensor design (i.e. reduction of the plate thickness), by increasing the concentrated mass or by increasing the delay of the SAW-delaying line.

SAW vibration sensors can be cascaded which offers many system designs what in turn offers an increase in the sensitivity and a reduction of the cross sensitivity of sensors. This is not possible with a MEMS sensor and a piezoelectric sensor. The high measuring frequency of SAW-comprising sensors allows designing wireless sensor versions.

In terms of the design and the method of measuring MEMS and SAW-VS acceleration sensors have a lot in common. In a MEMS sensor a vibrating plate of silicon changes the capacity of a capacitor. Mechanical properties of silicon (density 2.330kg/m$^{-3}$, and an equivalent Young's modulus of 106 Gpa) are similar to mechanical properties of quartz. Therefore the parameters of MEMS and SAW-VS must be similar. In our opinion SAW-VS sensors have their place in the measuring technology.

This work presents an example of using SAW-sensors in an electronic warning system. Vibration sensors placed at selected points record vibrations within areas they cover. An alert central station registers signals from a vibration sensor. This is a typical system to be applied for perimeter protection systems.

All sensors included in a system can be placed within one area; then this system will perform the role of an analyzer of vibrations within this area. This is a second prospective application of the system under discussion. The possibility of preparing high-sensitivity vibration sensors of resonance frequencies in the order of a few Hz is a prospective area of application for monitoring vibrations of bridges and buildings, where the frequency of free vibrations is in the order of fractions of Hz up to a dozen or so Hz. The presented application examples apply to working on the pulse response of an SAW-sensor. This kind of a sensor can be used to measure one component of the acceleration vector. Then, the pulse response means for a measurement of acceleration a parasitic signal, and should be eliminated. Therefore thru the use of sensor plates, where the decay of a pulse response is fast, acceleration sensors can be developed. The movement of a plate shows then a character of acceleration changes in time. Works [38] [39] show development of this kind of sensors.

## Author details

Jerzy Filipiak and Grzegorz Steczko
*Institute of Electronic and Control Systems, Technical University of Czestochowa, Częstochowa, Poland*

## Acknowledgement

This work was supported by the Polish Ministry of Science and Higher Education as a project "Vibration Warning System with SAW Vibration Sensors" included in the scientific budget of 2009–2011.

## 6. References

[1] Pakhamov A, Pisano D, Sicignano A and Golburt T (2005) High Performance Seismic Sensor Requirements for Military and Secuirity Applications Proceedings of SPIE, j. 5796: 117-124.
[2] Kaajakari V, (2009) Practical MEMS, Small Gear Publishing
[3] Lee I.G, Yoon G.H, Park J, Seok S, Chun K, Lee K (2005) Development and analysis of the vertical capacitive accelerometer. Sensors and Actuators A j.119: 8-18.
[4] www.metrozet.com
[5] Wohltjen H, Dessy R (1979) Surface Acoustic Waves Probe for Chemical Analysis I. Introductcion and Instrument Design. Analytical Chemistry. j. 9: 1458-1475.
[6] Nakamoto T, Nakamura K, Moriizumi T (1996) Study of Oscillator- Circuit Behavior for QCM Gas Sensor. Proc. Ultrasonics Symposium.j. 1: 351-354.
[7] Gizeli E, Liley M, Love C.R, Vogel H (1997) Antibody Binding to a Functionalized Supported Lipid Layer, A Direct Acoustic Immunosensor, Analytical Chemistry. j. 69: 4808-4813.
[8] Urbańczyk M, Jakubik W, Kochowski S (1994) Investigation of sensor properties of cooper phtalocyanine with the use of surface acoustic waves. Sensors and Actuators B j.22: 133-137.
[9] Cullen C, Reeder T, (1975) Measurement of SAW Velocity Versus Strain for YX and ST Quartz. Proc. Ultrasonics Symposium: 519-522.
[10] Cullen C, Montress T (1980) Progress in the Development of SAW Resonator Pressure Transducers. Proc. Ultrasonics Symposium. j. 2: 696-701.
[11] Pohl A, Ostermayer G, Reindl L, Seifert F (1997) Monitoring the Tire Pressure of Cars Using Passsive SAW Sensors. Proc. Ultrasonics Symposium.j.1: 471-474.
[12] Clayton L.D, EerNisse E.F (1998) Quartz thicknes-shear mode pressure sensor design for enhanced sensitivity. IEEE Transactions on Ultrasonics, Ferroelectrics, and Frequency Control. J.45: 1196-1203.
[13] Jiang Q, Yang X.M, Zhou H.G, Yang J.S (2005) Analysis of surface acoustic wave pressure sensor. Sensors and Actuators A j.118: 1-5.
[14] Drafts B (2000) Acoustic wave technology sensors. Sensors j.10: 1-9.
[15] Seifert F, Bulst W, Ruppel C (1994) Mechanical sensor based on surface acoustic waves. Sensors and Actuator A. j. 44: 231-239.
[16] Hauden D (1991) Elastic waves for miniaturized piezoelectric sensors: applications to physical quantity measurements and chemical detection. Archives of Acoustics. j. 16: 91-106.

[17] Filipiak J, Solarz L, Steczko G (2007) Surface acoustic wave stress sensors. Desinger Analysis. Molecular and Quantum Acoustics. j. 28: 71-80.

[18] Filipiak J (2006) Surface acoustic wave acceleration sensors. Technical University of Czestochowa. Monograph 121, p.198. (in Polish).

[19] Filipiak J, Kopycki C (1999) Surface acoustic waves for the detection of small vibrations. Sensors and Actuators. j. 76: 318-322.

[20] Filipiak J, Solarz L, Steczko G (2009) Surface acoustic wave vibration sensors for linear electronic warning systems. Acta Physica Polonica A j.116: 302-306.

[21] Filipiak J, Solarz L, Steczko G (2011) Surface Acoustic Wave (SAW) Vibration Sensor. Sensors. j.11: 11809-11832. http://www.mdpi.com/journal/sensors

[22] Filipiak J, Zubko K (2005) Determination of damping in piezoelectric crystals. Molecular and Quantum Acoustics j. 26: 75-80.

[23] Kopycki C (1999) Effect of substrates on the parameters of piezoelectric vibration sensor with a surface acoustic wave. Ph.D. Thesis (in Polish), Military University of Technology, Warsaw,

[24] Nye J.F (1957) Physical Properties of Crystals; Clarendon Press: Oxford, GB.

[25] Mason W.P (1958) Physical Acoustics and the Properties of Solids. Van Nostrand,

[26] Bogusz W, Dżygadło Z, Rogula D, Sobczyk K, Solarz L (1992) Vibrations and Waves A. Elsevier: Amsterdam

[27] Zubko K (2006) Applying of the Rayleigh method to determination of elastic and viscoelastic parameters of piezoelectric crystals. Ph.D. Thesis (in Polish), Military University of Technology, Warsaw,

[28] Filipiak J, Solarz L, Zubko K (2004) Analysis of Acceleration Sensor by the discrete model. Molecular and Quantum Acoustics j. 25: 89-99.

[29] Filipiak J (1993) Problems of synthesis of components of a surface acoustic wave signal processing to complex type "chirp". Ph.D. Thesis (in Polish), Military University of Technology, Warsaw,

[30] Matthews H L (1997) Surface Wave Filters. John Wiley and Sons, New York.

[31] Morgan D P (1985) Surface Wave Devices for Signal Processing. Academic

[32] Ruppel C C W, Fieldly T A (2001) Advances in Surface Acoustic Waves Technology, Systems and Applications (vol 2). Word Scientific Pub. Co. Inc,.

[33] www.saw-devices.com

[34] Danicki E, Filipiak J (1982) Bridging method for elimination of the direct signal, Electronics ( in Polish), j.10-12: 22-26

[35] Filipiak J, Solarz L, Steczko G(2011) Electronic Warning System Based on SAW Vibration Sensors, Acta Physica Polonica A j.120: 593-597

[36] Filipiak J, Solarz L, Steczko G(2009) Surface acoustic wave vibration sensors for linear electronic warning systems. Acta Physica Polonica A j.116: 302-306

[37] Filipiak J, Solarz L, Steczko G (2010) Analysis of Experimental Stand for SAW Vibrations Sensor, Acta Physica Polonica A j.118: 1118-1123

[38] Filipiak J, Kopycki C, Solarz L, Ostrowski J (1998) The SAW Acceleration Sensor. Proceedings European Frequency and Time Forum j.I :229-232.

[39] Filipiak J, Kopycki C, Solarz L, Ostrowski J (1997) Lithium niobiate as the substratum for the SAW acceleration sensor. Proceedings SPIE, The International Society for Optical Engineering j. 3179: 256-260.

Chapter 20

# Acoustics and Vibro-Acoustics Applied in Space Industry

Rogério Pirk, Carlos d´Andrade Souto,
Gustavo Paulinelli Guimarães and Luiz Carlos Sandoval Góes

Additional information is available at the end of the chapter

http://dx.doi.org/10.5772/49966

## 1. Introduction

During flight, Expandable Launch Vehicles (ELV) are excited by severe acoustic loads in three phases of flight: lift off, transonic flight and maximum dynamic pressure instant [1]. As such, principles to make onboard equipment compatible with the mission environments must be adopted. At lift off, the highly intense acoustic loads occur; and these levels are usually adopted to qualify payloads and equipments. However, during the transonic flight and maximum dynamic pressure phase, acoustic excitation is also present and such characteristics are also significant for performance evaluation as well as for specific system dynamic qualification/acceptance programs. In this way, noise control treatments (NCT) shall be adopted to alleviate internal vibro-acoustic environments, in view of decreasing costs and developments.

The hostile in-flight environments can damage sensors/conditioners as well as make measurements unreliable. In this way, installation adapters must be designed to protect the sensors. The acoustics of such protective cavities influence the measured sound pressure level (SPL) As such, the cavities must be analyzed and their amplitude-frequency characteristics evaluated. Finally, the measurement corrections, necessary to obtain the actual external SPL, are determined.

Concerning the internal environment found during flights, important launcher subsystems as payload fairing (PLF) and equipment bays shall be investigated and vibro-acoustic analysis can be done, as pointed by [2], [3] and [4]. The PLF is the structural compartment of a launcher where the payload is placed during the flight mission. PLF inner acoustics and its attenuation designs, using virtual prototypes are analyzed using deterministic and statistic techniques. However, when in-flight loading are not characterized, the accounted external air-borne excitation can be those described in [5]. In a similar way, SPL along the launcher structure at lift-off can also be estimated [6]. Furthermore, an alternative procedure to characterize external SPL during flight can also be adopted as described by [7]. Passive vibration control techniques can be used to attenuate structure-borne vibration and the use of viscoelastic materials adding structural damping to reduce the magnitude of vibrations is a well-known solution, usually applied in space and aeronautical industries. On the other hand, the use of active vibration control (AVC) is still considered difficult to be implemented in space industry.

For acoustic noise attenuation, the standard practice is to use passive techniques like blankets ([8] and [9]), which attenuate sound by trapping the energy in the blanket material and dissipating it as heat [10] and Helmholtz resonators tuned to absorb acoustic energy at one or some specific frequencies, typically the cavity frequencies as done by [11].

Another acoustic crucial subject in space industry is combustion instability, since it can severely impair the operation of Liquid Propelled Rocket Engine (LPRE) [12]. In this way, solutions for instability problems in combustion chambers of LPRE as well as solid rocket motors (SRM) are of large interest. In [13], it is described that combustion instability can be verified when the power spectrum of the acoustic pressure measured during tests is analyzed. When an oscillation is observed, i.e., combustion instability, well-defined sound pressure peaks, summed to the background noise are present. Such peaks are correlated to the resonance frequencies of the combustion chamber. In this way, the coupling of acoustic natural frequencies and burning oscillations of the combustion chamber occurs, which can cause instabilities and consequent unexpected behavior as efficiency loss or even the explosion of the engine. In the early developing phases of liquid rocket engines, it is usually proposed the investigation of different combustion chamber configurations [14]. This is usually done in two steps as follows: using theoretical calculation and through experimental measurements. In this way, theoretical and experimental natural frequencies of the acoustic cavity are obtained. Further studies must be performed, applying devices and techniques to attenuate pressure oscillations inside combustion chambers and devices as Helmholtz Resonators, baffles and ¼ wave filters are largely used ([15] and [16]).

This chapter describes three case studies applied on space industry. Firstly, analytical and numerical modelings of in-flight external microphone protection devices are described. Testing procedures and the SPL measurement correction factors determination are also presented. As a second case study, deterministic and statistical coupled vibro-acoustic analysis techniques are used to estimate PLF internal SPL at lift-off as well as to assess the effect of including NCT (blanket materials) on its skin. The modeling procedures and experimental ground test are described. Finally, in the third case, the acoustic characterization of combustion chambers is presented. Cold tests are described as well as the theoretical modeling procedures. The pressure attenuation control technique using Helmholtz Resonators are also presented. In all three case studies, theoretical x experimental results are depicted.

## 2. External on board microphones installation devices

At lift off, the source of the acoustic noise is the gas stream ejected by the motors (Fig. 1). Such acoustic pressure lies in the range of 140 to 180 dB near the rocket and is very close to an acoustic diffuse field (ADF) noise. At transonic flight the launcher is excited by the turbulent boundary layer (TBL) in the neighborhood of the shock waves. According to [17], when the maximum dynamic pressure occurs, the unsteady pressure field applied to the launcher is due to aerodynamic noise. The characteristics of such noise are very different from those at lift off. Non-attached flows increase the pressure in low-frequencies, which excite the launcher first structural modes. A simulation of the VLS-1 flight aerodynamics was done by Academician V. P. Makeyev State Rocket Centre (SRC- Makeyev), as shown in Fig. 2. Notice that the upper nose and 1st stage noses are the most exposed regions to aero-acoustic noise.

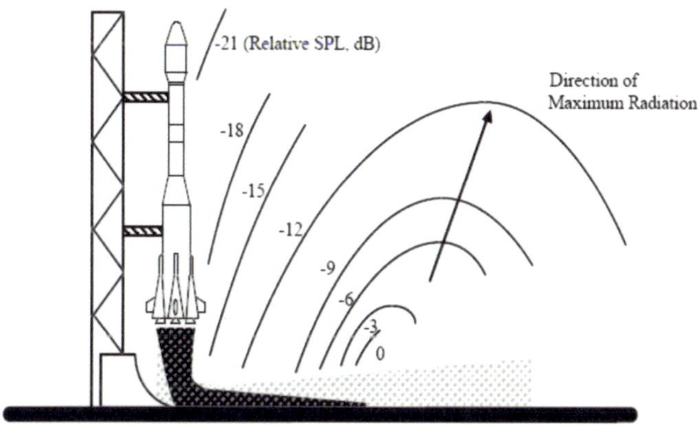


**Figure 1.** Acoustic noise at lift off

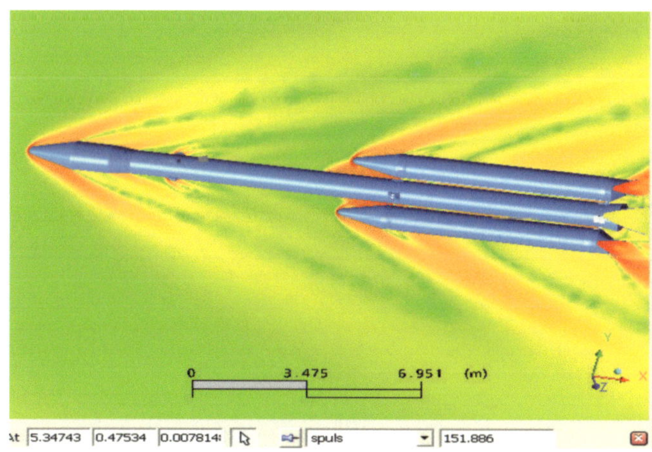


**Figure 2.** In-flight aero-acoustic noise

In view of having a good characterization of the in-flight acoustic loads acting upon the launcher structure, external acoustic measurements are required. Due to the high SPL and hostile environments found during flight, special microphones and adapters are specified. Such adapters must be designed in order to provide appropriate microphone/pre-amplifier installation and protection. Besides, when necessary, measurement correction procedures must be adopted. In this way, measurement programs for in-flight external acoustic characterization shall be developed, which may take into account three main phases as: preparation for experimental studies and acoustic testing sensors, ground development testing of acoustic sensors and methodology for reading acoustic pressures during flight.

Three different adapters for ¼" microphones were conceived, as described by [18]. On the upper parts of the launcher, two different configurations can be adopted. Firstly, at the PLF, heating and propellant dust effects are not significant; therefore, a structure flush installation (Fig. 3a) can be used. In this case, the measured SPL can be read directly. The second configuration, straight adapter (Fig. 3b), is applied for microphones installed near the equipment bays, where one has temperature and dust influences and, therefore, the sensor/conditioner must be protected. For such an assemblage, the protection channel dynamics directly affect the sensor response and, as a result, a measurement correction must be done.

On the bottom, the intense SPL at lift off generate a severe acoustic excitation of the first stage back modules region. Highly hostile dust, hot gas flow, heat flux and temperature environments are present during the motors operations. Nevertheless, the angular adapter must be used to install acoustic microphone/pre-amplifier, as shown in Fig. 3c. Notice in Figs. 3b and 3c, that the adapters were designed with small acoustic straight and angular cavities, respectively. When acoustically excited, the acoustical responses of such cavities directly influence the measured SPL, since the external pressure excitation profile and the measured signal are related by the cavity transfer function.

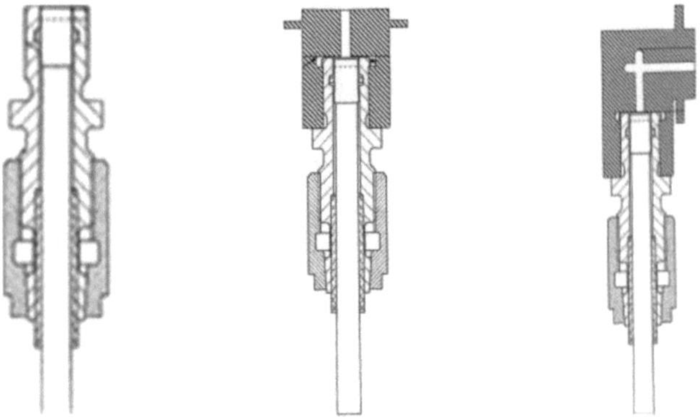

(a (b) (c)

**Figure 3.** (a) Flush adapter; (b) Straight adapter; (c) Angular adapter

In order to determine straight and angular adapters' dynamics, analytical and numerical calculations are done. The transfer functions of these channels are evaluated during ground acoustic tests, on which acoustic excitation with SPL close to those expected during flight is used to excite the cavities. Consequently, the measured SPL as well as the channels transfer functions are determined. Finally, the measurement corrections are determined, which may be applied when these adapters are used.

## 2.1. Mathematical models

### 2.1.1. Analytical model

In view of describing the dynamical behavior of the protective channels, one can assume the straight and angular channels as Helmholtz Resonators, which the channels and the space for microphone installation are accounted as the resonator throat and volume, as described by Eq. (1) [19].

$$f_0 = \frac{c_0}{2\pi}\sqrt{\frac{S}{V_0(l+l_c)}} \tag{1}$$

where: $f_0$ : natural frequency, $c_0$ : sound speed, $S$ : cross section of the resonator throat, $V_0$ : volume of the resonator cavity, $l$: length of the resonator throat, $l_c = 0.8r$ : end correction and $r$ : radius of the resonator throat cross section.

By considering the dimensions of the adapters into Eq. (1), one can calculate the natural frequencies of the straight and angular channels shown in Figs. 3b and 3c. These calculations are the starting point to assess the accuracy of the numerical models, built by using the Finite Element Method, once analytical x numerical natural frequencies can be compared.

### 2.1.2. Numerical model by Finite Element Method (FEM)

In a similar way as in structural dynamics, an acoustic cavity FEM model will have an acoustic stiffness matrix $[K_a]$, an acoustic mass matrix $[M_a]$, acoustic excitation vectors $\{F_{ai}\}$ and an acoustic damping matrix $[C_a]$. The combination of these components yields the acoustic finite element model, which can be solved for the unknown nodal pressure values $p_i$ [20].

$$\left([K_a] + j\omega[C_a] - \omega^2[M_a]\right).\{p_i\} = \{F_{ai}\} \tag{2}$$

Acoustic finite element models of the three adapters cavities are built. All cavities' surfaces were considered as rigid walls but the openings that are in direct contact with the external acoustic environment. In such cases, opened surfaces approximated using prescribed nodal pressures (equal to 0 for the eigenanalysis) were considered. The fluid inside the cavities is assumed as air at 15° C (c=340 m/s, ϱ=1.225 Kg/m$^3$, values used in this entire chapter).

Linear tetrahedral fluid elements are used in all three meshes. In order to have good prediction accuracy in the frequency range of interest, the general rule of thumb that requires at least 6 elements per wavelength is adopted. The main meshing characteristics are described in Table 1.

The acoustic load generated at lift off is simulated as an acoustic diffuse field (ADF). According to [21], an ADF is defined as an acoustic field in which the SPL is equal at any location and have an identical energy distribution in all directions. Such ADF can be obtained in an acoustic reverberant chamber, where the reflections along the rigid walls lead to this field. A formal way to describe an ADF consists on superimposing an infinite number of uncorrelated plane waves through different directions. In a FEM model, a finite number of uncorrelated plane waves can be generated and the pressure due the superposition of all the uncorrelated plane waves can then be applied as prescribed nodal pressures on the cavity's open surface (see [22]).

| Parameters | Mesh | | |
|---|---|---|---|
| | Flush adapter | Straight adapter | Angular adapter |
| Number of elements | 4,419 | 3,986 | 10,266 |
| Number of nodes | 1,009 | 980 | 2,367 |
| Maximum frequency (6 elements/wavelength) | 45,374 Hz | 44,593 Hz | 44,199 Hz |

**Table 1.** Meshes data

The FEM models of the three adapters are shown in Figs. 4a, 4b and 4c. The cavity's transfer functions are calculated by imposing prescribed nodal pressures in the nodes marked with small green arrows. In order to save time and computational efforts, a modal solution method is adopted using the first 14 modes. A modal damping of 5% is considered in these calculations.

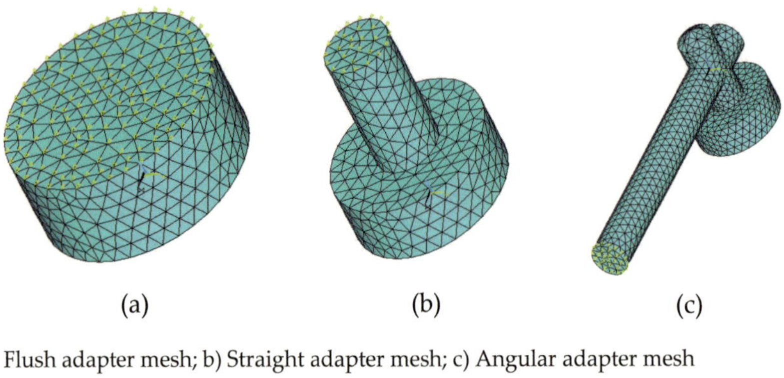

**Figure 4.** a) Flush adapter mesh; b) Straight adapter mesh; c) Angular adapter mesh

## 2.2. Experimental set up

An experimental unit is conceived to characterize all three adapters, as shown in Fig. 5. The experimental unit is placed into an acoustic reverberant chamber and submitted to an ADF, with a frequency profile shown in Fig. 6, which impinges the plate where the adapters and microphones/conditioners are installed. Care is taken to assure that the plate has structural response similar to that found along the launcher skin. Accelerometers are installed on the plate to measure the acoustically induced structural vibration.

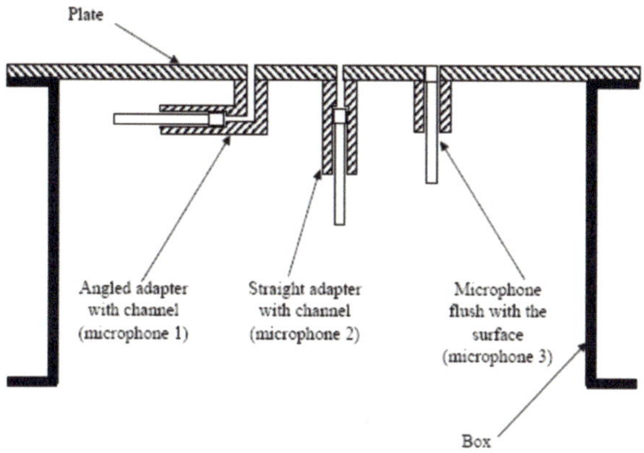


**Figure 5.** Experimental unit

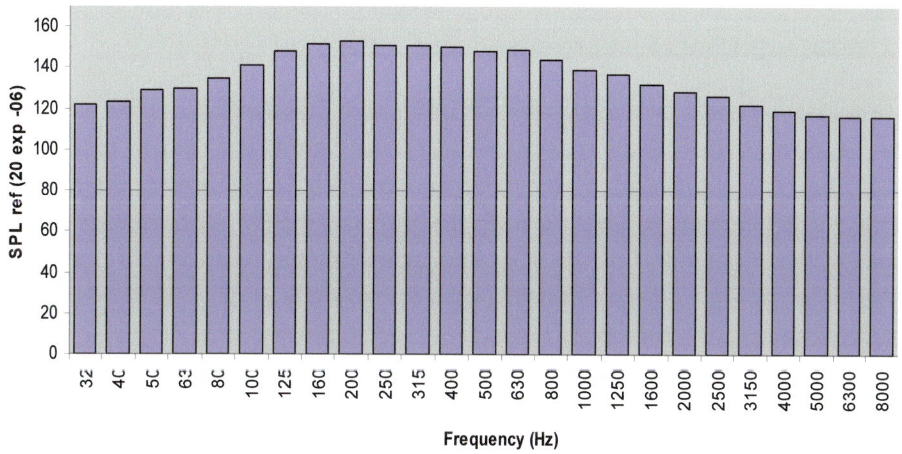


**Figure 6.** Excitation profile Overall Sound Pressure Level (OSPL ) – 160 dB

## 2.3. Methodology for the external measured SPL correction

### *2.3.1. Characterization of the adapters´ cavities*

The amplitude-frequency characteristics of such adapters must be accounted to determine the acoustic levels measured during the flight mission, since the device channels operate like filters. The SPL measured by the microphone installed with the flush adapter (Fig. 4a) is considered as the reference. Then the adapters transfer functions relating the input signal given by the microphone with the flush adapter and the output signal given by the microphones with the protective adapters (angular or straight) can be obtained by Eqs. (3) and (4)

$$M1(\omega) = H_{M1}(\omega)M3(\omega) \tag{3}$$

$$M2(\omega) = H_{M2}(\omega)M3(\omega) \tag{4}$$

where: $H_{M1}(\omega)$ and $H_{M2}(\omega)$: angular and straight adapter transfer functions, respectively. $M1$, $M2$, $M3$: measured SPL into angular, straight and flush adapters.

### 2.3.2. Results

Analytical and numerical natural frequencies results are compared with those obtained by acoustic testing in Table 2. As pointed out before, the measured SPL with the flush adapter can be read directly. In this way, only the adapters shown in Figures 4b and 4c are considered.

| | Straight Adapter | | | | Angular Adapter | | | |
|---|---|---|---|---|---|---|---|---|
| Mode | Analytical | FEM | Testing | FEM-Testing Difference (%) | Analytical | FEM | Testing | FEM-Testing Difference (%) |
| 1 | 5,000 | 5,396.6 | 5,202 | 3.74 | 2,300 | 2,204.7 | 1,953 | 12.88 |
| 2 | - | 27,032 | - | - | - | 8,660.3 | 7,831 | 10.58 |

**Table 2.** Theoretical x Experimental natural frequencies (in Hz)

Disagreements between test and predicted resonance frequencies can be explained by possible inaccuracy in microphone installation and complicated shape of the angular channel. The characterization of the adapters′ cavities is performed numerically, by calculating $H_{M1}(\omega)$ and $H_{M2}(\omega)$ (Eqs. (3) and (4)) using FEM. The calculated response functions of the angular and the straight adapter models are compared to those obtained experimentally in Fig. 7 (experimental frequency resolution Δf = 18.78 Hz; numerical frequency resolution Δf = 20.00 Hz).

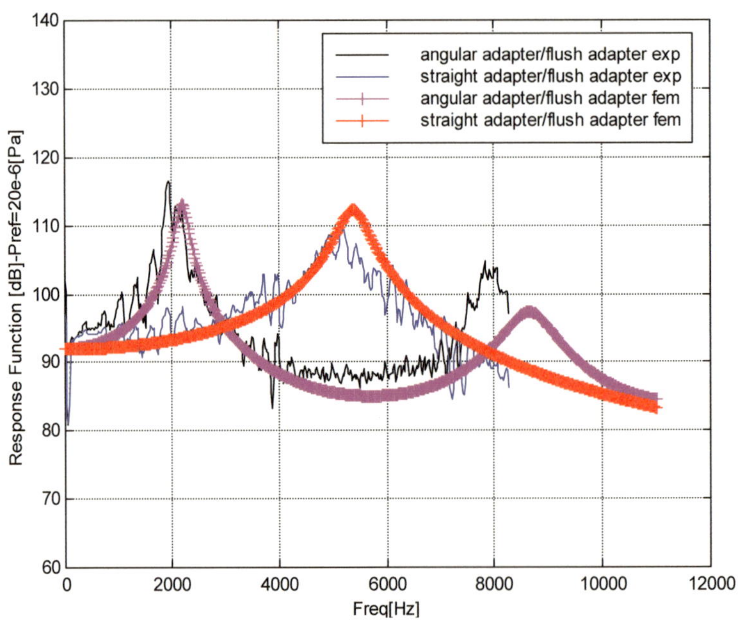


**Figure 7.** Microphones with adapters transfer functions (numerical and experimental)

The theoretical and experimental response functions show good agreement, with a shift in the second resonance peak for the microphone with angular adapter. The errors can be caused by: bad characterization of the ADF spectral distribution into the FEM model; adapter's geometry complexity; microphone installation inaccuracy or a combination of some of these factors.

Equations 3 and 4 show that the external noise (given by $M3(\omega)$) can be identified by knowing the internal noise and the inverses of the transfer functions for the angular and straight adapters.

## 3. Vibro-acoustic modelling of payload fairings (PLF)

A complete survey of PLF vibro-acoustic environment must be carried out, in order to determine its inner SPL. In this respect, it is important to have reliable numerical tools that can predict the responses of ELV systems, subjected to in-flight acoustic loads and that enable NCT design.

Low-frequency coupling techniques are used to estimate a PLF dynamic behavior. The fairing body and its inner acoustic domain are analyzed by using Finite Element Method (FEM) and Boundary Element Method (BEM). Structural FEM/fluid FEM and structural FEM/fluid BEM modeling techniques are then applied. In order to simulate the lift off acoustic excitation, an ADF of 145 dB OSPL is applied on the fairing body and coupled calculations are done from 5 to 150 Hz, which yielded the acoustic and skin responses for both models. Modal expansion and semi-modal expansion model techniques are applied, respectively.

For the high-frequency analysis, it is applied the Statistical Energy Analysis (SEA) technique, for a frequency range from 5 to 8,000 Hz. The 145 dB OSPL excitation is applied to the structural panels of the fairing and the acoustic and structural mean responses are calculated.

In view of validating the numerical predictions for the fairing, acoustic test is done to measure the acoustics inside the PLF. The PLF is submitted to 145 dB OSPL in a 1,200 $m^3$ acoustic reverberant chamber and microphones are positioned in its inner domain.

The implementation of sound absorption blankets is applied as a control technique to attenuate acoustic noise from medium- to high-frequency bands. SEA is a technique for high-frequency analysis; therefore, adequate to assess the influence of blankets on space systems. The generated SEA fluid-structure model is used to calculate internal SPL, with single-, double-, and multi-layered noise control treatments (NCT). Two NCT modeling approaches are used to simulate the effect of blanketing the fairing cavity:

i. acoustic materials Biot´s parameters, given by the manufacturer;
ii. material samples absorption coefficient, measured in a Kundt Tube.

### 3.1. Model description

The analyzed fairing is hammerhead type geometry and is composed of the body structure, functional components as electric and pyrotechnic components of the ejection system, mechanisms as well as the exterior cork liner. Figures 8a and 8b show the Brazilian VLS fairing structure.

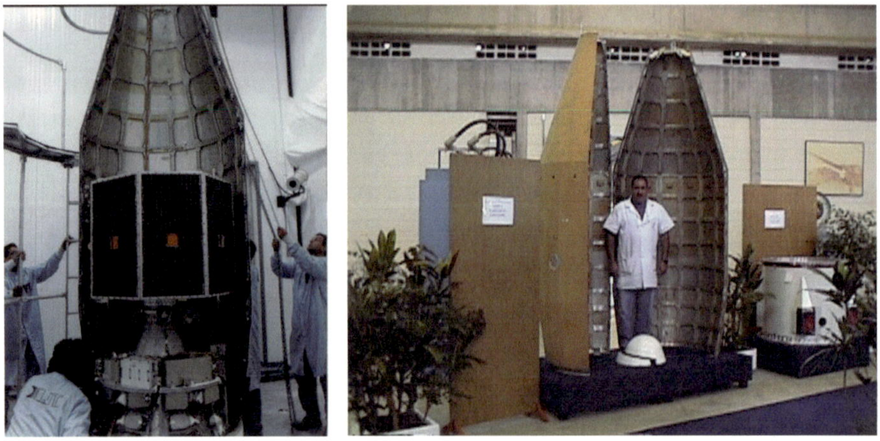

(a) (b)

**Figure 8.** a) PLF structure; b) PLF structure

## 3.2. Modelling methodology

### *3.2.1. Low-frequency modeling (deterministic) techniques*

In view of predicting the operational fairing cavity SPL, both the dynamic displacements of the fairing structure as well as the acoustic pressure fields at the interior and the exterior side of the fairing should be considered. In this study, however, the fluid-structure coupling interaction between the structural displacements and the exterior acoustic pressure field is neglected. The exterior acoustic pressure is assumed to be a known external excitation for the vibro-acoustic system, consisting of the fairing body and the internal acoustic cavity. The FEM and BEM are the most appropriate numerical techniques for the (low-frequency) dynamic analysis of this type of vibro-acoustic system.

FEM based models for coupled vibro-acoustic problems are most commonly described in an Eulerian formulation, in which the fluid is described by a single scalar function, usually the acoustic pressure, while the structural components are described by a displacement vector. The resulting combined FEM/FEM model in the unknown structural displacements and acoustic pressures at the nodes of, respectively, the structural and the acoustic FEM meshes are [20],

$$\left( \begin{bmatrix} K_S & K_C \\ 0 & K_A \end{bmatrix} + j\omega \begin{bmatrix} C_S & 0 \\ 0 & C_A \end{bmatrix} - \omega^2 \begin{bmatrix} M_S & 0 \\ -\rho K_C^T & M_A \end{bmatrix} \right) \begin{Bmatrix} w_i \\ p_i \end{Bmatrix} = \begin{Bmatrix} F_{Si} \\ F_{Ai} \end{Bmatrix} \tag{5}$$

Where:

$K_S$ – Structural stiffness matrix $M_S$ – Structural mass matrix $C_S$ – Structural damping matrix $K_C$ – Fluid – structure coupling matrix

$M_A$ – Fluid mass matrix $K_A$ – Fluid stiffness matrix $C_A$ – Fluid damping matrix
$w_i$ – Structural displacements nodal vector $p_i$ – Acoustic pressures nodal vector
$F_{Si}$ – Force nodal vector $F_{Ai}$ – Acoustic sources nodal vector

In comparison with a purely structural or purely acoustic FEM model, the coupled stiffness and mass matrices (Eq. (5)) are no longer symmetrical due to the fact that the force loading of the fluid on the structure is proportional to the pressure, resulting in a cross-coupling term $K_C$ in the coupled stiffness matrix, while the force loading of the structure on the fluid is proportional to the acceleration, resulting in a cross-coupling term $M_C = -\rho K_C^T$ in the coupled mass matrix.

Low-frequency vibro-acoustic problems can also be modeled by describing the structural behavior in a FEM model and the fluid behavior in a BEM model. In the same way as in the FEM/FEM technique, deterministic FEM/BEM models are usually described by acoustic double layer potential and structural displacement, which are the field variables. Equation (6) presents the resulting combined FEM structural displacements and BEM acoustic pressure differences at the nodes, for a coupled FEM/BEM mesh [23].

$$\begin{pmatrix} K_S + j\omega C_S - \omega^2 M_S & L_C \\ L_C & \dfrac{D}{\rho_0 \omega^2} \end{pmatrix} \begin{Bmatrix} w_i \\ \mu_i \end{Bmatrix} = \begin{Bmatrix} F_{Si} \\ F_{Ai} \end{Bmatrix} \tag{6}$$

Where $L_C$ is the fluid-structure coupling matrix, $D$ is the BEM acoustic matrix of coefficients and $\mu_i$ is the nodal vector of double layer potentials.

In deterministic models, the dynamic variables within each element are expressed in terms of nodal shape functions, usually based on low-order (polynomial) functions. Since these low-order shape functions can only represent a restricted spatial variation, a large number of elements is needed to accurately represent the oscillatory wave nature of the dynamic response. A general rule of thumb states that for fluid-structure interactions, at least 6 (linear) elements per wavelength are required to get reasonable accuracy. Since wavelengths decrease for increasing frequency, the FEM model sizes, computational efforts and memory requirements increase also with frequency. As a result, the use of FEM and BEM models is practically restricted to low-frequency applications. In comparison with uncoupled structural or acoustic problems, this practical frequency threshold becomes significantly smaller for coupled vibro-acoustic problems, since a structural and an acoustic problem must be solved simultaneously. Moreover, the matrices in a coupled model are no longer symmetrical, so that less efficient non-symmetrical solvers must be used. As a consequence, the computational effort, involved with the use of coupled FEM/FEM and FEM/BEM models for real-life vibro-acoustic engineering problems, becomes large at very low frequencies.

In order to obtain coupled vibro-acoustic response predictions within reasonable computational efforts, the dimensions of the FEM/FEM problem (Eq. (5)) have to be reduced. The most applied technique for model reduction is the modal superposition technique, which expresses the unknowns of the system in terms of a modal basis, resulting in a set of unknown modal participation factors, whose size is much smaller than the size of the original set of unknowns. A modal expansion in terms of uncoupled structural and acoustic modes is performed by using computationally efficient symmetric eigenvalue algorithms and requires much less computational effort than the use of vibro-acoustic (coupled) modes. However, a large number of high-order uncoupled acoustic modes is required to accurately represent the normal displacement continuity along the fluid-structure interface.

In a FEM/FEM virtual prototype, a modal expansion in terms of uncoupled structural and uncoupled acoustic modal bases is used, in order to keep the computational efforts within reasonable limits. On one hand, structural wavelengths are much smaller than acoustic wavelengths, so that the structural FEM mesh of the fairing must be finer than the acoustic FEM mesh of the inner cavity. On the other hand, due to the continuity of the normal structural and fluid displacements along the fluid-structure coupling interface, both meshes must be compatible in this region. In this framework, the following modeling methodology is adopted: 1) A fine FEM mesh of the fairing is used for the construction of the uncoupled structural modal data basis. 2) The resulting modes are then projected onto a FEM coarse mesh of the fairing structure. 3) For the acoustic cavity FEM mesh, the same mesh density is used along the fluid-structure coupling interface as the PLF structural coarse mesh, while the mesh density is slightly decreased towards the central axis of the cavity. 4) The uncoupled modes, resulting from this acoustic FEM mesh, together with the projected structural modal basis are used in a coupled FEM/FEM model. It is important to highlight that the coarse structural mesh has only the shells of the fairing structure, while all reinforcing beams are omitted, since it is assumed that these stiffeners have no significant effect on the fluid-structure coupling interaction, while their presence would increase the computational load of the modeling process.

For the case of the FEM/BEM problem (Eq. (6)), the modal expansion cannot be used, since the frequency dependency of the matrix coefficients in the acoustic part prohibits a standard eigenvalue calculation. Such as, the semi-modal approach, which uses only the expansion of the structural modal data basis, is applied. As mentioned above, BEM drawbacks as fully population of the matrices, complex and frequency dependent models result in a coupled FEM/BEM model less efficient than coupled FEM/FEM model. Therefore, the rule of thumb of 6 (linear) elements per wavelength becomes prohibitive for the actual fairing fluid-structure study. Such a way, a coarsest structural mesh may be generated and the same adopted frequency range for FEM/FEM model is kept for this FEM/BEM model, even considering that the structure has not enough discrete density. However, the modal data basis calculated using the structural fine mesh can assure good results for the structural displacements, in this fluid-structure model, since the expansion in terms of such a data basis is used on these Frequency Response Analysis (FRA) computations. The displacement continuity of the structural and acoustic meshes (same density in the fluid-structure interface) is considered to perform the link and calculate the coupled dynamic skin displacements and acoustic cavity pressure responses, for both FEM/FEM and FEM/BEM models.

### 3.2.2. *High-frequency modeling (statistical) technique*

A characteristic of high-frequency analysis is the uncertainty in modal parameters. The resonances and mode shapes show great sensitivity to small variations of geometry, construction and material properties. In addition, programs used to evaluate mode shapes and frequencies are known to be inaccurate for higher modes. In light of these uncertainties, a statistical model of the dynamic parameters seems natural and appropriate. As an alternative method for higher frequency analysis of the inner cavity of fairings, Statistical Energy Analysis (SEA) approach is proposed. This approach is the description of the dynamic system as a member of a statistical population or ensemble, whether or not the temporal behavior is random. SEA emphasizes the aspects of this field dynamical study.

The SEA equations express the energy balance of different subsystems in a model [24]. Some subsystems have direct power input of an independent source, e. g. an excitation force on a structural component, a sound power source in an acoustic medium etc. In general, subsystems can receive power (input power from external sources), dissipate power (internal losses due to damping) and exchange power with other subsystems to which they are coupled (losses due to coupling). SEA fundamental hypothesis as dissipation losses in relation to the energy variable and modal energy proportionality from connected subsystems are used to yield the SEA matrix equation of complex structures. The distribution of the dynamical response in the system due to some excitation is obtained from the distribution of the energy among the mode groups, based on a set of power balance equations for the mode groups.

## 3.3. FEM structural meshes

The fairing body is divided in five surfaces. The surfaces are discretized by using 4-noded quadrilateral shell elements, while 2-noded beam elements are used for the circumferential and the axial stiffeners. To account for the mass of the cork blanket on the exterior fairing surface, a distribution of concentrated mass elements are attached to the fairing nodes.

A total of 174 structural modes in a frequency range up to 220 Hz have been identified. Table 3 describes the main structural modes calculated by using FEM in the range up to 150 Hz.

| mode | frequency (Hz) |
|---|---|
| first bending | 38.637 |
| first breathing | 77.821 |
| second breathing | 92.694 |
| first longitudinal | 108.749 |
| first torsion | 125.772 |
| second bending | 150.735 |

**Table 3.** Fairing body structural modes

Table 3 shows that the first structural bending mode of the fairing is identified at 38.6 Hz, while the second mode is at 150.7 Hz. Figures 9a and 9b show the referred structural modes.

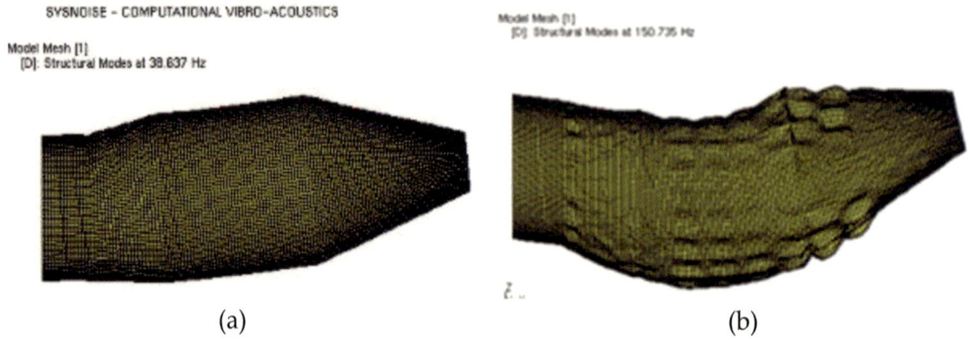


**Figure 9.** a) First structural bending mode; b) Second structural bending mode

### 3.4. FEM and BEM acoustic meshes

The acoustic FEM mesh consists of 119,577 nodes and 110,238 elements (106,050 8-noded hexahedral elements and 4,188 6-noded pentahedral elements). The cavity is considered filled with air at 15° C. The cavity's bottom and top faces are assumed to be acoustically closed (rigid walls). The acoustic mesh generation takes into account the meshes compatibility on the fluid-structure interface.

A total of 80 acoustic modes in a frequency range up to 566 Hz were identified. Acoustic wavelengths are bigger than structural wavelengths. Such that, a large number of high-order uncoupled acoustic modes is required to accurately represent the normal displacement continuity along the fluid-structure interface. That is why higher frequency range is used to describe the acoustic modal behavior of the fairing. Table 4 describes the acoustic modes in the frequency range up to 150 Hz.

| Mode | Frequency (Hz) |
|---|---|
| rigid body | 0.000 |
| first longitudinal | 63.491 |
| second longitudinal | 112.129 |

**Table 4.** Fairing cavity acoustic modes

It can be noticed in table 4 that the first and second acoustic modes of the fairing cavity are identified at 63.5 Hz and 112.1 Hz, respectively. Figures 10a and 10b show the referred acoustic modes.

The BEM acoustic mesh is a 2-D coarsest mesh. Therefore, as the coupled FEM/BEM equation is frequency dependent (Eq. (6)), the acoustic modes are not considered in the acoustic pressure calculations (semi-modal reduction model).

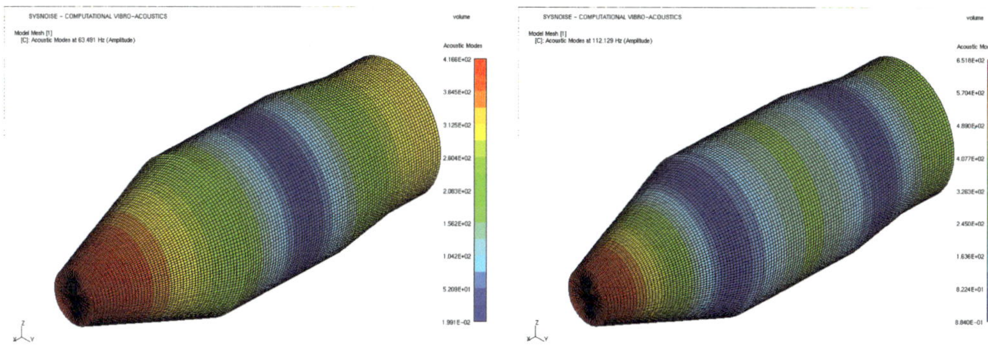


(a) (b)

**Figure 10.** a) First acoustic longitudinal mode; b) Second acoustic longitudinal mode

## 3.5. Model excitation

A uniform pressure loading is simulated by applying a normal point force varying harmonically on all nodes of the fairing shell elements. The force value is defined such that the total load is equivalent to a uniform pressure of 145 dB OSPL. Link of the acoustic and structural parts is done as well as the structural modal data basis is projected to the coarse and coarsest meshes.

In this way, all the meshes, modal data bases and excitation, needed to perform low-frequency calculations, using coupling fluid-structure techniques are ready. Next step is to perform FRA calculations for both models.

## 3.6. SEA fairing vibro-acoustic model

The fairing body is divided in four surfaces, as shown in Fig. 11a. To account for the rib-stiffened plates of the surfaces 2, 3 and 4, the SEA structural fairing model considers connected plates and beams (longitudinal and circular). The plate structural subsystems are generated as singly curved shells and uniform plates. Shell surface 1 has a thickness of 3mm and is modeled as a simple plate of aluminum (E=72 GPa, $\nu$=0.29, $\varrho$=2750 kg/m$^3$), while the other three surfaces are 0.8 mm thick and made of an aluminum alloy (E=72 GPa, $\nu$=0.29, $\varrho$=7000 kg/m$^3$). The circular and longitudinal beams are modeled by assigning the same material as the shells of the surfaces 2, 3 and 4 (Figure 11a). Damping loss factors of 1% (for flexure, extension and shear propagating waves) are assigned to the plates and beams subsystems, in order to account for the internal loss factors.

A total of 72 beams (44 longitudinal and 28 circular) and 8 shells (02 singly curved shells of the adaptor, 02 singly curved shells of the lower cone, 02 singly curved shells of the main cylinder and 02 singly curved shells of the upper cone) compose the structural SEA model. Figure 11a shows the SEA plates and beams generated to model the VLS-1 fairing structure. The external blanketed treatment of cork on the surfaces 2, 3 and 4, was simulated in this model as material addition. The layered area and the density of the cork were considered to assign this mass.

The acoustic environment inside the fairing was generated by starting from the structural model. This acoustic cavity was created considering air at 15° C as the fluid as well as the dimensional parameters of the fairing. The top and bottom face of the cavity were assumed to be acoustically closed. Figure 11b presents the 3D acoustic cavity of the fairing.

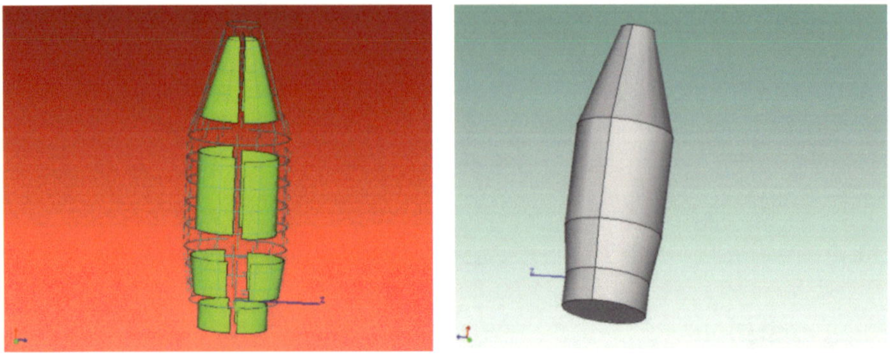

(a) (b)

**Figure 11.** a) Structural subsystems (shells, circular beams and longitudinal beams) b) Acoustic cavity of the fairing

The coupling boundary between all the structural and acoustic subsystems is modeled to consider the transmission of power across the junctions. A junction is comprised of connections to any number of coincident subsystems. As a result, all the subsystems that share common nodes are connected by point, line and area junctions and all the appropriate wavefields are connected as well as the corresponding coupling loss factors (CLF) between subsystems are created.

The estimated sound pressure levels at the lift off are assigned to the SEA model. Only elements with large surface areas, as plates and panels, are considered to be susceptible to acoustic excitation ([11], [24] and [25]). An ADF of 145 dB OSPL (Fig. 12) is applied to the plates of the SEA fairing model, which simulates the power input into a structural plate or shell element.

### 3.7. Analysis results

In view of having a complete knowledge of the fairing dynamic vibro-acoustic behavior, the fairing structural skin as well as its inner acoustic domain responses should be presented. However, since this chapter concerns acoustics, the body structural displacements are not presented here. Below, the obtained results of the acoustic behavior applying vibro-acoustic low-frequency and high-frequency analysis techniques predictions are presented.

#### *3.7.1. Low-frequency techniques*

##### *3.7.1.1. FEM/FEM response calculations*

A modal expansion in terms of 174 uncoupled structural and 80 uncoupled acoustic modes is used for the coupled calculations. A modal damping of 1% is assigned to all structural modes. All calculations are performed with a frequency resolution of 1 Hz. Figure 13 shows the low-frequency acoustic pressure spectra of the PLF for the case of a uniform exterior pressure loading, using FEM/FEM coupling analysis. It can be seen that the low-frequency pressure is dominated by the first longitudinal mode around 63.5 Hz and the second longitudinal mode around 112.1 Hz.

#### *3.7.1.2. FEM/BEM response calculations*

The same structural modal expansion as used for FEM/FEM is used for this FEM/BEM response calculations. Due to the frequency dependency of the boundary integral equation, the acoustic modal basis can not be used. A damping of 1% is assigned to all structural modes. All calculations are performed with a frequency resolution of 2 Hz. Figure 13 presents a comparison of the computed inner cavity space averaged acoustic pressure using FEM/FEM and FEM/BEM techniques.

### *3.7.2. High-frequency technique*

The energy balance (levels and interactions) between different subsystems of the SEA model is calculated. The interest frequency range is 5 to 8,000 Hz, by third octave bandwidth. As mentioned before, SEA technique is more effective in higher frequencies, where dynamic systems present higher modal density. The vibro-acoustic responses of the fairing, using SEA technique, are shown in Fig 13.

It is important to highlight that for the low-frequency range, SEA analysis results are not reliable, since the accuracy of the SEA technique is proportional to the modal density [24].

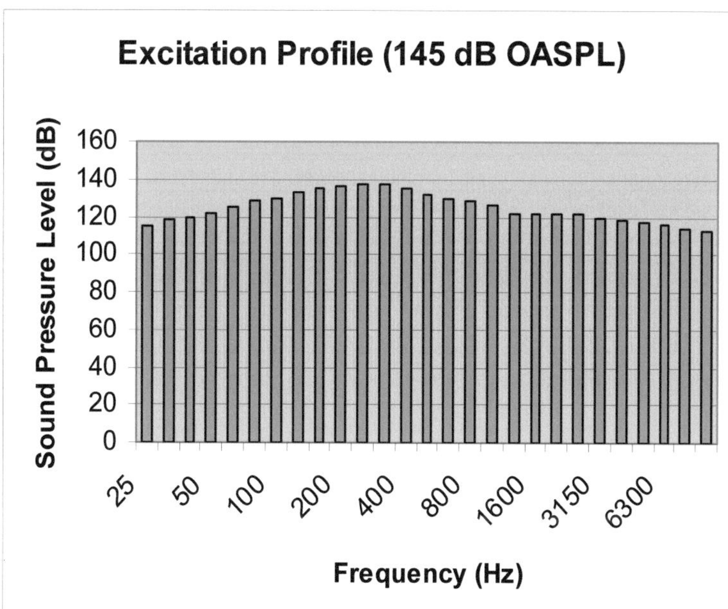


**Figure 12.** Acoustic diffuse field at lift off

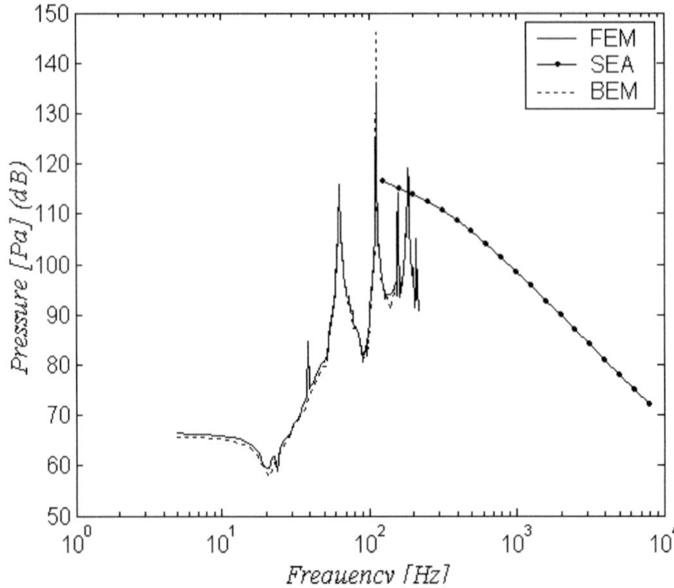


**Figure 13.** Acoustic Response Inside PLF

Considering the accuracy, advantages and drawbacks of the deterministic and statistical techniques, each of them is successfully applied in different frequency ranges. Such that, for the analyzed PLF, valid response results using deterministic techniques are assumed up to 150 Hz, while valid SEA results are assumed from 300 up 8,000 Hz. It is important to mention that in the "twilight zone" or medium frequency bandwidth (from 150 to 300 Hz), where deterministic models are inaccurate and present prohibitive computation time for the calculations and where the high modal density requirement is not yet accomplished for SEA, both results may be considered, as shows figure 13.

### 3.8. Model validation

The fairing structure was positioned inside an acoustic chamber and excited with an ADF of 145 dB OSPL. Eight control microphones were positioned inside the reverberant chamber, which feedback the control system. Four measurement microphones were located in the acoustic cavity of the PLF. The measured space averaged SPL is compared with the theoretical acoustic responses, computed using the virtual prototypes (FEM and SEA models) (Fig. 14).

The calculated internal acoustic frequency response function shown in figure 13 may be transformed into 1/3 octave band responses to be compared with the experimental (measured) results. Figure 14 presents the 1/3 octave comparisons for the frequency bandwidths ranging from 31.5 up to 8,000 Hz. It can be noticed, that experimental and calculated low-frequency responses have good agreement, presenting more significant differences only on the 1/3 octave bands 31.5, 40 and 50 Hz. This is because the low-frequency modes of the acoustic chamber are not well excited. However, in the regions where the cavity response is dominant (63 Hz and 112 Hz), differences are pretty small.

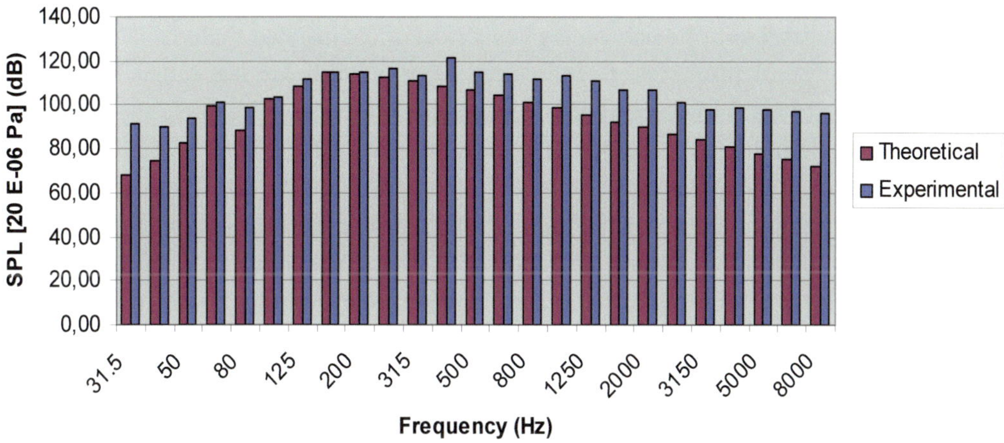


**Figure 14.** High-frequency theoretical X experimental comparison

For the higher frequencies, a more reasonable comparison should be done using Power Spectral Density (PSD) [28], since SEA calculations may be interpreted as mean values of energetic response functions when averaged at a given frequency over an ensemble of similar systems, differently of peak values resulting from deterministic approaches. However, a qualitative comparison can be presented for 1/3 octave bands from 160 up to 8,000 Hz, since one keeps in mind that mean values and the predicted magnitudes yielded by SEA should under estimate the dynamic response with a certain (acceptable) variance. Valid SEA results are assumed from 300 up 8,000 Hz, since the minimum five modes by bandwidth (modal density) requirement becomes true starting from 300 Hz.

At the beginning phases of space projects, the assessment of the effect of using different passive techniques for acoustic environment alleviation to be applied to PLF is an important issue. One of the main applications of numerical control prediction is the decision, still in the early product development phase, which design version is the most appropriate from the noise control point of view. By introducing the concept of sensitivity analysis, product development can be performed in a more systematic way. In order to predict the efficiency of a NCT, one compares the effects of design modifications.

In this framework, different blanket layers are implemented on the PLF elasto-acoustic virtual prototype and the effects of these NCT implementations are assessed. Since blankets acoustic absorption depend on certain material parameters, two blanket modeling approaches are assessed as follows: material physical Biot´s parameters as density, porosity, resistivity, tortuosity, viscous and thermal characteristic lengths, given by the blanket manufacturer and measured normal incidence absorption coefficients of material samples.

Furthermore, the influence of the NCT thicknesses and the presence of air-gaps between blankets are analyzed. Biot´s parameters and absorption coefficient approaches are implemented in the coupled elasto-acoustic SEA model of the PLF.

For the Biot´s parameters approach, an explicit model of the inserted material is considered, based on the physical properties of individual layers, which are accounted in the SEA model. Six types of glass wools are analyzed and the SPL inside the fairing are calculated. The wools' densities are given in pounds per cubic feet (pcf - 1 $lb/ft^3$ = 16.02 $Kg/m^3$). A thickness of 7,62 cm is adopted for almost all glass wools, but the two 1.2pcf glass wools, that presents particular behavior, which adopted thicknesses were 0.19 cm and 0.38 cm. The best performing material is chosen and a comparison between different thicknesses and percentages of layered surfaces of the fairing is done, considering the final weight of the applied NCT. The materials used were glass wools described in Table 5. The wools' Biot's parameters can be found in [4].

| Material characteristcs | | | | | | |
|---|---|---|---|---|---|---|
| Density (pcf) | 0.34 | 0.42 | 0.60 | 1.20 | 1.20 | 1.50 |
| Thickness (inches) | 3.00 | 3.00 | 3.00 | 0.75 | 1.50 | 3.00 |

**Table 5.** Glass wools used

On the other side, the measured absorption coefficient of multi- and single-layered samples of glass wools of 0.42 and 1.0 pcf were considered. According to [26], air gaps between materials increase the acoustic absorption at low-frequencies. For this case, samples with two different air gaps are positioned into a Kundt tube. The single-layered samples are 3.50 cm thick, while combinations are done with samples of 1.75 cm thick. Other configurations were assembled with air gaps of 1.0 and 3.0 cm between samples. Figure 15 shows the sample combinations. All the measured absorption coefficients are shown in Fig. 16. These absorption coefficients are assigned on the fairing vibro-acoustic model and SPL are calculated.

The PLF acoustic responses for different NCT configurations are shown in Fig. 17. Notice that the insertion of the 0.34pcf glass wool – 7.62 cm yields almost 20 dB of attenuation (chosen as the best performing material). The assessment of the thickness influence is done by assigning 0.34pcf glass wools of 7.62, 10.16 and 12.7 cm thicknesses, with total NCT weights of 3.90, 5.30 and 6.60 Kg, respectively. Figure 18 shows the internal SPL one-third octave distribution, as well as the OSPL.

Figure 19 shows SPL and OSPL from 50 to 8,000 Hz, for the NCT described in Fig. 16, without air gaps. A 3.50 cm double-layered blanket (0.42pcf/1.0pcf) is compared with two single-layered NCT. Notice in this figure that NCT decrease the internal OSPL from 132 dB to 128 dB. Figure 20 shows that single-layered treatment with 1.0 pcf and air gaps presented better results. One can see the air gap effect, since the SPL close to 100, 315 and

500 Hz are higher mainly when the 1.0 pcf material with 3.0 cm air gap is applied. The calculations yielded 127.5 dB OSPL inside the fairing cavity. This means that a gain of approximately 3.0 dB at 100 Hz bandwidth can be obtained, yielding an overall gain of 1.0 dB, approximately. However, air gap installation can be limited, due to fairing internal space. In this case, it is preferable to install the blanketed treatment distant from the panels by small air gaps, instead of bonded, once this installation configuration presents higher transmission loss [10].

**Figure 15.** Double-layered 0.42pcf (yellow)/1.0pcf (orange), 1.75 cm thick each.

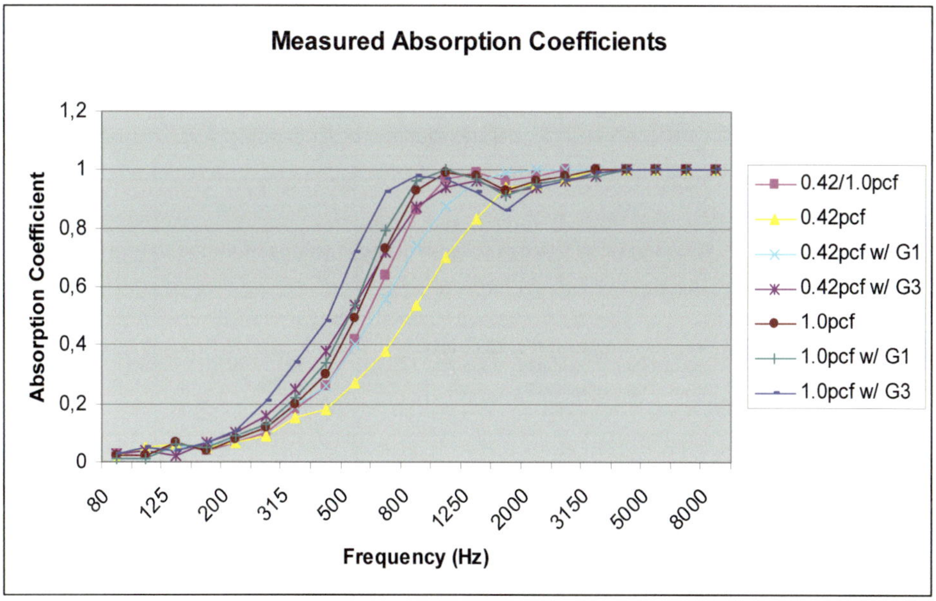


**Figure 16.** Measured absorption coefficients*G1 and G3: air gaps of 1.0 cm and 3.0 cm

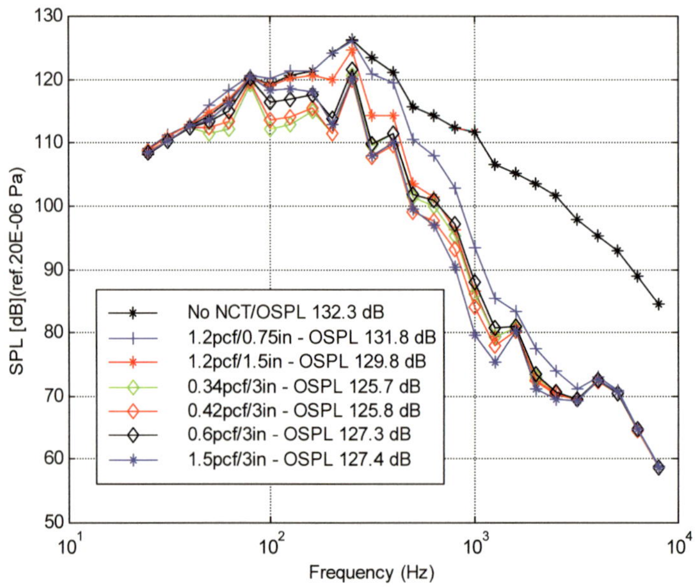


**Figure 17.** SPL for different blankets

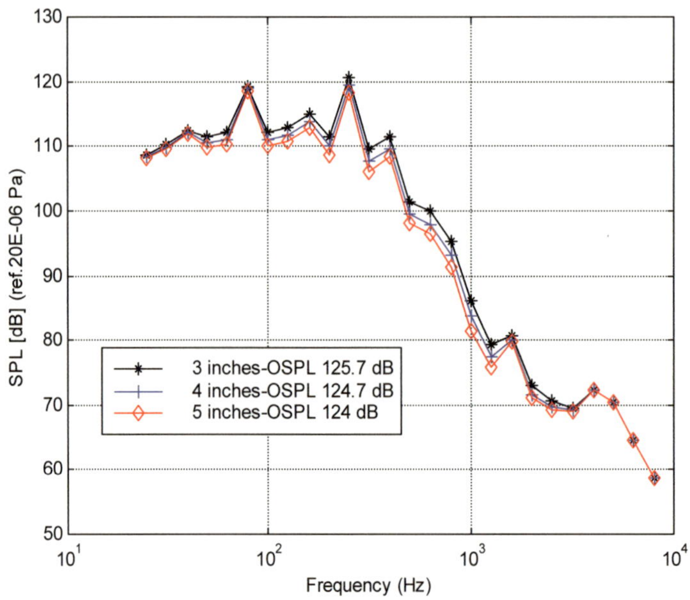


**Figure 18.** SPL for different thicknesses

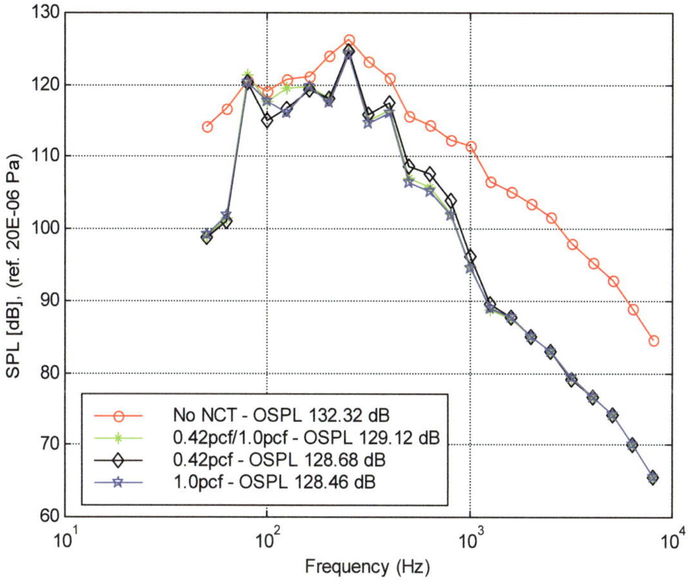


**Figure 19.** SPL for NCT without air gap.

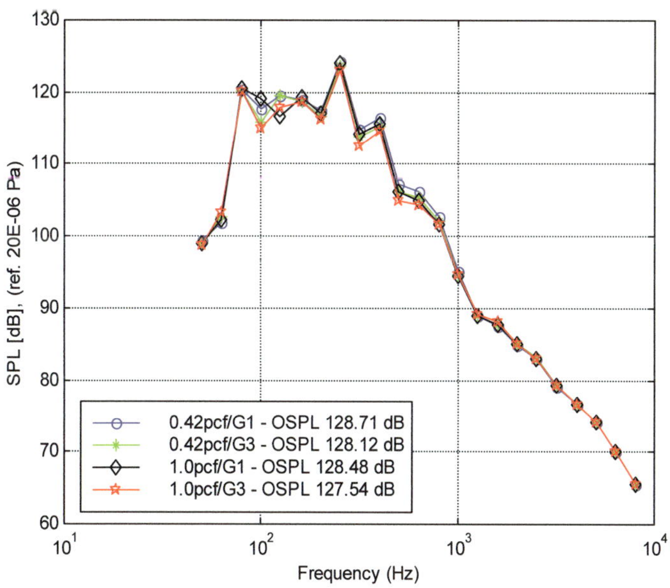


**Figure 20.** SPL for NCT with air gaps (*G1 and G3: air gaps of 1.0 cm and 3.0 cm)

## 4. Combustion instabilities of liquid propelled rocket engines due to chamber combustion acoustics

Combustion instabilities have been present in the development of LPRE over the last decades. There are basically three types of combustion instabilities: low-frequency (LF), medium-frequency (MF) and high-frequency (HF). LF instabilities, also called chugging, are caused by pressure interactions between the propellant feed system and the combustion chamber. MF instabilities, also called buzzing, are due to coupling between the combustion process and the propellant feed system flow. The HF instabilities are the most potentially dangerous and not well-understood ones. It occurs due to coupling of the combustion process and the chamber acoustics [27].

The presence of acoustic combustion instabilities must be considered still in development phase, although combustion instabilities can be clearly identified only during firing tests. In [13], it was described that instability can be verified when the power spectrum of the acoustic pressure levels, measured during burning tests, is analyzed. When an oscillation is observed, i.e., combustion instability, sound pressure peaks with well-defined magnitudes summed to the background noise are present. These peaks are correlated to the resonance frequencies of the combustion chamber. This phenomenon can cause instabilities and consequent unexpected behavior such as efficiency loss or even explosion of the engine. In this framework, the engine acoustic cavity characterization becomes an important issue to be investigated.

Acoustic behaviour of chambers is usually determined by doing cold tests measurements (without combustion). Acoustic dynamics in combustion environments are obtained by shifting the cold test resonant frequencies by a scalar factor defined by the ratio of sound velocity at the cold test temperature and at real operation temperature [15].

In view of attenuating acoustic pressure oscillations inside combustion chambers, reactive techniques as Helmholtz Resonators (HR), among others, are widely used ([13] and [16]). These devices are specially designed to attenuate oscillations at discrete resonance frequencies (pure tones). HR have been applied as combustion stabilization devices for solid motors and liquid rocket engines, with success. It could be noted in literature that they are used in a set of dozens or even hundreds in each chamber cavity, distributed along the walls or in a single row along the injector periphery [28].

This item describes a procedure for cold test acoustic characterization of LPRE combustion chambers. Firstly, the acoustical dynamic characterization of a combustion chamber is done and a typical longitudinal resonant frequency is chosen to be attenuated. A HR is designed (tuned at the chosen frequency) and applied to the mock up face plate. A LPRE mock-up [14] was used as experimental model. This test rig faithfully represents the internal acoustic cavity of the original engine. This procedure is followed by doing virtual prototypes of the combustion chamber. The acoustic natural frequencies and mode shapes are numerically calculated by a FEM model and validated through acoustic experimental modal analysis [29].

## 4.1. Experimental Acoustic Modal Analysis (EAMA)

Experimental Modal Analysis (EMA) is a well-applied technique in structure dynamics. However, due to the development of commercial acoustic sources, EAMA can be a suitable choice in view of extracting the acoustical Frequency Response Functions (FRF). In addition, the mathematical approach of structures modal parameters extraction can be applied to acoustic systems. [29].

In order to check the mutual orthogonality among modes from a modal model and to compare modes between different modal models (i.e., experimental and numerical solutions), the Modal Assurance Criterion (MAC) was used. This criterion indicates the degree of linear dependence between two eigenvectors and can be described as Eq. (7) [30].

$$MAC_{ijm} = 100 \cdot \frac{\left| \{\phi_{im}\}^T \{\phi_{jm}^*\} \right|^2}{\{\phi_{irm}\}^T \{\phi_{im}^*\} \{\phi_{jm}\}^T \{\phi_{jm}^*\}} \tag{7}$$

where: indexes $i$ and $j$ denotes modes obtained by different methods.

## 4.2. Helmholtz resonator

Helmholtz Resonators are widely applied in order to suppress or attenuate the acoustic pressure inside cavities, rooms and other volumes. A HR consists of a small volume connected to a bigger cavity (the combustion chamber, in this case) through an orifice by a flanged neck. The dimensions of the HR must be much smaller than the acoustic wavelength of interest, in order to consider the resonator as lumped elements coupled to a geometric discontinuity. The coupling condition is that the oscillatory volume flow in the neck is equal to that imposed on the fluid inside the cavity, neglecting the elastic property of the air in the neck [30].

A typical HR is shown in Fig.21 (left), being $d$ the neck diameter, $D$ the cavity diameter, $V_c$ the volume cavity, $l$ the neck length and $L$ the cavity length. $P_1$ is the incident acoustic pressure and $P_2$ is the cavity pressure. The gas motion in the HR coupled in an acoustic cavity can behave equivalently to a mass-spring-dashpot system (Fig.21, centre). The system can be divided into three distinct elements. The fluid enclosed in the neck behaves as an uncompressible gas, and its mass correspond to the $m$ element of the mechanical system. The air inside the cavity is compressible and stores potential energy, representing the mechanical stiffness $k$. The mechanical damping element ($c$) is represented by two factors: (i) the open-end of the neck radiates sound, introducing a radiation resistance and (ii) the gas movement in the neck introduces a viscous resistance. Considering the electrical analogue (Fig.21 right), the acoustic compliance C (analogous to electrical capacitance) is related to the stiffness of the air in the cavity, the acoustic inertance $M$ (analogous to electrical inductance) is associated to the inertia element (mass) and the acoustic resistance $R$ (analogous to electrical resistance) is related to the dissipative components stated above.

Considering that the gas beyond the end of the neck moves as a whole with the gas inside the neck, it is necessary to use an effective length $l_{eff}$ which is bigger than the true length $l$ of the neck [19]. The effective length $l_{eff}$ is obtained by adding a mass end correction $\delta$, which is empirically determined. In [28], it was presented a complete set of recommended equations for mass end correction, depending on the adopted considerations. For the purpose of this work, the appropriated equation is defined by:

$$\delta = 0.85d\left(1 - 0.7\sqrt{AR}\right) \text{for } AR < 0.16 \quad (8)$$

where $AR$ is the Area Ratio ($A_n$ / $A_c$), being $A_n$ and $A_c$ the neck cross-sectional area and the cavity cross-sectional area, respectively. The effective length is calculated as $l_{eff} = l + \delta$.

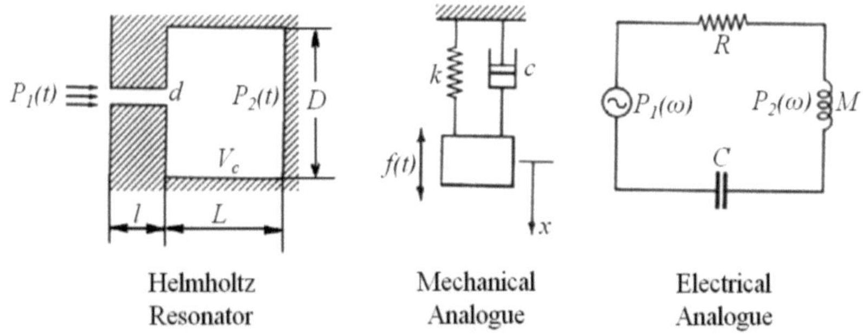


**Figure 21.** Helmholtz Resonators scheme and its analogues

The definition of acoustic inertance ($M$) applied to the Helmholtz resonator gives:

$$M = \frac{m}{A_n^2} = \frac{\rho l_{eff} A_n}{A_n^2} = \frac{\rho l_{eff}}{A_n} \quad (9)$$

where $\rho$ is the air density and $m$ is the effective mass.

The acoustic compliance C is defined as the volume displacement that is produced by the application of unit pressure [19]. By applying this definition to HR, one obtains:

$$C = \frac{V_c}{\rho c^2} \quad (10)$$

where $c$ is the velocity of sound.

The acoustic resistance in the neck ($R$) was approximated as the dissipation associated with viscous forces, considering the dynamic viscosity $\mu$ [28]:

$$R = \frac{8\pi\mu l}{A_n} \quad (11)$$

The acoustic impedance Z of the HR is:

$$Z = R + j\left(\omega M - \frac{1}{\omega C}\right) \quad (12)$$

As can be seen, the acoustic impedance is determined by the geometric and mechanical properties of the resonator. The resonance will occur when the acoustic reactance equals zero:

$$\omega M - \frac{1}{\omega C} = 0 \tag{13}$$

The resonance frequency can be determined by considering that the dimensions of the resonator are much smaller than the wavelength of interest:

$$\omega_0 = \sqrt{\frac{1}{MC}} = c\sqrt{\frac{A_n}{l_{eff}\, V_c}} \tag{14}$$

The resonance's sharpness of a HR can be quantified by its quality factor $Q$, given by:

$$Q = \frac{\omega_0 M}{R} \tag{15}$$

## 4.3. Finite Element Model

The cavity of a LPRE combustion chamber was analyzed using FEM in configurations without and with resonators. The first was modeled using 11,136 linear solid hexahedral elements, 12,510 nodes (12,093 degrees of freedom) and the second was modeled using 38,052 linear solid tetrahedral elements, 7,493 nodes (7,399 degrees of freedom). Both meshes are shown in Fig. 22. The fluid is air at 15° C. The eigenfrequencies were calculated from 0 to 2,400 Hz. Nodal pressures on the openings were assigned to zero.

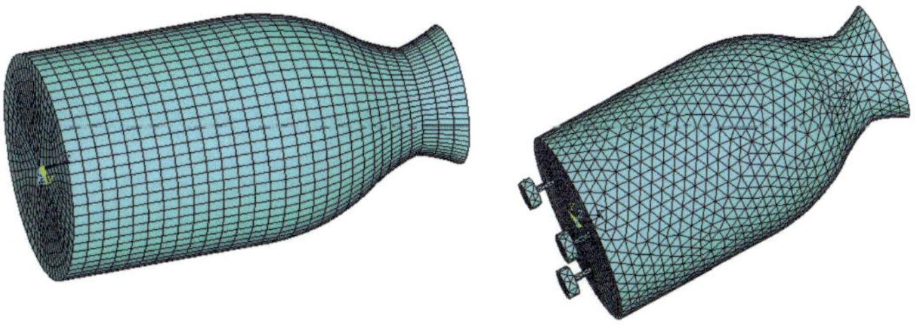

(a) (b)

**Figure 22.** a) Cavity without resonators; b) Cavity with resonators

## 4.4. Experimental setup

Figure 23 shows the experimental setup. The mid-high frequency volume acceleration source is composed by a driver, a tube and a nozzle, where it is installed a volume acceleration sensor. This source produces a voltage signal proportional to the volume acceleration [m$^3$/s$^2$] variation, with a nominal frequency range of 200 up to 8,000 Hz. This source nozzle was installed in the mock-up plane surface as shown in Fig. 23.

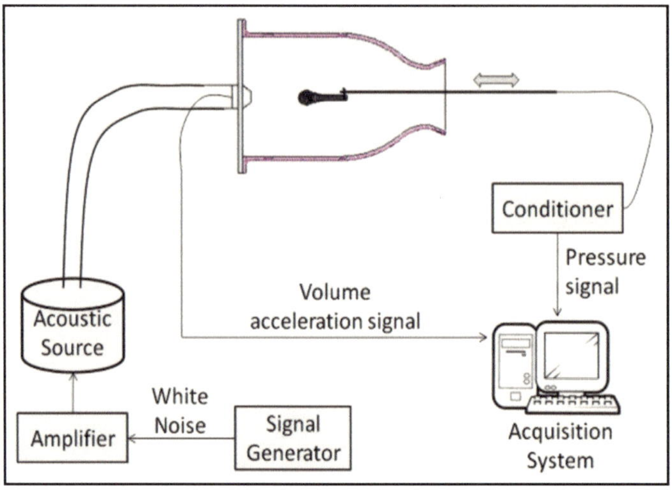


**Figure 23.** Measurement setup

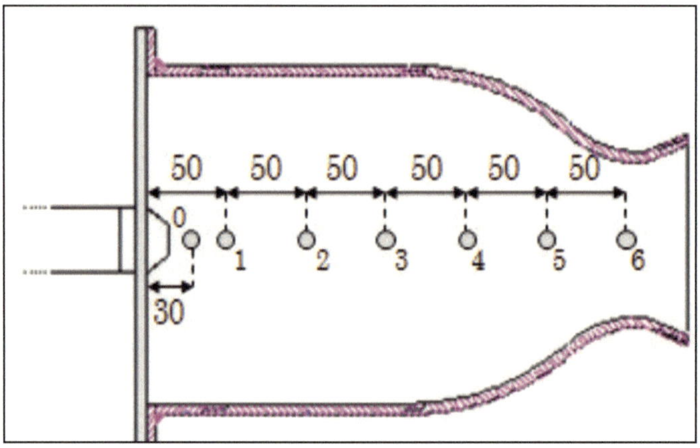


**Figure 24.** Source and microphone positions (distances in mm)

The chamber excitation was provided by a signal generator, a power amplifier and the source. The microphone was supported by a thin rod, placed in each measurement point inside the cavity. The pressure oscillations inside the cavity were captured by the microphone and registered by the data acquisition system. The volume acceleration source channel was settled as the reference channel. A white noise was used as excitation signal and the FRF were acquired at 7 points along the longitudinal axis (Fig. 24), being the point 0 the FRF driving point. The FRF were obtained by considering the volume acceleration as the excitation and the sound pressures as the responses. In order to make compatible theoretical x experimental comparisons, the volume velocity was assessed (instead of volume acceleration).

### 4.5. HR design

The objective is to tune the HR resonance as the same frequency that must be attenuated. Due to construction facility; it was chosen a cylindrical shape to develop the HR. Not only the resonance frequency must be observed during the design process, but also several factors that influence directly the behavior of the HR:

- Resonance frequency of interest = 730 Hz (second longitudinal mode to be attenuated);
- Relation HR dimensions and wavelength $\lambda$ (must be at least 10 times smaller than $\lambda$);
- Area Ratio (AR) must be smaller than 0.16, in order to assure an end mass correction
- Quality Factor (Q): monitored to be used in future designs (compare with other HR shapes);
- Constructive factors. Define such dimensions that can be feasible constructively.

The parameters used in the design were updated considering the room temperature (28 °C) observed during the experiment:: sound speed: $c = 348.3\ m/s$; air density: $\rho = 1.1839\ kg/m^3$; air dynamic viscosity: $\mu = 1.983x10^{-5}\ kg/ms$.

Considering the geometric dimensions in Eq. (14), one obtains the tuned resonance frequency as *726 Hz*, with a quality factor of $Q = 372.8$. The acoustic parameters were calculated $M = 1019\ kg/m^4$, $C = 4.72x10^{-11}\ s^2m^4/kg$, $R = 12468\ kg/sm^4$, using the Eq. (9), (10) and (11), respectively.

Three HR were manufactured in nylon. The presented measurement methodology was repeated in order to acquire the same FRF, considering the new configuration, with the resonators.

### 4.6. Results

The identified natural frequencies are summarized in Table 6. The transversal modes frequencies were identified only numerically. Considering the first four longitudinal modes, the maximum error comparing the numerical and experimental estimation of natural frequencies was 6.6%.

Figures 25a, 25c and 25e present the three first longitudinal numerical modes.

| Mode | Experimental | Numerical | Error (%) |
|---|---|---|---|
| Longitudinal 1st | 164.50 | 170.28 | 3.51 |
| Longitudinal 2nd | 730.85 | 721.90 | 1.23 |
| Transversal 1st | - | 1126.00 | - |
| Transversal 2nd | - | 1126.00 | - |
| Longitudinal 3rd | 1272.64 | 1296.00 | 1.84 |
| Transversal 3rd | - | 1382.00 | - |
| Transversal 4th | - | 1382.00 | - |
| Longitudinal 4th | 1689.52 | 1801.00 | 6.60 |

**Table 6.** Experimental and Numerical Natural Frequencies (Hz)

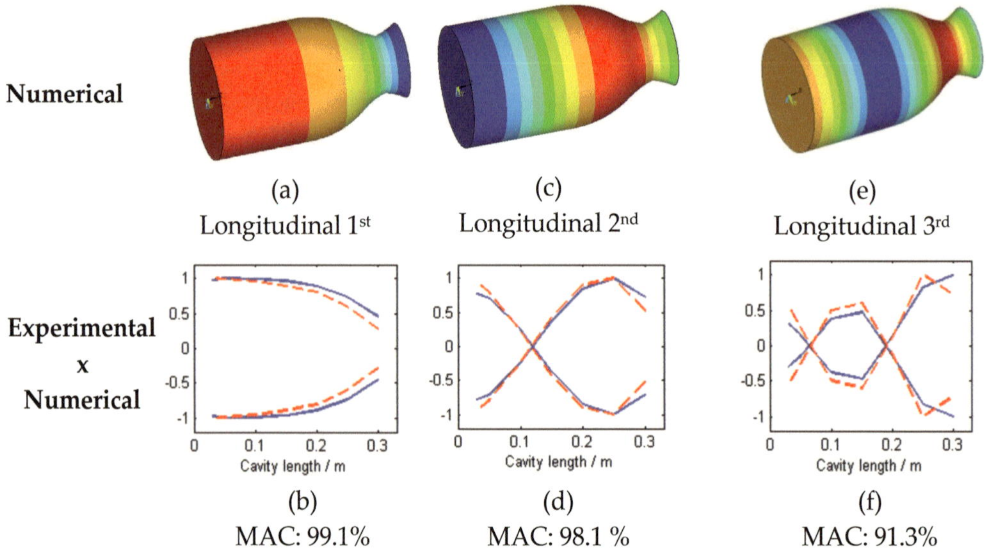


**Figure 25.** a) to f) Numerical modes and comparison between experimental (blue line) and numerical (red line) longitudinal mode shapes

Experimental versus numerical modes comparison, considering the normalized amplitude is also shown (Figs. 25b, 25d and 25f). In the numerical modes represented by the collor maps pictures (Figures 25a, 25c and 25e) the nodal regions are in green. The MAC (Eq. (7)) is also presented.

Notice in Fig. 25 that MAC values are bigger than 91% for the three first modes. For the first and second modes MAC values reach about 99%. Figures 25d and 25f show that the nodes in these modes are almost at the same point. After the introduction of the HR, the attenuation of the second mode is clearly noted in Fig. 26a, when comparing measurement results of the original cavity. At least 9 dB of attenuation can be observed in the new configuration. The FRF with and without HR are almost the same, but the second mode region (about 730 Hz), where the HR is tuned.

Figure 26 depicts the behavior of the chamber with HR. Numerical mode shapes of the configuration with HR were plotted and correlated to each part of the experimental FRF. The Fig. 26b was zoomed from the squared highlight in Fig.26a (point 1). This allows visualize the entire system behavior.

In Fig. 26 it can be noticed that mode shapes *(c)* and *(d)* have similar behavior, but different natural frequencies: it can be realized that the pressures inside all resonators vary in phase. In this case, the pressure inside the chamber remains almost unchanged. In the mode shapes *(e)* and *(f)*, the resonators act close its tuned frequency. In these modes, the whole chamber behaves as a nodal region and the pressure inside the resonators varies out of phase.

As a result of the movement of the air mass inside the neck in the four modes represented in Figs. 26, the acoustic energy on the resonators behaves as expected, reducing the energy inside the chamber.

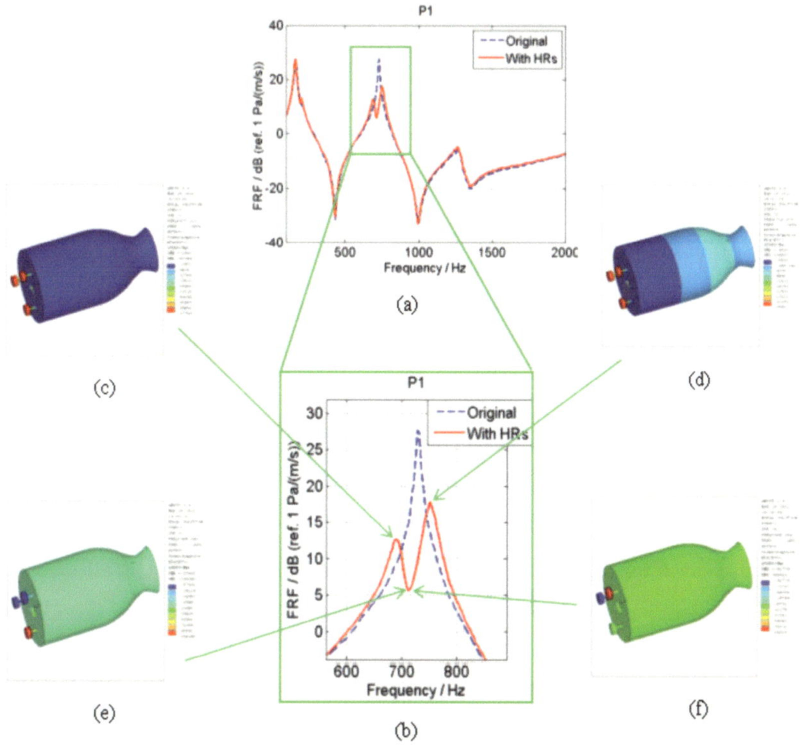


**Figure 26.** Numerical mode shapes near the resonances of the HR

## 5. Conclusions

This chapter presents three acoustic case studies applied on rocketry design.

Firstly, microphone protection devices design procedures for in-flight measurements are described. The modeling techniques using analytical and FEM numerical tools are presented as well as the validation acoustic testing procedures are presented. Good agreement among numerical and experimental results was obtained. A procedure to asses the SPL outer the launcher's structure by using the adapters' acoustic transfer functions and internal SPL measurements was also described.

Vibro-acoustic virtual prototypes were used to predict the acoustic response of a PLF cavity when excited by an ADF of 145 dB OSPL ranging from 5 to 8,000 Hz, generated at lift off.

Coupled deterministic techniques, using FEM/FEM and FEM/BEM, were applied to the fairing problem in a low-frequency band, considering accurate and efficient modeling techniques. The modal and semi-modal superposition techniques were applied to perform a FRA. In the higher frequencies, SEA coupling technique was applied to obtain the fairing acoustic responses in 1/3 octave bands.

The fairing was submitted to the lift off excitation in acoustic reverberant test and the internal acoustic pressure levels were measured. Experimental and numerical results show good agreement, except for the frequencies below 50 Hz and above 4,000 Hz.

The sensitivity analysis of acoustic blankets showed to be an effective tool for the development of the fairing NCT design. The effectiveness of a NCT considering its weight and performance can easily be evaluated using SEA, still in the development phase, when detailed subsystems are not required.

By analyzing many NCT configurations one can provide a library of performances and weights, important parameters that describe the ELV performance book. As in space industry the cost of a mission is a major issue, a trade-off between NCT weight and efficiency must be accounted. Acoustic testing in reverberant chamber may be conducted to validate the presented results and other porous-elastic materials may be investigated to complement the fairing NCT design library.

Finally, for the third case study, the use of a volumetric source in Experimental Acoustic Modal Analysis has important role in the process, once allows the accurate measurement of acoustic FRF.

The numerical model results were used as the basis for the HR design, in a first moment. In addition, numerical and experimental models were used to identify and localize, with a level of security, the node and maximum amplitude regions of each mode. The HR design seemed to be adequate, once it was verified an attenuation of 9 dB or bigger, depending of the location inside the chamber.

## Author details

Rogério Pirk*, Carlos d'Andrade Souto and Gustavo Paulinelli Guimarães
*Institute of Aeronautics and Space (IAE) and*
*Technological Institute of Aeronautics (ITA), São José dos Campos, Brazil*

Luiz Carlos Sandoval Góes
*Technological Institute of Aeronautics (ITA), São José dos Campos, Brazil*

## 6. References

[1] Jorge P. Arenas, Ravi N. Margasahayam, 2006, Noise and Vibration of Spacecraft Structures, Ingeniare. Revista Chilena de Ingeniería, vol. 14 Nº 3, pp. 251-264

[2] Pirk, R, Sas, P., Desmet, W. and Góes, L. C. S., "Vibro-Acoustic Analysis of the Vehicle Sattelite Launcher's (VLS) Fairing", Ph. D. Thesis, 2003, Technological Institute of Aeronautics (ITA), São José dos Campos, Brazil.

[3] Pirk, R. and Góes, L. C. S., 2006, "Acoustic Theoretical x Experimental Comparison of the Brazilian Satellite Launcher Vehicle (VLS) Fairing", Proceedings of the ISMA 2006, Leuven, Belgium.

* Corresponding Author

[4] Pirk, R. and Souto, Carlos, d´A, 2009, Implementation of Acoustic Blankets to the VLS Fairing – A Sensitivity Analysis Using SEA, Proceedings of the ISMA 2008, Leuven, Belgium.
[5] Scott, J. M., Jay, B. G., Robert, C. B., 1996, General Environmental Specification for STS and ELV – Payloads, Subsystems and Components, GEVS-SE, Rev. June 1996, NASA Goddard Space Flight Center, Maryland, 20771.
[6] Souto and Pirk, 2009, Estimation of the external sound pressure levels generated during VLS – 1 lift off, Institute of Aeronautics and Space, Technical Report RT 001/AIE/R/2009).
[7] Pirk et al. 2010, Alternative External Acoustic Loads Estimation of the Brazilian Satellite Launcher (VLS-1), Using In-Flight Experimental Data, International Congress on Sound and Vibration, Rio de Janeiro, Brazil
[8] Glaese, R. M. and Anderson, E. H., 2003, "Initial Structural-Acoustic Modeling and Control Results for a Full-Scale Composite Payload Fairing for Acoustic Launch Load Alleviation". http://www.csaengineering.com/techpapers/technicalpaperspdfs/CSA1999_Initial Structural
[9] Weissman, K., McNelis, M. E. and Pordan, W. D., 1994, "Implementation of Acoustic Blankets in Energy Analysis Methods with Application to the Atlas Payload Fairing, Journal of the Institute of Environmental Sciences – IES, July/August 1994.
[10] Bolton, J. S., 2005, "Porous Materials for Sound Absorption and Transmission Control", Proceedings of the 2005 Congress and Exposition on Noise Control Enginnering, INTERNOISE 2005, Rio de Janeiro, Brazil.
[11] H. Defosse, M.A. Hamdi, "Vibro-acoustic study of Ariane V launcher during lift-off", Proc. of Inter-Noise 2000, vol 1, 9-14 (Nice, 2000).
[12] Culick, F.E.C. Combustion Instabilities in Liquid Rocket Engines: Fundamentals and Control. California Institute of Technology, 2002]
[13] Burnley, V.S.; Culick, F.E.C., The Influence of Combustion Noise on Acoustic Instabilities, Air Force Research Laboratory, OMB N° 0704-0188, 1997.
[14] R. Pirk, C. d´A. Souto, D. D. Silveira and C. M. Souza, 2010, Liquid rocket combustion chamber acoustic characterization, Journal of Aerospace Technology and Management.
[15] Laudien, E. and Others 1994, Experimental Procedures Aiding the Design of Acoustic Cavities, DASA- Deutsche Aerospace AG, Liquid Rocket Engine Combustion Instability, Progress in Astronautics and Aeronautics, Chapter 14, volume 169
[16] Natanzon, M. S., 1999, Edited by Culick, F. E. C., California Institute of Technology
[17] Troclet, B., Analysis of Vibro-acoustic Response of Launchers in the Low-Frequency and High-Frequency Range, Proc. of NOVEM 2000, (Lyon, France, 2000).
[18] Khlybov, V. I. and Mak´hankov, S. A., 2009, The Preparation to Experimental Studies and Acoustic Sensors Testing Program Development, Technical Report.
[19] Kinsler, Lawrence E.; Frey, Austin R. Fundamentals of Acoustics, John Wiley & Sons, 2nd edition, 1962
[20] W. Desmet, D. Vandepitte, 2001, Finite Element Method in Acoustics, Course Graduate School in Mechanics – Advanced Acoustics, Katholieke Universiteit Leuven

[21] Coyette, J-P, Lecomte, C. & Meerbergen, K., 1997, Treatment of Random Excitations using SYSNOISE Rev. 5.3.1 – Documentation Theoretical Manual, Users Manual and Validation Manual, LMS Numerical Technologies NV.

[22] Klos, J. et al ,2003, Sound Transmission Through a Curved Honeycomb Composite Panel, AIAA Conference Paper AIAA-2003-3

[23] Desmet, W, Vandepite, D., Boundary Elements in Acoustics, Katholieke Universiteit Leuven (KUL), 2001.

[24] Lyon and DeJong, Theory and Application of Statistical Energy Analysis, Second Edition, 1995

[25] Thinh, T. Do, Vibro-acoustic Modeling Study of the Delta II 10-Foot Composite Fairing, Journal of the IEST, November/December 1999.

[26] Allard, J. F., 1993, Propagation of Sound in Porous Media – Modeling Sound Absorbing Materials, Elsevier Applied Science, London and New York.

[27] Sutton, G. P., Biblarz, O., 2001, "Rocket Propulsion Elements", New York, John Wiley & Sons.

[28] Santana Jr., A. Investigation of Passive Control Devices to Suppress Acoustic Instability in Combustion Chambers. Thesis of doctor in science. Aeronautics Institute of Technology, São José dos Campos, Brazil, 2008.

[29] Guimarães, G.P.; Pirk, R.; Souto C.A.; Góes, L.C.S. Acoustic Modal Analysis of Cylindrical-Type Cavities. Proceedings of the 8th Intern. Confer. on Structural Dynamics – EURODYN, Leuven, Belgium, 2011, p. 3160-3167.

[30] Fahy, F. Fundamentals of Engineering Acoustics. Academic Press. London, UK. 2001.

Section 6

# Acoustic Waves in Microfluidics

Chapter 21

# Acoustic Wave Based Microfluidics and Lab-on-a-Chip

J. K. Luo, Y. Q. Fu and W. I. Milne

Additional information is available at the end of the chapter

http://dx.doi.org/10.5772/56387

## 1. Introduction

Microfluidics refers to a set of technologies that control the flow of minute amounts of liquids, typically from a few picolitres (*pl*s) to a few microlitres (*μl*s) in a miniaturized system [1,2]. Lab-on-a-chip (LOC) systems typically consist of a set of microfluidics and sensors with dimensions from a few square millimetres ($mm^2$) to a few square centimetres ($cm^2$). Microfluidics handles liquids through droplet generation, transportation and mixing of liquid samples, chemical reactions etc. Sensors may include biochemical sensors, gas sensors and physical sensors such as humidity and temperature sensors, flow meter and viscometers etc. Therefore, LOCs are microsystems with a much broader meaning, and generally perform single or multiple laboratory processes and functions on a chip-scale.

Microfluidic and LOC systems have distinctive advantages:

- Low volume fluidic consumption (low reagents costs and less required sample volumes for analysis and diagnostics, less waste);
- Fast analysis and short response times due to short diffusion distances, high surface to volume ratios, small thermal capacity and fast heating rate;
- Better process control because of a faster response of the systems (e.g. thermal control for exothermic chemical reactions);
- Compactness of the systems owing to integration of many functionalities and the small dimensions of each component;
- Massive parallelization due to compactness, which allows high-throughput analysis;
- Low fabrication costs, allowing cost-effective disposable chips fabricated in mass production;
- Safe platform for chemical, radioactive or biological studies because of integration of functionality within the small systems and often on-board power generator.

Owing to their outstanding properties and great potential applications, microfluidics and LOCs have received tremendous interests from engineering, healthcare, medical research, drug-development sectors. They are regarded as the technologies of the future for great value-added manufacturing. So far, most of the LOCs and microfluidics are single function systems, the trend and demands are to develop LOCs and microfluidics with multi-functions.

Microfluidics and LOC systems based on acoustic waves generated through the piezoelectric effect have recently received a great attention, as acoustic waves can be utilized not only for actuation/microfluidics, but also for sensing/detection, allowing integration of various acoustic devices for LOCs to perform multi-functions. In this chapter, we will thus focus on the acoustic wave-based LOC systems.

## 1.1. Acoustic wave based microfluidics

Many microfluidic technologies have been explored and developed [1,3], including two major classes of devices: active devices such as micropumps, micromixers and droplet generators, and passive components such as microchannels, valves and microchambers. The mechanisms of micropumps vary widely. Based on the mechanisms, designs and applications, micropumps can be categorized into two main groups: mechanical and kinetic pumps. Mechanical micropumps typically represent miniaturized version of macro-sized pumps that typically consist of a microchamber, check valves, microchannels and an active diaphragm to induce displacements for liquid transportation. Thermal bimorph, piezoelectric, electrostatic, magnetic forces and shape memory mechanisms have been utilized to actuate the diaphragm in an oscillation mode [2,3]. These mechanical micropumps are complicated, expensive, typically made by multi-wafer processes, and are therefore difficult to be integrated with other microsystems such as integrated circuits (IC) for control and signal processing due to incompatible processes and structures [1,2,4]. They generally have a large dead volume, leading to an excessive waste of biosamples and reagents which are normally expensive and precious in biological analysis, especially for forensic investigations. These micropumps typically have moving parts (diaphragm and check-valves) which lead to low production yields in fabrication, high failure rates and poor reliability in operation.

The new trend is to develop non-mechanical or moving-part-free micropumps by utilizing electrokinetic forces and surface tension (or surface energy-related forces) such as the electroosmotic (EO) effect [5], electrophoresis (EP) [6], dielectrophoresis (DEP) [7,8], asymmetric electric field, electrowetting-on-dielectrics (EWOD) [9,10], electrostatic pumps [11,12] etc. Electrokinetic force based micropumps typically require electric/magnetic fields to mobilize ionic, or polarisable particles and species in a liquid which can drag the liquid through friction forces to form a continuous flow. The surface tension-based micropumps are typically droplet-based systems, which manipulate discrete droplets through modification of surface tension/energy by external stimuli [13]. Electric field, thermal and concentration gradients generated by localized heating or optical beams, photosensitization and capillary forces are used for these micropumps. Their key characteristic is that they can transport discrete droplets, acting as the so-called digital micropumps, on channel-less (or wall-less) planar surfaces without check valves [13]. The fabrication process is simple and requires no special substrate. Therefore, these pumps can be easily integrated with electronics for control and signal processing etc.

Acoustic waves utilized in microfluidics and lab-on-a-chip systems include ultrasound, bulk acoustic waves and surface acoustic waves (SAW). Ultrasound generated by external large piezoelectric (PE) transducers has been widely utilized for microfluidics which is very effective in mixing microfluidics, and has recently been utilized for transportation of liquid [14] and particle sorting etc through novel design of the systems [15,16]. Due to the different mechanism, ultrasonic wave based microfluidics and LOCs will not be discussed here.

SAW-based microfluidics and LOCs are one of the latest technologies. SAW can be used as an actuation force for pumping and mixing liquids, and for generating droplets and mist [17, 42,18,19]. SAW micropumps can not only manipulate discrete liquid droplets from *pl*s to a few tens of *μl*s, but can also pump a continuous fluid. SAW devices are also excellent sensors for monitoring physical parameters and detecting biochemical entities with high sensitivity. Furthermore the development of thin film SAW technology has opened the way for integration of SAW-based microfluidics and sensors with Si-based electronics on the same chip for LOCs with better functions and applications [20]. This chapter will mainly focus on acoustic wave technologies for microfluidics and lab-on-a-chip applications.

## 1.2. Acoustic wave resonant sensors

Sensors are a type of transducer that convert some physical stimulation such as temperature, pressure etc into electronic, optical, magnetic, or acoustic signals for quantitative measurement, while biosensors are devices that convert biological information into measurable electronic signals. Various technologies such as optical sensors, electrochemical sensors, field effect transistor based sensors, microcantilevers and acoustic resonators have been developed for sensing, particularly for biochemical sensing.

Compared with other biosensing technologies, acoustic wave based technologies have several advantages including simple operation, high sensitivity, small size, fast response and low cost, [21]. Another distinctive advantage of acoustic wave sensors over others is that they can be simply integrated with acoustic wave based microfluidics to form a LOC system driven by a single mechanism, which makes the system fabrication and operation much simpler.

There are four types of acoustic wave resonators: the quartz crystal microbalance (QCM), surface acoustic wave device (SAW), film bulk acoustic resonator (FBAR or BAW) and flexural plate wave (FPW) resonator. Acoustic wave sensors are able to detect not only mass/density changes, but also viscosity, elastic modulus, conductivity and dielectric properties. They have many applications in monitoring of pressure, moisture, temperature, force, acceleration, shock, viscosity, flow, pH levels, ionic contaminants, odour, radiation and electric fields [22,23]. By using specific gas absorbents and biological markers, the acoustic resonators can be made into gas sensors and biosensors. As the latter is the main focus of this chapter, we will only discuss acoustic wave biosensors here.

QCMs with a structure shown in Fig. 1 have a long history, and probably are the only one currently being commercialized and practically used. By cutting a quartz crystal with proper orientations, it is possible to make QCM sensors operate in a longitudinal mode as well as in a transverse (also called thickness shear) mode. The standing waves in the thickness shear-mode devices are parallel to the surface of the QCM plate, and the wave energy is largely preserved in the presence of liquids, therefore the shear mode QCM is suitable for sensing in liquid environments, a pre-request and necessary condition for most biosensing and physiological monitoring. QCM sensors have the advantages of simplicity in design and operation and a mature technology.

**Figure 1.** Typical QCM resonator.

The sensitivity of acoustic wave resonators is determined by the square of the resonant frequency, $f_r$, and base mass. The frequency shift $\Delta f$ of an acoustic wave resonator induced by a mass loading, $\Delta m$, is described by [24]

$$\Delta f = \frac{2\Delta m f_r^2}{A\sqrt{\rho\mu}} \tag{1}$$

where $A$, $\rho$, $\mu$ and $f_r$ are the area, density, shear modulus and intrinsic resonant frequency (sometimes defined as the operating frequency, $f_0$), respectively. QCM sensors have a fundamental limitation of low sensitivity due to the thickness of wafer and the large active area, hence low $f_0$ and large base mass.

SAW sensors consist of a pair of interdigitated transducers (IDTs) (Fig. 2). When a series of radio frequency (R.F.) signals are applied to one of the IDTs, surface acoustic waves are generated through the piezoelectric effect, and travel along the surface of the substrates, received by the IDT on opposite. The strength of the waves decays exponentially with depth into the substrate. Depending on the nature of the travelling acoustic waves, SAW can be longitudinal or transverse mode. The operating frequencies of SAW devices are typically in the range of 100-300 MHz, and the active base mass is much smaller than that of QCMs owing to the one wavelength depth of active area. Therefore, the sensitivity of SAW sensors could be much higher than that of the QCMs. The acoustic waves of shear mode SAW devices travel parallel to the surface with no longitudinal component with no acoustic energy dissipated into the liquid in contact. Therefore, the shear mode SAW devices are suitable for sensing in liquids. Furthermore rapid advances in thin film deposition technologies allow fabrication of high quality thin film SAW devices, resulting in the possibility for integration of SAW sensors with electronics on the same chip. Advantages of the SAW sensors are: simplicity in device structure and process, high sensitivity, small size compared to QCM, availability of thin film SAW, and the possibility for integration of SAW sensors with Si-electronics. However similar to QCM, the SAW devices are difficult to scale down, and other concerns include weak signal, relatively low quality factor and relatively low sensitivity.

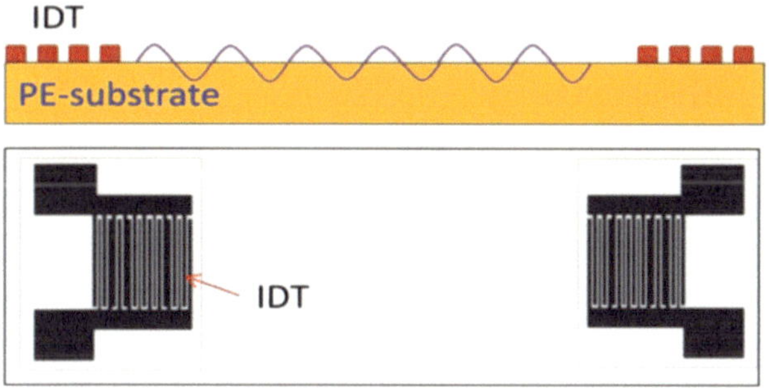


**Figure 2.** Typical structure of a SAW resonator.

FBARs are newly emerging acoustic nanodevices [25,26,27], with a structure similar to that of QCM as shown in Fig. 3, but thousand times smaller with a typical size from 100 μm × 100 μm to 30 μm × 30 μm and a thickness of a few micrometers. The operating frequency of FBARs is in the range from sub-gigahertz to a few GHz. The base mass of a FBAR is much smaller than those of the QCM and SAW devices owing to the much reduced area and thickness, and the sensitivity of FBAR sensors can be dramatically increased compared with other acoustic sensors as shown in Fig. 4 [28]. FBARs have the highest sensitivity, and the SAW devices are in the middle with the QCM ones the lowest.

FBARs have three basic structures: the Bragg acoustic mirror type (Fig. 3a), the back-trench type (Fig. 3b), and the air-bag type (Fig. 3c&3d). The Bragg reflector based FBARs are normally made on PE-films deposited on a solid substrate. The acoustic mirror is composed of many quarter-wavelength layers with alternating high and low acoustic impedances. Due to the high acoustic impedance ratio of the acoustic mirror, the acoustic energy is reflected and confined within the top piezoelectric layer, thus maintaining an excellent resonant bandwidth even on a solid substrate. This structure has a better mechanical robustness and a simpler fabrication process. Also cheap substrates such as glass or plastics may be used for the FBAR fabrication to reduce the cost [29]. The shortages of the Bragg reflector FBARs are long film deposition process and the difficulty in precise control of the layers thickness which may lead to poor quality factor. The back trench mode FBARs are typically made on a thin membrane with a thickness about 1-2 μm to reduce the base mass and increase the robustness. The back trenches are made either by anisotropic wet etching or by deep reactive ion etch (DRIE). The air bag types of FBARs are another type of back-trench type of FBARs, but with the large back-trench replaced by a small gaps formed through etching processes to remove the supporting material underneath. It has either a "standing-out" structure formed by a thin sacrificial layer or a "digging-in" structure.

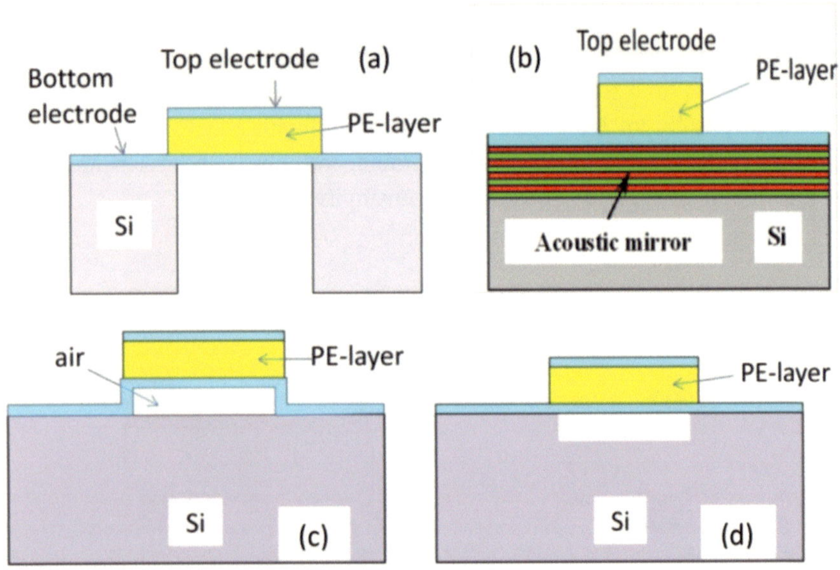


**Figure 3.** Various structures of FBARs. a) Back trench mode, b). Bragg reflector mode, c). Standing-out air bag mode and d). Digging-in air bag mode.

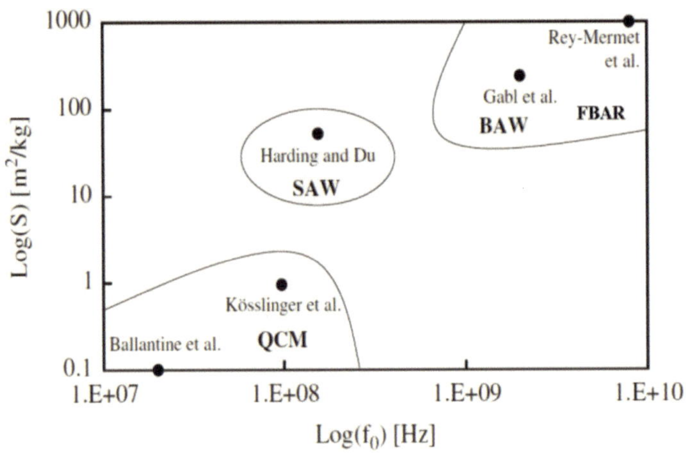


**Figure 4.** Comparison of sensitivity of QCM, SAW and FBAR biosensors. FBAR resonators have the highest sensitivity owing to its much reduced base mass and high operating frequency [28]. Reprinted with permission from Elsevier, Sensors and Actutars, B 114, 2006, 681.

FBARs also have extremely small size which allows the use of an array of FBARs for multiple-sensing in parallel. Also because they are based on thin film technology, the FBARs can be simply integrated with other acoustic microfluidics, microsystems and Si-electronics for LOC applications. Thickness shear mode FBARs have been tried by using off c-axis crystal materials with some success [30,31]. It is still not a simple task to obtain c-axis inclined film materials for low cost FBARs.

In a manner similar to a SAW device, flexural plate wave resonators (also called Lamb wave, and often regarded as the ultrasonic wave [32,33]) (Fig. 5) on a membrane have been developed for biochemical sensing in liquid [34]. The Lamb wave velocity in the FPW resonator is much smaller than that of the acoustic waves, and the dissipation of wave energy into the liquid is minimized, therefore it can be used for liquid sensing directly. The sensing mechanism is based on the detection of a relative change in wave magnitude induced by the perturbation on the membrane, rather than the resonant frequency shift as they would be very small. The sensitivity of the FPW devices increases as the membrane thickness becomes thinner [35,36]. The main drawback of the FPW biosensors is that there is a practical limit on the minimum film thickness due to the fragility. Compared with the other three types of acoustic wave sensors, FPW devices as sensors still need much further development.

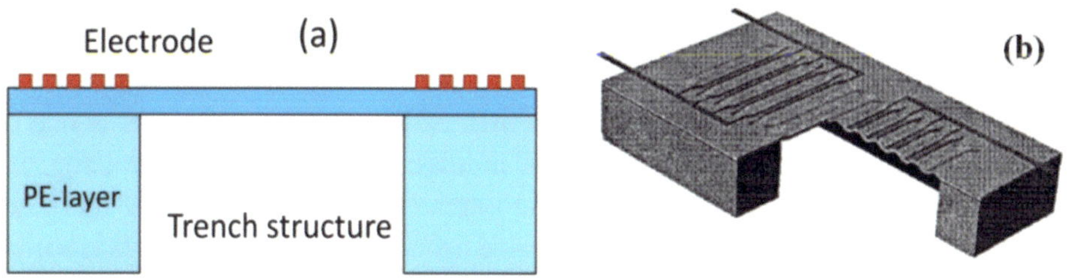


**Figure 5.** Typical structure of a flexural plate wave resonator.

## 1.3. Piezoelectric thin film technologies

QCM and SAW devices are typically fabricated using bulk materials which are expensive and cannot be integrated with electronics, microfluidics and sensors on the same substrate for applications. Various thin film-based QCMs [37,38], SAW devices [39,40,41] and FBARs [25,26,27] have been developed. PZT, ZnO and AlN piezoelectric thin films have good piezoelectric properties and high electro-mechanical coupling coefficient, $k^2$, thus they have been studied intensively for this purpose. They can be grown in thin film form on a variety of substrates such as Si, making these materials very promising for integration with electronic circuitry, particularly for devices aimed at low-cost and mass production for one-time use. PZT has the highest piezoelectric constant and $k^2$, but PZT films have very high acoustic attenuation, lower sound wave velocities, poor biocompatibility and worst of all, the requirement for extremely high temperature sintering and high electric field poling, making them unsuitable for integration with electronics. AlN and ZnO are the most common thin films used for SAW, FBARs and FPW devices. AlN is chemically inert and stable, with high acoustic velocity, but AlN thin films are relatively difficult to deposit, requiring stringent optimization for the process to obtain high quality thin films with smooth surfaces and right crystal orientations. On the other hand, ZnO PE films with a high PE quality are much easier to obtain using various deposition technologies such as sputtering, laser-ablation, chemical vapour deposition (CVD) and molecular beam epitaxy (MBE) etc, therefore, it is the most widely used PE material for thin film acoustic wave devices. The acoustic velocity of ZnO thin films is about 2700 m/s, smaller than that of AlN (11050 and 6090 m/s for longitudinal and transverse modes, respectively); hence ZnO SAW has a lower operating frequency than that of the AlN SAW devices. For wider applications, high operating frequency ZnO SAW devices have been developed by using non-PE supporting layers with high acoustic velocity such as sapphire and diamond [39, 40, 41]. Acoustic waves generated by ZnO SAW travel inside the supporting layer with high velocity, resulting in higher frequencies. Besides PZT, AlN, and ZnO, many other PE thin films have been developed, mostly for SAW device applications. Gallium arsenide, gallium nitrides, polyvinylidene fluoride (PVDF) and its copolymers are a few that have been investigated for piezoelectric applications.

## 2. Modelling of surface acoustic wave microfluidics

Microscale mixing and pumping are essential processes for microfluidic and LOC applications for biochemical analysis, disease diagnosis, DNA sequencing and drug development etc [42,43]. Various technologies have been developed. SAW based microfluidics is one of the most advanced technologies, which utilize SAW induced forces for pumping, mixing, droplet generating and ejecting etc. In this section, the interaction between the SAW and liquid is theoretically analysed, to show that the complicated acoustic streaming and particle sorting etc are physical phenomena which can be well addressed by theoretical model.

### 2.1. Navier-Stoke equation for fluid motion

It is well known that radiation of a high-intensity beam of acoustic waves into a liquid can result in acoustic streaming. Absorption of the acoustic energy results in significant attenuation of acoustic energy and moment in the fluid within a short range, meanwhile consumption of the energy leads to fluid motion. This phenomenon was observed by Lord Rayleigh in 1884 [44] and was then further studied in detail by Westervelt in 1951 [45] and Nyborg [46] in the 1960s'.

Under an external body force, $F_j$, induced by acoustic waves, the fundamental hydrodynamics of a steady viscous fluid is governed by Navier-stoke equation [47],

$$(V_i.\nabla)V_j = F_j - \frac{1}{\rho}\nabla p + \eta\nabla^2 V_j \tag{2}$$

where $V$ is the acoustic streaming velocity (or the particle velocity), $p$ the pressure, $\rho$ and $\eta$ are the fluid density and shear viscosity coefficient, respectively. It is generally assumed that the fluid flow exhibits viscous and incompressible laminar flow. Here the subscripts $i$ and $j$ = 1, 2, 3 represent the x, y and z coordinates respectively for a 3-dimentional (3D) phenomenon. The nonlinear body force is correlated to the Reynolds' stress, σ, induced by the acoustic wave in the fluid with spatial variation in all three coordinates [46],

$$\sigma_{ij} = \rho\overline{v_i v_j} \tag{3}$$

where $v_i$ and $v_j$ are the velocity fluctuations in x, y and z directions. The stress is a mean velocity fluctuation and density product represented by the upper bar. The relation for the force and the Reynold's stress is expressed as,

$$F_j = -\sum_{i=1}^{3} \frac{\partial \overline{v_i v_j}}{\partial x_i} \tag{4}$$

According to the continuity equation, the following equation applies,

$$\frac{\partial \rho}{\partial t} + \nabla.\rho \mathrm{V} = 0 \tag{5}$$

For a steady flow, the first term is zero and we obtain,

$$\nabla.\rho V = 0$$

This leads to zero for the right side of eq.(2),

$$F_j - \frac{1}{\rho}\nabla p + \eta \nabla^2 V_j = 0 \tag{6}$$

The governing equation of the acoustic streaming force has been derived by Nyborg [46] for an incompressible fluid, and is given by the following equation;

$$-F_j = < V_{aj}.\nabla V_{aj} > + V_{aj} < \nabla.V_{aj} > \tag{7}$$

where $V_a$ represents the acoustic wave velocity (different from the streaming velocity V), and the brackets < > indicate the time averaged value [46,48]. Therefore, the nonlinear acoustic streaming force $\Gamma_j$ can be calculated, once the wave velocity is known.

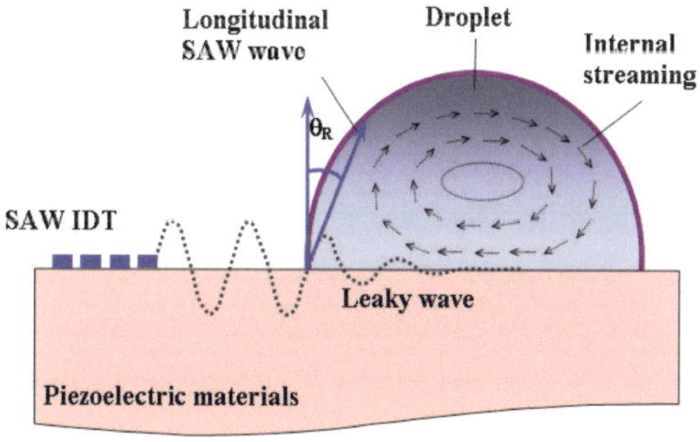


**Figure 6.** Schematic drawing of acoustic streaming in liquid fluid induced by SAW. The direction of SAW induced streaming is determined by the Rayleigh angle.

Acoustic waves travel in a solid medium with a relatively small attenuation. The attenuation becomes large once the waves enter the fluid medium and decays exponentially. Therefore the acoustic wave generated force is a short range force. Figure 6 schematically shows SAW induced acoustic streaming in a droplet.

The SAW travelling along the surface of the PE substrate has a small surface displacements typically less than one nanometre. The SAW changes its mode into a leaky SAW once it interacts with any liquid medium in its path. This leaky longitudinal SAW continuously travels within the liquid medium with the streaming angle determined by the Rayleigh angle $\Theta_R$ [49,50,51], as depicted in Fig. 6 according to the following equation;

$$\theta_R = \sin^{-1}\frac{V_L}{V_S} \tag{8}$$

where $V_S$ is the Rayleigh SAW velocity on the piezoelectric substrate, and $V_L$ is the acoustic velocity in the liquid. The leaky SAW has been well studied, and the displacement, ($u_x$, $u_y$), of the acoustic wave in the liquid is sinusoid in form with an exponential decay with distance. For a 2D case, the displacement can be expressed by [52],

$$u(x,t) = A\exp(-l(x)/l_S)\sin(\omega t - k.x + \varphi) \tag{9}$$

where $A$ is equivalent to the SAW amplitude at the point of entry into the liquid, $\omega = 2\pi f$, the angular frequency imposed by the SAW, $l$(x) is the path length along the wave path, and $\mathbf{k}$ is the wave vector of the acoustic wave. Streaming is a 3D phenomenon, especially for streaming inside a droplet; therefore displacements in all directions have to be considered. The displacements in the x and y directions are expressed by [47,53];

$$u_x = A\exp.(j\omega t).\exp(-jk_L x).\exp(-\alpha k_L y) \tag{10}$$

$$u_y = -j\alpha A\exp.(j\omega t).\exp(-jk_L x).\exp(-\alpha k_L y) \tag{11}$$

Here, $\alpha$ represents the attenuation constant;

$$\alpha^2 = 1 - (\frac{V_S}{V_L})^2 \tag{12}$$

where $V_S$ and $V_L$ are the leaky (Rayleigh) SAW velocity and the sound velocity in liquid respectively. $K_r=2\pi/\lambda$ is a real number, where $\lambda$ is the wave length and the leaky SAW wave number ($k_L=k_r+jk_i$) is complex with the imaginary part, $jk_i$, representing the SAW energy dissipation within the liquid. The leaky SAW wave number can be obtained by extending the method of Campbell and Jones [54, 47] into the solid-liquid structure assuming both stress and displacement to be continuous at $y$=0, and $V_L$=1500 m/s for water (Most of the biofluid is water-based, $V_L$ changes if other liquid is used.). If the wave displacements ($u_x$, $u_y$) are replaced by the wave velocities using $V_j = \frac{\partial u_j}{\partial t}$ and substituting into eq.(7), the two components of streaming force can be obtained for an incompressible fluid as follows [47,53];

$$F_x = -(1+\alpha_1^2)A^2\omega^2 k_i \exp 2(k_i x + \alpha_1 k_i y) \tag{13}$$

$$F_y = -(1+\alpha_1^2)A^2\omega^2 k_i \alpha_1 \exp 2(k_i x + \alpha_1 k_i y) \tag{14}$$

where, $\alpha = j\alpha_1$. The total SAW streaming force $F$ can be calculated by $F = \sqrt{F_x^2 + F_y^2}$, which is then given by;

$$F = -(1+\alpha_1^2)^{\frac{3}{2}} A^2\omega^2 k_i \exp 2(k_i x + \alpha_1 k_i y) \tag{15}$$

The SAW force $F$ acts on the main fluid volume as a body force, but the exponential decay of the leaky SAW limits the influence of this force within the whole fluid within the decay distance. This leads to a complete diminishing of the acoustic force within a few hundreds of micrometers from the interaction point between the SAW and the liquid.

The numerical simulation of acoustic streaming is a complicated process and should be treated case by case due to the different designs of the systems. One must solve the full set of nonlinear hydrodynamic equations consisting of the Navier-Stokes equations, the continuity equation and an equation of state for an incompressible fluid driven by the time-dependent boundary condition. The majority of researchers have implemented the modelling by using existing software such as COMSOL or ANSYS. The numerical simulation requires a few software modules to implement the multiphysics modelling for the solid components as well as the viscous liquid body, and requires the consideration of coupling between the modules. For the viscous liquid, a Finite Volume Method (FVM), OpenFOAM-1.6 CFD code (OpenCFD LTD) and Surface Evolver are often used [55, 53,].

## 2.2 Modelling of acoustic streaming induced by SAW

The interaction of SAW with liquid depends on the position of the SAW relative to the liquid. Figures 7a&7b show two cases of the interaction of a SAW with a droplet: (1) the droplet is in the path of surface acoustic wave with droplet size smaller than the aperture of the IDT of the SAW device; and (2) the droplet is partially on the SAW path.

The interaction of the SAW with the liquid is a dynamic process involving a transition of the flow. Figure 8 shows a comparison of the transitional and steady streaming velocities induced by a leaky SAW obtained numerically and experimentally with RF power as a function [53]. As can be seen, the streaming velocity increases with time rapidly, and becomes stable after the transitional period. The transition time is in the range up to a few hundreds of milliseconds, depending on the RF power applied.

The streaming pattern depends on the entry angle of the SAW into the liquid. Figure 9 demonstrates the different streaming patterns generated by a SAW entering from the centre

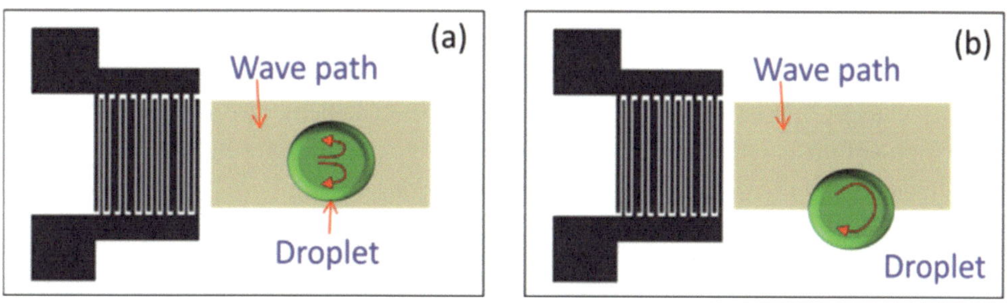


**Figure 7.** (a) Illustration of a droplet positioning symmetrically on surface of a SAW device; (b) asymmetric positioning of water droplet on the SAW device.

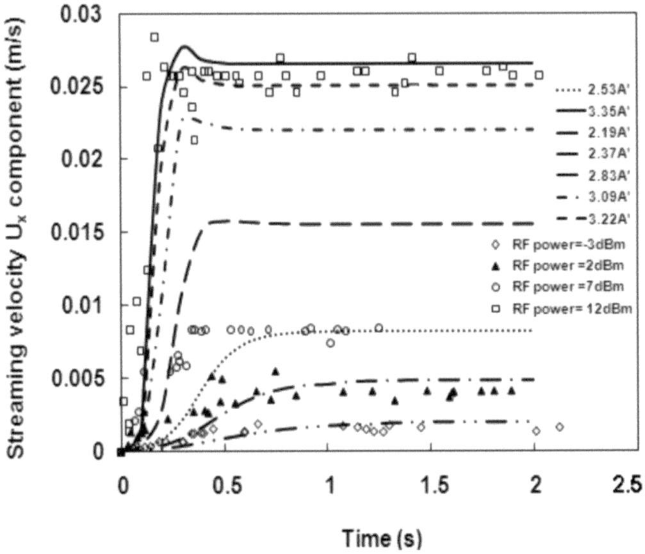


**Figure 8.** Streaming velocity at top centre of a 30μl droplet. SAW device has f=62MHz, an aperture of 2mm; Solid lines represent numerical results at different wave amplitudes; the symbols are experimentally measured data at different RF powers [53]. Reprinted with permission from Institute of Physics, J. Microeng. Micromech. 21, 2011, 015005.

and edge of the droplet [52]. A SAW entering in the centre of the droplet results in a symmetric shaped streaming pattern, with the highest velocity at the edge of the droplet where the SAW enters. When the SAW meets the droplet off-the central wave path, the streaming patterns is asymmetric and the degree of symmetry depends on the entry position of the SAW. A SAW entering into the droplet with a large off-centre line generates a large asymmetric flow pattern. As can be seen from Fig. 9, the numerical simulation can successfully reproduce the experimental results.

Modelling predicts that the streaming velocity is proportional to the RF power applied to the IDT electrode, in agreement with the experimental results, regardless as to whether 2D or 3D simulation was used. Figure 10 shows one of the results obtained by Alghane et al [53], demonstrating that the steady streaming velocity is approximately linearly correlated to the RF power applied.

For better understanding, 3D simulation is necessary to study the streaming phenomenon induced by leaky wave [44]. Figure 11 shows the simulated streamlines with a 3D circular flow pattern for a 30 μl droplet. The simulation results show that the highest value of a streaming velocity is located at the interaction area between the droplet and SAW as the highest momentum is delivered at this point. It attenuates rapidly as it enters the liquid droplet, in agreement with the experimental observation.

It is clear that the liquid initially moves at the Rayleigh angle before reaching the top of the droplet. It moves backward due to the constraint of the droplet boundary, forming a back flow on the two sides and the bottom of the droplet. Figure 12 is a comparison of the streaming patterns obtained experimentally and numerically, showing a good matching of

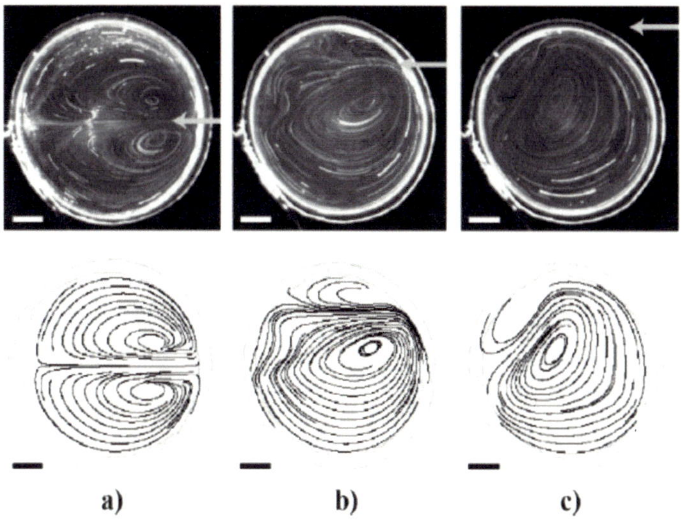


**Figure 9.** Comparison of streaming patterns for experiments and modelling. The narrow is SAW entering position from: (a) centre, (b) intermediate, and (c) outer part. The upper images are experimental results, while the bottom ones are the corresponding simulation results [52]. Reprinted with permission from IEEE, Transact. On Ultrasonics, Ferrelectrics and Freq. Control, 55, 2008, 2298. 1.

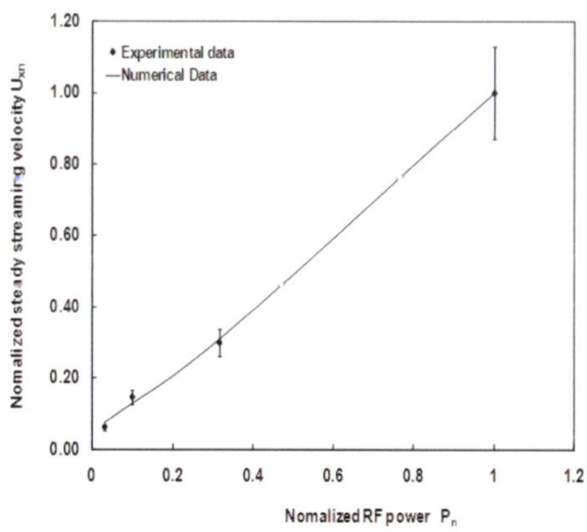


**Figure 10.** Flow normalized streaming velocity as a function of normalized RF power for a 30μl droplet size using a 128º YX-LiNbO$_3$ SAW device (IDT with 60 fingers) [53]. Reprinted with permission from Institute of Physics, J. Microeng. Micromech. 21, 2011, 015005.

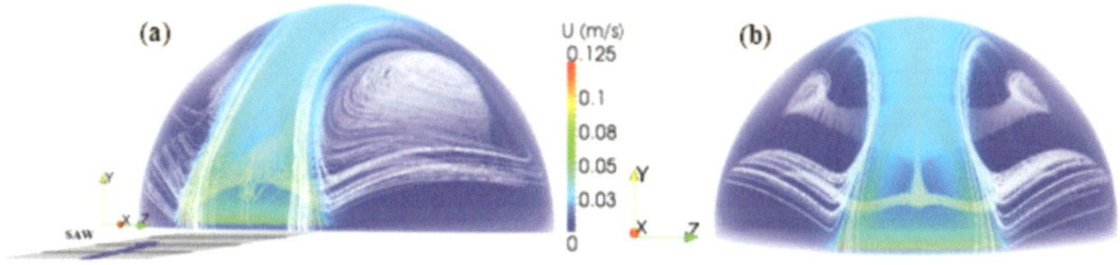


**Figure 11.** Acoustic streaming in 3D reproduced by numerical simulation in tilted view (a) and direct view (b). The droplet has a volume of 30ul, and the SAW enters through droplet centre [53]. Reprinted with permission from Institute of Physics, J. Microeng. Micromech. 21, 2011, 015005.

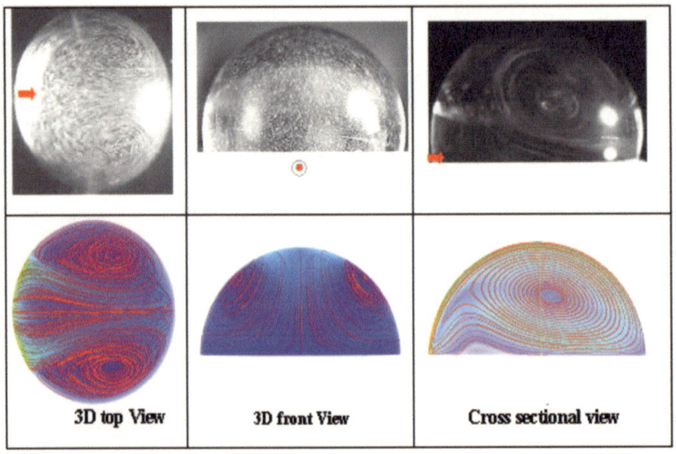


**Figure 12.** Comparison of experimental (upper row) and numerical modeling (bottom row) results for a 30μl droplet. SAW enters in a centre position of the droplet as shown by red arrow [53]. Reprinted with permission from Institute of Physics, J. Microeng. Micromech. 21, 2011, 015005.

the patterns. These numerical simulation results have clearly demonstrated that the chaotic acoustic streaming can be well explained by the physical laws.

## 3. Acoustic wave micropumps and micromixers

Although SAW devices have been commercialized over 60 years, applications in microfluidics and LOCs are only recent event. They are found to be very effective and efficient in microfluidics and LOCs [56]. SAW-based micropumps are one of them, and can transport liquid in a droplet form as well as in a continuous mode through proper design of a fluidic system. Acoustic streaming is also the most effective method of mixing liquids in small dimensions as it is quick and efficient, typically taking less than a few seconds to reach >95% mixing [57, 58]. In this section, SAW pumping and mixing are discussed.

### 3.1. IDT and SAW device structures

For microfluidics, one IDT electrode is enough for most pumping and mixing applications. The acoustic streaming velocity depends on the power output of the SAW devices and the amplitude of the RF signal applied to the IDT electrode. To obtain efficient acoustic streaming, it is desirable to have IDT electrodes with high and efficient power output. The SAW IDT design is important for delivering efficient SAW power output. The conventional bidirectional IDT may not be the most efficient for pumping and mixing, as the waves propagate in two opposite directions with half of the acoustic energy wasted. The simplest way is to reflect back some of the waves (Fig. 13a) by using reflector IDT. More sophisticated IDT designs include [59,60]: (a) split IDTs (Fig. 13b); (b) a SPUDT (single phase unidirectional transducer, Fig. 13c) which has the internally tuned reflectors within the IDT to form a unidirectional SAW propagation from one side of the IDT. These unidirectional acoustic wave transmission are essential for SAW microfluidics and sensors as they not only improve the performance, but maintain the SAW devices at the best operating conditions.

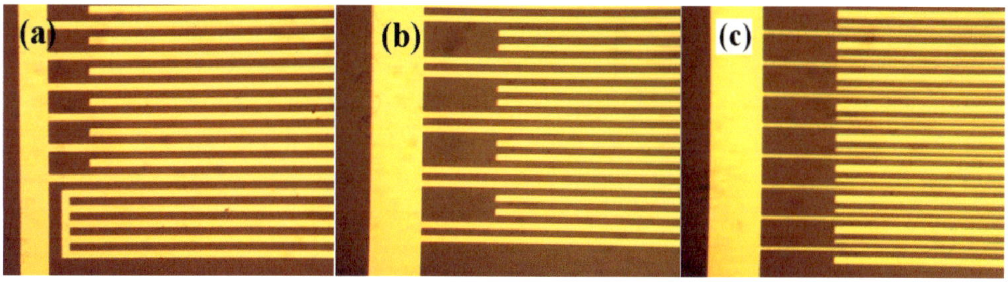


**Figure 13.** Common designs of IDT electrodes used for SAW devices. IDT with reflector (a), slitting electrode (b), and single phase unidirectional transducer (c).

Generally SAW IDT electrodes are so designed that the transmission spectra of the SAW devices have the highest quaility factor, Q, i.e. with a single well defined resonant peak with the highest amplitude. For microfluidic applications, the resonant frequency of a SAW device is changed slightly once liquid is loaded on the surface of the SAW device due to mass loading effect. It would be difficut to apply an RF signal with the exact frequency match to that of the SAW device, leading to the operation of SAW-microfluidics under mismatching conditions. This could be detrimental to the SAW devices as a much higher RF power is required to obtain the acoustic streaming effect, leading to overheating etc. Therefore a SAW device with a certain bandwidth of frequency would be better for SAW-microfluidic applications.

Thin film SAW devices have different transmission properties from those of SAW devices made on bulk substrates. Since the acoustic velocity of the thin film is different from that of the substrate, the acoustic waves generated by the top PE film layer may disperse into the substrate, i.e. some of waves propagate inside the substrate, resulting in two acoustic wave modes related to the PE layer and the substrate. A Sezawa mode acoustic wave is generated when the acoustic velocity in the substrate is higher than that in the PE-layer, and has been intensively studied as it can obtain SAW devices with high resonant frequencies using

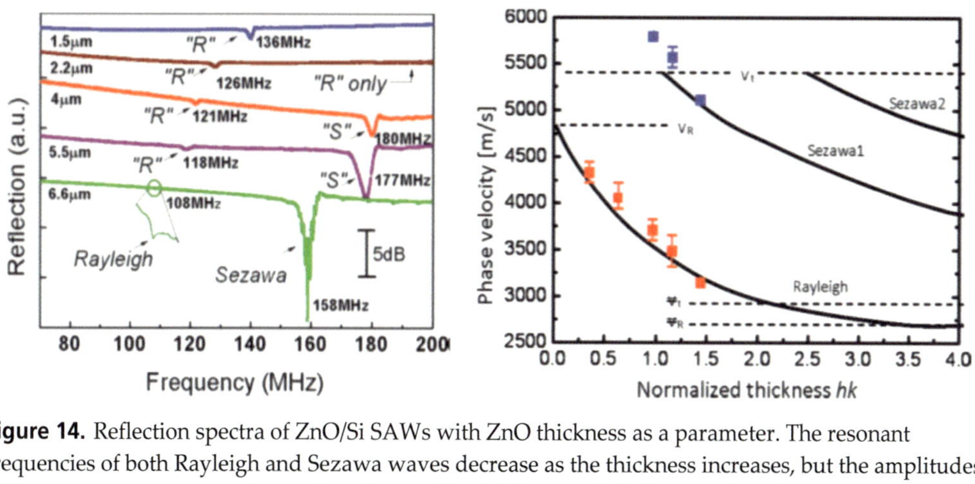


**Figure 14.** Reflection spectra of ZnO/Si SAWs with ZnO thickness as a parameter. The resonant frequencies of both Rayleigh and Sezawa waves decrease as the thickness increases, but the amplitudes of Sezawa waves are much larger than those of Rayleigh wave [61]. Reprinted with permission from AIP, Appl. Phys. Lett. 93, 2008, 094105.

conventional photolithography technology. Examples include ZnO films on Si, sapphire and diamond films. ZnO has a slow acoustic velocity of ~2700m/s, but ZnO SAW devices on a diamond layer can achieve a velocity over 10,000 m/s, close to those of the diamond substrate [39,40,41]. The acoustic velocity of the Sezawa waves depends on the thickness of the PE-layer, and can be well explained by the model of a layered structure. Figure 14 shows the dependence of the acoustic velocity of Rayleigh and Sezawa waves in ZnO/Si structures [61]. Sezawa waves often have much higher signal amplitude and resonant frequencies, particularly suitable for microfluidics and sensing applications.

### 3.2. SAW micropumps and micromixers

When the liquid is in the SAW path, acoustic moment and energy can be coupled into the liquid to induce acoustic streaming and flow. This has been utilized for fabrication of acoustic microfluidic systems, and many devices have been developed [62,63,64]. Generally speaking, both experimental and modelling results showed that acoustic waves can induce significant acoustic streaming in the liquid and result in mixing, pumping, ejection and atomization [65,66]. It was found that if the force is large enough, it can generate a significant acoustic streaming within the droplet (Fig. 15b). If the SAW device is immersed in a liquid container and an RF signal is applied to the IDT electrode, a steady flow pattern with a butterfly or quadrupolar streaming patterns can be obtained as shown in Fig. 15c. Significant acoustic streaming can facilitate internal agitation, which can speed up biochemical reactions, minimize non-specific bio-binding, and accelerate hybridization reactions in protein and DNA analysis which are commonly used in proteomics and genomics.

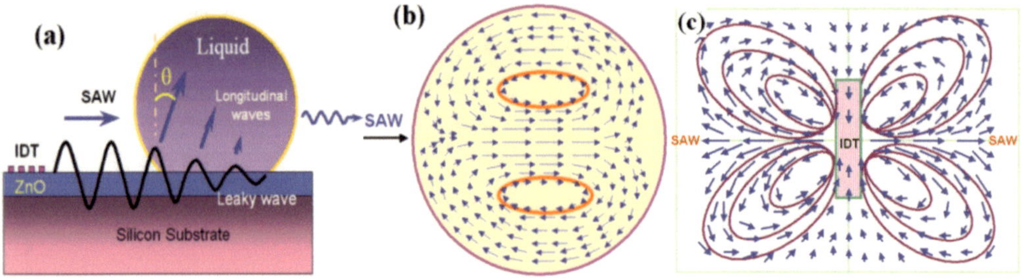


**Figure 15.** Patterns of acoustic streming in a droplet and in bulk liquid.

The SAW microfluidics have distinct advantages over other microfluidics, such as a simple device structure, no moving-parts, electronic control, high speed, programmability, manufacturability, remote control, compactness and high frequency response [67, 68,69]. The streaming velocity is proportional to the RF power applied, and could reach tens of centimetres per second. This is several orders of magnitude larger than other microfluidics, which are typically in the range of hundreds of micrometres to several millimetres per second [1,2,13]. Figure 16 shows the streaming velocity as a function of the amplitude of RF signal for LiNbO3 SAW devices with the IDT structure as a parameter [70]. It shows that the SAW device with a shorter wavelength has more power to induce streaming with higher velocity, and more IDT fingers are beneficial for coupling more RF power into the liquid. Du et al. also demonstrated that the third harmonic resonant waves can also be used for acoustic streaming though the streaming velocity is reduced to a third of the fundamental mode wave induced streaming [70] as higher harmonic modes have much lower RF output.

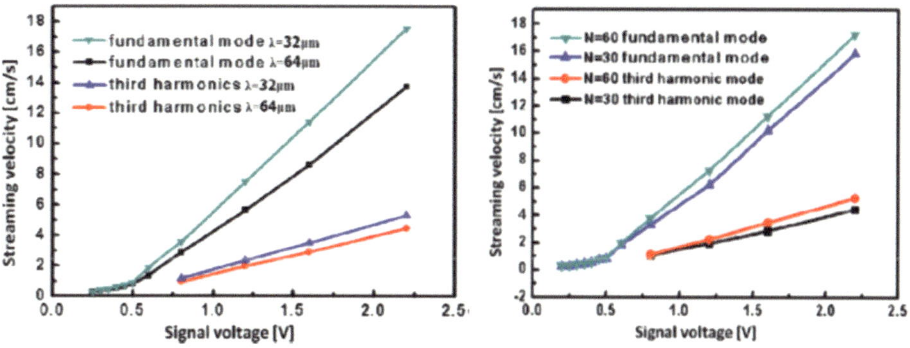


**Figure 16.** Acoustic streaming velocity as a function of RF signal voltage with finger pair and wave mode as parameters. [70]. Reprinted with permission from Institute of Physics, J. Microeng. Micromech. 19, 2009, 035016.

The streaming velocity depends on the RF power (or the signal voltage) applied to the IDT electrode. If the power is small enough and the droplet size is larger than the apperture of the IDT electrode, then the streaming velocity is proportional to the power as shown in Fig. 16, in agreement with the theoretical prediction as shown in Fig. 10. Once the power is over a certain range, velocity saturation is observed [71], and is believed that acoustic heating is responsible for the deviation from the linearity. Also the streaming velocity depends on the relative dimensions of the droplet and IDT aperture. At a fixed IDT aperture, a small doplet will have a low streaming velocity as the RF power can not be fully coupled into the liquid, and the streaming velocity increases with the droplet size, then reduces as the droplet size becomes larger than the aperture of the IDT. Therefore it is necessry to use an optimumal size/aperture ratio for a high performance SAW micromixer.

For streaming in a droplet, the boundary of the droplet becomes the boundary of the vortex, and the flow is confined within the droplet. This can be used for droplet-based mixing as shown in Fig. 17 by Xia et al on a $LiNbO_3$ SAW device [72]. Two drops with different contents are merged by a SAW, and immediate mixing occurs partially due to the kinetic energy involved and partially due to the acoustic streaming. This demonstrates the effective and efficient mixing of water and red dye droplets. The mixing process is completed in tens of ms, much shorter than those of most other micromixers.

Acoustic streaming has circulating flow patterns due to the back flow of the streaming as schematically shown in Fig.15 regardless as to whether it is a droplet or bulk liquid. This is caused by the short range of the acoustic force as discussed above. Directional flow by acoustic streaming, or liquid pumping, might be difficult due to this back flow. For pumping liquid in a specific direction, the SAW device and microchannels have to be arranged properly. Figure 18 is a schematic drawing of a SAW-based micropump. When the SAW-induced vortex size is smaller than the channel width, the SAW induces a circulating

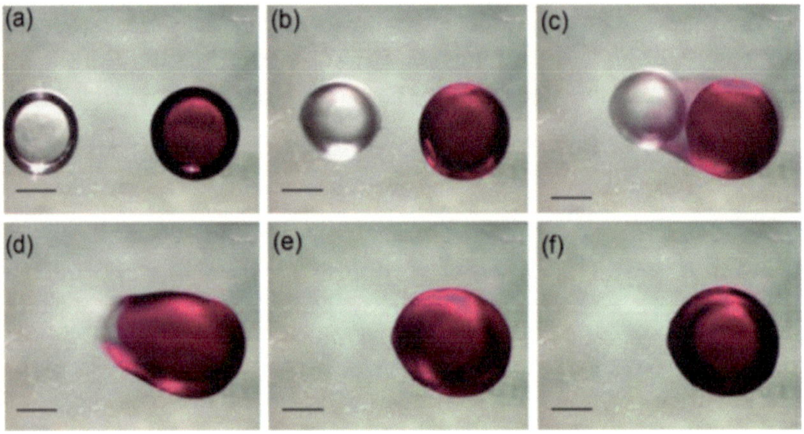


**Figure 17.** A water droplet and a red dye droplet mixture at (a) t=0s, (b) t = 4.066 s, (c) t = 4.4 s, (d) t = 4.466 s, (e) t = 5.000 s, and (f) t = 6.133 s [72]. Reprinted with permission from Elsevier, Talants, 84, 2011, 293.

vortex within the channel, and no net flow along the channel is produced. This can be utilized for in-channel mixing with high efficiency, but not pumping. When the size of vortex is larger than the width of the microchannel, the back flow is restricted by the channel wall, leading to a net directional flow along the channel. By using a channel with the width less than 200 μm, Yeo et al realized a SAW-based micropump [19].

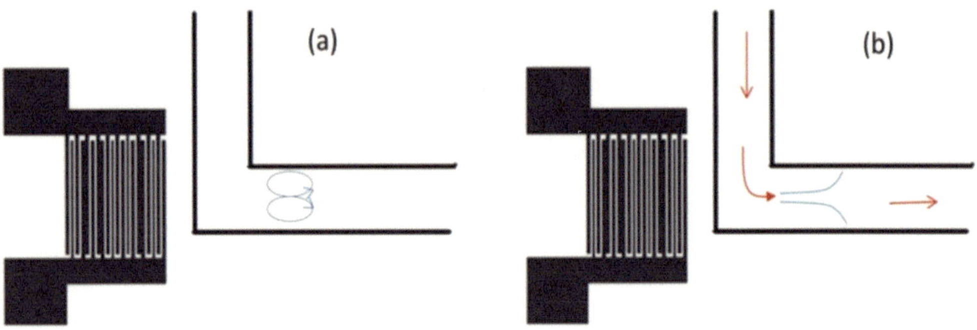


**Figure 18.** Principle of SAW-based micropumps with a channel. When the vortex is smaller than the channel width, the back flow occurs, suitable for in-channel mixing (a). When the vortex is larger than the channel width, a unidirectional flow is formed (b).

SAW has also been explored for in-channel mixing. Figure 19 shows two schemes used by Nguyen et al for in-channel mixing [73]. The microchannel is perpendicular to the SAW propagation direction. When a travelling acoustic wave encounters the liquid in the channel, streaming can be very effective mixing mechanism, and this mixing can be further enhanced by using curved IDT structures as shown in Fig. 19b. An alternative micromixing using SAW is to couple surface acoustic waves into liquids in a container to introduce agitation for mixing. As the streaming angle is determined by the Rayleigh angle, when a container is placed on a SAW device with a soft coupler on the surface, the SAW device can be an effective mixer [74].

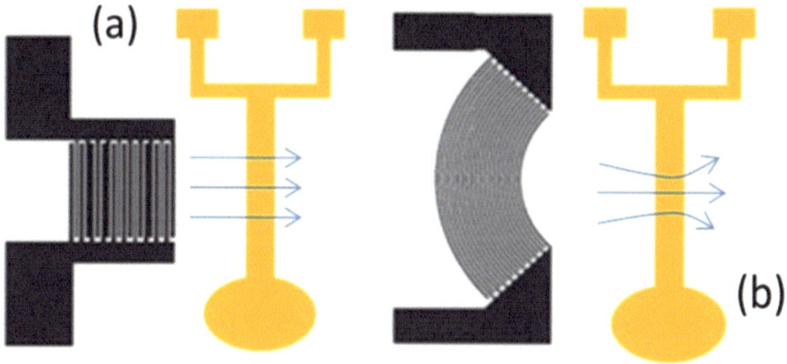


**Figure 19.** Schematic drawing of SAW-based in-channel micromixer.

SAW can be utilized for moving liquid droplets. If the liquid is on a hydrophobic surface with a contact angle larger than 90°, then the acoustic force may move the droplet along the wave path if the RF power is sufficiently large. Droplet-based microfluidics and LOCs are attractive as it provides the foundation for digital analysis and digital medicine with better and accurate results. The surface of most PE materials, however, is hydrophilic with water contact angle less than 90°. Surface acoustic waves are not strong enough to move the liquid on hydrophilic surfaces. Figure 20 shows the shape change of a droplet when it is subjected to an acoustic pressure on a hydrophilic surface. Water droplets cannot be moved freely on the surface under the stimulation of the SAW, but simply spreads on the wave path. The solution is to reduce the surface energy to move the droplets freely on the surface. Various surface coating technologies have been developed to increase the contact angle of the water droplet. CFx chemical vapour treatment was found to form contact angles larger than 90° and reduce the surface energy significantly [75]. Solution-based Teflon can be applied to the surface of PE substrates to form a hydrophobic surface to provide contact angle larger than 110°. The thickness of the Teflon and CFx layer has to be carefully controlled to reduce attenuation caused by the large acoustic absorption of the layer. Octadecyltrichlorosilane (OTS) was found to form a compact and hydrophobic self-assembled monolayer (SAM) with a thickness less than 10 nm. The OTS layer has a contact angle with water larger than 100° and does not damp the SAW amplitude visibly; therefore it is a good hydrophobic coating for microfluidic applications [70,71].

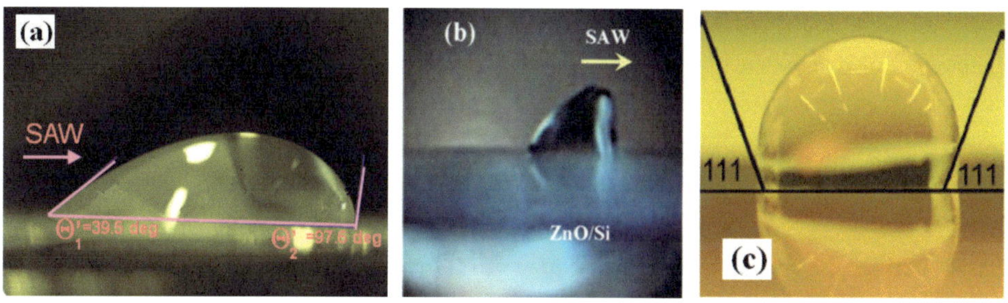


**Figure 20.** The shape of a droplet under acoustic pressure. The droplet deforms with the leading angle becomes larger and the trailing angle becomes smaller [70]. Reprinted with permission from Institute of Physics, J. Microeng. Micromech. 19, 2009, 035016 for (a) and (c).

Similar to streaming velocity, the droplet moving velocity driven by SAW is found to be proportional to the RF signal amplitude as shown in Fig. 21, and the velocity depends on the droplet size as well as the aperture of the IDT. However, when the droplet diameter is larger than the aperture, further increase in the power does not increase the velocity visibly as the acoustic wave does not fully coupled into the droplets as discussed in [76]. When the aperture is larger than the droplet size, it was found that the velocity decreases with increase of the droplet size as more power is needed to move a larger droplet. As can be seen, the droplet motion velocity is in the range up to 1-2 cm/sec, much larger than the velocities obtained by other methods. Although a SAW can be used to drive droplets at very high speeds, delivery of droplets to required locations with precision is more important for applications in biological analysis than driving the droplets at a high speed. This can be realized by using modulated pulsed RF signal to drive the SAW devices in a controlled manner. By adjusting the period of the on-off RF signal and the signal amplitude, it is possible to move droplets with precision distance. A moving rate of 100 μm/pulse for a droplet of 0.5 $\mu l$ on a $LiNbO_3$ substrate was obtained by using the pulsed RF signal [77].

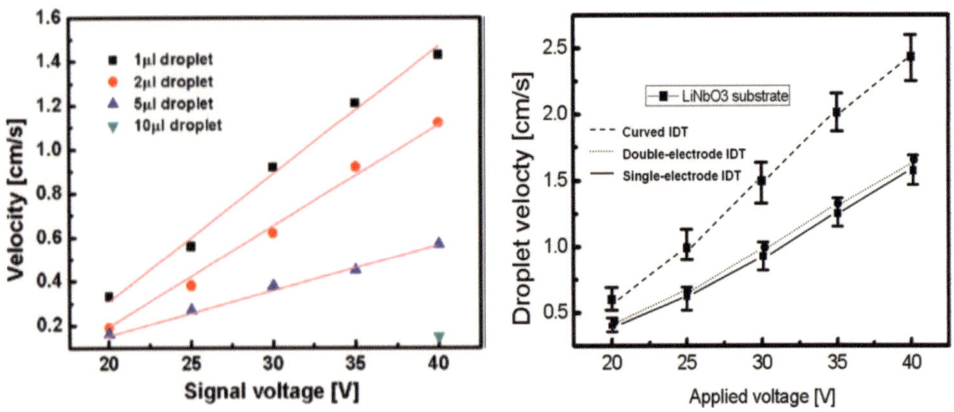


(a) (b)

**Figure 21.** Droplet moving velocity vs. RF signal amplitude (a) and droplet moving capability vs. IDT structures [70]. Reprinted with permission from Institute of Physics, J. Microeng. Micromech. 19, 2009, 035016 for (a).

As discussed before, the IDT structure significantly affects the energy output, and hence the acoustic streaming and droplet motion owing to its acoustic energy distribution. Generally, a unidirectional IDT has a higher acoustic energy density than a bi-directional IDT, and the curved IDT has the highest energy density. Figure 21b shows a comparison of the droplet motion velocity for SAW devices with different IDT structures. The curved IDT SAW has the highest droplet motion velocity as expected.

### 3.3. Flexural plate wave micropump and micromixer

Lamb waves have also been utilized for pumping and agitating of minute volumes of liquids [32, 33] and enhancing biochemical reactions [78], based on the fluid motion induced by the travelling flexural wave in a ZnO or AlN piezoelectric membrane [79]. A novel valveless pump based on a Lamb wave was proposed by Ogawa et al [80]. The liquid in the microchannel was transported by generating a travelling wave on the channel wall, which was composed of a piezoelectric PZT thin film actuator array. A mean flow velocity of about 118 and 172 μm/s was obtained for the 200 and 500 μm wide channels, respectively.

However, due to the low frequency resulting from the thin membrane, the agitation and pumping are too small and insufficient energy is coupled into the liquid. Furthermore, the microfluidic systems contain a membrane vibrating at a high speed, and the yield for fabrication is low, and the reliability of the systems during operation is poor compared to SAW micropumps and mixers. Therefore, research and application of FPW-based microfluidics are currently very limited.

# 4. Acoustic droplet generator and atomizer

## 4.1. SAW droplet generator and manipulator

Generation of droplets with volumes from a few *pls* to a few *μls* is extremely important for modern biotechnology, life science, medical research and diagnosis. For quantitative analysis in a small volume, it is essential to measure the small volume of reagents and biosamples precisely; otherwise false results can be easily obtained. Although a few technologies have been developed to generate droplets in volumes from a few *nls* to a few *μls*, they are difficult to be integrated with microfluidic systems. By changing the acoustic force and hydrophilic and hydrophobic patterns on the surface of a SAW device, it is possible to generate droplets from a few *pls* to *μls* on a free surface. Figure 22a is a schematic drawing of a SAW-based droplet generator, consisting of SAW devices with a reservoir and hydrophilic spots surrounded by a hydrophobic surface [49]. When the liquid in the reservoir is pushed forward under the acoustic force, it makes contact with the hydrophilic

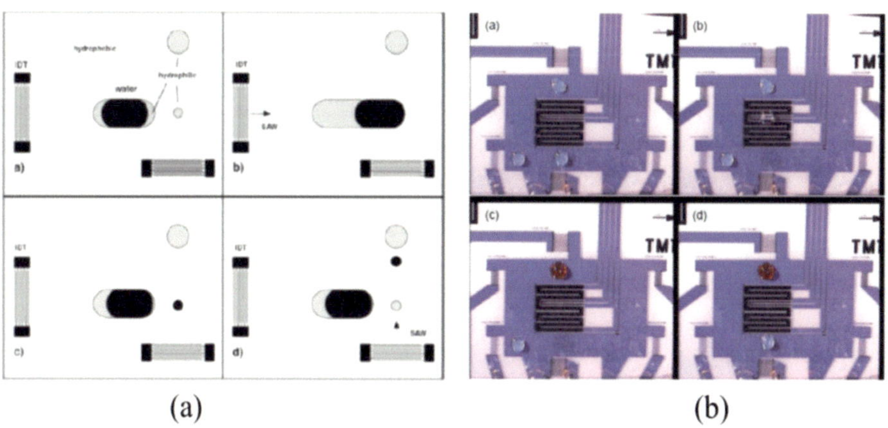


**Figure 22.** (a) the principle of acoustically driven nano dispenser by selective chemical modification of the wettability of parts of the chip surface and employing two SAW propagating at a right-angle to each other. (b) is a SAW driven microfluidic processor [49]. Reprinted with permission from Elsevier, Superlattice & Microstruct. 33, 2004, 389.

spot and fills it. Once the RF signal is off, the bulk liquid withdraws back to the reservoir as the acoustic force diminishes, leaving the spot filled with the liquid of a fixed volume determined by the size of the spot area and the contact angle. In this way, Woxforth et al. developed a SAW-based droplet generator [49]. Furthermore, by using the virtual containers and tracks formed by hydrophilic surfaces surrounded by hydrophobic surfaces for liquids on a chip, they have developed droplet manipulator and mixer as demonstrated in Fig. 22b.

A combination of a SAW pump with other droplet generators can realize new functions of microfluidics. Electrowetting on dielectrics (EWOD) has been combined with SAW devices to fabricate the EWOD-SAW microfluidics as shown in Fig. 23 [81]. EWOD is used to generate and separate the droplets, while the SAW device is used to move the droplets along the track. The EWOD force is employed to guide and position microdroplets precisely which can then be actuated by SAW devices for particle concentration, acoustic streaming, mixing and ejection, as well as for sensing using a shear-horizontal wave SAW device [82].

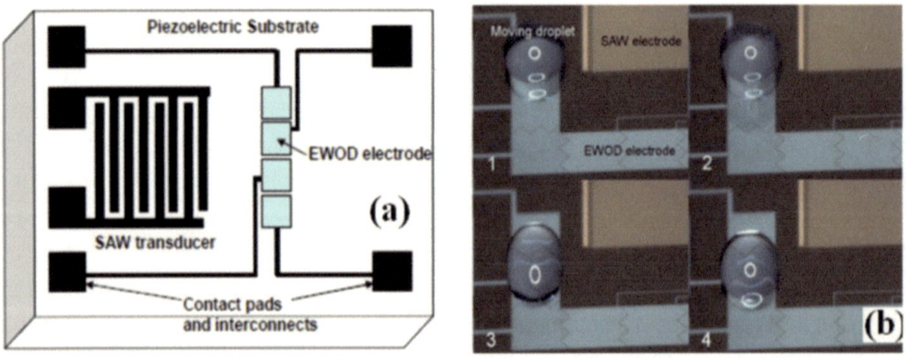


**Figure 23.** Schematic of integrated EWOD and SAW test structure and the droplet generation by EW and manipulation by SAW [82]. Reprinted with permission from AIP, Biomicrofluidics, 6, 2012, 012812.

## 4.2. SAW atomizer

If the RF power coupled to the liquid on a hydrophilic surface is sufficiently large, tiny droplets with volumes in the range of a few femtolitres (*fl*s) to *pl*s can be generated and escape from the surface of the host liquid, forming a continuous mist of droplets as shown in Fig. 24. This has been utilized for the development of SAW-based atomizers and droplet generators. Ejection of small particles and liquids has many applications ranging from inkjet printing, fuel and oil injection sprayers and propellers.

The height of the mist is dependent of the RF power applied, and could be up to 6 cm [20,71]. The ejection angle of tiny droplets escaping from the host liquid is determined by the Rayleigh angle, but affected by the RF power and height of the mist [83]. In order to generate a continuous mist on demand, there must be a continuous supply of liquid which can be realized by using a porous structure such as a filter paper linked to a large liquid reservoir as shown in Fig. 25 [65].

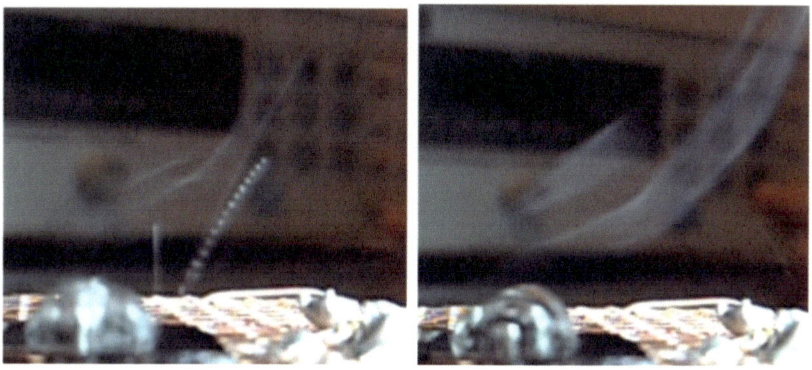

**Figure 24.** Acoustic streaming induced mist by a ZnO SAW device, which may reach 6mm height.

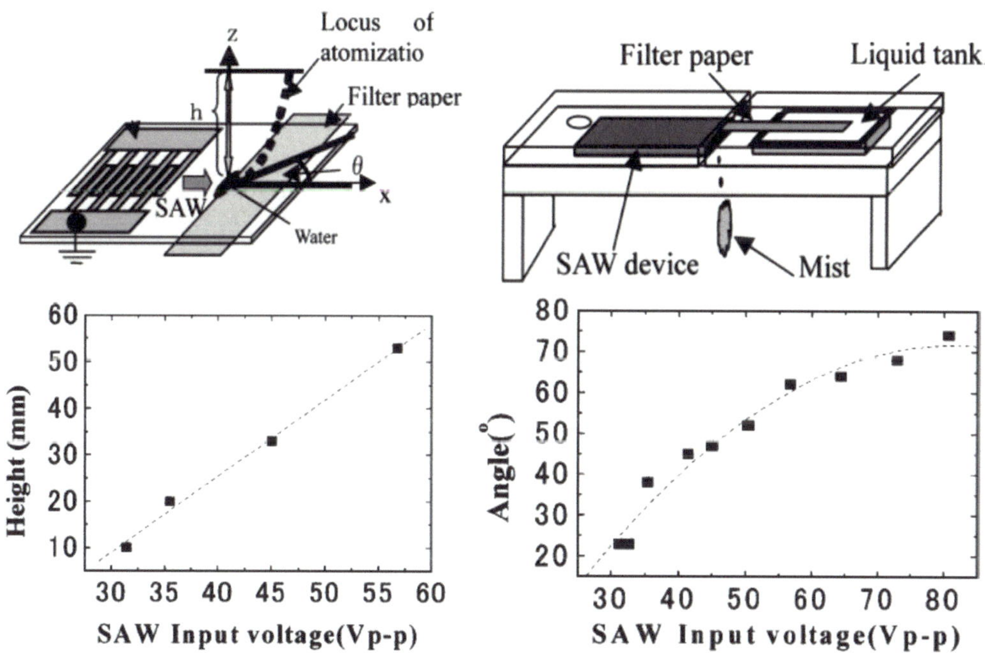


**Figure 25.** SAW atomization mechanism and setup for the SAW atomizer, and the mist height and angle as a function of SAW input voltage [65]. Reprinted with permission from Japan J. Appl. Phys. Lett. 43, 2004, 2987.

A proper design of SAW ejector can be utilized for nozzle-free ink applications. Tan *et al.* [84] demonstrated another method utilizing two opposite IDTs to converge the acoustic energies at point with a liquid drop ejected perpendicular to the surface, similar to the normal nozzle-based droplet ejector as schematically shown in Fig. 26. The radiation from two sides of the droplet resulted in an elongated liquid column with an angle of about 90°. These SAW ejectors do not have a nozzle head and offer a more cost effective solution when compared to the current ink ejector.

Atomization has been widely applied in pulmonary drug delivery as a promising technology to transport drug formulations directly to the respiratory tract in the form of inhaled particles. The most common methods employed for this application are jet atomization and ultrasonic atomization with difficulties to produce monodispersed particles, i.e. droplets with sizes in the range of 1~5μm in diameter. SAW atomizers are able to produce aerosol droplets with a good particle size distribution. By controlling the RF power applied to the SAW IDT, the droplet sizes can be less than 5 μm [85], suitable for the pulmonary drug delivery application.

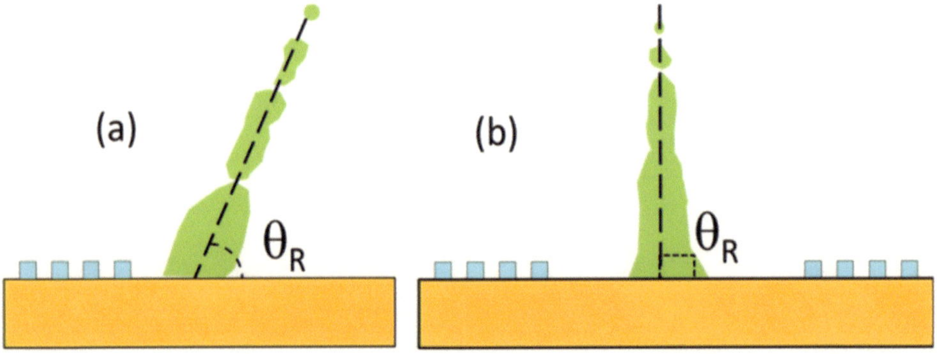


**Figure 26.** Droplet jetting induced by a single IDT SAW device (a), and droplet jetting induced by a pair of IDT electrodes (b).

Not just bulk SAW devices, but also thin film SAW devices can also achieve a similar atomization effect by using sufficient RF power. Figure 24 shows images of tiny liquid droplets ejected from the surface of a ZnO SAW device obtained by the authors, and the height of the mist is similar to those observed from the $LiNbO_3$ bulk SAW devices, even though thin film SAW devices have lower power delivered compared with the bulk devices.

## 5. Acoustic wave based biosensors

Most acoustic wave resonators (QCM, SAW and FBAR) can be used as sensors because all of them are sensitive to mechanical, chemical, optical or electrical perturbations on the surface of the devices [86,87]. They are versatile, sensitive and reliable, being able to detect not only mass/density changes, but also viscosity, elastic modulus, conductivity and dielectric properties etc. They have many applications such as sensing pressure, humidity, temperature, strain (stress), acceleration force, vibration, flow, pH values, radiation, electric fields etc. They are sensitive gas sensors once combined with specific gas absorption layers [88,89,90]. Development of acoustic wave based biosensors is relatively new but has a huge potential as they can detect tiny traces of biomolecules [22,23]. This can be utilized to detect viruses and genetic disorders, diagnose early stage diseases and cancers [91,92]. The principle of acoustic biosensors is similar to those of other biosensors, which are based on a specific interaction between biomarkers (also called probe molecules) deposited on the surface of the sensors with target molecules in the biosamples. Compared with other common biosensing technologies, such as surface plasmon resonance (SPR), optical fibres, and field effect transistors or cantilever-based sensors, acoustic wave based sensors have the combined advantages of simple operation, high sensitivity, small size, compact and low cost. In the following section, highlight the acoustic wave sensor technologies.

## 5.1. QCM sensors

Although the piezoelectric effect was discovered in the late 19th Century, quartz crystal resonators only found widespread applications in electronics, material and biological researches when it was demonstrated that there was a linear relationship between mass adsorbed on the surface and the resonant frequency of the crystal in 1959 by Sauerbrey [93]. Biosensing became possible when suitable oscillator circuits for operation in liquids were developed [94]. The QCM is one of the most developed biosensors that can be operated in a liquid environment using the thickness shear-mode. A QCM consists of a bulk piezoelectric material with typical dimensions of 1 cm in diameter and 500-1000 μm in thickness, sandwiched between two metal electrodes. When an A.C. electrical signal is applied to the two electrodes, it excites a standing wave between the two electrodes through the PE effect. The operating frequency of the QCM is determined by the thickness of the PE-layers, and is typically 5, 10 and 20MHz. As shown by eq.(1), the sensitivity of acoustic resonators is determined by the square of the frequency and the base mass. With decreasing thickness of the Quartz layer, the frequency of the QCM has been increased significantly, thus its sensitivity has been dramatically increased. There has been intensive research recently to develop thin film based QCMs with operating frequencies of several hundreds of MHz [37,38] which demonstrated their great potential for biosensing with better sensitivity.

For biosensing, QCM biosensors have been used to detect the interaction between protein-protein, DNAs, protein-DNA, viruses, bacteria etc, and demonstrated their usefulness, versatility and robustness with high sensitivity. For practical applications, more QCM sensors have been integrated with other structures and devices such as molecular imprint polymers [95], sensors [96] and microfluidics [97] for multi-task detection and monitoring with better accuracy and more functionality. There are hundreds of published papers on QCM biosensors; a review of QCM biosensors is beyond the scope of this book chapter. Readers can find more information in refs. [98,99].

## 5.2. SAW sensors

SAW devices are not only used for microfluidics, but are very good sensors. Since SAW devices have a much smaller active mass and much higher operating frequency than those of QCMs, the sensitivity of the SAW devices increases dramatically as shown in Fig. 3. Furthermore by using advanced photolithograph technology, especially e-beam writing techniques, it is now possible to fabricate SAW devices with operating frequencies up to the Gigahertz [100,101], and the sensitivity of SAW sensors can be further increased.

The longitudinal mode SAW device has a substantial surface-normal displacement that rapidly dissipates the acoustic wave energy into a liquid, leading to excessive damping, and hence poor sensitivity and noise for biodetection in liquid. Shear horizontal (SH-) mode SAW devices have substantially reduced coupling of acoustic energy into the liquid [102,103], hence they can maintain a high sensitivity in liquids. Consequently SH-SAW devices are suitable for biodetection, especially for "real-time" monitoring of physiological conditions of a patient. To further reduce the base mass of a SAW device to improve the sensitivity, Love wave SAW devices have been developed which consist of a normal SAW device and a thin wave guide layer (typically sub-micrometers) such as $SiO_2$ and polymers on top of the SAW surface. Since the acoustic velocity is slow in the wave guide layers, acoustic waves are trapped in the thin wave guide layer, resulting in a drastically reduced base mass and significantly improved sensitivity and quality of the sensors [104,105]. They are therefore frequently employed to perform biosensing in liquid conditions [106,103].

For LOC applications, integratable thin film SAW sensors are more attractive and desirable. A lot of efforts have been made to develop AlN and ZnO thin film SAW sensors. A ZnO/Si SAW device has been successfully used in the detection of aminohexanoic acid succinimidyl ester (DNP-X) and anti-DNP-KLH antibody [107]. The resonant frequency of the ZnO SAW devices was found to shift to lower frequencies as the PSAs are specifically immobilized on the surface-modified ZnO SAW device. A linear dependence has been measured between the resonance frequency change and the anti-DNP concentration over a range from 2 to 1000 ng/ml, and saturated as the concentration increases further due to the reduction of binding sites [107].

To realize biosensing in liquids with better sensitivity, Love wave SAWs have been studied intensively. ZnO has a shear wave velocity of ~2600 m/s, whereas that of ST-cut-quartz is about 4996 m/s. Therefore, it is reasonable to use ZnO as a guiding layer on substrates of ST-cut quartz to form Love mode biosensors. The other potential substrate materials for Love-mode ZnO sensors include $LiTaO_3$, $LiNbO_3$ and sapphire. A ZnO Love mode device of ZnO/ST-cut quartz has a maximum sensitivity up to $\sim 18.77\times10^{-8}$ $m^2$ s $kg^{-1}$, much higher than that of a $SiO_2$/quartz Love mode SAW device [108,109]. Mchale et al recently reported ZnO/$SiO_2$/Si SAW Love mode sensors with a sensitivity of 8.64 $\mu m^2$/mg [110], which is about 2 to 5 times that of ZnO/$LiTaO_3$ [111] and $SiO_2$/quartz Love sensors [112]. Another promising approach for making a ZnO based Love mode sensor is to use a polymer film (such as PMMA, polyimide, SU-8 or parylene C) on top of the ZnO layer as the guiding layer. However, this layered structure uses a polymer waveguide and has a relatively large attenuation compared with those of solid waveguide layers.

AlN SAW devices have higher acoustic velocities, for example, about 6000 m/s shear velocity for an AlN/Si SAW device, and thus it is desirable to use AlN SAW devices for sensors for high sensitivity. However there are not many reports on AlN based SAW biosensors. The reason could be the difficulties in the deposition of the thick AlN film (>4 μm) required for high quality SAW device fabrication. They normally have a large film stress and poor adhesion with the substrate. Although AlN films deposited to deposit on a $LiNbO_3$ substrate have been reported to form a highly sensitive Love mode sensing devices [113,114].

## 5.3. FBAR sensors

FBARs were initially developed as high frequency resonators for applications in electronics as filters, duplexer etc in 1980s' [25,26]. FBARs have been considered for biosensor application since 2000, and are considered as one of the most advanced sensors with extremely high sensitivity and very small dimensions. As the wavelength of the bulk resonators is determined by the thickness of the PE-layer, it is normal to fabricate FBARs with thin PE-layers. FBARs typically have frequencies of a few GHz, and ones with $f_r$ of 8 GHz or higher have been demonstrated [115,116]. Owing to the small base mass and high operation frequency, attachment of a small amount of target mass is able to induce a large frequency shift – typically a few MHz. This improves the sensitivity and makes the signal easily to be detected using simple electronic circuitry. Although the sensitivity of FBARs is not as good as predicted by eq.(1), it is still about three and two orders of magnitude higher than those of QCMs and SAWs respectively as shown in Fig.4 [28,117].

A ZnO-based label free FBAR biosensor with an operating frequency of 2 GHz was used to detect DNA and protein molecules [115]. It showed a sensitivity of 2400 Hz·cm$^2$/ng, which is approximately 2500 times higher than a conventional QCM device could achieve. A recent Al/ZnO/Pt/Ti FBAR design also showed a sensitivity of 3654 Hz·cm$^2$/ng with a better thermal stability than that of ZnO-based FBARs [118,119].

Based on eq.(1), an increase in $f_r$ would make FBARs of higher sensitivity. However it should be pointed out that the limitation to the sensitivity of acoustic resonators, especially the FBARs, is not only the frequency, but also the quality factor. It is easy to make FBARs of high $f_r$ by using a thin PE layer, but the quality factor of the FBARs was found to decrease dramatically with decrease of the thickness of the PE-layer, mostly due to the relatively poor crystal quality, small grain size, poor thickness uniformity, rough surfaces and the existence of a thick transition layer (20-80 nm), resulting in a severely reduced quality factor when a thin PE layer is used. Also it was found that the electrode shape and material properties also significantly affect the quality factor. Electrodes with 90° regular angles reflect the surface travelling wave and deteriorate the Q-factor. Electrodes with high acoustic impedance and low mass are preferred for fabrication of high performance FBARs. Aluminium is one of the most popular electrode materials in the microelectronics industry. It has a low mass density, but the acoustic impedance is low and thus is not the best material for the fabrication of FBARs. Au, Pt and W have high acoustic impedance but their mass densities are high, leading to large mass loading effects. A carbon nanotube (CNT) layer was found to be the best electrode material for FBARs as it has a high elastic modulus and low density (hence a high acoustic impedance and low mass loading) and CNTs have thus been used for fabrication of FBARs with much improved quality factor. An improvement of Q-value by 5 times was demonstrated simply by using a CNT layer on top of existing metal electrodes [120]. Garcia-Gancedo et al recently fabricated FBARs with CNT layer as the top electrode and demonstrated FBARs with a quality factor over 2000 [121,122], one of the best values reported. Also the spurious ripples round a resonant peak found in Au electrode FBARs disappeared when CNTs were used as shown in Fig. 27. The FBARs with CNT electrodes showed a better sensitivity than the Au electrode ones with a mass detection limit down to $10^{-13}$g, with the potential to go down to $10^{-15}$g, suitable for detecting a single molecule.

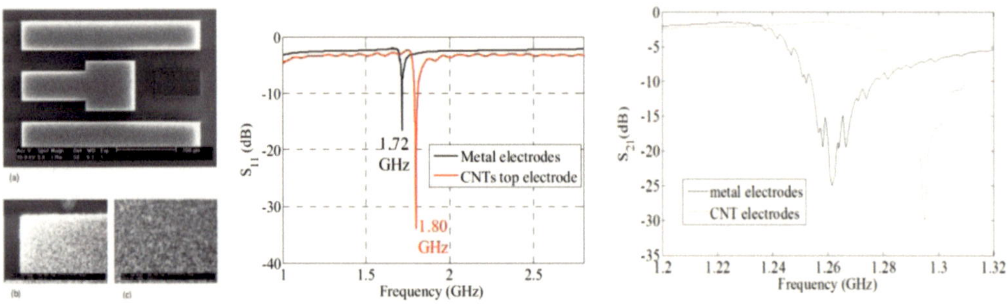


**Figure 27.** (a) SEM images of a fabricated SMR with CNTs layer top electrode. The middle is a typical frequency response, showing the main resonance at 1.75 GHz. The ripples for FBARs made of Au and CNTs top electrode. It disappears in CNTs-FBARs due to high acoustic impedance [122]. Reprinted with permission from Elsevier, Sensors and Actuators, B 160, 2011, 1386.

The majority of the PE thin films have crystal orientation (0002) normal to the surface of the substrates. They are suitable for fabrication of FBARs with longitudinal mode and appropriate for gas phase detection. FBARs with protein functionalized surfaces have been used as a gas phase biosensors, and demonstrated their feasibility for sensing an odorant binding protein of AaegOBP22 using N,N-diethyl-meta-toluamide (DEET) as the ligand to the odorant binding protein [123]. For sensing in liquids, novel structures, especially integrated with microfluidics, is needed. Zhang et al performed biodetection in liquids by taking advantage of the back trench structure. The trench was used as a container in which the bioreaction could take place, while a FBAR on the other side of the thin membrane was used for sensing. This demonstrated its feasibility for biodetection [124]. Similarly Wingqvist et al. used the surface of a FBAR for sensing with a built-in microchannel on the back which allows continuous flow of the biosamples or buffer solution to pass for a continuous measurement [31]. A schematic drawing is shown in Fig. 28.

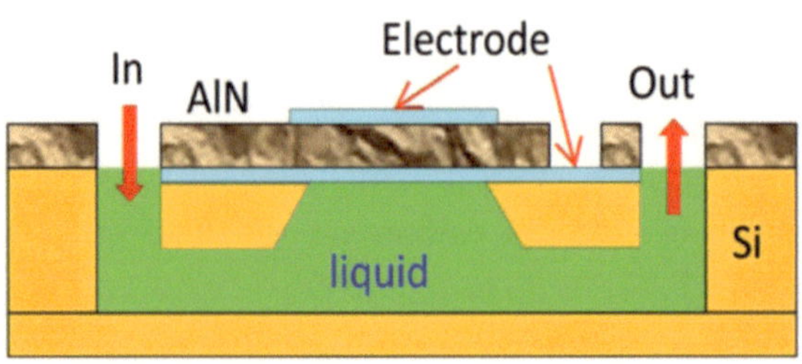


**Figure 28.** Typical structure of lateral field excited inclined AlN film FBAR.

For direct liquid contact sensing, it is necessary to develop ZnO or AlN films with crystal orientation inclined relative to the surface normal, allowing generation of shear waves to be used for detection in a liquid [125]. A (1120) textured ZnO film exhibits pure shear mode waves which can propagate in a liquid with little damping effect. A ZnO shear mode FBAR device has been used in a water-glycerol solution, with a high $f_r$ of 830 MHz and a sensitivity of 1000 Hz·cm²/ng [126]. Weber et al. fabricated ZnO FBAR devices using a ZnO film with 16° off c-axis crystal orientation, which operated in a transversal shear mode [125]. For an avidin/anti-avidin system, the fabricated devices had a high sensitivity of 585 Hz·cm²/ng and a mass detection limit of 2.3 ng/cm². The shear wave FBAR devices also showed a more stable temperature coefficient of frequency [30,31].

Another method for liquid phase biodetection is to use the lateral field excitation (LFE) for FBARs as shown in Fig. 29. This structural FBAR requires both signal and ground electrodes being in-plane and parallel on the exposed surface of the PE films [127,128,129]. Since the excitation is parallel to the surface and perpendicular to the normal c-axis crystal orientation, the FBARs exhibit a thickness shear mode operation. The devices are stable in biologically equivalent environments [130,128]. However, lateral structural FBARs normally have a low quality factor and the mechanism of the exciting resonance is not fully understood yet. A lot of research is needed before they can be effectively used as sensors. Furthermore, lateral FBARs have a very narrow active area, and it is difficult to incorporate microfluidics within the narrow channel for sensing.

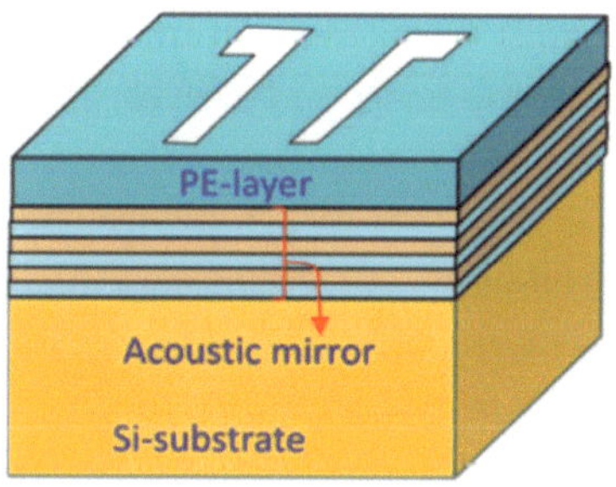


**Figure 29.** Schematical structure for the lateral field excited FBAR devices, which generates a thickness shear mode resonance.

## 5.4. Flexural plate resonant sensors

Lamb wave devices on a membrane structure have been used for biosensing in liquid [34]. Since the propagation velocity of the Lamb wave in the membrane is slower than that in the fluids on the surface, the acoustic energy is not easily dissipated, thus the Lamb wave sensors can be used for thus applications [131]. Since the resonant frequency of the FPW devices is small, it is not sensible to use the frequency as the parameter for sensing due to poor sensitivity as indicated by eq.(1), the amplitude of the resonant wave is normally used for sensing in liquid. Therefore, the sensitivity of these devices increases as the membrane thickness becomes thinner [35, 15].

A ZnO based FPW device has been used to monitor the growth of bacterium "Pseudomonas putida" in a boulus of toluene and the reaction of antibodies in an immunoassay for an antigen present in breast cancer patients [132]. Si/$SiO_2$/$Si_3N_4$/Cr/Au/ZnO FPW devices have been used for detecting human IgE based on the conventional cystamine SAM layer technology with a sentivity of $8.52 \times 10^7$ $cm^2/g$ at a wave frequency of 9 MHz [133]. However, the FPW biosensor has not been widely reported because of the low sensitivity, difficulty of fabrication and high temperature sensitivity of the thin film.

## 6. Other SAW-based functions and lab-on-a-chip

Since the acoustic wave mechanism can be utilized for fabrication of various microfluidic devices and sensors, it would be very attractive to fabricate single acoustic wave mechanism-based lab-on-chip systems [117]. Development of such systems has been rather limited so far, as the individual acoustic technologies are yet to be fully explored, developed and optimized. Current activities have been focused mainly on the development of individual acoustic wave based devices and systems such as acoustic microheater, SAW-based polymerase chain reaction (PCR), SAW-based particle concentrator, sorting and delivery devices etc. These will be highlighted in this section.

### 6.1. SAW microheater

For SAW devices, an input of high RF power will induce acoustic heating through crystal vibration and absorption of acoustic energy by defects in the substrates. For sensing, the input RF signal normally has a low power and acoustic heating is not a problem. For acoustic microfluidics, especially droplet-based pumps, acoustic heat may increase the surface temperature of the SAW device over 100 °C which will damage most of the cells and bio-molecules and reduce their biological integrity. Acoustic heating can be suppressed by using a pulsed RF signal to maintain the temperature below 40 °C. Although acoustic heating has many negative effects for biological and electronic applications, controlled acoustic heating can be utilized as a remote microheater for many applications, such as in polymerase chain reaction (PCR) to amplify DNA concentration for detection or to accelerate bioreaction.

Figure 30 shows the surface temperature as a function of RF signal voltage measured for a ZnO thin film SAW device. The temperature at a position 5 mm away from the IDT on the wave path was monitored. The temperature increases with the signal amplitude and the duration of the RF signal, and decreases with the distance from the IDT [77]. The maximum temperature can reach at 140 °C for a signal voltage of 60 V. The temperature was found to have a distribution along the wave path. It is higher near the IDT, and decreases away from the IDT due to attenuation of the wave. Acoustic heating has been utilized to construct PCR as will be discussed later [74,134]. For normal gas sensing, initialization of sensors requires a high temperature to remove all absorbed substances. Acoustic heating could be utilized for self-initialization for SAW gas sensors without need of an additional microheater, which greatly simplifies the system and reduces the cost and fabrication process. Effective utilization of acoustic heating would generate compact, useful microsystems with many functions, and many applications are yet to be explored.

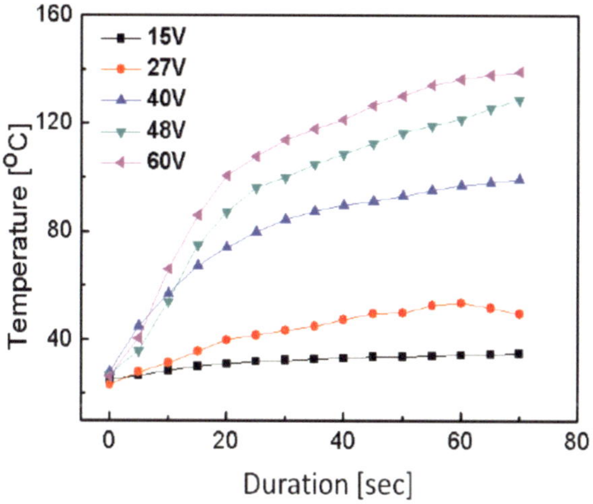


**Figure 30.** Acoustic heating induced temperature rise as a function of RF signal duration by a SAW microheater [77]. Reprinted with permission from AIP, J. Appl. Phys, 105, 2009, 024508.

## 6.2. SAW particle concentrator

Transportation and concentration of particles or bio-substances such as cells are important issues for biological applications. As above, the SAW acting on the edge of a droplet can generate a shear force within the droplet. This shear force can generate circulating streaming, moving particles towards the centre as shown in Fig. 31. From a side view, the fluid can be observed to be pushed upward just above the SAW propagation area which results in primary azimuthal rotation within the droplet periphery. Raghavan et al [135] reported that the flow phenomenon within liquid droplets due to SAW asymmetric positioning are similar to that obtained by the flow field between stationary and rotating disks. This azimuthal rotation phenomenon has been utilized for particle concentration.

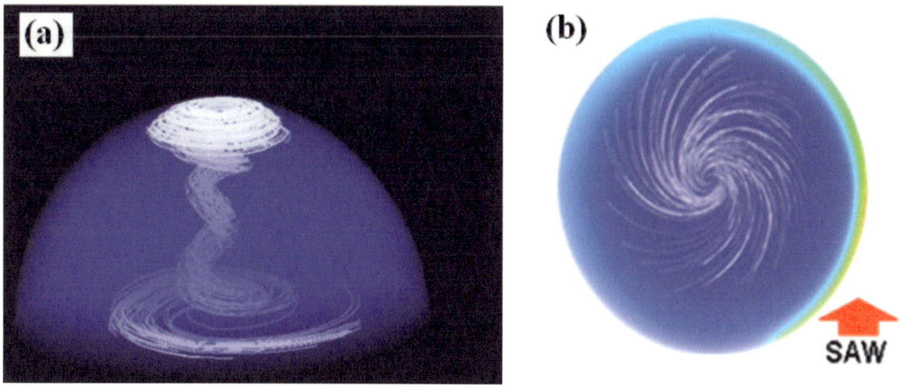


**Figure 31.** Side view (a) and overview (b) of simulated circulating streaming patterns induced by acoustic wave for a 10μl droplet [53].

Figure 32 shows the frame images of starch particles captured during the concentration process within a 20 *μl* droplet after applying an RF signal to the IDT electrode for a few seconds. Initially, the particles are uniformly dispersed in the water, and circulating streaming is induced once the RF signal is applied. The flow circulation rapidly establishes a particle cluster towards the centre of vortex, in the form of a conical shape similar to that depicted theoretically. The shear force induced particle migration is due to the gradient in the azimuthal streaming velocity in the droplet, resulting in particle motion from a higher shear force area at the droplet periphery to a lower shear force area at the bottom of the droplet centre. The shear velocity is large on the edge of the droplet, and gradually decreases on approaching the centre of the droplet. The particles circulate with the liquid in the droplet and simultaneously migrate from the high to the low shear velocity regions [53,136]. The concentration effect is dependent of the RF power as well as the properties of the particles. At a low power, it is not sufficient to generate a gradient in the azimuthal streaming velocity, whereas a high power produces a strong turbulent streaming, dispersing particles within the droplet randomly without any concentration effect [53]. The particle size is also critical for the efficient concentration. Particles with certain sizes can be easily agglomerated, whereas small particles can flow inside the liquid for very long time before being forced into the central region. Therefore it is possible to utilize this to separate particles with different sizes using a SAW device.

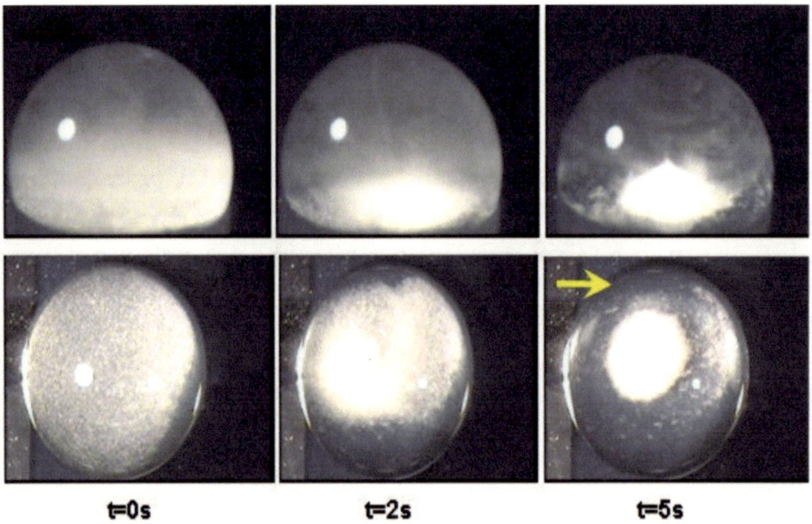


**Figure 32.** Captured video images illustrating the rapid starch particles concentration process for a 30 μl droplet. The first row shows a side view, while the second row a top view. The yellow arrow is the SAW propagation direction [53]. Reprinted with permission from Institute of Physics, J. Microeng. Micromech. 21, 2011, 01500581.

## 6.3. Particle sorting and manipulator

Particle sorting, separation and counting are frequently used in biological and medical analysis. Biological samples contain various cells such as blood plasma, and red and white cells. For analysis, these cells are to be separated and counted. Cell separation and counting are powerful tools being used for quantitative analysis. Acoustic waves can be utilized for particle sorting, separation, counting etc, mostly relying on the nodes and antinodes generated by standing waves [137]. Figure 33 shows the principle of a SAW-based particle sorting device. A pair of IDTs is arranged face to face with a distance between them equal to an integer of the half wavelength, $n\lambda/2$. Upon application of an RF signal, standing waves between the two IDT electrodes can be formed. If a channel is fabricated perpendicular to the wave path of one wavelength wide, the pressure node in the channel will confine the particles within the pressure node, generating streaming with particles confined within the line.

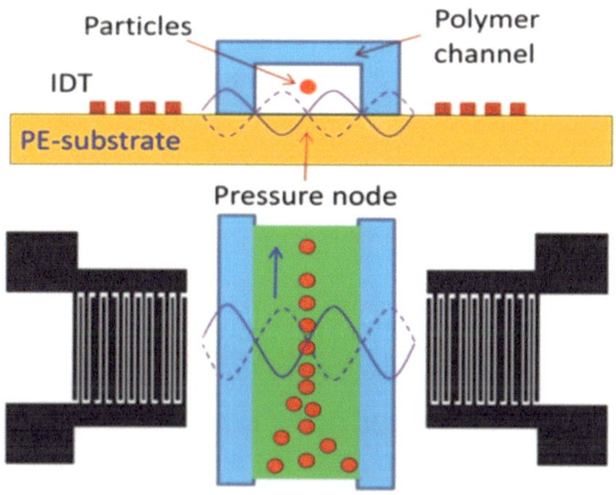


**Figure 33.** Formation of pressure node and anti-node by standing waves used for focusing particles inside a channel. A single wave node in a PDMS/SAW device and focus of the particles in a line.

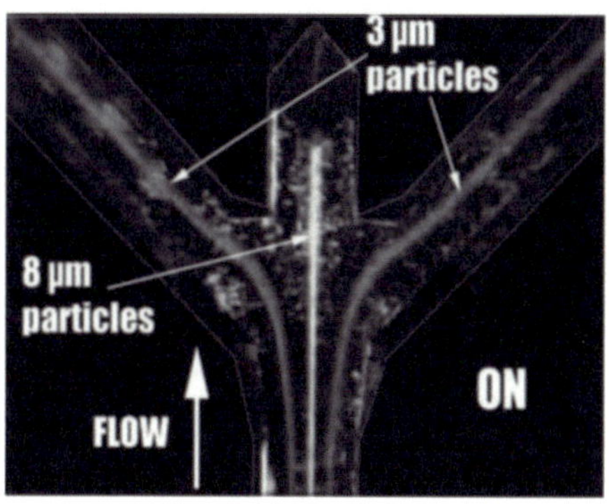


**Figure 34.** Microscope image of microparticle line induced by acoustic wave node [137]. Reprinted with permission from RSC Publishing, Chem. Soc. Rev. 36, 2007, 492.

Shi *et al.* employed this scheme to obtain focused particle lines using PDMS as the microchannel wall [138]. They have demonstrated fluorescent polystyrene particles sorting for a solution with a density of $1.176 \times 10^7$ beads/*ml*. The particle size is around 1.9 μm, and the flow containing the particles passes through a channel, and eventually forms a stable particle flow. By incorporating channels with different exits, particles with different sizes can be separated. Fig. 34 is the photo image of the particles captured by a camera [138].

It should be pointed out that standing waves have a long range force compared with the sizes of the microparticles; therefore it should be considered as a coarse manipulator or tweezers, suitable for simultaneous handling of a large group of particles or cells. It would be difficult to manipulate particles precisely down to the micrometre scale individually. For precision manipulation of particles, it would be best to integrate acoustic wave devices with other mechanisms. Wiklund et al [139] has integrated dielectrophoretic microfluidics mechanisms with ultrasonic particles concentrators. The ultrasonic standing wave delivers a long range force for high through-put particle manipulation, while the short-range dielectrophoretic forces are used for precision control to realize individual cell manipulation for bioanalysis.

Acoustic forces can be utilized to drive particles to places in order to realize scaffolding for cell growth. An in vitro cell culture is a technique used to grow cells in extra-cellular matrices and potentially for organ farming. Successful growth of cells depends on uniform distribution of seed cells into the scaffold of the matrices and the efficiency of the seeding process. The moving of a cell suspension into the scaffold material, typically a polymer, in the absence of external driving forces is exceptionally slow due to the large capillary resistance which may take from hours to days. SAW device can move a droplet with strong internal streaming and agitation as discussed, and can be used to deliver particles into the polymer matrices for cell growth. Yeo at al. have investigated the effect of SAW agitation on the efficiency of suspended fluorescence particles in a polycaprolactone (PCL) scaffold [140]. They demonstrated that efficiency up to 90% can be achieved on a SAW device within a few seconds, and the particles are uniformly distributed within the polymer matrix [140].

### 6.4. Other acoustic wave based functions

The acoustic wave technique has been used for biodegradable polymeric nanoparticle generation [141]. A polymeric incipient was dissolved into a solvent drop, and then atomized by a SAW. The solidified polymeric particles left behind are monodispersed. With this technique, 150–200 nm polymer spherical clusters were formed with sub-50 nm particulates. Periodically ordered polymer has also been patterned on a substrate by SAW atomization [142]. When a polymer solution was spread over the surface of a SAW device with two IDTs perpendicular to each other, the surface displacements induced by the standing waves displace the polymer film, breaking up the film in both the transverse and longitudinal directions producing evenly spaced solidified polymer microstructures. The spaces between the nanostructures in the X and Y-directions are approximately half the SAW wavelength in both directions with a pattern as shown in Fig. 35(left) schematically.

This can also be utilized to fabricate fine microwires of soft matter with nanoparticles uniformly dispersed in a solvent.

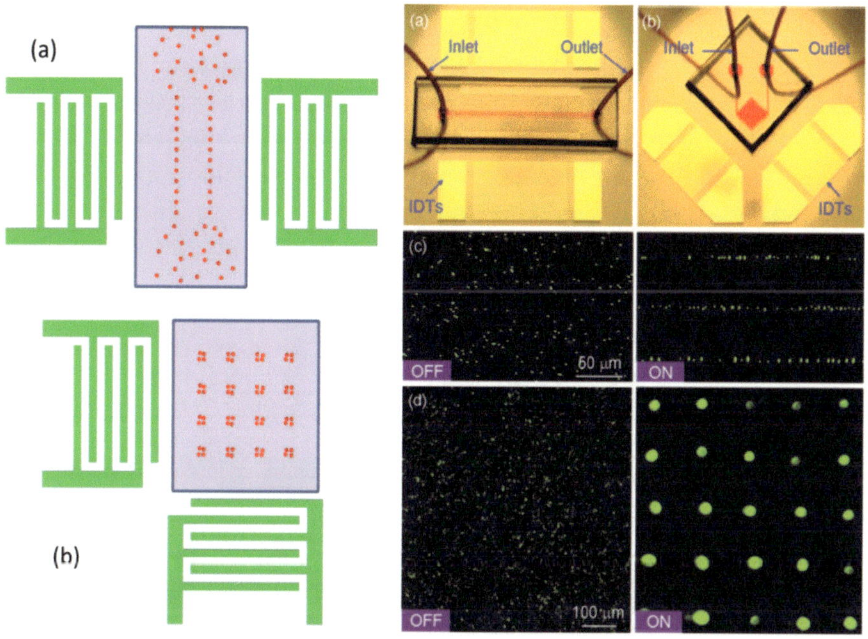


**Figure 35.** (Left) Particle lines formed in a fluid induced by standing waves from a pair of SAW IDTs; (Right) 2D array of particles form by a pair of IDTs in perpendicular [138]. Reprinted of (b) with permission from RSC Publishing, Lab on Chip, 9, 2009, 2890.

## 6.5. SAW based lab-on-a-chip

By using a combination of streaming induced mixing, enhanced biochemical reaction, droplet delivery and cell sorting etc by acoustic waves with other microfluidic and sensing functions, people have realized SAW based LOCs for various applications. A PCR system based on the combination of a SAW and resistance microheaters has been developed [74,134]. A schematic of the suggestion is shown in Fig. 36. The SAW-PCR system is a droplet-based DNA amplifier with droplets of the sample embedded in oil. SAW devices are used for moving and aligning the droplets between the zones with different temperatures through virtual tracks formed by chemical modification of the surface hydrophobicity. Once PCR amplification is completed, the droplet is moved to another heater for hybridization.

The chip is able to perform a fast and specific PCR with a small volume of 200 *nl* within 10 min. A single nucleotide polymorphism (SNP) responsible for the Leiden Factor V syndrome from human blood was successfully amplified by the PCR system and detected [134]. A SAW-based chip has several advantages over microfluidic channel systems. It can avoid the problems of clogging, large pressure drop and vaporization of liquid from the solid surface. Furthermore, the SAW streaming can also help to speed up the binding reaction and to get a more homogeneous fluorescence in hybridization.

Furthermore, attempts have been made to develop stand alone biodetection systems with integrated SAW microfluidics and sensors. In these cases, SAW microfluidics is mainly used for transporting liquid and agitating to minimize non-specific binding and speed-up the reaction. Due to the small dimensions, liquid in microchannels is dominated by a laminar flow, and the biochemical reaction is limited to mass transportation. The process is very slow, and the reaction is incomplete, thus additional agitation to speed up the reaction is therefore required. A SAW device is an ideal planar device to be integrated in the system for these purposes. Figure 37 is a schematic drawing of a SPR detection system with integrated SAW microfluidics [143]. By utilizing the abilities of moving droplets and acoustic streaming by SAW device, a droplet based SPR system for real-time sensing was realized.

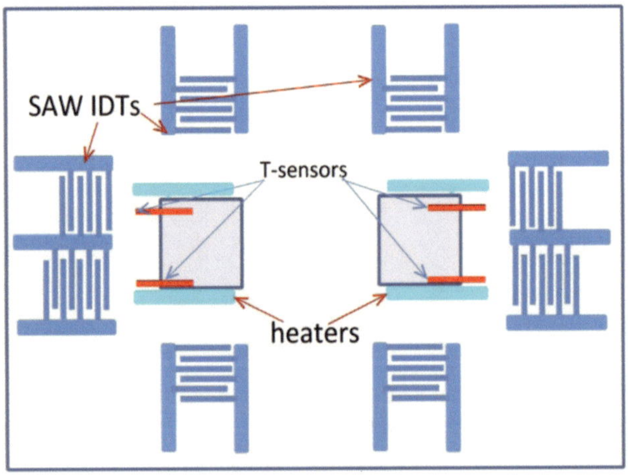


**Figure 36.** A schematic drawing the SAW based PRC system.

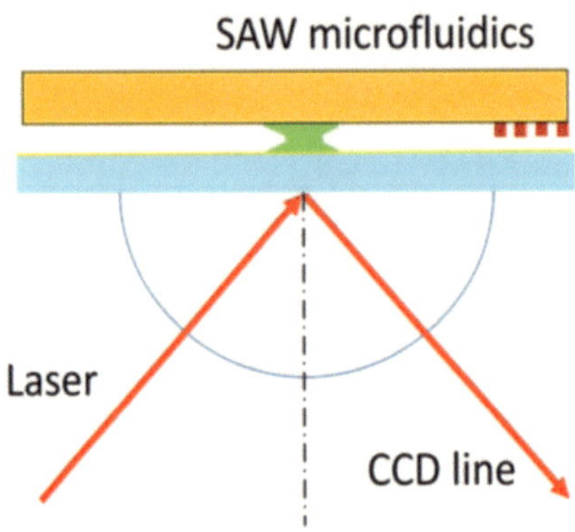


**Figure 37.** Schematic drawing of the set-up of the droplet-based SPR with SAW integrated for streaming to allow the real time monitoring of interactions.

## 7. Conclusions

Lab-on-a-chip systems for medical research, drug development and healthcare etc typically consist of a set of microfluidics and sensors. In most cases, the mechanisms for microfluidics and sensors are different and these devices are mostly assembled together to form LOCs, making the LOCs big and difficult to operate. Acoustic wave based lab-on-a-chip systems are a miniaturized microsystems with typical sizes of few square centimetres. LOCs may provide a single or multi-function such as transporting biosamples and sensing on a single chip.

Bulk and surface acoustic waves have found tremendous applications in LOCs. Surface acoustic waves have strong forces and can be utilized for fabrication of micropumps, mixers, droplets and mist generators for handling liquids and biosamples effectively and efficiently. On the other hand, bulk and surface acoustic wave based resonators (QCMs, SAWs and FBARs) are extremely sensitive to traces of absorbed mass and hence can be utilized for development of high sensitivity biosensors. Also acoustic waves can be utilized for generating many other functions such as remote heating, cell concentration and delivery. These unique functions of acoustic waves make it possible to develop lab-on-a-chip systems with a single actuation and sensing mechanism. Furthermore all these acoustic devices can be realized by using thin film technology, hence opening the way for integration of acoustic wave based LOCs with Si-based electronics on the same substrate.

## Author details

J. K. Luo
*Dept. Info. Sci. & Electron. Eng., Zhejiang University, China*
*Inst. Of Renew. Energy & Environ. Technol., Bolton University, UK*

Y. Q. Fu
*Thin Film Centre, University of West of Scotlant, Paisley, Scotland*

W. I. Milne
*Dept. of Eng. University of Cambridge, UK*

## Acknowledgement

The authors would like to acknowledge the partial financial support from the Leverhulme Trust under the grant No. of F/01431, The Knowledge Centre for Materials Chemistry under Grant No. of X00680PR, Canergie Trust Funding and Royal Society of Edinburgh, the Natural Science Foundation of China (No. 61274037) and Zhejiang Provincial Natural Science Foundation, China (No. Z1110168).

## 8. References

[1] Nguyen N T, Huang X and Chuan T K; J. Fluids Eng. 124; 2002; 384.
[2] Laser D J and Santiago J G; J. Micromech. Microeng. 14; 2004; R35.
[3] Woias P; Sens. Actuat. B105, 2005, 28.
[4] Truong T Q and Nguyen N T; J. Micromech.Microeng. 14, 2003, 632.
[5] Cho S K, Moon H and Kim C-J; J. Microelectromech. Syst. 12, 2003, 70.

[6] Lacharme F. and Gijs M A; SensorsActuators A 117, 2006, 384.
[7] Darabi J, Rada M, Ohadi M and Lawler J; J. Microelectomech. Syst. 11, 2002, 684.
[8] Jones T B; J. Electrostat. 51/52, 2001, 290.
[9] Pollack M G, Fair R B and Shenderov A D; Appl. Phys. Lett. 77, 2000, 1725.
[10] Pollack M G, Shenderov A D and Fair R B; Lab Chip 2, 2002, 96.
[11] Torkkeli A, Saarilahti J, Haara A, Harma H, Soukka T and Tolonen P; Proc. IEEE MEMS. pp.475–78, 2001.
[12] Mugele F and Herminghaus S; App. Phys. Lett. 81, 2002, 2303.
[13] Luo J K, Fu Y Q, Li Y, Du X-Y, Flewitt A J, Walton A J and Milne W I, J. Micromech. Microeng. 19, 2009, 054001.
[14] Chao C, Cheng C H, Liu Z B, Yang M, Leung W F; Proc. IEEE Int. Ultrason. Symp. 2008, pp.521.
[15] Luginbuhl P, Collins S D, Racine G A, Gretillat M A, De Rooij N F, Brooks K G and Setter N; J. Microelectromech. Syst. 6, 1997, 337.
[16] Kwon J W, Yu H, Zou Q, Kim E S; J. Micromech. Microeng. 16, 2006, 2697.
[17] Wixforth A, Strobl C, Gauer C, Toegl A, Scriba J and Guttenberg Z V; Anal. Bioanal. Chem. 379, 2004, 982.
[18] Friend J R and Yeo L Y, Rev. Modern Phys. 83, 2011, 647.
[19] Yeo L Y and Friend J R, Biomicrofludics, 33, 2009, 012002.
[20] Fu Y Q, Luo J K, Du X Y, Flewitt A J, Li Y, Markx G H, Walton A J and Milne W I, Sens. Actuat. B143, 2010, 606.
[21] Lange K, Rapp B E and Rapp M; Anal. Bioanal. Chem. 391, 2008, 1509.
[22] Shiokawa S, Kondoh J; Jap. J. Appl. Phys., 43, 2004, 2799.
[23] Ballantine D S Jr, White R M, Martin S J, Ricco A J, Frye G C, Zellars E T, Wohltjen H; 1997. Acoustic Wave Sensor—Theory, Design, and Physico-Chemical Applications, Academic Press, San Diego.
[24] Buttry D A and Ward M D; Chem. Rev. 92, 1992, 1355.
[25] Lakin K M and Wang J S; Appl. Phys. Lett. 38, 1981, 125.
[26] Lakin K M; IEEE Trans. Ultrason. Ferroelectr. & Freq. Control, 52, 2005, 707.
[27] Ruby R; Proc. IEEE Ultrasonics Symp., 1-6, pp.1029-1040, 2007
[28] Rey-Mermet S, Bjurstrom J, Rosen D and Petrov I; IEEE Trans. Ultrason. Ferroelectric and Freq. Control; 51, 2004, 1347.
[29] Kang Y R, Kang S R, Paek K K, Kim Y K, Kim S W and Ju B K, Sens. & Actuat. A117, 2005, 62.
[30] Link M, Schreiter M, Weber J, Gabl R, Pitzer D, Primig R, Wersing W, Assouar M B and Elmazria O; J. Vac. Sci. Technol. A24, 2006, 218.
[31] Wingqvist G, Bjurstrom J, Liljeholm L, Yantchev V and Katardjiev I; Sens. & Actuat. B123, 2007, 466.
[32] Nguyen N T and White R MSens. & Actuat. 77, 1999, 229.
[33] Meng A H, Nguyen N T and White R M, Biomed. Microdevice, 2:3, 2000, 169.
[34] Muralt P, Ledermann N, Baborowski J; IEEE Trans. Ultrason. Ferroelectr. Freq. Control. 52, 2005, 2276.

[35] Nguyen N T and White R M; IEEE Trans. Ultrason. Ferroelect. Freq. Control. 47, 2000, 1463.
[36] Lucklum R and Hauptmann P; Meas. Sci. Technol. 14, 2003, 1854.
[37] Kao P, Doerner S, Schneider T, Allara D, Hauptmann P and Tadigadapa S, J. Microelectromech. Syst. 18, 2009, 522.
[38] Kao P and Tadigadapa S, Sens. & Actuat. A149, 2009, 189.
[39] Wu T T and Wang W S; J. Appl. Phys. 96, 2004, 5249
[40] Benetti M, Cannatà D, Pietrantonio F D and Verona E, Proc. IEEE Ultrason. Symp. pp.1738, 2003.
[41] Hachigo A, Nakahata H, Itakura K, Fujii S and Shikata S, Proc. IEEE Ultrason. Symp. pp.325, 1999.
[42] Wixforth A; J. Associat. Lab. Automat., 11, 2006, 399.
[43] Josse F, Bender F and Cernosek R W; Anal. Chem. 73, 2001, 5937.
[44] Rayleigh L; On the circulation of air observed in Kundt's tubes, and on some allied acoustical problems, Trans. R. Soc. 25, 1884, 224.
[45] Westervelt P J; J. Acoust. Soc. Am. 25, 1953, 60.
[46] Nyborg W L; Acoustic streaming Physical Acoustics, Edited by W P Mason (New York: Academic) 1965, pp.265–331
[47] Shiokawa S, Matsui Y and Ueda T; Proc. Ultrasonic Symp. pp. 645, 1989.
[48] Nyborg W L Acoustic streaming Nonlinear Acoustics, Edited by M F Hamilton and D T Blackstock (New York: Academic), pp.207–331, 1998
[49] Wixforth A; Superlattices & Microstruct. 33, 2004, 389.
[50] Shiokawa S, Matsui Y and Morizum T, Jpn. J. Appl. Phys. 28, 1989, 126.
[51] Shiokawa S, Kondoh J; 2004. Jap. J. Appl. Phys. 43 2004, 2799.
[52] Frommelt T, Gogel D, Kostur M, Talkner P, Hanggi P; IEEE Transact. Ultrasonics, Ferroelectrics & Freq. Control. 55, 2008, 2298.
[53] Alghane M, Fu Y Q, Li Y, Luo J K, Bobbili B, Feng Y, Liu Y F, Hao Z C, Chen B X, Markx G, Wang C H and Walton A J; J. Microeng. Micromech. 21, 2011, 015005.
[54] Campbell J C and Jones W R; IEEE Trans. Sonics Ultrason. 15, 1968, 209.
[55] Trujillo F J and Knoerzer K, Proc. 7th Int. Conf. CFD in minerals & Process Ind. CSIRO, Melbourne, Australia, 9, Dec. 2009.
[56] Luo J K, Fu Y Q, Li Y F, Du X Y, Flewitt A J, Walton A J, Milne W I; J. Micromech. Microeng.19, 2009, 54001.
[57] Sritharan K, Strobl C J, Schneider M F, Wixforth A and Guttenburg Z; Appl. Phys. Lett. 88, 2006, 054102.
[58] Shilton R, Tan M K, Yeo L Y and Friend J R; J. Appl. Phys.104, 2008, 014910.
[59] Nakamura H, Yamada T, Ishizaki T and Nishimura K; IEEE Trans. On Ultrasonics. Ferroelectric. Freq. Control. 49, 2004, 761.
[60] Lehtonen S, Plessky V P, Hartmann C S and Salomaa M; IEEE Trans. Ultras. Ferroelectri. Freq. Control. 51, 2004, 1697.
[61] Du X Y, Fu Y Q, Tan S C, Luo J K, Flewitt A J, Milne W I, Lee D S, Maeng S, Kim S H, Park N M, Park J and Choi Y J; Appl. Phys. Lett. 93, 2008, 094105.
[62] Toegl A, Kirchner R, Gauer C, Wixforth A; J. Biomed. Technol. 14, 2003, 197.

[63] Wixforth A, Strobl C, Gauer C, Toegl A, Sciba J, Guttenberg Z V; Anal. Biomed. Chem. 379, 2004, 982.
[64] Newton M I, Banerjee M K, Starke T K, Bowan S M, McHale G; Sensor & Actuat. 76, 1999, 89.
[65] Chono K, Shimizu N, Matsu Y, Kondoh J, Shiokawa S; Jap. J. Appl. Phys. 43, 2004, 2987.
[66] Murochim N, Sugimoto M, Matui Y, Kondoh J; Jap. J. Appl. Phys. 46, 2007, 4754.
[67] Renaudin A, Tabourier P, Zhang V, Camart J C, Druon C; Sensor & Actuat. B113, 2006, 387.
[68] Toegl A, Scribe J, Wixforth A, Strobl C, Gauer C, Guttenburg Z V; Anal. Bioanal. Chem. 379, 2004. 69.
[69] Franke T and Wixforth A; Chem Phys Chem. 9, 2008, 2140.
[70] Du X Y, Swanwick M, Fu Y Q, Luo J K, Flewitt A J, Lee D S, Maeng S, Milne W I; J. Micromech. Microeng. 19, 2009, 035016.
[71] Fu Y Q, Garcia-Gancedo L, Pang H F, Porro S, Gu Y W, Luo J K, Zu X T, Placido F, Wilson J I B, Flewitt A J and Milne W I; Biomicrofluidics, 6, 2012, 024105.
[72] Zhang A L, Wu Z Q, Xia X H; Talanta, 84, 2011, 293.
[73] Luong T D, Phan V N and Nguyen N T; Microfluidic Nanofluid.10, 2011, 619.
[74] Wixforth A; JALA, 18, 399, 2006,
[75] D.Beyssen, L.Le.Brizoual, O.Elmazria and P.Alnot; Sens. & Actuat. B118, 2006, 380.
[76] M. Alghane, Y. Q. Fu1, B. X. Chen1, Y. Li, M. P. Y. Desmulliez and A. J. Walton, Microfluid. & Nanoflud. Submitted.
[77] Du X Y, Fu Y Q, Luo J K, Flewitt A J and Milne W I; J. Appl. Phys. 105, 2009, 024508.
[78] Nguyen N T and White R T; Sens. & Actuat. 77, 1999, 229.
[79] Moroney R M, White R M, Howe R T; Appl. Phys. Lett. 59, 1991, 774.
[80] Ogawa J, Kanno I, Kotera H, Wasa K, Suzuki T; Sens. & Actuat. A52, 2009, 211.
[81] Li Y, Flynn B W, Parkes W, Liu Y, Feng Y, Ruthven A D, Terry J G, Haworth L I, Bunting A, Stevenson J T M, Smith S, Bobbili P, Fu Y Q and Walton A J; Proc. European Solid State Device Res. Conf. pp.14-18, 2009.
[82] Li Y, Fu Y Q, Brodie S, Mansuor R and Walton A, Biomicrofluid. 6, 2012, 012812.
[83] Bennès J, Alzuaga S, Ballandras S, Chérioux F, Bastien F and Manceau J F; Proc. IEEE Ultrason. Symp. Rotterdam, pp.823-826, 2005.
[84] Tan M K, Friend J R and Yeo L Y; Phys. Rev. Lett. 103, 2009, 024501.
[85] Qi A, Friend J R and Yeo L Y; Proc. 2nd Micro/nanoscale heat & mass Transf. Int Conf. Shanghai, pp.1-8, 2009.
[86] Lucklum R and Hauptmann P; Meas. Sci. Technol. 14, 2003, 1854.
[87] Grate W J, Martin S J, White R M; Anal Chem. 65, 1993, 940.
[88] Cote G L, Lec R M, Pishko M V; IEEE Sens. J. 3, 2003, 251.
[89] Kuznestsova L A and Coakley W T; Biosens. & Bioelectron. 22, 2007, 1567.
[90] Teles F R R and Fonseca L P; Talanta, 77, 2008, 606.
[91] Vellekoop M J; Ultasonics. 36, 1998, 7.
[92] Gizeli E; Smart. Mater. Struct. 6, 1997, 700.
[93] Sauerbrey G. Verwendung von Schwingquarzen zur Wagung dunner Schichten und zur Microwagang. Z. Phys. 155, 1959, 206.

[94] Nomura T, Okuhara M; Anal. Chim. Acta. 142, 1982, 281.
[95] Lin T Y, Hu C H and Chou T C; Biosens. & Bioelectron. 20, 2004, 75.
[96] Maturos T, Wong-ek K, Sangworasil M, Pintavirooj C, Wisitsora-at A and Tuantranont A; Proc. Int. Conf. Electric/Electron. Eng. Comput, Telecom. & Info Tech. Vol.1, 478, 2009.
[97] Han J H, Zhang J P, Xia Y T, Li S H and Jiang L; Colloid. & Surf. A: Physicochem. Eng. Aspects 379, 2011, 2.
[98] Dickert F L, Hayden o, Blindeus R, Mann K J, Blaas D and Waigmann E, Anal Bioanal Chem, 378, 2004, 1929.
[99] Wolfbeis O S, Editor, Springer Series on Chemical Sensors and Biosensors, Methods and applications, 2007, Volume 5, Part C, 371-424,
[100] Wang Q J, Pflugl C, Andress W F, Ham D, Capasso F and Yamanishi M; J. Vac. Sci. Technol. B26, 2008, 1848.
[101] Kirsch P, Assouar M B, Elmazria O, Mortet V and Alnot P; Appl. Phys. Lett. 88, 2006, 223504.
[102] Barie N and Rapp M; Biosensors & Bioelectron. 16, 2001, 978.
[103] Kovacs G and Venema M; Appl. Phys. Lett. 61, 1992 639.
[104] Josse F, Bender F, Cernosek R W; Anal. Chem. 73, 2001, 5937.
[105] Mchale F G; Meas. Sci. Technol, 14, 2003, 1847.
[106] Lindner G; J. Phys. D. 41, 2008, 123002.
[107] Lee D S, Lee J H, Luo J K, Fu Y Q, Milne W I, Maeng S and Jung M Y; J. Nanosci. Nanotechnol. 9, 2009, 7181.
[108] Jian S J, Chu S Y, Huang T Y and Water W; J. Vac. Sci. Technnol. A22, 2004, 2424.
[109] Krishnamoorthy S and Iliadis A A; Solid-State Electronics, 50, 2006, 1113.
[110] Mchale G, Newton M I, Martin F; J. Appl. Phys. 91, 2002 9701.
[111] Powell D A, Zadeh K K, Wlodaiski W; Sens. & Actuat. A115, 2004, 456.
[112] Du J and Harding G L ; Sens. & Actuat. A65, 1998, 152.
[113] Kao K S, Cheng C C, Chen Y C and Lee Y H; Appl. Phys. A76, 2003,1125.
[114] Kao K S, Cheng C C, Chen Y C, et al., Appl. Surf. Sci. 230, 2004, 334.
[115] Gabl R, Feucht H D, Zeininger H, Eckstein G and Wersing W; Biosens. Bioelectron. 19, 615 (2004).
[116] Lanz R and Muralt P; IEEE Trans. Utralson, Ferroelectric & Freq. Control, 52, 2005, 936.
[117] Luo J K, Ashley G M, Garcia-Gancedo L, Kirby P B, Flewitt A J and Milne W I; Int. J. Nanomanufact. 7, 2011, 448.
[118] Lin R C, Chen Y C, Chang W T, Cheng C C, Koo K S; Sens.& Actuat. A147, 2008, 425.
[119] Yan Z, Song Z, Liu W, Appl. Surf. Sci. 253, 2007, 9372.
[120] Muller A, Neculoiu D, Vasilache D, Konstantinidis G, Grenier K, Dubuc D, Bary L, Plana R and Flahaut E; Appl. Phys. Lett. 89, 2006, 143122.
[121] García-Gancedo L, Al-Naimi F, Flewitt A J, Milne W I, Ashley G M, Luo J K, Zhao X B and Lu J R; IEEE Trans Ultrason. Ferroelectric & Freq. Control. 58, 2011, 2438.
[122] García-Gancedo L, Zhu Z, Iborra E, Clement M, Olivares J, Flewitt A J, Milne W I, Ashley G M, Luo J K, Zhao X B and Lu J R; Sens. & Actuat. B160, 2011, 1386.

[123] Zhao X B, Ashley G M, Garcia-Gancedo L, Jin H, Luo J K, A J Flewitt and Lu J R; Sens. & Actuat. B163, 2012, 242.

[124] Zhang H, Marma M S, Kim E S, McKenna C E and Thompson M E; J. Micromech. Microeng. 15, 2005, 1911.

[125] Weber J, Albers W M, Tuppurainen J, Link M, Gabl R, Wersing W, Schreiter M; Sensors & Actuat. A128, 2006, 84.

[126] Link M, Webber J, Schreiter M, Wersing W, Elmazria O, Alnot P; Sensors & Actuat. B121, 2007, 372.

[127] Ho G K, Abdolv R and Ayazi F; Digest of MEMS 2007, Kobe, Japan, pp.791-795, 2007.

[128] Corso C D, Dickherber A and Hunt W D; J. Appl. Phys. 101, 2007, 054514.

[129] Ho G K, Abdolv R, Sivapurapu A, Humad S and Ayazi F; J. Microelectromech. Syst. 17, 2008, 512.

[130] Dickherber A, Corso C D and Hunt W D; Sens. & Actuat. A144, 2008, 7.

[131] Wenzel S and White R; Sens. & Actuat. A21–23, 1990, 700.

[132] White R M; Faraday Discuss. 107, 1997, 1.

[133] Huang I Y and Lee M C; Sens.& Actuat. B132, 2008, 340.

[134] Guttenberg Z, Muller H, Habermuller H, Geisbauer A, Pipper J, Felbel J, Kielpinski M, Scriba J and Wixforth A, Lab Chip. 5, 2005, 308.

[135] Raghavan R V, Friend J and Yeo L Y; Microfluid. Nanofluid. 8, 2009, 0452.

[136] Wood C D, Evens S D, Cumingham J E, O'Rorke R, Walti C and Davies A G, Appl. Phys. Lett. 92, 2008, 044104.

[137] Laurell T, Petersson F and Nilsson A; Chem. Soc. Rev. 36, 2007, 492.

[138] Shi J J, Ahmed D, Mao X L, Lin S C S, Lawit A and Huang T J; Lab Chip. 9, 2009, 2890.

[139] Wiklund M, Gunther C, Lemor R, Jager M, Fuhr G and Hertz H M; Lab Chip. 6, 2006,1537.

[140] Li H Y, Friend J R and Yeo L Y; Biomaterials, 28, 2007, 4098.

[141] Friend J R, Yeo Y L, Arifin D R and Mechler A; Nanotechnol. 19, 2008, 145301.

[142] Alvarez M, Friend J R and Yeo L Y; Langmuir, 24, 2008, 10629.

[143] Galopin E, Beaugeois M, Pinchemel B, Camart J, Bouazaoui M and Thomy V; Biosens. & Bioelectron. 23, 2007, 746.

Section 7

# Acoustics in the Oceans

Chapter 22

# Underwater Acoustics Modeling in Finite Depth Shallow Waters

Emerson de Sousa Costa, Eduardo Bauzer Medeiros and João Batista Carvalho Filardi

Additional information is available at the end of the chapter

http://dx.doi.org/10.5772/56132

## 1. Introduction

Ocean Acoustics is the science which studies the sound in the sea and covers not only the study of sound propagation, but also its masking by the phenomena of acoustic interference [1].

Recent developments in underwater acoustic waves modeling have been influenced by changes in global geopolitics. These changes are evidenced by strategic shifts in military priorities as well as by efforts to transfer defense technologies to non-defense applications.

Despite the restrictiveness of military security, an extensive body of relevant research accumulated in the open literature, and much of this literature addressed the development and refinement of numerical codes that modeled the ocean as an acoustic medium [2].

One of the most important properties of the oceans as far practical applications are concerned lies in their high sensitivity to the propagation of acoustic signals with frequencies in the range of 1Hz to 20kHz that, different types of electromagnetic radiation, bring together a significant amount of information on the marine environment [3]. Another reason for the practical interest in acoustic propagation in the ocean is the distance the sound can spread, reaching several hundred kilometers.

Some properties of the seabed, such as the propagation velocities and compressional attenuation, density, among others, contribute to the spread in shallow waters significantly, making it interesting to perform a quantitative estimation of their values.

Underwater acoustic models are designed to simulate in detail the acoustic wave characteristics, thus enabling the prediction of the of the relevant phenomena behaviour. However a number of limitations are inherent to these models and often have to do with the medium characteristics, e.g., depth variation, number of degrees of freedom, just to mention a few. Other effects such as dispersion, are influenced by a different set of conditions such as surface irregularities, presence of substances derived from natural or artificial, and others.

## 2. Shallow waters

Acoustic wave propagation in conditions differing from the ideal infinite conditions for wave propagation, normally described as shallow water environments will be discussed in the following text.

### 2.1. What is shallow water

The term "shallow waters" is used when the ocean environment model is restricted by its surface at the top and by the seabed at the bottom. An important feature of this configuration is to allow the trapping of sound energy between these two interfaces which also favours the propagation of sound over long distances.

The existing criteria for defining the regions of what is "shallow" is based not only on the properties of sound propagation in the medium, but mainly by the frequency of the sound source and the interactions of sound with the background, resulting in a ratio linking the wavelength with the dimensions of the waveguide. Moreover, according to the hypsometric criterion [2], related to the depths, we define "shallow" as the waters of the continental shelf[1]. Since the average depth of the platform along the slope is usually found to be around 200m, the regions of "shallow" are defined as having depths less than 200m.

Additionally, ocean areas beyond the continental shelf can be considered to be "shallow" when the propagation of a signal with very low frequencies is accompanied by numerous interactions with the surface and the bottom Also, in practical terms, for a given frequency, " water regions are considered to be "shallow when the "shallow boundaries and reflective effects have a majour effect on the propagation and the energy is distributed in the form of a cylindrical divergence, getting trapped between the surface and the bottom.

### 2.2. Model of a shallow-water sound channel

The shallow-water acoustic communication channel can be classified as a multipath fading channel. It generally exhibits a long multipath delay spread, which can lead to intersymbol interference (ISI) if the spread exceeds the symbol time of communication system.

The main characteristic of sound propagation in the "shallow" is the profile setting the speed of sound, which usually has a negative or approximately constant gradient along the depth. This means that the spread over long distances due almost exclusively to the interactions of sound with the bottom and surface. Because each reflection at the bottom there is a large attenuation, spread over long distances is associated with large losses of acoustic energy [2].

[1] The ocean shelf is the zone around a continent, stretching from the low-water line to depths at which there is a sharp increase in the slope of the bottom in the direction of great depths [5].

The emission frequency of the source is also a crucial parameter. As in most regions of the ocean the bottom is composed of acoustic energy absorbing material, this will become more transparent to the energy in waves of low frequencies, which reduces the energy trapped in the waveguide Thus, for the lower frequencies, greater penetration of sound in the background is observed and therefore, exhibiting a greater dependence of propagation in relation to the parameters geoacoustics. At high frequencies (> 1kHz), sensitivity to the roughness of the interfaces and the marine life is greater, resulting in a greater spread, that is, a lower penetration of the bottom and a larger volume attenuation [4]. Spread over long distances therefore occurs in the range of intermediate frequencies (100 Hz to about 1 kHz) and is strongly dependent on the depth and the mechanisms of attenuation. Figure 1 shows the attenuation of sound absorption in seawater as a function of frequency. According to [2], the dependence with frequency can be categorized into four major regions, in increasing order of frequency: absorption in the background, the boric acid relaxation, relaxation of magnesium sulfate and viscosity.

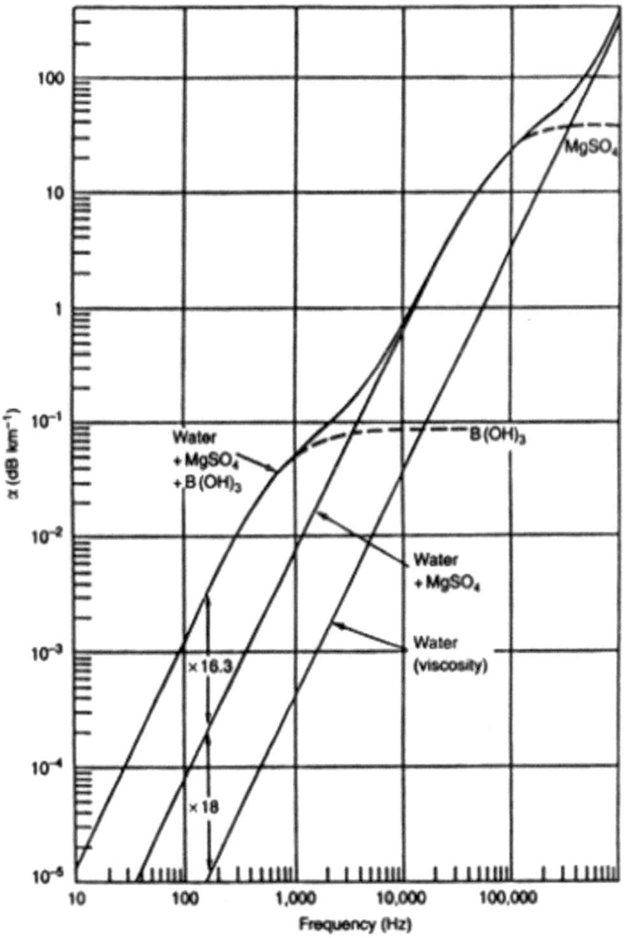


**Figure 1.** Absorption Coefficients for sea water [2].

### *2.2.1. Water layer*

In seawater, the sound speed is measured using special devices or computed from special empirical formulae, using measured values of the temperature, salinity and hydrostatic pressure.

In summer, the sound channel is mainly near the bottom, so that the propagation of sound takes place by sequential reflections of refracting rays from the bottom, causing high propagation losses.

The propagation of sound in winter, takes place either in a channel with a constant sound speed, when it is described by bottom-surface rays, or in a near-surface channel.

The propagation of low-frequency sound is most affected by random inhomogeneities with characteristic sizes ≥ 1m. Random onhomogeneities with vertical scales of 1-10 m and horizontal scales of 100-1000 m are mainly due to the fine thermohaline stricture an to internal waves [5].

In shallow water, the field of internal waves has a number of very specifics characteristics:

- A considerable inhomogeneity and non-stationarity, due to the characteristic trains of intense soliton-like internal waves against a relatively weak background;
- Clearly expressed anisotropy, determined by the bottom relief, when waves propagate mainly in the direction towards the coast, perpendicularly to the outside edge of the shelf;
- Synchronicity of the vertical fluctuations of all fluid layers, attesting to a predominance of the first gravitational mode.

As far the fine thermohaline structure is concerned, according to [5], the ocean is a finely stratified medium, in which there exist layers with thickness from tens of centimeters to tens of meters, with comparatively homogeneous properties, separated from each other by thin boundary layers with sharp changes in the thermodynamic characteristics (the vertical gradients of the physical properties in these layers may be 10-100 times greater than their average values). Using special sound-speed meters with a large resolution, the existence of sharp fine-scale changes in the vertical dependence of the sound speed was established. Roughnesses of the water layer boundaries (roughnesses of the bottom and a disturbed surface) may also have a marked effect on the propagation of sound in shallow water. Losses on the propagation of low-frequency (up to a few kHz) sound are due to various mechanisms (absorption, scattering, geometric divergence). The absorption of sound in clear water, propagation at low frequencies, is physically mainly due to the conversion of the sound energy to heat and is the result of the chemical composition of seawater, which is a complex electrolyte. A change in sound pressure leads to a periodic change in its ionic composition, which affects the volume viscosity. The absorption relaxation mechanism in this case is well described by a formula [5], which yields the absorption coefficient β

$$\beta = \frac{0{,}11 f^2}{1+f^2} + \frac{44 f^2}{4100+f^2} \tag{1}$$

where the frequency is measured in kHz, and the absorption coefficient is given in dB/km. In particular, in the region of interest to us, namely approximately from 100 to 1000 Hz, the coefficient increases monotonically from $10^{-3}$ to 0,06 dB/km.

### 2.2.2. *Layer of sediments*

This layer consists mainly of bottom deposits of the mud type, denser sedimentary rocks or basic rocks (granite, basalt, etc.). Its parameters, like the parameters of the other layers, depend on the geographical region.

In the layer of unconsolidated sediments, one characteristic property is the existence of abrupt random inhomogeneities. These include layered (intermittent and tapered) structures of length up to tens of kilometers, and vertical channels, associated with the venting of gases and diapers (dome-shaped folds, in which rocks with a high plasticity are extruded from below).

In the layer of semi-consolidated sediments, the speed of longitudinal waves is (2 – 3) x $10^3$ m/s, where a small positive gradient with respect to depth is possible. This layer is also absorptive and the absorption coefficients for longitudinal and transverse waves differ and are distinguished by a large spread.

In the layer of consolidated sediments basement is characterized by a high speed of both longitudinal ($c \approx (4 - 6) \times 10^3$ m/s) and transverse ($c_s \approx (1 - 3) \times 10^3$ m/s) waves where the attenuation coefficients are estimated as $\alpha \approx 0{,}1$ dB/(km.Hz), $\alpha_s \approx 0{,}01 - 0{,}1$ dB/(km.Hz).

In the theory of the propagation of sound sediments are seen as a two-component medium, consisting of a solid skeleton and a fluid component. These theories use a large number of parameters (porosity, average grain size, mean-square deviation from the average size, etc.), which determine the acoustic properties of a porous medium. This model allows one to consider the speed and attenuation coefficients of different types of waves. One of the most important characteristics of sediments, which can be calculated, is the frequency dependence of the attenuation coefficient of a longitudinal or transverse wave.

## 2.3. The sound field

Sound propagation in the ocean is conveniently described by the wave equation, having parameters and boundary conditions which are able to describe the ocean environment. There are essentially four types of computational models (computer solutions to the wave equation) normally used to describe sound propagation in the sea: Ray Theory, Fast Field Program (FFP), Normal Mode (NM) and Parabolic Equation (PE). All of these modes enable the ocean environment to vary with the depth. A model that also allows horizontal variations in the environment, i.e., sloping bottom or spatially variable oceanography, is termed "range dependent". For high frequencies (few kilohertz or above), ray theory, the infinite frequency approximation, is still the most practical whereas the other three model types become more and more applicable and useable below, say, a kilohertz [7].

The wave equation for an acoustic field of angular frequency $\omega$ is

$$\nabla^2\phi(r,z) + K^2(r,z)\phi(r,z) = -\delta^2(r - r_s)\delta(z - z_s);\ K^2(r,z) = \frac{\omega^2}{c^2(r,z)}, \tag{2}$$

where the subscript "s" denotes the source coordinates. The range dependent environment manifests itself as a coefficient, $K^2(r,z)$, of the partial differential equation for the sound speed profile and the range dependent bottom type and topography appears as both coefficients (elasticity effects are an added complication) and complicated boundary conditions.

Throughout the theoretical development of these five techniques, the potential function $G$ normally represents the acoustical field pressure. When this is the case, the Transmission Loss (TL) can easily be calculated as:

$$TL = 10\log_{10}\left[\phi^2\right]^{-1} = 20\log_{10}\left[\phi\right]. \tag{3}$$

## 3. Approximate methods in shallow-water acoustics

The various physical and mathematical models all have inherent limitations in their applicability. These limitations are usually manifested as restrictions in the frequency range or in your specification of the problem geometry. Such limitations are collectively referred to as "domains of applicability," and vary from model to model. The model selection criteria are provided to guide potential users to those models most appropriate to their needs [2].

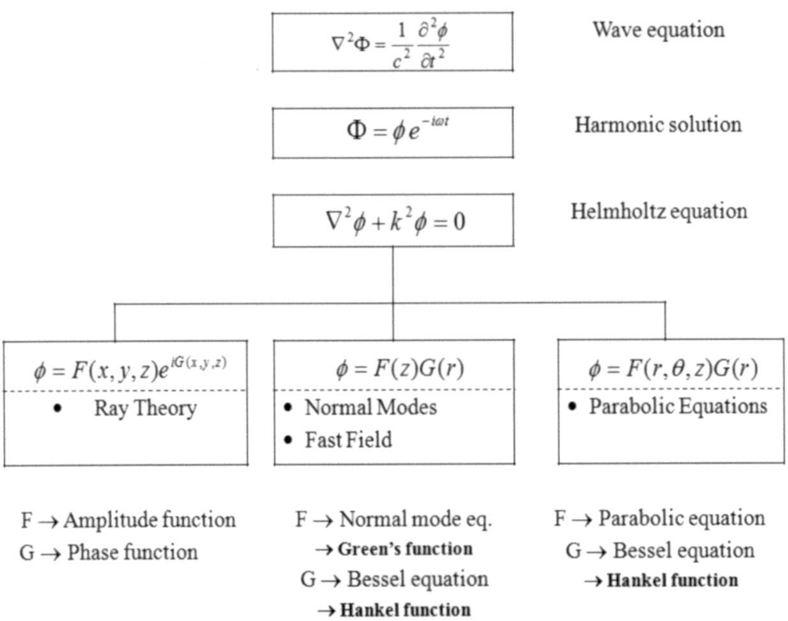


**Figure 2.** Summary relationship among theoretical approaches for propagation modeling.

A further subdivision can be made according to range-independent and range-dependent models. Range independence means that the model assumes a horizontally stratified ocean in which properties vary only as a function of depth. Range dependence indicates that some properties of the ocean medium are allowed to vary as a function of range ($r$) and azimuth ($\theta$) from the receiver, in addition on a depth ($z$) dependence. Such range-varying properties commonly include sound speed and bathymetry, although other parameters such as sea state, absorption and bottom composition may also vary. Range dependence can further be regarded as two dimensional (2D) for range and depth variations or three dimensional (3D) for range, depth and azimuthal variations [2].

### 3.1. Ray-theory models

Ray-theoretical models, a geometrical approximation, calculate TL on the basis of ray tracing. Ray theory starts with the Helmholtz equation. The solution for $\phi$ is assumed to be the product of a pressure amplitude function $A = A(x,y,z)$ and a phase function $P = S(x,y,z)$: $\phi = Ae^{iS}$, where the exponential term allows for rapid variations as a function of range and $A(x,y,z)$ is a more slowly varying "envelope" which incorporates both geometrical spreading and loss mechanisms. Substituting this solution into the Equation (2) and separating real and imaginary terms yields:

$$\frac{1}{A}\nabla^2 A - \left[\nabla S\right]^2 + K^2 = 0 \tag{4}$$

and

$$2\left[\nabla A \cdot \nabla S\right] + A\nabla^2 S = 0. \tag{5}$$

Equation 4 contains the real terms and defines the geometry of the rays. Equation 5, also known as the transport equation, contains the imaginary terms and determines the wave amplitudes. The separation of functions is performed under the assumption that the amplitude varies more slowly with position than does the phase (geometrical acoustics approximation). The geometrical acoustics approximation is a condition in which the fractional change in the sound-speed gradient over a wavelength is small compared to the gradient $c/\lambda$, where $c$ is the speed of sound and $\lambda$ is the acoustic wavelength [2]. Specifically

$$\frac{1}{A}\nabla^2 A << K^2. \tag{6}$$

In other words, the sound speed must not change much over one wavelength. Under this approximation, Equation 4 reduces to

$$\left[\nabla S\right]^2 = K^2. \tag{7}$$

Equation 7 is referred to as the eikonal equation. Surfaces of constant phase ($S$ = constant) are the wavefronts, and the normal to these wavefronts are the rays. Eikonal refers to the acoustic path length as a function of the path endpoints. Such rays are referred to as eigenrays when the endpoints are the source and receiver positions. Differential ray equations can then be derived from the eikonal equation.

The ray trajectories are perpendicular to surfaces of constant phase, $S$, and may be expressed mathematically as follow:

$$\frac{d}{dl}\left[K\frac{dR}{dl}\right]=\nabla K, \tag{8}$$

where $l$ is the arc length along the direction of the ray and $R$ is the displacement vector. One can determine that the direction of average flux (energy) follows that the trajectories and the amplitude of the field at any point can be obtained from the density of rays. Once $S$ is obtained, the Equation 5 yields the amplitude. We mention here, also, that "corrected" ray theory assumes that $A$ can be expanded in powers of inverse frequency-the leading term is the infinite-frequency result with the additional terms being frequency corrections [7].

The ray theory method is computationally rapid, extends to range dependent problems and the ray traces give a very physical picture of the acoustic paths. It is helpful in describing how noise redistributes itself when propagating long distances over paths that include shallow and deep environments and/or mid latitude to Polar Regions. The disadvantage of ray theory is that it does not include diffraction and such effects that describe the low frequency dependence ("degree of trapping") of ducted propagation.

### 3.2. Fast Field Program (FFP)

In the underwater acoustics, fast-field theory is also referred to as "wavenumber integration." Range independent wave theory solves the wave equation exactly when the ocean environment does not change in range. One of the possible derivations of the solution technique is to Fourier decompose the acoustic field an infinite set of horizontal waves,

$$\phi(r,z)=\frac{1}{2\pi}\int_{-\infty}^{\infty}d^2kg(k,z,z_s)e^{ik(r-r_s)}. \tag{9}$$

and from Equation 2, the depth dependent Green's function, $g(k, z, z_s)$, satisfies

$$\frac{d^2g}{dz^2}+\left(K^2(z)-k^2\right)g=-\frac{1}{2\pi}\delta\left(z-z_s\right). \tag{10}$$

Assuming azimuthal symmetry, we can integrate Equation 9 over the angular variable to Hankel functions and their asymptotic form reduces Equation 9 to (for simplicity, we take $r_s = 0$)

$$\phi(r,z)=\frac{e^{-i\pi/4}}{\left(2\pi r\right)^{1/2}}\int_{-\infty}^{\infty}dk(k)^{1/2}g(k,z,z_s)e^{ikr}. \tag{11}$$

Note that the factor $r^{-1/2}$ arises from cylindrical spreading. We now discretize the above integral and transform to a form amenable to the FFT technique by setting $k_m = k_0 + m\Delta k$; $r_n = r_0 + n\Delta r$ where $n, m = 0, 1, \ldots, N-1$. The additional condition $\Delta r\Delta k = 2\pi/N$ and $N$ is an integral power of two. The discretization scheme limits the solution to outgoing waves and Equation 10 becomes

$$\begin{aligned}\phi(r_n,z)&=\frac{\Delta k e^{i(k_0r_n-\pi/4)}}{\left(2\pi r\right)^{1/2}}\sum_{m=0}^{N-1}X_m e^{2\pi imn/N},\\ X_m&=\left(k_m\right)^{1/2}g(k_m,z,z_s)e^{imr_0\Delta k}.\end{aligned} \tag{12}$$

The above equation is now easily evaluated using the FFT algorithm with the bulk of the effort going into evaluating $g$ by solving Equation 10. Although the method is labeled "fast field" it is fairly slow because of the time required to calculate the $g$'s. However, it has advantages when one wishes to calculate the "near field" region or to include shear wave effects in elastic media. Because of this latter capability, it can be used as the propagation component of a description of (micro) seismic noise. The FFP method is often used as a benchmark for others less exact techniques. One such technique, not applicable to the near field but exact for a large class of range independent far-field problems is the computationally faster normal mode method [7].

### 3.3. Normal Mode Model (NM)

Normal-mode solutions are derived from an integral representation of the wave equation. In order to obtain practical solutions, however, cylindrical symmetry is assumed in a stratified medium (i.e. the environment changes as a function of depth only). The solution for the potential function $\phi$ can be written in cylindrical coordinates as the product of a depth function $F(z)$ and a range function $S(r)$:

$$\phi(z,\, r)=F(z)\cdot S(r). \tag{13}$$

Next, a separation of variables is performed using $\xi^2$ as the separation constant. The two resulting equations are:

$$\frac{d^2F}{dz^2}+(k^2-\xi^2)F=0 \tag{14}$$

$$\frac{d^2S}{dr^2}+\frac{1}{r}\frac{dS}{dr}+\xi^2S=0 \tag{15}$$

Equation 14 is the depth equation, better known as the normal mode equation, which describes the standing wave portion of the solution. Equation 15 is the range equation, which describes the traveling wave portion of the solution. Thus, each normal mode can be viewed as traveling wave in the horizontal *(r)* direction and as a standing wave in the depth *(z)* direction [2].

The normal-mode Equation 14 poses an eigenvalue problem. Its solution is known as the Green's function. The range Equation 15 is the zero-order Bessel equation. Its solution can be written in terms of a zero-order Hankel function $\left(H_0^{(1)}\right)$. The full solution for $\phi$ can be expressed by an infinite integral, assuming a monochromatic (single-frequency) point source:

$$\phi = \int_{-\infty}^{\infty} G(z, z_0; \xi) \cdot H_0^{(1)}(\xi r) \cdot \xi \, d\xi \tag{16}$$

where $G$ is the Green's function, $H_0^{(1)}$ a zero-order Hankel function of the first kind and $z_0$ the source depth. Note that $\phi$ is a function of the source depth $(z_0)$ and the receiver $(z)$. To obtain what is known as the normal-mode solution to the wave equation, the Green's function is expanded in terms of normalized mode functions.

The advantages of the Normal Modes procedure are: that once value problem is solved one has the solution for all source and receiver configurations, and, that is easily extended to moderate range dependent conditions using the adiabatic approximation.

### 3.4. Parabolic Equation Model (PE)

The parabolic approximation method was successfully applied to microwave waveguides, laser beam propagating, plasma physics, seismic wave propagation and underwater acoustic propagation.

The PE is derived by assuming that energy propagates at speeds close to a reference speed – either the shear speed or the compressional speed, as appropriate [2].

The PE method factors an operator to obtain an outgoing wave equation that can be solved efficiently as an initial-value problem in range. This factorization is exact when the environment is range independent. Range-dependent media can be approximated as a sequence of range-independent regions from which backscattered energy is neglected. Transmitted fields can then be generated using energy-conservation an single-scattering corrections [2].

The basic acoustic equation for acoustic propagation can be rewritten as:

$$\nabla^2 \phi + k_0^2 n^2 \phi = 0 \tag{17}$$

where $k_0$ is the reference wavenumber ($\omega/c_0$), $\omega(=2\pi f)$ the source frequency, $c_0$ the reference sound speed, $c(r, \theta, z)$ the sound speed in range ($r$), azimuthal angle ($\theta$) and depth ($z$), $n$ the refraction index ($c_0/c$), $\phi$ the velocity potential and $\nabla^2$ the Laplacian operator.

Equation 17 can be rewritten in cylindrical coordinates as:

$$\frac{\partial^2\phi}{\partial r^2}+\frac{1}{r}\frac{\partial\phi}{\partial r}+\frac{\partial^2\phi}{\partial z^2}+k_0^2n^2\phi=0 \tag{18}$$

where azimuthal coupling has been neglected, but the index of refraction retains a dependence on azimuth. Further, assume a solution of the form:

$$\phi=\Psi(r,z)\cdot S(r) \tag{19}$$

and obtain:

$$\Psi\left[\frac{\partial^2 S}{\partial r^2}+\frac{1}{r}\frac{\partial S}{\partial r}\right]+S\left[\frac{\partial^2\Psi}{\partial r^2}+\frac{\partial^2\Psi}{\partial z^2}+\left(\frac{1}{r}+\frac{2}{S}\frac{\partial S}{\partial r}\right)\frac{\partial\Psi}{\partial r}+k_0^2\Psi\right]=0 \tag{20}$$

Using $k_0^2$ as a separation constant, separate Equation 20 into two differential equations as follows:

$$\left[\frac{\partial^2 S}{\partial r^2}+\frac{1}{r}\frac{\partial S}{\partial r}\right]=-Sk_0^2 \tag{21}$$

and

$$\left[\frac{\partial^2\Psi}{\partial r^2}+\frac{\partial^2\Psi}{\partial z^2}+\left(\frac{1}{r}+\frac{2}{S}\frac{\partial S}{\partial r}\right)\frac{\partial\Psi}{\partial r}+k_0^2n^2\Psi\right]=\Psi k_0^2 \tag{22}$$

Rearrange terms and obtain:

$$\frac{\partial^2 S}{\partial r^2}+\frac{1}{r}\frac{\partial S}{\partial r}+k_0^2S=0 \tag{23}$$

which is the zero-order Bessel equation, and:

$$\frac{\partial^2\Psi}{\partial r^2}+\frac{\partial^2\Psi}{\partial z^2}+\left(\frac{1}{r}+\frac{2}{S}\frac{\partial S}{\partial r}\right)\frac{\partial\Psi}{\partial r}+k_0^2n^2\Psi-k_0^2\Psi=0. \tag{24}$$

The solution of the Bessel equation 24 for outgoing waves is given by the zero-order Hankel function of the first kind:

$$S=H_0^{(1)}\left(k_0r\right) \tag{25}$$

For $k_0r >> 1$ (far-field approximation):

$$S \approx \sqrt{\frac{2}{\pi k_0 r}} e^{i(k_0 r - \pi/4)} \tag{26}$$

which the asymptotic expansion for large arguments. The equation for $\Psi(r,z)$ (Equation 24) can be simplified to:

$$\frac{\partial^2 \Psi}{\partial r^2} + \frac{\partial^2 \Psi}{\partial z^2} + 2ik_0 \frac{\partial \Psi}{\partial r} + k_0^2(n^2 - 1)\Psi = 0. \tag{27}$$

Further assume that:

$$\frac{\partial^2 \Psi}{\partial r^2} << 2k_0 \frac{\partial \Psi}{\partial r} \tag{28}$$

which is the paraxial approximation. Then, Equation 27 reduces to:

$$\frac{\partial^2 \Psi}{\partial z^2} + 2ik_0 \frac{\partial \Psi}{\partial r} + k_0^2(n^2 - 1)\Psi = 0. \tag{29}$$

which is the parabolic wave equation. In this equation, $n$ depends on depth ($z$), range ($r$) and azimuth ($\theta$). This equation can be numerically solved by "marching solutions" when the initial field is known [2]. The computational advantage of the parabolic approximation lies in the fact that a parabolic differential equation can be marched in the range dimension whereas the elliptic reduced wave equation must be numerically solved in the entire range-depth region simultaneously Typically, a Gaussian field or a normal-mode solution is used to generate the initial solution.

## 4. Final considerations

A brief description of underwater acoustic propagation models for the "shallow environment has been considered. The choice of appropriate model depends on the simplifications needed for the environment in question. All the discussed models are well known and have been successfully developed by several authors for a variety of conditions.

The application of underwater acoustics is mostly sensor-based, including ocean sampling networks, environmental monitoring, undersea explorations, disaster prevention, assisted navigation, speech transmission between divers, distributed tactical surveillance, and mine reconnaissance.

Acoustical transmission is more flexible than others approaches, as it can be deployed in a wide variety of configurations, including networks consisting of both mobile and stationary nodes. It is not, however, free of complexity, In fact, certain aspects of underwater acoustic communications are more difficult than those RF terrestrial networks, especially high propagation delay. In general, underwater acoustic communications are influenced by transmission loss, bubbles, stratification, multipath propagation, Doppler spread, noise, and high propagation delay.

Transmission loss describes the weakening intensity of sound over a distance and is comprised of losses from both spreading and attenuation. Spreading loss is a geometrical effect that represents the weakening of sound as the wave moves outward from the source. It can be further classified as spherical spreading, cylindrical spreading, or a variant with properties somewhere between the two. Attenuation loss encompasses the effects of absorption, scattering, and leakage out of a sound channel. Absorption, a true loss of acoustic energy that results from the conversion of that energy into heat, accounts for the majority of attenuation. Bubbles produced by breaking waves at the surface can influence the propagation of high frequency signals. No bubble-induced losses were discovered for waves produced with wind speeds of 6 m/s or less [6].

The propagation of sound waves in the ocean is a somewhat complex process, particularly when there are multiple interactions with the seabed, which is often difficult to model. The theory of wave propagation is physical basis for the study of underwater acoustics and descriptive, and this we considered that the proper domain of the theory in simple environments is essential for proper understanding in solving problems realistic sound propagation in shallow water.

## Author details

Emerson de Sousa Costa
*CEFET-MG – Federal Center of Technological Education of Minas Gerais – Brazil*
*Postgraduate Program in Department of Mechanical Engineering – UFMG Brazil*

Eduardo Bauzer Medeiros and João Batista Carvalho Filardi
*Postgraduate Program in Department of Mechanical Engineering –UFMG-Brazil*

## 5. References

[1] Maia, L. P., Acoustic inversion and passive source location in shallow waters Master Thesis, Ocean Engineering. COPPE/UFRJ, Brazil, 2010. *(in Portuguese).*

[2] Etter, P. C., Underwater Acoustic Modeling and Simulation, Spon Press, 2003.

[3] Rodríguez, O. C., Submarine Acoustic Propagation Models: Comparison with the results of the with the analytical solution of the 3 layer problem. Signal processimg laboratory, Universidade do Algarve, Portugal, 1995 *(in Portuguese).*

[4] Xavier, B. C., Shallow Water Acoustic Propagation Models, MSc Dissertation, Ocean Engineering, COPPE/UFRJ, Brazil, (2005). *(in Portuguese).*

[5] Katsnelson, B. G. and Petnikov, V.G., Shallow water acoustics, Springer-Praxis books in geophysical sciences, 2002.

[6] Borowski, B. S., Doctoral Dissertation of Faculty of the Stevens Institute of Technology, 2010.

[7] Kerman, B. R., Sea Surface Sound, 253-272, Kluwer Academic Publishers, 1988.

Chapter 23

# Ray Trace Modeling of Underwater Sound Propagation

Jens M. Hovem

Additional information is available at the end of the chapter

http://dx.doi.org/10.5772/55935

## 1. Introduction

Modeling acoustic propagation conditions is an important issue in underwater acoustics and there exist several mathematical/numerical models based on different approaches. Some of the most used approaches are based on ray theory, modal expansion and wave number integration techniques. Ray acoustics and ray tracing techniques are the most intuitive and often the simplest means for modeling sound propagation in the sea. Ray acoustics is based on the assumption that sound propagates along rays that are normal to wave fronts, the surfaces of constant phase of the acoustic waves. When generated from a point source in a medium with constant sound speed, the wave fronts form surfaces that are concentric circles, and the sound follows straight line paths that radiate out from the sound source. If the speed of sound is not constant, the rays follow curved paths rather than straight ones. The computational technique known as ray tracing is a method used to calculate the trajectories of the ray paths of sound from the source.

Ray theory is derived from the wave equation when some simplifying assumptions are introduced and the method is essentially a high-frequency approximation. The method is sufficiently accurate for applications involving echo sounders, sonar, and communications systems for short and medium short distances. These devices normally use frequencies that satisfy the high frequency conditions. This article demonstrates that ray theory also can be successfully applied for much lower frequencies approaching the regime of seismic frequencies.

This article presents classical ray theory and demonstrates that ray theory gives a valuable insight and physical picture of how sound propagates in inhomogeneous media. However, ray theory has limitations and may not be valid for precise predictions of sound levels, especially in situations where refraction effects and focusing of sound are important. There exist corrective measures that can be used to improve classical ray theory, but these are not discussed in detail here. Recommended alternative readings include the books. [1-4] and the articles [5-6].

A number of realistic examples and cases are presented with the objective to describe some of the most important aspects of sound propagation in the oceans. This includes the effects of geographical and oceanographic seasonal changes and how the geoacoustic properties of the sea bottom may limit the propagation ranges, especially at low frequencies. The examples are based on experience from modeling sonar systems, underwater acoustic communication links and propagation of low frequency noise in the oceans. There exist a number of ray trace models, some are tuned to specific applications, and others are more general. In this chapter the applications and use of ray theory are illustrated by using Plane Ray, a ray tracing program developed by the author, for modeling underwater acoustic propagation with moderately range-varying bathymetry over layered bottom with a thin fluid sedimentary layer over a solid half with arbitrary geo-acoustic properties. However, the discussion is quite general and does not depend on the actual implementation of the theory.

## 2. Theory of ray acoustics

The theory of ray acoustics can be found in most books and [1-4] will not be repeated here, but instead we follow a heuristic approach based on Snell's law, which is expressed by.

$$\xi = \frac{\cos\theta(z)}{c(z)} = \frac{\cos\theta_0}{c_0}. \tag{1}$$

Figure 1 shows a small segment of a ray path and the coordinate system. The segment has horizontal and vertical components *dz* and *dr*, respectively, and has the angle $\theta$ with the horizontal plane. When the speed of sound varies with depth the ray paths will bend and the rays propagate along curved paths. The radius of curvature $R$ is defined as the ratio between an increment in the arc length and an increment in the angle

$$R = \frac{ds}{d\theta}. \tag{2}$$

Figure 1 shows that the radius of curvature is

$$R = \frac{1}{\sin\theta}\frac{dz}{d\theta}. \tag{3}$$

When the sound speed varies with depth the ray angle $\theta$ is a function of depth according to Snell's law. Taking the derivative of Eq. (3) with respect to gives the ray's radius of curvature at depth $z$ expressed as

$$R(z) = -\frac{c(z)}{\cos\theta(z)}\frac{1}{g(z)} = -\frac{1}{\xi g(z)}. \tag{3}$$

The ray parameter $\xi$ is defined in Eq. (1) and $g(z)$ is the sound speed gradient.

$$g(z)=\frac{dc(z)}{dz}. \tag{4}$$

At any point in space, the ray curvature is therefore given by the ray parameter $\xi$ and the local value of the sound speed gradient $g(z)$. The positive or negative sign of the gradient determines whether the sign of $R$ is negative or positive, and thereby determines if the ray path curves downward or upward.

A ray with horizontal angle $\theta_{in}$ strikes a plane with inclination $\alpha$, the reflected ray is changed to $\theta_{out}$.

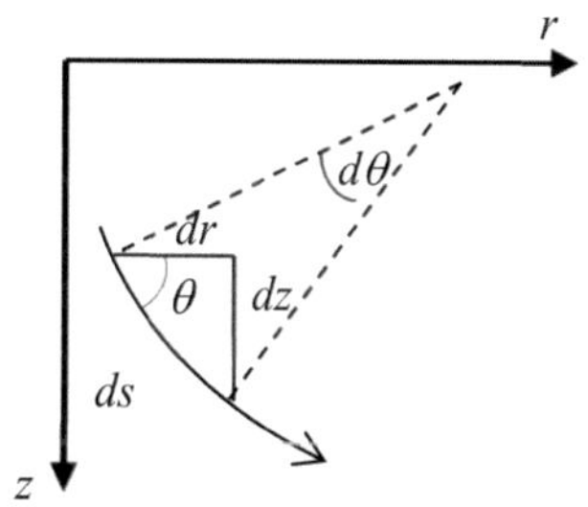


**Figure 1.** A small segment of a ray path in a isotropic medium with arc length ds.

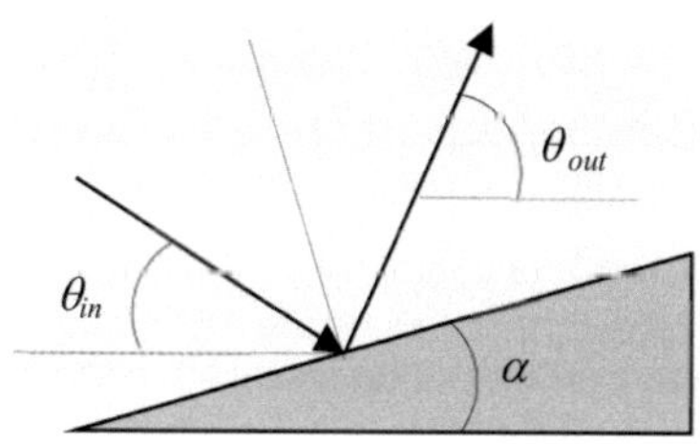


**Figure 2.** A ray with horizontal angle $\theta_{in}$ strikes a plane with inclination $\alpha$, the reflected ray is changed to $\theta_{out}$.

The ray parameter is not constant when the bathymetry varies with range. The change in ray direction is illustrated in Figure 2 showing that after reflection the angle $\theta_{in}$ of an incoming ray is increased by twice the bottom inclination angle $\alpha$.

$$\theta_{out}=\theta_{in}+2\alpha. \tag{5}$$

Consequently, after the ray is reflected, its ray parameter must change from $\xi_{in}$ to $\xi_{out}$, which is expressed as

$$\begin{aligned}\xi_{out}&=\frac{\cos(\theta_{out})}{c}=\frac{\cos(\theta_{in}+2\alpha)}{c}\\&=\xi_{in}\cos(2\alpha)\pm\frac{\sqrt{1-\xi_{in}^2c^2}}{c}\sin(2\alpha).\end{aligned} \tag{6}$$

The coordinates of a ray, starting with the angle $\theta_1$ at the point ($r_1$,$z_1$), where the sound speed is $c_1$, as shown in Figure 3. For the coordinates of the running point at ($r_2$, $z_2$) along the ray path, the horizontal distance is

$$r_2 - r_1 = \int_{z_1}^{z_2} \frac{dz}{\tan\theta(z)} = \int_{z_1}^{z_2} \frac{\cos\theta(z)dz}{\sqrt{1-\cos^2\theta(z)}} = \int_{z_1}^{z_2} \frac{\xi c(z)dz}{\sqrt{1-\xi^2 c^2(z)}}. \tag{7}$$

The travel time between the two points is obtained by integrating the quantity 1/$c$, the slowness, along the ray path:

$$\tau_2 - \tau_1 = \int_{z_1}^{z_2} \frac{ds}{c(s)} = \int_{z_1}^{z_2} \frac{dz}{c(z)\sin\theta(z)} = \int_{z_1}^{z_2} \frac{dz}{c(z)\sqrt{1-\xi^2 c^2(z)}}. \tag{8}$$

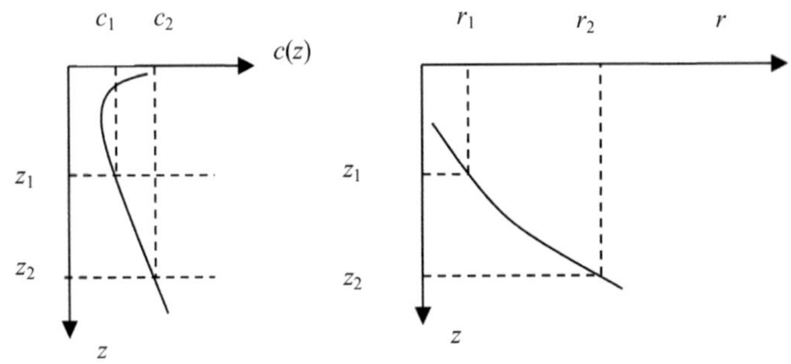


**Figure 3.** Left: The sound speed profile. Right: A portion of a ray traveling from point ($r_1$, $z_1$) to ($r_2$, $z_2$).

The acoustic intensity of a ray can, according to ray theory, be calculated using the principle that the power within a ray tube remains constant within that ray tube. This is illustrated in Figure 4 showing two rays with a vertical angle separation of $d\theta_0$ that define a ray tube centered on the initial angle $\theta_0$. At a reference distance $r_0$ from the source, the intensity is $I_0$. Taking into consideration the cylindrical symmetry about the $z$ axis, the power $\Delta P_0$ within the narrow angle $d\theta_0$ is

$$\Delta P_0 = I_0 2\pi r_0^2 \cos\theta_0 d\theta_0 . \tag{9}$$

At horizontal distance $r$, the intensity is $I$. In terms of the perpendicular cross section $dL$ of the ray tube, the power is

$$\Delta P = I 2\pi r dL . \tag{10}$$

Since the power in the ray tube does not change, we may equate Eq.(9) and eq. (10) , and solve for the ratio of the intensities:

$$\frac{I}{I_0} = \frac{r_0^2}{r}\cos\theta_0 \left|\frac{d\theta_0}{dL}\right|. \tag{11}$$

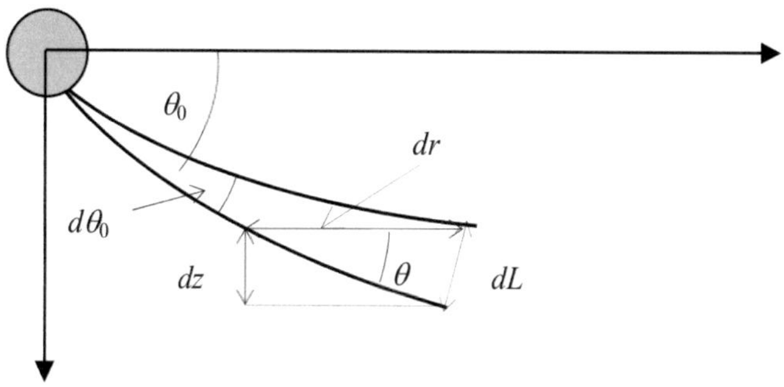


**Figure 4.** The principle of intensity calculations: energy radiated in a narrow tube remains inside the tube; $r_0$ represents a reference distance and $\theta_0$ is the initial ray angle at the source; $d\theta_0$ is the initial angular separation between two rays; *dr* is the incremental range increase; $\theta$ is the angle at the field point; *dz* is the depth differential; and *dL* is the width of the ray tube.

Instead of using Eq.(11) , it may be more convenient to use the vertical horizontal ray *dr*, which is

$$dr = \left| \frac{dL}{\sin\theta} \right|, \tag{12}$$

resulting in

$$\frac{I}{I_0} = \frac{r_0^2}{r} \left| \frac{\cos\theta_0}{\sin\theta} \frac{d\theta_0}{dr} \right| = \left( \frac{r_0^2}{r} \right) \left( \frac{c_0}{c} \right) \left| \frac{\cos\theta}{\sin\theta} \frac{d\theta_0}{dr} \right|. \tag{13}$$

The last expression in Eq.(13) is obtained by assuming that the ray parameter is constant and by using Snell's law. The absolute values are introduced to avoid problems with regard to the signs of the derivatives and of $\sin\theta$.

With respect to the reference distance $r_0$, the transmission loss *TL* is defined as

$$TL = -10\log(I / I_0). \tag{14}$$

By inserting Eq.(13) into Eq.(14) the transmission loss becomes

$$TL = 10\log\left(r / r_0^2\right) + 10\log\left| \frac{dz}{d\theta_0} \right| + 10\log\left( \frac{c}{c_0} \right). \tag{15}$$

The term $c_0/c$ is close to unity in water and can be ignored in most cases.

In this treatment the transmission loss includes only the geometric spreading loss. Therefore bottom and surface reflection losses and sea water absorption loss must be included separately.

The geometric transmission loss in Eq.(15) consists of two parts. The first term represents the horizontal spreading of the ray tube and results in a cylindrical spreading loss. The second and third terms represent the vertical spreading of the ray tube and are influenced by the depth gradient of the sound speed.

Eq.(13) predicts infinite intensity under either of two conditions: when $\theta = 0$ or when $dr / d\theta_0 = 0$. The first condition signifies a turning point where the ray path becomes horizontal; the second condition occurs at points where an infinitesimal increase in the initial angle of the ray produces no change in the horizontal range traversed by the ray. The locus of all such points in space is called a caustic. In both cases there is focusing of energy by refraction and where classical ray theory incorrectly predicts infinite intensity. Caustics and turning points will be discussed further in section 8.2.

## 3. A recipe for tracing of rays

A simple receipt for a ray tracing algorithm is to divide the whole water column into a large number of layers, each with the same thickness Δz. Within each layer, the sound speed profile is approximated as linear so that, in the layer $z_i < z < z_{i+1}$, the sound speed is taken to be

$$c(z) = c_i + g_i (z - z_i). \tag{16}$$

where $c_i$ is the speed at depth $z_i$, and $g_i$ is the sound speed gradient in the layer. From Eq. (7) and Eq. (8) the range and travel time increments in the layer are given by

$$r_{i+1} - r_i = \frac{1}{\xi g_i}\left[\sqrt{1-\xi^2 c^2(z_i)} - \sqrt{1-\xi^2 c^2(z_{i+1})}\right], \tag{17}$$

and

$$\tau_{i+1} - \tau_i = \frac{1}{|g_i|}\ln\left(\frac{c(z_{i+1})}{c(z_i)}\frac{1+\sqrt{1-\xi^2 c^2(z_i)}}{1+\sqrt{1-\xi^2 c^2(z_{i+1})}}\right). \tag{18}$$

When $\xi^2 c^2(z_{i+1}) \geq 1$, the ray path turns at a depth between $z_i$ and $z_{i+1}$, and Eq.s (17) and (18) must be replaced by the following expressions:

$$r_{i+1} - r_i = \frac{2}{\xi g_i}\sqrt{1-\xi^2 c^2(z_i)}, \tag{19}$$

and

$$\tau_{i+1} - \tau_i = \frac{2}{|g_i|}\ln\left(\frac{1+\sqrt{1-\xi^2 c^2(z_i)}}{\xi c(z_i)}\right). \tag{20}$$

These equations give the trajectories and the travel times for any ray's path to the desired range. By applying Eqs. (13) and (14), the geometrical transmission loss is also determined.

The simplicity of this method lies in the approximation of the sound speed profiles with straight-line segments and the ray path's subsequent decomposition into circular segments. The method's accuracy is determined by how well the linear fit matches the actual profile. In practice, the sound speed profile is often given as measured sound speeds at relatively few depth points. It is therefore advisable to use an interpolation scheme that is consistent with the usual behavior of the sound speed profile to increase the number of depth points to an acceptable high density.

The examples in this article are generated using the ray trace program PlaneRay that has been developed by the author [7-8]. However, any other ray programs with similar capabilities could have been use and the discussion is therefore valid for ray modeling in general. Other models frequently used and are the Bellhop model [9], and the models [10-11].

Figure 5 shows an example of ray modeling. The sound speed profile is shown at the left panel and the rays from a source at 50 m depth is shown in the right panel, which also shows the bathymetry and the thickness of the sediment layer over the solid half space.

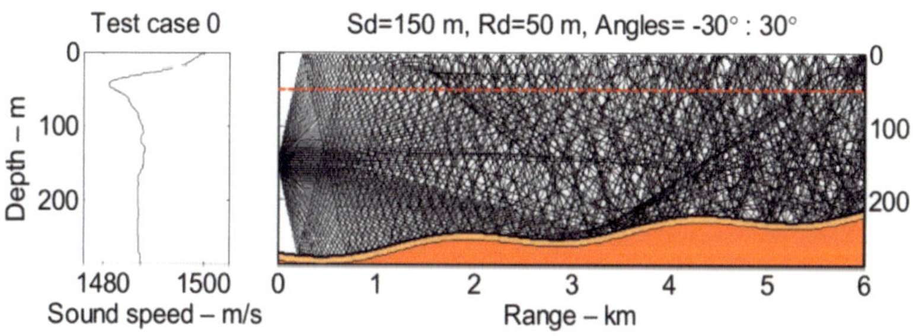


**Figure 5.** Sound speed profile and ray traces for a typical case. The source depth is 150 m and the red dotted line indicates a receiver line at a depth of 50 m. The initial angles of the rays at the source are from –30º to 30º.

## 4. Eigenray determination

To calculate the acoustic field it is necessary to have an efficient and accurate algorithm for determination of eigenrays. An eigenray is defined as a ray that connects a source position with a receiver position. In most case with multipath propagation there are many eigenrays for a given source/receiver configuration, which means that finding all eigenrays is not a trivial task.

The PlaneRay model uses a unique sorting and interpolation routine for efficient determination of a large number of eigenrays in range dependent environments. This approach is described by the two plots in Figure 6, which displays the ray history as function of initial angle at the source. All facts and features of the acoustic fields such as the transmission loss, transfer function and time responses are derived from the ray traces and their history The two plots show the ranges and travel times to where the rays cross the receiver depth line (marked by the red dashed line in Figure 5). A particular ray may intersect the receiver depth line, at several ranges. For instance at the range of 2 km, there are 11 eigenrays and from Figure 6 the initial angles of these rays are approximately found to be 5.9°, 9.6°, 22°, 24° for the positive (down going) rays and–2.0°,–3.6, °– 7-4°– 15.0°– 17.0°– 25.0°,–27.0°, for the negative (up-going waves). However, the values found in this way are often not sufficiently accurate for the determination of the sound field. Further processing may therefore be required to obtain accurate results.

The graphs of Figure 6 are composed of independent points, but it is evident that the points are clustered in independent clusters or groups. This property is used for sorting the points into branches of curves that represents different ray history. These branches are in most case relatively continuous and therefore amenable to interpolation. An additional advantage of this method is that the contribution of the various multipath arrivals can be evaluated separately, thereby enabling the user to study the structure of the field in detail.

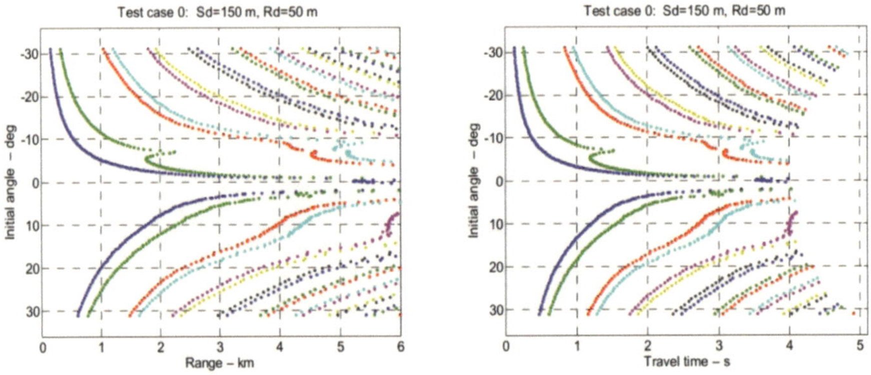


**Figure 6.** Ray history of the initial ray tracing in Figure 5 showing range (left) and travel time (right) to the receiver depth as function of initial angel at the source.

In most cases the eigenrays are determined by one simple interpolation yields values that are sufficiently accurate for most application, but the accuracy increases with increasing density of the initial angles at the cost of longer computation times.

Figure 7 shows examples of eigenrays traces with rays a receiver located at 2.5 km from the source for the scenario shown in Figure 5. To this receiver there are a total of 12 eigenrays, spanning the range of initial angles from -30° to 29°.

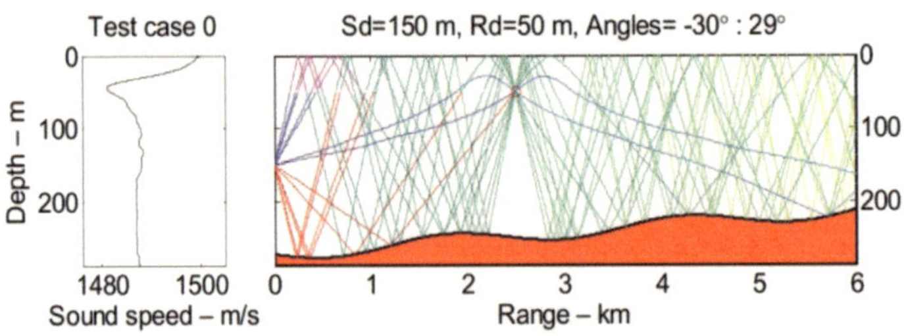


**Figure 7.** Eigenrays from a source at 150 m depth to a receiver at 50 m depth and distance of 2.5 km from the source.

# 5. Acoustic absorption in sea water

Sound absorption is important for long range propagation especially at higher frequencies. The absorption increases with frequencies and is dependent on temperature, salinity, depth and the pH value of the water. There exists several expressions for acoustic absorption in sea water; one of the preferred options is the semi-empirical formulae by Francoise and Garrison [12]. Figure 8 shows sound absorption as function of frequency in sea water using this expression for the values given in the figure caption.

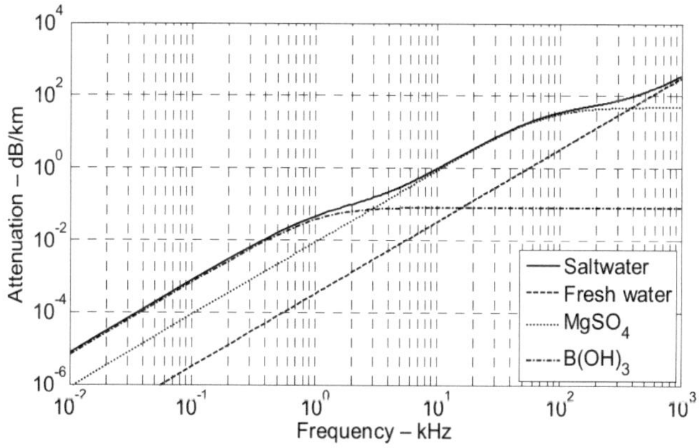


**Figure 8.** Acoustic absorption (dB/km) for fresh water and saltwater, plotted as a function of frequency (kHz) for water temperature of 10°C, atmospheric pressure of one atmosphere (surface), salinity of 35 pro mille, and pH value of 7.8. The various contributions to the absorption are also indicated.

# 6. Boundary conditions at the surface and bottom interfaces

Ray tracing is greatly simplified when no rays are traced into the bottom, but stops at the water-bottom interface. This avoids tracing of multiple reflections in layered bottoms. Instead the boundary conditions at the sea surface and the bottom can be approximately satisfied by the use of plane wave reflection coefficient.

A simple and useful bottom model is assuming a fluid sedimentary layer over a homogeneous solid half space. The reflection coefficient of a bottom with this structure is

$$R_b = \frac{r_{01} + r_{12}\exp\left(-2i\gamma_{p1}D\right)}{1 + r_{01}r_{12}\exp\left(-2i\gamma_{p1}D\right)}, \tag{21}$$

where $\gamma_{p1}$ is the vertical wave number for sediment layer and $D$ is the thickness of the sediment layer. The reflection coefficient between the water and the sediment layer, $r_{01}$, is given as

$$r_{01} = \frac{Z_{p1} - Z_{p0}}{Z_{p1} + Z_{p0}}, \tag{22}$$

and $r_{12}$ is the reflection coefficient between the sediment layer and the solid half space,

$$r_{12} = \frac{Z_{p2}\cos^2 2\theta_{s2} + Z_{s2}\sin^2 2\theta_{s2} - Z_{p1}}{Z_{p2}\cos^2 2\theta_{s2} + Z_{s2}\sin^2 2\theta_{s2} + Z_{p1}}. \tag{23}$$

In Eq (15) and (16) $Z_{ki}$ is the acoustic impedance for the compressional ($k = p$) and shear ($k = s$) waves in water column ($i = 0$), sediment layer ($i = 1$) and solid half-space ($i = 2$), respectively. The grazing angle of the transmitted shear wave in the solid half-space is denoted $\theta_{s2}$.

Figure 9 shows an example of the bottom reflection loss as function of angle and frequency for a bottom with a sediment layer with the thickness $D = 2$ m with sound speed of 1700 m/s and density 1800 kg/m$^3$ over a homogenous solid half space with compressional speed 3000 m/s, shear speed 500 m/s and density 2500 kg/m$^3$. The wave attenuations are 0.5 dB/ wavelength. The critical angle changes from 60° at very low frequencies to about 28° at high frequencies, the two angles are given by the sound speed in the water and the two bottom sound speed of 3000 m/s and 1700 m/s. The small, but significant, reflection loss at lower angles is caused by shear wave conversion and bottom absorption In this case the attenuation is about 1 dB in the frequency band around 50 Hz to 100 Hz.

The reflection coefficient of a flat even sea surface is −1 for. For a sea surface with ocean waves there will be diffuse scattering to all other direction than the specular direction, which result in a reflection loss that in the first approximation can be modeled by the coherent rough surface reflection coefficient

$$R_{coh} = \exp\left[-2\left(\frac{2\pi}{\lambda}\sigma_h \sin\theta\right)^2\right]. \tag{24}$$

In this expression $\theta$ is the grazing angle and $\sigma_h$ is the rms. wave height and $\lambda$, is the acoustic wavelength, both in meters.

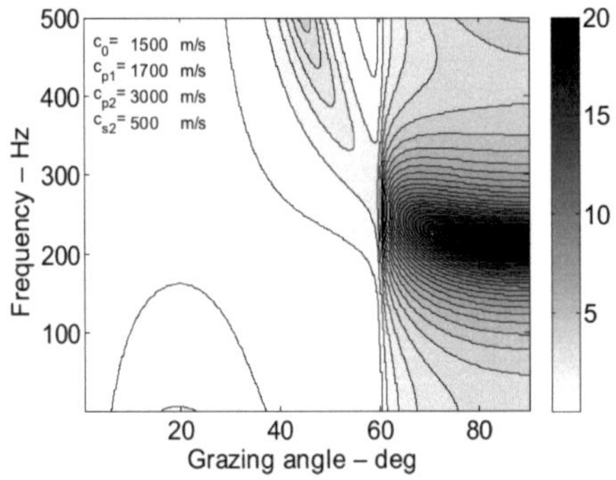


**Figure 9.** Bottom reflection loss (dB) as function of frequency and incident angle for a 2 m sediment layer over solid rock. The parameters are given in the text.

The reflection loss associated with reflection from a rough sea surface is

$$RL = -20\log10\left(R_{coh}\right). \tag{25}$$

The same rough surface reflection coefficient may also be applied to a rough bottom.

Figure 10 shows the rough surface reflection loss as function of grazing angle, calculated for a wave height of 0.5 m and the frequencies of 50 Hz, 100 Hz, 200 Hz and 400 Hz.

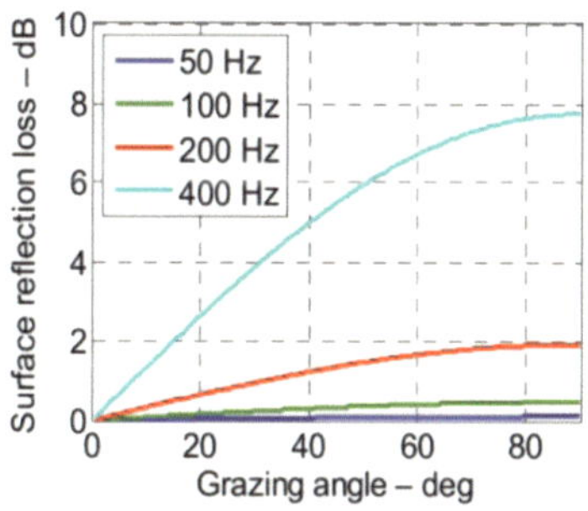


**Figure 10.** Reflection loss (dB) of rough surface with rms. wave height of 0.5 m as function of grazing angle, for the frequencies in the legend

# 7. Synthesizing the frequency domain transfer function and the time responses

The total wave field at any receiving point is calculated in the frequency domain by coherent summation of all the eigenray contributions. The first step in the calculation is to determine the geometrical transmission loss of each of the multipath contributions by applying Eq. (13) and Eq.(14) to the sorted and interpolated range-angle values. The frequency domain transfer function and the transmission loss are obtained by adding the multipath contributions coherently in frequency domain taken into account the phase shifts associated the travel times from the interpolated history of the travel times. The frequency dependent acoustic absorption of sound in water is included at this point in the process. The transfer function $H(\omega, r)$ can be expressed as

$$H(\omega,r) = \sum_{n} A_n B_n(\omega) S_n(\omega) T_n \exp(i\omega\tau_n). \tag{26}$$

Eq. (26) expresses the transfer function $H(\omega, r)$ to a distance $r$ from the source at the at angular frequency $\omega$ as a sum over the $n$ eigenrays that are included in the synthesis. $A_n$ is the geometrical spreading loss factor, defined as the square root of the expression in Eq. (13). $B_n$, and $S_n$, are the combined effects of all bottom reflections and surface reflections, respectively, $T_n$, is –90° phase shift associated with caustics and turning points, and $\tau_n$ is the travel time.

The synthesis of the received signals is performed in the frequency domain by multiplying the frequency spectrum of the source signal with the transfer function of each of the eigenrays and summing the contributions. The time domain response is obtained after multiplication with the frequency function of a source signal followed an inverse Fourier transform of the product. This requires the choice of a source signal, a sampling frequency ($f_s$) and a block length ($N_{fft}$) of the Fourier transform.

The total duration of the time window ($T_{max}$) after Fourier transform is

$$T_{max} = \frac{N_{fft}}{f_s}. \tag{27}$$

It is important to select the values of $N_{fft}$ and $f_s$ such that Fourier time window, $T_{max}$, is larger than the actual length or duration of the signal. In reality the real time duration of the received signal is often not known in advanced and therefore the user may have to experiment with different values to find appropriate values for of $N_{fft}$ and $f_s$.

Figure 11 shows an example where the transmission loss (in dB) as function of range has been calculated for the frequencies of 100 Hz and 200 Hz. The dashed black line indicates the geometrical spreading loss, which is added for comparison and given by,

$$TL_{geo}(r) = 10\log_{10}\left[ r^2 \left( 1 + \frac{r^2}{r_t^2} \right)^{-1/2} \right]. \tag{28}$$

This expression yields a transmission loss proportional to 20log($r$) when r< $r_t$ and proportional to 10 log(r) for $r> r_t$ . This approximation to the geometrical transmission loss may be used for approximate calculations of transmission loss for flat bottom and simple sound speed profiles. In the case shown in Figure 11 $r_t$ is set equal to the water depth at source location, which in this case is 200 m.

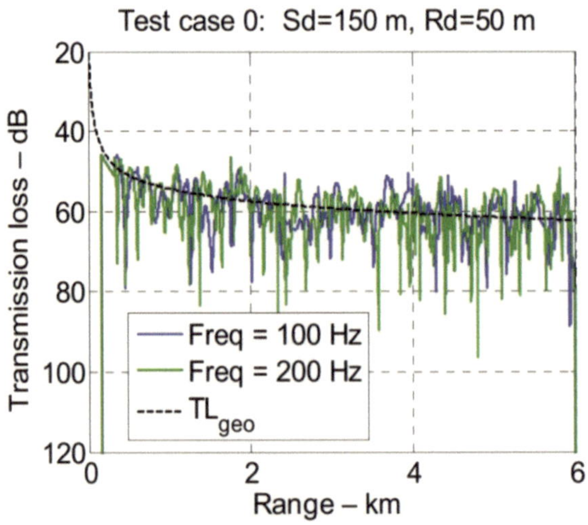


**Figure 11.** Transmission loss as function of range calculated for 100 Hz and 200 Hz The dashed black line is values of Eq.(28)

Figure 12 shows the synthesized time response at receivers spaced at 200 m separations in range up to 6 km. The sound speed and bathymetry is the same as in Figure 5 with the source at 150 m and all receivers at 50 m depth. The time scale is in reduced time to remove the gross transmission delay between the source and receiver. The reduced time is defined as

$$t_{red} = t_{real} - \frac{r}{c_{red}}. \tag{29}$$

In Eq. (29), $t_{real}$ and $t_{red}$ are the real and reduced times, respectively, $r$ is range and $c_{red}$ is the reduction speed. The actual value of $c_{red}$ is not important as long as the chosen value results in a good display of the time responses.

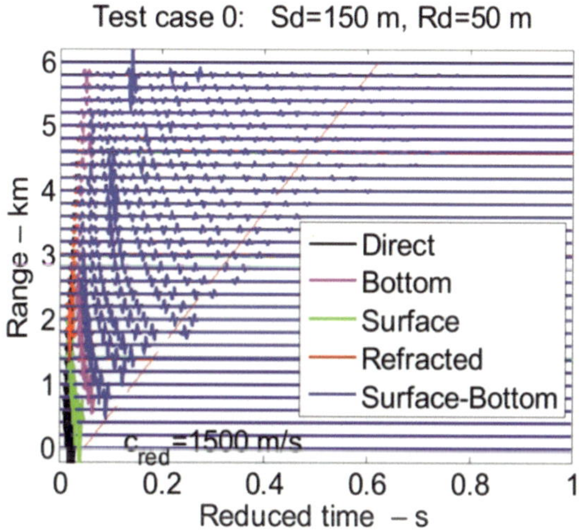


**Figure 12.** Received time signals as function of range and reduced time.

In the example shown above, the time signal and calculated assuming a narrow band-limited source signal in the form of a Ricker pulse. An example of a Ricker pulse and its frequency spectrum are shown in Figure 13.

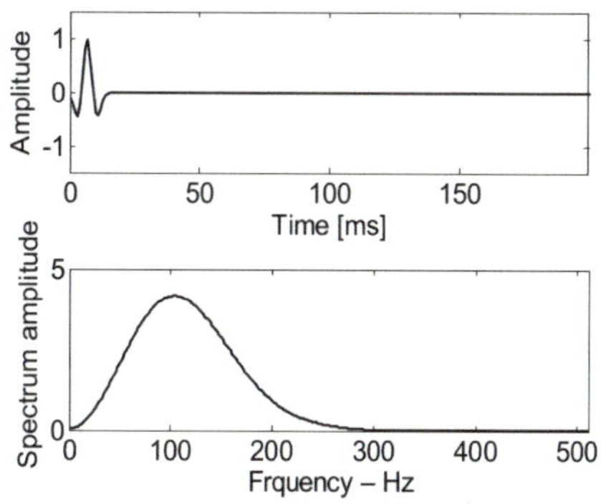


**Figure 13.** Ricker time pulse and frequency function the center frequency of 100 Hz

The time responses in Figure 12 are sorted according to the history of their eigenrays and color coded to allow for studying the various multipath contributions. This is particularly useful when dealing with transient signal and broad band signal, especially when knowledge of the multipath structure is important. In many such situations only the direct arrival or the refracted arrivals in the water column may carry the useful signals and all the other arrivals represent interference. In this case there are direct arrivals, followed by surface reflected and refracted arrivals at the turning points. Notice the high sound pressure values caused by the caustics at 3 km, 6, km and 7 km, which are apparent in both plots, this issue is discussed in section 8.2.

The red dotted line in Figure 12 represents an estimate of the duration of the cannel impulse response. This time duration is mainly given by the bottom reflection coefficient and the critical angle. Rays that propagate at angles closer to the horizontal plane than the critical angle experience almost no bottom reflection loss and may therefore propagate to long distances. Rays with steeper angles will experience higher reflection losses and die out more rapidly with range. Thus the time duration of the impulse is directly determined by the ratio of sound speeds in the water and the bottom as

$$t_{red} = \frac{r}{c_0}\left(\frac{1}{\cos\theta_{crit}} - 1\right) = r\left(\frac{c_b - c_0}{c_0^2}\right) \tag{30}$$

This estimate of the time duration of the channel impulse response assumes that the bottom is fluid, homogenous and flat, but the estimate may also be useful in other cases with moderately range dependent depth and with solid or layered bottom.

## 8. Special considerations

### 8.1. Frequency of applications

Ray tracing is a high frequency approximation to the solution of the wave equation and in principle more valid for high than for low frequency applications. However, high resolution prediction of higher frequency acoustic fields is difficult both for numerical and physical reasons. Principally most important is the physical limitation caused by the fact that the sound speed and the environment are generally not known in sufficient detail. This can be illustrated by a simple example. Consider coherent communication using a frequency of 10 kHz with wavelength of 10 mm. The required accuracy in order to be correct at a distance of 1 km is that the sound speed is known and stable with a relative error less $10^{-5,}$ an impossible requirement to satisfy in practice regardless of the numerical accuracy of the computer model.

### 8.2. Caustics and turning points

As mentioned before, the locations where $dr/d\theta_0=0$ are called caustics where the ray phase is decreased by 90° and the where the intensity, according to ray theory, goes to infinity. In reality the intensity is high, but finite, and the basic ray theory breaks down at these points. There exists theories to amend and repair the defects of ray theory at these points [1, 2, 13], but that is not discussed here.

Figure 14 shows details of the field at a showing the rays with initial angles in the range of −6° to −1°. The scenario is the same as in of Figure 5, but for clarity the tracing of rays have been stopped after the first bottom reflection and the figure concentrates on the details the field at the caustic at 1760 m range for a ray with initial angle of −5.6°. Figure 15 shows the time responses for ranges in the interval from 1.6 km to 1.9 km. In this case, the source signal is a Ricker pulse with a peak frequency of 200 Hz. There is a first direct arrival (black color) at all ranges. From the range 1760 there is also a refracted arrival a little later than the direct, but with higher amplitude, in particular near the range of 1760 m. Notice the effect of the 90° phase shift for ranges beyond the caustic at 1760 m and that the amplitude at this range is considerable higher than at the other ranges.

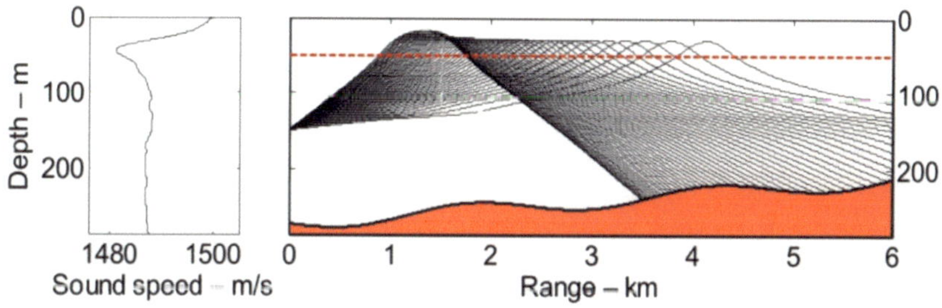


**Figure 14.** Rays through a caustic

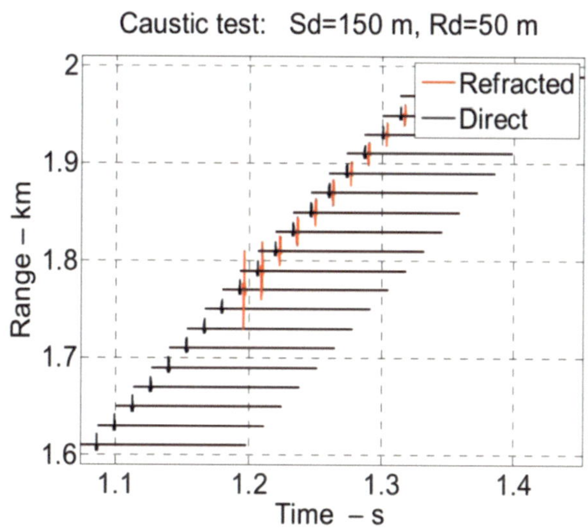


**Figure 15.** Time responses around the caustic at 1.76 km. The transmitted signal is a Ricker pulse with peak frequency of 200 Hz.

## 8.3. The principle of reciprocity and its validity in ray modeling

The principle of reciprocity is an important and useful property of linear acoustics and systems theory. The principle is very general and valid also in cases where the wave undergoes reflections at boundaries on its path from source to receiver [14]. The reciprocity principle is correctly represented in ray modeling, as can easily be understood from the eigenray plots of Figure 7. The eigenrays from a source position to the receiver position are the same as when source and receiver changes positions. The reflections at the bottom and at the sea surface are also symmetric in angles and consequently the acoustic fields are the same. However, it should be noted that the reciprocity principle applies to a point-to-point situation. This means that, for instance, that the development of the transmission loss as function source-receiver separation is generally not the same for the two directions.

## 8.4. The validity of using plane wave reflection coefficients

The accuracy of any ray model depends on the validity and limitation of ray theory and the implementation. A fundamental assumption of model is that the interactions with the boundaries are adequately described by plane wave reflection coefficient. In this section the validity of this assumption is investigated.

The general expression for the reflected field is given in text books, for instance in [13], over horizontal wave numbers $k$, as

$$\Phi_{ref}\left(r,z_r,\omega\right)=\frac{S(\omega)}{8\pi i}\int_0^\infty \Re\left(k\right)\frac{\exp\left(i\gamma\left|z_r+z_s\right|\right)}{\gamma}kH_0^{(1)}\left(kr\right)dk \ . \tag{31}$$

$\Phi_{ref}(r, z_r\ \omega)$ is the reflected field due to point source with frequency $\omega$ and source strength $S(\omega)$. $\Re\left(k\right)$ is the reflection coefficient, $H_0^1\left(kr\right)$ is the Hankel function of first kind, which represents a wave progressing in the positive r-direction. The horizontal wave number $k$ and the vertical wave number $\gamma$ are related to the sound speed, frequency and the angle $\theta$ by

$$k=\frac{\omega}{c}\cos\theta,$$
$$\gamma=\frac{\omega}{c}\sin\theta. \tag{32}$$

Eq.(31) states that the field is given as an integral over all horizontal wave numbers, or as consequence of Eq.(32), integration over all angles both real and the imaginary.

Consider now the situation where $\Re\left(k\right)=\Re$ is constant and independent of $k$ or the angle. The integral in Eq.(31) becomes a standard integral and

$$\Phi_{ref}\left(r,z_r,\omega\right)=\frac{S(\omega)}{4\pi R}\Re\exp(ikR). \tag{33}$$

$$R=\left[r^2+\left(z_s+z_r\right)^2\right]. \tag{34}$$

According to Eq. (33) the reflected wave is the same as the outgoing spherical wave from the image of the source in the mirror position of the real source and modified by the constant reflection coefficient $\Re$ . The situation with a constant reflection coefficient is valid for perfectly flat sea surface where the reflection coefficient is equal to -1 for all angles of incidence. Thus the reflection from a smooth sea surface is accurately described plane wave reflection coefficients.

In the general case, and for reflections from the bottom, the reflection coefficient $\Re(k)$ is not constant and the integral can only be solved approximately or numerically. In order to obtain an approximation of the integral in Eq.(31) , the Hankel function is expanded in a power series with the first terms being

$$H_0^1(kr)\approx\sqrt{\frac{2}{\pi kr}}\exp\left[i\left(kr-\pi/4\right)\right]\left[1+\frac{1}{8ikr}+\cdots\right]. \tag{35}$$

Restricting the integral of Eq.(31) to the first term yields

$$\begin{aligned}&\Phi_{ref}\left(r,z_r,\omega\right)=\\&=\frac{S(\omega)}{4\pi}\frac{1}{\sqrt{2\pi r}}\int_{-\infty}^{\infty}\Re(k)\frac{\sqrt{k}}{\gamma}\exp\left[ikr+i\gamma\left(z_r+z_s\right)\right]dk.\end{aligned} \tag{36}$$

The exponential in the integrand will normally be a rapid varying function and therefore the value of the integral will be small except when the phase term of Eq.(36) is nearly constant. The phase term of Eq.(36) is

$$\alpha=i\gamma\left(z_r+z_s\right)+ikr. \tag{37}$$

The stationary points are defined to the values of the horizontal wave number $k$ where the derivative of the phase with respect to $k$ is equal to zero, that is where $d\alpha/dk=0$, giving the stationary point as

$$r=\frac{\left(z+z_s\right)}{\tan\left(\theta_0\right)}. \tag{38}$$

The interpretation of this result is quite simple; the reflected wave field is equal to that of the image source multiplied with the reflection coefficient at the specular angle $\theta_0$.

There are however situations where this approximation is not sufficient in practice. This is discussed in [13] and in the following their results are cited without proof. The accuracy of the approximation depends on the source or receiver distance from the bottom interface. The result of the analysis is that the distance $z$ from the bottom must satisfy

$$z >> \frac{\lambda}{2\pi} \frac{\frac{\rho_b}{\rho_w}}{\sqrt{\left(\frac{c_b}{c_w}\right)^2 - 1}}. \tag{39}$$

With the water parameters of $\rho_w$ = 1000 kg/m³ and $c_w$=1500 m/s, and the bottom parameters of $\rho_b$ = 1500 kg/m³ and $c_b$=1700 m/s. Equation (39) requires than the distance from the bottom satisfy z>> 0.5 λ for the validity of using plane wave reflection coefficient at the bottom interface. A harder bottom with $\rho_b$ = 1800 kg/m³ and $c_b$=3000 m/s, gives the requirement that z>> 1.0 λ. Hence the condition for validity is somewhat easier to satisfy for a soft bottom than for a hard bottom.

## 8.5. Bench marking ray modeling

The wave number integration model OASES [15] has been used to validate the accuracy and the limitation of the ray trace model using the simple case with constant water depth of 100 m and constant sound speed of 1500 m/s.

Figure 16 show the calculated transmission loss for the frequencies of 25 Hz, 50 Hz, 100 Hz and 200 Hz. The agreements between the results are very god for the higher frequency, but with some discrepancies for the lower frequencies, in particular for 25 Hz. The discrepancy is mainly a phase shift in the interference patterns of the two results, most pronounced for low frequencies and long ranges. This observation agrees with the theory outlined earlier. The seriousness of this discrepancy or errors may not very important in practice since the mean level is nearly the same as shown by the comparison with the OASES model.

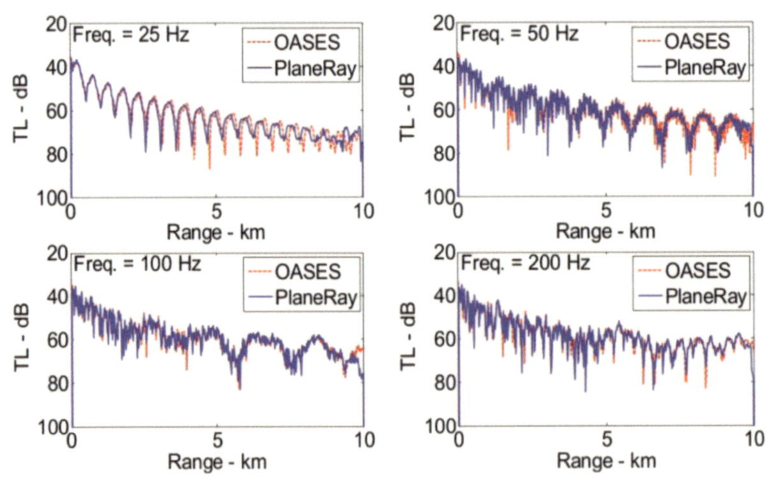


**Figure 16.** Comparison of the transmission loss as function of range for selected frequencies by PlaneRay (solid blue line) and OASES (dotted red line) for Pekeris' waveguide with a homogenous solid bottom with compressional wave speed of 3000 m/s and shear wave speed 500 m/s. Both wave attenuations have the values of 0.5 dB/wavelength.

# 9. Case studies

In the following we present two case studies that are relevant application of the modeling techniques descried in this article. The first if these is in connection with acoustic underwater communication and the transmission of digital information. In this case the multipath communication may be a significant problem causing intersymbol interference and significant degradation of reliability and performance. The second case is related to studies on the propagation of low frequency sound and the effect such noise may affect marine life, sea mammals and fish.

## 9.1. Seasonal variations of communication links

In connection with a study of underwater acoustic communication the propagation over a 6 km track has been modeled for the various seasonal sound speed profiles.

The sound speed some months are shown in Figure 17. The sound speed profiles depend on the sea water temperature, the salinity and the depth. In the present case the sea water temperature variation with depth and the seasons is the main reason for changes in sound speed profile. During winters the surface water is cold and the sound speed is low, in the summer the surface water temperature and the sound speed is higher. The seasonal heating and cooling of the surface water propagates also to deeper depths, but with diminishing temperatures changes. At very large depths the water temperature is nearly the same at all seasons and the sound speed increases linearly and slowly with depth.

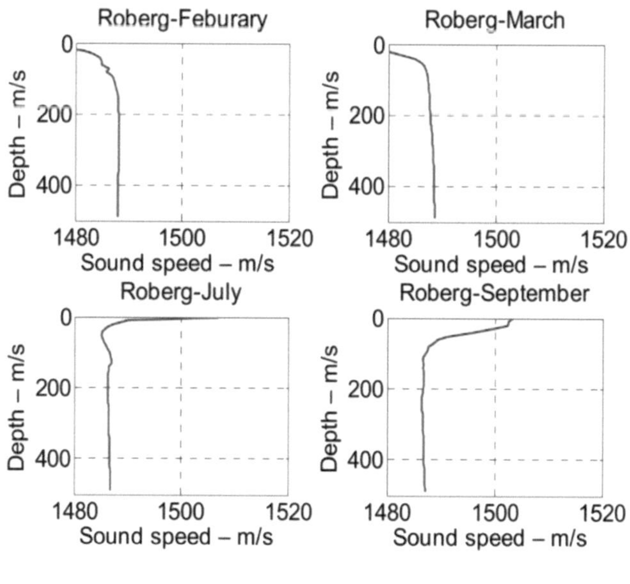


**Figure 17.** Sound speed profiles measured at specific dates for the months given in the figures

Figure 18 shows ray tracing results are for the same profiles as displayed in Figure 17. The purpose of the study was to investigate the possibility of communication to positions beyond a sea mount and to study the multipath arrival structure as function of range and depth.

There is a seamount with a peak at about 3 km from the transmitting station. In order to simplify the interpretation ray tracings in these plots have been terminated after 6 bottom reflections, but all rays are included in the calculation of the acoustic field, but rays with so many bottom reflections, or more, will in most case not be useful for data communication because of the reflection loss and reduced coherence.

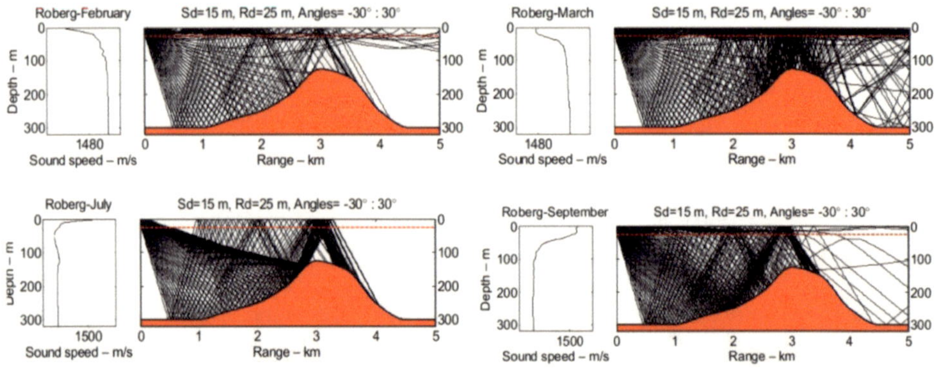


**Figure 18.** Ray tracing plots assuming a source depth of 15 meter for four monthly conditions at the Roberg test site. The sound speed profiles are the same as shown in Figure 17.

Figure 19 shows examples of received time responses at 25 m depth using a Ricker pulse as source signal. The different multipath contributions are color coded for clarity. At distances from the source over 1.5 km the first arrivals is follow paths surface reflected and upward refracted paths

Figure 20 shows the channel responses at a fixed range as function of depth down to 50 m. This figure shows the total response after adding all the individual multi path contributions. The plots demonstrate that the surface channel consists of deep refracted path and a number of paths reflected from the surface and deeper upwards refractions. The stability of these paths may be uncertain and subject to rapid changes in the environmental conditions near the surface due to temperature wind and current.

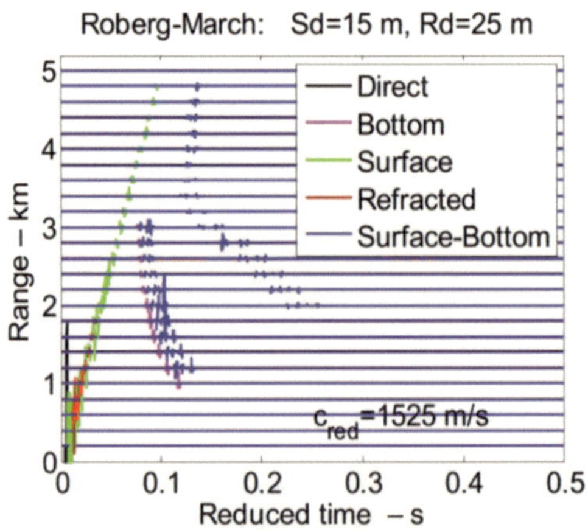


**Figure 19.** Time responses as function of range for receivers at depths of 25 m with a source at 15 m.

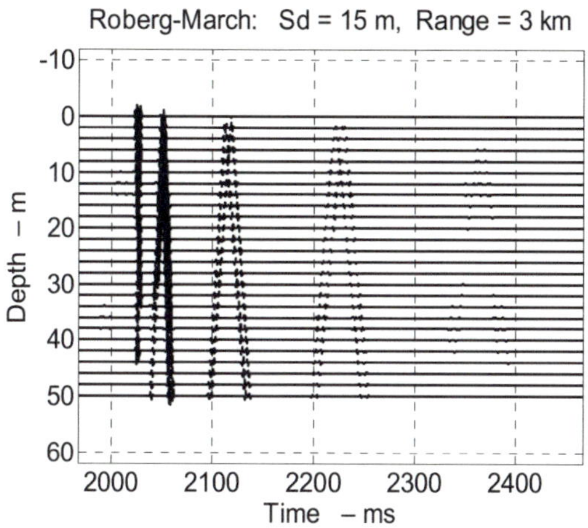


**Figure 20.** *Time responses as function of receiver depth at a fixed horizontal distance of 3 km from a source at 15 m depth.*

## 9.2. Seismic noise propagation

In many areas of the world anthropogenic noise often dominates over the natural ambient noise, especially in the low frequency band from approximately from 10 Hz and upwards to 1000 Hz, or more. This frequency band coincides approximately with the frequencies of perception of sea mammals and fish and may therefore be harmful to their natural activities, or even cause physical damages. An example is the case of the seismic exploration for oil and gas in certain areas where there is important commercial fishing interest. The propagation and distribution of acoustic noise depends the environmental conditions, in particular the oceanographic parameters, the topography of the seafloor and the acoustic properties of the bottom. In this section some of examples are presented to illustrate how the environment may affect the distribution of sound and noise. This study and discussion is also relevant for passive sonar applications to detect and track submerged vehicles and objects base emitted acoustic noise

The effects of bathymetric are illustrated in Figure 21 showing ray traces of upslope and downslope conditions for typical summer conditions at the Halten Bank in the Norwegian Sea. With downslope propagation there is a thinning the ray density with distance and upslope propagation gives a concentration of rays as the water depth diminishes.

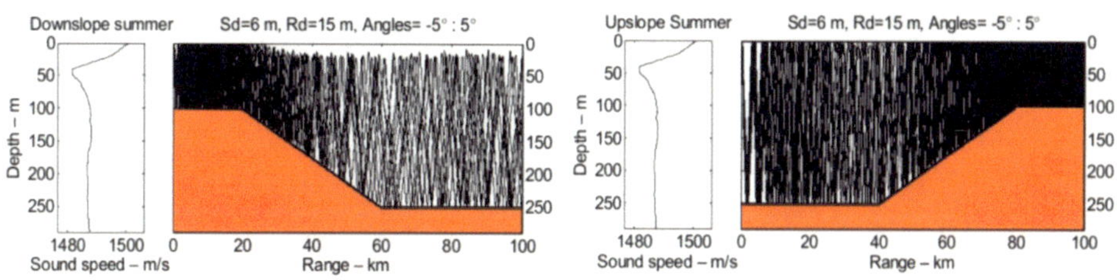


**Figure 21.** The effect of up and down sloping bottoms on the acoustic field distribution calculated for the typical summer condition in the month of July.

Figure 22 and Figure 23 show the calculated sound pressure level as function of range for the downslope and upslope propagation. The sound pulse from an airgun array is modeled as s a Ricker pulse with a peak pressure of 260 dB rel. 1μPa, centered on the frequency of 50 Hz, The horizontal dashed line is the assumed threshold value for fish reaction to sound. The bottom is modeled with a 2 m thick sedimentary layer over solid rock. The sound speed in the sediment layer is 1700 m/s and the density is 1800 kg/m$^3$. The compressional sound speed in the rock is 3000 m/s, and density is 2500 kg/m$^3$. The results in Figure 22 and Figure 23 are obtained under two conditions: (a) with a shear speed of 500 m/s, and (b) with no shear wave in the rock, i.e. the shear speed is zero. The absorptions are assumed to be 0.5 dB per wavelength for all the waves in the sediment layer and the rock. In the first case (a) the bottom reflection loss is as shown in Figure 9 with a significant low frequency reflection loss at angles lower than the critical angle caused by absorptions and conversion to shear wave in the bottom, which draws energy for the reflected wave. In the case of Figure 22 this results in a low-frequency and low-angle reflection loss of about 1 dB. For long ranges and many reflections this adds up to a significant total propagation loss. With no shear conversion the reflection loss is considerably reduced and the sound propagates easier to long ranges. The difference between the sound level at 50 Hz and 100 Hz is partly a result of increase attenuation at the higher frequency and partly that the source level in this case is higher for 50 Hz than for 100 Hz.

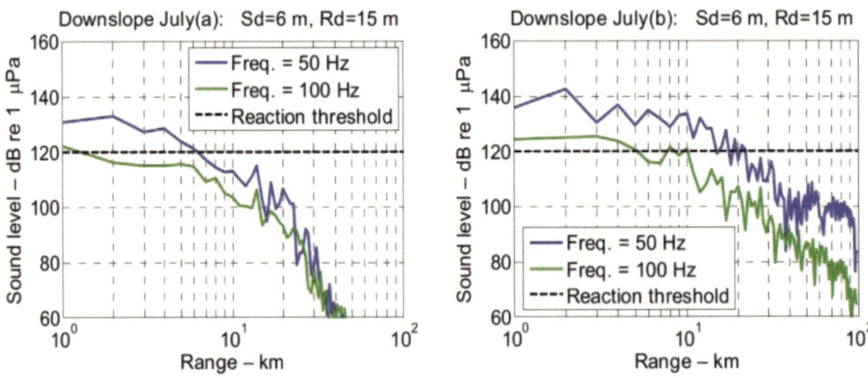


**Figure 22.** Sound pressure level as function of range for downslope propagation and July conditions. Left: With shear wave conversion (500 m/s). Right: No shear wave conversion.

Figure 24 and Figure 23 show similar results for downslope and upslope propagation for typical winter conditions represented by a sound speed profile measured in the month of February. For downslope conditions the sound level decrease rapidly with increasing depth and much more rapidly with shear wave conversion (Figure 24a) than without shear (Figure 24b). With upslope propagation (Figure 23) the sound levels are near independent of shear conversion except at the very long rages where the water depth becomes constant. The examples demonstrate that sound propagation in the ocean is strongly influenced by both by the oceanographic conditions and the geophysical properties of the bottom. Reliable prediction of acoustic propagation condition requires modeling tool that can that can handle both bottom and water properties.

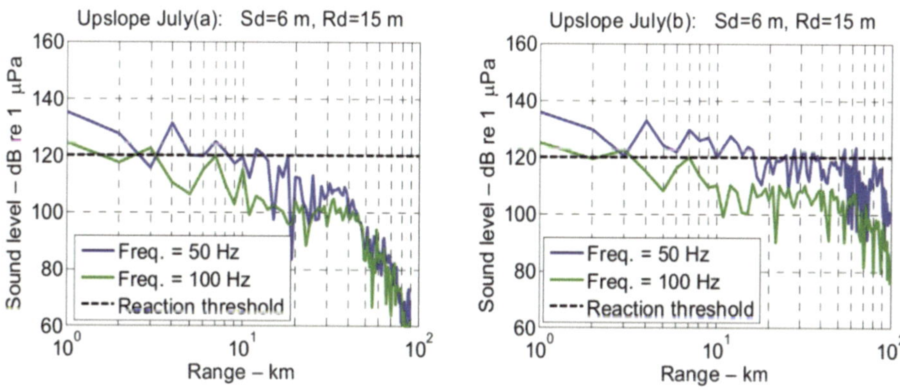


**Figure 23.** Sound pressure level as function of range for upslope propagation and July conditions. Left: With shear wave conversion (500 m/s). Right: No shear wave conversion.

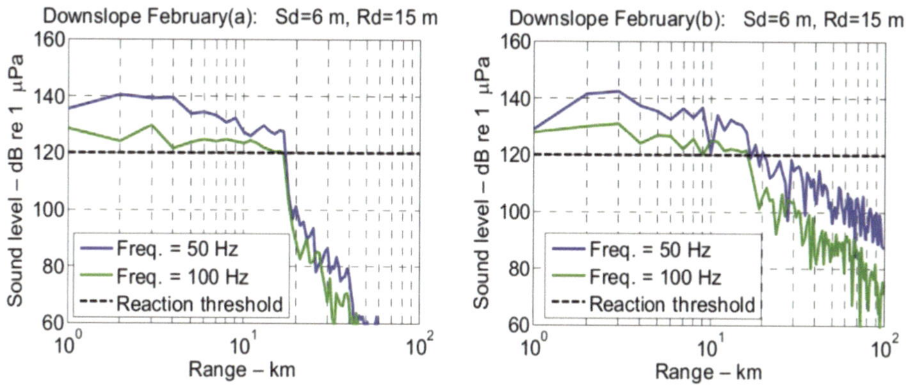


**Figure 24.** Sound pressure level as function of range for downslope propagation and February conditions. Left: With shear wave conversion (500 m/s). Right: No shear wave conversion.

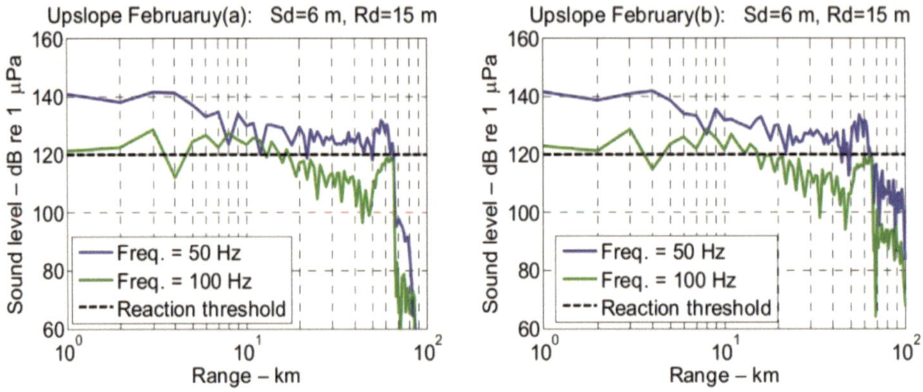


**Figure 25.** Sound pressure level as function of range for upslope propagation and February conditions. Left: With shear wave conversion (500 m/s). Right: No shear wave conversion

## 10. Summary

The article has outlined the theory of ray modeling and described how the theory can be applied to study acoustic wave propagation in the ocean. The complete acoustic fields are calculated by coherent addition of the contributions of a large number of eigenrays. In this method no rays are traced into the bottom, but the bottom interaction is modeled by plane wave reflection coefficients. Ray tracing is, by definition, frequency independent and therefore the ray trajectories through the water column are valid for all frequencies. Frequency dependency is introduced by reflections from the sea surface and the bottom, including loss associated with absorption and diffuse scattering of a rough ocean and bottom interfaces. Ray tracing is therefore a computational effective method for modeling broad of frequency band wave fields and for calculation of time responses.

Ray tracing is high-frequency approximation to the solution of the wave equation and the accuracy and validity at lower frequencies may be questioned, in particular the use of plane ray reflection coefficient to represent the bottom effects. This problem has been considered both theoretically and by simulations and comparison with more accurate model. The results of this study shows that source and receiver should be at a height above the bottom of at least half a wavelength, but there is no similar requirement to the distance from the sea surface. Less fundamental is the limitation of the numerical accuracy of the determination of the eigenrays, which is most serious in the calculation of the ray amplitude and the transmission loss. These inaccuracies are of more practical nature and can be reduced by refinements in the calculations.

Examples relevant for application in acoustic underwater communication and active sonar have been presented. The propagation of low frequency sound to large distances has been presented showing the effect of the bathymetry and the acoustic properties of the bottom. An important conclusion is the effect of bathymetry and the sound speed structure interacts and that accurate modeling of sound propagation requires information about the oceanography, the bathymetry and the geology of the bottom.

## Author details

Jens M. Hovem
*Norwegian University of Science and Technology, Norway*

## 11. References

[1] Officer, C. B. (1958). *Introduction to the theory of sound transmission*. McGraw-Hill, New York City.

[2] Jensen F. B., W. A. Kuperman, M. B. Porter, and H. Schmidt, (2011). *Computational Acoustics*, Second edition, Springer, New York.

[3] Kinsler, L. E., A. R. Frey, A. B. Coppens, and J. V. Sanders. 2000. *Fundamentals of acoustics*, 4th ed. New York: John Wiley & Sons.

[4] Hovem, J. M. (2010). *Marine Acoustics: The Physics of Sound in Underwater Environments.* Peninsula Publishing, Los Altos, California, USA. ISBN 9780932146656

[5] Westwood, E. K and P. J. Vidmar. (1987). *Eigenray finding and time series simulation in a layered-bottom ocean*, J. Acoust. Soc. Am. 81, pp. 912-924.

[6] Westwood, E. K and C. T. Tindle. (1987). *Shallow water time simulation using ray theory*. J. Acoust. Soc. Am. 81, pp. 1752 -1761.

[7] Hovem Jens M. (2008). *PlaneRay: An acoustic underwater propagation model based on ray tracing and plane wave reflection coefficients*, in Theoretical and Computational Acoustics 2007, Edited by Michael Taroudakis and Panagiotis Papadakis, Published by the University of Crete, Greece, pp. 273-289, (ISBN: 978-960-89785-4-2).

[8] Hovem, Jens M. *Ray trace modeling of underwater sound propagation.* SINTEF Report A21539, 2011.11.23, ISBN978 82 14 04997-8.

[9] Porter, M. B. (1991). *The Kraken normal mode program*. Rep.SM-245. (Nato Undersea Research Centre), La Spezia, Italy. (1991).

[10] Abrahamsson, L. (2003). *RAYLAB---a ray tracing program in underwater acoustics*, Sci. Rep. FOIR--1047--SE, Division of Systems Technology, Swedish Defence Research Agency, Stockholm, Sweden.

[11] Ivansson, S. (2006). *Stochastic ray-trace computations of transmission loss and reverberation in 3-D range-dependent environments*, ECUA 2006, Carvoeiro, Portugal, pp. 131-136.

[12] Francois, R. E., and G. R. Garrison. 1982. Sound absorption based on ocean measurements: Part II. Boric acid contribution and equation for total absorption. *J. Acoust. Soc. Am.* 72(6), 1879–1890.

[13] Brekhovskikh, L. M., and Yu. Lysanov. (2003). *Fundamentals of ocean acoustics*, 3rd ed. Springer-Verlag, New York City.

[14] Landau, L. D., and E. M. Lifshitz. (1959). *Fluid Mechanics*. Pergamon Press, Oxford UK.

[15] Schmidt H. (1987). *SAFARI:* Seismo-acoustic fast field algorithm for range independent environments. User's guide, SR-113, SACLANT Undersea Research Centre, La Spezia, Italy.